Nitroxide Mediated Polymerization
From Fundamentals to Applications in Materials Science

RSC Polymer Chemistry Series

Editor-in-Chief:
Professor Ben Zhong Tang, *The Hong Kong University of Science and Technology, Hong Kong, China*

Series Editors:
Professor Alaa S. Abd-El-Aziz, *University of Prince Edward Island, Canada*
Professor Stephen Craig, *Duke University, USA*
Professor Jianhua Dong, *National Natural Science Foundation of China, China*
Professor Toshio Masuda, *Shanghai University, China*
Professor Christoph Weder, *University of Fribourg, Switzerland*

Titles in the Series:
1: Renewable Resources for Functional Polymers and Biomaterials
2: Molecular Design and Applications of Photofunctional Polymers and Materials
3: Functional Polymers for Nanomedicine
4: Fundamentals of Controlled/Living Radical Polymerization
5: Healable Polymer Systems
6: Thiol-X Chemistries in Polymer and Materials Science
7: Natural Rubber Materials: Volume 1: Blends and IPNs
8: Natural Rubber Materials: Volume 2: Composites and Nanocomposites
9: Conjugated Polymers: A Practical Guide to Synthesis
10: Polymeric Materials with Antimicrobial Activity: From Synthesis to Applications
11: Phosphorus-Based Polymers: From Synthesis to Applications
12: Poly(lactic acid) Science and Technology: Processing, Properties, Additives and Applications
13: Cationic Polymers in Regenerative Medicine
14: Electrospinning: Principles, Practice and Possibilities
15: Glycopolymer Code: Synthesis of Glycopolymers and their Applications
16: Hyperbranched Polymers: Macromolecules in-between of Deterministic Linear Chains and Dendrimer Structures
17: Polymer Photovoltaics: Materials, Physics, and Device Engineering
18: Electrical Memory Materials and Devices
19: Nitroxide Mediated Polymerization: From Fundamentals to Applications in Materials Science

How to obtain future titles on publication:
A standing order plan is available for this series. A standing order will bring delivery of each new volume immediately on publication.

For further information please contact:
Book Sales Department, Royal Society of Chemistry, Thomas Graham House, Science Park, Milton Road, Cambridge, CB4 0WF, UK
Telephone: +44 (0)1223 420066, Fax: +44 (0)1223 420247
Email: booksales@rsc.org
Visit our website at www.rsc.org/books

Nitroxide Mediated Polymerization
From Fundamentals to Applications in Materials Science

Edited by

Didier Gigmes
Aix-Marseille University, CNRS, Marseille, France
Email: Didier.gigmes@univ-amu.fr

ROYAL SOCIETY OF CHEMISTRY

THE QUEEN'S AWARDS FOR ENTERPRISE: INTERNATIONAL TRADE 2013

RSC Polymer Chemistry Series No. 19

Print ISBN: 978-1-78262-061-7
PDF eISBN: 978-1-78262-263-5
ISSN: 2044-0790

A catalogue record for this book is available from the British Library

Published by The Royal Society of Chemistry,
Thomas Graham House, Science Park, Milton Road,
Cambridge CB4 0WF, UK

Registered Charity Number 207890

For further information see our web site at www.rsc.org

Printed in the United Kingdom by CPI Group (UK) Ltd, Croydon, CR0 4YY, UK

Preface

Among the various reversible-deactivation radical polymerization (RDRP) techniques, nitroxide mediated polymerization (NMP) was the first to be developed in the mid-1980s. This book traces the major developments of NMP from its discovery to date. Through intense research efforts made in the field of NMP, significant progresses have been made in recent years, including the understanding of the main parameters influencing the control quality of a NMP process, the synthesis of highly potent nitroxides and corresponding alkoxyamines and the preparation of polymer and/or hybrid materials with advanced properties. NMP stands out from other techniques in its simplicity because it requires only a thermal or photochemical trigger without adding an external radical source or metal catalyst. This book collates the contributions of renowned experts in the NMP process in order to provide a comprehensive and general overview of this technique.

The book commences with the history of NMP (Chapter 1) followed by a detailed description of the kinetics (Chapter 2) associated with this technique. Chapter 3 is dedicated to the synthesis of nitroxides and alkoxyamines while Chapters 4, 5 and 7 are focused on the NMP process in homogenous, dispersed media and the controlled polymerization of methyl methacrylate, respectively. The latest developments in NMP are given in Chapter 6. Chapter 8 offers a picture of the various complex macromolecular architectures achievable by NMP. Chapter 9 highlights the most relevant results in the field of bioconjugates and biomaterials prepared by NMP. Chapter 10 demonstrates the high potential of NMP for surface functionalization, while Chapter 11 is dedicated to the synthesis of different materials for a variety of emerging and next generation applications.

RSC Polymer Chemistry Series No. 19
Nitroxide Mediated Polymerization: From Fundamentals to Applications in Materials Science
Edited by Didier Gigmes

Published by the Royal Society of Chemistry, www.rsc.org

This book will be of major interest to industry and academic researchers as a reference guide for using the nitroxide mediated polymerization technique. It also aims at being a valuable document for graduate students in the broad field of polymer chemistry.

Contents

RSC Polymer Chemistry Series No. 19
Nitroxide Mediated Polymerization: From Fundamentals to Applications in Materials Science
Edited by Didier Gigmes

Published by the Royal Society of Chemistry, www.rsc.org

CHAPTER 1

The History of Nitroxide-mediated Polymerization

GRAEME MOAD* AND EZIO RIZZARDO

CSIRO Manufacturing Flagship, Bag 10, Clayton South, Vic 3169, Australia
*Email: graeme.moad@csiro.au

1.1 Introduction

This chapter traces the early history of nitroxide-mediated polymerization (NMP) during its first ~15 years. It begins with a short prehistory of observations made during studies on defining initiation mechanisms using nitroxide radical trapping that can be seen to have inspired the initial experiments. The main part of the chapter is devoted to an account of the discovery of NMP in the early 1980s and research carried out in the period through to 1993, which saw most aspects of the mechanism defined and the process exploited at CSIRO, mainly in the synthesis of acrylic block copolymers. Many nitroxides for NMP were evaluated, including TEMPO (**1.1**). However, those found to be more effective and which were most used were 1,1,3,3-tetraethylisoindolin-*N*-oxyl (**1.2**) and di-*t*-butyl nitroxide (**1.3**) (Figure 1.1). We also provide a brief summary of developments in both the patent and open literature during the period 1993–2000 when the process came to the attention of the wider polymer community, mainly for the TEMPO-mediated polymerization of styrenics. The end of this period saw the discovery of nitroxides such as SG1 (**1.4**) and TIPNO (**1.5**) (Figure 1.2), which provided utility and versatility to NMP.

RSC Polymer Chemistry Series No. 19
Nitroxide Mediated Polymerization: From Fundamentals to Applications in Materials Science
Edited by Didier Gigmes

Published by the Royal Society of Chemistry, www.rsc.org

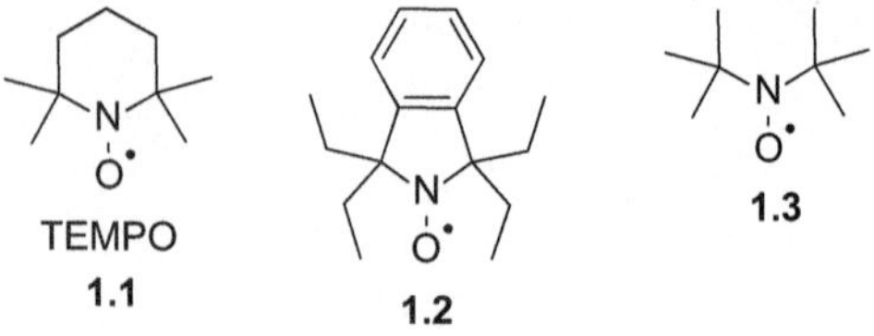

Figure 1.1 Structures of nitroxides favored in early studies of nitroxide-mediated polymerization.

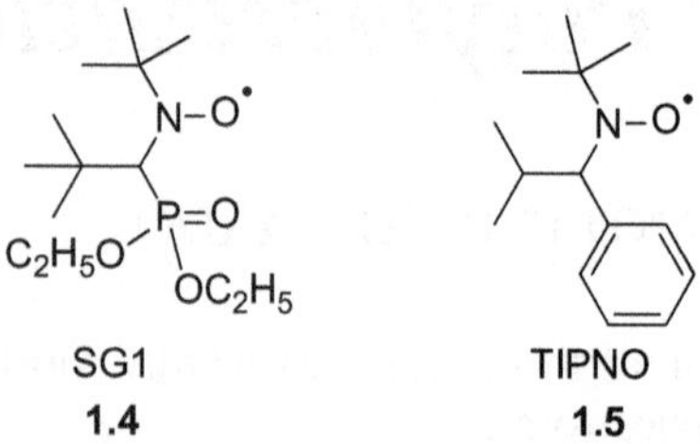

Figure 1.2 Structures of α-hydrido nitroxides commonly known as SG1 and TIPNO.

The term "nitroxide" is discouraged in IUPAC nomenclature, which instead recommends the term "aminoxyl". The IUPAC recommended term for "nitroxide-mediated polymerization" (NMP) is "aminoxyl-mediated radical polymerization" (AMRP).[1] However, in keeping with the historical context, the terms in common use are used throughout this chapter.

1.2 Radical Polymerization

Over the past 20 years, radical polymerization has proved to be one of the most active and fertile fields for research into polymer synthesis. The growth of interest in radical polymerization over this period can be largely attributed to the development of techniques for reversible deactivation radical polymerization (RDRP),[1] which impart living character to the process. These techniques include nitroxide-mediated polymerization (NMP—*vide infra*), atom transfer radical polymerization (ATRP),[2–5] reversible addition-fragmentation chain transfer (RAFT),[6–13] and tellurium-mediated radical polymerization (TERP).[14] Papers on these methods now account for more than two-thirds of all papers on radical polymerization.[15]

NMP was discovered at CSIRO, in the then Division of Applied Organic Chemistry, and lodged as an Australian patent application *Alkoxyamines useful as initiators* in 1984. The first publication on NMP was a European patent application, *Free radical polymerization and the produced polymers*, which appeared in 1985.[16] In these documents, the process was described as a method for controlled-growth radical polymerization. The introduction[16] states "The present invention relates generally to improved processes for free radical polymerization, particularly to improved processes in which it is possible to control the growth steps of the polymerization to produce relatively short chain length homopolymers and copolymers, including block and graft copolymers, and further relates to new initiators which find particular application in the improved processes," and goes on to define suitable alkoxyamine initiators. The process was said to have living character and be particularly suited for the preparation of well-defined short-chain or oligomeric chains with M_n in the range 500–5000. However, the preparation of higher molecular weight polymers was also outlined.

NMP was briefly described in our review entitled *Other initiating systems*, which appeared in *Comprehensive Polymer Science* in 1989.[17] However, the origins of the process at CSIRO are given little mention in most reviews of NMP.[18–29] Some details of the discovery are revealed in articles by Solomon[30] and Rizzardo and Solomon.[7]

In the 1980s, radical polymerization was possibly the most widely used processes for the commercial production of high molecular weight polymers.[15] Radical polymerization provides the ability to polymerize a vast array of monomers. This versatility can be attributed to the technique's tolerance of unprotected functionality in monomer and solvent, its compatibility with a variety of reaction conditions, and the relative simplicity and low cost of implementation. However, use of the conventional process imposes severe limitations on the degree of control that can be asserted over features such as molecular weight distribution, copolymer compositions and macromolecular architecture.

Conventional radical polymerization is a chain reaction (Scheme 1.1).[31] Chains are initiated by radicals, formed from an initiator, adding to monomer. These chains propagate by sequential addition of monomer units. Chain termination occurs when the propagating radicals self-react by combination or disproportionation. Continuous initiation and termination

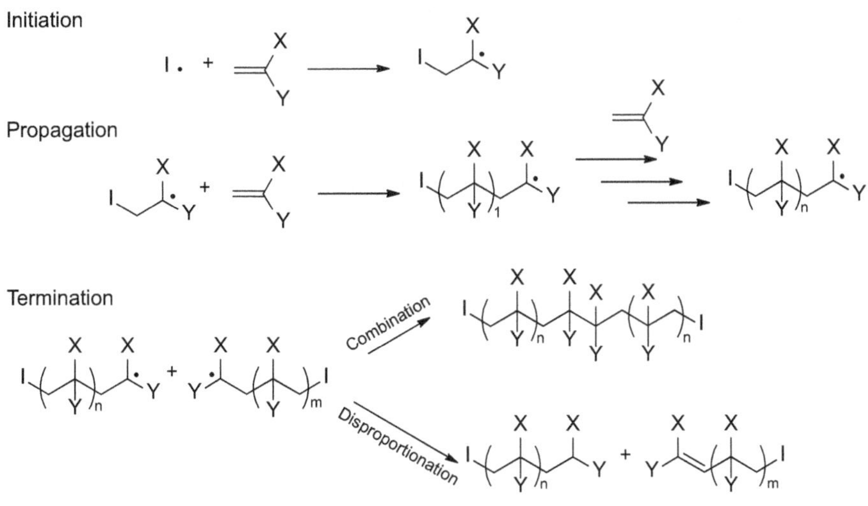

Scheme 1.1 Ideal reaction scheme for conventional radical polymerization.

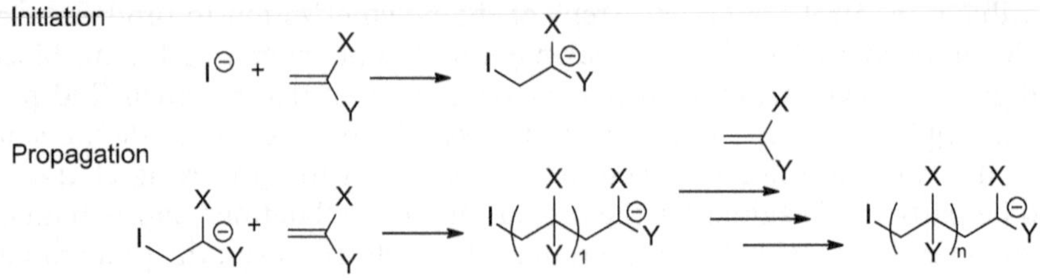

Scheme 1.2 Reaction scheme for living anionic polymerization.

provides a steady-state radical concentration of only ~10^{-7} M and the lifetimes of individual chains are typically ~5–10 seconds within a reaction span that may be many hours. In the absence of chain transfer, the lengths of the chains formed during the early stages of polymerization are high. The breadth of the molecular weight distribution is governed by statistical factors. The dispersity ($Đ$), the ratio of weight to number average molecular weights (M_w/M_n), is ideally 2.0 if termination is by disproportionation or chain transfer, or 1.5 if termination is by combination.[31]

In marked contrast, in a living polymerization, all chains are initiated at the beginning of the process, grow at a similar rate and all survive the polymerization (Scheme 1.2). By definition, there is no termination or irreversible chain transfer.[32–34] If initiation is rapid with respect to propagation, the molar mass distributions should be very narrow, approaching a Poisson distribution. In a living polymerization, chains can be extended indefinitely with the provision of monomer and conditions to support polymerization. In a conventional radical polymerization, the propensity of radicals to undergo self-termination means that all chains cannot be simultaneously active.

The first examples of what we now recognize as stable radical-mediated polymerization (SRMP) mediated by dithiocarbamyl radicals were reported by Otsu[35,36] in 1956–1957. Further examples of what may be considered SRMP mediated by diarylmethyl radicals were described by Braun and colleagues in the period 1970–1990.[37] Braun termed these processes "resuscitatable" radical polymerizations.

However, the concept of living radical polymerization (now known as RDRP)[1] was only introduced by Otsu and coworkers in 1982.[38–40] They recognized that radical polymerizations might display living attributes in the presence of reagents that are capable of reversibly deactivating active chains (propagating radicals, $P_n^•$) such that the majority of living chains are maintained in a dormant form (P_n-X). A further requirement is reaction conditions that support an equilibrium between active and dormant chains that is rapid with respect to propagation. The terms "initer" (for *ini*tiator-*ter*minator) and "iniferter" (for *ini*tiater-trans*fer* agent-*ter*minator) were introduced to describe the reagents used. A similar terminology ["inifer" (for *ini*tiater-trans*fer* agent)] had already been used by Kennedy[41,42] in describing cationic polymerizations with reversible deactivation.

1.3 Initiation Mechanisms in Radical Polymerization by Radical Trapping

Even though there are clear differences in the mechanism of radical (Scheme 1.1) and other processes for chain polymerization, for example, anionic polymerization (Scheme 1.2), polymers formed are usually represented by the same structure (Figure 1.3). This structure defines only the dominant repeat unit and ignores connectivity, end-groups and side reactions during propagation.

This idealized structure has deficiencies when it comes to rationalizing certain polymer properties. For example, poly(methyl methacrylate) (PMMA) synthesized by anionic polymerization can be more stable than the (apparently) same polymer made by radical polymerization.[43] The stability of PMMA made by radical polymerization depends on the initiator used and other details of its preparation. Similar observations have been made for other polymers, including poly(vinyl chloride) (PVC) and polystyrene (PSt).[44] These observations led to a recognition that chain polymers contain structural irregularities, which include the structures formed by chain initiation and termination (Scheme 1.1).[45–47]

At CSIRO, these issues prompted the application and development of methods for probing the detailed chemistry of initiation of radical polymerization. Amongst these was the radical-trapping method making use of nitroxides [the IUPAC recommended term for nitroxide is aminoxyl].[48] A now well-known feature of the chemistry of nitroxides (*e.g.* **1.1**–**1.8**) is that they combine with carbon-centred radicals to give alkoxyamines at close to diffusion controlled rates.[49–53] This property led to the use of nitroxides in the so-called inhibitor method for the determination of initiator efficiency[54] and was the basis of the radical-trapping method.

The radical trapping method using nitroxides had been developed at CSIRO in the late 1970s in response to a need to be able to quantitatively characterize radical reactions.[48] The use of spin-trapping using nitroso-compounds and nitrones had already been explored by several research groups in this context[55] and had been used to study the initiation of polymerization.[56–61] However, that method was not generally regarded as quantitative due to the complication of various side reactions. One side reaction is the formation of stable alkoxyamines by the further reaction of the nitroxides formed to scavenge carbon-centred radicals.[55] There was other literature to indicate that the nitroxides were able to selectively scavenge carbon-centred radicals[62] and limited kinetic data to indicate that the rate of reaction was extremely rapid (k_c 10^7–10^9 M^{-1} s^{-1}).[63] Nitroxides were also known to be effective inhibitors of radical polymerization.[62] Thus a stable

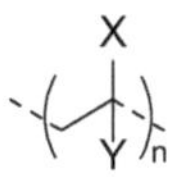

Figure 1.3 Generic polymer structure.

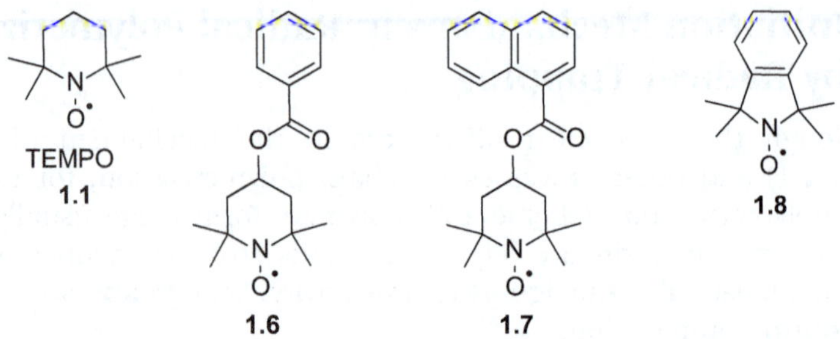

Figure 1.4 Nitroxides used in radical trapping experiments.

nitroxide, in particular 2,2,6,6-tetramethylpiperidin-1-oxyl (TEMPO, **1.1**) (Figure 1.4), was explored as a radical trap initially to study initiation pathways in methyl acrylate (MA) polymerization.[48] This and subsequent studies confirmed that nitroxides selectively scavenge carbon-centred radicals to yield stable alkoxyamines (under the conditions used and when isolated) while oxygen-centred radicals either did not react or reacted reversibly with the nitroxide.

Largely over the period 1979–2000, the radical trapping technique was successfully used to define the initiation pathways for the reactions of mainly oxygen-centred (*t*-butoxy,[48,64–81] cumyloxy,[82] other *t*-alkoxy,[83–87] isopropoxy,[88,89] ethoxy,[89] benzoyloxy,[70–72,77,79,90–92] isopropoxycarbonyloxy,[70] hydroxy,[93,94]), more reactive carbon-centred radicals (methyl,[81] *t*-butyl,[81,95] phenyl[71,72,77,78,92]) and cyanoisopropyl radicals[85,96] with a range of monomers. The nitroxides used in these studies included 2,2,6,6-tetramethylpiperidin-1-oxyl (TEMPO, **1.1**) and its derivatives (**1.6**,[71] **1.7**[73]), and the isoindoline nitroxide, 1,1,3,3-tetramethylisoindolin-2-oxyl (**1.8**).[71,97] The nitroxides **1.6–1.8** (Figure 1.4) possess a UV chromophore facilitating chromatographic detection.

A number of observations were made when applying the radical-trapping method with nitroxides that led to the development of NMP (note that most of the references cited below postdate the invention of on NMP—the original observations were not published or were published later).

- Certain alkoxyamines appeared unstable during isolation or subsequent handling.[77,98] This was sometimes indicated by color development or the appearance of the characteristic absorbance of the corresponding nitroxide.
- Certain alkoxyamines were observed to equilibrate to a mixture of isomers on heating or on standing for prolonged periods. For example, the isolated "styrene dimer" alkoxyamines shown in Scheme 1.3[99] undergo *cis–trans* isomerization to form the same mixture of isomers on heating. The allylic alkoxyamines shown in Scheme 1.4 isomerize with a 1,3-shift of the nitroxide functionality.[100] In both examples, the findings can be understood if the alkoxyamines undergo reversible homolytic dissociation as shown.

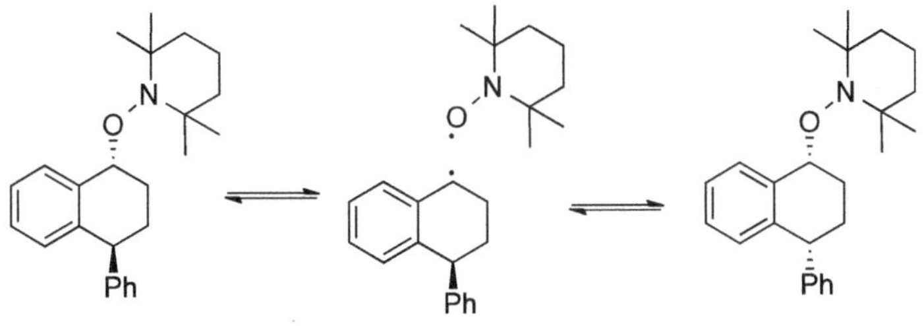

Scheme 1.3 Mechanism of *cis–trans* isomerization of "styrene dimer" alkoxyamines.

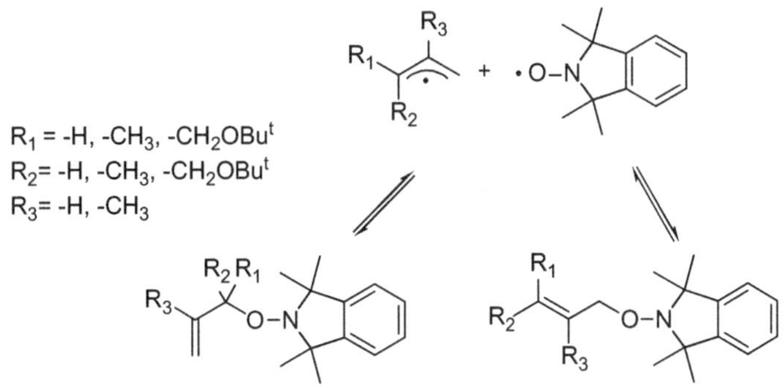

Scheme 1.4 Mechanism of isomerization of allylic alkoxyamines.

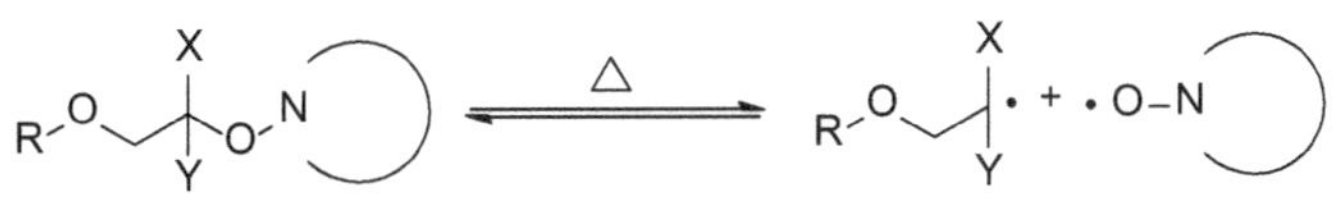

Scheme 1.5 Activation–deactivation equilibrium in nitroxide-mediated polymerization.

- In some trapping experiments, the formation of small amounts of oligomers was observed.[72,82,92] This was attributed to the rate of trapping of the initiating radical being slow with respect to propagation. The yield of propagation products was consistent with the known kinetics of propagation and trapping, and this remains the likely explanation. NMP was unlikely under the conditions used for the trapping experiments because of the nitroxide used (**1.1** or **1.8**), the significant excess of nitroxide (~10% at complete initiator consumption), and the low reaction temperatures used (60 °C). Nonetheless, the observations led us to consider other possibilities.

These and similar observations suggested that the alkoxyamines formed were thermally labile, perhaps undergoing reversible dissociation as shown in Scheme 1.5. This in turn suggested the possibility of a method for controlled-growth radical polymerization based on alkoxyamine chemistry with the potential to open the door to structures that were not readily approached by conventional radical polymerization.

1.4 First Examples of Nitroxide-mediated Polymerization

The concept of NMP as a method for living radical polymerization was first disclosed in a Plenary lecture by Ezio Rizzardo at the 14th Australian Polymer Symposium, which was held in Ballarat in February 1984[101] only a few days after the first patent application had been lodged.

The alkoxyamines based on the nitroxides used in the trapping work (**1.1** and **1.6–1.8**) were not very effective in NMP and required the use of relatively high reaction temperatures for homolysis, particularly, with acrylates and styrene. It was reasoned that the rate of homolysis should be lowered by steric congestion. Therefore, the first successful NMP experiment (Scheme 1.6) made use of the tetraethylisoindoline nitroxide **1.2**, an analog of the tetramethylisoindoline nitroxide **1.8** that had been used in the trapping experiments. The patent[102] generically covered the use of nitroxides of general structure **1.9** in NMP with specific examples being **1.1–1.3** (Figure 1.1) and **1.10–1.13** (Figure 1.5).

1.4.1 Homopolymer Synthesis

The first successful NMP experiment carried out in 1982, detailed as Example 23 of the original patent application,[102] is summarized in Scheme 1.6. It was found that heating the alkoxyamine **1.19** with MA in benzene (50% v/v) solution at 80 °C provided complete conversion of the alkoxyamine to the single-unit monomer insertion product **1.20**). When the same alkoxyamine **1.19** was heated at 80 °C in bulk MA, ~7 units of MA were inserted to give the heptamer **1.21**. In both cases, no further reaction (oligomerization) was observed when the mixtures were heated at 80 °C for longer times.

These observations are explained as follows. At 80 °C, the alkoxyamine **1.19** dissociates and MA units add; the number is dictated by the relative concentrations of MA and nitroxide. Combination of the propagating species with nitroxide gives alkoxyamine **1.20** or **1.21**. These insertion products **1.20** and **1.21** are stable at 80 °C, so no NMP was observed.

NMP required higher reaction temperatures, sufficient to allow reversible dissociation of the alkoxyamine products. When the heptamer **1.21** was heated at 100 °C, slow NMP was observed such that a 14-mer was obtained after 1.5 h. Faster NMP took place at 120 °C providing a 70-mer after 1.5 h. NMP of styrene (Scheme 1.7) was also successful at 100 °C. A 4.5-mer macroalkoxyamine was obtained after 1 h and a 12-mer after 2 h at 100 °C. The mechanism proposed for NMP is shown in Scheme 1.8.

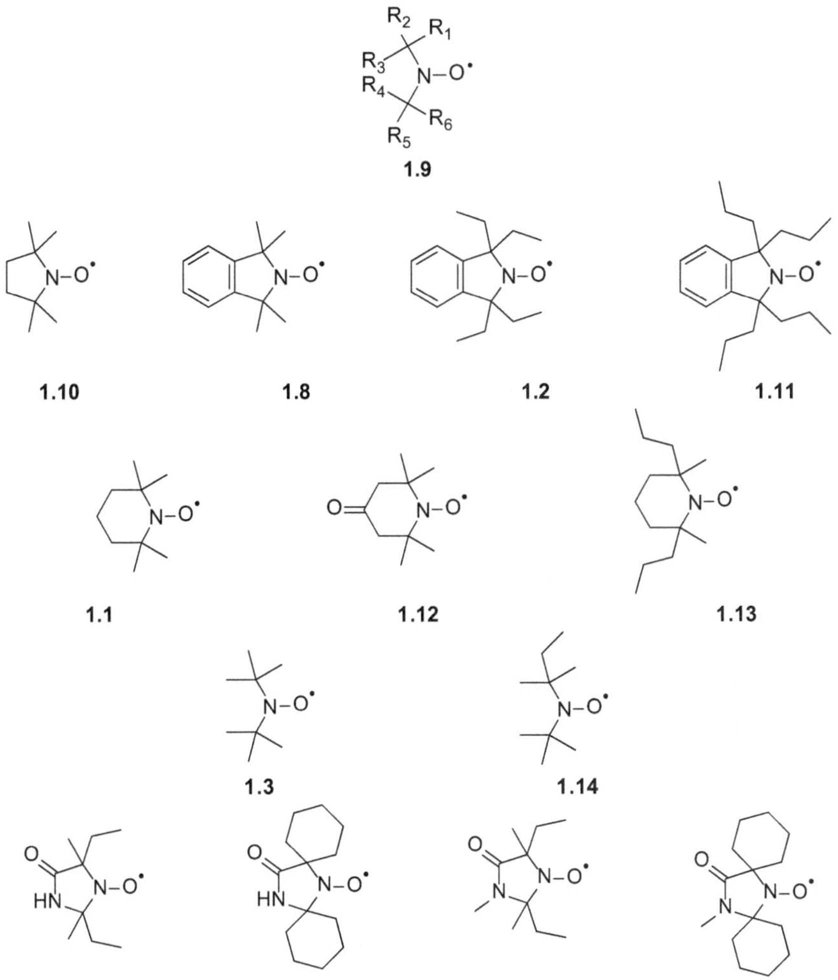

Figure 1.5 Nitroxides used in nitroxide-mediated polymerization (NMP).

NMP of methyl methacrylate (MMA) provided a low dispersity macromonomer rather than a macro-alkoxyamine as the isolated product (Scheme 1.9, Scheme 1.10). Two explanations for this finding were proposed:[102] (a) that the tertiary alkoxyamine underwent thermal elimination to provide the macromonomer and hydroxylamine and/or (b) that the PMMA propagating species gave some disproportionation with nitroxide to give the same products (Scheme 1.11).

A MMA-based macro-alkoxyamine was prepared by NMP of MMA in the presence of styrene (Scheme 1.12). Since the macro-alkoxyamine with a terminal styrene is stable at 60 °C, under these reaction conditions, the molecular weight is determined by the styrene concentration and the

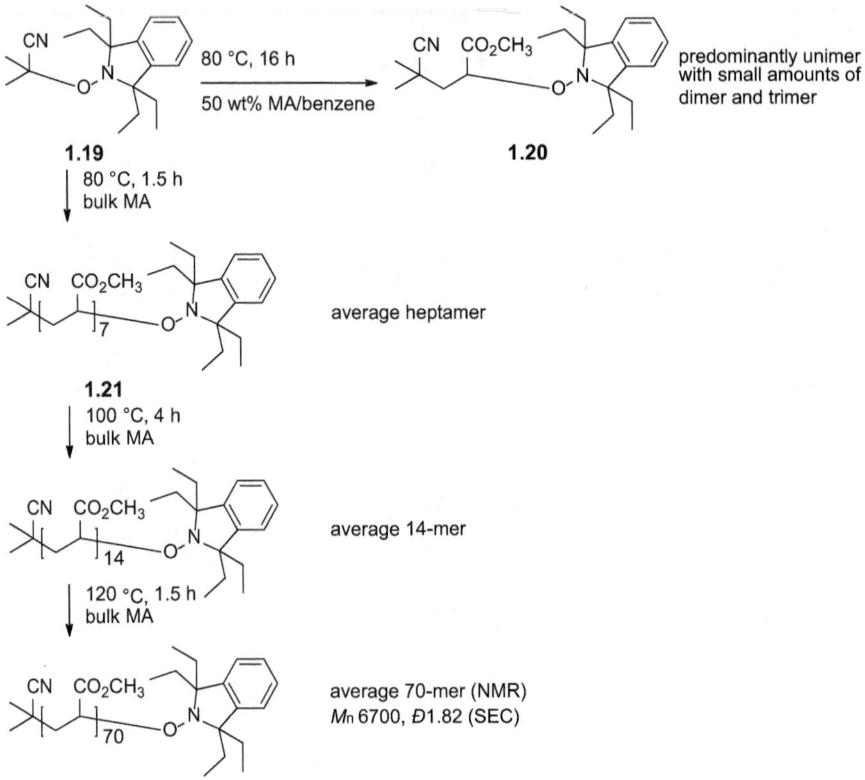

Scheme 1.6 Nitroxide-mediated polymerization of methyl acrylate (MA). Example 23 of the original patent application.[102]

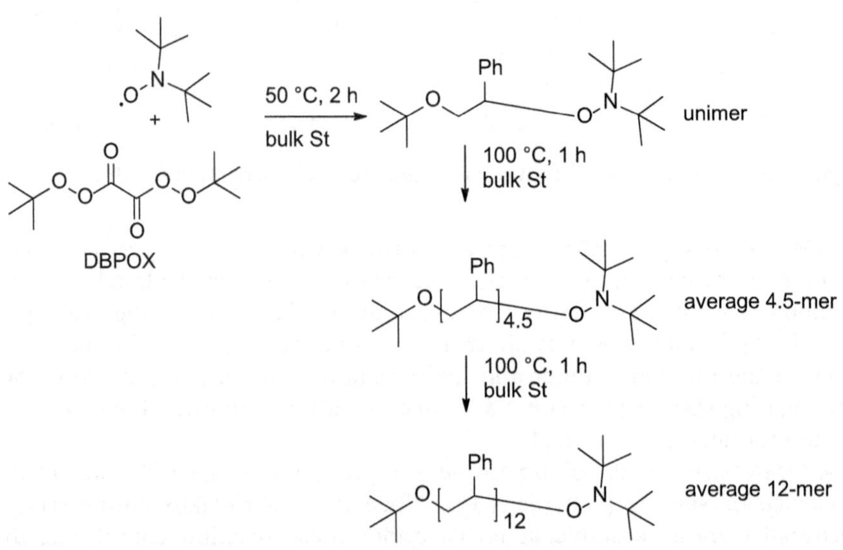

Scheme 1.7 Nitroxide-mediated polymerization of styrene (St). Example 15, St unimer preparation, and Example 27 of the original patent application.[102]

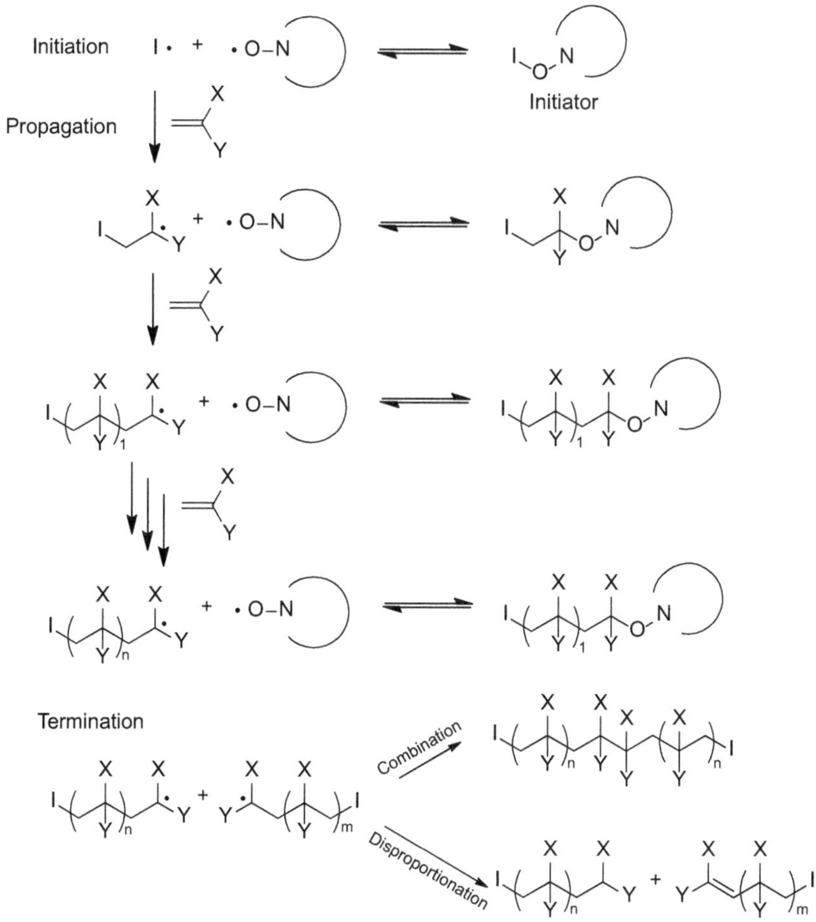

Scheme 1.8 Mechanism for nitroxide-mediated polymerization (NMP) proposed in the CSIRO patent application.[102]

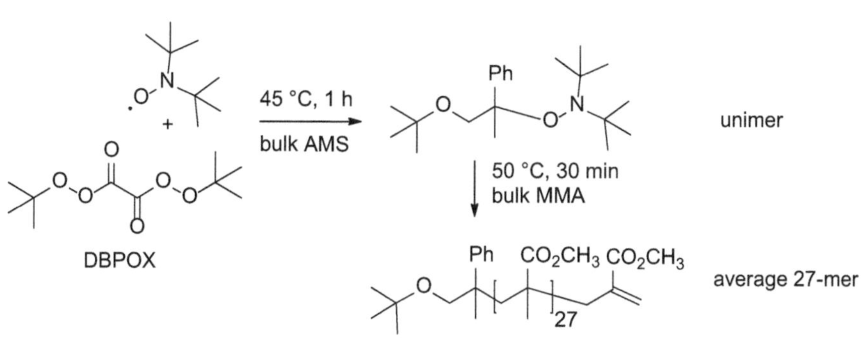

Scheme 1.9 Nitroxide-mediated polymerization of methyl methacrylate (MMA). Example 13, α-methylstyrene (AMS) unimer preparation, and Example 33 of the original patent application.[102]

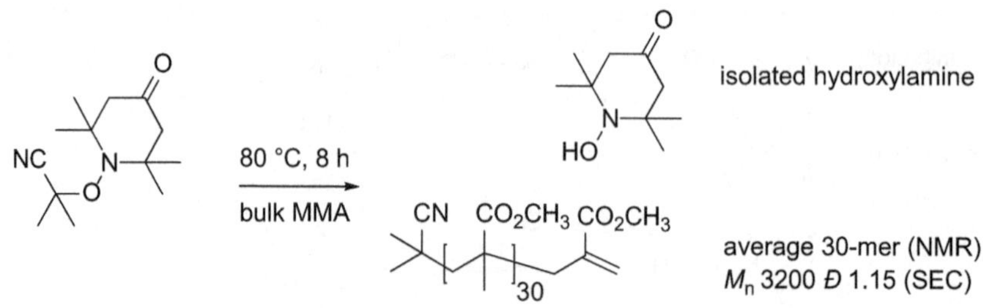

Scheme 1.10 Nitroxide-mediated polymerization of methyl methacrylate (MMA). Example 34 of the original patent application.[102]

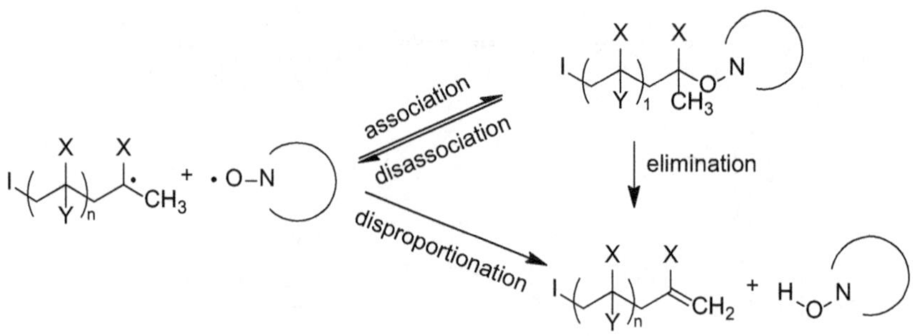

Scheme 1.11 Possible mechanisms for end-group loss during NMP of α-methyl vinyl monomers.[102]

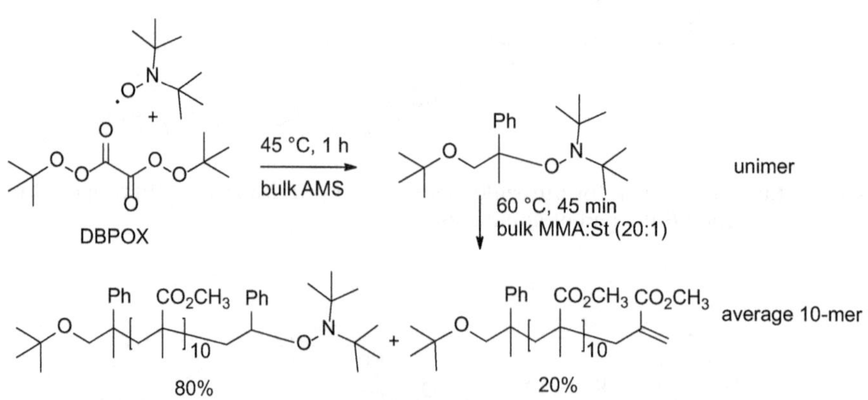

Scheme 1.12 Nitroxide-mediated polymerization of methyl methacrylate (MMA) in the presence of styrene. Example 13, α-methylstyrene (AMS) unimer preparation, and Example 31 of the original patent application.[102]

monomer reactivity ratios. Nitroxide-mediated copolymerization of MMA and EA was also successful in providing a macro-alkoxyamine with a terminal EA unit (Scheme 1.13).

In later work,[103,104] a comparative study of a series of nitroxides in controlling MMA polymerization at 90 °C was performed in experiments where

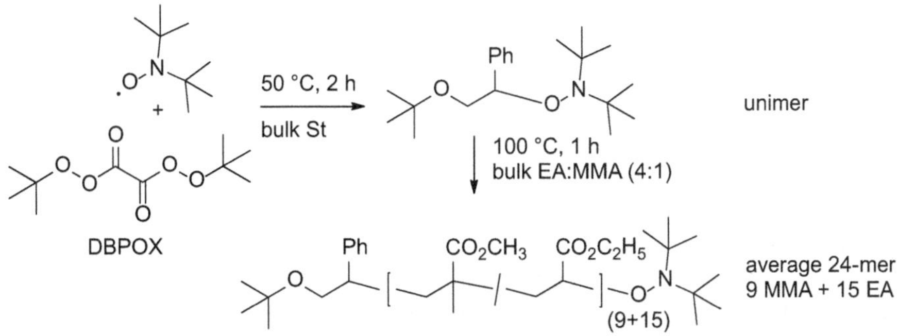

Scheme 1.13 Nitroxide-mediated copolymerization of ethyl acrylate (EA) and methyl methacrylate (MMA). Example 32 of the original patent application.[102]

Table 1.1 Molecular weight/conversion data for polymerizations (90 °C, bulk) of methyl methacrylate in the presence of various nitroxides and with azobis(2,4-dimethyl-2-pentanenitrile) as initiator.[a,103,104]

Nitroxide	M_n[b]	Đ	Conv. %	[nitroxide] M	M_n(calc)[c]
1.15	35 700	1.57	38	0.0077	36 800
	33 800	1.65	36	0.0069	34 500
	20 400[d]	1.70	19	0.0077	18 500
1.17	31 021	1.63	17	0.0077	18 200
1.16[e]	6500	1.44	6	0.0077	6100
1.18[e]	28 800	1.89	21	0.0077	22 500
1.2	18 300	1.71	18	0.0079	17 400
	5600	1.68	7	0.0115	5900
1.8	10 500	2.24	11	0.0076	10 800
1.12	17 000	3.30	18	0.0077	16 900
1.1	22 100	3.11	14	0.0076	13 800
1.14	19 200	4.10	18	0.0076	17 200

[a][Initiator] 0.0054 M.
[b]Number average molecular weight in polySt equivalents.
[c]Calculated molecular weight based on an initiator efficiency of 90%. Values shown have been rounded to the nearest hundred.
[d]Reaction temperature 100 °C.
[e][initiator] 0.0049 M.

the alkoxyamine was formed *in situ*.[103,104] Azobis(2,4-dimethyl-2-pentanenitrile) was used as the initiator since it has a very short half-life at 90 °C (<20 sec).[105] The results (conversions, dispersities, molecular weights) obtained in these experiments are summarized in Table 1.1. The lowest dispersities, in the range 1.4–1.7, were obtained with the 5-membered ring nitroxides. These dispersities are not exceptional but compare very favorably with the values in the range 3.1–4.1 obtained with the 6-membered ring nitroxides, TEMPO (**1.1**) and its 4-oxo derivative **1.12**, or the open chain nitroxide **1.14**. In these experiments, polymerization ceased after less than one hour reaction time. Best conversions (up to 38%) were obtained with the imidazolidinone nitroxide **1.17**. All other experiments in the series gave monomer conversions <20%.

The findings are consistent with NMP of MMA proceeding only until the nitroxide concentration builds to a level so as to inhibit further polymerization. After this time, reversible dissociation of the alkoxyamine continues but little propagation occurs and the propagating radicals ultimately disproportionate with nitroxide. Kinetic simulations[103] supported this hypothesis and indicated that the dependence on nitroxide type could be attributed to the combination : disproportionation ratio for the reaction the MMA propagating species and nitroxide, which is lower in the case of the 5-membered ring nitroxides than it is with 6-membered ring or open chain nitroxides. The 'worst case' combination : disproportionation ratio consistent with the experimental results was <0.01.[103]

It is worth noting that macromonomers have wide utility in the synthesis of graft copolymers,[106] as transfer agents,[107] and as so-called macromonomer RAFT agents.[108,109]

The molecular weight dispersities achieved in most of the examples of NMP in the patent are high with respect to what we have come to expect from effective RDRP. In part, this can be attributed to the polymerization conditions and to the relatively low molecular weights that were targeted. Nonetheless, the patent contains an example of polyMMA with $Đ$ of 1.15 and several polymers with $Đ<1.5$. The dispersity of chains formed by conventional radical polymerization under ideal conditions with termination by disproportionation is 2.0.

1.4.2 Block Copolymer Synthesis

The experiments shown in demonstrated that the NMP had living characteristics in that chain extension, by re-subjecting the macro-alkoxyamine to the polymerization conditions, was possible. Nonetheless, it was important to provide further evidence and demonstrate block copolymer synthesis. In the patent, this was exemplified with the synthesis of polyEA-*block*-polyMA (Scheme 1.14) and polyMA-*block*-polyEA-*block*-polyMMA (Scheme 1.15). Note that the polyMMA block must be made last since the end-group is lost during NMP and the product is a macromonomer.

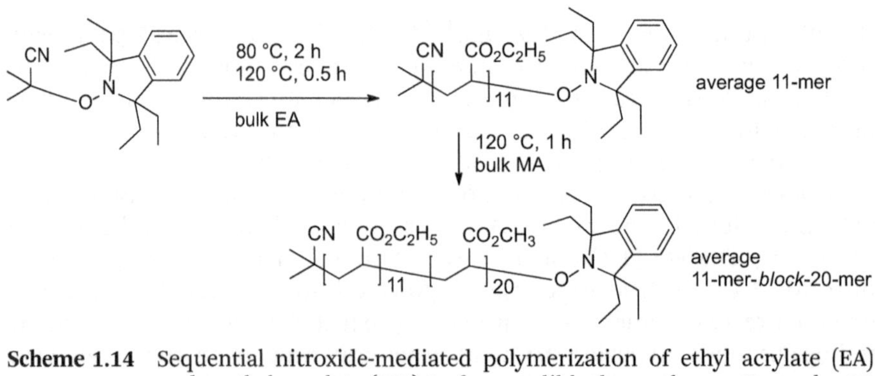

Scheme 1.14 Sequential nitroxide-mediated polymerization of ethyl acrylate (EA) and methyl acrylate (MA) to form a diblock copolymer. Examples 29 and 38 of the original patent application.[102]

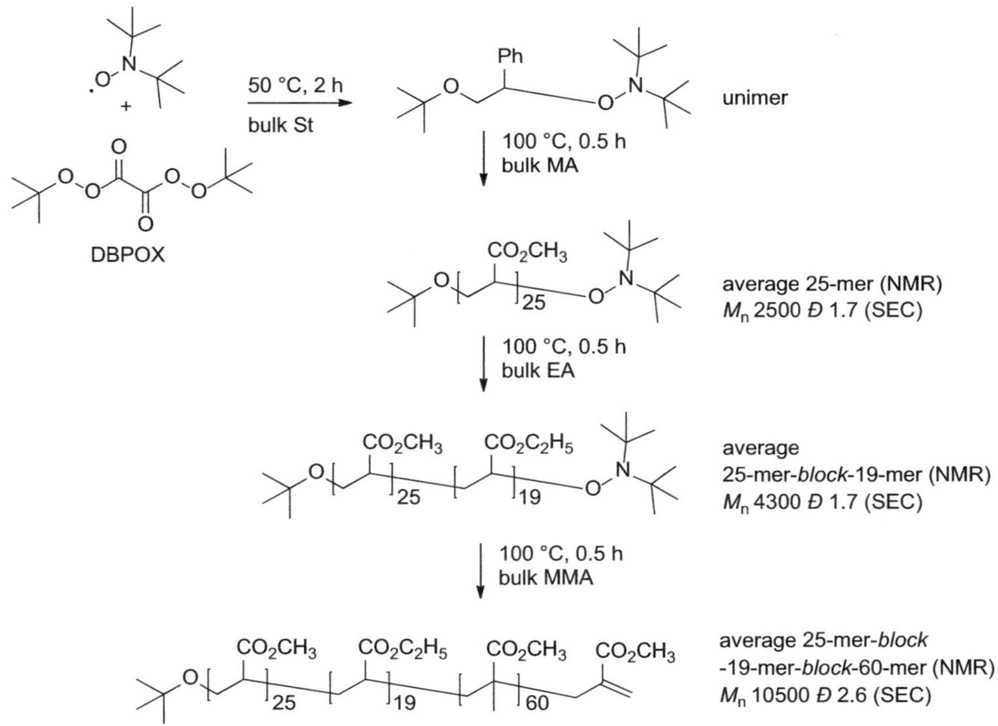

Scheme 1.15 Sequential nitroxide-mediated polymerization of methyl acrylate (MA), ethyl acrylate (EA) and methyl methacrylate (MMA) to form a triblock copolymer. Examples 26, 36 and 39 of the original patent application.[102]

A major application interest for NMP at CSIRO, dictated in part by our commercial partners, lay with the synthesis of low molecular weight acrylic acid–alkyl acrylate block copolymers. This research was briefly reported in two papers.[110,111] The methodology was applied in the multi-gram synthesis of block copolymers useful as pigment dispersants. For example, poly(BA-*co*-MA)-*block*-polyAA, as shown in Scheme 1.16.[112]

1.4.3 Graft Copolymer Synthesis

The CSIRO patent also demonstrated the utility of NMP in forming branched or graft copolymers from copolymers containing alkoxyamine functionality. The latter polymers could be formed in various ways, of which two were exemplified: (a) copolymerization of an appropriate functional monomer, and (b) trapping radicals formed on a polymer substrate. The process (a) is illustrated in Scheme 1.17.

1.4.4 Effects of Alkoxyamine Structure and Polymerization Conditions on Homolysis Rates

Following the success of the experiments shown in Scheme 1.6, much early work revolved around ascertaining the dependence of NMP on nitroxide structure. In 1995,[113] we reported a study on the kinetics of alkoxyamine decomposition based in large part on the experimental data from the patent[102] and two earlier papers.[114,115] The half-lives of alkoxyamines formed between various nitroxides and radicals were measured by a nitroxide-exchange process that involved following the rate of consumption of the alkoxyamine (or appearance of product) when it was heated in the presence an excess of nitroxide **1.1** or **1.8** in ethyl acetate solution (Scheme 1.18, Table 1.2).[102,113] The reaction between the carbon-centred radicals and **1.1** or **1.8** is essentially irreversible under the measurement conditions such that the exchange could be conveniently followed by HPLC. The concentration of nitroxides **1.1** or **1.8** was chosen to be in excess such that the rate of disappearance of alkoxyamine is a direct measure of the homolysis rate constant (*i.e.* there was no significant likelihood of recombination to form the original alkoxyamine). It was later found that similar data could be obtained simply by heating the alkoxyamines in air when the released radicals were scavenged by oxygen.

The first NMP experiments showed that homolysis rates increased with an increase in steric bulk of the substituents α to the nitroxide nitrogen. The nitroxide exchange experiments confirmed this hypothesis. The experiments also showed that alkoxyamine homolysis rates increased with increasing ring size, for the cyclic nitroxides, and that alkoxyamines based on the open chain nitroxide (**1.3**) offered the fastest homolysis rates amongst those studied at the time. These findings were in accord with bond dissociation energies and activation energies for C–O bond homolysis estimated with

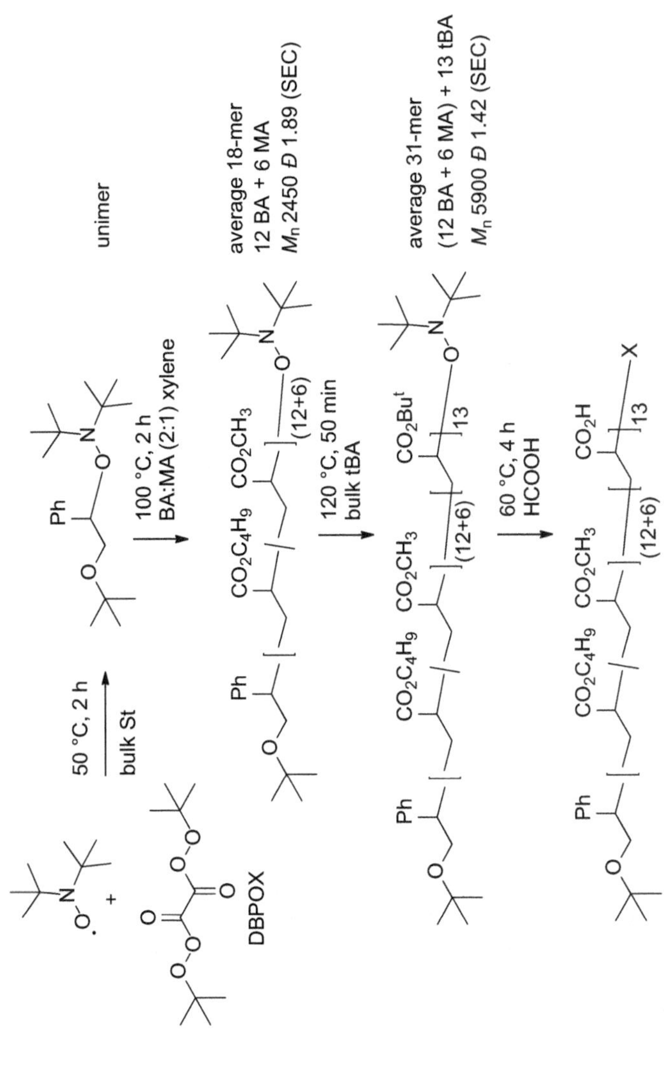

Scheme 1.16 Sequential nitroxide-mediated copolymerization of methyl acrylate (MA) and *n*-butyl acrylate (BA) followed by polymerization of *t*-butyl acrylate (tBA) to form a diblock copolymer. The tBA units were converted to acrylic acid units by heating with formic acid. The end-groups X in the final polymer were not determined.[112]

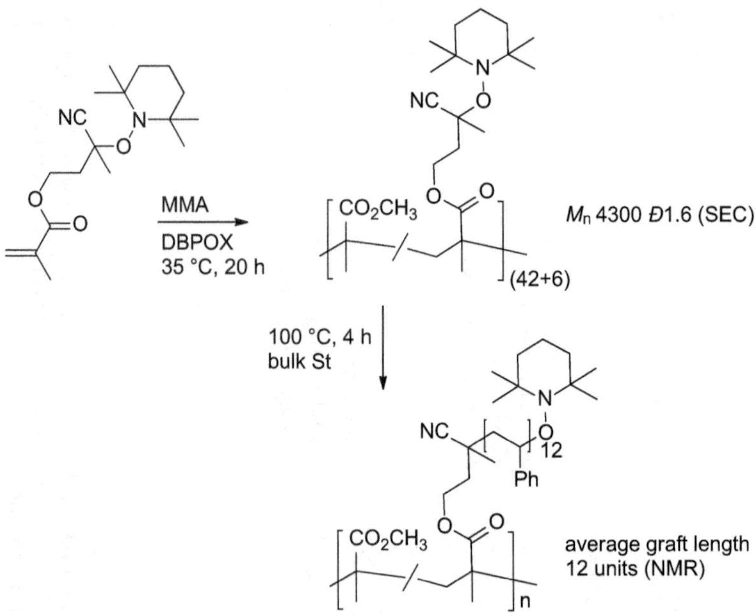

Scheme 1.17 Synthesis of alkoxyamine functional methyl methacylate (MMA) copolymer by conventional radical copolymerization at low temperature initiated by di-*t*-butyl peroxyoxalate (DBPOX) and nitroxide-mediated graft copolymerization of styrene (St). Examples 21 and 44 of the original patent application.[102]

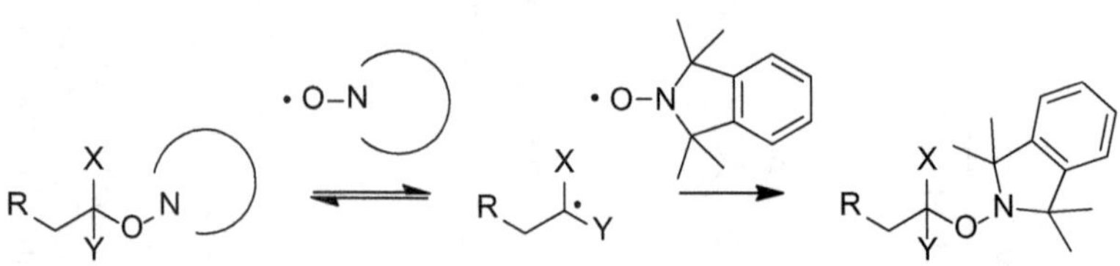

Scheme 1.18 Nitroxide exchange experiment used to determine alkoxyamine half-lives.

molecular orbital calculations. Later work showed that rates for combination exhibit the inverse trend, *i.e.*, 5-membered ring > 6-membered ring > open chain, but are somewhat smaller.[53]

The effect of alkyl group structure on homolysis rates was also examined.[113] Alkoxyamines based on tertiary radicals with an α-methyl substituent gave substantially faster homolysis rates than analogous secondary radicals (Figure 1.6).

1.4.4.1 *Effect of Polymerization Conditions on Homolysis Rates*

The reaction medium was also found to have a significant effect on the rate constant for C–O bond homolysis. Half-life data at 60 °C for the alkoxyamine

Table 1.2 Half-lives (min) of alkoxyamines[a] (R–X) in ethyl acetate solution.[102,113]

R	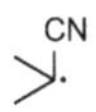	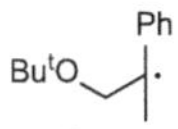	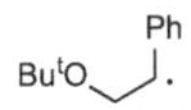	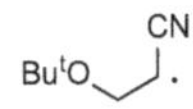	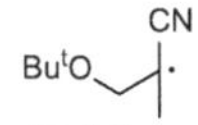	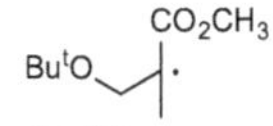
X	60 °C	60 °C	80 °C	90 °C	60 °C	60 °C
1.10	280	—	—	—	—	—
1.8	780	—	—	—	—	—
1.2	33	75	>1000	—	55	123
1.11	31	—	—	—	—	—
1.1	65	—	—	—	—	—
1.13	—	10	400	—	—	—
1.3	<2[b]	~2[c]	70[d]	105	—	—
1.14[297]	—	—	25	—	—	—
1.15[297]	—	—	180[e]	—	—	—

[a]Alkoxyamines (R–X) formed between nitroxide (X) and radical (R) indicated.
[b]Half-life 8.5 min at 40 °C.
[c]Half-life is 18 min at 40 °C.
[d]Half-life is 22 min at 90 °C.
[e]At 110 °C.

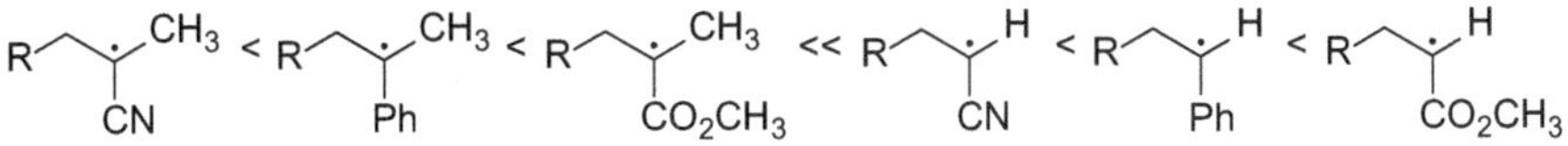

Figure 1.6 Effect of alkyl group structure on homolysis rates.

Table 1.3 Solvent dependence of homolysis and trapping reactions.

Solvent	Half-life[a] (min)	k_H[b] (s^{-1})	$k_T/10^7$[c] ($M^{-1}\ s^{-1}$)
Hexane	38	0.00062	48
Ethyl acetate	33	0.00071	17
Acetonitrile	22	0.00107	9.5
Dimethylformamide	20	0.00118	—
Methanol	17	0.00139	13
10% (v/v) Water in methanol	16	0.00147	—
10% (v/v) Acetic acid in methanol	15	0.00157	—

[a]Half-life of alkoxyamine **1.19** formed between 2-cyanoprop-2-yl and nitroxide **1.2** at 60 °C.[113]
[b]Rate constant for alkoxyamine homolysis at 60 °C derived from half-life value.
[c]Rate constant for trapping of benzyl radical at 18 °C with TEMPO (**1.1**).[116]

formed between nitroxide **1.2** and the cyanoisopropyl radical determined by the radical exchange method are presented in Table 1.3.[113] Rate constants for C–O bond homolysis (k_H) estimated on the basis of these half-lives and rate constants for the reverse trapping reaction (k_T)[116] are also provided. Clear trends for an increase in alkoxyamine dissociation rate (decrease in half-life) and increase in association or trapping with increasing solvent polarity were apparent. Significantly higher alkoxyamine dissociation rates were observed for NMP in acidic media (acetic acid in methanol).

1.4.4.2 *Kinetics of Nitroxide-mediated Polymerization*

The first full paper from CSIRO on NMP, published in 1990,[117] dealt with kinetic simulation of NMP of acrylates. The paper showed that high molecular weight polymers with very low dispersities were feasible and indicated some of the conditions that should be met to achieve this. The paper described the persistent radical effect, but did not attach a name to the phenomenon or associate it with the theory that had been previously expounded by Fischer. Fischer had described the persistent radical effect in radical reactions in a non-polymerization context in 1986.[118] A full description of the persistent radical effect in RDRP, in particular NMP, was not published by Fischer until 1997.[119] The kinetics of NMP will not be described in this chapter since they are detailed in another chapter of this volume.

1.4.5 End-group Removal

Three processes for end-group removal from macro-alkoxyamines were described in the original CSIRO patent application[102] and are summarized in Scheme 1.19. The first simply involves heating the macro-alkoxyamine in an α-methylvinyl monomer, such as MMA, to convert the macro-alkoxyamine to a macromonomer and a hydroxylamine. While the hydroxylamine is a good hydrogen atom donor, its low concentration dictates that it is inert under normal reaction conditions for NMP. The hydroxylamine is isolable but it is converted to the nitroxide on exposure to air. This process also provides for a relatively simple method of nitroxide recycling. For an example of the chemistry, see Scheme 1.10.[102]

A second method for end-group removal involves heating the macro-alkoxyamine in the presence of an excess of a good hydrogen atom-donor. An excess of the hydrogen atom-donor is required so that hydrogen atom-transfer can compete with other processes (Scheme 1.20)

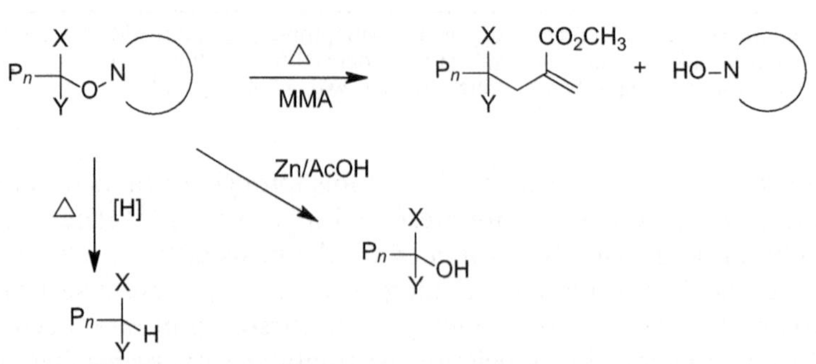

Scheme 1.19 Processes for end-group transformation of macro-alkoxyamines described in the CSIRO patent.[102]

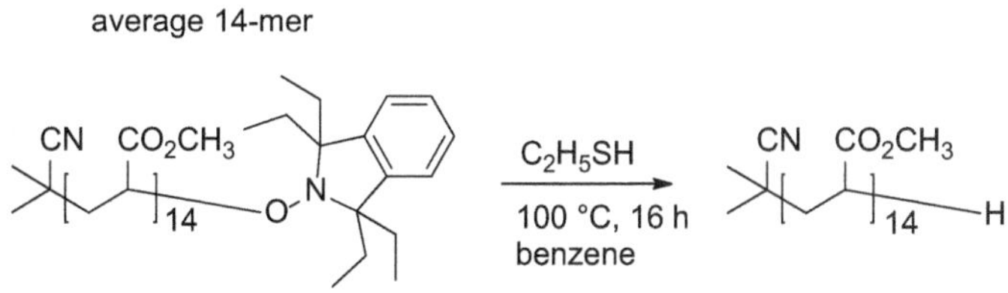

Scheme 1.20 Conversion of alkoxyamine end-group (of polymer shown in Scheme 1.6) to hydrogen end-group by heating in the presence of a thiol. Example 48 of original patent application.[102]

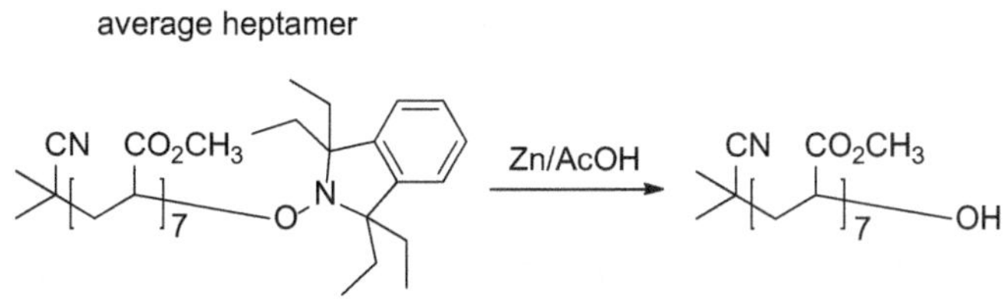

Scheme 1.21 Conversion of alkoxyamine end-group (of polymer shown in Scheme 1.6) to hydroxy end-group by zinc–acetic acid reduction. Example 48 of original patent application.[102]

The third method of end-group transformation involves reduction with zinc in acetic acid to provide a polymer with a hydroxyl chain end (Scheme 1.21).

1.5 Developments in NMP 1993–2000

In this section, we describe developments in NMP produced by various research groups over the period 1993–2000. We focus on reports in the open literature. However, it is informative to also examine the patent literature of the period. In Table 1.4, we provide a summary of 77 granted US Patents relating to NMP that were first filed during the period 1983–2000. The patents are listed in chronological order with respect to priority date. It will be seen that many publications in the open literature were preceded by patent filings. The list includes patents assigned to Xerox (28),[120–147] Elf Atochem/Atofina (8),[148–155] BASF (5),[156–160] Ciba (7),[161–167] Symyx (5),[168–172] Lubrizol (4),[173–176] IBM (3),[177–179] Bayer (3),[180–182] CSIRO/DuPont (2),[102,183] Carnegie-Mellon University (2),[184,185] Dow (2),[186,187] Melbourne University (2),[188,189] and others (BF Goodrich,[190] Kaneka,[191] Kansai Paint,[192] Shipley,[193] Sumitomo[194]; 1 each). There are additional patents in the field that were not filed in the US.

1.5.1 Xerox Group

Following on from the early CSIRO work, the next significant development in NMP stemmed from the publication in 1993 by Georges and coworkers[195] on

Table 1.4 US patents relating to nitroxide-mediated polymerization 1983–2000 (patents are listed in chronological order with respect to priority date).

Patent No.	Assignee	Priority date	US Patent Publication date	Comment
US4581429A[102]	CSIRO	11 Jul 1983	8 Apr 1986	Original CSIRO NMP patent
US5322912A[120]	Xerox	16 Nov 1992	21 Jun 1994	First Xerox NMP patent (BPO TEMPO in bulk or suspension)
US5401804A[121]	Xerox	16 Nov 1992	28 Mar 1995	First Xerox NMP patent divisional
US5549998AA[127]	Xerox	16 Nov 1992	27 Aug 1996	Xerox NMP patent divisional on toner compositions
US5412047A[122]	Xerox	13 May 1994	2 May 1995	Acrylate polymerization with 4-oxo-TEMPO
US6087451A[140]	Xerox	18 Aug 1994	11 Jul 2000	NMP with bis-alkoxyamines in for synthesis of telechelics
US6258911B1[144]	Xerox	18 Aug 1994	10 Jul 2001	Telechelics formed from use of bis-alkoxyamines in NMP
US5545504A[126]	Xerox	3 Oct 1994	13 Aug 1996	Toner compositions
US5714993A[131]	Xerox	3 Oct 1994	3 Feb 1998	Toner compositions for inkjet
US6320007B1[145]	Xerox	18 Nov 1994	20 Nov 2001	NMP
US5449724A[123]	Xerox	14 Dec 1994	12 Sep 1995	NMP of ethylene
US5530079A[125]	Xerox	3 Jan 1995	25 Jun 1996	SRMP with phenoxy radical mediator
US5610250A[130]	Xerox	3 Jan 1995	11 Mar 1997	Polymer from SRMP with phenoxy radical mediator
US5610249A[192]	Kansai Paint	24 Jan 1995	11 Mar 1997	NMP of styrene in presence of a phosphorus compound
US5498679A[124]	Xerox	30 Mar 1995	12 Mar 1996	NMP (multifunctional TEMPO alkoxyamines) to form star polymers
US5608023A[129]	Xerox	30 Mar 1995	4 Mar 1997	Use of sulfonic acids to enhance the rate of NMP
US5723554A[194]	Sumitomo	6 Sep 1995	3 Mar 1998	NMP of styrene
US5627248A[186]	Dow	26 Sep 1995	6 May 1997	bis-nitroxides in NMP
US5739229A[136]	Xerox	7 Nov 1995	14 Apr 1998	NMP with TEMPO plus an electron acceptor
US5552502A[128]	Xerox	16 Nov 1995	3 Sep 1996	NMP in supercritical SO_2
US6239226B1[156]	BASF	25 Jan 1996	29 May 2001	Acrylic block copolymer by NMP. No examples.
US5721320A[187]	Dow	29 Mar 1996	24 Feb 1998	NMP formation of polydiene block and graft polymers
US5723511A[132]	Xerox	17 Jun 1996	3 Mar 1998	Synthesis of branched polymer from unsaturated polymer substrate and macroalkoxyamine
US5919861A[138]	Xerox	17 Jun 1996	6 Jul 1999	Synthesis of branched polymer
US6114499A[141]	Xerox	17 Jun 1996	5 Sep 2000	NMP synthesis of star polymer from unsaturated polymer substrate

US5728747A[133]	Xerox	8 Aug 1996	17 Mar 1998	Photoinitiated NMP
US5910549A[184]	Carnegie-Mellon University	22 Aug 1996	8 Jun 1999	Conversion of ATRP catalyst to alkoxyamine
US6288186B1[185]	Carnegie-Mellon University	22 Aug 1996	11 Sep 2001	NMP rate enhancement with added initiator
US6075105A[139]	Xerox	26 Aug 1996	13 Jun 2000	NMP crosslinking polymerization
US6255422B1[149]	Elf Atochem	20 Nov 1996	3 Jul 2001	NMP with inifers
US6271340B1[183]	CSIRO/DuPont	10 Jan 1997	7 Aug 2001	NMP with imidazolinone nitroxodes
US6300443B1[188]	Melbourne University	15 Jan 1997	9 Oct 2001	Star microgel by NMP
US5744560A[134]	Xerox	21 Jan 1997	28 Apr 1998	NMP with TEMPO plus metal complex (Lewis acid)
US6281311B1[190]	BF Goodrich	31 Mar 1997	28 Aug 2001	Nitroxides for NMP
US6262206B1[161]	Ciba	15 Jul 1997	17 Jul 2001	Alkoxyamines from open chain nitroxides for NMP
US6111025A[174]	Lubrizol	24 Jun 1997	29 Aug 2000	Amphiphilic block copolymer by NMP
US6696533B1[160]	BASF	27 Jun 1997	24 Feb 2004	NMP in emulsion
US6156858A[143]	Xerox	25 Jun 1997	5 Dec 2000	NMP (BPO TEMPO) in presence of base
US6121397A[142]	Xerox	14 Jul 1997	19 Sep 2000	NMP in miniemulsion (TEMPO alkoxyamine)
US6365675B1[146]	Xerox	14 Jul 1997	2 Apr 2002	NMP in miniemulsion. Multiblock copolymer synthesis.
US6472485B2[147]	Xerox	14 Jul 1997	29 Oct 2002	NMP in miniemulsion (TEMPO alkoxyamine)
US6503983B1[159]	BASF	27 Jul 1997	7 Jan 2003	NMP in miniemulsion (TEMPO alkoxyamine)
US5817824A[135]	Xerox	1 Aug 1997	6 Oct 1998	Synthesis of 4-oxoTEMPO
US5891971A[137]	Xerox	15 Aug 1997	6 Apr 1999	NMP with reducing agent
US6911511B1[155]	Elf Atochem	30 Dec 1997	28 Jun 2005	NMP with a small amount of nitroxide
US6107425A[193]	Shipley	6 Feb 1998	22 Aug 2000	NMP of 4-acetoxystyrene
US6353107B1[162]	Ciba	9 Mar 1998	5 Mar 2002	Multialkylpiperidine-*N*-oxyl derivatives for NMP (alkoxyamines)
US6927295B2[163]	Ciba	9 Mar 1998	24 Apr 2003	Multialkylpiperidine-*N*-oxyl derivatives for NMP (alkoxyamines)
US6683142B2[165]	Ciba	9 Mar 1998	27 Jan 2004	Multialkylpiperidine-*N*-oxyl derivatives for NMP (NMP)
US7402639B2[167]	Ciba	9 Mar 1998	22 Jul 2008	Multialkylpiperidine-*N*-oxyl derivatives for NMP (NMP)
US6531547B1[176]	Lubrizol	25 Mar 1998	11 Mar 2003	Vinyl aromatic-co-acrylic block copolymers by NMP
US6545095B1[189]	Melbourne University	7 May 1998	8 Apr 2003	Star microgel by NMP and other RDRP

Table 1.4 (*Continued*)

Patent No.	Assignee	Priority date	US Patent Publication date	Comment
US6353065B1[148]	Elf Atochem	10 Jul 1998	5 Mar 2002	NMP in emulsion (SG1 used)
US6521710B1[164]	Ciba	3 Sep 1998	18 Feb 2003	Graft copolymers by NMP (in extruder)
US6734269B1[154]	Atofina	6 Oct 1998	11 May 2004	NMP of ethylene
US6369162B1[175]	Lubrizol	26 Oct 1998	9 Apr 2002	Star microgel by NMP
US6479603B1[157]	BASF	16 Dec 1998	12 Nov 2002	Polymerizable nitroxides
US6649701B1[191]	Kaneka	27 Mar 1998	18 Nov 2003	Chain-extending NMP with non-conjugated dienes (example is ATRP)
US6538141B1[150]	Atofina	8 Jan 1999	25 Mar 2003	Preparation of SG1 (**1.4**) and related nitroxides
US6569967B1[151]	Atofina	18 Feb 1999	27 May 2003	SG1 (**1.4**) and related nitroxides in NMP
US6624322B1[152]	Atofina	8 Jan 1999	23 Sep 2003	Preparation of SG1 (**1.4**) and related nitroxides
US6472486B2[168]	Symyx	9 Mar 1999	29 Oct 2002	NMP with "unstable" Nitroxides (TIPNO, **1.5**), NMP in seeded emulsion
US6559255B2[169]	Symyx	9 Mar 1999	6 May 2003	NMP in seeded emulsion
US6355756B1[177]	IBM	18 May 1999	12 Mar 2002	Electroactive polymers
US 6433115B2[178]	IBM	18 May 1999	13 Aug 2002	Devices from electroactive polymers
US6512070B2,[179]	IBM	18 May 1999	28 Jan 2003	Electroactive polymers
US6657043B1[298]	Atofina	19 May 1999	2 Dec 2003	Multi-alkoxyamines based on SG1 (**1.4**)
US6716948B1[170]	Symyx	31 Jul 1999	6 Apr 2004	Multi-alkoxyamines in NMP
US7259217B2[172]	Symyx	31 Jul 1999	21 Aug 2007	Multi-alkoxyamines in NMP
US6639033B[158]	BASF	18 Aug 1999	28 Oct 2003	Nitroxides for NMP
US6573346B1[180]	Bayer	7 Sep 1999	2 Jun 2003	NMP synthesis of end-functional polymers
US6632895B1[181]	Bayer	7 Sep 1999	14 Oct 2003	Functional alkoxyamine synthesis
US6800708B2[182]	Bayer	7 Sep 1999	5 Oct 2004	End-functional polymers made by NMP
US6706832B2[153]	Atofina	26 Jan 2000	16 Mar 2004	NMP of ethylene
US7067586B2[171]	Symyx	3 Oct 2000	27 Jun 2006	Blocks based on amphiphilic copolymer
US6844405B2[166]	Ciba	13 Nov 2000	18 Jan 2005	NMP of vinyl chloride

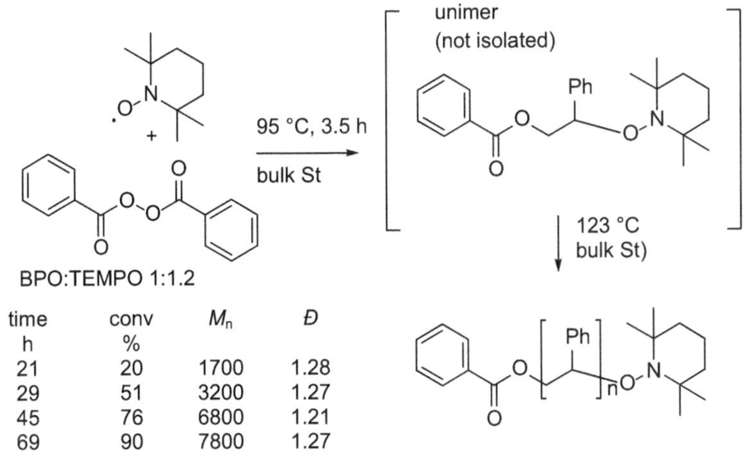

time h	conv %	M_n	$Đ$
21	20	1700	1.28
29	51	3200	1.27
45	76	6800	1.21
69	90	7800	1.27

Scheme 1.22 Nitroxide-mediated polymerization of styrene with TEMPO as nitroxide, benzoyl peroxide (BPO) as initiator, and *in situ* generation of alkoxyamine.[195]

the use of NMP for the synthesis of poly(styrene-*co*-butadiene) and low dispersity polystyrene (Scheme 1.22). A brief overview of the history of the research in NMP conducted at Xerox is provided in the 2010 review of Szkurhan *et al.*[23] The studies by Georges and co-workers focused narrowly on the use of TEMPO as nitroxide and benzoyl peroxide (BPO) as initiator with *in situ* generation of the alkoxyamine. The *in situ* production of alkoxyamine was an option that was mentioned in the text of the CSIRO patent[102] but was not exemplified.

A large number of papers[18,195–219] and patents (see Table 1.4)[120–147] from the Xerox group followed the initial report[195] with topics over the period 1993–2000 including the mechanism of the reaction between BPO and TEMPO,[196] aspects of the kinetics of NMP,[198,202,203,207,208,212,214–217,219] the effect of added acid to accelerate NMP,[198,208] other additives for accelerating NMP,[201,211] the role of excess nitroxide in providing a controlled process,[207,214] controlling NMP with small amounts of nitroxide,[210] NMP of styrenesulfonic acid,[200] chloromethylstyrene[209] and *n*-butyl acrylate,[204] styrene–butadiene block copolymers[213] and NMP in miniemulsion.[218]

Some attention was paid to nitroxides other than TEMPO for use in NMP. Semi-empirical MO calculations were used to rank nitroxides on the basis of predicted C–O bond dissociation energies.[197,199] Use of di-*t*-butyl nitroxide (**1.3**) was found to give improved control and faster polymerization rates than TEMPO in the case of the Xerox study for styrene polymerization.[199] However, this finding was not followed up.

The induced decomposition of BPO by TEMPO had been examined at CSIRO[220] in the context of a study of the initiation of styrene polymerization by BPO using the radical trapping method. The mechanism proposed (Scheme 1.23)[220] involved the nitroxide being transformed to the

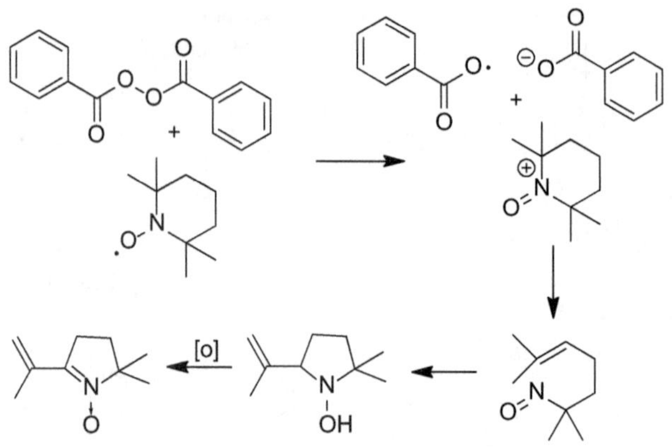

Scheme 1.23 Mechanism of reaction between TEMPO and benzoyl peroxide (BPO).[220]

oxammonium benzoate or an equivalent covalent adduct with formation of an equivalent of benzoyloxy radicals. The oxammonium benzoate then decomposed to benzoic acid and an unsaturated nitroso-compound, which underwent an intramolecular 'ene' reaction to give the hydroxylamine, oxidation of which afforded the nitrone shown in Scheme 1.23.

1.5.2 Hawker Group

The research of Hawker and coworkers is detailed in the review by Hawker *et al.* published in 2001.[20] The *in situ* method of alkoxyamine formation used by Xerox, while convenient, adds a degree of complexity and uncertainty to the process. In 1994, Hawker and colleagues published their first paper on NMP.[221] Claiming to have been inspired by the Xerox work[195] and being aware of the CSIRO patent,[102] they re-examined NMP of styrene initiated by a preformed alkoxyamine based on TEMPO (Scheme 1.24), which they called a unimer. The alkoxyamine was prepared using the methodology developed by CSIRO.[102] They demonstrated that alkoxyamine-initiated NMP could be used to form high molecular weight low dispersity polystyrene. Lower than anticipated molecular weights, when higher molecular weights were targeted (Scheme 1.24), can be understood in terms of the contribution of thermal initiation to the process.

In subsequent work, Hawker and coworkers examined the relative merits of the unimer *vs.* the *in situ* "initiator plus nitroxide" method,[221,222] methods for alkoxyamine synthesis,[223] and aspects of the kinetics and mechanism of NMP.[224,225] They also demonstrated the utility of NMP in preparing a variety of statistical copolymers,[226] star, graft and hyperbranched copolymers,[227–231] block copolymers[232] and end functional polymers.[19,228] In 1999, Benoit *et al.* published on the screening of various nitroxides for NMP. This study showed that alkoxyamines based on α-hydrido species such as TIPNO (**1.5**) provided

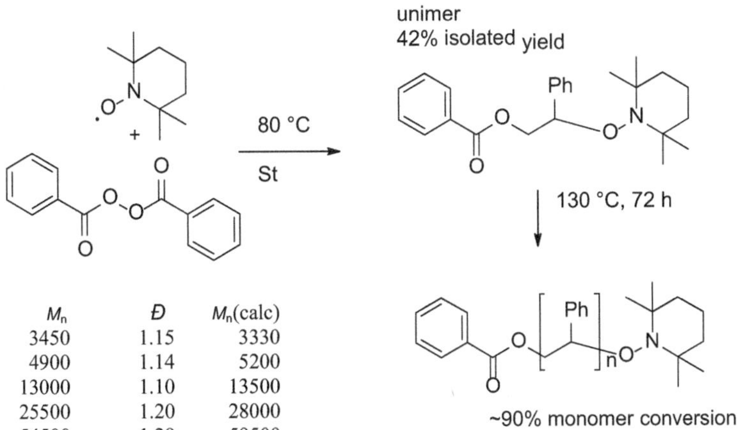

M_n	$Đ$	M_n(calc)
3450	1.15	3330
4900	1.14	5200
13000	1.10	13500
25500	1.20	28000
54500	1.29	59500
82000	1.36	90500
110000	1.41	123000

Scheme 1.24 Nitroxide-mediated polymerization of styrene with TEMPO-based unimer as initiator.[221] All experiments were carried out at 130 °C in bulk styrene for 72 h and gave ∼90% monomer conversion. Calculated molecular weights were based on the alkoxyamine concentration and the monomer conversion.

very much improved outcomes in NMP (lower dispersities, faster rates of polymerization at lower temperatures).[232–234] These alkoxyamines were also claimed in a patent filed by Symyx.[168]

1.5.3 Fukuda Group

Fukuda and coworkers' initial work focused on analyzing the kinetics and mechanism of NMP.[235–245] They developed methods for determining homolysis (activation) rate constants based on GPC analysis.[236,237,245] They also demonstrated the effect of added initiator to control the excess nitroxide in NMP[238,246] and used this knowledge to obtain enhanced rates in butyl acrylate polymerization mediated by di-*t*-butyl nitroxide.[246] NMP was exploited in block copolymer synthesis[247,248] crosslinking polymerization,[249] and in the preparation of glycopolymers.[250–252]

1.5.4 Fischer Group

The kinetics of NMP and other SRMP were studied by Fischer *et al.*,[119,253,254] who demonstrated the applicability of what he termed the persistent radical effect to NMP. Fischer had first described the persistent radical effect in radical reactions in a non-polymerization context in 1986.[118]

In the reversible deactivation processes, the transient and the persistent species are formed and disappear with equal rates. If reversible deactivation was the only reaction then the concentrations of the two species would increase equally with time to reach steady state values. However, transient radicals also decay by irreversible bimolecular self-termination. Hence, the concentration of the transient species will reach a maximum and decrease while the concentration of the persistent radical will steadily increase. The excess concentration must balance the self-termination loss of the transient species. No steady state will exist. As bimolecular self-termination occurs, the cross reaction will be increasingly favored.[255,256] Bimolecular self-termination is self-suppressing.

1.5.5 Other Groups

In a 1994 publication on possibilities for living radical polymerization, Greszta *et al.*[257] postulated an alternative mechanism for NMP based on degenerative chain transfer. This[257,258] prompted work by several groups to prove (or disprove) the original mechanism.[224,239]

Much effort was devoted to improving NMP, particularly with respect to increasing the rate of polymerization and enhancing end-group fidelity. Many studied the kinetics of NMP.[259–262] Catala *et al.* showed that the rate of NMP of styrene at higher temperatures was independent of the concentration of alkoxyamine initiator.[259] This result was soon confirmed and it was then shown that the rate of NMP of styrene with TEMPO at higher reaction temperatures was determined by the rate of thermal initiation by the Mayo mechanism.[119,235,260,263]

Various rate accelerating additives were devised to facilitate, in particular, TEMPO-mediated NMP.[255] These reagents include organic acids,[208,264] acid salts,[201,211] anhydrides,[225,243,264] added initiator[238,246,264–266] and reducing agents such as ascorbic acid.[215] Many of these are thought to work by removing the excess of nitroxide that builds during the course of polymerization as a side effect of termination.

New nitroxides/alkoxyamines for use in NMP were also reported by those outside of the major groups. Mention should be made of the work of Yamada and coworkers,[267] and Puts and Sogah.[268] The latter authors also reported a multifunctional initiator suitable for both cationic ring-opening polymerization and NMP.[269] However the most significant developments in this context were the reports of SG1 (**1.4**) and TIPNO (**1.5**) (Hawker Group), and related species (Figure 1.2). The discovery of SG1 (**1.4**) was reported in a series of papers in 1998–2000[233,270–273] and appears in patents filed by Elf Atochem.[148,150–152]

The initial reports of NMP in heterogeneous media appeared in the patent literature with filing in the period from Xerox,[142,146,147] BASF,[159,160] Symyx[168,169] and Elf Atochem.[148] The Early attempts at NMP emulsion,[274] miniemulsion[218,274,275] and dispersion polymerization[276,277] were complicated by the high reaction temperatures then necessary for NMP making use of TEMPO derivatives and by poor colloidal stability. The development is summarized in reviews of RDRP in heterogeneous media.[278,279]

Many were active in exploring synthesis of block copolymers, functional copolymers and other structures mainly using TEMPO-based NMP. For example:

- Synthesis of a polybutadiene block copolymer based on transformation of anionic polymerization to NMP[280–282] with TEMPO. Various approaches for transforming anionic polybutadiene to the macro-alkoxyamine were explored.
- polySt-*block*-poly(*para*-chlorostyrene),[283] polySt-*block*-poly(St-*co*-BMA).[284] The polySt macro-alkoxyamine was formed by BPO-initiated NMP with TEMPO.
- polySt-*block*-poly(EA, MA or BA).[285] (The polySt TEMPO macro-alkoxyamine was formed anionic polymerization and end-group transformation)
- Li *et al.*[286] examined the mechanism of thermal decomposition of alkoxyamines. The end-group fidelity in TEMPO-mediated styrene polymerization appears >90% for low molecular weights but drops off rapidly for higher molecular weights (>10 000), both due to the contribution of autopolymerization (*i.e.* thermal initiation by the Mayo mechanism) and the poor thermal stability of the alkoxyamine chain end.[287]
- Blocks by a combination of conventional radical polymerization and NMP.[288]
- polySt-*block*-polyEA, polySt-*block*-polyMMA.[289] The polySt macro-alkoxyamine was formed by AIBN-initiated NMP with TEMPO. Poor reinitiation efficiencies were observed, which were improved with addition of camphorsulfonic acid. This led to the development of alternative agents for SRMP[290,291] and a proposal that inbuilt instability of the stable radical could be used as a process for regulating the formation of excess stable radical during SRMP.[292,293]
- polyTHF-*block*-PolySt.[294] The polyTHF TEMPO macro-alkoxyamine was formed from a macro azoinitiatior.
- PolySt-*block*-poly(4-methylstyrene)-*block*-polySt.[295] Synthesis from a bis-alkoxyamine. Alkoxyamine synthesis based on hydrogen atom abstraction from an ethylbenzene in the presence of TEMPO.[296]
- Moroni *et al.*[205] prepared a polymer with pendent phenylenevinylene units by TEMPO-mediated NMP.

1.6 Conclusions

The early CSIRO work demonstrated NMP as a viable method for controlled polymer synthesis by RDRP. The method allowed the synthesis of polymers of lowered dispersity and made available block and graft copolymers that were not previously possible by radical polymerization. However, little attention was paid to NMP by the wider polymer community until the publications of Georges, Hawker, Fukuda and others in 1993–1995, a decade after the original invention. The period 1993–2000 saw many papers on kinetics and mechanism, and the exploitation of the method mainly for polymerization of styrene and derivatives in a range of architectures. A step change in the versatility of NMP was realized following the discovery of SG1 (**1.4**) and TIPNO (**1.5**) (Figure 1.2) in 1998–2000, but that further development is outside the scope of this chapter.

References

1. A. D. Jenkins, R. I. Jones and G. Moad, *Pure Appl. Chem.*, 2010, **82**, 483-491.
2. K. Matyjaszewski and J. Xia, *Chem. Rev.*, 2001, **101**, 2921–2990.
3. K. Matyjaszewski, *Macromolecules*, 2012, **45**, 4015–4039.
4. M. Kamigaito, T. Ando and M. Sawamoto, *Chem. Rev.*, 2001, **101**, 3689-3745.
5. M. Ouchi, T. Terashima and M. Sawamoto, *Chem. Rev.*, 2009, **109**, 4963-5050.
6. G. Moad, E. Rizzardo and S. H. Thang, *Chem. - Asian J.*, 2013, **8**, 1634-1644.
7. E. Rizzardo and D. H. Solomon, *Aust. J. Chem.*, 2012, **65**, 945–969.
8. G. Moad, E. Rizzardo and S. H. Thang, *Aust. J. Chem.*, 2012, **65**, 985-1076.
9. G. Moad, E. Rizzardo and S. H. Thang, *Aust. J. Chem.*, 2009, **62**, 1402-1472.
10. G. Moad, E. Rizzardo and S. H. Thang, *Polymer*, 2008, **49**, 1079–1131.
11. G. Moad, E. Rizzardo and S. H. Thang, *Acc. Chem. Res.*, 2008, **41**, 1133-1142.
12. G. Moad, E. Rizzardo and S. H. Thang, *Aust. J. Chem.*, 2006, **59**, 669–692.
13. G. Moad, E. Rizzardo and S. H. Thang, *Aust. J. Chem.*, 2005, **58**, 379–410.
14. S. Yamago, *Chem. Rev.*, 2009, **109**, 5051–5068.
15. G. Moad and D. H. Solomon, *The Chemistry of Radical Polymerization*, Elsevier, Oxford, 2006.
16. D. H. Solomon, E. Rizzardo and P. Cacioli, *Free radical polymerization and the produced polymers*, Commonwealth Scientific and Industrial Research Organization, Australia., EP135280A2, 1985.
17. G. Moad, E. Rizzardo and D. H. Solomon, in *Comprehensive Polymer Science*, ed. G. C. Eastmond, A. Ledwith, S. Russo and P. Sigwalt, Pergamon, Oxford, 1989, p. 141.
18. M. K. Georges, R. P. N. Veregin, P. M. Kazmaier and G. K. Hamer, *Trends Polym. Sci.*, 1994, **2**, 66–72.
19. C. J. Hawker, *Trends Polym. Sci.*, 1996, **4**, 183–188.
20. C. J. Hawker, A. W. Bosman and E. Harth, *Chem. Rev.*, 2001, **101**, 3661-3688.
21. A. Studer and T. Schulte, *Chem. Rec.*, 2005, **5**, 27–35.

22. P. Nesvadba, *Chimia*, 2006, **60**, 832–840.
23. A. R. Szkurhan, J. Lukkarila and M. K. Georges, in *Stable Radicals: Fundamentals and Applied Aspects of Odd-Electron Compounds*, ed. R. Hicks, Wiley, Hoboken, N.J., 2010, pp. 407–431.
24. D. Bertin, D. Gigmes, S. R. A. Marque and P. Tordo, *Chem. Soc. Rev.*, 2011, **40**, 2189–2198.
25. J. Nicolas, Y. Guillaneuf, D. Bertin, D. Gigmes and B. Charleux, in *Polymer Science: A Comprehensive Reference*, ed. K. Matyjaszewski and M. Möller, Elsevier, Amsterdam, 2012, pp. 277–350.
26. P. Nesvadba, in *Fundamentals of Controlled/Living Radical Polymerization*, The Royal Society of Chemistry, 2013, pp. 112–167.
27. J. Nicolas, Y. Guillaneuf, C. Lefay, D. Bertin, D. Gigmes and B. Charleux, *Prog. Polym. Sci.*, 2013, **38**, 63–235.
28. G. Audran, P. Bremond and S. R. A. Marque, *Chem. Commun.*, 2014, **50**, 7921–7928.
29. E. G. Bagryanskaya and S. R. A. Marque, *Chem. Rev.*, 2014, **114**, 5011–5056.
30. D. H. Solomon, *J. Polym. Sci., Part A: Polym. Chem.*, 2005, **43**, 5748–5764.
31. G. Moad and D. H. Solomon, *The Chemistry of Radical Polymerization*, Elsevier, Oxford, 2006, pp. 1–9.
32. M. Szwarc, *J. Polym. Sci., Part A: Polym. Chem.*, 1998, **36**, IX–XV.
33. M. Szwarc, *Nature*, 1956, **178**, 1168–1169.
34. S. Penczek and G. Moad, *Kem. Ind.*, 2012, **61**, 215–236.
35. T. Otsu, *J. Polym. Sci.*, 1957, **26**, 236–239.
36. T. Otsu, *J. Polym. Sci.*, 1956, **21**, 559–561.
37. D. Braun, *Macromol. Symp.*, 1996, **111**, 63–71.
38. T. Otsu and M. Yoshida, *Makromol. Chem., Rapid Commun.*, 1982, **3**, 127–132.
39. T. Otsu, M. Yoshida and A. Kuriyama, *Polym. Bull.*, 1982, **7**, 45–50.
40. T. Otsu, *J. Polym. Sci., Part A: Polym. Chem.*, 2000, **38**, 2121–2136.
41. J. Kennedy, V. C. Chang, R. Smith and B. Iván, *Polym. Bull.*, 1979, **1**, 575–580.
42. J. P. Kennedy and R. A. Smith, *J. Polym. Sci., Polym. Chem. Ed.*, 1980, **18**, 1523–1537.
43. G. Moad and D. H. Solomon, *The Chemistry of Radical Polymerization*, Elsevier, Oxford, 2006, pp. 417–420.
44. G. Moad and D. H. Solomon, *The Chemistry of Radical Polymerization*, Elsevier, Oxford, 2006, pp. 414–421.
45. D. H. Solomon, *J. Macromol. Sci., Chem.*, 1982, **A17**, 337–367.
46. D. H. Solomon, P. Cacioli and G. Moad, *Pure Appl. Chem.*, 1985, **57**, 985–992.
47. G. Moad and D. H. Solomon, *Aust. J. Chem.*, 1990, **43**, 215–239.
48. E. Rizzardo and D. Solomon, *Polym. Bull.*, 1979, **1**, 529–534.
49. A. L. J. Beckwith, V. W. Bowry and G. Moad, *J. Org. Chem.*, 1988, **53**, 1632–1641.

50. A. L. J. Beckwith, V. W. Bowry, M. O'Leary, G. Moad, E. Rizzardo and D. H. Solomon, *J. Chem. Soc., Chem. Commun.*, 1986, 1003–1004.
51. G. Moad, E. Rizzardo, D. Solomon and A. J. Beckwith, *Polym. Bull.*, 1992, **29**, 647–652.
52. J. Chateauneuf, J. Lusztyk and K. U. Ingold, *J. Org. Chem.*, 1988, **53**, 1629–1632.
53. V. W. Bowry and K. U. Ingold, *J. Am. Chem. Soc.*, 1992, **114**, 4992–4996.
54. T. Fukuda, Y.-D. Ma and H. Inagaki, *Macromolecules*, 1985, **18**, 17–26.
55. E. G. Janzen, *Acc. Chem. Res.*, 1971, **4**, 31–40.
56. G. R. Chalfont, M. J. Perkins and A. Horsfield, *J. Am. Chem. Soc.*, 1968, **90**, 7141–7142.
57. T. Kunitake and S. Murakami, *Polym. J.*, 1972, **3**, 249–251.
58. T. Kunitake and S. Murakami, *J. Polym. Sci., Polym. Chem. Ed.*, 1974, **12**, 67–81.
59. T. Sato and T. Otsu, *Polymer*, 1975, **16**, 389–391.
60. T. Sato and T. Otsu, *Makromol. Chem.*, 1977, **178**, 1941–1950.
61. J. C. Bevington, B. J. Tabner and P. F. Fridd, *Rev. Roumaine De Chim.*, 1980, **25**, 947–952.
62. E. G. Rozantsev and V. D. Sholle, *Synthesis*, 1971, **1971**, 401–414.
63. K. U. Ingold, in *Landoldt-Bornstein. Reaction Kinetics Radical Reaction Rates in Liquids Radicals Centered on N, S, P and Other Heteroatoms. Nitroxyls*, ed. H. Fischer, Springer-Verlag, 1983, pp. 181–182.
64. R. D. Grant, E. Rizzardo and D. H. Solomon, *Makromol. Chem.*, 1984, **185**, 1809–1817.
65. P. G. Griffiths, E. Rizzardo and D. H. Solomon, *J. Macromol. Sci., Chem.*, 1982, **A17**, 45.
66. P. G. Griffiths, E. Rizzardo and D. H. Solomon, *Tetrahedron Lett.*, 1982, **23**, 1309–1312.
67. R. D. Grant, P. G. Griffiths, G. Moad, E. Rizzardo and D. H. Solomon, *Aust. J. Chem.*, 1983, **36**, 2447–2454.
68. M. J. Cuthbertson, E. Rizzardo and D. H. Solomon, *Aust. J. Chem.*, 1985, **38**, 315–324.
69. M. J. Cuthbertson, E. Rizzardo and D. H. Solomon, *Aust. J. Chem.*, 1983, **36**, 1957–1973.
70. M. J. Cuthbertson, G. Moad, E. Rizzardo and D. H. Solomon, *Polym. Bull.*, 1982, **6**, 647–651.
71. G. Moad, E. Rizzardo and D. H. Solomon, *Aust. J. Chem.*, 1983, **36**, 1573–1588.
72. G. Moad, E. Rizzardo and D. H. Solomon, *Macromolecules*, 1982, **15**, 909–914.
73. M. J. Jones, G. Moad, E. Rizzardo and D. H. Solomon, *J. Org. Chem.*, 1989, **54**, 1607–1611.
74. W. K. Busfield, I. D. Jenkins, S. H. Thang, E. Rizzardo and D. H. Solomon, *Tetrahedron Lett.*, 1985, **26**, 5081–5084.
75. W. K. Busfield, I. D. Jenkins, S. H. Thang, E. Rizzardo and D. H. Solomon, *Aust. J. Chem.*, 1985, **38**, 689–698.

76. W. K. Busfield, I. D. Jenkins, S. H. Thang, E. Rizzardo and D. H. Solomon, *J. Chem. Soc., Perkin Trans. 1*, 1988, 485–490.
77. D. Bednarek, G. Moad, E. Rizzardo and D. H. Solomon, *Macromolecules*, 1988, **21**, 1522–1528.
78. S. Bottle, W. K. Busfield, I. D. Jenkins, S. Thang, E. Rizzardo and D. H. Solomon, *Eur. Polym. J.*, 1989, **25**, 671–676.
79. S. Bottle, W. K. Busfield, I. D. Jenkins, S. Thang, E. Rizzardo and D. H. Solomon, *Eur. Polym. J.*, 1989, **25**, 671–676.
80. W. K. Busfield, D. I. Grice and I. D. Jenkins, *Polym. Int.*, 1992, **27**, 119–123.
81. T. Nakamura, W. K. Busfield, I. D. Jenkins, E. Rizzardo, S. H. Thang and S. Suyama, *J. Am. Chem. Soc.*, 1996, **118**, 10824–10828.
82. E. Rizzardo, A. K. Serelis and D. H. Solomon, *Aust. J. Chem.*, 1982, **35**, 2013–2024.
83. T. Nakamura, W. K. Busfield, I. D. Jenkins, E. Rizzardo, S. H. Thang and S. Suyama, *Macromolecules*, 1997, **30**, 2843–2847.
84. T. Nakamura, W. K. Busfield, I. D. Jenkins, E. Rizzardo, S. H. Thang and S. Suyama, *J. Org. Chem.*, 1997, **62**, 5578–5582.
85. W. K. Busfield, I. D. Jenkins, T. Nakamura, M. J. Monteiro, E. Rizzardo, S. Sayama, S. H. Thang, P. V. Le and C. I. Zayas-Holdsworth, *Polym. Adv. Technol.*, 1998, **9**, 94–100.
86. T. Nakamura, S. Suyama, W. K. Busfield, I. D. Jenkins, E. Rizzardo and S. H. Thang, *Polymer*, 1999, **40**, 1395–1401.
87. T. Nakamura, W. K. Busfield, I. D. Jenkins, E. Rizzardo, S. H. Thang and S. Suyama, *J. Org. Chem.*, 1999, **65**, 16–23.
88. W. K. Busfield, I. D. Jenkins, E. Rizzardo, D. H. Solomon and S. H. Thang, *J. Chem. Soc., Perkin Trans. 1*, 1991, 1351–1354.
89. W. K. Busfield, I. D. Jenkins, S. H. Thang, E. Rizzardo and D. H. Solomon, *Eur. Polym. J.*, 1993, **29**, 397–400.
90. G. Moad, E. Rizzardo and D. H. Solomon, *Makromol. Chem., Rapid Commun.*, 1982, **3**, 533–536.
91. G. Moad, E. Rizzardo and D. H. Solomon, *Tetrahedron Lett.*, 1981, **22**, 1165–1168.
92. G. Moad, E. Rizzardo and D. H. Solomon, *J. Macromol. Sci., Chem.*, 1982, **A17**, 51–59.
93. R. D. Grant, E. Rizzardo and D. H. Solomon, *J. Chem. Soc., Chem. Commun.*, 1984, 867.
94. R. D. Grant, E. Rizzardo and D. H. Solomon, *J. Chem. Soc., Perkin Trans. 2*, 1985, 379–384.
95. T. Nakamura, Y. Watanabe, H. Tezuka, W. K. Busfield, I. D. Jenkins, E. Rizzardo, S. H. Thang and S. Suyama, *Chem. Lett.*, 1997, 1093–1094.
96. K. Ewing, W. K. Busfield, I. D. Jenkins and P. B. Zetterlund, *Polym. Int.*, 2003, **52**, 1671–1675.
97. P. Griffiths, G. Moad and E. Rizzardo, *Aust. J. Chem.*, 1983, **36**, 397–401.

98. S. Bizilj, D. Kelly, A. Serelis, D. Solomon and K. White, *Aust. J. Chem.*, 1985, **38**, 1657–1673.
99. G. Moad, E. Rizzardo and D. H. Solomon, *Polym. Bull.*, 1982, **6**, 589–593.
100. M. J. Cuthbertson, E. Rizzardo and D. H. Solomon, *Aust. J. Chem.*, 1983, **36**, 1957–1973.
101. E. Rizzardo, in *14th Australian Polymer Symposium*, ed. G. B. Guise, Polymer Division, Royal Australian Chemical Insititute, Ballarat, Victoria, Australia, 1984, p. 36.
102. D. H. Solomon, E. Rizzardo and P. Cacioli, *Polymerization Process and Polymers Produced Thereby*, CSIRO, US4581429, 1986.
103. G. Moad, F. Ercole, C. H. Johnson, J. Krstina, C. L. Moad, E. Rizzardo, T. H. Spurling, S. H. Thang and A. G. Anderson, *ACS Symp. Ser.*, 1998, **685**, 332–360.
104. Y. K. Chong, F. Ercole, G. Moad, E. Rizzardo, S. H. Thang and A. G. Anderson, *Macromolecules*, 1999, **32**, 6895–6903.
105. G. Moad, *Prog. Polym. Sci.*, 2015, submitted.
106. P. Cacioli, D. G. Hawthorne, R. L. Laslett, E. Rizzardo and D. H. Solomon, *J. Macromol. Sci., Chem.*, 1986, **A23**, 839–852.
107. E. Rizzardo, D. Harrison, R. L. Laslett, G. F. Meijs, T. C. Morton and S. H. Thang, *Prog. Pac. Polym. Sci.*, 1991, **2**, 77–88.
108. J. Krstina, G. Moad, E. Rizzardo, C. L. Winzor, C. T. Berge and M. Fryd, *Macromolecules*, 1995, **28**, 5381–5385.
109. J. Krstina, C. L. Moad, G. Moad, E. Rizzardo, C. T. Berge and M. Fryd, *Macromol. Symp.*, 1996, **111**, 13–23.
110. E. Rizzardo, *Chem. Aust.*, 1987, **54**, 32.
111. E. Rizzardo and Y. K. Chong, in *2nd Pacific Polymer Conference, Preprints*, Pacific Polymer Federation, Tokyo, 1991, pp. 26–27.
112. J. Krstina, G. Moad, G. R. Morrow, E. Rizzardo, S. H. Thang and G. Van Diepen, in *Engineered Resins Strategic Alliance Technical Report - February 1993*, CSIRO Chemicals and Polymers, *1993*, pp. 35–37.
113. G. Moad and E. Rizzardo, *Macromolecules*, 1995, **28**, 8722–8728.
114. G. Moad and E. Rizzardo, in *Third Pacific Polymer Conference*, Polymer Division, Royal Australian Chemical Institute, Brisbane, Queensland, 1993, pp. 651–652.
115. G. Moad and E. Rizzardo, in *MacroAkron'94*, IUPAC, Akron OH, 1994, p. 9.
116. A. L. J. Beckwith, V. W. Bowry and K. U. Ingold, *J. Am. Chem. Soc.*, 1992, **114**, 4983–4992.
117. C. H. J. Johnson, G. Moad, D. H. Solomon, T. H. Spurling and D. J. Vearing, *Aust. J. Chem.*, 1990, **43**, 1215–1230.
118. H. Fischer, *J. Am. Chem. Soc.*, 1986, **108**, 3925–3927.
119. H. Fischer, *Macromolecules*, 1997, **30**, 5666–5672.
120. M. K. Georges, R. P. N. Veregin, P. M. Kazmaier and G. K. Hamer, *Polymerization processes and toner compositions therefrom*, Xerox Corp., USA, US5322912A, 1994.

121. M. K. Georges, R. P. N. Veregin, P. M. Kazmaier and G. K. Hamer, *Polymerization processes and toner compositions therefrom*, Xerox Corp., USA, US5401804A, 1995.
122. M. K. Georges, M. D. Saban, P. M. Kazmaier, R. P. N. Veregin, G. K. Hamer and K. A. Moffat, *Homoacrylate polymerization processes with oxonitroxides*, Xerox Corp., USA, US5412047A, 1995.
123. K. A. Moffat, M. D. Saban, R. P. N. Veregin, M. K. Georges, G. K. Hamer and P. M. Kazmaier, *Stable free radical polymerization process and thermoplastic materials produced therefrom*, Xerox Corp., USA, US5449724A, 1995.
124. K. A. Moffat, P. M. Kazmaier, G. K. Hamer and M. K. Georges, *Process for producing telechelic, branched and star thermoplastic resin polymers*, Xerox Corp., USA, US5498679A, 1996.
125. R. P. N. Veregin, P. M. Kazmaier, M. K. Georges and G. K. Hamer, *Polymerization Processes*, Xerox Corp., USA, US5530079A, 1996.
126. B. Keoshkerian, M. K. Georges and S. V. Drappel, *Ink jettable toner compositions and processes for making and using*, Xerox Corp., USA, US5545504A, 1996.
127. M. K. Georges, R. P. N. Veregin, P. M. Kazmaier and G. K. Hamer, *Polymerization processes and toner compositions therefrom*, Xerox Corp., USA, US5549998A, 1996.
128. P. G. Odell and G. K. Hamer, *Polymerization process and compositions thereof*, Xerox Corp., USA, US5552502A, 1996.
129. P. G. Odell, R. P. N. Veregin, G. K. Hamer and M. K. Georges, *Rate-enhanced pseudoliving polymerization process for producing polymers having narrow polydispersity*, Xerox Corp., USA, US5608023A, 1997.
130. R. P. N. Veregin, P. M. Kazmaier, M. K. Georges and G. K. Hamer, *Stable free radical in polymerization for narrow polydispersity of molecular weight*, Xerox Corp., USA, US5610250A, 1997.
131. B. Keoshkerian, S. Drappel and M. K. Georges, *Ink jettable toner compositions and processes for making and using*, Xerox Corp., USA, US5714993A, 1998.
132. P. M. Kazmaier, B. Keoshkerian, K. A. Moffat, M. K. Georges, G. K. Hamer and R. P. N. Veregin, *Processes for preparing telechelic, branched and star thermoplastic resins* Xerox Corporation, USA, US5723511A, 1998.
133. P. M. Kazmaier, B. Keoshkerian, R. O. Loutfy, K. A. Moffat, M. K. Georges, G. K. Hamer and R. P. N. Veregin, *Stable free radical polymerization processes for polymers of narrow molecular weight distribution*, Xerox Corp., USA, US5728747A, 1998.
134. D. A. Foucher, P. J. MacLeod, M. K. Georges and T. E. Burrow, *Metal-accelerated radical polymerization processes* Xerox Corporation, USA, US5744560A, 1998.
135. P. M. Kazmaier, M. D. Saban, K. A. Moffat and M. K. Georges, *Processes for stable free radicals* Xerox Corporation, USA, US5817824A, 1998.

136. B. Keoshkerian, M. K. Georges, P. M. Kazmaier and G. K. Hamer, *Polymerization processes*, Xerox Corp., USA, US 5739229A, 1998.
137. B. Keoshkerian, M. H. Quinlan, M. K. Georges and B. H. Goodbrand, *Polymerization processes*, Xerox Corporation, USA, US5891971A, 1999.
138. P. M. Kazmaier, B. Keoshkerian, K. A. Moffat, M. K. Georges, G. K. Hamer and R. P. N. Veregin, *Processes for the preparation of branched polymers*, Xerox Corporation, USA, US5919861A, 1999.
139. B. Keoshkerian, M. K. Georges, P. M. Kazmaier and G. K. Hamer, *Polymerization processes and resin particles formed thereby*, Xerox Corp., USA, US6075105A, 2000.
140. M. K. Georges, T. Inoue, G. K. Hamer, P. M. Kazmaier and R. P. N. Veregin, *Bifunctional macromolecules and toner compositions therefrom*, Xerox Corp., USA, US6087451A, 2000.
141. P. M. Kazmaier, R. P. N. Veregin, K. A. Moffat, M. K. Georges, M. D. Saban and G. K. Hamer, *Process for preparation of branched polymers*, Xerox Corp., USA, US6114499A, 2000.
142. P. J. MacLeod, P. G. Odell, F. E. Torres and M. K. Georges, *Polymerization processes using oligomeric compound, monomer and surfactant*, Xerox Corp., Japan, US6121397A, 2000.
143. B. Keoshkerian, M. K. Georges and P. M. Kazmaier, *Stable free radical polymerization processes*, Xerox Corp., USA, US6156858A, 2000.
144. M. K. Georges, T. Inoue, G. K. Hamer, P. M. Kazmaier and R. P. N. Veregin, *Bifunctional macromolecules and toner compositions therefrom*, Xerox Corp., USA, US6258911B1, 2001.
145. P. M. Kazmaier, M. K. Georges, R. P. N. Veregin, B. Keoshkerian, G. K. Hamer, K. A. Moffat and M. D. Saban, *Process for producing thermoplastic resin polymers*, Xerox Corp., USA, US6320007B1, 2001.
146. P. J. MacLeod, P. G. Odell, F. E. Torres and M. K. Georges, *Polymerization processes*, Xerox Corp., Japan, US6365675B1, 2002.
147. P. J. MacLeod, P. G. Odell, F. E. Torres and M. K. Georges, *Polymerization process for the preparation of a resin*, Xerox Corp., Japan, US6472485B2, 2002.
148. B. Charleux, M. Lansalot, R. Pirri, J.-P. Vairon and S. Denie, *Emulsion polymerization in the presence of a stable free radical* Elf Atochem S.A., Fr.; Atofina., US6353065B1, 2000.
149. D. Bertin, B. Boutevin and P. Nicol, *Polymerization in the presence of stable free radicals and inifers*, Elf Atochem S.A., Fr., US6255422B1, 2001.
150. J.-P. Gillet, O. Guerret and J.-P. Lascombe, *Method for preparing nitroxides*, Elf Atochem S.A., Fr., US6538141B1, 2003.
151. J.-L. Couturier, C. Henriet-Bernard, C. Le Mercier, P. Tordo and J.-F. Lutz, *Alkoxyamines derived from phosphorus-containing nitroxides and their use*, Elf Atochem S.A., Fr., US6569967B1, 2003.
152. J.-P. Gillet, O. Guerret and P. Tordo, *Method for preparing β-phosphorus nitroxide radicals*, Elf Atochem S.A., Fr., US6624322B1, 2003.

153. E. Minaux, L. Greci, M. Buback, P. Tordo, T. Senninger, P. Stipa, P. Carloni, E. Damiani and G. Tommasi, *Method for controlled free radical polymerization or copolymerization of ethylene under high pressure in the presence of indoline nitroxide radicals ATOFINA*, Fr., US6706832B2, 2004.
154. M. Buback, E. Minaux, T. Senninger and J.-M. Le Blevec, *Method for controlled free radical polymerization or copolymerization of ethylene under high pressure in the presence of an initiator controller Elf Atochem S.A.*, Fr., US6734269B1, 2004.
155. D. Bertin, *Controlled radical polymerization process using a small amount of a stable free radical Elf Atochem S.A.*, Fr.; ATOFINA., US6911511B1, 2005.
156. M. Fischer and J. Koch *Block copolymers* BASF A.-G., Germany, US6239226B1, 1997.
157. Y. Yang, J. He, J. Cao, J. Chen and C. Li, *Method for living free-radical polymerization* BASF Aktiengesellschaft, Germany, US6479603B1, 2002.
158. S. Brinkmann-Rengel, S. Haremza, H. F. Sutoris, D. Christie, R. B. Raether and J. Cao, *Preparation of novel N-oxyl radicals as stabilizers for ethylenically unsaturated monomers*, BASF Aktiengesellschaft, Germany, US6639033B1, 2003.
159. B. R. Morrison, K. Mathauer, M. Fischer and Y. Durant, *Stable N-oxyl radial assisted mini emulsion polymerization*, BASF A.-G., Germany, US6503983B1, 2003.
160. M. Fischer, B. R. Morrison, H. F. Sutoris and Y. Durant, *Stable N-oxyl radical assisted emulsion polymerization under pressure*, BASF A.-G., Germany, US6696533B1, 2004.
161. P. Nesvadba, A. Kramer, A. Steinmann and W. Stauffer, *Polymerizable compositions containing alkoxyamine initiators derived from nitroso- or nitrone compounds*, Ciba Specialty Chemicals Holding Inc., Switz., US6262206B1, 2001.
162. A. Kramer and P. Nesvadba, *1-Alkoxypolyalkylpiperidine derivatives and their use as polymerization regulators*, Ciba Specialty Chemicals Holding Inc., Switz., US6353107B1, 2002.
163. A. Kramer and P. Nesvadba, *1-Alkoxypolyalkylpiperidine derivatives and their use as polymerization regulators*, Ciba Specialty Chemicals Holding Inc., Switz., US6927295B2, 2003.
164. M. Roth, R. Pfaendner and P. Nesvadba, *Grafting of ethylenically unsaturated monomers onto polymers*, Ciba Specialty Chemicals Holding Inc., Switz., US6521710B1, 2003.
165. A. Kramer and P. Nesvadba, *1-Alkoxypolyalkylpiperidine derivatives and their use as polymerization regulators*, Ciba Specialty Chemicals Holding Inc., Switz., US6683142B2, 2004.
166. R. Pfaendner, T. Wannemacher and D. Braun, *Polymerization of vinyl chloride in the presence of a stable nitroxyl radical*, Ciba Specialty Chemicals Holding Inc., Switz., US6844405B2, 2005.

167. A. Kramer and P. Nesvadba, *1-Alkoxypolyalkylpiperidine derivatives and their use as polymerization regulators*, Ciba Specialty Chemicals Holding Inc., Switz., US7402639B2, 2008.
168. G. Klaerner, A. Safir, R. B. Nielsen, B. Jandeleit, P. Huefner and Y. Li, *Controlled stable free radical emulsion polymerization processes*, Symyx Technologies, Inc., USA, US6472486B2, 2002.
169. G. Klaerner, A. Safir, R. B. Nielsen, B. Jandeleit, P. Huefner and Y. Li, *Controlled free radical emulsion and water-based polymerizations and seeded methodologies*, Symyx Technologies Inc., USA, US6559255B2, 2003.
170. G. Klaerner, M. Petro, D. Benoit and D. Charmot, *Controlled-architecture polymers, their preparation, and use as separation media*, Symyx Technologies, Inc., USA, US6716948B1, 2004.
171. M. Liu, R. B. Nielsen, D. Charmot, D. Hadjuk, J. M. J. Frechet and E. Khosdel, *Method of making ABA-type block copolymers having a random block of hydrophobic and hydrophilic monomers*, Symyx Technologies, Inc., USA, US7067586B2, 2006.
172. G. Klaerner, M. Petro, D. Benoit and D. Charmot, *Controlled-architecture polymers and use thereof as separation media*, Symyx Technologies, Inc., USA, US7259217B2, 2007.
173. D. C. Visger and R. M. Lange, *Vinyl aromatic-(vinyl aromatic-co-acrylic) block copolymers prepared by stabilized free radical polymerization*, The Lubrizol Corporation, USA, EP945474A1, 1999.
174. D. C. Visger, K.-W. Damm, R. M. Lange and B. J. Schober, *Block copolymer surfactants prepared by stabilized free-radical polymerization*, The Lubrizol Corporation, USA, US6111025A, 2000.
175. D. C. Visger and R. M. Lange, *Radial polymers prepared by stabilized free-radical polymerization*, The Lubrizol Corporation, USA, US6369162B1, 2002.
176. D. C. Visger and R. M. Lange, *Vinyl aromatic-(vinyl aromatic-co-acrylic) block copolymers prepared by stabilized free radical polymerization*, The Lubrizol Corporation, USA, US6531547B1, 2003.
177. C. J. Hawker, G. Klaerner, J.-I. Lee, V. Y.-W. Lee, R. D. Miller and J. C. Scott, *Dual purpose electroactive copolymers, preparation thereof, and use in opto-electronic devices*, International Business Machines Corporation, USA; The Board of Trustees of the Leland Stanford Junior University, US6355756B1, 2002.
178. C. J. Hawker, G. Klaerner, J.-I. Lee, V. Y.-W. Lee, R. D. Miller and J. C. Scott, *Opto-electronic devices fabricated with dual purpose electroactive copolymers*, International Business Machines Corporation, USA, The Board of Trustees of the Leland Stanford Junior University, US6433115B2, 2002.
179. C. J. Hawker, G. Klaerner, J.-I. Lee, V. Y.-W. Lee, R. D. Miller and J. C. Scott, *Dual purpose electroactive copolymers*, International Business Machines Corporation, USA, The Board of Trustees of the Leland Stanford Junior University., US6512070B2, 2003.

180. M. Melchiors, H. Hoecker, H. Keul and D. Achten, *Oligomeric and polymeric telechelics*, Bayer A.-G., Germany, US6573346B1, 2003.
181. M. Melchiors, H. Hoecker, H. Keul and D. Achten, *Functionalized alkoxyamine initiators*, Bayer A.-G., Germany, US6632895B1, 2003.
182. M. Melchiors, H. Hoecker, H. Keul and D. Achten, *Oligomeric and polymeric telechelics*, Bayer A.-G., Germany., US6800708B2, 2004.
183. A. G. Anderson, A. Gridnev, G. Moad, E. Rizzardo and S. H. Thang, *Method for controlling polymer molecular weight and structure*, E. I. Du Pont de Nemours & Co., USA, Commonwealth Scientific and Industrial Research Organization, US6271340B1, 2001.
184. K. Matyjaszewski and D. Greszta, *Method for preparation of alkoxyamines from nitroxyl radicals*, Carnegie-Mellon University, USA, US5910549A, 1999.
185. K. Matyjaszewski and D. Greszta, *Rate enhancement of nitroxyl radical-mediated polymerization*, Carnegie-Mellon University, USA, US6288186B1, 2001.
186. R. A. Koster, D. B. Priddy and I. Li, *Difunctional living free radical polymerization initiators for vinyl aromatic monomers*, Dow Chemical Co., USA, US5627248A, 1997.
187. D. B. Priddy and I. Q. Li, *In-situ block copolymer formation during polymerization of a vinyl aromatic monomer*, Dow Chemical Co., USA, US5721320A, 1998.
188. D. H. Solomon, S. Abrol, P. A. Kambouris and M. G. Looney, *A process for preparing polymeric microgels*, The University of Melbourne, Australia, US6300443B1, 2001.
189. D. H. Solomon, G. G. Qiao and S. Abrol, *Process for microgel preparation*, The University of Melbourne, Australia, US6545095B1, 2003.
190. J. T.-y. Lai, W. F. Masler, P. P. Nicholas, N. Pourahmady, R. D. Puts and S. Tahiliani, *Controlled free radical polymerization process*, B.F. Goodrich Co., USA, US6281311B1, 2001.
191. Y. Nakagawa and N. Fujita, *Preparation of olefin polymers by coupling polyolefins in radical polymerization*, Kaneka Corporation, Japan, US6649701B1, 2003.
192. T. Ogawa, *Manufacture of styrene polymers with narrow distribution of molecular weight*, Kansai Paint Co Ltd, Japan, US 5610249A, 1997.
193. R. F. Sinta, G. G. Barclay, J. H. Georger, Jr., M. D. Denison and S. L. Ablaza, *Narrow molecular weight distribution polymers and use of same for negative-acting photoresists*, Shipley Company, L.L.C., USA, US6107425A, 2000.
194. M. Fujita and H. Kihara, *Styrene polymers, their preparation and their use*, Sumitomo Chemical Co., Ltd., Japan, US5723554A, 1998.
195. M. K. Georges, R. P. N. Veregin, P. M. Kazmaier and G. K. Hamer, *Macromolecules*, 1993, **26**, 2987–2988.
196. R. P. N. Veregin, M. K. Georges, P. M. Kazmaier and G. K. Hamer, *Macromolecules*, 1993, **26**, 5316–5320.

197. M. K. Georges, R. P. N. Veregin, G. K. Hamer and P. M. Kazmaier, *Macromol. Symp.*, 1994, **88**, 89–103.
198. M. K. Georges, R. A. Kee, R. P. N. Veregin, G. K. Hamer and P. M. Kazmaier, *J. Phys. Org. Chem.*, 1995, **8**, 301–305.
199. P. M. Kazmaier, K. A. Moffat, M. K. Georges, R. P. N. Veregin and G. K. Hamer, *Macromolecules*, 1995, **28**, 1841–1846.
200. B. Keoshkerian, M. K. Georges and D. Boils-Boissier, *Macromolecules*, 1995, **28**, 6381–6382.
201. P. G. Odell, R. P. N. Veregin, L. M. Michalak, D. Brousmiche and M. K. Georges, *Macromolecules*, 1995, **28**, 8453–8455.
202. M. D. Saban, M. K. Georges, R. P. N. Veregin, G. K. Hamer and P. M. Kazmaier, *Macromolecules*, 1995, **28**, 7032–7034.
203. R. P. N. Veregin, M. K. Georges, G. K. Hamer and P. M. Kazmaier, *Macromolecules*, 1995, **28**, 4391–4398.
204. N. A. Listigovers, M. K. Georges, P. G. Odell and B. Keoshkerian, *Macromolecules*, 1996, **29**, 8992–8993.
205. M. Moroni, A. Hilberer and G. Hadziioannou, *Macromol. Rapid Commun.*, 1996, **17**, 693–702.
206. R. P. N. Veregin, P. G. Odell, L. M. Michalak and M. K. Georges, *Macromolecules*, 1996, **29**, 3346–3352.
207. R. P. N. Veregin, P. G. Odell, L. M. Michalak and M. K. Georges, *Macromolecules*, 1996, **29**, 2746–2754.
208. R. P. N. Veregin, P. G. Odell, L. M. Michalak and M. K. Georges, *Macromolecules*, 1996, **29**, 4161–4163.
209. P. M. Kazmaier, K. Daimon, M. K. Georges, G. K. Hamer and R. P. N. Veregin, *Macromolecules*, 1997, **30**, 2228–2231.
210. P. J. MacLeod, R. P. N. Veregin, P. G. Odell and M. K. Georges, *Macromolecules*, 1997, **30**, 2207–2208.
211. P. G. Odell, R. P. N. Veregin, L. M. Michalak and M. K. Georges, *Macromolecules*, 1997, **30**, 2232–2237.
212. R. P. N. Veregin, P. M. Kazmaier, P. G. Odell and M. K. Georges, *Chem. Lett.*, 1997, **26**, 467–468.
213. M. K. Georges, G. K. Hamer and N. A. Listigovers, *Macromolecules*, 1998, **31**, 9087–9089.
214. M. K. Georges, R. P. N. Veregin and K. Daimon, in *Controlled Radical Polymerization*, ed. K. Matyjasewski, ACS Symposium Series, vol. 685, American Chemical Society, 1998, pp. 170–179.
215. B. Keoshkerian, M. Georges, M. Quinlan, R. Veregin and B. Goodbrand, *Macromolecules*, 1998, **31**, 7559–7561.
216. P. J. MacLeod, R. P. N. Veregin, P. G. Odell and M. K. Georges, *Macromolecules*, 1998, **31**, 530–531.
217. K. A. Moffat, G. K. Hamer and M. K. Georges, *Macromolecules*, 1999, **32**, 1004–1012.
218. P. J. MacLeod, R. Barber, P. G. Odell, B. Keoshkerian and M. K. Georges, *Macromol. Symp.*, 2000, **155**, 31–38.

219. W. G. Skene, J. C. Scaiano, N. A. Listigovers, P. M. Kazmaier and M. K. Georges, *Macromolecules*, 2000, **33**, 5065–5072.
220. G. Moad, E. Rizzardo and D. H. Solomon, *Tetrahedron Lett.*, 1981, **22**, 1165–1168.
221. C. J. Hawker, *J. Am. Chem. Soc.*, 1994, **116**, 11185–11186.
222. C. J. Hawker, G. G. Barclay, A. Orellana, J. Dao and W. Devonport, *Macromolecules*, 1996, **29**, 5245–5254.
223. J. Dao, D. Benoit and C. J. Hawker, *J. Polym. Sci., Part A: Polym. Chem.*, 1998, **36**, 2161–2167.
224. C. J. Hawker, G. G. Barclay and J. Dao, *J. Am. Chem. Soc.*, 1996, **118**, 11467–11471.
225. E. Malmström, R. D. Miller and C. J. Hawker, *Tetrahedron*, 1997, **53**, 15225–15236.
226. C. J. Hawker, E. Elce, J. Dao, W. Volksen, T. P. Russell and G. G. Barclay, *Macromolecules*, 1996, **29**, 2686–2688.
227. C. J. Hawker, *Angew. Chem., Int. Ed. Engl.*, 1995, **34**, 1456–1459.
228. C. J. Hawker and J. L. Hedrick, *Macromolecules*, 1995, **28**, 2993–2995.
229. C. J. Hawker, D. Mecerreyes, E. Elce, J. Dao, J. L. Hedrick, I. Barakat, P. Dubois, R. Jérôme and W. Volksen, *Macromol. Chem. Phys.*, 1997, **198**, 155–166.
230. M. R. Leduc, C. J. Hawker, J. Dao and J. M. J. Fréchet, *J. Am. Chem. Soc.*, 1996, **118**, 11111–11118.
231. E. E. Malmström and C. J. Hawker, *Macromol. Chem. Phys.*, 1998, **199**, 923–935.
232. D. Benoit, C. J. Hawker, E. E. Huang, Z. Lin and T. P. Russell, *Macromolecules*, 2000, **33**, 1505–1507.
233. D. Benoit, V. Chaplinski, R. Braslau and C. J. Hawker, *J. Am. Chem. Soc.*, 1999, **121**, 3904–3920.
234. B. Didier, H. Eva, H. Brett, R. Ian, V. Robert, R. Marlene and J. H. Craig, in *Controlled/Living Radical Polymerization*, ed. K. Matyjasewski, ACS Symposium Series, vol. 768, American Chemical Society, 2000, pp. 123–137.
235. T. Fukuda, T. Terauchi, A. Goto, K. Ohno, Y. Tsujii, T. Miyamoto, S. Kobatake and B. Yamada, *Macromolecules*, 1996, **29**, 6393–6398.
236. A. Goto, T. Terauchi, T. Fukuda and T. Miyamoto, *Macromol. Rapid Commun.*, 1997, **18**, 673–681.
237. T. Fukuda and A. Goto, *Macromol. Rapid Commun.*, 1997, **18**, 683–688.
238. A. Goto and T. Fukuda, *Macromolecules*, 1997, **30**, 4272–4277.
239. A. Goto and T. Fukuda, *Macromolecules*, 1997, **30**, 5183–5186.
240. K. Ohno, Y. Tsujii and T. Fukuda, *Macromolecules*, 1997, **30**, 2503–2506.
241. K. Ohno, M. Ejaz, T. Fukuda, T. Miyamoto and Y. Shimizu, *Macromol. Chem. Phys.*, 1998, **199**, 291–297.
242. T. Fukuda, A. Goto, K. Ohno and Y. Tsujii, in *Controlled Radical Polymerization*, ed. K. Matyjasewski, ACS Symposium Series, vol. 685, American Chemical Society, 1998, pp. 180–199.
243. A. Goto, Y. Tsujii and T. Fukuda, *Chem. Lett.*, 2000, 788–789.

244. T. Fukuda, A. Goto and K. Ohno, *Macromol. Rapid Commun.*, 2000, **21**, 151–165.
245. A. Goto and T. Fukuda, *Macromol. Chem. Phys.*, 2000, **201**, 2138–2142.
246. A. Goto and T. Fukuda, *Macromolecules*, 1999, **32**, 618–623.
247. T. Fukuda, T. Terauchi, A. Goto, Y. Tsujii, T. Miyamoto and Y. Shimizu, *Macromolecules*, 1996, **29**, 3050–3052.
248. H. Okamura, N. Ide, M. Minoda, K. Komatsu and T. Fukuda, *Macromolecules*, 1998, **31**, 1859–1865.
249. N. Ide and T. Fukuda, *Macromolecules*, 1997, **30**, 4268–4271.
250. K. Ohno, Y. Tsujii, T. Miyamoto, T. Fukuda, M. Goto, K. Kobayashi and T. Akaike, *Macromolecules*, 1998, **31**, 1064–1069.
251. K. Ohno, T. Fukuda and H. Kitano, *Macromol. Chem. Phys.*, 1998, **199**, 2193–2197.
252. K. Ohno, Y. Izu, S. Yamamoto, T. Miyamoto and T. Fukuda, *Macromol. Chem. Phys.*, 1999, **200**, 1619–1625.
253. T. Kothe, S. Marque, R. Martschke, M. Popov and H. Fischer, *J. Chem. Soc., Perkin Trans. 2*, 1998, 1553–1560.
254. H. Fischer, *J. Polym. Sci., Part A: Polym. Chem.*, 1999, **37**, 1885–1901.
255. M. Souaille and H. Fischer, *Macromolecules*, 2000, **33**, 7378–7394.
256. S. Marque, C. Le Mercier, P. Tordo and H. Fischer, *Macromolecules*, 2000, **33**, 4403–4410.
257. D. Greszta, D. Mardare and K. Matyjaszewski, *Macromolecules*, 1994, **27**, 638–644.
258. K. Matyjaszewski, S. Gaynor, D. Greszta, D. Mardare and T. Shigemoto, *J. Phys. Org. Chem.*, 1995, **8**, 306–315.
259. J. M. Catala, F. Bubel and S. O. Hammouch, *Macromolecules*, 1995, **28**, 8441–8443.
260. S. A. F. Bon, F. A. C. Bergman, J. J. G. S. van Es, B. Klumperman and A. L. German, in *Controlled Radical Polymerization*, ed. K. Matyjasewski, ACS Symposium Series, vol. 685, American Chemical Society, 1998, pp. 236–255.
261. W. G. Skene, S. T. Belt, T. J. Connolly, P. Hahn and J. C. Scaiano, *Macromolecules*, 1998, **31**, 9103–9105.
262. M. Drache, A. Koch and G. Schmidt-Naake, *Angew. Makromol. Chem.*, 1999, **265**, 47–54.
263. D. Greszta and K. Matyjaszewski, *Macromolecules*, 1996, **29**, 5239–5240.
264. S. Butz, H. Baethge and G. Schmidt-Naake, *Angew. Makromol. Chem.*, 1999, **270**, 42–48.
265. D. Greszta and K. Matyjaszewski, *J. Polym. Sci., Part A: Polym. Chem.*, 1997, **35**, 1857–1861.
266. H. Baethge, S. Butz, C.-H. Han and G. Schmidt-Naake, *Angew. Makromol. Chem.*, 1999, **267**, 52–56.
267. B. Yamada, Y. Miura, Y. Nobukane and M. Aota, in *Controlled Radical Polymerization*, ed. K. Matyjasewski, ACS Symposium Series, vol. 685, American Chemical Society, 1998, pp. 200–213.

268. R. D. Puts and D. Y. Sogah, *Macromolecules*, 1996, **29**, 3323–3325.
269. R. D. Puts and D. Y. Sogah, *Macromolecules*, 1997, **30**, 7050–7055.
270. D. Benoit, S. Grimaldi, J. P. Finet, P. Tordo, M. Fontanilleand Y. Gnanou, in *Controlled Radical Polymerization*, ed. K. Matyjasewski, ACS Symposium Series, vol. 685, American Chemical Society, 1998, pp. 225–235.
271. S. Grimaldi, J.-P. Finet, F. Le Moigne, A. Zeghdaoui, P. Tordo, D. Benoit, M. Fontanille and Y. Gnanou, *Macromolecules*, 2000, **33**, 1141–1147.
272. D. Benoit, S. Grimaldi, S. Robin, J. P. Finet, P. Tordo and Y. Gnanou, *J. Am. Chem. Soc.*, 2000, **122**, 5929–5939.
273. R. Sophie and G. Yves, in *Controlled/Living Radical Polymerization*, ed. K. Matyjasewski, ACS Symposium Series, vol. 768, American Chemical Society, 2000, pp. 334–346.
274. S. A. F. Bon, M. Bosveld, B. Klumperman and A. L. German, *Macromolecules*, 1997, **30**, 324–326.
275. C. Marestin, C. Noël, A. Guyot and J. Claverie, *Macromolecules*, 1998, **31**, 4041–4044.
276. L. I. Gabaston, R. A. Jackson and S. P. Armes, *Macromolecules*, 1998, **31**, 2883–2888.
277. M. Hölderle, M. Baumert and R. Mülhaupt, *Macromolecules*, 1997, **30**, 3420–3422.
278. P. B. Zetterlund, Y. Kagawa and M. Okubo, *Chem. Rev.*, 2008, **108**, 3747–3794.
279. M. F. Cunningham, *Prog. Polym. Sci.*, 2008, **33**, 365–398.
280. S. Kobatake, H. J. Harwood, R. P. Quirk and D. B. Priddy, *Macromolecules*, 1997, **30**, 4238–4240.
281. S. Kobatake, H. J. Harwood, R. P. Quirk and D. B. Priddy, *Macromolecules*, 1998, **31**, 3735–3739.
282. S. Kobatake, H. J. Harwood, R. P. Quirk and D. B. Priddy, *Macromolecules*, 1998, **32**, 10–13.
283. S. Butz, J. Hillermann, G. Schmidt-Naake, J. Kressler, R. Thomann, B. Heck and B. Stühn, *Acta Polym.*, 1998, **49**, 693–699.
284. S. Butz, H. Baethge and G. Schmidt-Naake, *Macromol. Rapid Commun.*, 1997, **18**, 1049–1055.
285. E. Yoshida, T. Ishizone, A. Hirao, S. Nakahama, T. Takata and T. Endo, *Macromolecules*, 1994, **27**, 3119–3124.
286. I. Li, B. A. Howell, K. Matyjaszewski, T. Shigemoto, P. B. Smith and D. B. Priddy, *Macromolecules*, 1995, **28**, 6692–6693.
287. Z. Yucheng, I. Q. Li, B. A. Howell and D. B. Priddy, in *Controlled Radical Polymerization*, American Chemical Society, 1998, pp. 214–224.
288. I. Q. Li, B. A. Howell, M. T. Dineen, P. E. Kastl, J. W. Lyons, D. M. Meunier, P. B. Smithand D. B. Priddy, *Macromolecules*, 1997, **30**, 5195-5199.
289. M. Steenbock, M. Klapper, K. Müllen, N. Pinhal and M. Hubrich, *Acta Polym.*, 1996, **47**, 276–279.

290. D. Colombani, M. Steenbock, M. Klapper and K. Müllen, *Macromol. Rapid Commun.*, 1997, **18**, 243–251.
291. M. Steenbock, M. Klapper and K. Müllen, *Macromol. Chem. Phys.*, 1998, **199**, 763–769.
292. M. Steenbock, M. Klapper, K. Müllen, C. Bauer and M. Hubrich, *Macromolecules*, 1998, **31**, 5223–5228.
293. M. Steenbock, M. Klapper and K. Müllen, *Acta Polym.*, 1998, **49**, 376–378.
294. Y. Yağci, A. Başkan Düz and A. Önen, *Polymer*, 1997, **38**, 2861–2863.
295. I. Q. Li, B. A. Howell, R. A. Koster and D. B. Priddy, *Macromolecules*, 1996, **29**, 8554–8555.
296. B. A. Howell, D. B. Priddy, I. Q. Li, P. B. Smith and P. E. Kastl, *Polym. Bull.*, 1996, **37**, 451–456.
297. F. Ercole, J. Krstina, N. Le, L. Hutson, G. Moad, E. Rizzardo, S. H. Thang, C. L. Winzor and J. Chiefari, in *Engineered Resins Strategic Alliance Technical Report – January 1995*, CSIRO Chemicals and Polymers, 1995, pp. 45–49.
298. O. Guerret, J.-L. Couturier, J.-F. Lutz, C. Le Mercier, S. Robin and B. Vuillemin, *Alkoxyamine phosphonates and their use as polymerization catalysts*, ATOFINA, Fr., US6657043B1, 2003.

CHAPTER 2

Kinetic Aspects of Nitroxide Mediated Polymerization

ELENA G. BAGRYANSKAYA*[a,b] AND SYLVAIN R. A. MARQUE*[c]

[a] N. N. Vorozhtsov Novosibirsk Institute of Organic Chemistry SB RAS, Pr. Lavrentjeva 9, 630090 Novosibirsk, Russia; [b] Novosibirsk State University, Pirogova 2, 630090 Novosibirsk, Russia; [c] Aix Marseille Université, CNRS, ICR UMR 7273, 13397 Marseille cedex 20, France, *Email: egbagryanskaya@nioch.nsc.ru; sylvain.marque@univ-amu.fr

2.1 Nitroxide Mediated Polymerization (NMP) and Key Reactions that Determine its Kinetics

During the last decades, a number of polymerization techniques making it possible to obtain polymers with the desired and well defined properties have been developed,[1] among which are ionic controlled polymerization,[2,3] nitroxide mediated polymerization,[4–11] reversible addition-fragmentation chain transfer,[1,12–17] and atom transfer radical polymerization.[17–19]

The main features of ionic controlled polymerizations are polymers exhibiting very narrow mass distribution (polydispersity index, Figure 2.1a) and re-activable polymer chain ends allowing the re-initiation of the polymerization (Figure 2.1b).

Such properties are now available using controlled radical polymerization mediated by nitroxide, *aka* nitroxide mediated polymerization, invented in 1985,[20,21] as well as using other controlled radical polymerization techniques. Hence, polymers exhibiting complex architectures (Figure 2.1c) are now readily available for a moderate cost using the robust and versatile

RSC Polymer Chemistry Series No. 19
Nitroxide Mediated Polymerization: From Fundamentals to Applications in Materials Science
Edited by Didier Gigmes

Published by the Royal Society of Chemistry, www.rsc.org

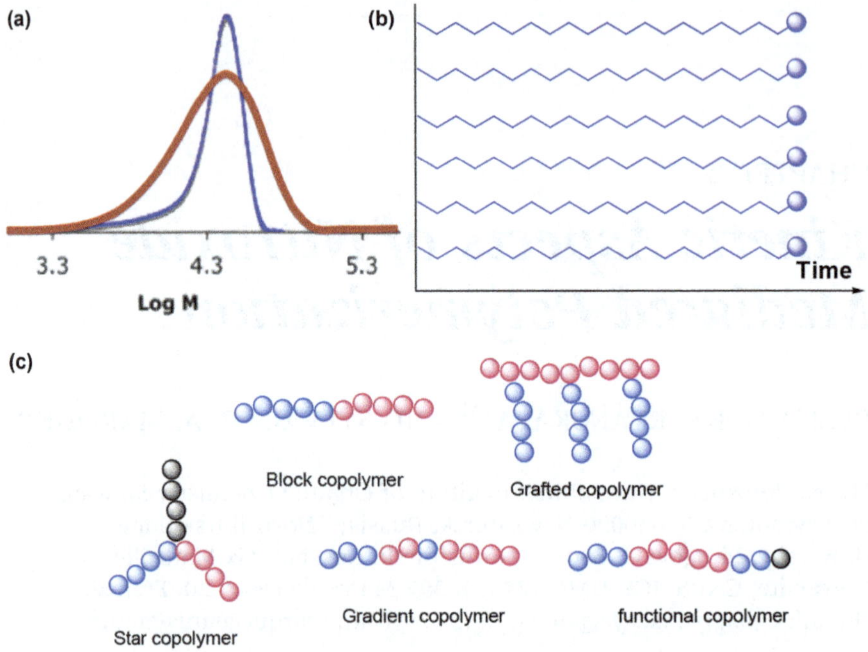

Figure 2.1 (a) Difference in chain length distribution between conventional radical polymerization (red bold line PDI = 2.5) and NMP or ionic polymerizations (blue thin line, PDI = 1.15). (b) Polymer with re-activable chain ends. (c) Example of complex architectures for polymers afforded by NMP. Reproduced from ref. 10.

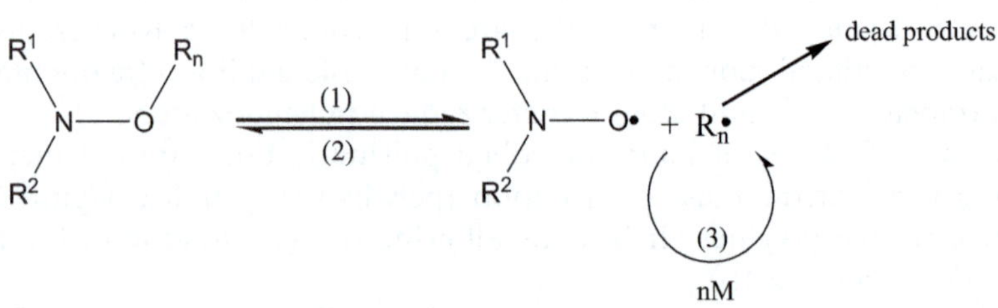

Scheme 2.1 Simplified scheme for NMP.

radical polymerizations under less drastic conditions. The NMP process—as described in the simplified Scheme 2.1—relies on the homolytic decomposition of an alkoxyamine[22–29] $R_1R_2NOR_n$ (reaction 1, Scheme 2.1) into the propagating alkyl radical $R_n^{\bullet}$ and the controlling nitroxide $R_1R_2NO^{\bullet}$. The recombination of $R_n^{\bullet}$ with $R_1R_2NO^{\bullet}$ (reaction 2, Scheme 2.1) leads to a very low concentration of $R_n^{\bullet}$, limiting the self-termination reaction of the alkyl radical, and allowing the controlled growth of the macro-radical (reaction 3, Scheme 2.1).

As soon as NMP was discovered, the conventional 3-stage scheme—initiation, propagation, and termination—of radical polymerization was proposed by Johnson[30] and analysed numerically (Scheme 2.2). However, the role of the initiation stage in the fate of NMP was disregarded, as long as the homolysis of the initiator was not slower than the homolysis of the dormant species ($k_{d,ds}$). Later, the importance of the alkoxyamine and the role of the homolysis of its C–ON bond in the fate of NMP were delineated by the work of Fukuda[31] and Matyjazewsky.[32] Then, Fischer[33–36] and Fukuda[37,38] developed independently the analytical kinetic equations linking the features of the polymer (chain length M_n, polydispersity index PDI, conversion) to the values of the rate constants for the propagation and termination stages. However, both approaches were performed assuming the independence of the rate constants from the chain length and the high persistence of the nitroxide. The complete kinetic scheme (Scheme 2.2) is composed of 3 stages, with k_d and $k_{d,ds}$ the homolysis rate constants of the initiator R^1R^2NOR and of the dormant species ds, respectively; k_c and $k_{c,ds}$ the rate constants for the cross-coupling between the alkyl radicals (the initiating one $R^•$ and the propagating one $RM_n^•$) and the controlling nitroxide $R^1R^2NO^•$, respectively; k_{add} the rate constant of addition of the initiating alkyl radical $R^•$ onto monomer M; k_p the propagation rate constant of the propagating radical $RM_n^•$; k_t the self-termination rate constants for the C-centred radicals; k_{dD} the rate constants for the intramolecular proton-transfer (Cope-type elimination) between the initiator and the dormant species; and k_{cD} the rate constant for the intermolecular H-transfer between nitroxides and C-centred radicals.

In this chapter, the peculiarities of these key reactions for NMP and the main factors determining their rate constants are viewed and discussed.

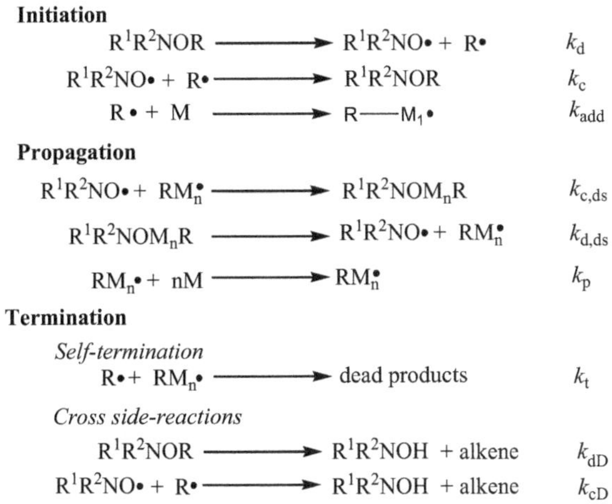

Scheme 2.2 Three-stage scheme for NMP. Reproduced from ref. 10.

2.1.1 Persistent Radical Effect (PRE) and its Application for NMP

When two radicals are generated at the same time, at the same rate, and one is more persistent than the other, the self-termination reactions are dramatically lowered, leading to an unusually high selectivity for the cross-coupling reaction (k_c in Scheme 2.3).[41] This effect was first observed for pentaarylethane by Bachmann in 1936.[39] However, it was "forgotten" and "re-discovered" regularly[34,40] until the kinetic analysis developed by Hans Fischer[41] in 1986 and the pioneer work on the NMP by Rizzardo and Solomon in 1985[20], and was named persistent radical effect (PRE) by Daikh and Finke.[35] PRE can be described by means of two kinetic schemes (Scheme 2.3) as long as the conditions mentioned above are fulfilled. Hence, the PRE occurs for radicals generated either from the decomposition of two different initiators (Scheme 2.3a) affording a transient radical $T^\bullet$ and a persistent radical $P^\bullet$ or from the decomposition of an initiator PT generating simultaneously a persistent radical $P^\bullet$ and a transient radical $T^\bullet$ (Scheme 2.3b). That is, the kinetics of PRE are installed when the self-termination rate constant for the "persistent" radical[42] $P^\bullet$ is at least 4 orders of magnitude lower than the k_t value for the transient radical $T^\bullet$. Another important, but often omitted, requirement underlying the occurrence of PRE is the absence of side reactions.[34]

Resolving the equations for these two systems leads to eqn (2.1) showing that the concentration of persistent species obeys a $t^{1/3}$ law (with $[PT]_0$ the initial concentration of alkoxyamine, k_d the rate constant of alkoxyamine homolysis, k_t the recombination rate constant of transient radicals $T^\bullet$, and k_c the recombination rate constant for radical $T^\bullet$ and persistent radical $P^\bullet$).

$$[P^\bullet] = [PT]_0^{2/3} \cdot \left(\frac{3k_d^2 \cdot 2k_t}{k_c^2}\right)^{1/3} \cdot t^{1/3} \tag{2.1}$$

Only a few examples displaying the perfect kinetic behaviour are known, mainly, due to the difficulty to observe the $t^{1/3}$ law over a reasonable time scale.[29,35,43,44]

One of the best examples highlighting the occurrence of PRE was provided by Fischer's group in 1998,[29] with the investigation of the decomposition of

(a)

(a1) $A \longrightarrow P^\bullet$

(a2) $B \longrightarrow T^\bullet$ $\quad k_d$

(b) $P^\bullet + T^\bullet \longrightarrow P\text{—}T$ $\quad k_c$

(c) $T^\bullet + T^\bullet \longrightarrow$ self-termination products $\quad k_t$

(b)

(a) $P\text{—}T \longrightarrow P^\bullet + T^\bullet$ $\quad k_d$

(b) $P^\bullet + T^\bullet \longrightarrow P\text{—}T$ $\quad k_c$

(c) $T^\bullet + T^\bullet \longrightarrow$ self-termination products $\quad k_t$

Scheme 2.3 (a) Bi-components system and (b) monocomponent system with $\mathbf{P}^\bullet$ and $\mathbf{T}^\bullet$ for persistent and transient radicals, respectively. D. Gigmes and S. R. A. Marque, Nitroxide Mediated Polymerization and its Applications, in: *Encyclopedia of Radicals in Chemistry, Biology, and Materials*, ed. C. Chatgilialoglu and A. Studer, Wiley, Chichester, U.K.[59]

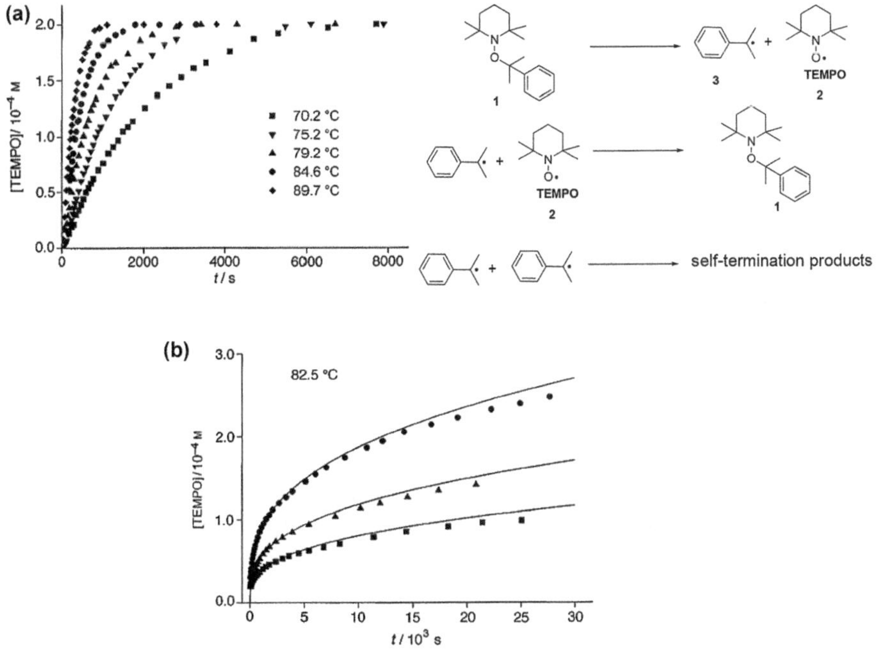

Figure 2.2 Thermolysis of **1** (a) in the presence of galvinoxyl as alkyl radical scavenger and (b) under PRE conditions.
Reproduced from ref. 29. (http://dx.doi.org/10.1039/a802773k).

the TEMPO-based alkoxyamine **1** complying with Scheme 2.3. Indeed, in **1** as PT, the C–ON bond homolysis (k_d in Scheme 2.3b) afforded a persistent nitroxide **2** and a transient alkyl radical $PhMe_2C^\bullet$ **3**. The latter either recombined with **2** to yield the starting material (cross-coupling reaction k_c in Scheme 2.3b) or self-terminated (k_t in Scheme 2.3b) to afford termination products (*aka* dead products). Indeed, the decay of **1** was dramatically accelerated in the presence of a free alkyl radical scavenger such as galvinoxyl (100% of **2** released in 3000 s at 79.2 °C, Figure 2.2a) whereas it was extremely slow in the absence of a scavenger (2% of **2** released in 30 000 s at 82.5 °C, Figure 2.2b) and obeyed a $t^{1/3}$ law [eqn (2.1)], as expected. Thus, experiments with scavengers showed faster homolysis of **1** whereas experiments without scavengers showed that the recombination reaction was favoured over the termination reactions.

Solving the kinetic equations afforded the evolution of each species over time until complete dissociation of **1** in the absence of alkyl radical scavenger. The main parameters determining the character of the kinetics are the rate constants for each elementary reaction. The kinetic behaviour of **1** exhibited three different regimes:

- A very short regime I (Figure 2.3) for less than 0.1 second where a transient and a persistent radical are generated at the same rate, obeying a law in *t*. Hence, during this regime, the self-termination reactions occur, affording an excess of persistent radical required for setting the PRE.

- A second regime II (Figure 2.3) lasting around *175 days* ($t \approx 10^7$ s) before significant decay of **1** (*ca.* 20%) can be observed. During that time, the pre-equilibrium is set, and the growth of the nitroxide obeys a $t^{1/3}$ law [eqn (2.1)] and the decay of the transient cumyl radical obeys a $t^{-1/3}$ law.
- The third regime III (Figure 2.3), corresponding to the time when 99% conversion of **1** is reached, lasts around *4 centuries* (!). During that time, the decays of alkoxyamine **1** and of the cumyl radical **3** obey a $1/t$ law.

More details on the development of the kinetic equations and the subsequent consequences are available in reviews by Fischer,[34,45–47] Fukuda[37,48] and the recent update[49] made by Fukuda and Matyjaszewski. It must be mentioned that some models had been proposed by Solomon and colleagues[30] and Smirnov[50] before the kinetics of PRE were recognized as governing the kinetics of NMP. The kinetic evidence given by **1** exemplifies nicely the importance of PRE in controlling the reactivity and selectivity of highly reactive radical species. Consequently, in an idealized case, such a long decay for an alkoxyamine makes it possible to perform polymerizations

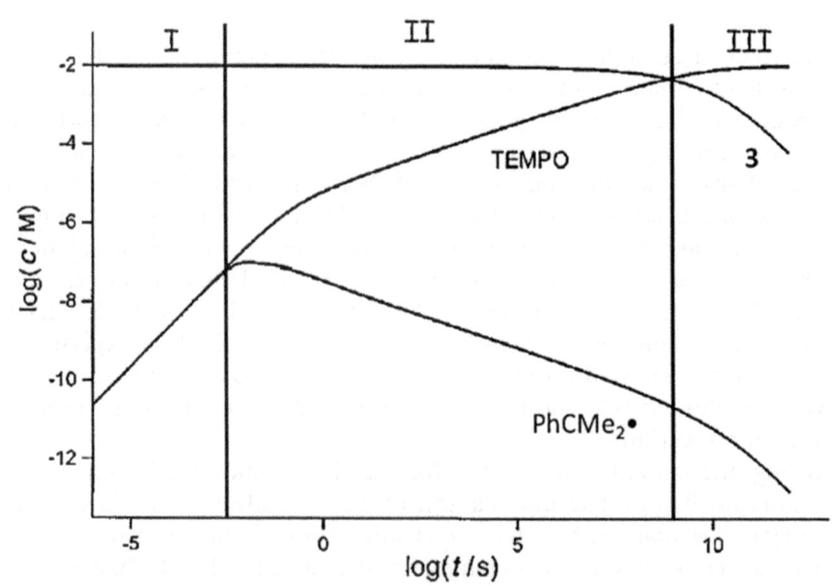

Figure 2.3 Evolution of **1** and of its subsequent radical species in log plot. $k_d = 2.3 \times 10^{-3}$ s^{-1}, $k_c = 9.1 \times 10^7$ M^{-1} s^{-1}, and $2k_t = 2.3 \times 10^9$ M^{-1} s^{-1} were given at 83 °C.
Reproduced from ref. 29. (http://dx.doi.org/10.1039/a802773k).

before significant loss of livingness is noticed. Besides, the significant role of PRE was also highlighted in recent applications of alkoxyamines in organic chemistry and biology.[51–55]

2.1.2 Equations for NMP and the Phase Diagram Approach

Taking into account the aforementioned requirements, Fischer[33,34,45–47,56,57] developed a set of equations using $k_{c,ds}$ and $k_{d,ds}$ as parameters. Indeed, the success of NMP depends on the occurrence of the quasi-equilibrium between dormant and active species. Thus, the PRE depends on the quasi-equilibrium constant $K = k_{d,ds}/k_{c,ds}$, and on a few kinetic requirements [eqn (2.2)]. Hence, it was possible to develop equations affording the control [PDI, eqn (2.3)], the livingness [ϕ, eqn (2.4)], and the polymerization time [$t_{conv.}$, eqn (2.5)] as function of several rate constants. From this set of equations, it is possible to build the so-called phase diagram (*vide infra*) and to predict the fate of NMP. As initiator concentrations are never higher than 0.1 M, $k_{t,ds}$ values are higher than 10^6 M^{-1} s^{-1} (except at very high conversions), and $k_{c,ds}$ values are often higher than 10^4 M^{-1} s^{-1}, the quasi-equilibrium condition is fulfilled for most of the alkoxyamines used in NMP.

$$K \ll \frac{k_{c,ds}[\text{initiator}]_0}{k_t}, \quad K \ll [\text{initiator}]_0, \quad k_{d,ds} \leq k_{t,ds}[\text{initiator}]_0 \tag{2.2}$$

$$\text{PDI}_\infty = 1 + \frac{[\text{initiator}]_0}{[\text{monomer}]_0} + \left(\frac{\pi k_p^3[\text{initiator}]_0}{k_{d,ds}k_{c,ds}k_{t,ds}}\right)^{1/2} \tag{2.3}$$

$$\phi = \left(\frac{2k_{d,ds}k_{t,ds}\ln(1/(1-\text{conversion}))}{k_{c,ds}k_p[\text{initiator}]_0}\right)^{1/2} \tag{2.4}$$

$$t_{conv.} = \left(\frac{(2\ln(1/(1-\text{conversion})))^3 \cdot k_{t,ds}}{9 \cdot k_p^3 \cdot [\text{initiator}]_0} \cdot \frac{k_{c,ds}}{k_{d,ds}}\right)^{1/2} \tag{2.5}$$

Using eqn (2.2)–(2.5), it is possible to plot a phase diagram that shows different sections leading to NMP experiments of various performances (Figure 2.4).[46,47] The line QE borders the zone where quasi-equilibrium is achieved, that is, above this line there can be no living and controlled polymerization. The bold lined zone, where point A lies, corresponds to the zone for which the polymerization exhibits an ideal control and a high livingness in the required polymerization time. The zone below the bolded line T and above the dashed line δ ($\delta = \text{PDI} - 1$) will provide a living and controlled polymerization in a longer time than aimed. The zone below the dashed line δ and the thick line ϕ will provide a poorly controlled but living polymer in a time depending on line T. On the other hand, the zone below the dashed line δ and above the thick line ϕ will provide a poorly living and

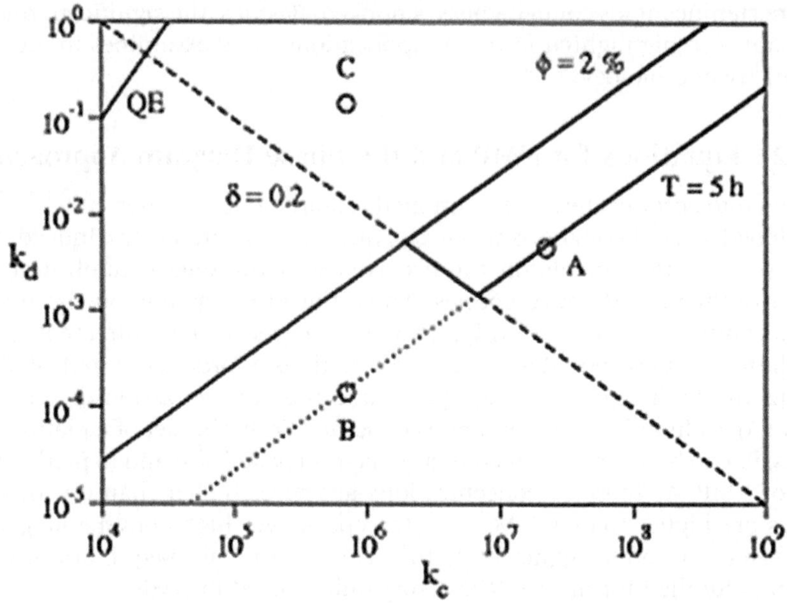

Figure 2.4 Range of rate constants $k_{d,ds}$ and $k_{c,ds}$ for optimized NMP experiments in $t_{90\%} = 5$ hours (eqn (2.5)) for 90% conversion with less than 2% unreactive polymer (eqn (2.4)) and PDI (δ = PDI-1, Equation X.3) smaller than 1.2. $k_p = 5000\ M^{-1}\ s^{-1}$, $k_t = 10^8\ M^{-1}\ s^{-1}$, and $[\text{initiator}]_0 = 0.1$ M. The Persistent Radical Effect in Living Radical Polymerization – Border Cases and Side-Reactions, H. Fischer and M. Souaille, *Macromol. Symp.*, 2001, **174**, 231–240.[47] Copyright Wiley-VCH Verlag GmbH & Co. KGaA.

poorly controlled polymer, that is, an unsuccessful NMP experiment. Interestingly, going from A to B, the product $k_{d,ds} \times k_{c,ds}$ decreases, K being constant, which implies that the same livingness and conversion at the same time is reached for B as for A, except that the polymer obtained for B will exhibit poor control (below the line δ). Indeed, the PDI of 1.1 for A increases to 3 for B, keeping the dead fraction smaller than 2%. Going from A to C, the K value increases, keeping the product $k_{d,ds} \times k_{c,ds}$ constant, implying that the same PDI and the same conversion are reached for C in a shorter time than for A, except that the polymer for C will exhibit good control but poor livingness. Indeed, the dead fraction of 2% for A increases to 50% for C keeping the PDI close to 1.1. "Obviously for NMP the properties of livingness and control do not mutually imply each other," as quoted by Fischer.[46,47] Importantly, it must be stressed that the construction of this diagram is based on an ideal case with no additional initial nitroxide, no additional alkyl radical generation, no side-reactions, and assuming the non-importance of the initiation stage. The effect of the last two terms on the fate of NMP will be discussed in the next section. The effect of the addition of extra alkyl radicals was thoroughly discussed by Fukuda.[37,48,58] It can originate from intentionally added initiators, residual peroxide impurities, or from the auto-initiation of the monomer.

$$A_2 \xrightarrow{r} 2R^{\bullet} \tag{2.6}$$

In the case when the additional alkyl radical $R^{\bullet}$ generation rate r is slow (eqn (2.6)), only the kinetics of polymerization are affected while eqn (2.2)–(2.5) still hold. When r is fast, the polymerization is faster, the $t^{1/3}$ law is not obeyed and the conversion follows the more conventional t law (eqn (2.7)), implying that eqn (2.2)–(2.5) need to be modified.[34] When extra initial nitroxide is added, the kinetics are also modified. The $t^{1/3}$ law is not obeyed any longer and is replaced by a t law. The polymerization slows down, which helps to control the monomer exhibiting high k_p values.[34,57]

$$\ln\frac{[\text{monomer}]_0}{[\text{monomer}]} = k_p\sqrt{\frac{r}{k_t}}t \tag{2.7}$$

These aspects as well as their consequences have been discussed in more detail in several reviews.[1,9,10,34,45–48,59] Moreover, the kinetics have also been extended to more complicated systems, such as copolymerization, compartimentalized polymerization, and emulsion.[7,9,60–64]

2.1.3 Evidence of the Occurrence of PRE in NMP

One of the first pieces of evidence was provided by Ohno *et al.*,[65] who described the polymerization of VLA **4** (*N*-(*p*vinylbenzyl)-[*O*-β-D-galactopyranosyl-(1→4)]-D-gluconamide) initiated by **5** and controlled by DBNO **6** (di*tertio*butyl-*N*-oxyl). As expected, the ($\ln([M]_0/[M])$ conversion obeyed the $t^{2/3}$ law (eqn (2.5) rearranged into eqn (2.8)), as displayed in Figure 2.5. Indeed, the most striking evidence was observed for the monomer/initiator ratio 20/1 (**4**/**5**, ▲ in Figure 2.5).[66] A few years later, investigating the NMP of styrene initiated by 7, Lutz *et al.*[67] showed that the conversion varied monotonically with $t^{2/3}$, as expected for various amounts of initiator 7 (eqn (2.8) and Figure 2.5b). These two examples confirmed nicely the approaches developed by Fischer[34] and Fukuda.[37]

$$\ln\frac{[M]_0}{[M]} = \frac{3}{2}\cdot k_p\cdot\left(\frac{k_d}{3k_c}\cdot\frac{[I]_0}{k_t}\right)^{1/3}\cdot t^{2/3} \tag{2.8}$$

However, it should be mentioned that the typical kinetic behaviour due to PRE might be spoiled by side-reactions such as thermal monomer self-initiation,[68] and also by the use of an extra amount of nitroxide,[49,69] of conventional initiators,[70] or of nitroxide reducing agents.[71]

2.1.4 Importance of k_c

The cross-coupling reaction between an alkyl radical and a nitroxide, also named the alkoxyamine re-formation reaction, is merely the back-reaction of

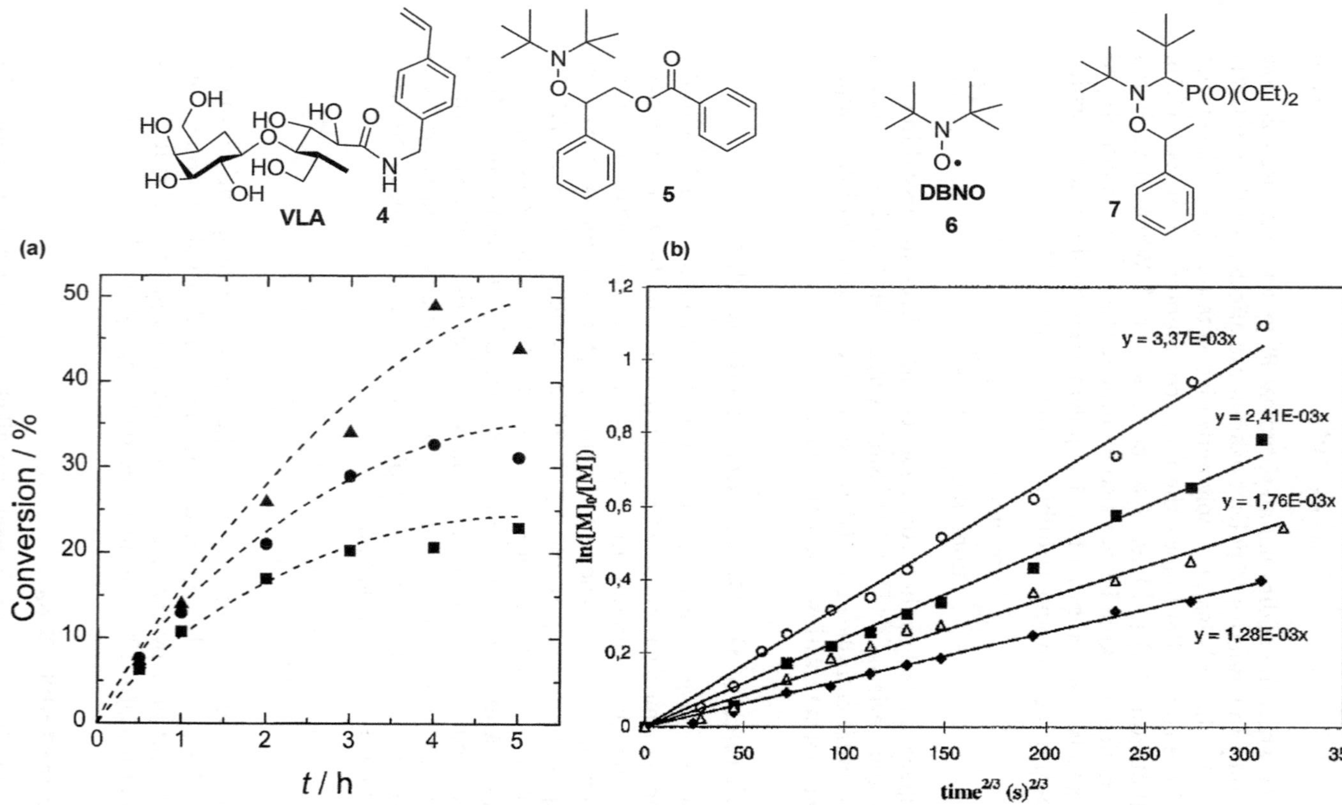

Figure 2.5 (a) Plots conversion *vs.* t for the NMP of VLA **4** carried out with the monomer/initiator ratios of 20/1 (▲), 40/1 (●), and 70/1 (■). (b) Plots conversion *vs.* $t^{2/3}$ for the NMP of styrene carried out with various amounts of 7: 0.188 M (○), 0.044 M (■), 0.023 M (△), and 0.009 M (♦).
(a) Reprinted with permission from K. Ohno, Y. Tsujii, T. Miyamoto, T. Fukuda, M. Goto, K. Kobayashi and T. Akaike, *Macromolecules*, 1998, **31**, 1064–1069. Copyright (1998) American Chemical Society.[65] (b) J. F. Lutz, P. Lacroix-Desmazes and B. Boutevin, *Macromol. Rapid Commun.*, 2001, **22**, 189.[67]

the alkoxyamine C–ON bond homolysis. Due to very high values of k_c (in general, higher than 10^6 M^{-1} s^{-1}), its impact is very difficult to observe and only a few nice examples are available in the literature. As reported in eqn (2.2)–(2.5), $k_{c,ds}$ (or k_c) plays an important role in the time of polymerization, on the control (PDI), and on the livingness of the polymer. In our opinion, the most striking evidence of its importance was provided by both Guillaneuf *et al.*[72] and Studer and Siegenthaler.[73]

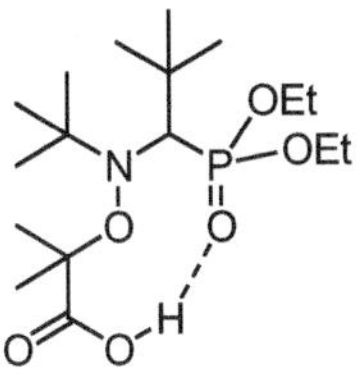

8

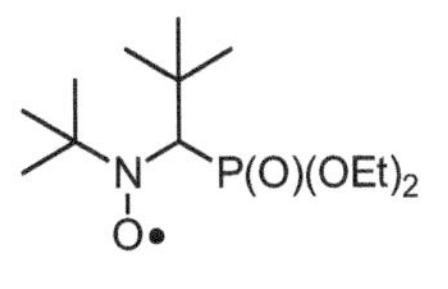

SG1
9

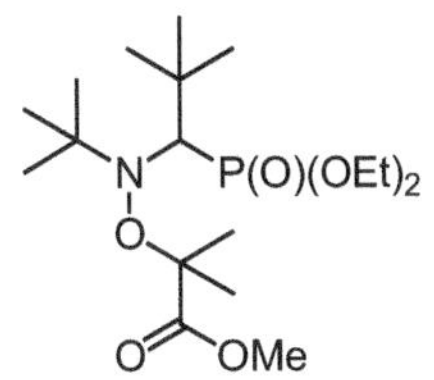

10

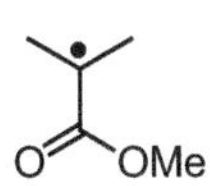

11

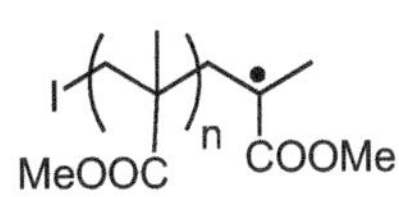

12

The analysis[46,74] of the phase diagram for the NMP of methyl methacrylate (MMA) initiated by **8** and controlled by SG1 **9**, taking into account the known k_d and k_c values of the corresponding molecular species **10**, predicted successful NMP (Figure 2.6a).[72] However, the authors found a dramatic effect of the polymer chain length on k_c ($k_c = 5.4 \times 10^6$ M^{-1} s^{-1} for the cross-coupling reaction between **9** and **11** and $k_{c,ds} = 6.0 \times 10^4$ M^{-1} s^{-1} for the cross-coupling reaction between **9** and the polymethylmethacrylyl radical **12**) and, when this fact was included in their simulations, it became obvious that the NMP of MMA under these conditions could only fail (red dotted arrow in Figure 2.6a).[72,75]

The other example provided by Siegenthaler and Studer[73] reported the NMP of styrene with alkoxyamines **13**, **14**, and **15** as initiators. Interestingly, while both polymerizations exhibited good control, the polymerization was much faster with **16** as the controlling agent (6 h for 50% conversion, and PDI = 1.15) than with **17** (29 h for 50% conversion and PDI = 1.15) although **13** and **14** exhibited close k_d values ($k_d = 6.7 \times 10^{-4}$ s^{-1} at 90 °C). On the other hand, the polymerization controlled by **18** (for **15**, $k_d = 4.9 \times 10^{-3}$ s^{-1} at 90 °C) was faster (2 h to reach 67% conversion) with poor control (PDI = 1.76). Such striking differences in the polymerization fate were ascribed to dramatic variations in $k_{c,ds}$ values. The Fischer's diagram (Figure 2.6b) using the k_c values estimated with an empirical multiparameter relationship predicts a polymerization controlled by **18** very close to the zone of poorly controlled and poorly living NMP.[76] On the other hand, NMPs controlled by **16** and by **17** are predicted to be living and controlled, with an experimental time shorter for **16** than for **17**.

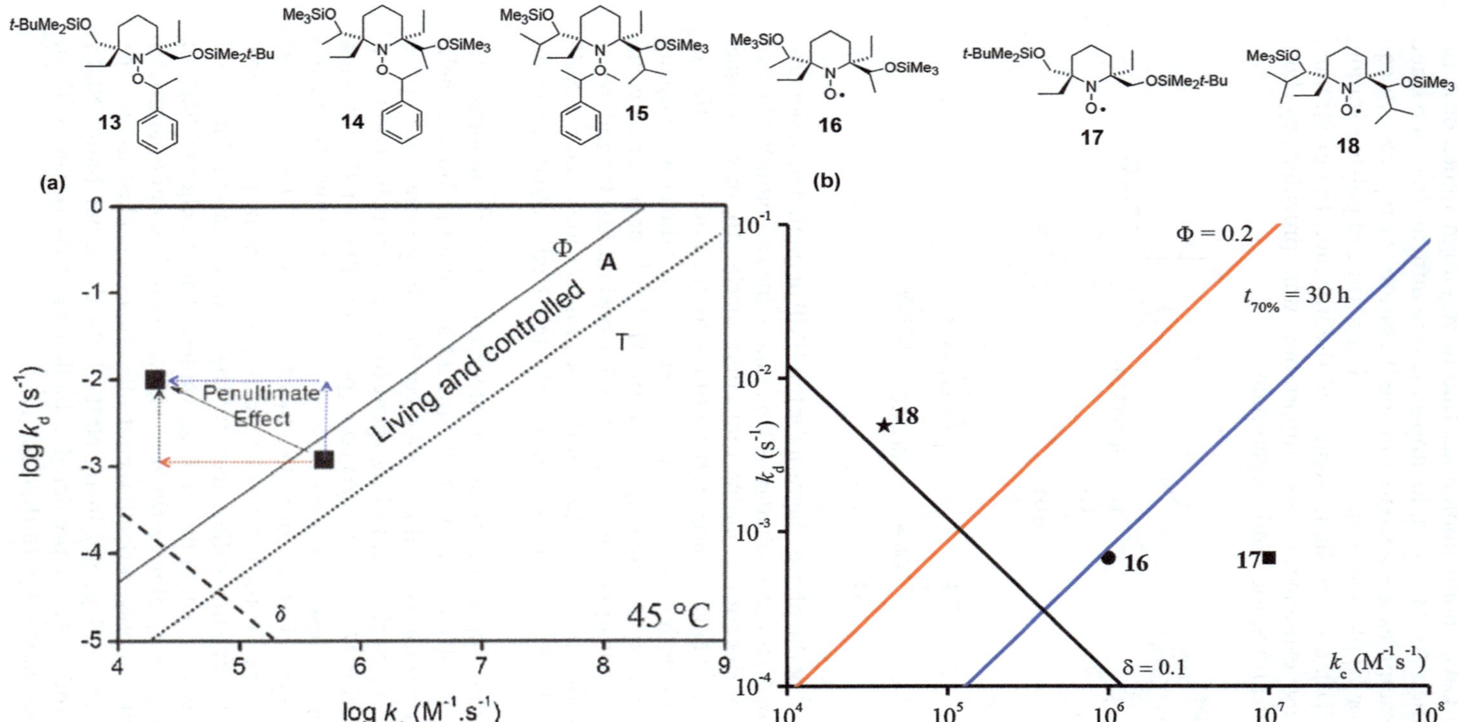

Figure 2.6 Fischer phase diagram for NMP: (a) of MMA, controlled by **9** at 45 °C (bottom square for **10** and top square for the polymeric species); (b) of styrene, controlled by **16** (●), **17** (■), and by **18** (★) at 90 °C.
(a) "Nitroxide Mediated Polymerization of Methyl Methacrylate using an SG1-Based Alkoxyamine. How Penultimate Effect could Lead to Uncontrolled and Unliving Polymerization" Y. Guillaneuf, D. Gigmes, S. R. A. Marque, P. Tordo and D. Bertin, *Macromol. Chem. Phys.*, 2006, **207**(6), 1278.[72] Copyright 2006 WILEY-VCH Verlag GmbH & Co. KGaA, Weinheim. (b) Reprinted with permission from E. G. Bagryanskaya, S. R. A. Marque and Y. P. Tsentalovich, *J. Org. Chem.*, 2012, **77**(11), 4996–5005.[76] Copyright 2012 American Chemical Society.

There is no doubt that these two examples highlight very well the importance of k_c for the fate of NMP experiments. A more detailed discussion and more examples of the importance of k_c for NMP have been provided in recent reviews.[9,10,59]

2.1.5 Importance of k_d

In general, radical reactivity involves the cleavage/formation of bonds and, thus, the occurrence of the reaction can readily be predicted using the bond dissociation energy (BDE) which mainly governs the reaction enthalpy and the activation barrier (Bell–Evans–Polanyi–Semenov relationship).[77,78] Consequently, the reactivity of alkoxyamines will be dramatically dependent on the strength of the C–ON bond and, thus, on the value of the C–ON bond homolysis rate constant k_d. Hence, the conversion C depends on the cubic root of k_d [eqn (2.8)] and the livingness ϕ depends on the square root of k_d [eqn (2.4)], whereas the polymerization time t depends on the reciprocal of the square root of k_d [eqn (2.5)], as does PDI [eqn (2.3)]. Many examples can be found in the literature to highlight how k_d can change the fate of NMP experiments.[10,34,59,79–81] We will consider two such examples related to the challenging NMP of MMA. Indeed, in the example displayed in Figure 2.6a, NMP of MMA controlled by **9** failed, as highlighted by the shift of the polymeric species from zone A to the zone above the line Φ (Figure 2.6a). In fact, the success of NMP was first estimated using the k_d and k_c values of the molecular model **10** (square in zone A of successful NMP, Figure 2.6a). However, when $k_{d,ds}$ ($k_{d,ds}$[82,83] $= 10^{-2}$ s^{-1} for **19** and $k_d = 3.1\times10^{-4}$ s^{-1} for **10**[84,85] at 45 °C) and $k_{c,ds}$ (*vide supra*) for the expected macroalkoxyamine were implemented in the phase diagram equations, the square in zone A moved first upward out of the zone A and then on the left, cancelling any chance of successful NMP of MMA controlled by **9** (blue dotted arrows in Figure 2.6a).

Nevertheless, although the bulk NMP of MMA controlled by **9** was shown to be unsuccessful (Figure 2.6a), Charleux and colleagues[86] showed that NMP of MMA controlled by **9** at 78 °C and at 83 °C were successful when performed in the presence of 4–8% of styrene as co-monomer (Figure 2.7a), as highlighted by the linear evolution of M_n with the conversion. Indeed, when the Fischer diagram is plotted, NMP of bulk MMA is not controlled at 75 °C, as expected (square for **19** in Figure 2.7b). On the other hand, when the ultimate unit is assumed to be a styrenic unit as for model **20** (circle in Figure 2.7b), a nicely successful NMP of MMA targeting livingness larger than 70% is predicted for such a dormant species, assuming k_d[87,88] $= 10^{-3}$ s^{-1} and k_c[89] $= 60\,000$ M^{-1} s^{-1}. In fact, due to the lesser steric hindrance of the ultimate styrenic moiety, $k_{c,ds}$ is slightly larger for **20** than for **19** and $k_{d,ds}$ is dramatically smaller for **20** than for **19**, insuring successful NMP of MMA (Figure 2.7). Any other combination between styrene and MMA as ultimate and penultimate units (MMA–MMA–, S–S–, S–MMA–) afforded unsuccessful NMP experiments.[86] Some monomers other than styrene were used to achieve the same efficiency in the NMP of MMA.[90]

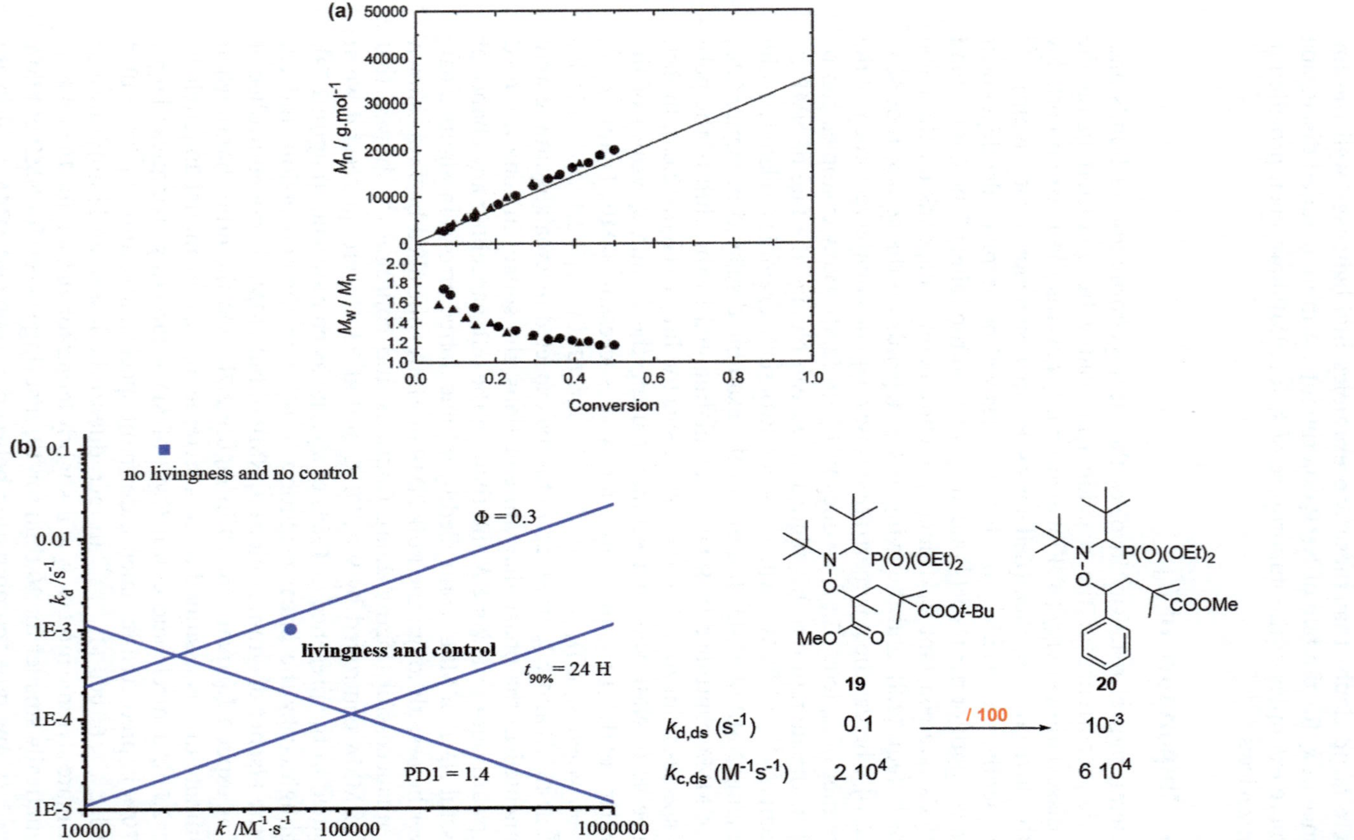

Figure 2.7 (a) NMP of MMA with 8% of styrene controlled by **9** (● at 83 °C and ▲ 78 °C, $M_{n,th} \approx 35\,000$ g mol^{-1}), (b) NMP of MMA with 8% of styrene controlled by **9** at 75 °C (square for **19** and circle for **20**).
(a) Reprinted with permission from J. Nicolas, C. Dire, L. Mueller, J. Belleney, B. Charleux, S. R. A. Marque, D. Bertin, S. Magnet and L. Couvreur, *Macromolecules*, 2006, **39**, 8274–8282.[86] Copyright 2006 American Chemical Society. (b) Reproduced from ref. 10.

These two examples highlight nicely how dramatic the effect of k_d can be on the fate of NMP and how it is important to own tools able to predict these values with one glance at the structure of the nitroxyl and alkyl fragments.

2.1.6 Importance of the Initiation Stage

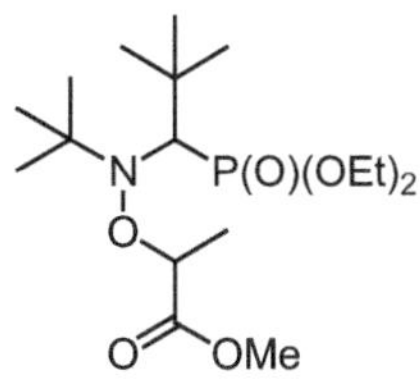

21

In the three main models developed by Johnson,[30] Fukuda[91] and Fischer,[34] the initiation stage was assumed to play a minor role, as long as k_d was not smaller than $k_{d,ds}$. Despite this assumption, the theoretical models nicely described the experimental observation concerning the kinetics of the polymerization, and the evolutions of the mass, polydispersity index and living fraction. Otherwise, for $k_d \ll k_{d,ds}$, poor control or inhibition would be observed.[34,45–48] Nevertheless, in some cases, depending hardly on the experimental conditions, the initiation stage may play a pivotal role.[92] For example, the NMP of styrene at 120 °C, controlled by **9** and initiated by both **21** and **8**, afforded the same mass evolution (Figure 2.8a) whereas when initiated at 90 °C by **7**, **8**, and **21** the control of the polymerization increased from **21** to **8** (Figure 2.8b). Consequently, the faster the homolysis of the initiating alkoxyamine, the better the control of the polymerization.[92]

At first glance, the first radical addition should not play any role. However, Bagryanskaya *et al.*[80] reported a very surprising kinetic behaviour when the NMP of styrene was initiated by either alkoxyamine **23** or **22**, both based on nitroxide **24**. While the NMP of styrene at 120 °C was controlled by **24** (Figure 2.9a), the expected kinetic behaviour was observed only for the experiment initiated with **23**, whereas the polymerization initiated by **22** seemed inhibited despite faster homolysis for **22**[93] than for **23**.[80] Edeleva[93] showed that during the cross-coupling reaction of **25** and **24** around 3% of the intermolecular H-transfer occurred (k_{cD} in Scheme 2.2). Consequently, as the addition of **25** onto styrene ($k_{add} \approx 10^4$ M^{-1} s^{-1}) was slower than that of **26** ($k_{add} \approx 3\ 10^4$ M^{-1} s^{-1}) and the cross-coupling of **24** with **25** ($k_c \approx 3\ 10^8$ M^{-1} s^{-1}) was clearly faster than that of **26** with **24** ($k_c = 2\ 10^7$ M^{-1} s^{-1}), the intermolecular H-transfer reaction (HAT) becomes preponderant over the addition onto styrene when **22** was used as initiator.

Using the series of TEMPO-based alkoxyamines **27–31**, Zubenko *et al.*[79] observed a very different evolution of the conversion with time depending

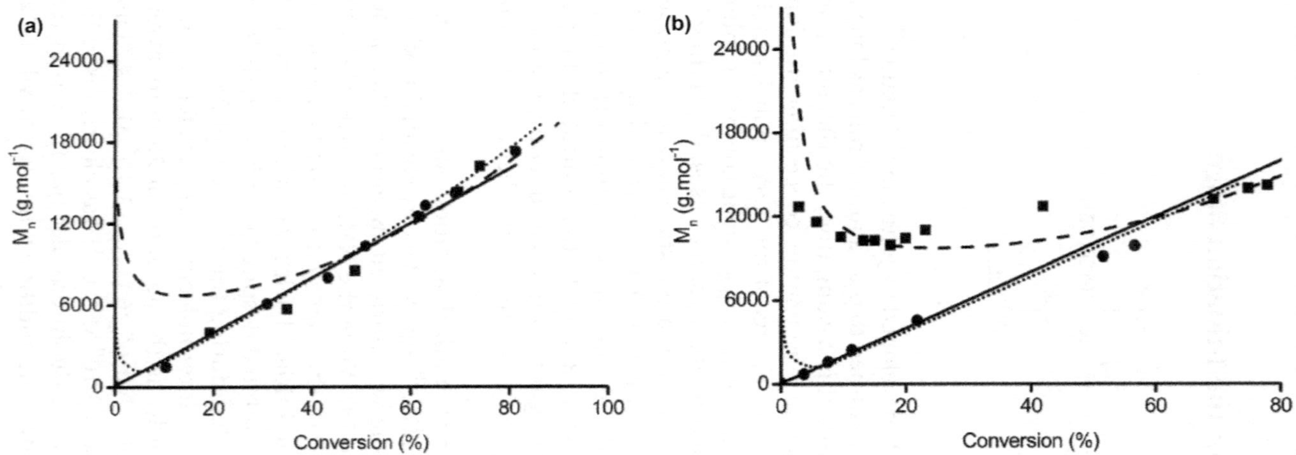

Figure 2.8 Effects of k_d of **7** (▲), **8** (●) and **21** (■) on the evolution of M_n with conversion for the polymerization of styrene by NMP (a) at 120 °C ($k_{d,8} = 0.28\ s^{-1}$, and $k_{d,21} = 10^{-3}\ s^{-1}$), and (b) at 90 °C ($k_{d,7} = 2.3 \times 10^{-4}\ s^{-1}$, $k_{d,8} = 1.7 \times 10^{-2}\ s^{-1}$, and $k_{d,21} = 3.6 \times 10^{-5}\ s^{-1}$). Reprinted with permission from F. Chauvin, P.-E. Dufils, D. Gigmes, Y. Guillaneuf, S. R. A. Marque, P. Tordo and D. Bertin, *Macromolecules*, 2006, **39**, 5238–5250.[92] Copyright 2006 American Chemical Society.

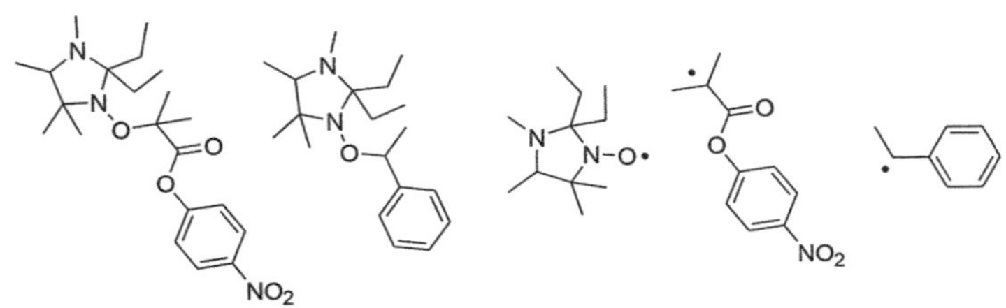

22 **23** **24** **25** **26**

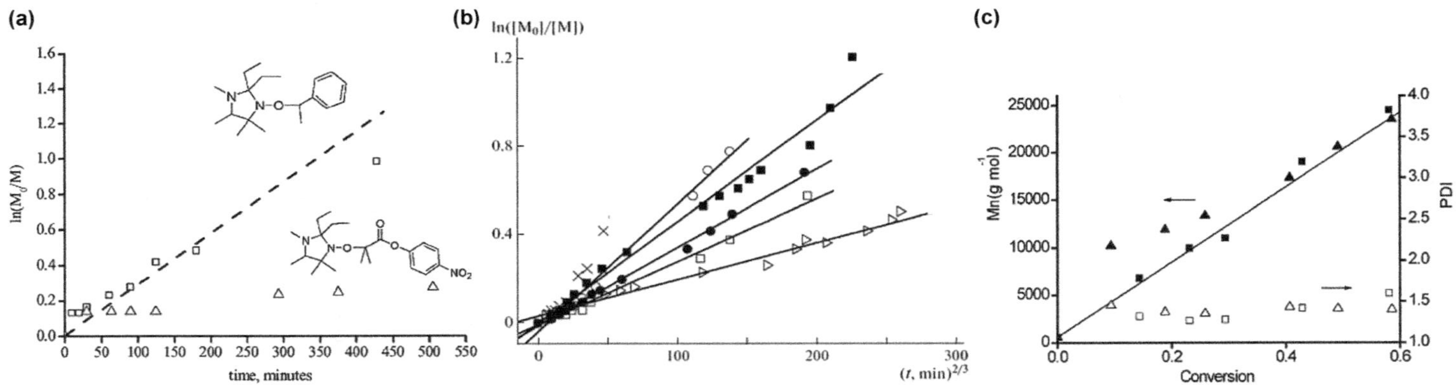

Figure 2.9 Plots conversion *vs.* time for NMP of styrene at 120 °C initiated by (a) **23** and **22**, and by (b) **27** (■), **28** (●), **29** (□), **30** (▷), **31** (○), and crosses are for its self-initiated polymerization. (c) NMP of MMA initiated by **33** (■) and **35** (▲) controlled by **34**. (a) "Can the first addition of alkyl radical play a role in the fate of NMP?" E. Bagryanskaya, D. Bertin, Gigmes, D. I. Kirilyuk, S. R. A. Marque, V. Reznikov, G. Roshchupkina, I. Zhurko and D. Zubenko, *Macromol. Chem. Phys.*, 2008, **209** (13) 1345.[80] Copyright 2008 WILEY-VCH Verlag GmbH & Co. KGaA, Weinheim. (b) "Role of the Alkyl Fragment of Initiating Alkoxyamine in Pseudoliving Polymerization of Styrene" D. P. Zubenko, I. A. Kirilyuk, S. R. A. Marque, I. P. Chuikov, E. G. Bagryanskaya, *Polymer Sci., Ser. B*, 2010, **52** (5–6), 327. Russian Version: *Vysokomol. Soedin., Ser. B*, 2010, **52** (6), 1030.[79] Copyright 2010 Springer. Reprinted with kind permission from Springer Science and Business Media. (c) Reprinted with permission from Y. Guillaneuf, D. Gigmes, S. R. A. Marque, P. Astolfi, L. Greci, P. Tordo and D. Bertin, *Macromolecules*, 2007, **40**, 3108.[97] Copyright 2007 American Chemical Society.

on the initiation fragment. Amazingly, the initiator **30** exhibiting the highest k_d value ($k_d = 0.13$ s^{-1} at 120 °C) was less efficient than the initiator **27** exhibiting a much lower k_d value ($k_d = 5.2 \times 10^{-4}$ s^{-1} at 120 °C). Indeed, the lack of efficiency of **30** during the initiation stage is due to the combination of 3 factors: (i) a faster cross-coupling between **2** and **32** ($k_c = 29.0 \times 10^8$ M^{-1} s^{-1} at 27 °C)[94] than between **2** and **26** ($k_c = 2.2 \times 10^8$ M^{-1} s^{-1} at 27 °C);[76] (ii) HAT ($k_{cD} = 5.6 \times 10^5$ M^{-1} s^{-1}) occurring between **2** and **32**; and (iii) a low rate constant of addition of **32** onto styrene ($k_{add} = 1.9 \times 10^4$ M^{-1} s^{-1}).

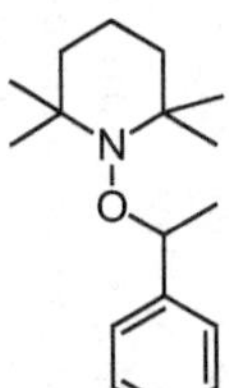

27

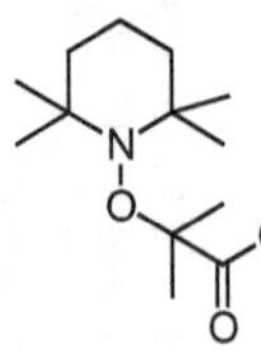

28

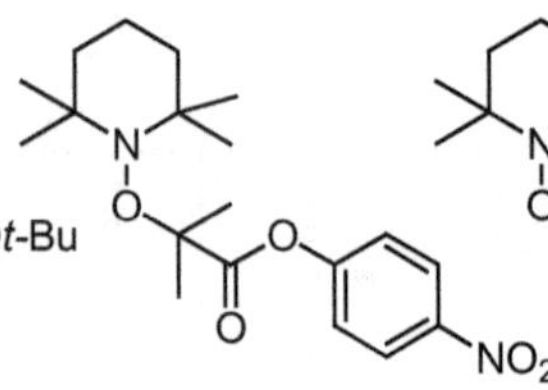

29

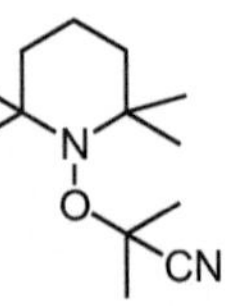

30

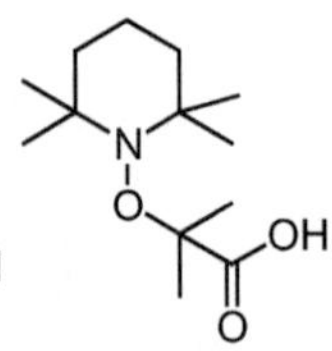

31

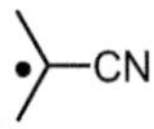

32

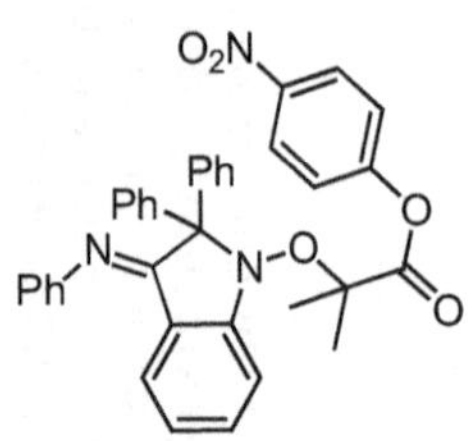

33

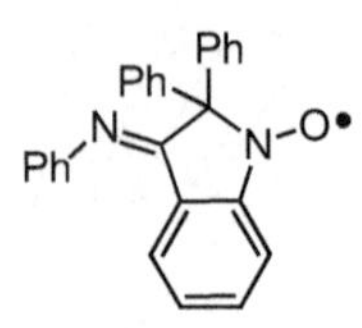

34

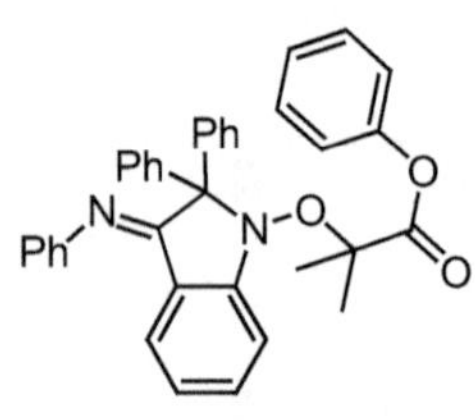

35

Recently, Harisson *et al.*[95,96] showed that the NMP of isoprene is more efficient when a secondary or primary SG1-based alkoxyamine initiators is used instead of a tertiary one.

Another nice example of the importance of the initiation stage was provided by Guillaneuf *et al.*,[97] who presented the first effective polymerization of methyl methacrylate (MMA) using **33** as initiating alkoxyamine, **34** as controlling agent and **25** as initiating alkyl radical, whereas the experiment was of lesser quality when **35** was used as initiating alkoxyamine (Figure 2.9c). Recently, using homologues of **34** and **35**, Astolfi *et al.* prepared a highly living block-copolymer of MMA and styrene.[98]

Hence, alkoxyamines with high k_d are recommended for initiation as long as the H-transfer reaction (k_{cD} in Scheme 2.2) cannot compete with the first addition of the initiating alkyl radical onto the monomer (k_{add} in Scheme 2.2). Such a kinetic behaviour was not observed with a SG1-based initiator because H-transfer reactions are very slow[99,100] in the case of SG1 (less than 0.5%).[101]

2.1.7 Importance of the Side-reactions on the Fate of NMP

The termination stage in Scheme 2.2 includes self-termination reactions and cross side-reactions. The self-termination reactions are needed to install the PRE because their occurrence helps to generate an excess of nitroxide in the very early stages of the polymerization.[34,48] The dimerization and disproportionation self-termination reactions of alkyl radicals were intensively investigated and reviewed in the late 1980s.[102] In most cases, the values of the rate constants for small alkyl and oligomeric radicals are close to diffusion control and are $\sim 10^9$ M^{-1} s^{-1}. Recently, Buback and colleagues[103] reviewed in details the dimerization and disproportionation self-termination reactions for polymeric radicals, which are very important for the least two stages of NMP (stage 2 and 3 in Scheme 2.2). It was shown that k_t values for alkyl and oligomeric radicals are several orders of magnitude larger than those for polymeric radicals. Thus to build a phase diagram, it is recommended to use the k_t values for the polymeric species when available.

The key cross side-reactions are the intramolecular proton transfer reaction, IPT (Cope type elimination in Scheme 2.4),[104] which occurs during the thermal decomposition of the alkoxyamine, and the intermolecular (Scheme 2.4) H-transfer reaction, HAT,[105] which occurs during the cross-coupling between C-centred radicals and nitroxides.[1,34,48,106,107]

Based on the preliminary investigations of Ananchenko *et al.*,[99,108] Edeleva *et al.*[93,109] provided a procedure to differentiate these processes (*vide infra*) and investigated the kinetics when both were occurring. Both side-reactions led to the same consequence: NMP of low performances.

Fischer described the effect of IPT and HAT on the fate of NMP.[56,99,108] When intermolecular H-transfer occurs, the decay of alkoxyamine is given by eqn (2.9), with $f_{cD} = k_{cD}/(k_{c,ds} + k_{cD})$ the disproportionation fraction, thus the maximal concentration C_{max} reachable, the fraction p of the maximum value of the polymerization index is reached, the optimal polymerization time t_p and the living fraction LF are given by eqn (2.10)–(2.12), respectively. Furthermore, the maximal f_{cD} tolerated to obtain a viable NMP experiment is given as a function of the conversion, LF, $k_{d,ds}$, $k_{c,ds}$, k_p, and k_t in eqn (2.13). As a rule of thumb, 0.5% H-transfer has no influence on NMP, as it can tolerate up to 2% of H-transfer (80% conversion, 40% livingness). For a larger percentage of H-transfer the NMP process is spoiled. These limits are modified of a few percent, in general, depending on the conversion and the livingness targeted and the values of k_p, $k_{c,ds}$ and $k_{d,ds}$.

$$[\mathrm{ds}]_t = [\mathrm{initiator}]_0 \cdot e^{-f_{cD} \cdot k_{d,ds} \cdot t} \tag{2.9}$$

$$C_{max} = 1 - \exp\left(-\frac{3}{2} \cdot k_p \left(\frac{[\mathrm{initiator}]_0}{3k_{t,ds} \cdot (k_{c,ds} + k_{cD}) \cdot k_{d,ds} \cdot f_{cD}^2} \right)^{1/3} \cdot 1.08 \right) \tag{2.10}$$

$$t_p = \frac{1}{f_{cD} k_{d,ds}} \ln\left(\frac{1}{2.59} \cdot \frac{2}{1-p} \right) \tag{2.11}$$

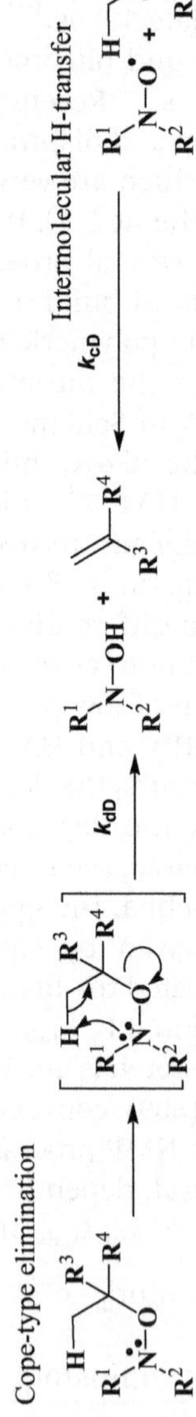

Scheme 2.4 Cope-type elimination and intermolecular H-transfer reactions.

$$\mathrm{LF} = \frac{[\mathrm{ds}]_{t_\mathrm{p}}}{[\mathrm{initiator}]_0} = \mathrm{e}^{-f_{\mathrm{cD}} \cdot k_{\mathrm{d,ds}} \cdot t_\mathrm{p}} \tag{2.12}$$

$$f_{\mathrm{cD}} < \frac{\ln(1/\mathrm{LF})}{\left(2\ln\frac{1}{1-C}\right)^{3/2}} \cdot 3k_\mathrm{p}^{3/2} \sqrt{\frac{[\mathrm{initiator}]_0}{k_{\mathrm{d,ds}} k_{\mathrm{c,ds}} k_{\mathrm{t,ds}}}} \tag{2.13}$$

It was shown that the effect of the intramolecular proton transfer (Cope-type elimination) was the same as for the intermolecular H-transfer and, consequently, the equations displayed above are suitable to predict the fate of NMP, except that the decay of the alkoxyamine (dormant species) is given by eqn (2.9) with $f_{\mathrm{dD}} = k_{\mathrm{dD}}/k_{\mathrm{d,ds}}$.

Due to very complicated kinetics, only few results are available to confirm the set of equations developed by Fischer and presented here.[65,107,110] Some of them have been discussed above and are displayed in Figures 2.6a and 2.9a,b.

Despite their importance, these side-reactions have not been extensively investigated and thus their occurrence is often considered as a random event and depends a lot on the structure of the nitroxide, the alkyl radical/fragment and the experimental conditions (*vide infra*). However, some trends can be drawn: (i) in general, in the case of the styryl radical, the H-transfer is negligible and does not hamper the polymerization;[1,34,48,108] (ii) the H-transfer is more favoured in the case of acrylyl radicals and not all nitroxides can afford controlled polymerization of this type of monomer;[34,106,108] (iii) H-transfer in the case of methacrylyl radicals occurs for most of the nitroxides making the NMP of MMA a challenge for chemists.[1,8,34,48,111]

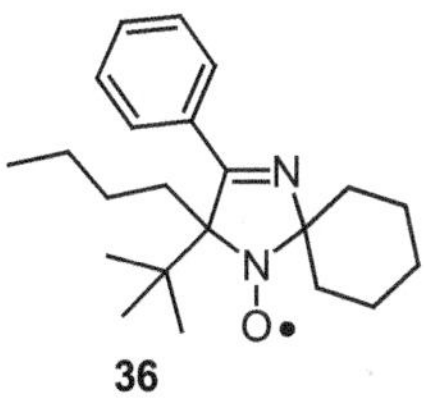

36

In good agreement with the theoretical predictions of Fischer *et al.*,[57] it was observed experimentally[69,71,112] that the decay of nitroxide due either to its self-decomposition or to the presence of an additive substantially shortened the polymerization time with no alteration of both control and livingness. The interesting outcome of this analysis is that a very unstable nitroxide (a several minutes of lifetime) can be used to control the polymerization. Amazingly, the occurrence of the (self-) decay of the nitroxide might balance the detrimental effect of the H-transfer reactions when suitable conditions are gathered. This very surprising aspect was confirmed and nicely highlighted by Edeleva *et al.*[93] with **36**.

The CO–N bond homolysis, another case of alkoxyamine decomposition, has not been thoroughly investigated. In general, it occurs for an alkoxyamine exhibiting BDE(C–ON) higher than 140 kJ mol^{-1}. Alkoxyamines based on **34** were successfully applied to the NMP of MMA, which exhibited very good control up to 80% conversion and good livingness.[97] Surprisingly, despite a rather low BDE(C–ON) of around 126.0 kJ mol^{-1},[109] such nitroxide did not undergo the conventional decay reactions affording the hydroxylamine.[109] In fact, the occurrence of the N–OC bond cleavage[113] is likely to moderate the occurrence of intermolecular H-transfer by decreasing the amount of nitroxide.[114]

2.2 Recombination of Nitroxides and Alkyl Radicals

The typical values of the rate constant of the cross-coupling between nitroxide and an alkyl radical, k_c, are close to the recombination rate constants of fast radical reactions and cover the range 10^6–10^9 M^{-1} s^{-1}. The peculiarities of recombination of nitroxides and C-centred radicals have recently been thoroughly and extensively reviewed.[115] All rate constants (k_c) measured during the last 20 years both for molecular and polymeric species have been gathered in the aforementioned review. In this section, the review of the main k_c measurement techniques is presented and the main factors determining k_c are discussed.

2.2.1 Techniques and Methods of Determination

During the last decades, different approaches for the measurement of k_c and $k_{c,ds}$ have been developed.[115] The direct measurement of k_c requires techniques with nanosecond time resolution, such as laser flash photolysis kinetic absorption spectroscopy (LFP-KAS), where the decay of the C-centred radicals is measured *via* their UV absorbance at different nitroxide concentrations.[116] Other approaches to the measurements of k_c are based on nondirect measurement methods: modification of LFP using an additional probe,[117] time-resolved chemically induced dynamic nuclear polarization (CIDNP),[29] CIDNP with fast switching of the external magnetic field,[118] the radical clock method,[119] C-centred radical nitroxide trapping (competitive method),[116] and radical nitroxide recombination-pulsed lamp photolysis-size exclusion chromatography (RNR-PLP-SEC).[120] The most popular techniques—LFP-KAS and RNR-PLP-SEC—are discussed hereafter.

2.2.1.1 Laser Flash Photolysis-Kinetic Absorption Spectroscopy (LFP-KAS)

LFP-KAS is the most widely used and convenient technique to study fast radical reactions, and has been accurately described in many books and articles (LFP-KAS experimental setup includes pulse laser and digital oscilloscope).[115] LFP-KAS involves the generation of radicals using laser irradiation and the monitoring of the short-lived species by UV-visible at the

$$\mathrm{I} \xrightarrow{h\nu} \mathrm{R}^\bullet \quad (1)$$

$$2\mathrm{R}^\bullet \xrightarrow{k_t} \mathrm{R_2} \quad (2)$$

$$\mathrm{R}^\bullet + \mathrm{R_1R_2NO}^\bullet \xrightarrow{k_c} \mathrm{R_1R_2NOR} \quad (3)$$

Scheme 2.5 Reactions required for the LFP-KAS method.

nanosecond time-scale (1–10 ns) set by the duration of the laser pulse and the response time of the oscilloscope.

The first example of the use of LFP-KAS to determine k_c was proposed by Ingold's and Beckwith's groups.[121,122] C-centred radicals were generated by photolysis (reaction 1 in Scheme 2.5) of symmetrical ketones (Norrish 1 type decomposition) or of symmetrical acyl peroxides, followed by monitoring of the C-centred radical UV-absorption A. For a large excess of nitroxide (reaction 3 in Scheme 2.5), a pseudo-first order decay was monitored[116,119,123] and k_{exp} is given by eqn (2.14)

$$\ln\frac{A_t}{A_0} = -\,k_{exp}\cdot t \tag{2.14}$$

$$\frac{A_t}{d} = \frac{k_{exp}}{\frac{2k_t}{\varepsilon}\left[\left(1+\frac{k_{exp}\;d}{2k_t/\varepsilon A_0}\right)e^{k_{exp}t}-1\right]} \tag{2.15}$$

$$k_{exp} = k_0 + k_c\cdot[\mathrm{R_1R_2NO^\bullet}] \tag{2.16}$$

When a large excess of nitroxide cannot be used (high absorbance of nitroxide at the wavelength of irradiating light; overlap of nitroxide and C-centred radical UV spectra; dimerization of nitroxides, *etc.*), k_{exp} is contributed by both first and second order decays and is given by eqn (2.15); (A_0 and A_t for the absorption at initial time and at time t, respectively, k_0 includes all reactions except for the coupling between C-centred radicals and nitroxides, d is the optical density, and k_t is the self-termination reaction of C-centred radicals).[117] In both cases, k_c is obtained by plotting k_{exp} against the nitroxide concentration $[\mathrm{R_1R_2NO^\bullet}]$ [(eqn (2.16)].

The main drawbacks of this method are sometimes the tedious preparation of elaborated radical initiators, the low extinction coefficients of some of the C-centred radicals, the overlap of the absorption spectra of the C-centred radical and of the reactants or products in some cases, and, up to now, this technique can only be applied to molecular radical species. Thus, other variants of the FLP-KAS method can be used or different methods can be applied.

2.2.1.2 *Radical Nitroxide Recombination-Pulsed Lamp Photolysis-size Exclusion Chromatography (RNR-PLP-SEC)*

As mentioned above, the LFP-KAS technique is only suitable for measuring the k_c of monomeric species because it requires the "instantaneous" formation of C-centred radicals.[124] The RNR-PLP-SEC technique was designed especially for measuring k_c of polymeric species.[120,125] The initiator (most often an arylketone) is irradiated in the presence of monomer and nitroxide. Hence, all the chains are generated instantaneously and are scavenged by the nitroxide before the next pulse. The solution is irradiated several times (in a row) to afford a concentration in polymer species large enough for analysis by SEC to afford the average values of M_n and M_w. Nevertheless, for analysis by SEC, the determination of $k_{c,ds}$ must comply with several requirements:

- To prevent homolysis of the formed alkoxyamine, the experimental temperature T must be low enough in comparison with the critical temperature of homolysis T_c [eqn (2.17)];
- 98% of the generated macro-radicals must decay between each pulse [eqn (2.18), where t_d is the interval time, corresponding to the reverse of the irradiation frequency and $[NO^\bullet]_0$ the initial concentration of nitroxide];
- The accuracy of $k_{c,ds}$ relies on the accuracy of the SEC technique, which is generally reliable for M_n greater than 1000 g mol^{-1} [eqn (2.19), where DP_{min} is the value determined accurately using the SEC method for the lower M_n values, A_p and E_{ap} are the pre-exponential and the activation energy of the propagation reaction, respectively, $[M]_0$ is the initial concentration in monomer, and ln10 is required for a large excess of nitroxide];
- The contribution of the self-termination reaction of propagating radicals is negligible [eqn (2.20), with $A_{t,ds}$ and $E_{a,t,ds}$ the pre-exponential and the activation energy of the self-termination reaction of propagating radicals].

$$\frac{1}{T} \geq \frac{1}{T_c} \tag{2.17}$$

$$\log[NO^\bullet]_0 \geq \log\left(\frac{4}{k_{c,ds} \cdot t_d}\right) \tag{2.18}$$

$$\frac{1}{T} \leq \frac{R\ln(10)}{E_{a,p}}\left[-\log\left(\frac{k_{c,ds} \cdot DP_{min}}{A_p \cdot [M]_0}\right) - \log[NO^\bullet]_0\right] \tag{2.19}$$

$$\frac{1}{T} \geq \frac{R\ln(10)}{E_{a,t,ds}}\left[-\log\left(\frac{10A_{t,ds} \cdot [P^\bullet]_0}{k_{c,ds}}\right) - \log[NO^\bullet]_0\right] \tag{2.20}$$

Using these equations, a phase diagram affords the zone that is suitable to perform reliable and accurate determination of $k_{c,ds}$, as exemplified for the cross-coupling rate constant $k_{c,ds}$ between **2** and polystyryl radical **37** (gray zone in Figure 2.10).

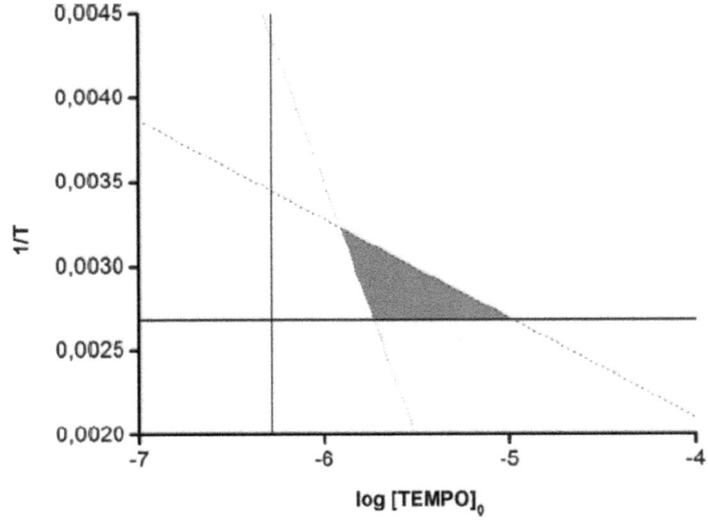

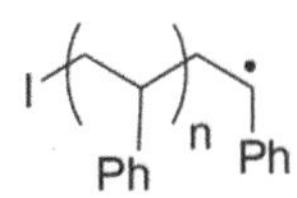

37

Figure 2.10 Phase diagram for the cross-coupling reaction between **2** and **37** (gray zone suitable for RNR-PLP-SEC experiment). Temperature requirement [bold line, eqn (2.17)], single pulse condition [thin line, eqn (2.18)], SEC requirement [dashed line, eqn (2.19)], and cross-coupling reaction as the main chain growth breaking event [dotted line, eqn (2.20)]. $k_p = 4.27\ 10^7 \exp(-32\,500/RT)$ L mol^{-1} s^{-1}, $k_t = 1.06\ 10^9 \exp(-6500/RT)$ L mol^{-1} s^{-1}; $[P^\bullet]_0 = 10^{-7}$ mol L^{-1}, $DP_{min} = 10$, $[M]_0 = 8.74$ mol L^{-1}, frequency = 10 Hz; $T_c = 100$ °C, preestimated $k_{c,ds} = 10^8$ L mol^{-1} s^{-1}. Reprinted with permission from Y. Guillaneuf, D. Bertin, P. Castignolles and B. Charleux, *Macromolecules*, 2005, **38**, 4638.[120] Copyright 2005 American Chemical Society.

Thus, k_{obs} was obtained by plotting either eqn (2.21), eqn (2.22) or eqn (2.23), each depending on a different assumption: the Mayo method,[126] linking the DP_n to the propagation rate constant and the chain growth breaking constants, the Flory–Schultz distribution,[127] linking the weight-mass average $DP_w/2$ (DP_w being generally more accurately determined than DP_n) to the propagation rate constant and the chain growth breaking constants, and the chain length distribution determination (x_i) method, as given by Gilbert,[128] respectively.

$$\overline{DP}_n = \frac{k_p[M]_0}{k_{c,ds}[NO^\bullet]_0}. \tag{2.21}$$

$$\frac{DP_w}{2} = \frac{k_p[M]_0}{k_{c,ds}[NO^\bullet]_0} \tag{2.22}$$

$$\frac{d[\ln(x_i)]}{di} = -\frac{k_{c,ds}[NO^\bullet]_0}{k_{c,ds}[NO^\bullet]_0 + k_p[M]_0} \tag{2.23}$$

The limitations of this technique are: (i) all chains are required to be initiated at the same time; (ii) very drastic kinetic conditions are required to measure reliable and accurate k_c values; (iii) reliable and accurate values of k_p are needed; (iv) the k_c values are averaged for short polymer chains; and (v) $k_{c,ds}$ size hints must be provided.

2.2.1.3 *Competitive Inhibition Method*

Recently Zaremski and colleagues[129] proposed an approach for the measurements of the rate constants of recombination of propagating radicals with nitroxides in NMP *via* the competitive inhibition method with the use of ESR spectroscopy. In this method, polymerization is performed in the presence of two nitroxides A and B. The rate constant of coupling between the nitroxide A or B is taken as reference and the monomer propagating radical, k_{c1}, should be known. If the rates of consumption of A and B, w_A and w_B, of both nitroxides can be measured separately, the rate constant of coupling of propagating radicals with the second nitroxide, k_{c2}, can be estimated through the ratio $w_A/w_B = d[A]/d[B] = k_{c1}[A]/k_{c2}[B]$. The advantage of this method is that there is no need to know the exact concentrations of A and B in the experiments because, in the calculations of k_{c2}, the concentrations of the reagents are in relative units ($d[A]/[A]$ and $d[B]/[B]$), so the values of [A] and [B] can be in arbitrary units. Using this approach, the authors showed that the values of recombination rate constants obtained for polymer radicals of styrene, acrylic acid, and methyl methacrylate turn out to be an order of magnitude lower than those known for low molecular mass model systems.

2.2.2 Arrhénius Parameters

For the nitroxide–alkyl radical cross-coupling reaction, only 5 Arrhénius parameters have been obtained from direct measurement by LFP-KAS although 13 of them have been determined.[115] These 5 parameters are in the expected range: 10^6–10^8 M^{-1} s^{-1} for the pre-exponential factor A_c and 3–10 kJ mol^{-1} for the activation energy $E_{a,c}$.[115] The cross-coupling reaction between alkyl radical and nitroxide can be described by either a one-step mechanism (Scheme 2.6 left) or a two-step mechanism (Scheme 2.6 right) involving the formation of a complex intermediate.

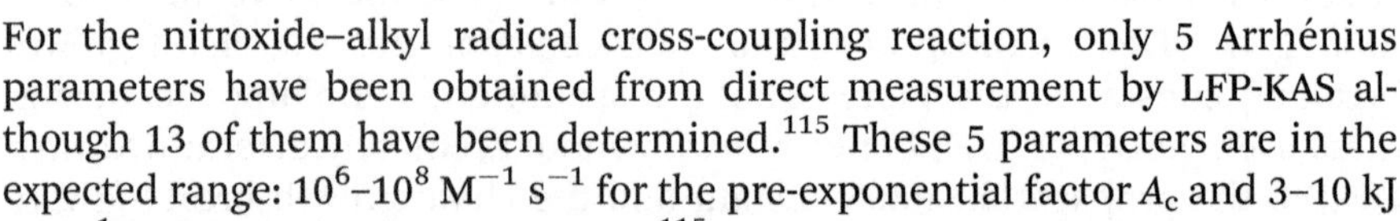

Scheme 2.6 One- (left hand side) and two-step (right hand side) mechanism for the cross-coupling reaction between an alkyl radical and a nitroxide.

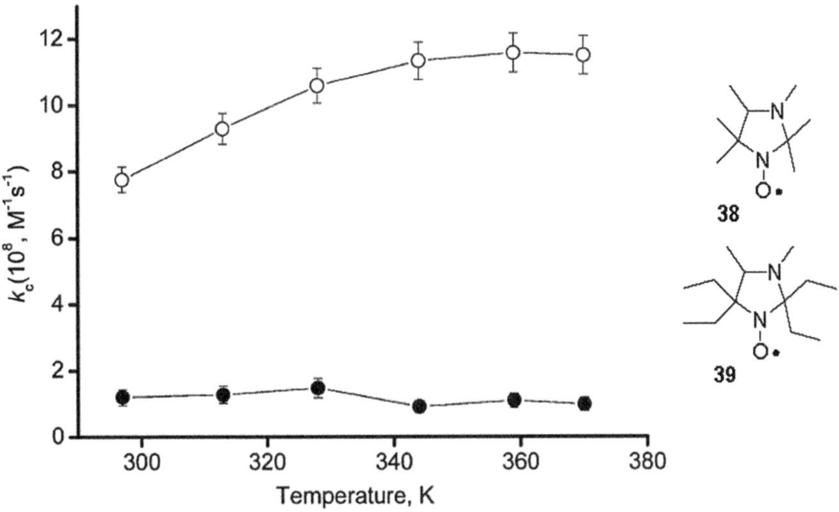

Figure 2.11 Arrhénius plots of T *vs.* k_c for the cross-coupling reaction between alkyl radical **11** and nitroxides **38** (○) and **39** (●).
Reprinted with permission from D. Zubenko, Y. Tsentalovich, N. Lebedeva, I. Kirilyuk, G. Roshchupkina, I. Zhurko, V. Reznikov, S. R. A. Marque and E. Bagryanskaya, *J. Org. Chem.*, 2006, **71**(16), 6044.[130] Copyright 2006 American Chemical Society.

Moreover, depending on the structures of the nitroxide and of the alkyl radical, non-Arrhénius behaviours have been reported, that is, k_c values decrease with increasing temperature.[115] Hence, the non-Arrhénius behaviour as displayed in Figure 2.11,[130] exemplified with the coupling reaction of **11** with **38** or **39**, can be due either to a two-step mechanism (Scheme 2.6)—as proposed by Ingold and colleagues[116] for high viscosity solvents—or to a one-step mechanism implying an activation entropy effect $\Delta S^{\neq}$ (negative values to the coupling of congested radicals) predominant over the activation enthalpic effect $\Delta H^{\neq}$ (low values due to the coupling of two stabilized radicals), as proposed by Fischer and colleagues,[117] implying a decrease in k_c with increasing temperature.

2.2.3 Effects Influencing k_c

Two different approaches to the analysis of numerous experimental data on k_c for different nitroxides and alkyl radicals were developed. The first approach based on the bond dissociation energy (BDE) of alkanes described the effect of the stabilization of the alkyl radicals, while the second one used molecular descriptors for the steric, polar and stabilization effects of the alkyl radicals and nitroxides.[115]

2.2.3.1 *Bond Dissociation Energy Approach to the Structural Effects in Alkyl Radicals*

The plots of BDE[115,131] *vs.* $E_{a,c}$ for 5 series of nitroxides **2**, **6**, **9**, **40**, and **41** (Figure 2.12) showing several correlations for the same nitroxide or scattered plots mean that this approach provides only very rough values of E_a or k_c.

2.2.3.2 *Multiparameter Approach to the Structural Effects in Alkyl Radicals*

To overcome the drawbacks of the plots BDE(C–H) *vs.* $E_{a,c}$, a multiparameter approach relying on the stabilization (σ_{RSE}), polarity (σ_I) and steric (υ) effects of the released alkyl radicals was developed affording, rather accurate values of k_c [eqn (2.24)]. Values of k_c are given by eqn (2.24) for 5 nitroxides and exemplified with **9** and **2** in Figure 2.13 and eqn (2.25),[132,133] and eqn (2.26),[134,135] respectively.

$$\log(k_c/\mathrm{M}^{-1}\,\mathrm{s}^{-1}) = \log k_{c,0} + \rho_{RS}\cdot\sigma_{RS} + \delta\cdot\upsilon + \rho_I\cdot\sigma_I \tag{2.24}$$

$$\log(k_c/\mathrm{M}^{-1}\cdot\mathrm{s}^{-1}) = 10.17(\pm 0.66) - 4.76(\pm 1.26)\cdot\sigma_{RS} - 3.25(\pm 0.31)\cdot\upsilon + 16.82(\pm 2.67)\cdot\sigma_I \tag{2.25}$$

$$\log(k_c/\mathrm{M}^{-1}\cdot\mathrm{s}^{-1}) = 9.65(\pm 0.09) - 4.64(\pm 0.25)\cdot\sigma_{RS} - 0.16(\pm 0.08)\cdot\upsilon + 7.17(\pm 0.83)\cdot\sigma_I \tag{2.26}$$

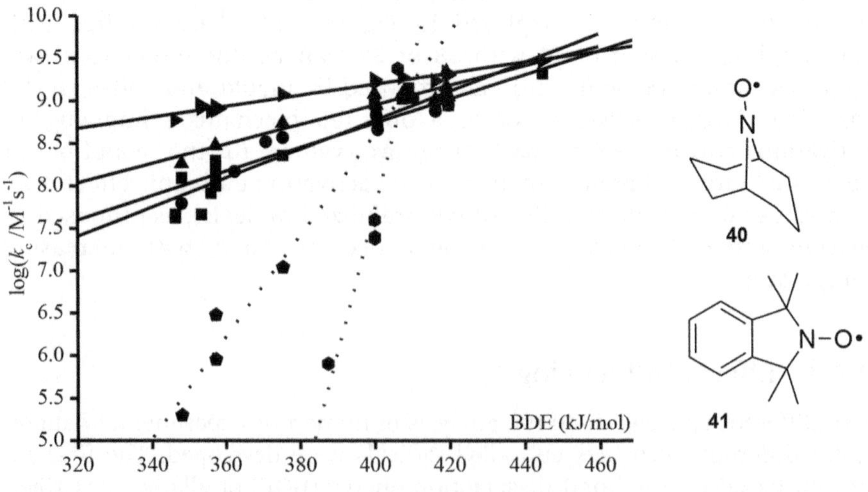

Figure 2.12 BDE(C—H) *vs.* E_a for **2** (■), **6** (●), **9** (⬢, ⬟), **40** (►), and **41** (▲). See ref. 115.
Reprinted with permission from E. G. Bagryanskaya and S. R. A. Marque, *Chem. Rev.*, 2014, **114**, 5011.[115] Copyright 2014 American Chemical Society.

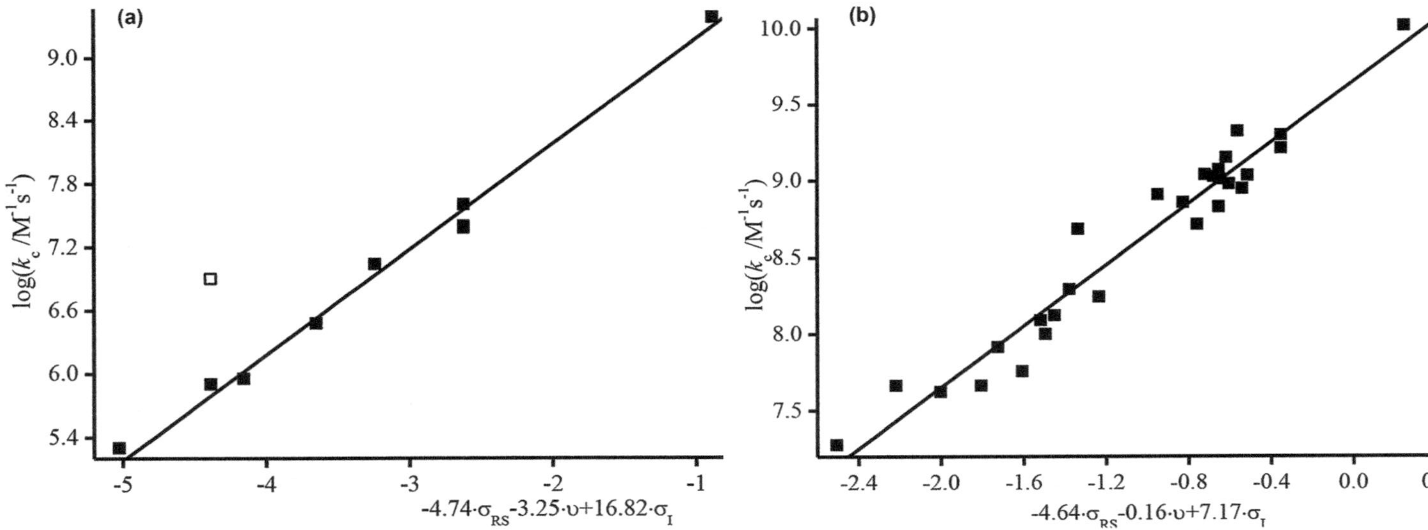

Figure 2.13 Log k_c at room temperature for the cross-coupling of nitroxide (a) **9** in *tert*-butylbenzene; (b) **2** in *iso*-octane with various C-centred radicals *vs.* $f(\sigma_{RS},\sigma_I,\upsilon)$ [eqn (2.24)]. Empty symbol is for the outlier.
Reprinted with permission from E. G. Bagryanskaya and S. R. A. Marque, *Chem. Rev.*, 2014, **114**, 5011.[115] Copyright 2014 American Chemical Society.

Thus, the correlations show that the stabilization, steric, and polar effects govern the cross-coupling reaction between a nitroxide and a C-centred radical (C–ON bond re-formation).

It was shown that stabilization and bulkiness of the alkyl radical lead to lower k_c values. In the case of bulky nitroxides, these effects are enhanced. An increase in the polarity of the alkyl radical leads to higher k_c.

2.2.3.3 Multiparameter Approach to the Structural Effects in Nitroxides

In sharp contrast to the case of alkyl radicals, since only a few BDE(NO–H)[136,137] are available, no BDE(NO–H) plot has been proposed.[138] The stabilization of the nitroxide and the effect of the EWG attached to the nitroxyl moiety are described by the electrical Hammet constant σ_I and the congestion around the nitroxyl moiety is described by the steric Taft–Fujita constant E_s, as displayed in eqn (2.27).[76,115]

$$\log(k_c/\mathrm{M}^{-1}\cdot\mathrm{s}^{-1}) = \log k_{c,0} + \delta' \cdot E_s + \rho_I' \cdot \sigma_I \qquad (2.27)$$

It must be mentioned that E_s is given by the sum of the bulkiness of R^1 and of R^2, which is obtained from a relationship based on the primary steric constant of each group carried by the carbon atom attached to the nitroxyl moiety. Importantly, the steric effect for groups *anti* to the N–O$^\bullet$ bond is disregarded.

This multiparameter approach was applied to 5 alkyl radicals and 43 nitroxides. To evaluate the steric effect, the nitroxides were classified into 10 main families. For the sake of simplicity, only the correlation [eqn (2.28)][139,140] for the largest series corresponding to the coupling of various nitroxides with alkyl radical **26** is reported (Figure 2.14).[141]

$$\log(k_c/\mathrm{M}^{-1}\cdot\mathrm{s}^{-1}) = 10.22(\pm 0.10) + 0.46(\pm 0.02)\cdot E_s + 0.41(\pm 0.17)\cdot \sigma_I \qquad (2.28)$$

Thus, the correlations show that the stabilization, polar and steric effects govern the cross-coupling reaction between a nitroxide and a C-centred radical (C–ON bond re-formation).

As mentioned above, polar and stabilization effects are aliased and described by the electrical Hammett constant σ_I. Hence, strong EWG destabilize the nitroxide by favouring canonic form B (Scheme 2.7) which, in turn, implies an increase in spin density on the oxygen atom, favouring the cross-coupling reaction. Moreover, strong EWG also favour the pyramidalization of the nitrogen, as required in the final alkoxyamine (sp^3 hybridized N atom). On the other hand, the congestion around the nitroxyl moiety due to the bulky groups attached to the N atom hampers the approach to the alkyl radical, leading to a decrease in k_c. This detrimental effect is enhanced for bulky and/or highly stabilized alkyl radicals whereas the presence of EWG has a beneficial effect as it increases the spin density on

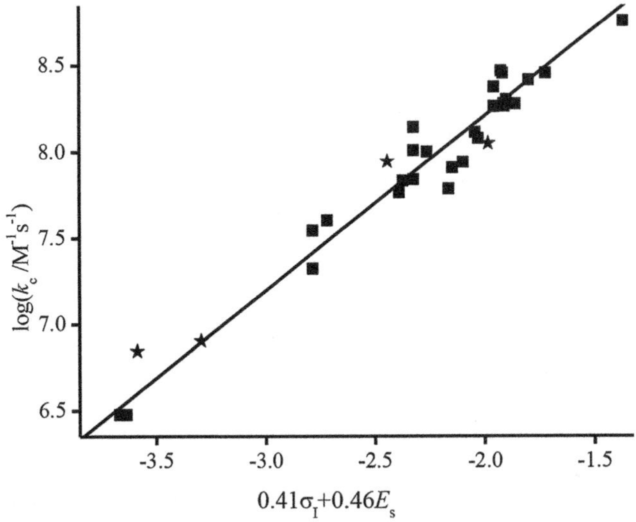

Figure 2.14 Log k_c at room temperature of the cross-coupling of alkyl radical 2.26 and various nitroxides in *tert*-butylbenzene *vs.* $f(\sigma_{RS},\sigma_I,\upsilon)$ [eqn (2.28)]. Star symbols are for nitroxides carrying a phenyl group.[141]
Reprinted with permission from E. G. Bagryanskaya and S. R. A. Marque, *Chem. Rev.*, 2014, **114**, 5011.[115] Copyright 2014 American Chemical Society.

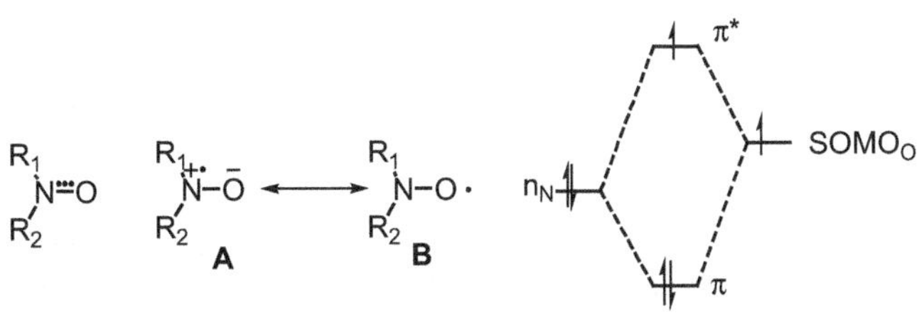

Scheme 2.7 Mesomeric forms A and B of the nitroxide.

the O atom. This effect is also enhanced when the coupling occurs with alkyl radicals carrying EWGs.

2.2.3.4 *Energy Profiles, Transition State (TS), and Calculations*

The first attempt to develop correlations based on computed parameters was made by Iwao *et al.*[142] using the Omega constant Ω to describe the steric effect of both the C-centred radical and nitroxide, and the spin density ρ to describe the delocalization (stabilization) effect on the C-centred radical. More than one decade later, Coote and colleagues[143] published preliminary investigations on correlations with k_c using other parameters to describe a larger set of C-centred radicals and nitroxides. They used: (i) the ionization potential and the radical stabilization energy to describe the nucleophilic/

electrophilic character and the stabilization of the C-centred radical; and (ii) the radical stabilization energy of the nitroxide to describe both the stabilization and the steric effect around the nitroxyl moiety.

The impact of the cross-coupling reaction and its subsequent side-reaction depends on small changes in the structure of both the C-centred radical and the nitroxide. Moreover, the calculations on the large molecules experiencing small changes cannot yet be treated at a sufficient level of theory within acceptable time and, thus, the cross-coupling reaction and its side-reaction were not thoroughly investigated by computational chemistry.

Nevertheless, the first calculations on the pathway of the cross-coupling reaction between **2** and **26** were done by Puts and Sogah.[144,145] Using the semi-empirical method PM3, they showed that TS should exhibit a O···C distance of 2.07 Å and a N–O···C angle of 123° for the cross-coupling reaction. In 2000, Zipse and colleagues[146] were not able to calculate the correct TS of the cross-coupling reaction between the methyl radical and the dihydronitroxide using UB3LYP and UCCSD(T)/cc-pVZ levels of theory because the energy barrier between the starting materials (C-centred radical and nitroxide) and TS was of a few kJ mol^{-1}.[146] However, with a simple model, they were able to determine some typical geometric parameters, such as the angle of attack that ranges between 110° and 117° for a distance around 2.4 Å. Later, Kaim *et al.*[147,148] investigated the cross-coupling reaction between **2** and the phenylethyl radical by DFT (UB3LYP/6-311 + G(3df,2p)/(U)AM1), and found reasonable values for distances (*ca.* 2.5 Å to 1.9 Å) and angles (*ca.* 118° to 117°) at TS.

As the cross-coupling reaction is strongly exothermic, the effects observed—polarity, steric, and stabilization of alkyl radicals and nitroxides—influence only the starting materials and TS (Figure 2.15). Therefore, as the stabilization of both the nitroxide (positive ρ'_I meaning increasing k_c with increasing σ_I values, and conversely) and the alkyl radical (negative values of ρ_{RS}) increases, the starting materials are more stabilized than TS, meaning a rise in the energy barrier. Moreover, the presence of EWG implies a slight rehybridization from sp^2 to sp^3 in the formed alkoxyamine for both the alkyl radical (positive ρ_I) and the nitroxide (positive ρ'_I) at the N atom. This should again afford a gain in energy by stabilizing TS and, hence, should decrease the energy barrier. The increase in the congestion around the alkyl radical (negative slope δ) and around the nitroxide (positive slope δ') leads to destabilization of TS in comparison with the starting materials as the approach to the two reactive centres is hampered, meaning that the energy barrier rises.

To summarize, the (de)stabilization effect for both the alkyl radical and the nitroxide acts on the starting materials, while the polar (rehybridization) effect for both the alkyl radical and the nitroxide, as well as the congestion around the nitroxyl moiety and the bulkiness of the alkyl radical, acts at TS.

The cross-coupling reaction between a nitroxide and a C-centred radical also depends on the interaction between their respective SOMOs (Figure 2.16), and more specifically on the energy gap between these two

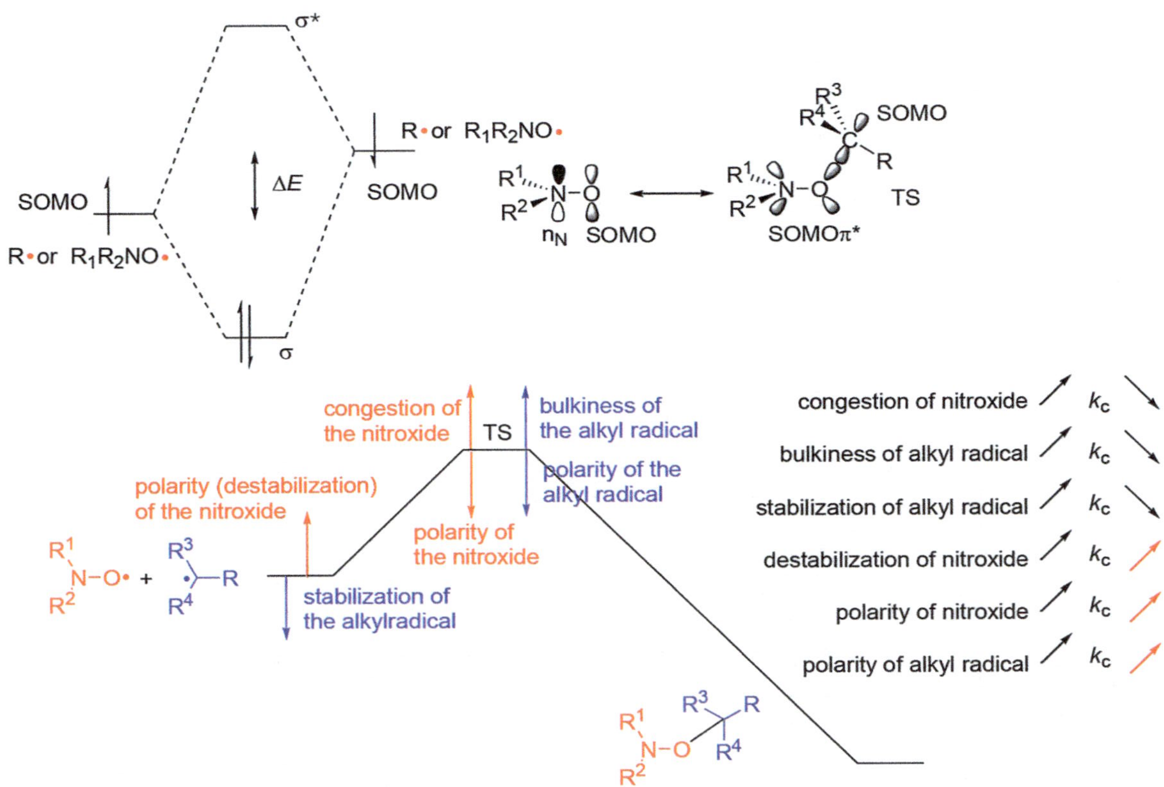

Figure 2.15 Orbital diagram for the SOMO interactions between nitroxide and C-centred radical, and the energy profile of the cross-coupling reaction.
Reprinted and adapted with permission from E. G. Bagryanskaya, S. R. A. Marque, Y. P. Tsentalovich, *J. Org. Chem.*, 2012, 77(11), 4996–5005.[76] Copyright (2012) American Chemical Society.

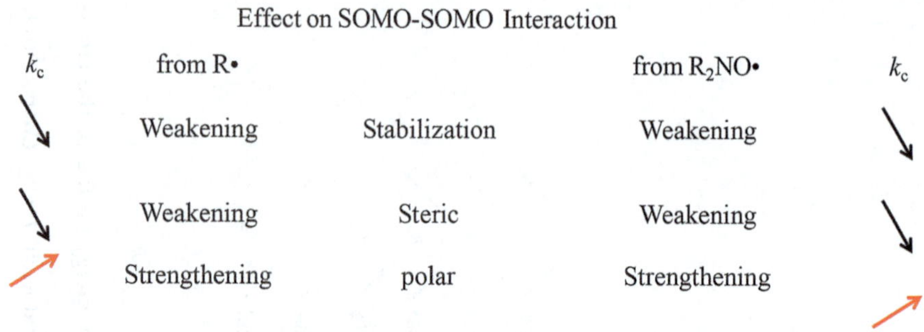

Figure 2.16 Influence of the effects in alkyl radicals and nitroxides on the SOMO–SOMO interaction.

orbitals (Figure 2.15),[149] which in turn depends on: (i) a good match between the orbitals; (ii) a high spin density in each orbital; and (iii) a good orbital overlapping. As displayed in Scheme 2.7, the orbital interaction between the nitrogen lone pair and the odd electron localized on the oxygen atom leads to anti-bonding π^* 3-electron orbital implying that the SOMO of the nitroxide exhibits a symmetry π that matches perfectly the SOMO of symmetry π, which describes the odd electron in a conventional alkyl radical (Figure 2.15). The cross-coupling reaction is favoured by the presence of a high electron density in the SOMO of both the alkyl radical and the nitroxide. In turn, this electron density depends dramatically on the (de)-stabilization of the starting materials, *i.e.*, more stabilized starting materials correspond to a lower the spin density in the SOMO, and a larger energy gap ΔE, and consequently, a lower k_c.

As the odd electron of the nitroxide is localized in the π^* orbital, it can be assumed that the cross-coupling reaction should occur through an approach obeying the Bürgi–Dunitz trajectory[150,151] (Figure 2.15), as for the addition of alkyl radical onto alkene.[152] Consequently, this approach depends dramatically on the congestion around the reactive centre in each reactant, *i.e.*, the orbital overlap is hampered for a congested nitroxide and a bulky alkyl radical implying an increase in ΔE and hence a decrease in k_c for increasing υ and $-E_s$ values for the alkyl radical and the nitroxide, respectively.

Moreover, the increase in polarity (presence of EWG) of the nitroxide implies a relocation of the odd electron on the oxygen atom favouring the $SOMO_{nitroxide}$–$SOMO_{alkyl}$ overlap at TS, hence stabilizing TS, and consequently decreasing the energy barrier. Another consequence of the pyramidalization of the radical centre in the alkyl radicals is a release of the congestion affording a less encumbered SOMO for the alkyl radical, favouring again the $SOMO_{nitroxide}$–$SOMO_{alkyl}$ overlap, and, hence, stabilizing TS, and consequently decreasing the energy barrier.

Each effect discussed above is ascribed to an enthalpic and an entropic contribution for both the alkyl radical and the nitroxide (Figure 2.17). For both the stabilization effect is better ascribed to an enthalpic effect, which stabilizes the starting materials and at the same time decreases the spin

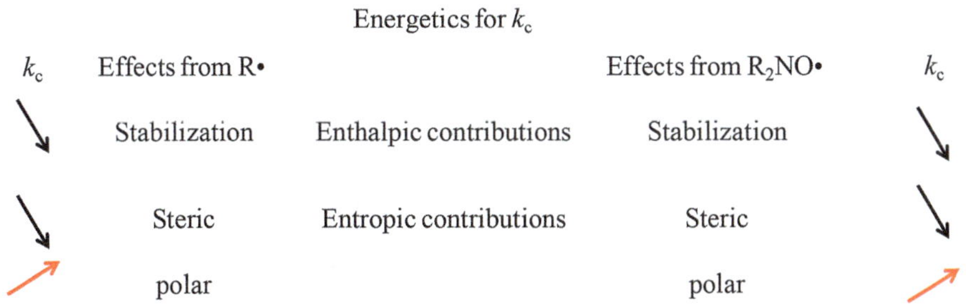

Figure 2.17 Enthalphic and entropic contributions of the various effects governing k_c values.

density at the radical centre, weakening the SOMO–SOMO interaction at TS between them (stereoelectronic effect). For both of them, the steric effect is better ascribed to an entropic effect, which destabilizes TS as the approach of the two radicals towards each other is impeded by the bulkiness around their radical centres. The polar effect is better ascribed to an entropic effect,[153] as it implies the rehybridization of the radical centre from sp^2 to sp^3 for the N and C atoms, as required in the alkoxyamines. Such rehybridization is expected to unshield the SOMO of the nitroxide and, hence, to favour the overlapping of the two SOMOs.

2.2.3.5 *Miscellaneous Effects*

In this section, the effects modifying the main trends described above will be briefly discussed. More details, and especially a collection of k_c values, can be found in the authors' review on this topic.[115]

2.2.3.5.1 Solvent Effects. The effect of the cybotactic region[154] on k_c has been investigated for several nitroxides and alkyl radicals in several solvents. Based on the results obtained, a chart for the correction of the k_c values shifting from a solvent to another solvent was developed (Table 2.1).

No mono- or multi-parameter correlations were observed between $\log k_c$ and the conventional descriptors of the bulk properties of the solvent, that is, the dipole moment μ[155] and the dielectric constant ε.[116] However, several correlations were observed with the following parameters: cohesive pressure c, normalized Dimroth–Reichardt polarity constant $E_T{}^N$, viscosity, and EPR nitrogen hyperfine coupling constant a_N—describing the solvent effect in the cybotactic region. A decrease in k_c with increasing cohesive pressure c of the solvent was observed with several alkyl radicals and nitroxides, meaning that some energy was required to create a hole in the solvent, so that it would accommodate a solute molecule.

The effect of the solvent viscosity was investigated for several nitroxides.[116] All the k_c values measured were clearly smaller than the diffusion-controlled rate constant.[102,156] It was noted that k_c decreases as the viscosity η increases. Nevertheless, the solvent viscosity effect is not larger than one order

Table 2.1 Ratios of k_c for various solvents.

Solvents[a]	*Iso*octane	*t*-Bubenzene	Benzene	Acetonitrile	Triethylamine	*Cyclo*hexane	Di*t*butyl peroxide	Toluene
*iso*octane	1							
t-Bubenzene	1.5	1						
Benzene	2.5	1.6	1					
Acetonitrile	4.5	3.0	1.8	1				
Triethylamine	1.3	0.9	0.5	0.3	1			
*Cyclo*hexane	1.1	0.7	0.5	0.3	0.8	1		
Di*t*butyl peroxide	1.1	0.7	0.4	0.2	0.8	1.0	1	
Toluene	2.2	1.5	0.9	0.5	1.7	2.0	2.1	1

[a]The ratios (column/row) are averaged values. See ref. 115. Reprinted with permission from E. G. Bagryanskaya and S. R. A. Marque, *Chem. Rev.*, 2014, **114**, 5011.[115] Copyright 2014 American Chemical Society.

$$\mathrm{R\bullet + R_1R_2NO\bullet} \underset{k_d'}{\overset{k_s}{\rightleftharpoons}} \underset{\text{encounter complex}}{\left[\mathrm{R_1R_2NO\bullet\bullet R}\right]^*} \xrightarrow{k_p'} \mathrm{R_1R_2NOR}$$

Scheme 2.8 Kinetic model describing the viscosity effect.

of magnitude. This effect was accounted for by the presence of an encounter complex (see Section 2.2.2 and Scheme 2.8), as proposed by Ingold and colleagues.[116] Indeed, the cross-coupling reaction would imply the formation of an encounter complex with a rate constant k_s (given by the Stoke–Einstein–Smoluchowski relationship), which depends on the solvent viscosity. This complex would either collapse (rate constant k_d') to give back the alkyl radical and the nitroxide or proceed to the coupling of the two radicals to yield the alkoxyamine (rate constant k_p').

The nitrogen hyperfine coupling constant a_N of the nitroxide is well known for probing the solvent effect at the molecular level and for encompassing several solvent effects.[157,158] Nevertheless, Ingold *et al.*[116] observed that k_c decreased with increasing solvent polarity. This effect was ascribed to two events in polar events: a stronger stabilisation of the nitroxide as form A (Scheme 2.7) is favoured, the solvation cage is "tight" around the nitroxyl moiety, implying that more energy is required for the C-centred radical to displace solvent molecules, and hence to get access to the reactive radical centre.[116] However, several outliers have been reported for protic solvents, such as MeOH, EtOH, *i*PrOH, and *t*BuOH, and highly viscous solvents, such as hexadecane and ethylene glycol (*vide supra*). In fact, in polar protic solvents, k_c is higher than expected because the intermolecular hydrogen bonding does not occur with the electron in the SOMO but with the oxygen lone pairs perpendicular to the π* SOMO. Hence, the access to the SOMO by the C-centred radical is less hampered than expected.[116] It is worth mentioning that the effect of the polarity is not larger than one order of magnitude for polar non-protic solvents, and no more than 4 times larger for protic polar solvents.

It is well known that a_N is well correlated with E_T^N for many nitroxides.[157,158] Thus, the observation of good correlations between $\log k_c$ and E_N^T is not surprising. It confirms the stabilization of the nitroxide in polar solvents as well as a strong solvation cage, both affording a decrease in k_c with increasing E_N^T values.

2.2.3.5.2 **Protonation Effects.** The presence of an acid is sometimes required during polymerizations[1,10,48,71,112,159,160] or during applications in organic synthesis[51,54,161–163] to remove the excess of nitroxide. However, the effect of an acidic solution on k_c cannot be disregarded. In fact, the effect of an acidic medium on k_c was investigated by Scaiano and colleagues[164] and Bagryanskaya and colleagues,[165] after a 15 year time gap. These two teams reported opposite trends, that is, Scaiano and colleagues[164] observed a two-fold decrease in k_c upon protonation under strong acidic conditions in sharp contrast with the two-fold increase of k_c

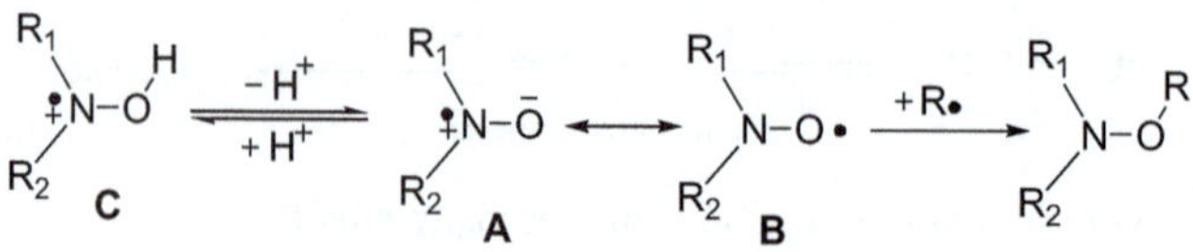

Scheme 2.9 Effect of the protonation on mesomeric forms.

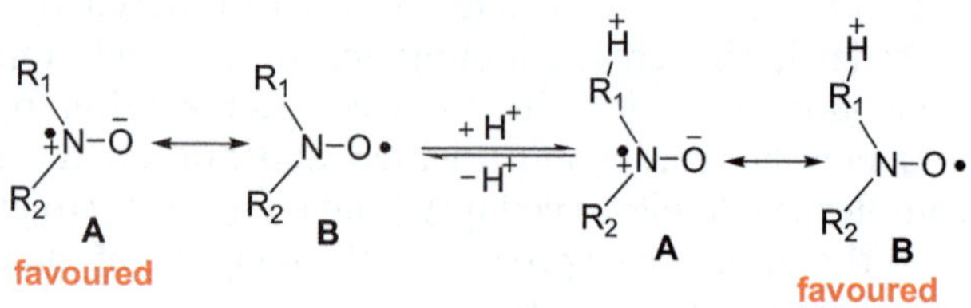

Scheme 2.10 Protonation effect on the weight of each mesomeric form.

reported by Bagryanskaya and colleagues.[165] In fact, two different reaction systems were investigated, respectively:

- The cross-coupling reaction between the benzyl radical and **2** for which only the nitroxyl moiety can be protonated. In this case, the oxygen atom is much more basic than the nitrogen atom affording form C (Scheme 2.9) which,[166] in turn, implies electron density only on the nitrogen atom, virtually suppressing the cross-coupling reaction. However, as form C is in equilibrium with form A, a cross-coupling reaction still occurs although with lower apparent k_c values.

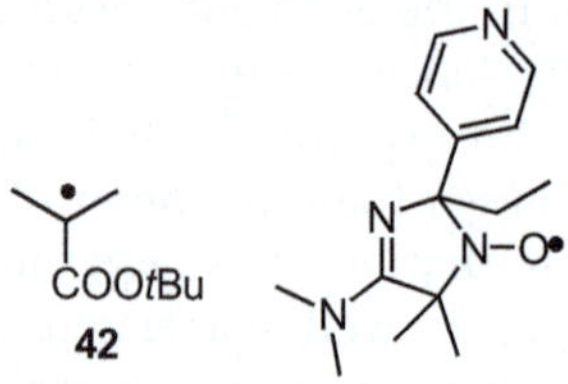

42 **43**

- The cross-coupling reaction between **42** and **43**—which is a nitroxide carrying a pyridinyl group capable of protonation. Upon protonation, the electron withdrawing effect is dramatically increased (*vide supra*) implying that form A is less stabilized and thus form B is favoured (Scheme 2.10). Consequently, from the (de)stabilization and the polar effects described above, an increase in k_c was expected upon protonation.

2.2.3.5.3 Penultimate Unit and Chain Length Effects

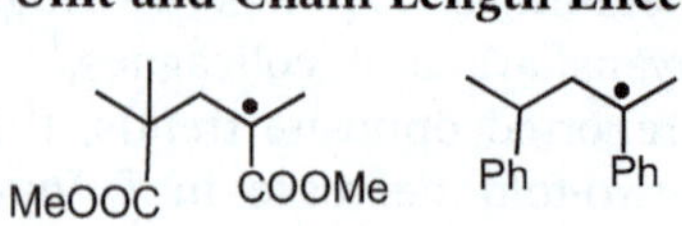

44 **45**

Although the rate constant $k_{c,ds}$ for cross-coupling between the nitroxide and the macro radical is crucial for the success of the NMP experiments, it has been little investigated compared to the cross-coupling reaction rate constants for molecular species. This is mainly due to the smaller number of monomers that are available and suitable for NMP, and the lack of reliable techniques to straightforwardly determine $k_{c,ds}$ values. In fact, all the techniques applied to determine k_c values for molecular species require the preparation of specific initiators, which turns out to be a chemical challenge in the case of macromolecular species. Indeed, RNR-PLP-SEC is the only technique affording rather accurate and reliable $k_{c,ds}$ values (*vide supra*). Otherwise, $k_{c,ds}$ can be determined using the equations related to PRE, provided the chain length dependence of k_p, $k_{d,ds}$, and k_t is accurately known, and the assumptions for PRE hold. Nevertheless, the few $k_{c,ds}$ values available show unambiguously a chain length dependence of $k_{c,ds}$, depending on the nature of the polymer chain and on the nitroxide. For example, for the cross-coupling reaction between **9** and **12**, $k_{c,ds}$ is given at 6.0×10^4 M^{-1} s^{-1} whereas for the corresponding molecular species **11** k_c is given as *ca.* 100-fold larger, at 5.0×10^6 M^{-1} s^{-1}. For the cross-coupling reaction between **2** and **37**, $k_{c,ds}$ is given as 1.8×10^8 M^{-1} s^{-1} whereas for the corresponding molecular species **26** k_c is the same. Unfortunately, the preparation of initiators suitable for LFP-KAS and carrying a penultimate unit is still a challenging task. Thus, the chain length effect cannot yet be matched with experimental values reporting the effect of the penultimate unit. As k_c and k_d are anti-correlated,[115,167] the multi-parameter relationships developed to describe k_c values are also suitable to describe the effect of the penultimate unit on k_c. Thus, the k_c values for the cross-coupling between **9** and **44** and between **2** and **45** are given at 2.2×10^4 M^{-1} s^{-1} and 1.7×10^8 M^{-1} s^{-1} by eqn (2.24) and (2.25), respectively. The estimated values match nicely the experimental values of $k_{c,ds}$ reported for the macromolecular species.

2.2.3.5.4 Intermolecular Hydrogen Atom Transfer (HAT) as Side-reaction. Although the fate of NMP depends dramatically on the occurrence of the intermolecular hydrogen atom transfer (HAT, k_{cD} in Scheme 2.11),[168] which is the main side-reaction of the cross-coupling reaction (k_c in Scheme 2.11), this occurrence has been little investigated.

As displayed in Scheme 2.11,[93] the HAT event affords an alkene and a hydroxylamine, which are the same side-products yielded by the intramolecular proton transfer event (IPT, the side-reaction occurring during the homolysis in Scheme 2.11). As HAT was mainly investigated during the thermal alkoxyamine decomposition during polymerization experiments, it was not always obvious to separate it from the IPT event (*vide infra*). Nevertheless, the authors developed an approach combining the separate measurements of k_d and k_c, and the kinetics of the alkoxyamine decomposition under the PRE conditions affording k_{cD}, as given in eqn (2.29); [k_{dD} the rate constant for IPT, $f_{dD} = k_{dD}/k_d$ and $f_D = k_{cD}/(k_{cD} + k_c)$].

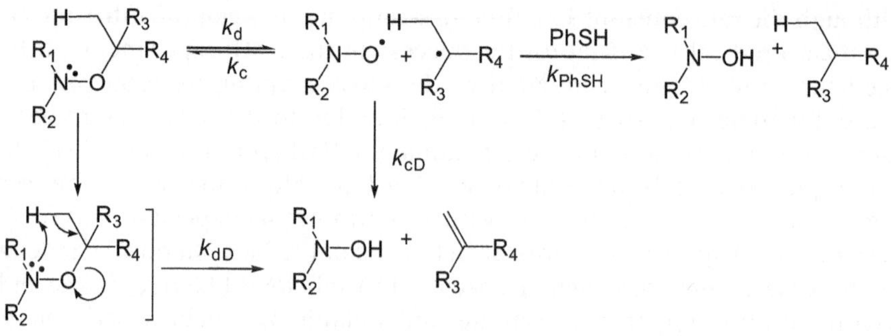

Scheme 2.11 Kinetic scheme of the procedure developed to determine unambiguously the rate constants for IPT (k_{dD}) and for HAT (k_{cD}). "Kinetic Study of H-atom Transfers in Imidazoline-, Imidazolidine-, and Pyrrolidine-Based Alkoxyamines: Consequences for Nitroxide Mediated Polymerization." M. V. Edeleva, I. A. Kirilyuk, D. P. Zubenko, I. F. Zhurko, S. R. A. Marque, D. Gigmes, Y. Guillaneuf and E. G. Bagryanskaya, *J. Polym. Sci., Part A: Polym. Chem.*, 2009, **47**(23), 6579–6595.[93] Copyright 2009 Wiley Periodicals, Inc.

$$k_{obs} = k_{dD} + \frac{k_{cD} \cdot k_d}{k_{cD} + k_c} = k_{dD} + k_d \cdot f_D = k_d(f_{dD} + f_D) \tag{2.29}$$

HAT has mainly been investigated for macromolecular alkoxyamine models. The main approach investigating the HAT event is to monitor the polymerization of monomers or the thermolysis of alkoxyamines by NMR, or to analyse end products with various techniques, such as NMR, HPLC, or MS. CIDNP can be used, except that in some cases the results obtained with experiments performed at room temperature can be misleading.[130,169] The occurrence of HAT is also expected to be dependent on the observation time, *i.e.*, too short time of observation combined with too low temperature might lead to the absence of HAT, and hence be misleading.[130,169]

Whatever the alkyl radical or the nitroxide, the HAT impact on NMP depends on the level of its occurrence: for f_D values lower than 1% the impact can be disregarded, for f_D values between 1% and 5%, the impact can be moderate depending on the experimental conditions, and for f_D values larger than 5% the NMP experiment fails.[56,99,108]

2.3 The C–ON Bond Homolysis of Alkoxyamines

Over recent years, the homolysis of alkoxyamines has been considered in several reviews.[9,10,34,48,59] Although they all discuss C–ON bond homolysis, none is specifically devoted to this event. They are more focused on the kinetics of polymerization,[1,7,10,34,48,59] on the development of new initiators/controllers,[1,5,51,54] on the applications in polymerization,[1,7,8,17,48,170] on designing new materials[4–11,170] and on tin-free radical chemistry[51,53,54].

The aim of this section is not to comprehensively cover the fundamental aspects of the C–ON bond homolysis but rather to highlight the main results and effects.

2.3.1 Techniques and Methods of Determination

Taking into account that the kinetics of the homolysis of alkoxyamines obey PRE, the values of the alkoxyamine C–ON bond homolysis rate constant k_d are obtained only when the re-formation (cross-coupling) reaction is suppressed by the presence of either an alkyl radical scavenger or a nitroxide reductant. Various types of radical scavengers/reductants—nitroxides, galvinoxyl, oxygen, quinone, and H-donors such as thiophenol and phenylhydrazine—can be used, as long as the corresponding signals do not overlap the signal that being monitored and as long as the excess of radical scavenger/reductant is large enough to suppress the back reaction.[27,28,171–173]

Thus, k_d values can be determined using the complete alkoxyamine decay as given by eqn (2.30), using the plateau method (concentration of nitroxide at infinite time), as given by eqn (2.31), when the nitroxide growth is monitored, or using the initial slope, as given by eqn (2.32).

$$\ln\frac{[\text{alkoxyamine}]_t}{[\text{alkoxyamine}]_0} = -\, k_d \cdot t \tag{2.30}$$

$$\ln\frac{[\text{nitroxide}]_\infty - [\text{nitroxide}]_t}{[\text{alkoxyamine}]_\infty} = -\, k_d \cdot t \tag{2.31}$$

$$-\frac{[\text{alkoxyamine}]_t}{[\text{alkoxyamine}]_0} = \frac{[\text{nitroxide}]_t}{[\text{nitroxide}]_\infty} = k_d \cdot t \tag{2.32}$$

Various techniques can be used to determine k_d values. The most frequently used is EPR. In this technique, only the signal of the nitroxide is observed, affording easy analysis and allowing kinetics to be measured *in situ*.[171] Moreover, for the preparation of EPR samples, less than 1 mg of material is required for each measurement, due to the very low threshold of detection. The k_d values are accurate and reliable as long as the nitroxide released is stable under heating.[171] Furthermore, any type of solvent is suitable for EPR. By HPLC, both the alkoxyamine (starting material) and the nitroxide released can be detected as long as they are separated on the column and can be detected by UV-visible or refractive index detectors. HPLC is not suitable for *in situ* monitoring and requires the analysis of several samples, which is time-consuming (no less than 5 minutes per sample). Moreover, the species are only identified by comparison with authentic samples. The k_d of polymeric species can be estimated with good accuracy and reliability using SEC or GPC.[174] The fluorescence technique[175] has also been applied to determine k_d values. A nitroxide carrying a quenched fluorescent probe is used as alkyl radical scavenger. Upon scavenging, the fluorescence is restored and its growth is used to determine k_d. NMR also can be used to investigate the kinetics of alkoxyamine homolysis. In particular, ^{1}H,[108,176] ^{31}P[172] and ^{15}N[177] NMR have been applied and there is no doubt that ^{13}C, ^{14}N, and ^{19}F[178] are also suitable, although they are not yet used. The main advantage of ^{1}H NMR is the detection of all alkoxyamine decomposition products in one experiment, which makes it possible not only to measure k_d, but also to obtain information about any side-reactions.[93,99,101,108,109,179] An important limitation is the small number of deuterated solvents available to investigate the solvent effect. The advantages of heteroatom NMR are: (i) the possibility to use any type of solvent as long as it is mixed with a few percent of deuterated solvent; and (ii) a very simple signal which Zaremski affords easy access to the kinetics of the diastereoisomers.[180,172] Moreover, heteroatom NMR is well suited for non-*in situ* investigations of the kinetics.

Zaremski[181] proposed another approach to the determination of quantitative kinetic parameters of NMP, in particular k_c, L (length of a single step, *i.e.* the number of monomeric units added to a macroradical over a time between reinitiation and "falling asleep") and the number of steps *n* that the form macromolecule. The method was based on the linearization of the distribution of macromolecule chain length. The distribution of polymers in length obtained by NMP was measured at different degrees of conversion and analyzed. The author verified the applicability of the method theoretically and experimentally using the NMP of styrene and butyl-acrylate mediated by TEMPO as an example.

Besides determining the k_d values, it is also valuable to estimate the strength (BDE) of the C–ON bond in alkoxyamines. This value can be estimated using two approaches: one relying on the direct measurement of the BDE(C–ON) using photoacoustic calorimetry (PAC) or very low pressure pyrolysis (VLPP)[182] and the second one based on the very simple thermodynamic relationship given in eqn (2.33) which, in turn, requires determining the reliable activation energies $E_{a,c}$ and $E_{a,d}$ for the cross-coupling reaction and the homolysis, respectively (Figure 2.18). Often, the values of $E_{a,c}$ are small enough (less than 10 kJ mol^{-1}) to be considered as negligible, leading to assume that the $E_{a,d}$ values are equivalent to the BDE.

$$\mathrm{BDE(C{-}ON)} = \Delta H_r = E_{a,d} - E_{a,c} \tag{2.33}$$

2.3.2 Arrhénius Parameters

As displayed in Figure 2.18, alkoxyamine C–ON bond homolysis is an exothermic reaction involving late or loose TS. Consequently, any effect stabilizing the products, *i.e.*, alkyl radical and nitroxide, will stabilize TS, will decrease $E_{a,d}$ and hence will increase k_d. Conversely, any effect stabilizing the reactant, *i.e.*, the alkoxyamine, will increase $E_{a,d}$, and hence will decrease k_d.

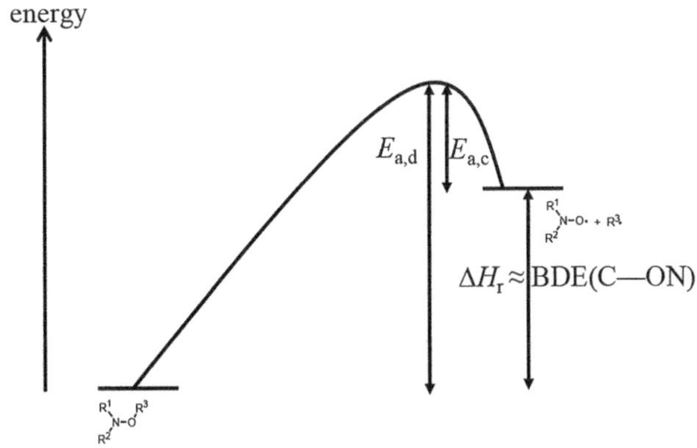

Figure 2.18 Homolysis of an alkoxyamine into an alkyl radical and a nitroxide and its reformation. ΔH_r for the reaction enthalpy, $E_{a,d}$ for the C–ON bond homolysis activation energy, and $E_{a,c}$ for the C–ON bond reformation activation energy.

As given by the transition state theory [eqn (2.34)], the frequency factor A_d in the Arrhénius relationship [eqn (2.35)] depends on the activation entropy $\Delta S^{\neq}$.[183,184] For $\Delta S^{\neq}=0$, the A_d value is around 10^{13} s^{-1}. It is commonly accepted that the fragmentation implies an increase in the freedom of motion, affording A_d between 10^{13} s^{-1} to 10^{15} s^{-1} for small molecules and larger values up to 10^{18} s^{-1} for large molecules.[183,184] Consequently, for alkoxyamines being large molecules, positive $\Delta S^{\neq}$ values and A_d values larger than 10^{13} s^{-1} are expected for their homolysis. On the other hand, A_d values lower than 10^{13} s^{-1} imply negative values of $\Delta S^{\neq}$ leading to question the reliability of these values.

$$k=\frac{k_B \cdot T}{h} \cdot \exp(\Delta S^{\neq}/R)\exp(-\Delta H^{\neq}/RT) \tag{2.34}$$

$$k=A_d \exp(\Delta E_{a,d}/RT) \tag{2.35}$$

Around 50 Arrhénius parameters (frequency factors A_d and activation energies $E_{a,d}$) have been determined during the last two decades. Only a few values of A_d are lower than 10^{13} s^{-1} and higher than 10^{15} s^{-1}. These values are likely erroneous due to too short ranges of temperature used for their determination, the use of techniques/methods that are non suitable, or a poor analysis of the data.[185] The range 10^{13}–10^{15} s^{-1} observed for most of the values of A_d means that the homolysis needs conformational changes costing some activation entropy $\Delta S^{\neq}$.

In general, high A_d are associated with high E_a and conversely, leading to ambiguous discussion on the Arrhénius parameters. However, great care must be given to molecules exhibiting A_d values higher than 10^{13} s^{-1}, and $E_{a,d}$ lower than 95 kJ mol^{-1} because, in such a case, the half-life time $t_{1/2}$

would be shorter than 2 h, making the preparation, handling, and storage of such molecules a challenge. The few cases reported are related to alkoxyamines considered as transient or prepared in very low yields.

2.3.3 Effects Influencing k_d

This section is intended to describe the interplay of the main effects—stabilization of the products and reactants, polarity, steric strain and stereoelectronic effects—governing the alkoxyamines C–ON homolysis and how these effects are enhanced or weakened by minor effects.

2.3.3.1 Bond Dissociation Energy Approach

As shown in Figure 2.18, the stabilization of the released alkyl radical and the nitroxide is expected to play a major role on the size of the activation barrier $E_{a,d}$. The common way to investigate this effect is to plot some BDE or E_a against BDE(C–H) of the alkane corresponding to the released alkyl radical, which accounts for the stabilization of the odd electron. The first plot BDE(C–ON) *vs.* BDE(C–H) was reported by Mulder and colleagues[182] for a short series of **2**-based alkoxyamines. It showed that for more stabilized alkyl radicals, lower BDE(C–ON) and higher k_d were observed. Later, Marque *et al.*[171,186] reported plots $E_{a,d}$ *vs.* BDE(C–H) and $\log k_d$ *vs.* BDE(C–H) for a large (13 data) series of **2**-based alkoxyamines. A plot performed for **9**-based alkoxyamines[85] revealed several correlations (Figure 2.19): one (6 data) gathering primary and secondary alkyl radicals carrying weak EWG ($\sigma_I < 0.10$); one (10 data) gathering the primary and secondary alkyl radicals carrying moderate/strong EWG ($\sigma_I > 0.10$); one gathering the tertiary alkyl radicals carrying moderate/strong EWG; and a last one (one data) gathering the tertiary alkyl radical carrying weakly EWG. Consequently, the stabilization of the released radical, and the steric strain due to the bulkiness and polarity of the alkyl fragment have a major influence on the value of k_d. Obviously, a plot E_a *vs.* BDE(C–H) or $\log k_d$ *vs.* BDE(C–H) was not reliable enough to predict k_d. Nevertheless, it was possible to propose an incremental averaged substituent scale (Figure 2.20) able to provide a rough estimate of k_d readily.[85] It must be mentioned that this scale was built by summing all data available for each alkyl radical, whatever the nitroxyl fragment. Hence, strong discrepancies between predicted and measured data might arise when synergetic or antagonistic effects occur between the effects due to the nitroxyl and alkyl fragments.

2.3.3.2 Multiparameter Approach to the Structural Effects in Alkyl Radicals

To circumvent the weaknesses of the $E_{a,d}$ *vs.* BDE(C–H) plots, a relationship [eqn (2.36)] based on the molecular descriptors[187] was developed using σ_{RS}

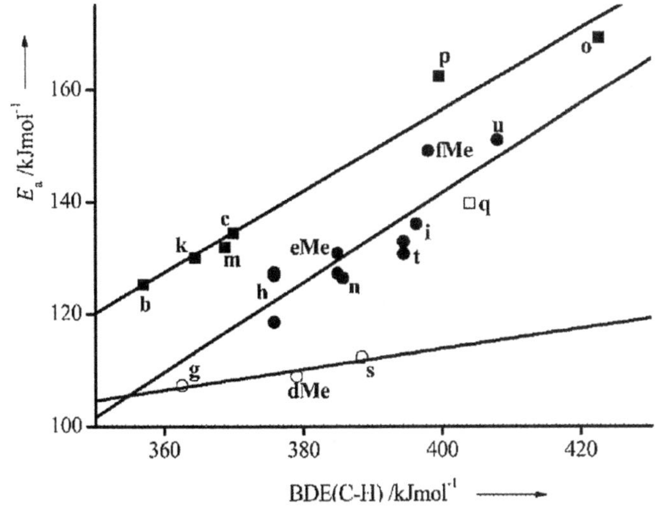

Figure 2.19 $E_{a,d}$ *versus* BDE(C–H) for SG1-based alkoxyamines. Filled symbols: primary and secondary alkoxyamines; empty symbols: tertiary alkoxyamines; squares: nonpolar alkoxyamines; circles: polar alkoxyamines. "Alkoxyamine C–ON Bond Homolysis: Stereoelectronic Effects". E. Beaudoin, D. Bertin, D. Gigmes, S. R. A. Marque, D. Siri and P. Tordo, *Eur. J. Org. Chem.*, 2006, 7, 1755.[85] Copyright 2006 Wiley-VCH Verlag GmbH & Co. KGaA, Weinheim.

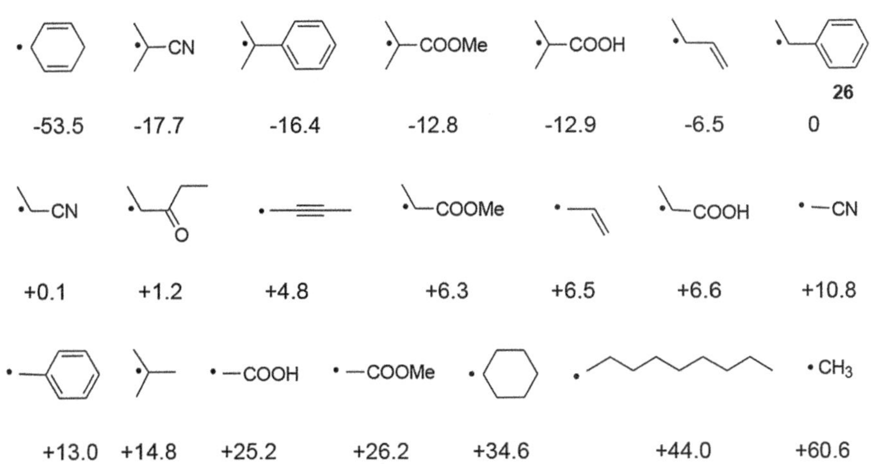

Figure 2.20 Incremental scale for various alkyl fragments with **2.26** as benchmark. Energies are given in kJ mol^{-1}.
"Alkoxyamine C–ON Bond Homolysis: Stereoelectronic Effects" E. Beaudoin, D. Bertin, D. Gigmes, S. R. A. Marque, D. Siri and P. Tordo, *Eur. J. Org. Chem.*, 2006, 7, 1755.[85] Copyright 2006 Wiley-VCH Verlag GmbH & Co. KGaA, Weinheim.

as the radical stabilization Hammett-type constant, σ_I as the electrical Hammett constant and υ as the steric constant to account for the stabilization, polar and steric effects of the alkyl fragment, respectively.

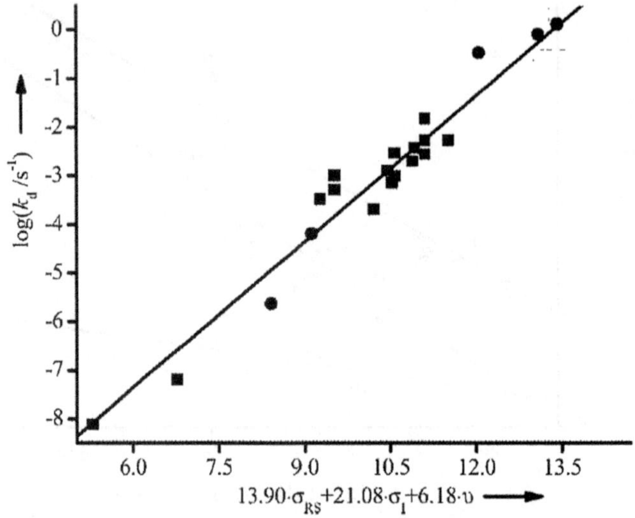

Figure 2.21 Log k_d for C–ON bond homolysis in SG1-based alkoxyamines *versus* a linear combination of molecular descriptors (σ_{RS}, σ_I and υ). Circles denote tertiary alkyl SG1-based alkoxyamines.
Reprinted with permission from D. Bertin, D. Gigmes, S. R. A. Marque and P. Tordo, *Macromolecules*, 2005, **38**(7), 2638.[187] Copyright 2005 American Chemical Society.

This approach was applied to **2**- and **9**-based alkoxyamines and provided nice correlations, as exemplified with SG1-based alkoxyamines [eqn (2.37) and Figure 2.21).[188,189] The positive sign for each effect denotes that k_d increases with increasing effect. Interestingly, for **9**-based alkoxyamines, all effects are evenly weighted.[187,189] In general, during the homolysis the occurrence of charge distribution is low, the polar effect is better ascribed to a destabilization of the alkoxyamine. Indeed, Pauling[190,191] described the BDE [eqn (2.38)] of a heteroatom bond A–B as the sum of an enthalpic term related to the bond strength of the corresponding homoatom bonds A–A and B–B and of a polar term related to the square of the difference in the electronegativity χ of the atoms in the A–B bond. Thus, an increase in χ on the carbon atom in the C–ON bond due to the presence of EWG implies a decrease in the difference $\chi_O - \chi_C$ and consequently a weaker BDE, which in turn implies a lower $E_{a,d}$ and, hence, a higher k_d.[186,187] Taking into account Figure 2.18, the radical stabilization is expected to stabilize TS, and, thus, to decrease $E_{a,d}$, leading to increasing k_d with increasing σ_{RS}. On the other hand, the steric effect is expected to be weak for a late or loose TS (endothermic reaction) and, then, is expected to destabilize the alkoxyamine through internal strains affording a decrease in $E_{a,d}$, and hence, an increase in k_d with increasing υ. The absence of charge at TS implies that the polar effect destabilizes the alkoxyamine by changing the electronegativity of the C–ON bond affording a decrease in E_a and, hence, an increase in k_d with increasing σ_I values.

$$\log(k_d/s^{-1}) = \log k_{d,0} + \rho_{RS} \cdot \sigma_{RS} + \delta \cdot \upsilon + \rho_I \cdot \sigma_I \tag{2.36}$$

$$\log(k_d/s^{-1}) = -14.04(\pm 0.83) + 14.30(\pm 1.41) \cdot \sigma_{RS} + 6.89(\pm 0.72) \cdot \upsilon + 21.44(\pm 1.91) \cdot \sigma_I \tag{2.37}$$

$$\mathrm{BDE}(\mathrm{A-B}) = \frac{1}{2}(\mathrm{BDE}(\mathrm{A-A}) + \mathrm{BDE}(\mathrm{B-B})) + 23 \cdot (\chi_A - \chi_B)^2 \tag{2.38}$$

2.3.3.3 Multiparameter Approach to the Structural Effects in Nitroxides

The multiparameter approach for correlation of k_d on the structure of nitroxides is very similar to the same for k_c (Section 2.2.3.3) leading to eqn (2.39). Indeed, as long as the two alkyl groups attached to the nitroxyl moiety are of the tertiary type, the steric effect is described in the same way for k_c as for k_d. If one of the alkyl groups is of the secondary type, the steric effect of the remaining tertiary group is disregarded.[193]

$$\log(k_d/M^{-1} \cdot s^{-1}) = \log k_{d,0} + \delta' \cdot E_s + \rho'_I \cdot \sigma_I \tag{2.39}$$

$$\log(k_d/M^{-1} \cdot s^{-1}) = -2.98(\pm 0.14) - 0.86(\pm 0.03) \cdot E_s + 5.83(\pm 0.10) \tag{2.40}$$

Up to now, the multiparameter analysis for k_d has only been applied with **26** as alkyl radical. As displayed in Figure 2.22, most data are nicely correlated with eqn (2.40),[192–196] except a few outliers. The negative slopes reported in eqn (2.40) denote an increase in k_d with increasing steric strain and a decrease in k_d with increasing the electron withdrawing properties of the groups attached to the nitroxyl moiety. In fact, going from alkoxyamine to TS implies the lengthening of the C–ON bond, affording a relief of the steric strain. Hence, the alkoxyamine is destabilized by bulky groups attached to the nitroxyl moiety. However, the impact of the steric effect is strongly modulated by the size and by the substituent attached to the ring in the cyclic nitroxyl moiety. As k_c and k_d are anticorrelated,[167] stabilization and polar effects for k_d are aliased and described by the same parameter σ_I, *i.e.*, the presence of strong EWG attached to the nitroxyl moiety will destabilize the nitroxide (form B in Scheme 2.7 is favoured) and hence TS, affording an increase in $E_{a,d}$, and will increase the χ_O character of the C–ON bond affording higher BDE and, consequently, higher $E_{a,d}$. However, it is not yet possible to disentangle these effects.

2.3.3.4 Energy Profiles, TS, and Calculations

Taking into account the progress made in calculations, only calculations performed with DFT or higher methods will be discussed hereafter. In

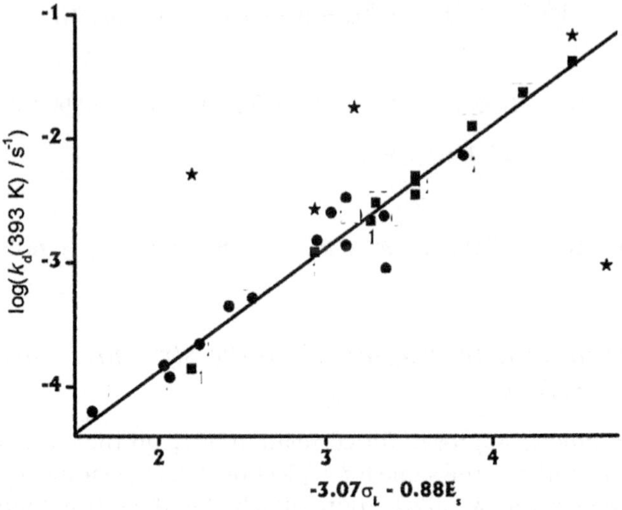

Figure 2.22 Log k_d for the C–ON bond cleavage of nitroxide-**26**-derived alkoxyamines *vs.* eqn (2.39). Circles are for cyclic nitroxyl moieties, squares for non-cyclic nitroxyl moieties, and stars for outliers.
Reprinted with permission from S. R. A. Marque, *J. Org. Chem.*, 2003, **68**(20), 7582.[193] Copyright 2003 American Chemical Society.

general, DFT yields BDE values accurate enough for a qualitative discussion.[197,198] However, Coote *et al.*[199,200] strongly recommend to use G3(MP2)-RAD. In general, calculations agree nicely with the above discussions and are very useful for investigating minor effects (*vide infra*). As for the cross-coupling reaction, it was not possible to reach TS because of the very low energy barrier $E_{a,c}$. Nevertheless, the calculations by Braslau *et al.*[146] and Megiel *et al.*[147,148] nicely support the discussions hereafter.

As C–ON bond homolysis is an endothermic reaction, TS is late or "loose" (Hammond postulate), implying that the structure of the activated complex at TS closely resembles that of the products (Figure 2.23). Thus, the orbital interactions at TS play a minor role, except that they must comply with a few requirements (Scheme 2.12): (i) the change in hybridization from sp^3 at the N atom in the alkoxyamine to sp^2 at TS aiming at favouring the interaction between the N lone pair and the antibonding orbital of the O–C bond $n_N \rightarrow \sigma^*_{O-C}$ which will ultimately lead to the generation of a nitroxide exhibiting a 3-electron orbital π^*; (ii) the change in hybridization from sp^3 at the C atom of the C–ON bond to sp^2 at TS to ultimately afford a π alkyl radical; and (iii) the C–C bond rotation affording the interaction between the bonding orbital σ of the NO–C bond and the antibonding orbital π^* of the unsaturated moiety C=X attached to the C–ON bond.

Thus, the steric strain in the alkyl and nitroxyl fragments is expected to favour items (i) and (ii) as the sp^2 hybridization relieves the steric strain, and, thus, increases k_d. On the other hand, this steric strain might impede the C–C bond rotation affording the $\sigma_{O-C} \rightarrow \pi^*_{C=X}$ interaction leading to a cost in

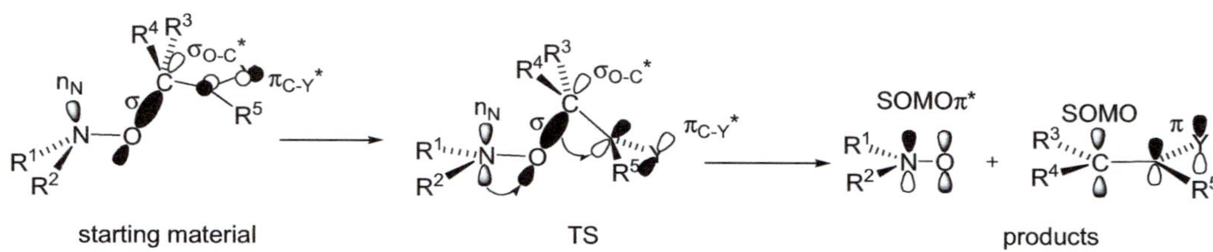

Scheme 2.12 Orbital interactions in alkoxyamine, TS, alkyl radical and nitroxide.

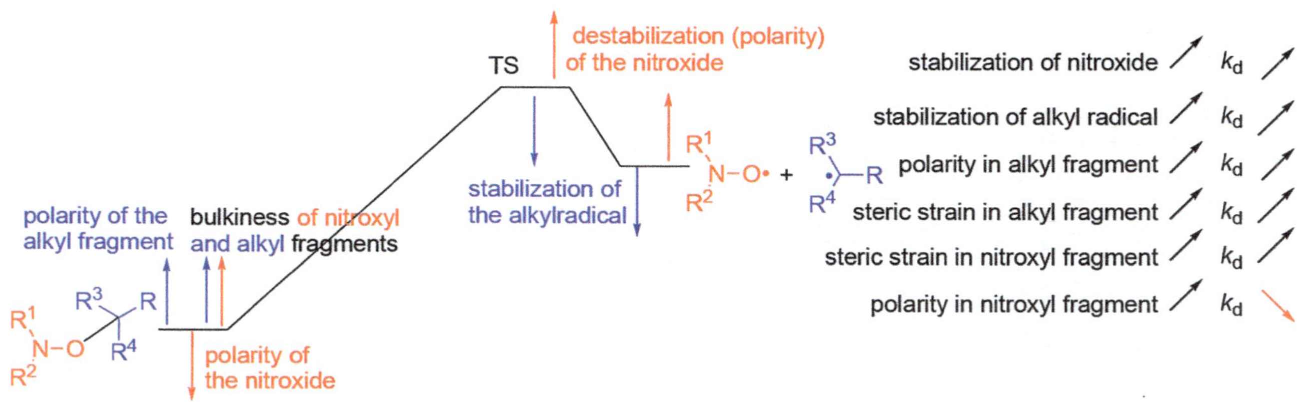

Figure 2.23 Energy profile for the C–ON bond homolysis.

activation entropy $\Delta S^{\neq}$ and, thus, to a decrease in k_d. It is therefore not surprising to observe A_d values lower than 10^{15} s^{-1}.

As TS is late (Figure 2.23), any stabilization of the products will stabilize TS. On the other hand, effects in alkoxyamine do not influence TS. Therefore, as the stabilization of both the nitroxide (negative ρ'_I, *i.e.*, meaning decreasing k_d with increasing σ_I values, and conversely) and the alkyl radical (positive values of ρ_{RS}) increases, TS is more stabilized, implying a smaller energy barrier $E_{a,d}$ and consequently higher k_d. The polarity has opposite effects in the alkyl and the nitroxyl fragments as a positive slope ρ_I is observed—meaning k_d increases with increasing σ_I—for the alkyl radical whereas a negative slope ρ'_I is observed—meaning k_d decreases with increasing σ_I—for the nitroxyl fragment. Indeed, taking into account eqn (2.38), the polarity effect (presence of EWG) in the alkyl fragment increases the electronegativity χ_C in the C–ON bond, reducing the difference $(\chi_O - \chi_C)^2$ and thus decreasing BDE, whereas the polarity effect (presence of EWG) in the nitroxyl fragment increases the electronegativity χ_O in the C–ON bond, increasing the difference $(\chi_O - \chi_C)^2$ and thus increasing BDE. As the bulkiness of the alkyl fragment (positive slope δ) and of the groups attached to the nitroxyl fragment (negative slope δ') increases, k_d increases, meaning that the alkoxyamine is more destabilized than TS, affording a smaller barrier $E_{a,d}$.[201]

To summarize, the (de)stabilization effect for both the alkyl radical and the nitroxide acts on TS, the polar (the electronegativity changes) effect as well as the steric effect of both the alkyl and the nitroxyl fragments acts on alkoxyamine (Figure 2.23).

All the effects discussed above are ascribed to enthalpic and entropic contributions (Figure 2.24). For both the alkyl radical and the nitroxide, the radical stabilization effect is better ascribed to an enthalpic contribution as it stabilizes the products and hence stabilizes TS, leading to a decrease in

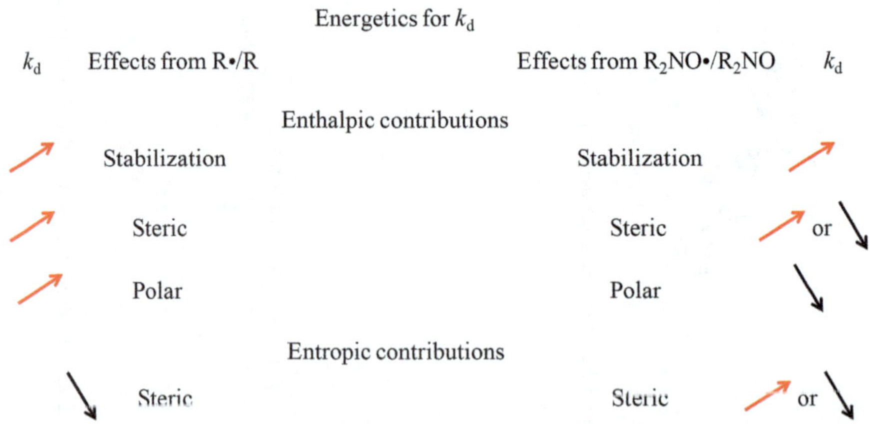

Figure 2.24 Enthalphic and entropic contributions of the various effects governing k_d values.

$E_{a,d}$. As the polar effect plays a role only in the alkoxyamines and modifies the strength of the C–ON bond through eqn (2.38), it is better ascribed to an enthalpic contribution. For both the nitroxyl and the alkyl fragment, a part of the steric effect—electrostatic repulsion between the two fragments—is better ascribed to the activation enthalpic contribution and the other part—ring conformation, bond rotation—is better ascribed to the activation entropic contribution.

2.3.3.5 Miscellaneous Effects

Besides the main effects discussed above, several minor effects modifying the values of k_d have been unveiled during the last two decades. Thus, only the effects playing a significant role in NMP are addressed hereafter. For other effects, such as the anomeric[182,186,202] and anchimeric[85] effects, the steric levelled effect,[186,193,203] the polar[204,205] and steric[206] remote effects,[207] and the counter-anion effect,[208,209] the readers are invited to refer to the references cited

2.3.3.5.1 Intramolecular Hydrogen Bonding (IHB)

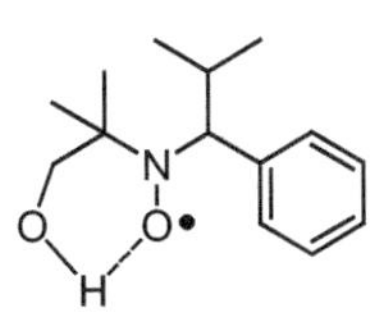

46

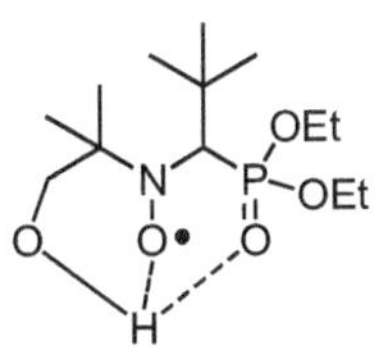

47

IHB[210] can affect k_d in two ways: (i) stabilization of the released nitroxide by interaction between the hydroxyl group and the nitroxyl moiety; and (ii) intramolecular hydrogen interaction between the alkyl (H-donor) and the nitroxyl (H-acceptor) fragment. The first one was shown to play a significant role in **46**[186] and in the phosphoryl[211] moiety of **47**. It has also been highlighted with other nitroxides.[186,196,212] This effect has been shown to stabilize the nitroxide and, hence, to increase k_d, which has been successfully applied to improve NMP.[211–213]

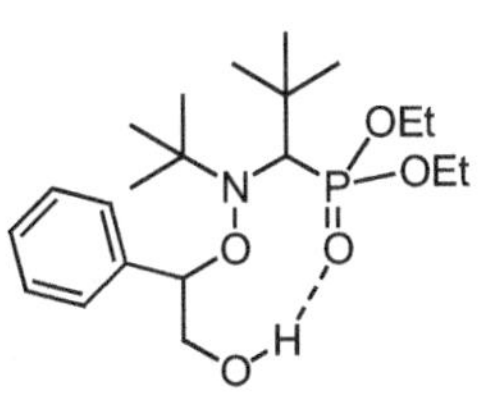

48

Brémond and colleagues[214] showed that such IHB, when occurring between the alkyl and nitroxyl fragments, significantly decreased k_d values in **48**. The existence of IHB was confirmed by X-ray analysis, IR, NMR and kinetic investigations. In this case, IHB stabilized the alkoxyamine because the homolysis involves two bonds: the *strong* C–ON bond and the weak hydrogen bond COH···O=P.

Nicolas and colleagues[95,96] proposed the occurrence of such IHB between the alkyl and nitroxyl fragments in **8**, which exhibited a detrimental effect on the NMP of isoprene, and this IHB was recently confirmed by calculations, NMR, IR and solvent investigations.[215]

2.3.3.5.2 Penultimate Unit and Chain Length Effects. The chain length effect was first investigated first with the polymeric species **49** whose k_d are at least twice as large as those of the molecular species **27**. Using dyads **50**, Georges *et al.*[176] showed that this difference is accounted for by the presence of the penultimate unit. A few years later, using DFT calculations, Blachon *et al.*[216] showed that the difference in k_d between the diastereoisomers of dyad **50** and between each diastereoisomers and **27** is due to the difference in steric strain for the two diastereoisomers. However, no significant difference was observed between the macromolecular species **51** and the molecular one **7**, that is, a 1.5-fold increase in k_d for the *poly*styryl chain.[83,217] Interestingly, a linear 3–4-fold increase in k_d was observed for the chain length of the *poly*-methyl acrylate based alkoxyamine **52**.[83] It was ascribed to the increase in the polarity, likely transmitted through space, of the *poly*-methyl acrylate chain. This effect was also observed by Zaremski *et al.*[173] with **53**.

49

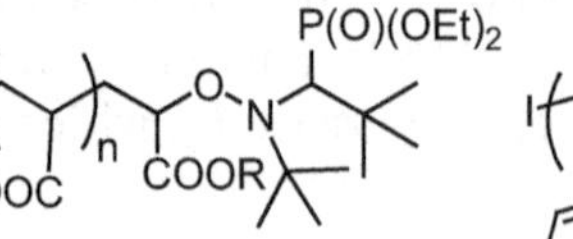

50

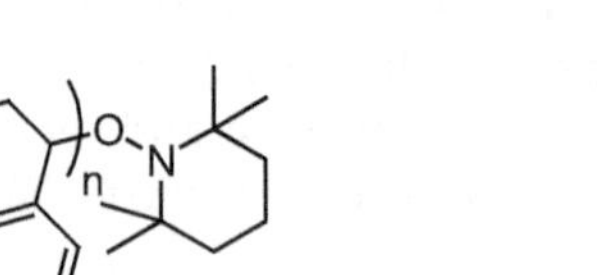

51

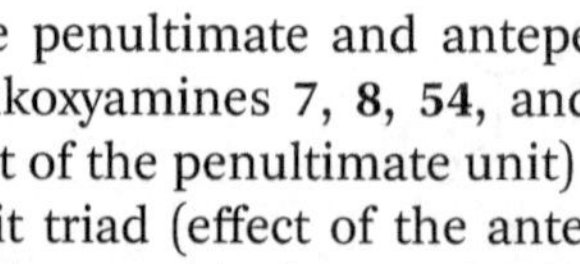

52

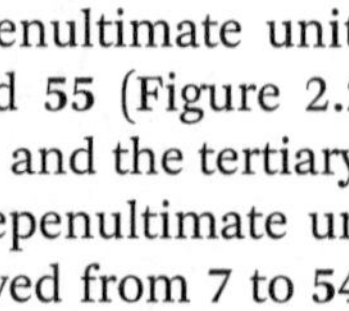

53

The effect of the penultimate and antepenultimate units on k_d was investigated using alkoxyamines **7**, **8**, **54**, and **55** (Figure 2.25).[87] When the styrenic dyad (effect of the penultimate unit) and the tertiary unit–secondary unit–secondary unit triad (effect of the antepenultimate unit) of **54**, a non significant 1.5-fold increase in k_d was observed from **7** to **54**. A clear 18-fold increase in k_d was observed from **7** to **55**, in which only the effect of the penultimate unit is observed due to the tertiary unit–secondary unit dyad. Consequently, the effect of the penultimate unit is significant only when it is a tertiary carbon. Importantly, the effect of the antepenultimate and

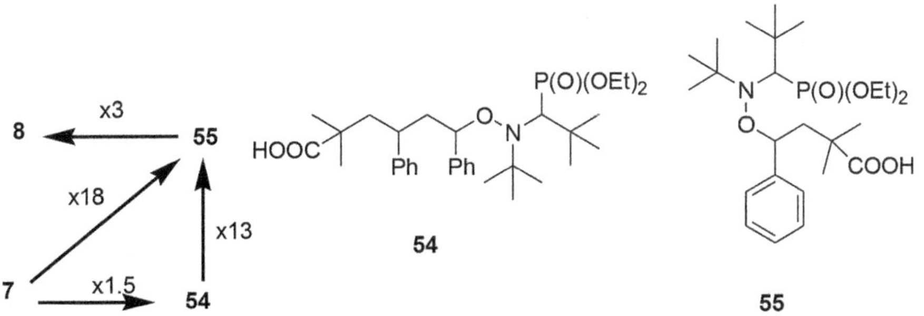

Figure 2.25 Effect of the antepenultimate and penultimate unit on k_d of alkoxyamines.
"Effect of the Penultimate Unit on the C–ON Bond Homolysis in SG1-Based Alkoxyamines" D. Bertin, P.-E. Dufils, I. Durand, D. Gigmes, B. Giovanetti, Y. Guillaneuf, S. R. A. Marque, T. Phan and P. Tordo, *Macromol. Chem. Phys.*, 2008, **209**, 220.[87] Copyright 2006 Wiley-VCH Verlag GmbH & Co. KGaA, Weinheim.

penultimate units is nicely described with eqn (2.36) by estimating steric, polar, and stabilization effects, as given in Section 2.3.3.2.[87,187]

2.3.3.5.3 Protonation Effects on the Alkyl and Nitroxyl Fragments. Recently, the authors showed that the protonation of the nitroxyl (Figure 2.26a)[165] and alkyl (Figure 2.26b)[218] fragment led to either a 3–4-fold decrease (de-activation) or a 20-fold increase (activation) in the C–ON bond homolysis constants k_d, respectively.

In fact, upon protonation of the nitroxyl fragment, k_d values decreased from **56** to **57** and eventually to **58** (Figure 2.26a). This was ascribed to an increase in the electron withdrawing properties of the nitroxyl fragment, which in turn slightly increased the electronegativity χ_O on the oxygen atom in the C–ON bond and, consequently, as given by eqn (2.38),[187,190,191] the C–ON bond was strengthened. However, the main effect of the protonation was to destabilize the released nitroxide by increasing the electron withdrawing properties of the substituents it carried.[76,158,186,196] Calculations nicely confirmed the strengthening of the C–ON, due to the destabilization of the nitroxide upon protonation,[219] occurring through a switch between the HOMO and SOMO orbitals.[220]

Indeed, upon protonation of the alkyl fragment, from **59** to **60**, k_d values increased. This was ascribed to an increase in the electron withdrawing properties of the alkyl fragment. Calculations[221] showed that the effect of the protonation on the stabilization of the released alkyl radical was very weak. Upon protonation of the alkyl fragment, the positive charge generated on the nitrogen atom at position *para* was delocalized on the *ipso* carbon of **60**, leading to an increase in electronegativity χ_C of the carbon atom of the C–ON bond, implying a weakening of this bond, as given by eqn (2.38) and as highlighted in Figure 2.26b with **59** and **60**.

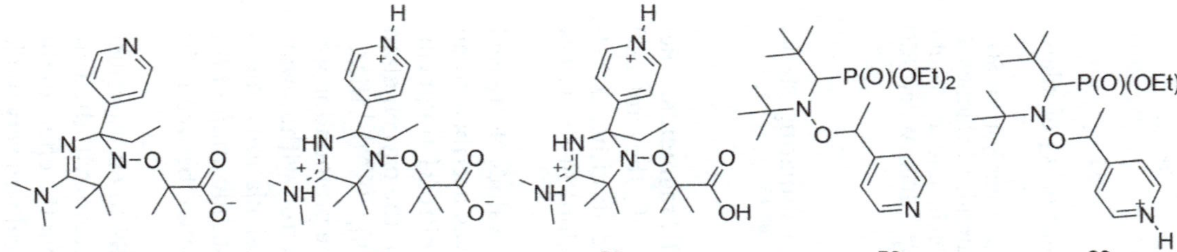

56 57 58 59 60

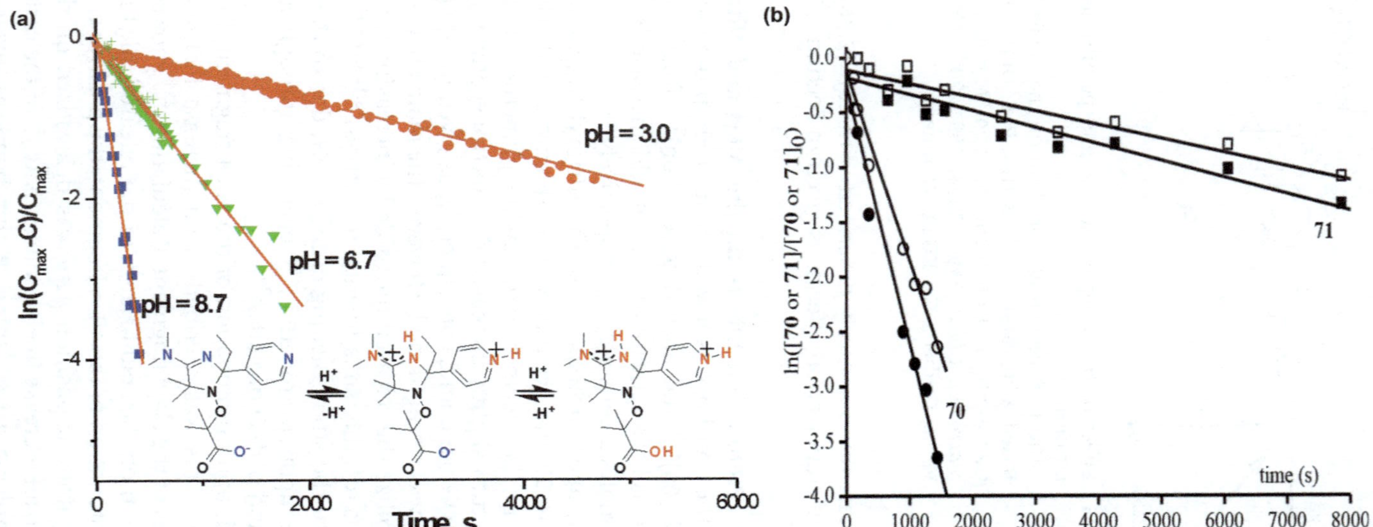

Figure 2.26 Protonation of nitrogen atoms carried by groups attached to (a) nitroxyl fragment and to (b) alkyl fragment. (a) Reprinted with permission from M. V. Edeleva, I. A. Kirilyuk, I. F. Zhurko, D. A. Parkhomenko, Y. P. Tsentalovich and E. G. Bagryanskaya, *J. Org. Chem.*, 2011, **76**, 5558.[165] Copyright 2011 American Chemical Society. (b) Reproduced from ref. 218.

The protonation effect was also investigated for **61**[222] and **62**.[223] As expected, the protonation at the *meta* position in **61**[222] afforded only a small 2-fold increase in k_d whereas the protonation at the *ortho* position in **62**[223] afforded the same increase as position *para* in **60**.[221]

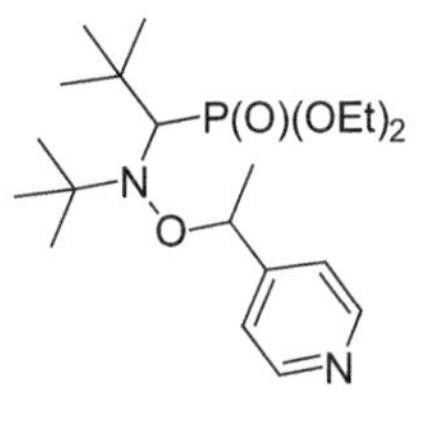

59

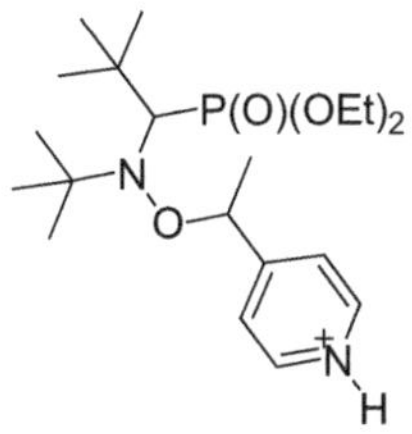

60

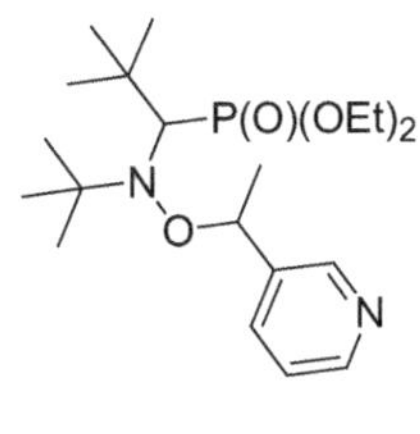

61

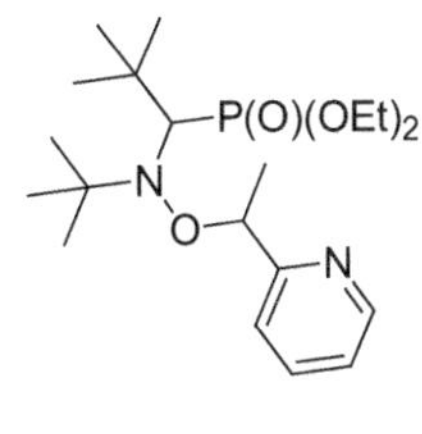

62

Recently Edeleva *et al.*[224] performed NMP of MMA and styrene using a series of imidazoline nitroxides with bulky spirocyclic moieties at positions 2 or 5 of the imidazole ring. MMA polymerization was found to proceed in a controlled regime up to 55% of monomer conversion and the polymer obtained was able to reinitiate the polymerization of styrene. The ratio of "living" chains was estimated at 90%. Homolysis rate constants were measured and possible contributions of side reactions were quantified.

Alkoxyamines **59** and **60** were successfully applied to the NMP of styrene and sodium styrene sulfonate, performing as well as the benchmark alkoxyamine **8** (BlocBuilder™).[225]

2.3.3.5.4 Solvent Effects. In 1995, Moad and Rizzardo[226] reported a small 3–4-fold increase in k_d with increasing the solvent polarity for **63**. Some years later, Marque and colleagues[83] also observed a small 2–3-fold increase in k_d for **7** and **21**. This has also been reported by other groups.[99,108,186,227–231] A small 2–4-fold increase in k_d was also observed even for alkoxyamines such as **60** and **64** capable of H-bonding with HBD solvents.[232] On the other hand, **65** exhibited a striking 20–70-fold increase in k_d whereas no H-bonding sites were available on the alkyl fragment.[233]

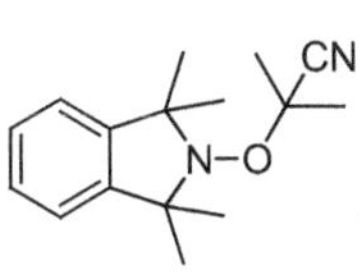

63

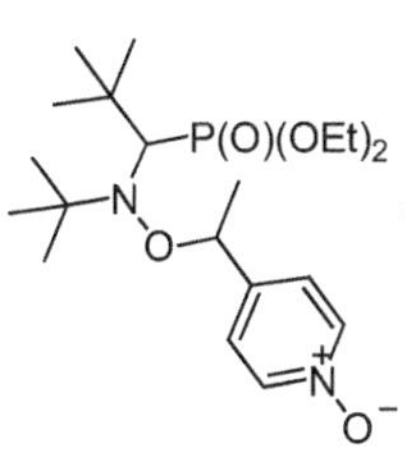

64

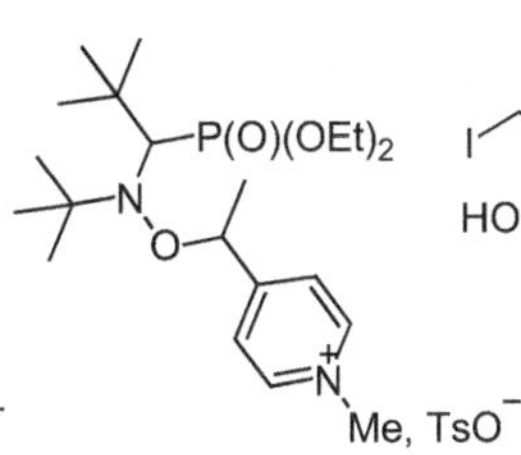

65

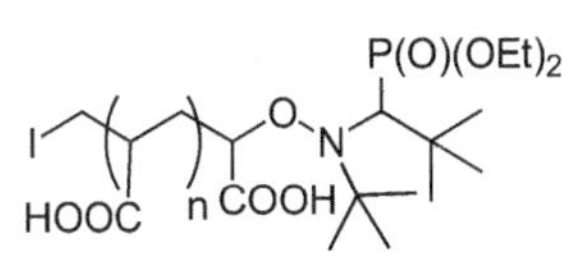

66

In general, this increase in k_d (Figure 2.27) was ascribed to the stabilization of the nitroxide by the polarity/polarizability effect (parameter π^*). It was shown that for **60** and **64** this effect was major (weight $w_{\pi^*} > 70\%$) while for **65** it was minor ($w_{\pi^*} < 40\%$). The HBD property of the solvent (parameter α)

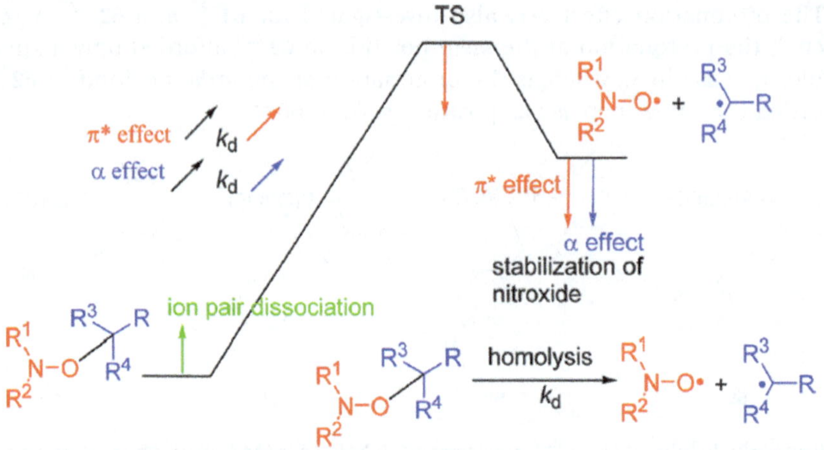

Figure 2.27 Energy profile for the solvent effect in **2.60**, **2.64**, and **2.65**, displaying the effects of the polarity/polarizability of the solvent given by parameter π* and of the HBD property α of the solvent.
Reprinted with permission from G. Audran, P. Brémond, S. R. A. Marque and G. Obame, *J. Org. Chem.*, 2012, 77(21), 9634.[233] Copyright 2012 American Chemical Society.

had a minor effect for **60** and **64** and a major effect for **65**. The polarizability/polarity effect also stabilized TS for all alkoxyamines investigated. The enhanced effect of the HBD property observed for **65** was due to the shift from intimate ion pair in poor HBD solvents—the effect of the positive charge is cancelled by the closeness of the counter-anion—to the dissociate ion pair in strong HBD solvents—the effect of the free positive charge is maximized.

The solvent effect on macro-alkoxyamines has been very little investigated.[48,234,235] Recently, Zaremski *et al.*[173] reported the absence (or non-significant effect) of the solvent on k_d values for **2**-based macroalkoxyamines **53** whereas a 2–20 fold increase in k_d was observed for **51** and **66** **9**-based macroalkoxyamines.

2.3.3.5.5 Intramolecular Proton Transfer (IPT) as Side-reaction

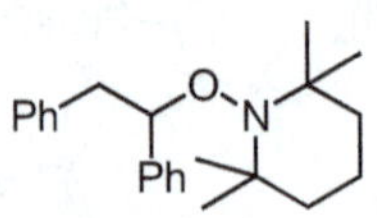

67

IPT (Cope-type elimination) is the side-reaction occurring during the homolysis of the C–ON bond and was disregarded as it affords the same products as the HAT reaction (Schemes 2.4 and 2.11). Its occurrence was probably reported for the first time by Scaiano and colleagues[236] for **67**. The authors recently investigated this reaction for a series of alkoxyamines and showed that its occurrence was randomly dependent on the structure.[93] DFT calculations showed that the energy barrier did not depend significantly on the structure of the alkoxyamine and IPT significantly occurred for a C–ON bond homolysis barrier higher than 130 kJ mol^{-1}.[197]

HAT and IPT afford the same reaction products (hydroxylamine and alkene) although they are chemically different reactions (Scheme 2.4): HAT is an intermolecular radical process while IPT is an intramolecular ionic process. Hence, the consequences of the occurrence of IPT on the fate of the NMP are the same as those of the occurrence of HAT.[56,57]

2.3.3.5.6 C–ON Bond Homolysis *vs.* CO–N Bond Homolysis. The CO–N bond homolysis side-reaction (Scheme 2.13) has been very little investigated, as CO–N bond homolysis is expected to compete with C–ON bond homolysis only when the latter exhibits a BDE(C–O) larger than 140 kJ mol^{-1}.[55] Nevertheless, CO–N bond homolysis was reported for **35** despite quite low E_a (*ca.* 130 kJ mol^{-1}) for the C–ON bond homolysis.[72,113,237] It was also reported for other derivatives of **34**.[113] In general, the EPR technique using the spin-trapping method is the approach to detecting the occurrence of the CO–N bond homolysis and the level of its occurrence.[113] This approach was applied to other types of indoline-based alkoxyamines.[98] Using calculations, Coote and colleagues[199] investigated the effects ruling the competition C–ON *vs.* CO–N bond homolysis and noted that, in general, C–ON bond homolysis was favoured by 50 kJ mol^{-1} over the CO–N bond homolysis, whatever the conditions. It was also noted that the free enthalpy ΔG for the CO–N bond homolysis is greater than 140 kJ mol^{-1} in the gas phase at 298 K, except for indoline-derivatives of type **34**. The absence of correlation between the ΔG values for C–ON and CO–N bond homolysis was also mentioned. In fact, the CO–N bond homolysis competes with the C–ON bond homolysis only when the released alkyl radical is not stabilized, anomeric interactions occurring in the alkoxyamine, and the released aminyl radical (CO–N bond homolysis) is strongly stabilized as, for indoline-derivatives.

2.3.4 Other Types of Activation

All the effects discussed in Section 2.3.3.5 were related to the thermal initiation of the C–ON homolysis. However, some other types of activation are known, although they are not too well developed, except the photochemical activation. Among the chemical activation processes available, only the protonation has been applied to NMP.[225]

Microwave triggered C–ON bond homolysis was proposed by Studer and colleagues[238] for applications in tin-free radical chemistry and NMP.[239–242]

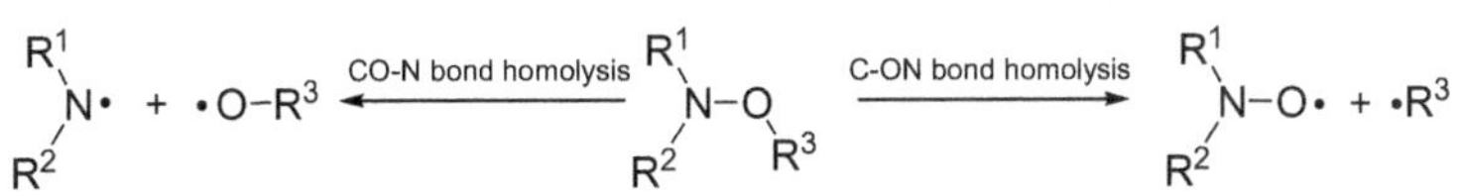

Scheme 2.13 C–ON bond homolysis *vs.* CO–N bond homolysis.

Microwave initiation is another type of thermal initiation, except that the heating of the solution is much faster and more efficient.[243] Nevertheless, it is reasonable to expect that the aforementioned discussions hold.

As far as the authors know, only one report on NMP under sonication was reported.[244] In this article, Kojima *et al.*[244] noticed that the NMP of styrene under sonication in the presence of **2** exhibited an inhibition time due to the formation of the dormant species. The polymerization occurred under sonication in a controlled fashion at a higher rate than for the conventional thermal NMP. The increase in the rate can be ascribed to either a higher homolysis rate constant k_d [eqn (2.8)] under sonication or an increase in the frequency of collision between the monomer and the growing radical. Nevertheless, whatever the event, the increase in reactivity is due to the release of thermal energy occurring when the cavitation bubbles collapse.

The ability of alkoxyamines to cleave under UV irradiation was first reported by Scaiano and colleagues[245] with **27** using an external photosensitizer. A decade later, the process was independently improved by the groups of Gigmes[246] and Yoshida[247] in order to develop UV-initiated nitroxide mediated polymerization.

2.4 Conclusion and Outlook

Nowadays, NMP has left the academic benches to be currently and routinely applied in factories to prepare various types of polymers for very different applications. This will be the scope of several chapters in this book. Nevertheless, we can claim today that the amazing reactivity of alkoxyamines observed in 1974 by Kovtun *et al.*[27] has found very nice and important applications in the preparation of polymers. Several aspect of NMP kinetics, such as the kinetics of copolymerization in (micro, mini)emulsion, as well as other NMP variants such as bicomponent systems (nitroxide + initiator, nitrone + initiator ....), have not been discussed in this chapter and are the topics of several chapters in this book.

In our opinion, the fundamentals of NMP are well known and this technique is mature enough to look to go further on some emerging directions, *e.g.*, its combination with non-radical polymerization to develop an orthogonal approach, polymerization on a surface for hybrid materials...

All along this chapter, the authors tried to provide the main guidelines for estimating and discussing k_c and k_d values. Nevertheless, we invite readers to refer to several reviews dedicated to each of the rate constants as well to the fundamental kinetics and the various systems derived from NMP.

Acknowledgements

SM is grateful to CNRS (ASR 21245 and 23961) and to Agence Nationale pour la Recherche (ANR-14-CE06-0023-01) for travelling grants and support. EB thanks for financial support RFBR-15-03-05250.

References

1. G. Moad and D. H. Solomon, *The Chemistry of Radical Polymerization*, Elsevier, Amsterdam, 2nd fully revised edn, 2006.
2. G. Odian, *Principles of Polymerization*, Wiley-Interscience, Hoboken, New Jersey, 4th edn, 2004.
3. S. Aoshima and S. Kanaoka, *Chem. Rev.*, 2009, **109**, 5245.
4. V. Sciannamea, R. Jerome and C. Detrembleur, *Chem. Rev.*, 2008, **108**, 1104.
5. R. B. Grubbs, *Polym. Rev.*, 2011, **51**, 104.
6. P. Nesvadba, *Chimia*, 2006, **60**, 832, and references cited therein.
7. B. Charleux and J. Nicolas, *Polymer*, 2007, **48**, 5813.
8. C. J. Hawker, A. W. Bosman and E. Harth, *Chem. Rev.*, 2001, **101**, 3661.
9. J. Nicolas, Y. Guillaneuf, C. Lefay, D. Bertin, D. Gigmes and B. Charleux, *Progr. Polym. Sci.*, 2013, **38**(1), 63.
10. D. Bertin, D. Gigmes, S. R. A. Marque and P. Tordo, *Chem. Soc. Rev.*, 2011, **40**, 2189.
11. D. F. Grishin and I. D. Grishin, *Russ. J. Appl. Chem.*, 2011, **84**(12), 2021, *Z. Priklad. Kim.*, 2011, **84**, 12, 1937.
12. *Handbook of RAFT Polymerization*, ed. C. Barner-Kowollik, Wiley-VCH, Weinheim, 2008.
13. C. Boyer, V. Bulmus, Th. P. Davis, V. Ladmiral, J. Liu and S. Perrier, *Chem. Rev.*, 2009, **109**, 5402.
14. M. Beija, J.-D. Marty and M. Destarac, *Prog. Polym. Sci.*, 2011, **36**, 845.
15. A. Gregory and M. H. Stenzel, *Prog. Polym. Sci.*, 2012, **37**, 38.
16. A. B. Lowe and C. L. McCormick, *Prog. Polym. Sci.*, 2007, **32**, 283.
17. W. A. Braunecker and K. Matyjaszewski, *Prog. Polym. Sci.*, 2007, **32**, 93.
18. D. J. Siegwart, J. Kwon and Oh, K. Matyjaszewski, *Prog. Polym. Sci.*, 2012, **37**, 19.
19. J. Ran, L. Wu, Z. Zhang and T. Xu, *Prog. Polym. Sci.*, 2014, **39**, 124.
20. D. H. Solomon, E. Rizzardo and P. Cacioli, 1985, *Eur. Pat. Appl.* 135280; 1986, *US Pat.* 4,581,429; 1985, *Chem. Abstr.*, 102, 221335q.
21. D. H. Solomon, *J. Polym. Sci., Part A: Polym. Chem.*, 2005, **43**(23), 5748.
22. The first report on alkoxyamines $R_1R_2NOR_3$ (trialkyl hydroxylamine) was given by Jones and Major[23] at the beginning of last century. That family of molecules was developed for medical purposes. The physical and chemical properties of this family as well as their application in ionic chemistry have been investigated and reviewed all along the last century.[24–26] However, the radical reactivity of this family was only discovered by Kovtun *et al.* in 1974.[27] Using cyclohexyl-**2** alkoxyamine in the presence of dioxygene O_2 as alkyl radical scavenger showed that the corresponding nitroxide **2** was generated upon heating.[29] A few years later, Howard *et al.*[28] re-investigated this reactivity with different **2**-derivatives and showed that the life-time of the alkoxyamine was much longer in the absence of oxygene than in its presence or in the presence of any other alkyl radical scavenger. Nevertheless, the kinetics

underlying the decomposition of alkoxyamines were not understood until the two seminal works of Fischer on the unusual selectivity of radical reactions[41] and on the radical reaction kinetics during homolysis of *N*-alkoxyamines.[29]

23. L. W. Jones and R. T. Major, *J. Am. Chem. Soc.*, 1927, **49**, 1527.
24. J. S. Roberts, *Comprehensive Organic Chemistry*, ed. D. H. R. Barton and W. D. Ollis, Series ed. I. O. Sutherland, Pergamon Press, Oxford, 1979, vol. 2, p. 185.
25. R. Askani, D. F. Taber, *Comprehensive Organic Synthesis*, ed. B. M. Trost and I. Fleming, Series ed. E. Winterfeld, Pergamon Press, Oxford, 1991, vol. 6, p. 104.
26. C. M. Marson and A. D. Hobson, *Comprehensive Organic Functional Groups*, ed. A. R. Katritzky, O. Meth-Cohn and C. W. Rees, Series ed. S. V. Ley, Pergamon Press, Oxford, 1995, vol. 2, p. 298.
27. G. A. Kovtun, A. L. Aleksandrov and V. A. Golubev, *Bull. Akad. Sci. USSR: Div. Ser.*, 1974, 2115, *Izv. Akad. Nauk SSSR: Ser. Khim.*, 1974, 2197.
28. J. A. Howard and J. C. Tait, *J. Org. Chem.*, 1978, **43**, 4279.
29. T. Kothe, S. Marque, R. Martschke, M. Popov and H. Fischer, *J. Chem. Soc. Perkin Trans. 2*, 1998, 1553.
30. C. H. J. Johnson, G. Moad, D. H. Solomon, T. H. Spurling and D. J. Vearing, *Aust. J. Chem.*, 1990, **43**, 1215.
31. T. Fukuda, T. Terauchi, A. Goto, K. Ohno, Y. Tsujii, T. Miyamoto, S. Kobatake and B. Yamada, *Macromolecules*, 1996, **29**, 6393.
32. D. Greszta and K. Matyjaszewski, *Macromolecules*, 1996, **29**, 7661.
33. H. Fischer, *Macromolecules*, 1997, **30**, 5666.
34. H. Fischer, *Chem. Rev.*, 2001, **101**, 3581, and references cited therein.
35. This term Persitent Radical Effect (PRE) was coined by: B. E. Daikh and R. G. Finke, *J. Am. Chem. Soc.*, 1992, **114**, 2938.
36. The kinetic equations for the theoretical model were developed by H. Fischer, see ref. 41.
37. T. Fukuda, A. Goto and K. Ohno, *Macromol. Rapid Commun.*, 2000, **21**, 151.
38. C. Yoshikawa, A. Goto and T. Fukuda, *Macromolecules*, 2002, **35**, 5801.
39. W. E. Bachmann and F. Y. Wiselogle, *J. Org. Chem.*, 1936, **1**, 354.
40. M. J. Perkins, *J. Chem. Soc.*, 1964, 5932.
41. H. Fischer, *J. Am. Chem. Soc.*, 1986, **108**, 3925.
42. D. Griller and K. U. Ingold, *Acc. Chem. Res.*, 1976, **9**, 13.
43. D. Ruegge and H. Fischer, *Int. J. Chem. Kinet.*, 1989, **21**, 703.
44. K.-S. Focsaneanu and J. C. Scaiano, *Helv. Chim. Acta*, 2006, **89**, 2473.
45. H. Fischer, *J. Polym. Sci., Part A: Polym. Chem.*, 1999, **37**, 1885.
46. H. Fischer and M. Souaille, *Chimia*, 2001, **55**, 109.
47. H. Fischer and M. Souaille, *Macromol. Symp.*, 2001, **174**, 231.
48. A. Goto and T. Fukuda, *Prog. Polym. Sci.*, 2004, **29**, 329, and the references therein.

49. W. Tang, T. Fukuda and K. Matyjaszewski, *Macromolecules*, 2006, **39**, 4332.
50. B. R. Smirnov, *Vysokomol. Soedin.*, 1990, **A32**, 583.
51. A. Studer, *Chem. Soc. Rev.*, 2004, **33**, 267.
52. P.-E. Dufils, N. Chagneux, D. Gigmes, T. Trimaille, S. R. A. Marque, D. Bertin and P. Tordo, *Polymer*, 2007, **48**, 5219.
53. T. Vogler and A. Studer, *Synthesis*, 2008, 1979.
54. L. Tebben and A. Studer, *Angew. Chem., Int. Ed.*, 2011, **50**(22), 5034.
55. G. Audran, P. Brémond and S. R. A. Marque, *Chem. Commun.*, 2014, **50**(59), 7921.
56. M. Souaille and H. Fischer, *Macromolecules*, 2001, **34**, 2830.
57. M. Souaille and H. Fischer, *Macromolecules*, 2002, **35**, 248.
58. A. Goto and T. Fukuda, *Macromol. Chem. Phys.*, 2000, **201**, 2138.
59. D. Gigmes and S. R. A. Marque, Nitroxide Mediated Polymerization and its Applications, in *Encyclopedia of Radicals in Chemistry, Biology, and Materials*, ed. C. Chatgilialoglu and A. Studer, Wiley, Chichester, U.K., 2012, vol. 4, p. 1813.
60. J. Qiu, B. Charleux and K. Matyjaszewski, *Prog. Polym. Sci.*, 2001, **26**, 2083.
61. P. B. Zetterlund, Y. Kagawa and M. Okubo, *Chem. Rev.*, 2008, **108**, 3747.
62. M. F. Cunningham, *Prog. Polym. Sci.*, 2002, **27**, 1039.
63. M. F. Cunningham, *Prog. Polym. Sci.*, 2008, **33**, 365.
64. J. M. Asua, *Prog. Polym. Sci.*, 2002, **27**, 1283.
65. K. Ohno, Y. Tsujii, T. Miyamoto, T. Fukuda, M. Goto, K. Kobayashi and T. Akaike, *Macromolecules*, 1998, **31**, 1064.
66. The plot of eqn (2.8) in log–log scale affords the slope for the monomer/initiator 20/1 ratio the closest to the expected 2/3 value.
67. J. F. Lutz, P. Lacroix-Desmazes and B. Boutevin, *Macromol. Rapid Commun.*, 2001, **22**, 189.
68. L. Catala, E. Rubel and S. O. Hammouch, *Macromolecules*, 1995, **28**, 8441.
69. D. Benoit, V. Chaplinski, R. Braslau and C. J. Hawker, *J. Am. Chem. Soc.*, 1999, **121**, 3904.
70. D. Greszta and K. Matyjaszewski, *J. Polym. Sci., Part A: Polym. Chem.*, 1997, **35**, 1857.
71. M. K. Georges, R. P. N. Veregin, P. M. Kazmaier, G. K. Hamer and M. Saban, *Macromolecules*, 1994, **27**, 7228.
72. Y. Guillaneuf, D. Gigmes, S. R. A. Marque, P. Tordo and D. Bertin, *Macromol. Chem. Phys.*, 2006, **207**, 1278.
73. K. O. Siegenthaler and A. Studer, *Macromolecules*, 2006, **39**, 1347.
74. D. Gigmes, D. Bertin, C. Lefay and Y. Guillaneuf, *Macromol. Theory Simul.*, 2009, **18**, 402.
75. $k_{d,ds}$ for the polymeric species was also different from that for the molecular model **10** and estimated using equations reported in ref. 87 and taking **19** as a model.
76. E. G. Bagryanskaya, S. R. A. Marque and Y. P. Tsentalovich, *J. Org. Chem.*, 2012, 77, 4996.

77. M. G. Evans, *Discuss. Faraday Soc.*, 1947, **2**, 271.
78. N. N. Semenov, *Some Problems in Chemical Kinetics and Reactivity* (Engl. Transl.), Princeton Press, Princeton, 1958, p. 29.
79. D. P. Zubenko, I. A. Kirilyuk, S. R. A. Marque, I. P. Chuikov and E. G. Bagryanskaya, *Polymer Sci., Ser. B*, 2010, **52**(5–6), 327, *Vysokomol. Soedin., Ser. B* 2010, **52**, 6, 1030.
80. E. Bagryanskaya, D. Bertin, D. Gigmes, I. Kirilyuk, S. R. A. Marque, V. Reznikov, G. Roshchupkina, I. Zhurko and D. Zubenko, *Macromol. Chem. Phys.*, 2008, **209**, 1345.
81. D. F. Grishin and L. L. Semyonycheva, *Russ. Chem. Rev.*, 2007, **70**, 425.
82. $E_a = 102.5$ kJ mol^{-1} as given in ref. 87 and assuming a chain length effect of a factor 3 as in ref. 83.
83. O. Guerret, J.-L. Couturier, F. Chauvin, H. El-Bouazzy, D. Bertin, D. Gigmes, S. Marque, H. Fischer and P. Tordo, *ACS Symp. Ser.*, 2003, **854**, 412.
84. $E_a = 108.9$ kJ mol^{-1} as given in ref. 85.
85. E. Beaudoin, D. Bertin, D. Gigmes, S. R. A. Marque, D. Siri and P. Tordo, *Eur. J. Org. Chem.*, 2006, 1755.
86. J. Nicolas, C. Dire, L. Mueller, J. Belleney, B. Charleux, S. R. A. Marque, D. Bertin, S. Magnet and L. Couvreur, *Macromolecules*, 2006, **39**, 8274.
87. $E_a = 116$ kJ mol^{-1} D. Bertin, P.-E. Dufils, I. Durand, D. Gigmes, B. Giovanetti, Y. Guillaneuf, S. R. A. Marque, T. Phan and P. Tordo, *Macromol. Chem. Phys.*, 2008, **209**, 220.
88. The chain length effect was disregarded.
89. k_c was estimated using equations reported in ref. 115 assuming no effect of the penultimate unit on the radical stabilization. $\sigma_{RS} = 0.34$, $\sigma_I = 0.05$ with $\sigma_{I,CH2CMe2COOMe} = 0.02$, and $\upsilon = 1.43$ with $\upsilon_1 = \upsilon_{i\text{-}Pr} = 0.76$, $\upsilon_2 = \upsilon_{CMe2COOMe} = 1.43$ and $\upsilon_3 = \upsilon_{Ph} = 0.57$.
90. J. Nicolas, S. G. N. Brusseau and B. Charleux, *J. Polym. Sci., A Polym. Chem.*, 2010, **48**, 34.
91. T. Fukuda, A. Goto and Y. Tsujii, *Handbook of Radical Polymerization*, ed. K. Matyjazewski and T. P. Davis, Wiley-Interscience, 2002, 407 and references cited therein.
92. F. Chauvin, P.-E. Dufils, D. Gigmes, Y. Guillaneuf, S. R. A. Marque, P. Tordo and D. Bertin, *Macromolecules*, 2006, **39**, 5238.
93. M. V. Edeleva, I. A. Kirilyuk, D. P. Zubenko, I. F. Zhurko, S. R. A. Marque, D. Gigmes, Y. Guillaneuf and E. G. Bagryanskaya, *J. Polym. Sci., Polym. Chem.: Part A*, 2009, **47**, 6579.
94. It was estimated from eqn (2.26) with $\sigma_{RS} = 0.22$, $\sigma_I = 0.14$, and $\upsilon = 1.20$ for **32** in ref. 187.
95. S. Harisson, P. Couvreur and J. Nicolas, *Macromolecules*, 2011, **44**, 9230.
96. S. Harisson, P. Couvreur and J. Nicolas, *Macromol. Rapid Commun.*, 2012, **33**, 805.
97. Y. Guillaneuf, D. Gigmes, S. R. A. Marque, P. Astolfi, L. Greci, P. Tordo and D. Bertin, *Macromolecules*, 2007, **40**, 3108.

98. P. Astolfi, L. Greci, P. Stipa, C. Rizzoli, C. Ysacco, M. Rollet, L. Autissier, A. Tardy, Y. Guillaneuf and D. Gigmes, *Polym. Chem.*, 2013, **4**, 3694.
99. G. S. Ananchenko, M. Souaille, H. Fischer, C. Le Mercier and P. Tordo, *J. Polym. Sci., Part A: Polym. Chem.*, 2002, **40**, 3264–3283.
100. For SG1-PMMA, k_{cD} was estimated at *ca.* 1000 M^{-1} s^{-1} C. Dire, J. Belleney, J. Nicolas, D. Bertin, S. Magnet and B. Charleux, *J. Polym. Sci., Part A: Polym. Chem.*, 2008, **46**, 6333.
101. M. Edeleva, S. R. A. Marque, K. Kabytaev, Y. Guillaneuf, D. Gigmes and G. Bagryanskaya, *J. Polym. Sci., Part A: Polym. Chem.*, 2013, **51**(6), 1323.
102. H. Fischer and H. Paul, *Acc. Chem. Res.*, 1987, **20**, 200.
103. M. Buback, M. Egorov, R. G. Gilbert, V. Kaminsky, O. F. Olaj, G. T. Russell, P. Vana and G. Zifferer, *Macromol. Chem. Phys.*, 2002, **203**, 2570.
104. R. D. Bach, D. Andrzejewski and L. R. Dusold, *J. Org. Chem.*, 1973, **38**, 1742.
105. A. A. Gridnev, *Macromolecules*, 1997, **30**, 7651.
106. G. Moad, A. G. Anderson, F. Ercole, H. J. Johnson, J. Krstina, C. L. Moad, E. Rizzardo, T. H. Spurling and S. H. Thang, *ACS Symp. Ser.*, 1998, **685**, 332.
107. A. Goto and T. Fukuda, *Macromolecules*, 1999, **32**, 618.
108. G. S. Ananchenko and H. Fischer, *J. Polym. Sci., Part A: Polym. Chem.*, 2001, **39**, 3604.
109. M. Edeleva, S. R. A. Marque, D. Bertin, D. Gigmes, Y. Guillaneuf, S. V. Morozov and E. G. Bagryanskaya, *J. Polym. Sci., Part A: Polym. Chem.*, 2008, **46**, 6828.
110. A. Ohno, Y. Tsujii and T. Fukuda, *Macromolecules*, 1997, **30**, 2503.
111. In recent years, the homopolymerization of MMA by NMP (refs. 97, 98 and 165) as well as its block-copolymerization (refs. 98 and 165) has been solved for the large part.
112. R. P. N. Veregin, P. G. Odell, L. M. Michalak and M. K. Georges, *Macromolecules*, 1996, **29**, 4161.
113. D. Gigmes, A. Gaudel-Siri, S. R. A. Marque, D. Bertin, P. Tordo, P. Astolfi, L. Greci and C. Rizzoli, *Helv. Chim. Acta*, 2006, **89**, 2312.
114. N–OC homolysis occurs in **34** derivatives as it affords a highly stabilized aminyl radical through the delocalization of the odd electron onto the aromatic ring.
115. E. G. Bagryanskaya and S. R. A. Marque, *Chem. Rev.*, 2014, **114**, 5011.
116. V. W. Bowry and K. U. Ingold, *J. Am. Chem. Soc.*, 1992, **114**, 4992.
117. J. Sobek, R. Martschke and H. Fischer, *J. Am. Chem. Soc.*, 2001, **123**, 2849.
118. N. V. Lebedeva, D. P. Zubenko, E. G. Bagryanskaya, R. Z. Sagdeev, G. S. Ananchenko, S. Marque, D. Bertin and P. Tordo, *Phys. Chem. Chem. Phys.*, 2004, **6**, 2254.
119. A. L. J. Beckwith, V. W. Bowry and K. U. Ingold, *J. Am. Chem. Soc.*, 1992, **114**, 4983.
120. Y. Guillaneuf, D. Bertin, P. Castignolles and B. Charleux, *Macromolecules*, 2005, **38**, 4638.

121. A. L. J. Beckwith and V. Bowry, *J. Org. Chem.*, 1988, **53**(8), 1632.
122. J. Chateauneuf, J. Lusztyk and K. U. Ingold, *J. Org. Chem.*, 1988, **56**, 1629.
123. W. G. Skene, J. C. Scaiano, N. A. Listigovers, P. M. Kazmaier and M. K. Georges, *Macromolecules*, 2000, **33**, 5065.
124. To the best of our knowledge, nobody has reported the preparation of ketones or any other radical initiator carrying a polymer chain and suitable for the generation of polymeric radicals by photolysis. Moreover, no reports based on the approach developed in ref. 115 involving polymer chains are available.
125. M. Buback, R. G. Gilbert, R. A. Hutchinson, B. Klumperman, F. D. Kuchta, B. G. Manders, K. F. O'Driscoll, G. T. Russell and J. Schweer, *Macromol. Chem. Phys.*, 1995, **196**, 3267.
126. F. R. Mayo, *J. Am. Chem. Soc.*, 1943, **65**, 2324.
127. J. P. A. Heuts, T. P. Davis and G. T. Russell, *Macromolecules*, 1999, **32**(2), 6019.
128. P. A. Clay and R. G. Gilbert, *Macromolecules*, 1995, **28**, 552.
129. O. V. Borisova, M. Yu. Zaremski, V. B. Golubev, A. V. Plutalova, O. V. Borisov and L. Billon, *Polym. Sci., Ser. B*, 2013, **55**(9–10), 508.
130. D. Zubenko, Y. Tsentalovich, N. Lebedeva, I. Kirilyuk, G. Roshchupkina, I. Zhurko, V. Reznikov, S. R. A. Marque and E. Bagryanskaya, *J. Org. Chem.*, 2006, **71**(16), 6044.
131. BDE of the C–H bond corresponding to the released alkyl radical.
132. $N=8$, $R^2=0.98$, Student-Fischer test $F_{0.01\%}=122$, Student t-test $t>99.6\%$.
133. The weight w of each effect was also estimated in %: $w_{\sigma RS}=48$, $w_{\upsilon}=27$ and $w_{\sigma I}=25$.
134. $N=28$, $R^2=0.95$, Student-Fischer test $F_{0.01\%}=140$, Student t-test $t>94\%$.
135. The weight w of each effect was also estimated in %: $w_{\sigma RS}=63$, $w_{\upsilon}=13$ and $w_{\sigma I}=24$.
136. Y.-R. Luo, *Comprehensive Handbook of Chemical Bond Energies*, CRC Press, Boca Raton, 2007.
137. J. R. Mahoney, G. D. Mendenhall and K. U. Ingold, *J. Am. Chem. Soc.*, 1973, **95**, 8610.
138. As far as we know, no article other than ref. 137 deals with the experimental determination of the BDE of NO–H.
139. Statistical output: $N=28$, $R^2=0.95$, Student-Fischer test $F_{0.01\%}=254$, Student t-test $t>99.7\%$.
140. The weight w of each effect was also estimated in %: $w_{Es}=89$ and $w_{\sigma I}=11$.
141. It is well known that the phenyl group is considered as a Janus group concerning its bulkiness, *i.e.*, it varies depending on the other groups in its vicinity. Consequently, these nitroxides were discussed separately. See ref. 76.
142. K. Iwao, K. Sakakibara and M. Hirota, *J. Comput. Chem.*, 1998, **19**, 215.

143. J. L. Hodgson, C. Y. Lin, M. L. Coote, S. R. A. Marque and K. Matyjaszewski, *Macromolecules*, 2010, **43**, 3728–3743.
144. R. D. Puts and D. Y. Sogah, *Macromolecules*, 1996, **29**, 3323.
145. Calculations of the thermodynamics of the homolysis reaction were first performed by M. K. Georges, R. P. N. Veregin, P. M. Kazmaier and G. K. Hamer, *Trends Polym. Sci.*, 1993, **2**, 66; O. Kysel, P. Mach and M. Micov, *Polym. Degrad. Stab.*, 1993, **40**, 31.
146. R. Braslau, N. Naik and H. Zipse, *J. Am. Chem. Soc.*, 2000, **122**, 8421.
147. A. Kaim and E. Megiel, *J. Polym. Sci., Part A: Polym. Chem.*, 2005, **44**, 914.
148. A. Kaim, *J. Polym. Sci., Part A: Polym. Chem.*, 2007, **45**, 232.
149. A. Rauk, *Orbital Interaction Theory of Organic Chemistry*, Wiley-Interscience, New York, 2nd edn, 2001.
150. J. D. Dunitz, J. M. Lehn and G. Wipf, *J. Am. Chem. Soc.*, 1974, **96**, 1956.
151. H. B. Burgi, J. D. Dunitz, J. M. Lehn and G. Wipf, *Tetrahedron*, 1974, **30**, 1563.
152. H. Fischer and L. Radom, *Angew. Chem., Int. Ed.*, 2001, **40**, 1340.
153. The presence of an EWG attached to the radical center, *i.e.*, increasing the polarity, is expected to destabilize the C-centred radical, and, consequently, to increase the spin density on the radical centre. Hence, it would be ascribed to an enthalpic effect. However, it is likely that such an effect on the distribution of spin density is already described by the stabilization parameter σ_{RS}.
154. As quoted by Reichardt: "A cybotactic region may be defined as the volume around a solute molecule in which the ordering of the solvent molecules has been influenced by the solute, including both the first solvation shell and the transition region." C. Reichardt, T. Welton, *Solvent and Solvent Effect in Organic Chemistry*, Wiley-VCH, Weinheim, 4th edn, 2011.
155. G. E. Zaikov, R. G. Makitra, G. G. Midyana and L. I. Bazylyak, *Influence of the Solvent of Some Radical Reactions*, Chemistry Research and Applications Series, Nova Science Publishers Inc., New York, 2010.
156. J. T. Edward, *J. Chem. Ed.*, 1970, **47**(4), 261.
157. B. Knauer and J. J. Napier, *J. Am. Chem. Soc.*, 1976, **98**, 4395.
158. G. Audran, P. Brémond, S. R. A. Marque and G. Obame, *ChemPhysChem*, 2012, **13**, 3542.
159. P. G. Odell, R. P. N. Veregin, L. M. Michalak and M. K. Georges, *Macromolecules*, 1997, **30**, 2232.
160. C. Li, J. He, L. Li, J. Cao and Y. Yang, *Macromolecules*, 1999, **32**(21), 7012.
161. A. Studer, *Angew. Chem., Int. Ed.*, 2000, **39**, 1108.
162. A. Studer and T. Schulte, *Chem. Rec.*, 2005, **5**, 27.
163. D. Bertin, D. Gigmes and S. R. A. Marque, *Recent Res. Dev. Org. Chem.*, 2006, **10**, 63, and references therein.
164. M. V. Baldoví, N. Mohtat and J. C. Scaiano, *Macromolecules*, 1996, **29**, 5497.

165. M. V. Edeleva, I. A. Kirilyuk, I. F. Zhurko, D. A. Parkhomenko, Y. P. Tsentalovich and E. G. Bagryanskaya, *J. Org. Chem.*, 2011, **76**, 5558.
166. V. Malatesta and K. U. Ingold, *J. Am. Chem. Soc.*, 1973, **95**(19), 6404.
167. S. Marque, J. Sobek, H. Fischer, A. Kramer, P. Nesvadba and W. Wunderlich, *Macromolecules*, 2003, **36**, 3440.
168. A general discussion on the various type of HAT is provided in J. J. Warren, T. A. Tronic and J. M. Mayer, *Chem. Rev.*, 2010, **110**, 6961.
169. D. Zubenko, I. Kirilyuk, G. Roshchupkina, I. Zhurko, V. Reznikov, S. R. A. Marque and E. Bagryanskaya, *Helv. Chim. Acta*, 2006, **89**(10), 2341.
170. M. Destarac, *Macromol. React. Eng.*, 2010, **4**, 165.
171. S. Marque, C. Le Mercier, P. Tordo and H. Fischer, *Macromolecules*, 2000, **33**(12), 4403.
172. D. Bertin, D. Gigmes, S. Marque and P. Tordo, *e-Polym.*, 2003, **2**, 1–9.
173. M. Zaremski, O. Borisova, C. Xin, V. B. Golubev and L. Billon, *J. Polym. Sci., Part A: Polym. Chem.*, 2012, **50**, 3437.
174. T. Fukuda and A. Goto, *Macromol. Rapid. Commun.*, 1997, **18**, 683.
175. O. G. Ballesteros, L. Maretti, R. Sastre and J. C. Scaiano, *Macromolecules*, 2001, **34**, 6184.
176. L. C. Li, G. K. Hamer and M. K. Georges, *Macromolecules*, 2006, **39**, 9201.
177. P. Kelemen, J. Lugtenburg and B. Klumperman, *J. Org. Chem.*, 2003, **68**, 7322–7328.
178. Personal communication of the authors.
179. G. Ananchenko, S. Marque, D. Gigmes, D. Bertin and P. Tordo, *Org. Biomol. Chem.*, 2004, **2**, 709.
180. *In situ* NMR experiments require the booking of NMR machines for a long time as well as an accurate and reliable temperature control device.
181. M. Yu. Zaremski, *Polym. Sc., Ser. A*, 2006, **48**(3), 234.
182. M. V. Ciriano, H.-G. Korth, W. B. van Scheppingen and P. Mulder, *J. Am. Chem. Soc.*, 1999, **121**, 6375.
183. S. W. Benson, *Thermochemical Kinetics*, John Wiley & Sons, New York, 1968, p. 55.
184. I. W. M. Smith, *Chem. Soc. Rev.*, 2008, **37**, 812.
185. S. R. A. Marque and D. Siri, *Macromolecules*, 2009, **42**(4), 1404.
186. S. Marque, H. Fischer, E. Baier and A. Studer, *J. Org. Chem.*, 2001, **66**(4), 1146.
187. D. Bertin, D. Gigmes, S. R. A. Marque and P. Tordo, *Macromolecules*, 2005, **38**(7), 2638.
188. $N = 21$, $R^2 = 0.94$; $\sigma = 0.48$; $F_{18,0.01\%} = 75$.
189. $w_{\sigma I} = 35\%$, $w_{\sigma RS} = 34\%$, and $w_{\upsilon} = 31\%$.
190. L. Pauling, *J. Am. Chem. Soc.*, 1931, **53**, 1367.
191. L. Pauling, *The Nature of the Chemical Bond*, Cornell University Press, Ithaca, NY, 1960.
192. $N = 25$, $R^2 = 0.95$, $\sigma = 0.17$, $F_{23,0.01\%} = 210$.

193. S. R. A. Marque, *J. Org. Chem.*, 2003, **68**(20), 7582.
194. Non-cyclic nitroxyl fragments were removed from the correlation.
195. $N = 48$, $R^2 = 0{:}96$; $\sigma = 0{:}14$; $F_{47,0.01\%} = 524$.
196. H. Fischer, A. Kramer, S. R. A. Marque and P. Nesvadba, *Macromolecules*, 2005, **38**, 9974.
197. D. Parkhomenko, E. G. Bagryanskaya, S. R. A. Marque and D. Siri, *Phys. Chem. Chem. Phys.*, 2013, **15**, 13862.
198. P. Marsal, M. Roche, P. Tordo and P. D. Claire, *J. Phys. Chem. A*, 1999, **103**, 2899.
199. J. L. Hodgson, L. B. Roskop, M. S. Gordon, C. Y. Lin and M. L. Coote, *J. Phys. Chem. A*, 2010, **114**, 10458.
200. B. Chan, M. L. Coote and L. Radom, *J. Chem. Theory Comput.*, 2010, **6**(9), 2647.
201. In fact, as TS is late, the distance between the alkyl and nitroxyl fragments is longer than in an alkoxyamine leading us to expect a weak influence of the steric effect.
202. A. Gaudel-Siri, D. Siri and P. Tordo, *ChemPhysChem*, 2006, 7, 430.
203. S. Acerbis, E. Beaudoin, D. Bertin, D. Gigmes, S. Marque and P. Tordo, *Macromol. Chem. Phys.*, 2004, **205**, 973.
204. D. Bertin, D. Gigmes, S. R. A. Marque, S. Milardo, J. Peri and P. Tordo, *Collect. Czech. Chem. Commun.*, 2004, **69**, 2223.
205. G. Audran, M. Bim Batsiandzy Ibanou, P. Brémond, S. R. A. Marque, G. Obame, V. Roubaud and D. Siri, *J. Phys. Org. Chem.*, 2014, **27**, 387.
206. D. Bertin, D. Gigmes, S. Marque, R. Maurin and P. Tordo, *J. Polym. Sci., Part A: Polym. Chem.*, 2004, **42**, 3504.
207. G. Ananchenko, E. Beaudoin, D. Bertin, D. Gigmes, P. Lagarde, S. R. A. Marque, E. Revalor and P. Tordo, *J. Phys. Org. Chem.*, 2006, **19**, 269.
208. D. Bertin, D. Gigmes, S. R. A. Marque, D. Siri, P. Tordo and G. Trappo, *ChemPhysChem*, 2008, **9**, 272.
209. G. Audran, P. Brémond, S. R. A. Marque and G. Obame, *J. Org. Chem.*, 2013, **78**, 7754.
210. The role in NMP of IHB in nitroxide was first proposed by K. Matyjaszewski, S. Gaynor, D. Greszta, D. Mardare and T. Shigemoto, *J. Phys. Org. Chem.*, 1995, **8**, 306
211. S. Acerbis, D. Bertin, B. Boutevin, D. Gigmes, P. Lacroix-Desmazes, C. Le Mercier, J.-F. Lutz, S. R. A. Marque, D. Siri and P. Tordo, *Helv. Chim. Acta*, 2006, **89**(10), 2119.
212. C. A. Knoop and A. Studer, *J. Am. Chem. Soc.*, 2003, **125**, 16327.
213. E. Harth, B. Van Horn and C. J. Hawker, *J. Chem. Soc., Chem. Commun.*, 2001, 823.
214. P. Brémond, T. Butscher, V. Roubaud, D. Siri and S. Viel, *J. Org. Chem.*, 2013, **75**, 10524.
215. E. G. Bagryanskaya, P. Brémond, T. Butscher, S. R. A. Marque, D. Parkhomenko, V. Roubaud, D. Siri and S. Viel, *Macromol. Chem. Phys.*, 2015, **216**(5), 475–488.

216. A. Blachon, S. R. A. Marque, V. Roubaud and D. Siri, *Polymers*, 2010, **2**, 353.
217. D. Bertin, F. Chauvin, S. Marque and P. Tordo, *Macromolecules*, 2002, **35**, 3790.
218. P. Brémond and S. R. A. Marque, *Chem. Commun.*, 2011, **47**, 4291.
219. D. A. Parkhomenko, M. V. Edeleva, V. G. Kiselev and E. G. Bagryanskaya, *J. Phys. Chem. B*, 2014, **118**(20), 5542.
220. G. Gryn'ova, M. Marshall, S. J. Blanksby and M. L. Coote, *Nat. Chem.*, 2013, **5**, 474.
221. P. Brémond, A. Koïta, S. R. A. Marque, V. Pesce, V. Roubaud and D. Siri, *Org. Lett.*, 2012, **14**, 358.
222. G. Audran, M. Bim Batsiandzy Ibanou, P. Brémond, S. R. A. Marque, V. Roubaud and D. Siri, *Org. Biomol. Chem.*, 2013, **11**, 7738.
223. G. Audran, L. Bosco, P. Brémond, S. R. A. Marque, V. Roubaud and D. Siri, *J. Org. Chem.*, 2013, **78**, 9914–9920.
224. M. V. Edeleva, D. A. Parkhomenko, D. A. Morozov, S. A. Dobrynin, D. G. Trofimov, B. Kanagatov, I. A. Kirilyuk and E. G. Bagryanskaya, *J. Polym. Sci., Part A: Polym. Chem.*, 2014, **52**, 929.
225. E. Bagryanskaya, P. Brémond, M. Edeleva, S. R. A. Marque, D. Parkhomenko, V. Roubaud and D. Siri, *Macromol. Rapid Commun.*, 2011, **33**, 152.
226. G. Moad and E. Rizzardo, *Macromolecules*, 1995, **28**, 8722–8728.
227. D. W. Grattan, D. J. Carlsson, J. A. Howard and D. M. Wiles, *Can. J. Chem.*, 1979, **57**, 2834.
228. K. H. Kuo, W. Y. Chiu and K. C. Cheng, *Polym. Int.*, 2008, **57**, 730.
229. M. Chenal, S. Mura, C. Marchal, D. Gigmes, B. Charleux, E. Fattal, P. Couvreur and J. Nicolas, *Macromolecules*, 2010, **43**, 9291.
230. Z. Ding, A. Fischer, S. W. Yang, P. Wu, A. Brembilla and P. Lochon, *Eur. Polym. J.*, 2001, **37**, 1561.
231. Z. Ding, A. Fischer, A. Brembilla and P. Lochon, *J. Polym. Sci., Part A: Polym. Chem.*, 2000, **38**, 3067.
232. G. Audran, P. Brémond, S. Marque and G. Obame, *Polym. Chem.*, 2012, **3**, 2901.
233. G. Audran, P. Brémond, S. R. A. Marque and G. Obame, *J. Org. Chem.*, 2012, 77(21), 9634.
234. S. A. F. Bon, G. Chambard and A. L. German, *Macromolecules*, 1999, **32**, 8269.
235. M. Yu. Zaremski, C. Xin, A. P. Orlova, V. B. Golubev, S. A. Kurochkin and V. P. Grachev, *Polym. Sci. Ser. B*, 2010, **52**, 528.
236. W. G. Skene, J. C. Scaiano and G. P. A. Yap, *Macromolecules*, 2000, **33**, 3536.
237. The driving force in the CO–N homolysis in **35** is the generation of a highly stabilized aminyl radical by the delocalization of the odd electron on the aromatic ring.
238. C. Wetter and A. Studer, *Chem. Commun.*, 2004, 174.
239. J. Li, X. Zhu, J. Zhu and Z. Cheng, *Radiat. Phys. Chem.*, 2007, **76**, 23.

240. M. Leenen, F. Wiesbrock, R. Hoogenboom and U. S. Schubert, *e-Polym.*, 2005, **71**, 1–9.
241. J. Rigolini, B. Grassl, S. Reynaud and L. Billon, *J. Polym. Sci., Part A: Polym. Chem.*, 2010, **48**, 5775.
242. Y. Kwak, R. T. Mathers and K. Matyjaszewski, *Macromol. Rapid Commun.*, 2011, **33**, 80.
243. L. Perreux and A. Loupy, *Tetrahedron*, 2001, **57**, 9199.
244. Y. Kojima, S. Koda, H. Nomura and S. Kawaguchi, *Ultrason. Sonochem.*, 2001, **8**, 81.
245. J. C. Scaiano, T. J. Connolly, N. Mohtat and C. N. Pliva, *Can. J. Chem.*, 1997, **75**, 92.
246. Y. Guillaneuf, D. Bertin, D. Gigmes, D. L. Versace, J. Lalevee and J. P. Fouassier, *Macromolecules*, 2010, **43**, 2204.
247. E. Yoshida, *Colloid Polym. Sci.*, 2010, **288**, 1639, However, the role of the irradiation as well of the dye is not clearly discussed in this article.

CHAPTER 3

Synthesis of Nitroxides and Alkoxyamines

JAMES P. BLINCO,* STEVEN E. BOTTLE, KATHRYN E. FAIRFULL-SMITH, EMILY SIMPSON AND KOMBA THOMAS

School of Chemistry, Physics and Mechanical Engineering, Queensland University of Technology (QUT), 2 George St, Brisbane, Queensland 4001, Australia
*Email: j.blinco@qut.edu.au

3.1 Introduction

The persistent radical effect (PRE) established in the presence of a nitroxide has been widely investigated for the preparation of macromolecules with a narrow molar mass distribution through radical polymerization (both in a bi-molecular fashion employing a nitroxide and radical initiator, and uni-molecularly using a thermally unstable alkoxyamine to supply both fragments). The technique provides telechelic polymers with end group functionality being governed by moieties incorporated onto the mediating nitroxide/alkoxyamine. The efforts of several groups have resulted in the development of the synthesis of many new nitroxides and alkoxyamines, which have aided in the construction of numerous functional and complex architectures. Some of these results are illustrated herein.

This chapter is divided into two sections. The first section focuses on the synthesis of nitroxides and their precursors, including methodologies for the introduction of functionality. The second provides an overview on

RSC Polymer Chemistry Series No. 19
Nitroxide Mediated Polymerization: From Fundamentals to Applications in Materials Science
Edited by Didier Gigmes

Published by the Royal Society of Chemistry, www.rsc.org

techniques for the subsequent transformation of these compounds into alkoxyamines—the key mediating species in unimolecular, nitroxide mediated polymerization.

In both sections, a number of literature examples will be highlighted, with the aim being to provide the reader with a workable selection of design tools for the modification of both nitroxide and R-group moieties.

3.1.1 General Considerations

3.1.1.1 Nitroxide Structure

With the advent of nitroxide mediated polymerization, the synthesis of novel nitroxide classes has dramatically increased over the past three decades. The most stable nitroxides occur in the form of hindered cyclic compounds with piperidines, pyrrolines, pyrrolidines, imidazolines and isoindolines being among the most common ring classes of cyclic nitroxides. However, a significant number of acyclic forms also exist.

Despite the diverse nature of many of these structures, each stable nitroxide molecule has two common structural characteristics that influence its properties within organic synthesis, biomedicine and material science: the effect of substituents in the immediate proximity of the nitroxide moiety and the resonance delocalization of the three-electron N–O bond (detailed in Section 3.1.1.2). Recently, Marque and co-workers demonstrated through the analysis of 10 classes of nitroxides by X-ray diffraction and electron paramagnetic resonance spectroscopy that the conformation and geometry of steric substituents around the radical centre influence the coupling reaction of the nitroxides with carbon centred radicals.[1] From this study, they deduced six conformational and geometrical configurations around the nitroxide moiety (Figure 3.1). All currently reported stable nitroxide compounds fall exclusively within one of these classes.

3.1.1.2 Nitroxide Stability

Like most radical species, nitroxide free radicals have an open-shell electronic configuration with an odd number (single unpaired) of higher energy valence electrons. However, while many radicals are quite unstable, highly reactive and short-lived, hindered nitroxide radicals are remarkably stable. This feature allows organic chemists to synthesize, characterize and store them under normal laboratory conditions over long periods of time. As a result, a number of nitroxide compounds are commercially available and have been used extensively as biophysical tools[2] and in materials science.[3] Such unique stability is partly due to the strong delocalized three-electron bond within the radical centre between the nitrogen and the oxygen atoms. The high delocalization energy within this bond confers substantial thermodynamic stability at the radical centre (Scheme 3.1).[4]

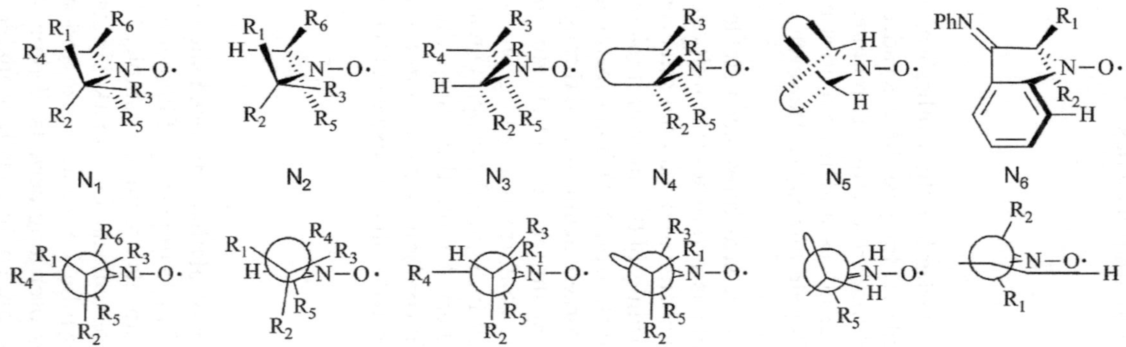

Figure 3.1 Six deduced conformations (N_1–N_6) around the nitroxide center and their corresponding Newman projections: N_1 (has a syn and an anti β-R group), N_2 (with anti β-hydrogen atom and a syn β-R group), N_3 (with an anti-R group and an anti-hydrogen atom), N_4 (with an anti-ring), N_5 (with two syn hydrogen atoms), and N_6 (with the radical centre fused to an aromatic ring). Reprinted (adapted) with permission from E. G. Bagryanskaya, S. R. Marque, Y. P. Tsentalovich, *J. Org. Chem.*, 2012, 77, 4996. Copyright (2012) American Chemical Society.

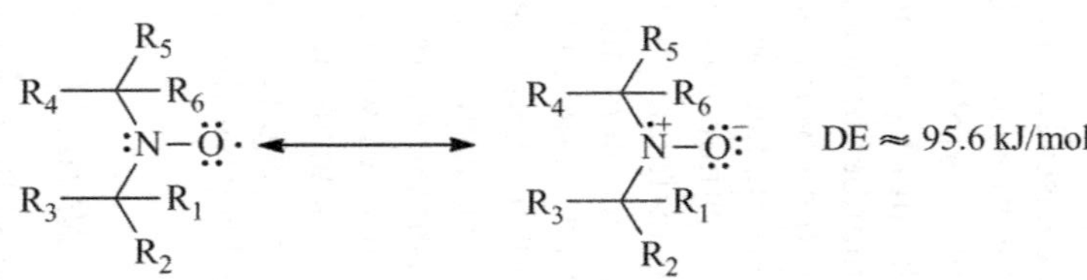

Scheme 3.1 Resonance structures of the nitroxide.

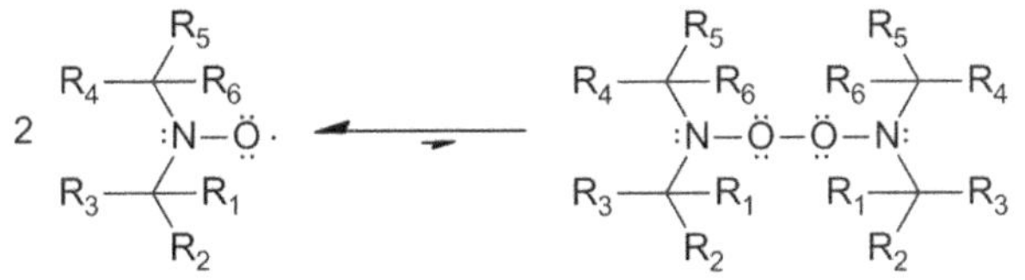

Scheme 3.2

This makes a pair of nitroxide radicals essentially unreactive toward one another. Thus, dimerization, a common radical-to-radical termination reaction, is highly unlikely to occur between nitroxide radicals (Scheme 3.2).[4,5]

Nitroxides are also stabilized significantly by steric hindrance and the polarity of substituents around the radical centre, particular at the α-carbon.[6–9] In general, saturating the α-carbon with bulky substituents confers kinetic stability at the radical center. As a result, the prospect of many common radical mediated degradation pathways, such as hydrogen abstraction, is significantly diminished. As a practical example, tetraethyl substituted nitroxides have been shown to be reduced at a slower rate by ascorbic acid in comparison to their tetramethyl counterparts.[10,11] This is also the case in biological environments. Moreover, having an electron withdrawing group at the α-carbon increases the electron density of the oxygen atom on the resonance structure.[12] This consequently increases the nucleophilicity of the oxygen atom, which in turn increases the ease of reduction of the nitroxide group.

3.1.2 Nitroxide Preparation

Nitroxides can be generated from a number of functional precursors (Scheme 3.3). However, most stable nitroxides are commonly prepared from their corresponding amines (secondary and tertiary) and hydroxylamines. These precursors undergo oxidation reactions in the presence of a number of oxidizing agents to form stable nitroxide radical-containing compounds.

3.1.2.1 Oxidation of Secondary Amines

Oxidation of hindered secondary amines is the most common approach used to generate nitroxides. There are a variety of reagents currently employed to achieve this transformation. Amongst them, aqueous hydrogen peroxide is the main reagent used to oxidise secondary amines to nitroxides.[13] The reaction is performed in the presence of sodium tungstate and an inorganic weak base (*e.g.* sodium bicarbonate) (Scheme 3.4). There are also examples within the literature where sodium tungstate has been successfully replaced with phosphotungstic acid.[14,15] In either case, the reactions are usually undertaken in water, aqueous methanol or acetonitrile at room temperature. Despite being the most cost-effective reagents of choice, the hydrogen peroxide/sodium tungstate oxidation reaction generally

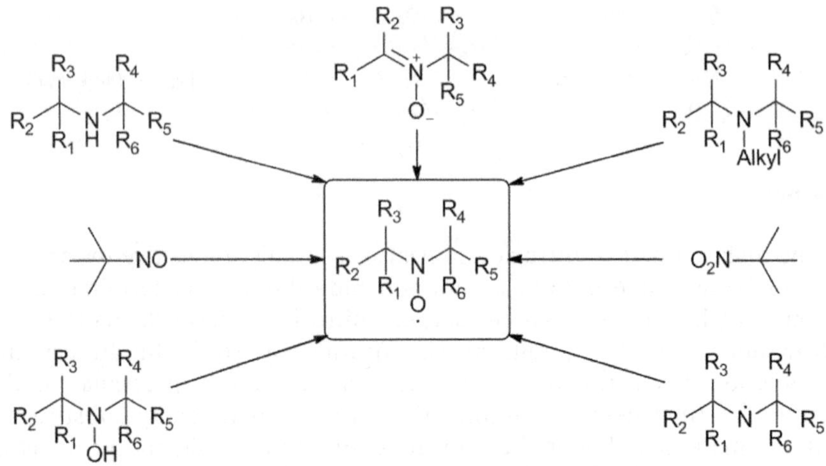

Scheme 3.3 Synthetic precursors for generating nitroxide radicals.

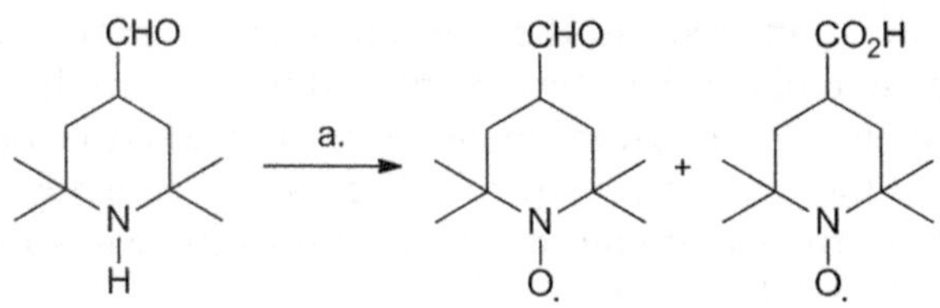

Scheme 3.4 Reagents and conditions: a. H_2O_2, Na_2WO_4, $NaHCO_3$, MeOH.

requires the use of excess reagents and longer reaction times (1 or more days). Such reaction conditions may also affect the stability of other labile functional groups present. Under such conditions, the oxidation of a compound bearing both a secondary amine and an aldehyde functionality resulted in the simultaneous oxidation of the amine to a nitroxide and the aldehyde to its acid derivative (Scheme 3.4).[16]

Such limitations can be avoided either by designing appropriate synthetic sequences in order to introduce the desired functionalities or by employing a protecting group strategy. Peracids, most commonly *m*-chloroperoxybenzoic (*m*CPBA) and sometimes *m*-nitroperoxybenzoic acid, are also widely used reagents for the oxidation of secondary amines to nitroxide moieties in high yields.[17–19] Unlike the hydrogen peroxide–tungstate system, peracid oxidation reactions are conducted in an organic solvent, with much shorter reaction times (30 mins to a few hours). The resulting *m*-chlorobenzoic acid by-product can be removed easily *via* extraction with base, or separated from the desired product using chromatographic techniques or trituration.

Specifically, this technique has been employed for the synthesis of phosphonate-containing nitroxides (*e.g.* SG1-type nitroxides). Treatment of the α-aminophosphonates with 1 mol equiv. of *m*-chloroperbenzoic acid (*m*CPBA) in dichloromethane gave the corresponding nitroxides in yields above 90%. If the secondary amine being oxidized is also chiral in nature, it is possible to produce optically active nitroxides using this same methodology.[20,21]

Generating nitroxides from secondary amines can also be achieved in high yields using 2 molar equivalents of dimethyldioxirane. This reagent was initially investigated by Murray and co-workers when they added a dimethyldioxirane stock solution to an ice cold reaction mixture of a variety of hindered secondary amines in acetone.[22] The authors proposed a reaction mechanism in which the amine was first oxidized to the hydroxylamine derivative by one molar equivalent of the reagent, with the second molar equivalent required to easily oxidize the resulting hydroxylamine to the desired nitroxide. A slight modification to the Murray procedure conducted by Brik involved the *in situ* generation of dimethyldioxirane by the reaction of peroxomonosulfate with acetone under phase-transfer conditions in biphasic media.[23] In each case, the authors claimed superiority of this reagent because of its versatility, short reaction time (30 min), convenience and high yields. Using this method, the aldehyde nitroxide shown in Scheme 3.4 was obtained from its secondary amine precursor in 95% yield with no evidence of over-oxidation.

3.1.2.2 Oxidation of Tertiary Amines

Hindered tertiary amines are another type of functional group that can be used as a direct precursor for generating nitroxides. Although tertiary amines are comparatively less common than their secondary amine counterparts, there are still a number of examples in the literature where tertiary amines have been successfully oxidized to the corresponding nitroxides in moderate to high yields. *N*-alkyl tertiary amines are usually employed as protecting groups under undesirable reaction conditions with respect to the either the nitroxide moiety, or other intermediate functionalities. As with secondary amines, the H_2O_2/Na_2WO_4 oxidation system has been employed to convert heterocyclic hindered tertiary amines into nitroxides.[24,25] One drawback to this method is that longer reaction times (days) are required. There are a number of literature examples where this methodology has failed to yield the desired nitroxide product. For example, Grigor'ev *et al.* used this method for the oxidation of an amide derivative of *N*-methyl-3-imidazoline to the corresponding nitroxide but were unsuccessful with the carboxylic acid and the ester derivatives (see Scheme 3.5).[24]

In the case of oxidation of *N*-benzyl type hindered tertiary amines to nitroxides, *m*CPBA is the reagent of choice. As shown in Scheme 3.6, it was possible to successfully oxidize the hindered *N*-benzyl azaphenalene derivatives to their corresponding nitroxides using an excess *m*CPBA.[26]

Even the extremely crowded tetraphenylisoindoline compound was completely oxidized in high yield with *m*CPBA after six days (Scheme 3.7 route a).[27] Alternatively, *N*-benzyl tertiary amines can be first easily converted to their secondary amines by hydrogenolysis and then easily oxidized to the

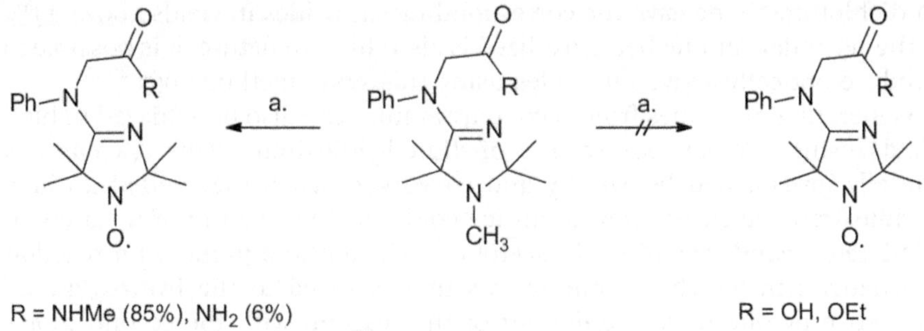

Scheme 3.5 Reagents and conditions: a. H_2O_2, Na_2WO_4, $NaHCO_3$.

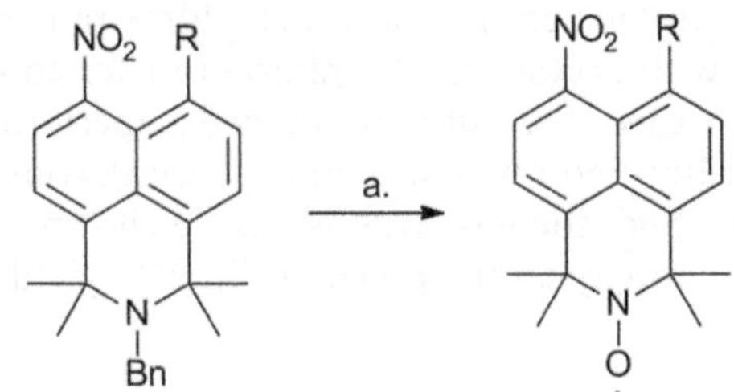

Scheme 3.6 Reagents and conditions: a. *m*CPBA, DCM, 1 d, R = H (90%), NO_2 (65%).

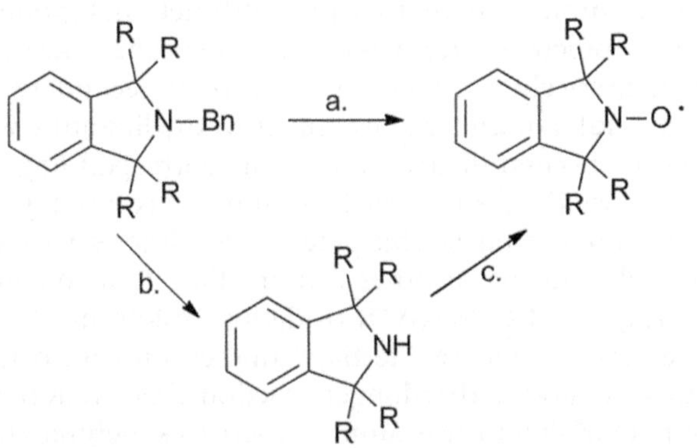

Scheme 3.7 Reagents and conditions: a. *m*CPBA, DCM, 1–6 d, R = Ph (71%), $(CH_2)_3Ph$ (54%); b. H_2, 10% Pd/C, R = Me (92%), Et (99%), *n*-Pr (83%), *n*-Bu (92%); *m*CPBA, DCM, 3 h, R = Me (91%), Et (90%), *n*-Pr (96%), *n*-Bu (99%).

desired nitroxides using the H_2O_2/Na_2WO_4 system or *m*CPBA (Scheme 3.7 route b).[28] This route gives a comparatively higher yield over the two steps and can significantly reduce reaction times.

3.1.2.3 *Oxidation of Hydroxylamines*

Hindered hydroxylamines are the most easily oxidized precursors to nitroxides. As a result, it is highly unlikely that they are isolated in their pure

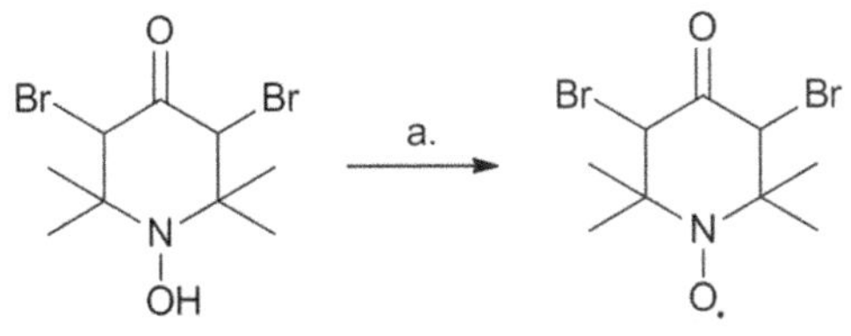

Scheme 3.8 Reagents and conditions: a. HNO_2.

form without nitroxide derivatives being found as impurities. In most cases, even brief exposure to oxygen in the atmosphere is enough to oxidize hydroxylamines to nitroxides.[29,30] There are some examples, however, where hydroxylamine derivatives are very stable and their oxidation requires more aggressive conditions. For example, attempts to oxidize the dibromo hydroxylamine derivative using common oxidizing agents such as PbO_2, MnO_2 or H_2O_2/Na_2WO_4 proved unsuccessful. The oxidation was only successful when HNO_2 was employed (Scheme 3.8).[31]

As a general rule, it becomes much harder to oxidize hydroxylamine compounds as the number of electron withdrawing substituents around the radical centre increases (see Section 3.1.1.1). Some Cu^{2+} salts have also been used as oxidizing agents for hydroxylamines.[32] Judicious choice of oxidizing agent is required for the oxidation of hydroxylamines as certain oxidizing agents can lead to over oxidation, such as Br_2, which is known to further oxidise the desired nitroxide to the oxoammonium tribromide salt.[33]

3.1.2.4 Oxidation of Alkoxyamines

Although there is a significant volume of literature demonstrating the homolytic cleavage of the NO–C bond to generate nitroxides in controlled radical polymerization systems,[34–36] synthetic methods for the preparation of nitroxides from alkoxyamines are less common.[37] Recently, Bottle *et al.* investigated the oxidative cleavage of a range of functionalized *N*-methoxy type alkoxyamines and the application of such systems as robust protecting group chemistry for the nitroxide moiety.[38] Interestingly, it was found that quantitative regeneration of the nitroxide compounds from their methoxyamine derivatives could be achieved using mild conditions with *m*CPBA as the oxidant in a Cope-type elimination (Scheme 3.9). Complete transformation to the nitroxide was observed at room temperature within 30 min. There are other literature examples where *m*CPBA inadvertently cleaved the NO–C bond of a range of alkoxyamines to give nitroxides as an unwanted biproducts during aldehyde synthesis.[35,39–45]

3.1.2.5 Other Methods for Nitroxide Synthesis

Nitroxides can also be prepared by the oxidation of nitrones. Like nitroxides, nitrone compounds are regularly employed as spin traps for the

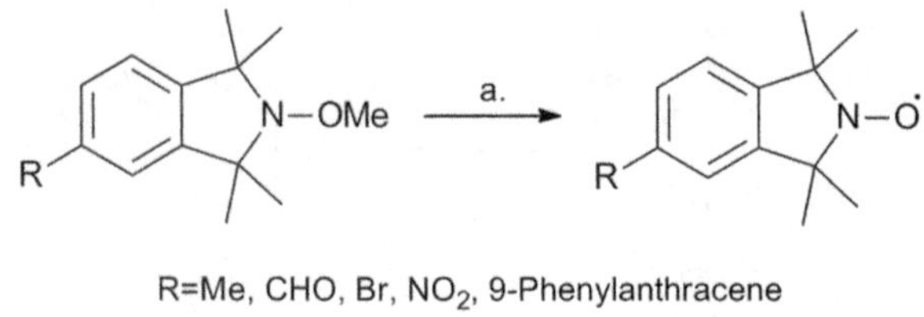

Scheme 3.9 Reagents and conditions: a. *m*CPBA, DCM, 0 °C, 30 min^{-1} h; 70–90% yield.

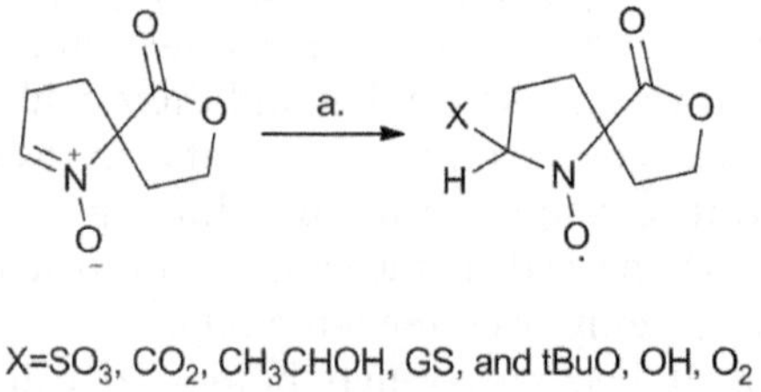

Scheme 3.10 Reagents and conditions: PBS, H_2O_2, UV, and respective radical sources $NaHCO_2$, Na_2SO_3, EtOH, GSSG, $(t\text{BuO})_2$, KO_2, DMSO.

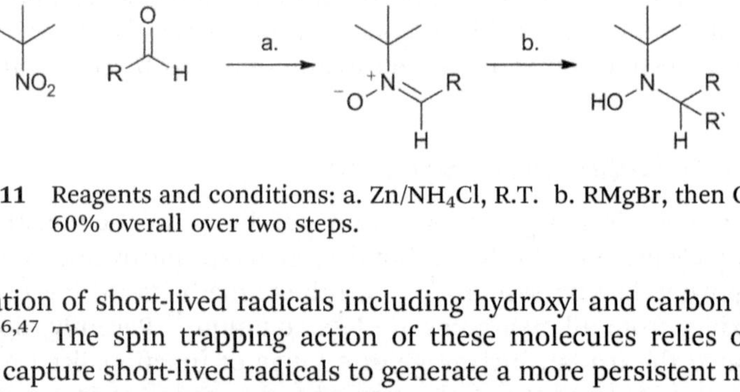

Scheme 3.11 Reagents and conditions: a. Zn/NH_4Cl, R.T. b. RMgBr, then Cu^{2+}/air, 60% overall over two steps.

identification of short-lived radicals including hydroxyl and carbon centred radicals.[46,47] The spin trapping action of these molecules relies on their ability to capture short-lived radicals to generate a more persistent nitroxide radical species that can be analysed by EPR spectroscopy. An example of such nitroxide generation is shown in Scheme 3.10 where a spirolactonyl nitroxide spin adduct is generated from a nitrone precursor.[48]

Alkylation of nitrones with Grignard reagents affords the corresponding hydroxylamine which, as discussed earlier, rapidly undergoes oxidation in air (or with other oxidizing agents) to give a nitroxide.[49] This route is particularly important for the synthesis of acyclic α-hydrogen-bearing nitroxides, which have found many uses within nitroxide mediated polymerization. Braslau and coworkers demonstrated that the reductive condensation of 2-methyl-2-nitropropane with a range of aldehydes gave the intermediate nitrone compounds in a single, high yielding step. Addition of a variety of aryl Grignards followed by copper(II) catalysed oxidation under ambient atmosphere resulted in the formation of the desired nitroxides in good yields (Scheme 3.11).[50]

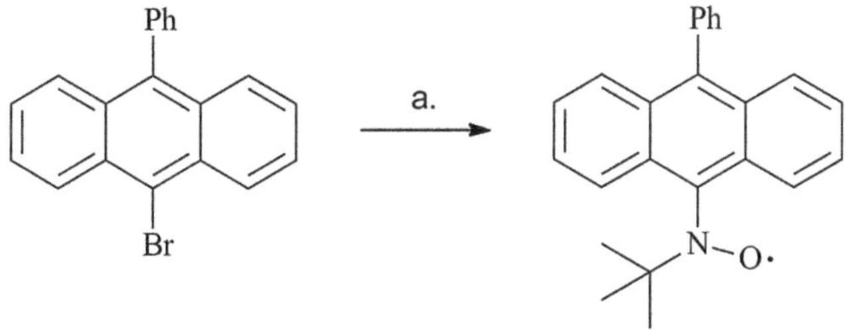

Scheme 3.12 Reagents and conditions: a. 1. *t*BuLi; 2. *t*BuNO; 3. Air, 52%.

Nitroso compounds, particularly tertiary aliphatic and aromatic nitroso compounds, are another functional group of compounds that can be transformed to stable nitroxide radicals. As shown in Scheme 3.12, upon lithiation of 9-bromo-10-phenylanthracene with *t*-butyllithium, the resulting anion can be quenched with *t*-butylnitroso to afford the hydroxylamine intermediate, which was subsequently oxidized in air spontaneously to the corresponding nitroxide in 52% yield.[51]

3.1.3 Nitroxide Design

There is a significant volume of literature that supports the robust nature of the nitroxide moiety under many chemical reaction conditions. With a number of complex free radical polymerization mediators being employed to incorporate various functionality onto the ends of polymer chains, the design of more nitroxides for NMP is required. This section is aimed at highlighting recent synthetic transformations undertaken to design functional nitroxide compounds within the past 10–15 years. However, it is important to note that there are certain reaction conditions in which the nitroxide moiety itself is known to react. Such undesirable reactions have the tendency to produce side products and to lower the yield of the desired product, or stop its formation completely. Thus, being aware of such conditions will help synthetic chemists to adequately design and undertake chemical transformations that involve other functional groups without any significant undesirable interference from the nitroxide radical center.

In addition to their ability to recombine with short-lived radical species, reactions at the nitroxide's radical center commonly involve an initial reversible redox triad between the nitroxyl radical and its hydroxylamine and oxoammonium ion derivatives (Scheme 3.13).[52]

In the presence of strong oxidizing agents, the nitroxide undergoes a single-electron oxidation to its oxoammonium salt. The oxidation potential and the stability of the resulting oxoammonium ion mainly depend of the nitroxide ring class as well as the surrounding substituents.

While the oxidation of a hydroxylamine is an efficient synthetic method for producing a nitroxide, the nitroxide moiety, a weak oxidant by itself, can also be easily reduced in a reverse reaction to its hydroxylamine derivative by

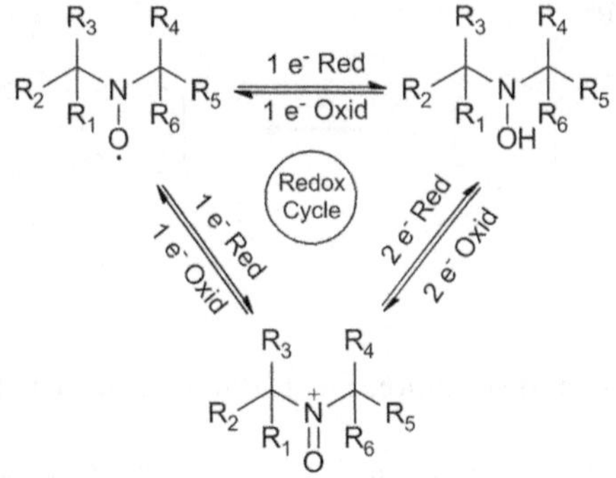

Scheme 3.13 Reversible redox cycle between nitroxide, hydroxylamine and *N*-oxoammonium cation.

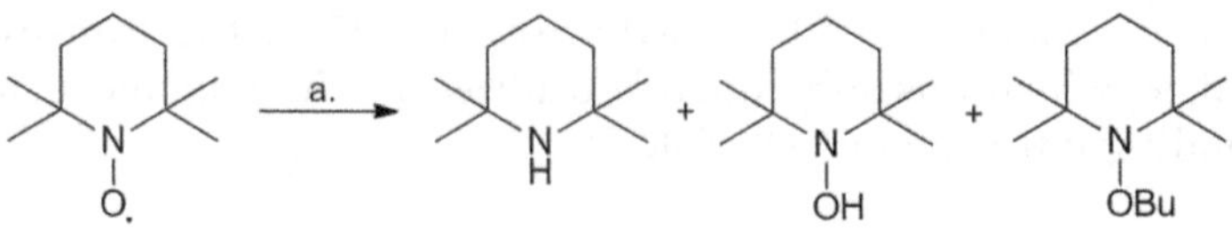

Scheme 3.14 Reagents and conditions: a. *n*BuLi.

a variety of reducing agents, including some metals and metal salts,[53] hydrazines,[54–56] ascorbic acid,[57–59] and hydrogen. Due to the efficient reducing power and the strong nucleophilicity of hydrazines, it is not recommended, and is highly unlikely, to have a compound that contains both nitroxide and hydrazine functionalities. Most commonly, metal hydrides have been successfully employed in chemical transformations in the presence of the nitroxide moiety. As noted earlier, the reduction is reversible. In some instances, oxidation of the hydroxylamine back to the nitroxide is so easy that even exposing it to air is enough to regenerate the nitroxide. Nitroxides are also reduced by organometallic reagents. The products isolated when the piperidine nitroxide was treated with *n*-butyllithium included the secondary amine and the hydroxylamine and butoxyamine derivatives (as depicted in Scheme 3.14).[60,61]

3.1.3.1 Functional Nitroxides

The concept of functionalizing a molecule containing the nitroxide moiety is not solely restricted to reactions conducted in the presence of the radical species. In many instances, functional groups are introduced in the molecular framework of nitroxide precursors, such as the amines or the hydroxylamines. In other instances convenient protecting groups, such as the acetyl protecting group, can be used to mask the radical centre if necessary. The following sections focus on reactions that can be conveniently performed in the presence of the nitroxide radical (or its immediate precursors) with little or no interference by the radical centre.

Aromatic substitution reactions are synthetic transformations commonly used in organic chemistry. Hence, functionalizing nitroxide compounds using a range of aromatic substitution reaction conditions offers a convenient method for the introduction of functionality onto the nitroxide compound.

In the synthesis of the terminal acetylenic sulfone linked to a pyrrolidine nitroxide, the intermediate nitroxide sulfonyl chloride (shown in Scheme 3.15) was obtained following direct electrophilic aromatic sulfonation of the pyrrolidine nitroxide derivative with chlorosulfonic acid at 0 °C.[62] Due to the high acidic content of the reaction media, PbO_2 was employed to oxidize any pre-formed hydroxylamine (formed due to the acid disproportionation of nitroxide) back to the desired nitroxide.

The isoindoline class of nitroxides is commonly functionalized on the aromatic ring, particularly at the 5 and 6 positions. Access to 5-amino isoindoline nitroxides is available *via* catalytic reduction of a nitro isoindoline derivative which can be sysnthesised by direct nitration of the parent nitroxide (both reactions are achieved in quantitative yield).[63,64] The amino functionality can be alkylated to give a water soluble trimethyl ammonium derivative or undergo diazotation to give an isolable diazonium salt used for the synthesis of phenol and diazo nitroxide derivatives.[63,65]

Like most aromatic compounds, introducing a halide group on the aromatic ring can increase the synthetic versatility of the substrate. While direct ring bromination of isoindoline nitroxide precursors (with either single or double substitution available at the 5 and 6 positions) has been demonstrated in a number of examples in the literature,[66] it is only recently that the direct iodination of aryl nitroxides has been established. Di-iodination of both tetramethyl and tetraethyl substituted isoindolines was successfully achieved using periodic acid and potassium iodide in concentrated sulfuric acid as the source of a strongly electrophilic iodonium ion (Scheme 3.16

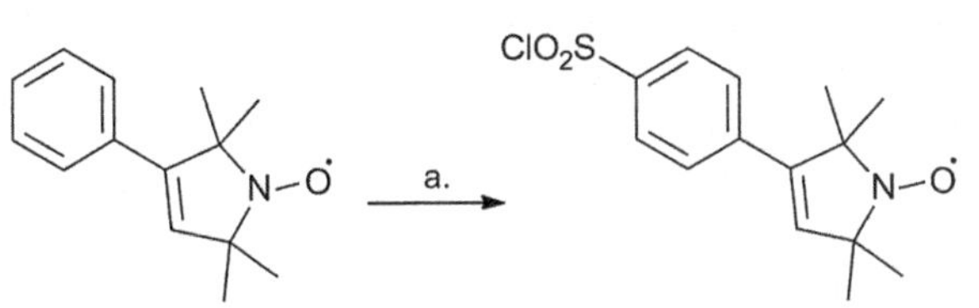

Scheme 3.15 Reagents and conditions: a. HSO_3Cl, 0 °C to rt, 2 h, then PbO_2, O_2, 10 min, 73% yield.

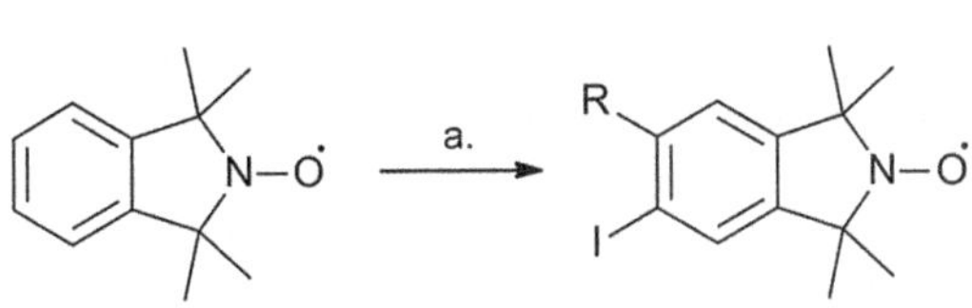

Scheme 3.16 Reagents and conditions: a. H_5IO_6, KI, conc. H_2SO4, 0 °C to rt. 3 h, R = H (48%), I (82%).

and 3.17).[67] In the case of the tetraethyl nitroxide reaction, the authors also reported the isolation of iodinated nitrones. This can be attributed to the formation and stability of the expected acid disproportionation product (oxoammonium ion and hydroxylamine) under these reaction conditions.

However, under the same reaction conditions, selective single iodidations were always accompanied by the formation of the di-iodo counterparts. The products were separated by flash column chromatography. The reaction conditions were also successfully employed on the secondary amine precursors. These halide containing nitroxides have subsequently been used in numerous palladium and copper catalysed coupling reactions.[68,69]

5-Bromo-1,1,3,3-tetramethylisoindoline is a versatile nitroxide precursor used in a variety of synthetic transformations. Scheme 3.18 illustrates some useful synthetic transformations that have been carried out using this brominated derivative, including formylation,[70] cyanation,[71,72] iodination,[73] carboxylation,[74,75] copper catalysed nucleophilic methoxylation[13] and palladium catalysed Heck coupling reactions (Scheme 3.18).[68] The intermediate dianion formed when the brominated isoindoline is treated with 2 molar equivalents of *n*-butyllithium can be quenched with dry DMF, dry CO_2, or iodine to give the benzaldehyde, benzoic acid, or iodobenzene derivatives, respectively. The benzaldehyde nitroxide could also be subjected to a Wittig reaction with triphenylmethylphosphonium iodide in the presence of K_2CO_3, KOH and 18-crown-6 in refluxing dioxane to give a styrene nitroxide derivative, which could also be further reacted in palladium catalysed Heck reactions (see Scheme 3.18 i and j).[76]

Piperidine type nitroxides are commonly functionalized at the 4 position. Two stable Schiff base piperidine nitroxide derivatives of benzothiazole and benzoxazole-based dyes were synthesised in high yields when 4-amino TEMPO was reacted in an alcohol solution with relevant aldehydes in the presence of a catalytic amount of acetic acid (see Scheme 3.19).[77]

Nitroxides have also been incorporated into Fischer-type carbene complexes to give novel organometallic nitroxide spin systems.[78] Scheme 3.20 demonstrates the aminolysis of a ferrocenyl methoxy carbene with excess 4-amino TEMPO to afford a TEMPO-linked ferrocenyl methoxy carbene complex. This complex is unique in that it contains two redox systems (both a nitroxide and ferrocenyl group) and still remains a remarkably stable compound leading to its investigation as a potential charge-storage moiety.

Organoselenium compounds are useful molecules that have found application as building blocks for materials and as potential pharmaceutical agents. In the synthesis of some functionalized selenium-nitroxide compounds, 4-amino TEMPO was successfully transformed to a range of selenoureas and selenocarbamates (Scheme 3.21).[79] The transformation proceeded *via* the reaction of the primary amine with excess ethyl formate to give the 4-formylamino derivative, which was in turn dehydrated with

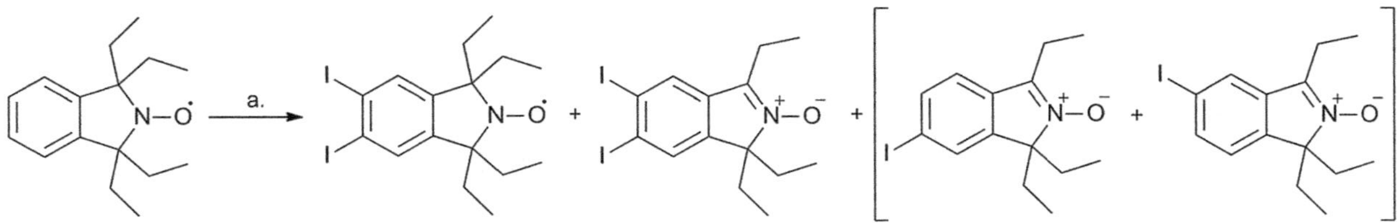

Scheme 3.17 Reagents and conditions: a. H_5IO_6, KI, conc. H_2SO_4, 0 °C to rt. 3 h; yields: 3a (37%), 3b (8%), 3c (13%).

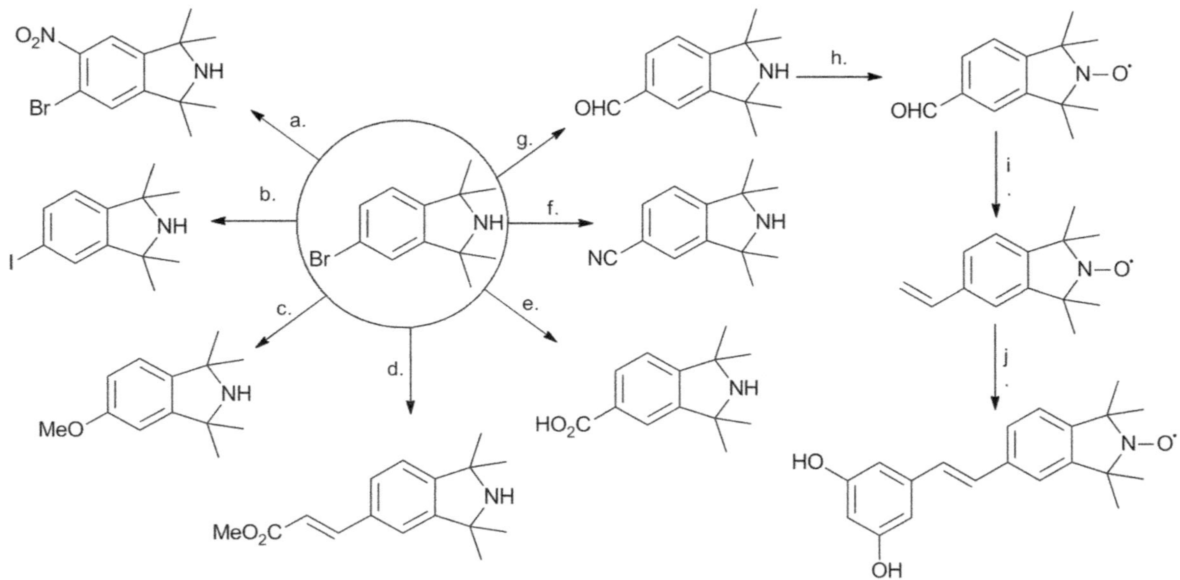

Scheme 3.18 Reagents and conditions: a. HNO_3, H_2SO_4, 0 °C to rt. 3 h, 56%; b. *n*BuLi, THF, −78 °C, then I_2, 70%; c. NaOMe, CuI, DMF, MeOH, reflux, 16 h, 88%; d. K_2CO_3, $Pd(OAc)_2$, PPh_3, DMF, 120 °C, 72 h, 62%; e. *n*BuLi, THF, −78 °C, 95%; f. CuI, *n*Bulmi, $K_4[Fe(CN)_6]$, *o*-xylene, 180 °C, 4 d, 78%; g. *n*BuLi, THF, −78 °C, then DMF, 100%; h. *m*CPBA, DCM, 72%; i. PCH_3Ph_3I, K_2CO_3, KOH, 18-crown-6, dioxane, reflux, 48 h, 62%; j. KOAc, $Pd(OAc)_2$, Bu_4NBr, DMF, 80 °C, 10 h, then NaOMe (0.1 equiv), MeOH, 1 h., rt, 78%.

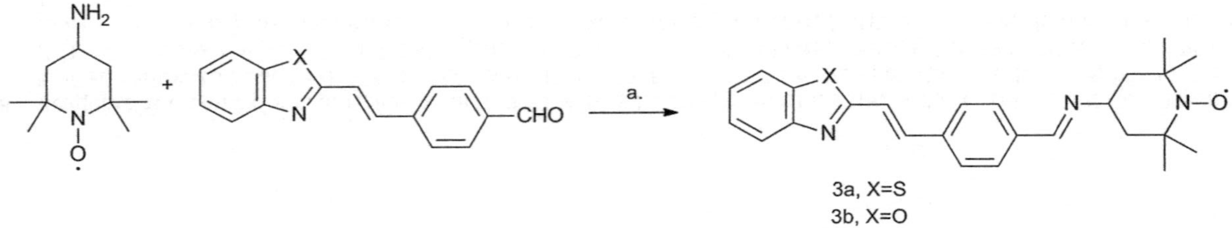

Scheme 3.19 Reagents and conditions: a. AcOH, EtOAc, reflux, 2 h; 3a (83%), 3b (81%).

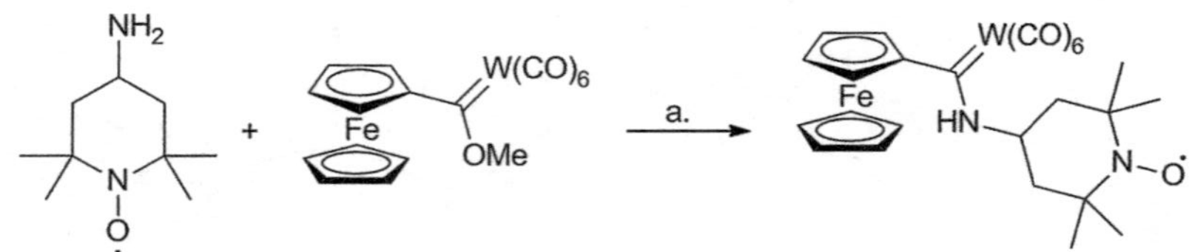

Scheme 3.20 Reagents and conditions: a. THF, rt, 2 d, 58%.

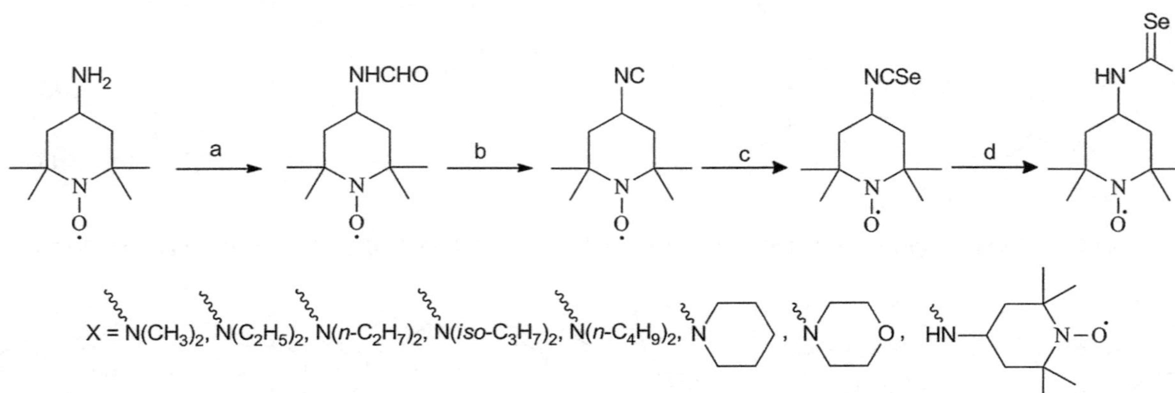

Scheme 3.21 Reagents and conditions; a. HCO_2Et, reflux, 5 h, (96%); b. $COCl_2$ or $ClCOOCCl_3$, TEA, DCM, 0 °C, (78%); c. Se, $CHCl_3$, 61 °C 70 h, 38%; d. HX/NaX, benzene, 19–87%.

gaseous phosgene to the 4-isocyano compound. This compound was then selenated with metallic selenium to give 4-isoselenocyanato derivative, which was reacted with a series of amines and alcohols to give the desired selenourea and selenocarbamate targets.

Turro and co-workers reported a synthetic procedure used to synthesise polynitroxides *via* the nucleophilic aromatic substitution of fluorides using the alkoxide anion of 4-hydroxy TEMPO (Scheme 3.22).[80] Using a range of polyfluorinated aromatic cores, they were able to selectively introduce mono-, di-, tri- and tetra-nitroxides under appropriate reaction conditions. In contrast to mononitroxide compounds, a polynitroxide compound has efficient electron spin–electron spin exchange between individual nitroxide moieties. This enables their use as efficient paramagnetic relaxation agents as well as electron spin agents for dynamic nuclear polarization experiments.

Copper catalysed azide alkyne cycloaddition (CuAAC) reactions are another synthetically useful approach currently employed to produce functionalized nitroxides. Scheme 3.23 gives examples of stable azides and acetylenes that have been synthesized and successfully used in click reactions for the synthesis of a range of triazole functionalized nitroxide compounds.[62,81–85]

The triazole coumarin-based nitroxide shown in Scheme 3.24 was targeted as a profluorescent nitroxide, *i.e.* a nitroxide with a significantly suppressed fluorescence when the nitroxide and chromophore are present together. It was synthesized *via* a copper-catalysed azide alkyne 1,3-dipolar cycloaddition reaction (CuAAC) between 3-azido-7-diethylamine coumarin and the alkyne isoindoline nitroxide derivative (Scheme 3.24).[82] The fluorescence of the molecule was restored when it was converted to a methoxyamine derivative *via* reaction with methyl radicals generated using Fenton chemistry.

A number of examples of nitroxides grafted to carbon based surfaces have also been reported, which if employed as "grafting from" NMP agents would give highly bonded carbon–polymer composite materials.

The first example of this was achieved in an attempt to design higher capacity next generation lithium ion batteries, wherein graphene was

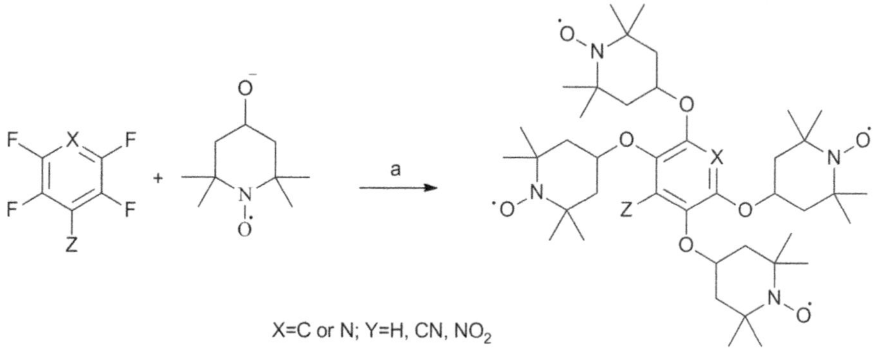

Scheme 3.22 Reagents and conditions; a. DMI, $NaNH_2$, 0 °C–rt–80 °C, 37–81%.

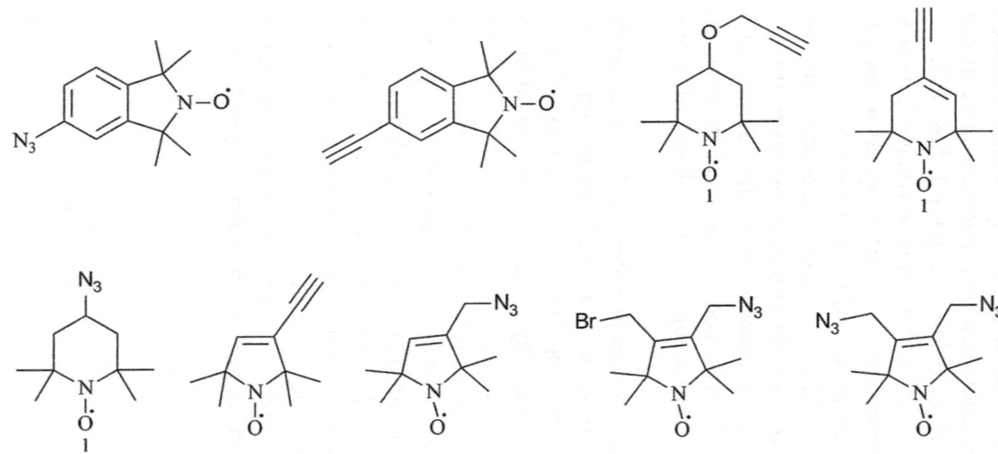

Scheme 3.23 Common azides and acetylene nitroxides.

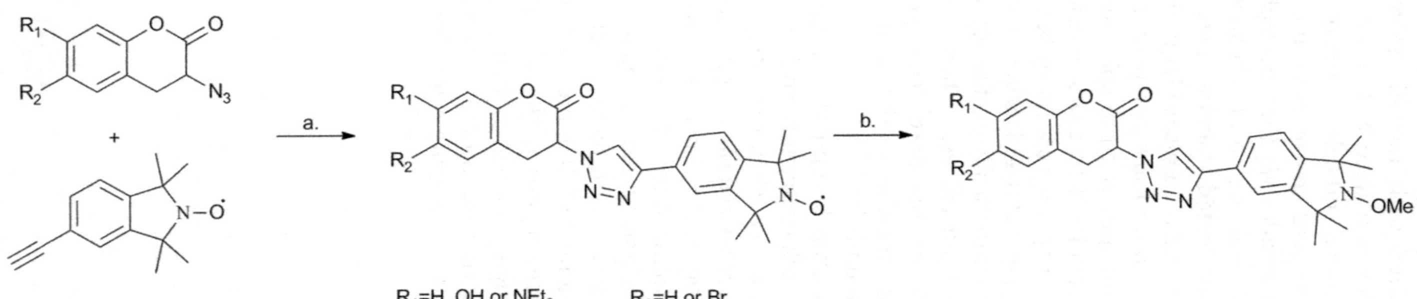

Scheme 3.24 Reagents and conditions: a. $CuSO_4$, NaAsc, $EtOH/H_2O$, 16 h, rt, 60–90%; b. H_2O_2, $FeSO_4.7H_2O$, DMSO, 30 min, 78–90%.

grafted with 4-hydroxy TEMPO *via* conventional esterification of its carboxylic acid groups.[86] The hybrid nitroxide–graphene anode conjugate material possesses superior lithium storage capacity as a result of its dual reservoir of graphene and the reversible redox cycling nitroxide (Scheme 3.25).

Yan and co-workers have also demonstrated grafting to carbon surfaces by attaching isoindoline nitroxides to a fullerene (C60 and C70) cage.[87] As shown in Scheme 3.26, the nitroxide-linked nanoparticle (TMIO-C_{60}) was prepared by refluxing fullerene and a benzaldehyde nitroxide derivative in toluene in the presence of sarcosine. Along the same lines, the Bingel–Hirsch reaction of fullerene and the malonate ester nitroxide afforded the nitroxide adorned methanofullerene derivative.[88]

Blinco *et al.* were also able to design a novel diazonium-based isoindoline nitroxide that could be grafted to a number of different carbon allotropes (single and multi-walled carbon nanotubes, graphene, glassy carbon and carbon fibre) using both electrochemical reduction of the diazonium and, in the case of graphene and carbon nanotubes, spontaneous reaction (Scheme 3.27)[65]

Incorporating nitroxides in complex biomolecules is an effective technique commonly used to obtain useful structural information about these systems. Recently, isoindoline based nitroxides were synthesized and incorporated into duplex DNA as spin labels. These systems could also be utilised to form polymer–bio hybrids. In Scheme 3.28, a CuAAC reaction of an azide functionalized isoindoline nitroxide with 5-ethynyl-2′-deoxyuridine oligomers afforded the rigid triazole-linked spin labelled oligonucleotide derivatives.[84]

Isoindoline-based nitroxide spin-labelled benzimidazole and benzoxazole nucleosides that display limited mobility in duplex DNA were also synthesized and incorporated into DNA oligonucleotides (Scheme 3.29).[89] Noncovalently binding nucleic acid probes based on a fused pyrido[1,2-a]benzimidazole (PyrImid) scaffold adorned with a nitroxide label have also been recently explored.[90]

3.1.3.2 Chromophore Containing Nitroxides

With the advent of nitroxide mediated photopolymerization the ability to tether a chromophore in close proximity to a nitroxide moiety is a very important synthetic technique. Chromophore containing nitroxides are usually designed by combining one or more nitroxide moieties with a known chromophore. In the simplest examples, the two components have been joined *via* ester,[91–98] amide[99–101] or sulfonamide linkers.[102–104] More elegant examples can also be found where the chromophore is incorporated into the carbon framework of the nitroxide ring cycle.[26,66,73,75,90,105–107]

Zhang *et al.* took advantage of the ability of nitronyl nitroxide radicals to coordinate to metal ions, act as chromophores that absorb in the visible

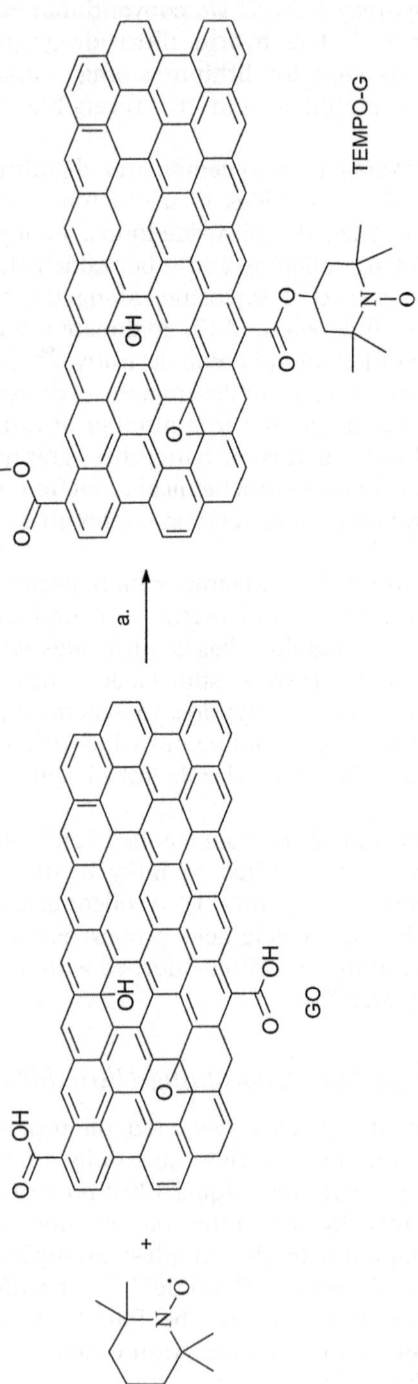

Scheme 3.25 Reagents and conditions: a. $SOCl_2$, DMF, 70 °C, 24 h; then TEA, 100 °C, 60 h, $NaHCO_3$.

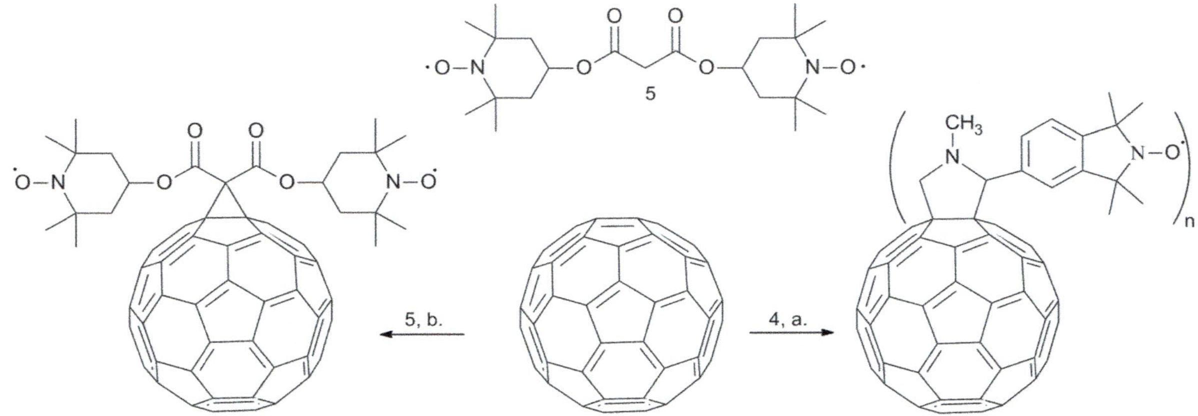

Scheme 3.26 Reagents and conditions: a. Sarcosine, toluene, reflux, 20 h; b. CBr_4, DBU, toluene, 25 °C.

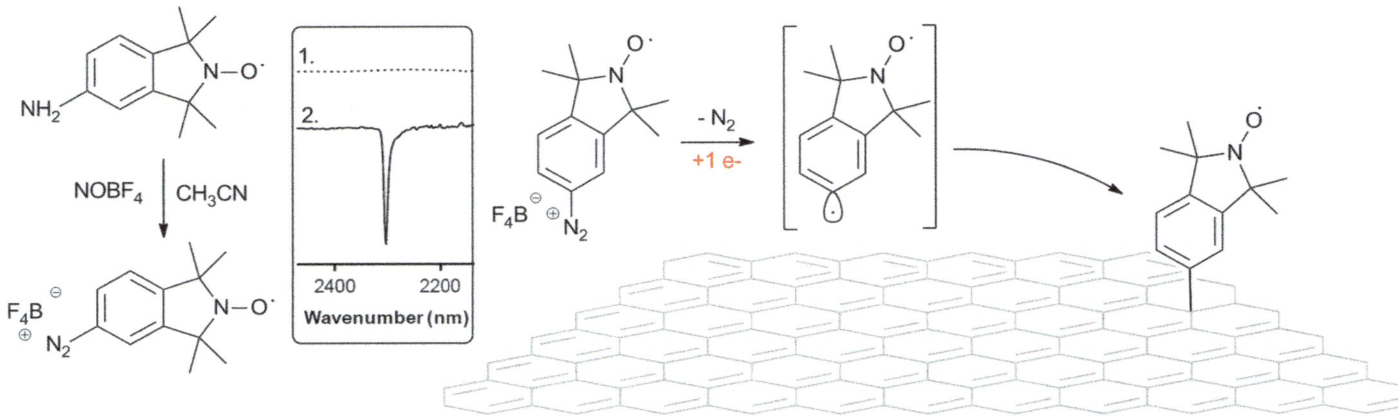

Scheme 3.27 Synthetic pathway for grafting of various carbon allotropes with isoindoline nitroxide free radicals. INSET: Confirmation of the diazonium moiety, observed by the characteristic N-N stretching visible at 2300 cm^{-1}, in the IR spectrum. Reproduced from J. P. Blinco, B. A. Chalmers, A. Chou, K. E. Fairfull-Smith, S. E. Bottle., *Chem. Sci.*, 2013, **4**, 3411.

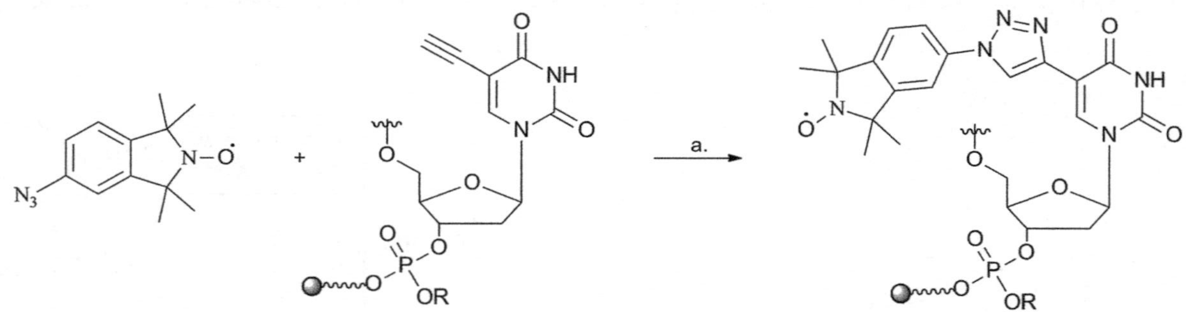

Scheme 3.28 Reagents and conditions: a. TBTA, CuBr, *t*BuOH/DMSO; then NH_4OH, 55 °C, quant.

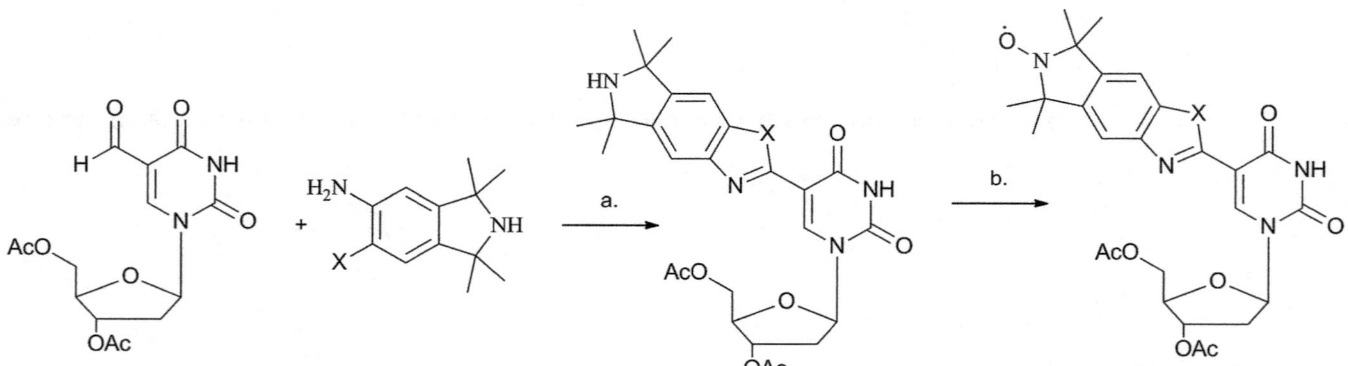

X= OH, NH_2

9A (X= O)
9B (X=NH)

10A (X= O)
10B (X=NH)

Scheme 3.29 Reagents and conditions: a. $K_3Fe(CN)_6$, MeOH, 40% (For X = OH); Iodobenzene diacetate, MeOH, 47% (For X = NH_2); b. *m*CPBA, NaN_3, 50% (X = OH), 56% (X = NH_2).

range, and exhibit reversible redox cycle characteristics, to design a bisnitronyl nitroxide (Scheme 3.30) sensor.[108]

Recently, Liu and co-workers synthesised and investigated the photochemical properties of three photosensitive chromophore-containing nitroxides.[109] The target chromophores, namely benzophenone, naphthalene and quinolone, were linked to TEMPOL using DCC coupling in high yields (Scheme 3.31). These ester linked dual chromophores were successfully used in a controlled photo-induced nitroxide mediated polymerization of methyl methacrylate (MMA). The authors proposed that the absorbed energy of the chromophore components served as the required energy transferred to cleave the NC–O bond in the polymerization process.

Turro *et al.* also used the same coupling conditions to synthesise the ^{15}N-labelled TEMPO-thioxanthone derivative (Scheme 3.32), which they encapsulated in a nanocapsule.[110] When photoexcited, this encapsulated thiozanthone derivative generated an electron spin polarization which could be transferred to a free nitroxide in the bulk solution. Using time-resolved EPR spectroscopy, they were able to investigate the efficiency of the spin–spin interaction between the two nitroxides. The nitroxide 2,2,5-trimethyl-4-phenyl-3-azahexane-3-nitroxide (TIPNO) was covalently merged with the highly fluorescent pyrene chromophore and the spectral properties of the resulting conjugate were determined both in solution and in polymer matrices.

The perylene-based profluorescent (Scheme 3.33), obtained by coupling 5-amino-1,1,3,3-tetramethyisoindolin-2-yloxyl with a perylene-bisimide fluorophore, has been shown to efficiently monitor radical mediated degradation of crosslinked melamine-formaldehyde polyesters under harsh weather conditions.[111]

3.2 Synthesis of Alkoxyamines

3.2.1 General Considerations

3.2.1.1 Alkoxyamine Structure

Alkoxyamines are a large family of compounds with a general formula of $R_1R_2NOR_3$. Alkoxyamines, also known as *o*-alkyl hydroxylamines, have been utilised in several applications, such as radical cyclisation,[112] intermolecular addition reactions[112] and in the formation of heterobivalent libraries.[113] These compounds can undergo fragmentation to form a transient alkyl $R_3^\bullet$ and a stable nitroxide radical, $R_1R_2NO^\bullet$. This concept is utilised in their major application to undergo nitroxide mediated polymerisation (NMP), using alkoxyamines as initiators.[114] Due to this application, extensive research into the synthesis of alkoxyamines containing bulky, sterically hindered R_3 groups has been employed. Investigations into the dissociation rate constants (k_d) of various alkoxyamine initiators found that control was governed by the alkyl fragment, whereas the nitroxide structure influenced the livingness of the polymerisation.

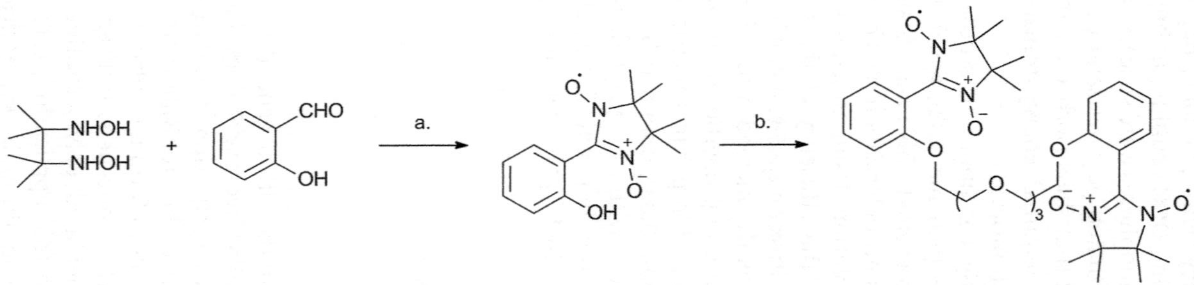

Scheme 3.30 Reagents and conditions: a. pCH$_3$C$_6$H$_4$SO$_3$H, MeOH, rt, 1 d, then PbO$_2$, 70%; b. Tetraethylene glycol, DEAD, PPh$_3$, THF, 75%.

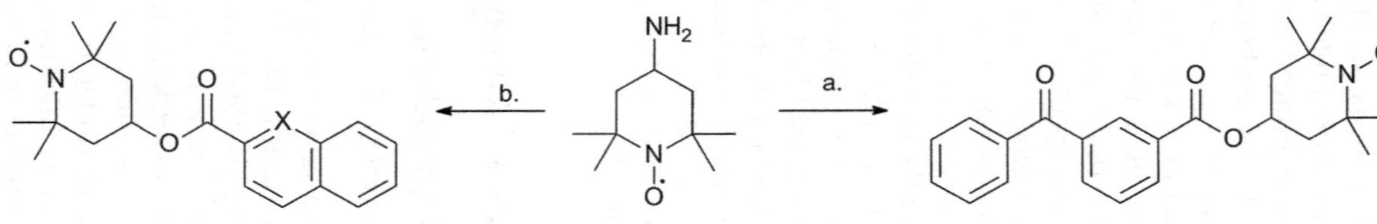

X= C, N

Scheme 3.31 Reagents and conditions: a/b. DCC, DMAP, DCM, 1 d. Each corresponding acid derivative was reacted with 4-amino TEMPO.

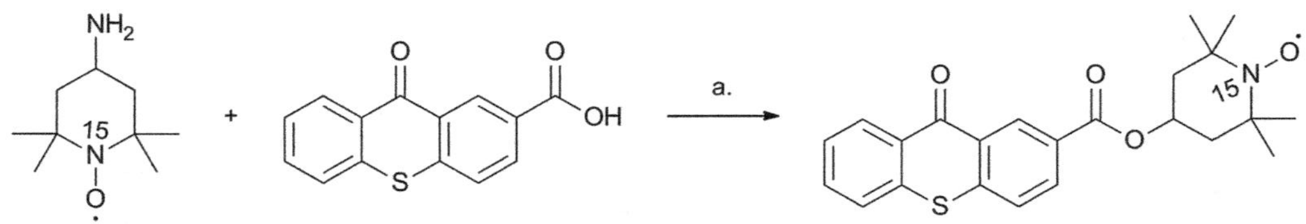

Scheme 3.32 Reagents and conditions: a. DCC, 4-pyrrolidinopyridine, 1 d, 60%.

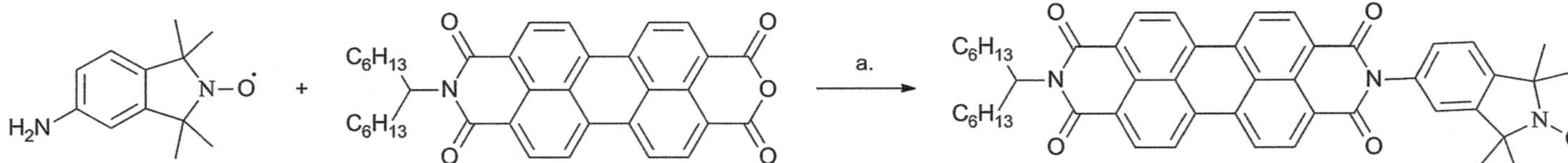

2

Scheme 3.33 Reagents and conditions: a. [a] H_2SO_4, HNO_3/AcOH, 40 °C, 95%.

3.2.1.2 *Alkoxyamine Stability*

Kovtun *et al.* were the first to report that the stability of alkoxyamines was surprisingly dependent on experimental conditions.[115] It was observed that the decomposition of alkoxyamines into alkyl radicals and nitroxides relied on the presence and amount of scavenging species in the system, including oxygen or iodine. Common alkoxyamines have been optimised for the ability of the nitroxide derivative to mediate radical polymerisation. The most important structural factor when considering this application is steric bulk. Largely, the greater steric bulk of the nitroxide leads to increased strain on the alkoxyamine, resulting in homolytic cleavage under certain conditions. Considering cyclic nitroxides, six-membered ring systems have been shown to cleave more readily than their five-membered ring counterparts. However, acyclic derivatives containing *t*-butyl moieties as the R groups cleave fastest of these bond types.[116]

3.2.2 Alkoxyamine Preparation

3.2.2.1 *Preparation from Hydroxylaminate Anion*

This method of producing alkoxyamines is established by the nucleophilic substitution of the hydroxylaminate anion on the alkyl halide derivative, requiring the production of the corresponding hydroxylamine and its anion. These compounds are both generally unstable and therefore this pathway has not been utilised to a great extent. The pathway is restricted to the production of some 2,2,6,6-tetramethylpiperidinyl-1-oxy (TEMPO) and di-*tert*-butyl nitroxide derivatives.[117] Using the same principles, a one-pot reaction was achieved to produce 2,2-diphenyl-3-phenylimino-2,3-dihydroindol-1-yloxyl (DPAIO) alkoxyamine derivatives by combining the reduction of the nitroxide by the addition of phenyl hydrazine, the preparation of the hydroxylaminate anion by the addition of *t*BuOK and the nucleophilc substitution by a desired alkyl halide (Scheme 3.34).[118]

3.2.2.2 *Preparation from Nitrone or* N*-oxides*

From the nitrone starting material, alkoxyamines can be synthesised *via* several potential routes. Studer *et al.*[119] and Grubbs *et al.*[120] both produced alkoxyamines from the utilisation of nitroso compounds as a radical trap. However, both methods differ significantly in the method applied for the

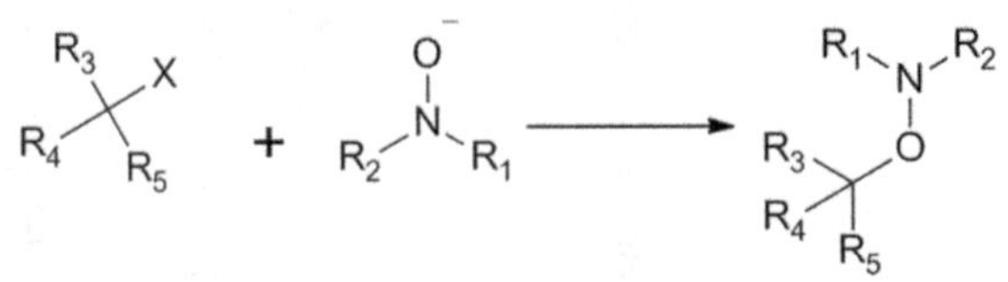

Scheme 3.34 Reaction of a hydroxylaminate anion to produce an alkoxyamine.

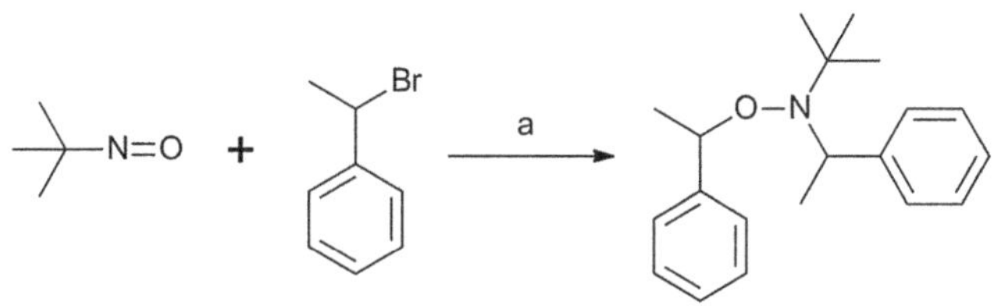

Scheme 3.35 An example of Grubbs *et al.* improved synthesis. a: PMDETA, CuBr, Cu(0), toluene, rt, 68%.

production of the radicals to be scavenged. Studer utilised the reaction between hydrazine and lead oxide to encourage alkyl radical formation while Grubbs and co-workers reacted an alkyl bromide with a copper(I) complex to produce the alkyl radical *via* an atom transfer radical addition reaction. Grubbs *et al.* were also able to improve on the earlier work of Studer by removing the need for lead oxide in excess as well as extending the scope of the reaction to include a number of commercial compounds (Scheme 3.35).[121]

This methodology is akin to the technique employed in atom transfer radical polymerisation. The copper(0) present serves to reestablish the copper(I) species *via* disproportionation with the copper (II) species formed during abstraction with the halide (bromide in this case) from the alkyl halide (1-bromoethylbenzene). This alkyl halide, in the presence of the Cu(I)/Br/Cu(0)/ligand catalytic system, is used as a prime source of required radicals at room temperature.[121]

3.2.2.3 *Preparation from* N-*Oxoammonium Cation*

Stable nitroxides can produce *N*-oxoammonium cations by a one-electron oxidation of the nitroxide. These are prime oxidants for various substrates, including alcohols and tertiary amines.[122] Functionalised alkoxyamines can be prepared by the 1,2 addition of oxoammonium salts on electron rich vinyl ethers and enamines (Scheme 3.36).[123]

A similar reaction with styrene as the olefin was also performed to produce styryl-based alkoxyamines containing a halogen in the β-position to the oxygen atom.[124] The reactions of oxoammonium salts and enolisable compounds to produce alkoxyamines have been reported[125] including a reaction of acetophenone or acetone with an oxoammonium salt by enolate addition to produce the resulting alkoxyamines.[126,127]

3.2.2.4 *Preparation from Nitroxide Radicals*

3.2.2.4.1 Metal-mediated Alkoxyamine Synthesis. Initially, metal complexes were reacted with alkenes to produce alkoxyamines from nitroxide derivatives. A manganese-based Jacobsen's catalyst was utilised to prepare a suite of functionalised alkoxyamines.[128,129] This reagent, first used as a reagent for the enantioselective epoxidation of alkenes,[130] required

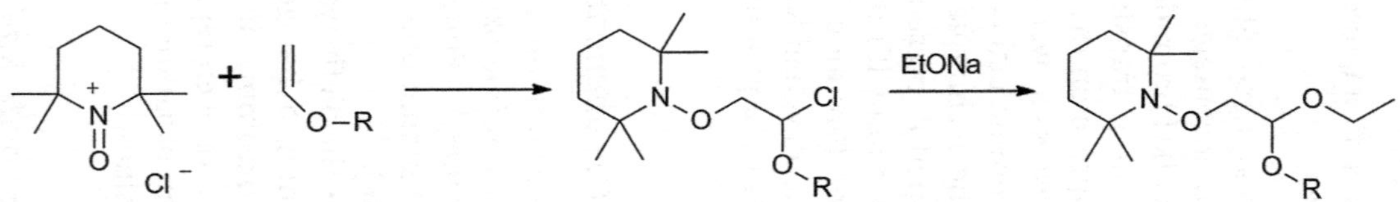

Scheme 3.36 Reaction of an oxoammonium salt with an alkene for the synthesis of alkoxyamines.

oxidation to the manganese–oxo species before reaction with an activated alkene, to produce the radical intermediate. This radical species was then trapped by the nitroxide and, upon reduction, the desired alkoxyamine was formed. These two reported methods are performed under mild conditions and produce good yields with few side products.

A low-cost alternative option to the Jacobsen's catalyst was investigated following this, using a Mn(salen)Cl complex.[131] This reaction also allowed, if stirring of the reaction mixture was undertaken in the presence of air, the removal of peroxide from the reagent requirements. This was originally developed utilising commercially available manganese(III) acetate to produce the TEMPO and TIPNO derivatives in high yield.[132] This reaction required a stoichiometric amount of the manganese complex as opposed to the catalytic amount required in the previous studies discussed. Further improvements of this reaction involved the use of ultrasound to acquire an improved homogenisation and the *in situ* generation of the active complex by a continuous addition of $Mn(OAc)_2$ in the presence of $KMnO_4$ in toluene/acetic acid.[133]

Activated olefins have also been reacted under conditions with metals such as mercury-based complexes. Using oxymercuration with mercuric acetate, TEMPO-based alkoxyamines have been synthesised.[134]

It has also been reported that TEMPO can be efficiently reacted with β-alkylcatecholboranes to produce alkyl radicals that can then be trapped by a second equivalent of TEMPO to produce the alkoxyamines derivatives. This route has also been expanded to ketone-derived catecholboron enolates by the reduction of the α-ketone with catecholborane.[135]

A simple copper-catalysed synthesis of alkoxyamines from nitroxide radicals was proposed by Nicolas *et al.*[136] This reaction was based on a nitroxide with an alkyl halide catalysed by Copper(0) as the reducing agent and a specific ligand, either *N*,*N*,*N*′,*N*″,*N*″-pentamethyldiethylenetriamine or 2,2′-bipyridine (PMDETA or bipy, respectively), as illustrated in Scheme 3.37.

The formation of the required alkoxyamines was observed, with the most rapid reaction and simplest purification occurring when acetonitrile was used as the solvent. A variety of nitroxides and alkyl halides were investigated to determine the scope of the methodology and to produce a suite of alkoxyamines.

3.2.2.4.2 Radical Addition onto an Olefin. The aminoxyl radical can be trapped *in situ* by alkyl radicals produced by the addition of a radical species $Y^•$ to an olefin (so $R_1 = CH_2–Y$). The radical precursor determines the nature of $Y^•$ and is generated by the photochemical or thermal decomposition of radical initiators (an example is shown in Scheme 3.38). Oxygen centred radicals such as alkoxy and/or acyloxy radicals can be generated by common radical initiators, such as dialkyl, diacyloxy or diaryloxy peroxide,[122] peroxalate or peresters.[137] Alkyl radicals have also been generated by azo derivatives.[138] Photochemical methods are a major advantage in that the reaction can be performed at room temperature precluding

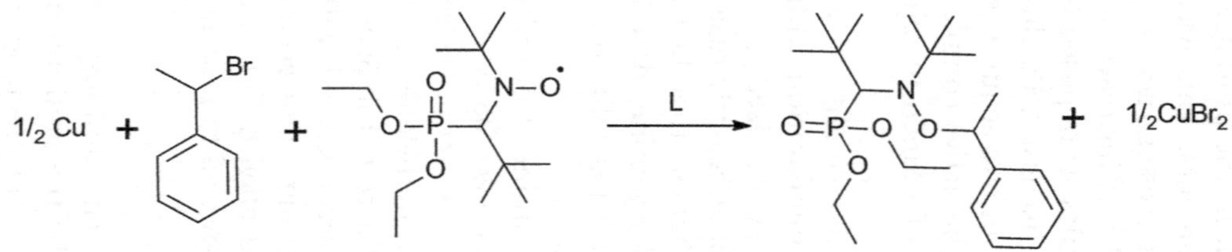

Scheme 3.37 Alkoxyamine synthesis *via* copper metal-mediated reaction of a nitroxide with an alkyl halide (L = bipy [89%] or PMDETA [55%]).

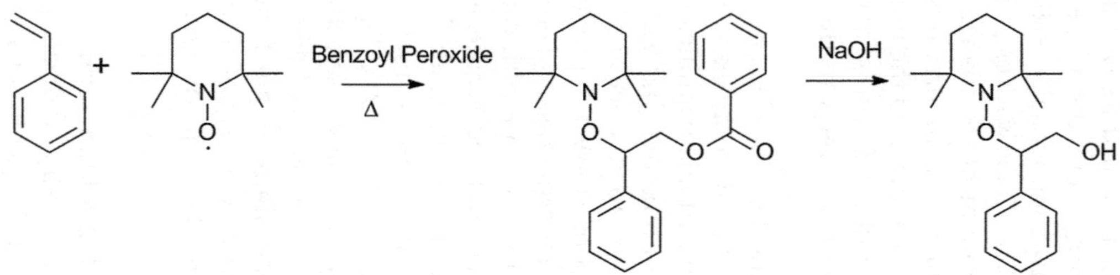

Scheme 3.38 Synthesis of a functionalized alkoxyamine by the radical addition of a acyloxy radical onto an olefin and subsequent ester hydrolysis.

decomposition of the alkoxyamine with only moderate yields due to side reactions including multiple addition or hydrogen abstraction. At increased temperatures, if no other radicals are produced, nitroxides such as isoindoline-based nitroxides, can react with activated olefins, such as methyl acrylate (MA) or styrene, to produce a bis–nitroxide adduct.[139]

3.2.2.4.3 Grignard Reagents. Alkoxyamines have also been prepared *via* utilisation of organometallic reagents, such as Grignard reagents, alkyl lithium, *etc.* These methods have been reported for TEMPO-based alkoxyamines.[140] The proposed mechanism for these reactions includes the oxidation of the organometallic compound by TEMPO to produce the alkyl radical, which is followed by radical trapping by the nitroxide. This reaction is not possible with aryl-, alkenyl-, or alkynyl-based Grignard reagents as the resulting radicals are destabilised and therefore prevent oxidation.[141,142] The mechanism was not effectively clarified, with only homocoupling of the reagent being observed in good yields (Scheme 3.39).

Similar procedures were utilised with lithiated 1,3-dithiane.[140] The reaction scope has been extended to use organomercury chloride or bromide in conjunction with O_2–TEMPO–$NaBH_4$ in dimethylformamide (DMF) at 0 °C.[143] This has an excellent advantage, as the nonactivated alkyl radicals can be produced and transformed into alkoxyamines in good yields.

Currently, the only alternative to prepare non-activated alkyl-based alkoxyamines relies on the β-fragmentation of an α-hydroxyl alkoxyl radical to release an alkyl radical.[144] Reaction of H_2O_2 with a ketone to give an α-hydroxyl hydroperoxide succeeded by a one-electron oxidation using CuCl has been utilised to produce the α-hydroxyl alkoxyl radical. This method was then extended and used amongst aldehydes and was found to be more selective and therefore to generate fewer side-products.[145]

3.2.2.4.4 Hydrogen Abstraction. If a targeted alkoxyamine is sufficiently stable, it can be generated by activated hydrogen abstraction through thermal decomposition of a dialkyl peroxide. For example, the styryl–TEMPO alkoxyamine can be produced in this method as oxygen-centred radicals can be used to abstract activated hydrogens from the carbon substrates to generate the required carbon centred radical.[146] This was prepared by reacting ethylbenzene with di-*tert*-butylperoxide in the presence of the nitroxide at 125 °C.[147] The use of photochemical and mild thermal oxygen-centred radical generating conditions enabled the temperature of

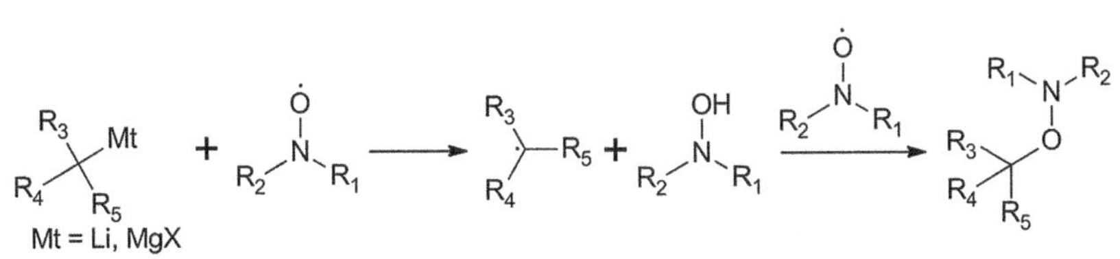

Scheme 3.39 The basis for the synthesis of alkoxyamines using Grignard reagents.

the hydrogen abstraction route to be decreased. The radical generation can be achieved by the photodecomposition of dialkyl peroxide[148,149] and at low temperatures from dialkylperoxyoxalate[150] and dialkyl hyponitrite.[148]

3.2.2.4.5 Oxidation of Ester Enolates. The development of copper(II)-mediated oxidation of lithium enolates in the presence of TEMPO has been investigated to prepare alkoxyamines containing carbonyl groups adjacent to the carbon attached to the aminoxyl group.[148] This method was expanded to a malonate-based TEMPO alkoxyamine, which was effectively used to undertake a 1,2 intermolecular addition onto non-activated olefins.[151] Ferrocenium hexafluorophosphate has been reported to have use as a mild, stable, recyclable, single electron transfer oxidant for these malonate esters, allowing α-ester radicals to be readily converted into the analogous alkoxyamines (Scheme 3.40).[152]

3.2.2.4.6 Decomposition of Azo-compounds. The direct reaction of azobisisobutyronitrile (AIBN) with TEMPO at 60 °C to synthesise alkoxyamines is limited to the production of alkoxyamines that are stable at the decomposition temperature of the azo initiator.[153,154] This methodology was furthered by the use of ultra-violet irradiation as opposed to thermal irradiation allowing thermally sensitive compounds to also be synthesised, such as tertiary SG1-based alkoxyamines.[155] This reaction required a high azo-compound/nitroxide molar ratio as the cage effect can decrease the efficiency of the azo-compound at room temperature (Scheme 3.41).

3.2.2.4.7 Reaction of Halogenoalkanes in the Presence of Metal Complexes. The preparation of several TEMPO derivative alkoxyamines has been reported based on atom transfer radical addition (ARTA).[156] The

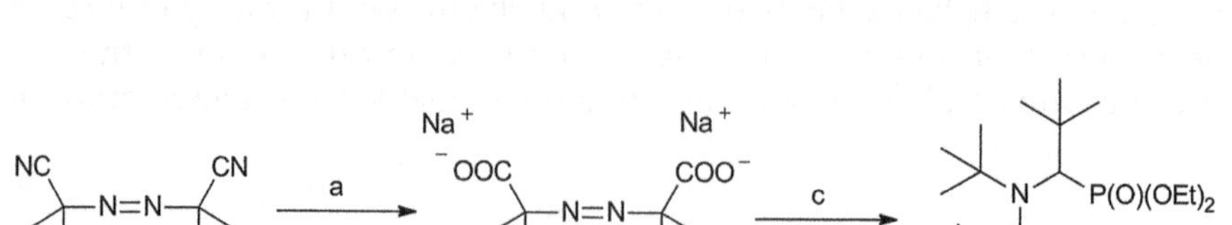

Scheme 3.40 Oxidation of ester enolates to produce alkoxyamines.

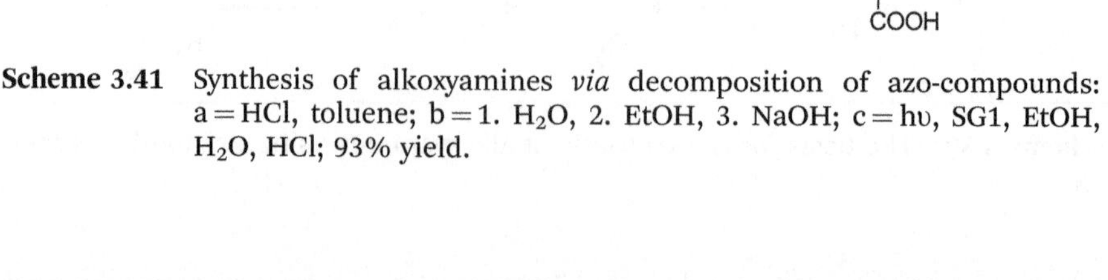

Scheme 3.41 Synthesis of alkoxyamines *via* decomposition of azo-compounds: a = HCl, toluene; b = 1. H_2O, 2. EtOH, 3. NaOH; c = hυ, SG1, EtOH, H_2O, HCl; 93% yield.

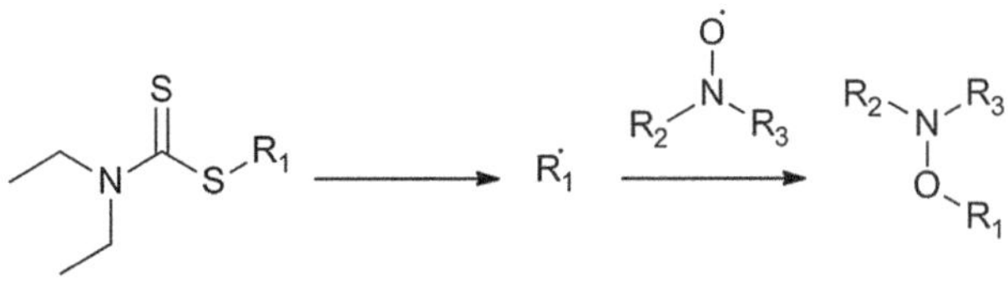

Scheme 3.42 Synthesis of Alkoxyamines by photodecomposition of alkyl dithiocarbamates.

reaction is based on the initial reaction of atom transfer radical polymerisation, producing alkyl radicals from an activated bromide with CuBr which was generated *in situ* by proportionation of Cu metal and $CuBr_2$. This reaction used TEMPO to trap the radicals to produce the desired alkoxyamines. This is a very versatile technique with high yields, simple purification and readily available reagents. This method, however, required high temperatures over several hours, which can result in decomposition of the more labile SG1-based alkoxyamines. When optimising the procedure, room temperature reactions became possible with good yields, however they require a 3-fold excess of copper as well as excess activated alkyl bromide.

3.2.2.4.8 Photodecomposition of Alkyl Dithiocarbamates. A successfully mild methodology facilitates the production of highly labile alkoxyamines based on the photodecomposition of alkyl dithiocarbamates.[155] Secondary and tertiary alkyl dithiocarbamates were previously known to undergo a reversible C–S bond dissociation, producing alkyl and dithiocarbamyl radicals.[157] Nucleophilic substitution of the halide with diethyldithiocarbamate salts[158] or radical addition onto tetraethylthiuram disulfide [159,160] were two methods utilised to produce these compounds. Unlike previously described photochemical methods, only a stoichiometric amount of dithiocarbamate was required and the disulfide was recoverable at the end of the reaction. As a result, a simple two-step process can be used for the photodecomposition of both azo and dithiocarbamate compounds which combines the advantages of using a stoichiometric amount of reagents and only producing sulfurous compounds as intermediates (Scheme 3.42).

3.3 Conclusions

In this chapter, the principal synthetic methodologies for the synthesis of nitroxide free radicals and subsequent alkoxyamine derivatives have been detailed. As outlined, the unique advantage of nitroxides compared to other classes of stable free radicals is that the nitroxide group is resistant to numerous experimental conditions needed to perform various organic transformations. As a consequence, the chemistry of nitroxides is very rich, offering access to an almost unlimited number of molecules in which the physical and chemical characteristics can be tailored.

Other chemical properties of the nitroxides, including their redox behaviour and ability to scavenge carbon centred radicals, also make them ideal precursors (either as starting material or as non-isolated intermediates) for the synthesis of alkoxyamines that can be used as mediating groups within controlled free radical polymerisation. Through select examples, this chapter has demonstrated many of the synthetic techniques that can be applied to produce nitroxide/alkoxyamines with wide ranging functionalities, installed at various positions on the molecule.

References

1. E. G. Bagryanskaya, S. R. Marque and Y. P. Tsentalovich, *J. Org. Chem.*, 2012, 77, 4996.
2. J. B. Mitchell, *Antioxid. Redox Signaling*, 2007, **10**, 1731.
3. J. P. Blinco, K. E. Fairfull-Smith, B. J. Morrow and S. E. Bottle, *Aust. J. Chem.*, 2011, **64**, 373.
4. A. Nilsen and R. Braslau, *J. Polym. Sci., part A: Polym. Chem.*, 2006, **44**, 697.
5. J. F. Keana, *Chem. Rev.*, 1978, **78**, 37.
6. T. Yamasaki, F. Mito, Y. Ito, S. Pandian, Y. Kinoshita, K. Nakano, R. Murugesan, K. Sakai, H. Utsumi and K.-I. Yamada, *J. Org. Chem.*, 2010, **76**, 435.
7. I. V. Tikhonov, V. D. Sen, L. I. Borodin, E. M. Pliss, V. A. Golubev and A. I. Rusakov, *J. Phys. Org. Chem.*, 2014, **27**, 114.
8. H. Fischer, A. Kramer, S. R. A. Marque and P. Nesvadba, *Macromolecules*, 2005, **38**, 9974.
9. S. R. A. Marque, *J. Org. Chem.*, 2003, **68**, 7582.
10. L. Marx, R. Chiarelli, T. Guiberteau and A. Rassat, *J. Chem. Soc., Perkin Trans. 1*, 2000, 1181.
11. A. P. Jagtap, I. Krstic, N. C. Kunjir, R. Hänsel, T. F. Prisner and S. T. Sigurdsson, *Free Radical Res.*, 2015, **49**, 78.
12. E. G. Bagryanskaya and S. R. A. Marque, *Chem. Rev.*, 2014, **114**, 5011.
13. E. Cote, L. Chafin, M. DiFazio, J. Robbins, J. Kotrola, T. Nocentini and K. Schoening, *Org. Process Res. Dev.*, 2014, **18**, 1843.
14. R. Briere, H. Lemaire and A. Rassat, *Bull. Soc. Chim. Fr.*, 1965, **11**, 3273.
15. J. Zakrzewski, *Beilstein J. Org. Chem.*, 2012, **8**, 1515.
16. H. Schlude, *Tetrahedron*, 1973, **29**, 4007.
17. K. Tokumaru, H. Sakuragi and O. Simamura, *Tetrahedron Lett.*, 1964, **5**, 3945.
18. H. C. Shi and Y. Li, *J. Mol. Catal. A: Chem.*, 2007, **271**, 32.
19. T. Toda, E. Mori and K. Murayama, *Bull. Chem. Soc. Jpn.*, 1972, **45**, 1904.
20. R. R. Shah, D. Merreceyes, M. Husemann, I. Rees, N. L. Abbott, C. J. Hawker and J. L. Hedrick, *Macromolecules*, 2000, **33**, 597.

21. D. Benoit, S. Grimaldi, S. Robin, J.-P. Finet, P. Tordo and Y. Gnanou, *J. Am. Chem. Soc.*, 2000, **122**, 5929.
22. R. W. Murray and M. Singh, *Tetrahedron Lett.*, 1988, **29**, 4677.
23. M. E. Brik, *Tetrahedron Lett.*, 1995, **36**, 5519.
24. M. A. Voinov, J. F. Polienko, T. Schanding, A. A. Bobko, V. V. Khramtsov, Y. V. Gatilov, T. V. Rybalova, A. I. Smirnov and I. A. Grigor'ev, *J. Org. Chem.*, 2005, **70**, 9702.
25. C. P. Sár, K. Tamás, J. Jekő and K. Hideg, *Synthesis of Deprenyl-like nitroxide free radicals and their diamagnetic derivatives*. MPublishing, University of Michigan Library, Ann Arbor, MI, 2011.
26. J. P. Blinco, J. L. Hodgson, B. J. Morrow, J. R. Walker, G. D. Will, M. L. Coote and S. E. Bottle, *J. Org. Chem.*, 2008, **73**, 6763.
27. H. Sato, V. Kathirvelu, A. Fielding, J. P. Blinco, A. S. Micallef, S. E. Bottle, S. S. Eaton and G. R. Eaton, *Mol. Phys.*, 2007, **105**, 2137.
28. K. S. Chan, X. Z. Li and S. Y. Lee, *Organometallics*, 2010, **29**, 2850.
29. E. M. Skoda, J. R. Sacher, M. Z. Kazancioglu, J. Saha and P. Wipf, *ACS Med. Chem. Lett.*, 2014, **5**, 900.
30. J. M. Bobbitt, M. C. F. Guttermuth, Z. Ma and H. Tang, *Heterocycles*, 1990, **30**, 1131.
31. I. Y. Zhukova, E. S. Kagan and V. A. Smirnov, *Khim. Geterotsikl. Soedin.*, 1992, 73.
32. P. Wipf, J. S. Greenberger, M. W. Epperly, M. M. Sprachman and J. Goff, Patent No. WO2013059651A1, 2013.
33. V. A. Golubev, G. N. Voronina and E. G. Rozantsev, *Izv. Akad. Nauk SSSR, Ser. Khim.*, 1972, 161.
34. Y. Guillaneuf, D. Bertin, D. Gigmes, D. L. Versace, J. Lalevee and J. P. Fouassier, *Macromolecules*, 2010, **43**, 2204.
35. Y. Guillaneuf, D. L. Versace, D. Bertin, J. Lalevée, D. Gigmes and J. P. Fouassier, *Macromol. Rapid Commun.*, 2010, **31**, 1909.
36. D. L. Versace, J. Lalevée, J. P. Fouassier, Y. Guillaneuf, D. Bertin and D. Gigmes, *Macromol. Rapid Commun.*, 2010, **31**, 1383.
37. T. Inokuchi and H. Kawafuchi, *Tetrahedron*, 2004, **60**, 11969.
38. B. A. Chalmers, J. C. Morris, K. E. Fairfull-Smith, R. S. Grainger and S. E. Bottle, *Chem. Commun.*, 2013, **49**, 10382.
39. X. H. Ho, W. J. Jung, P. K. Shyam and H. Y. Jang, *Catal. Sci. Technol.*, 2014, **4**, 1914.
40. T. Itoh, Y. Shimizu and M. Kanai, *Org. Lett.*, 2014, **16**, 2736.
41. Y.-X. Xie, R.-J. Song, Y. Liu, Y.-Y. Liu, J.-N. Xiang and J.-H. Li, *Adv. Synth. Catal.*, 2013, **355**, 3387.
42. B. Zhang and A. Studer, *Org. Lett.*, 2013, **15**, 4548.
43. M. Hartmann, Y. Li and A. Studer, *J. Am. Chem. Soc.*, 2012, **134**, 16516.
44. Y. F. Wang, K. K. Toh, J. Y. Lee and S. Chiba, *Angew. Chem., Int. Ed.*, 2011, **50**, 5927.
45. U. Jahn and E. Dinca, *Chem. – Eur. J*, 2009, **15**, 58.

46. F. Choteau, B. a. Tuccio, F. A. Villamena, L. Charles, B. Pucci and G. g. Durand, *J. Org. Chem.*, 2012, 77, 938.
47. N. Gruber, L. L. Piehl, E. R. d. Celis, J. E. Diaz, M. B. Garcia, P. Stipa and L. R. Orelli, *RSC Adv.*, 2015, **5**, 2724.
48. Y. Han, B. Tuccio, R. Lauricella, A. Rockenbauer, J. L. Zweier and F. A. Villamena, *J. Org. Chem.*, 2008, **73**, 2533.
49. P. Astolfi and P. Stipa, *J. Org. Chem.*, 2011, **76**, 9253.
50. D. Benoit, V. Champlinski, R. Braslau and C. Hawker, *J. Am. Chem. Soc.*, 1999, **121**, 3904.
51. R. Toba, H. Gotoh and K. Sakakibara, *Org. Lett.*, 2014, **16**, 3868.
52. V. A. Golubev and V. D Sen, *Russ. J. Org. Chem.*, 2011, **47**, 869–876.
53. C. P. Sár, T. Kálai, N. M. Bárácz, G. Jerkovich and K. Hideg, *Synth. Commun.*, 1995, **25**, 2929.
54. T. A. Prolla and R. J. Mehlhorn, *Free Radical Res.*, 1990, **9**, 135.
55. T. D. Lee and J. F. W. Keana, *J. Org. Chem.*, 1975, **40**, 3145.
56. P. M. Gannett, E. Darian, J. H. Powell and E. M. Johnson, *Synth. Commun.*, 2001, **31**, 2137.
57. C. M. Paleos and P. Dais, *Chem. Commun.*, 1977, 345.
58. Y. Kinoshita, K. Yamada, T. Yamasaki, F. Mito, M. Yamato, N. Kosem, H. Deguchi, C. Shirahama, Y. Ito and K. Kitagawa, *Free Radical Biol. Med.*, 2010, **49**, 1703.
59. A. A. Bobko, I. A. Kirilyuk, I. A. Grigor'ev, J. L. Zweier and V. V. Khramtsov, *Free Radical Biol. Med.*, 2007, **42**, 404.
60. G. M. Whitesides and T. L. Newirth, *J. Org. Chem.*, 1975, **40**, 3448.
61. J. F. W. Keana, K. Hideg, G. B. Birrell, O. H. Hankovszky, G. Ferguson and M. Parvez, *Can. J. Chem.*, 1982, **60**, 1439.
62. T. Kalai, M. R. Fleissner, J. Jeko, W. L. Hubbell and K. Hideg, *Tetrahedron Lett.*, 2011, **52**, 2747.
63. D. Reid and S. Bottle, *Chem. Commun.*, 1998, 1907.
64. A. M. Girroud and A. Rassat, *Bull. Soc. Chim. Fr.*, 1979, **1–2**, 48.
65. J. P. Blinco, B. A. Chalmers, A. Chou, K. E. Fairfull-Smith and S. E. Bottle, *Chem. Sci.*, 2013, **4**, 3411.
66. A. S. Micallef, R. C. Bott, S. E. Bottle, G. Smith, J. M. White, K. Matsuda and H. Iwamura, *J. Chem. Soc., Perkin Trans. 1*, 1999, 65.
67. K. E. Fairfull-Smith, E. A. Debele, J. P. Allen, M. C. Pfrunder and J. C. McMurtrie, *Eur. J. Org. Chem.*, 2013, **22**, 4829.
68. D. J. Keddie, T. E. Johnson, D. P. Arnold and S. E. Bottle, *Org. Biomol. Chem.*, 2005, **3**, 2593.
69. K. E. Fairfull Smith and S. E. Bottle, *Eur. J. Org. Chem.*, 2008, 5391.
70. S. E. Bottle, D. G. Gillies, D. L. Hughes, A. S. Micallef, A. I. Smirnov and L. H. Sutcliffe, *J. Chem. Soc., Perkin Trans. 2*, 2000, 1285.
71. K. Thomas, B. A. Chalmers, K. E. Fairfull-Smith and S. E. Bottle, *Eur. J. Org. Chem.*, 2013, **5**, 853.
72. A. G. Barrett, G. R. Hanson, A. J. White, D. J. Williams and A. S. Micallef, *Tetrahedron*, 2007, **63**, 5244.

73. K. E. Fairfull-Smith, J. P. Blinco, D. J. Keddie, G. A. George and S. E. Bottle, *Macromolecules*, 2008, **41**, 1577.
74. K. E. Fairfull-Smith, F. Brackmann and S. E. Bottle, *Eur. J. Org. Chem.*, 2009, **12**, 1902.
75. S. Bottle, D. Gilles, D. Hughes, A. Micallef, A. Smirnov and L. Sutcliffe, *Perkin 2*, 2000, 7, 1285.
76. T. Kálai, E. Borza, C. Antus, B. Radnai, G. Gulyás-Fekete, A. Fehér, B. Sümegi and K. Hideg, *Bioorg. Med. Chem.*, 2011, **19**, 7311.
77. H. Zhao, J. Wu, X. Meng, S. Zuo, W. Wang, H. Yuan and M. Lan, *J. Heterocycl. Chem.*, 2008, **45**, 371.
78. P. L. Poul, B. Caro, N. Cabon, F. R.-L. Guen and S. Golhen, *J. Organomet. Chem.*, 2013, **745–746**, 57.
79. J. Zakrzewski and M. Krawczyk, *Heteroat. Chem.*, 2008, **19**, 549.
80. O. Zeika, Y. Li, S. Jockusch, G. Parkin, A. Sattler, W. Sattler and N. J. Turro, *Org. Lett.*, 2010, **12**, 3696.
81. S. A. Shelke, G. B. Sandholt and S. T. Sigurdsson, *Org. Biomol. Chem.*, 2014, **12**, 7366.
82. J. C. Morris, J. McMurtrie, S. E. Bottle and K. E. Fairfull-Smith, *J. Org. Chem.*, 2011, **76**, 4964.
83. T. Kalai, W. L. Hubbell and K. Hideg, *Synthesis*, 2009, 1336.
84. U. Jakobsen, S. A. Shelke, S. Vogel and S. T. Sigurdsson, *J. Am. Chem. Soc.*, 2010, **132**, 10424.
85. E. E. Romanova, R. Akiel, F. H. Cho and S. Takahashi, *J. Phys. Chem. A*, 2013, **117**, 11933.
86. Z. Du, W. Ai, L. Xie and W. Huang, *J. Mater. Chem. A*, 2014, **2**, 9164.
87. G.-P. Yan, B. Zhao, S. E. Bottle, Q. Zhang and J.-L. Li, *Fullerenes, Nanotubes, Carbon Nanostruct.*, 2014, **23**, 734.
88. V. P. Gubskaya, L. S. Berezhnaya, A. T. Gubaidullin, I. I. Faingold, R. A. Kotelnikova, N. P. Konovalova, V. I. Morozov, I. A. Litvinov and I. A. Nuretdinov, *Org. Biomol. Chem.*, 2007, **5**, 976.
89. D. B. Gophane and S. T. Sigurdsson, *Chem. Commun.*, 2013, **49**, 999.
90. B. A. Chalmers, S. Saha, T. Nguyen, J. McMurtrie, S. T. Sigurdsson, S. E. Bottle and K.-S. Masters, *Org. Lett.*, 2014, **16**, 5528.
91. N. V. Blough and D. J. Simpson, *J. Am. Chem. Soc.*, 1988, **110**, 1917.
92. S. A. Green, D. J. Simpson, G. Zhou, P. S. Ho and N. V. Blough, *J. Am. Chem. Soc.*, 1990, **112**, 7337.
93. J. L. Gerlock, P. J. Zacmanidis, D. R. Bauer, D. J. Simpson, N. V. Blough and I. T. Salmeen, *Free Radical Res. Commun.*, 1990, **10**(1–2), 119.
94. C. Coenjarts, O. Garcia, L. Llauger, J. Palfreyman, A. L. Vinette and J. C. Scaiano, *J. Am. Chem. Soc.*, 2003, **125**(3), 620.
95. X.-F. Yang and X.-Q. Guo, *Analyst*, 2001, **126**, 1800.
96. G. Moad, D. A. Shipp, T. A. Smith and D. H. Solomons, *J. Phys. Chem. A*, 1999, **103**, 6580.
97. O. G. Ballasteros, L. Maretti, J. C. Scaiano and A. Aspee, *Photochem. Photobiol.*, 2008, **84**, 1535.

98. C. Bueno, L. Mikelsons, L. Maretti, J. C. Scaiano and A. Aspee, *Photochem. Photobiol.*, 2008, **84**, 1535.
99. C. Toniolo, M. Crisma and F. Formaggio, *Biopolymers*, 1998, **47**, 153.
100. B. Pispisa, A. Palleschi, L. Stella, M. Venanzi and C. Toniolo, *J. Phys. Chem. B*, 1998, **102**, 7890.
101. B. Pispisa, C. Mazzuca, A. Palleschi, L. Stella, M. Venanzi, M. Wakselman, J.-P. Mazaleyrat, M. Rainaldi, F. Formaggio and C. Toniolo, *Chem. – Eur. J.*, 2003, **9**, 4084.
102. E. Hideg, T. Kalai, K. Hideg and I. Vass, *Biochemistry*, 1998, **37**, 11405.
103. T. Kalai, E. Hideg, I. Vass and K. Hideg, *Free Radical Biol. Med.*, 1998, **24**, 649.
104. E. Lozinsky, V. V. Martin, T. A. Berezina, A. I. Shames, A. L. Weis and G. I. Likhtenshtein, *J. Biochem. Biophys. Methods*, 1999, **38**, 29.
105. J. P. Blinco, J. C. McMurtrie and S. E. Bottle, *Eur. J. Org. Chem.*, 2007, **48**, 4638.
106. B. J. Morrow, D. J. Keddie, N. Gueven, M. F. Lavin and S. E. Bottle, *Free Radical Biol. Med.*, 2010, **49**, 67.
107. D. J. Keddie, K. E. Fairfull-Smith and S. E. Bottle, *Org. Biomol. Chem.*, 2008, **6**, 3135.
108. Y. Yu, D. Zhang, G. Zhang, Z. Wang and D. Zhu, *Tetrahedron Lett.*, 2006, **47**, 3413.
109. J. Su, X. Liu, J. Hu, Q. You, Y. Cui and Y. Chen, *Polym. Int.*, 2015, 867–874.
110. S. Jockusch, O. Zeika, N. Jayaraj, V. Ramamurthy and N. J. Turro, *J. Phys. Chem. Lett.*, 2010, **1**, 2628.
111. P. D. Sylvester, H. E. Ryan, C. D. Smith, A. S. Micallef, C. H. Schiesser and U. Wille, *Polym. Degrad. Stab.*, 2013, **98**, 2054.
112. A. Studer and T. Schulte, *Chem. Rec.*, 2005, **5**, 27.
113. D. Crich, D. Grant and A. A. Bowers, *J. Am. Chem. Soc.*, 2007, **129**, 12106.
114. C. J. Hawker, A. W. Bosman and E. Harth, *Chem. Rev.*, 2001, **101**, 3661.
115. G. A. Kovtun, A. L. Aleksandrov and V. A. Golubev, *Izv. Akad. Nauk SSSR, Ser. Khim.*, 1974, 2115.
116. G. Moad and E. Rizzardo, *Macromolecules*, 1995, **28**, 8722.
117. J. M. Catala, F. Bubel and S. O. Hammouch, *Macromolecules*, 1995, **28**, 8441.
118. D. Gigmes, A. Gaudel-Siri, S. R. A. Marque, D. Bertin, P. Tordo, P. Astolfi, L. Greci and C. Rizzoli, *Helv. Chim. Acta*, 2006, **89**, 2312.
119. A. Studer, K. Harms, C. Knoop, C. Müller and T. Schulte, *Macromolecules*, 2004, **37**, 27.
120. R. B. Grubbs, J. K. Wegrzyn and Q. Xia, *Chem. Commun.*, 2005, 80.
121. J. M. Bobbitt and M. C. L. Flores, *Heterocycles*, 1988, **27**, 509.
122. T. Takata, Y. Tsujino, S. Nakanishi, K. Nakamura, E. Yoshida and T. Endo, *Chem. Lett.*, 1999, 937.
123. S. Kobatake, J. S. Harwood, P. Quirk and D. B. Priddy, *J. Polym. Sci., Part A: Polym. Chem.*, 1998, **36**, 2555.

124. D. H. Hunter, D. H. R. Barton and W. J. Motherwell, *Tetrahedron Lett.*, 1984, **25**, 603.
125. V. A. Golubev, R. V. Miklyush and Z. G. Rosantsev, *Izv. Akad. Nauk SSSR, Ser. Khim.*, 1972, 656.
126. T. Ren, Y. C. Liu and Q. X. Guo, *Bull. Chem. Soc. Jpn.*, 1996, **68**, 2935.
127. M. Takeuchi and K. Kano, *Bull. Chem. Soc. Jpn.*, 1994, **64**, 1726–1733.
128. J. Dao, D. Benoit and C. J. Hawker, *J. Polym. Sci., Part A: Polym. Chem*, 1998, **36**, 2161.
129. N. S. Finney, P. J. Pospisil, S. Chang, M. Palucki, R. G. Konsler, K. B. Hansen and E. N. Jacobsen, *Angew. Chem., Int. Ed.*, 1997, **39**, 1720.
130. M. Bothe and G. Schmidt-Naake, *Macromol. Rapid Commun.*, 2003, **24**, 609.
131. T. Krause, W. D. Habicher, M. Messerschmidt and B. I. Voit, *Des. Monomers Polym.*, 2004, **4**, 391.
132. Y. Thiessen and T. Wolff, *Des. Monomers Polym.*, 2005, **8**, 397.
133. C. Ollivier, R. Chuard and P. Renaud, *Synthesis*, 1999, 807.
134. J. L. Lukkarila, G. K. Hamer and M. K. Georges, *Tetrahedron Lett.*, 2004, **45**, 5317.
135. M. Pouliot, P. Renaud, K. Schenk, A. Studer and T. Vogler, *Angew. Chem., Int. Ed.*, 2009, **48**, 6037.
136. S. Harrisson, P. Couvreur and J. Nicolas, *Polym. Chem.*, 2011, **2**, 1859.
137. C. J. Hawker and J. L. Hedrick, *Macromolecules*, 1995, **28**, 2993.
138. D. Wang, X. Bi and Z. Wu, *Macromolecules*, 2000, **33**, 2293.
139. F. Aldabbagh, W. K. Busfield, I. D. Jenkins and S. H. Thang, *Tetrahedron Lett.*, 2000, **41**, 3673.
140. T. Nagashima and D. P. Curran, *Synlett*, 1996, 330.
141. M. S. Maji, T. Pfeifer and A. Studer, *Synthesis*, 2009, 2467.
142. M. S. Maji, T. Pfeifer and A. Studer, *Angew. Chem., Int. Ed.*, 2009, **47**, 9547.
143. P. Hayes, B. D. Suthers and W. Kitching, *Tetrahedron Lett.*, 2000, **41**, 6175.
144. A. Dichtl, M. Seyfried and K. U. Schoening, *Synlett*, 2008, 1877.
145. K. U. Schoening, W. Fischer, S. Hauck, A. Dichtl and M. Kuepfert, *J. Org. Chem.*, 2009, **74**, 1567.
146. C. Walling, *Pure Appl. Chem.*, 1967, **15**, 69–80.
147. C. J. Hawker, G. C. Barclay, A. Orellana, J. Dao and W. Devonport, *Macromolecules*, 1996, **29**, 5245.
148. R. Braslau, L. C. B. M. Siano, N. Naik, R. K. Howden and L. K. Mahal, *Macromolecules*, 1997, **30**, 6445.
149. T. J. Connolly, M. V. Baldovi, N. Mohtat and J. C. Scaiano, *Tetrahedron Lett.*, 1996, **37**, 4919.
150. Y. Miura, K. Hirota, H. Moto and B. Yamada, *Macromolecules*, 1999, **32**, 8356.
151. C. Wetter, K. Jantos, K. Woithe and A. Studer, *Org. Lett.*, 2003, **5**, 2899.
152. U. Jahn, *J. Org. Chem.*, 1998, **63**, 7130.
153. D. H. Solomon, E. Rizzardo and P. Cacioli, US Pat., 4581429, 1993.
154. D. Wang and Z. Wu, *Macromolecules*, 1998, **31**, 6727.

155. Y. Guillaneuf, J. L. Couturier, D. Gigmes, S. R. A. Marque, P. Tordo and D. Bertin, *J. Org. Chem.*, 2008, **73**, 4728.
156. K. Matyjaszewski, B. E. Woodworth, X. Zhang, S. G. Gaynor and Z. Metzner, *Macromolecules*, 1998, **31**, 5955.
157. T. Doi, A. Matsumoto and T. Otsu, *J. Polym. Sci., Part A: Polym. Chem*, 1994, **32**, 2241.
158. K. Ishizu, R. A. Kahn, Y. Ohta and M. Furo, *J. Polym. Sci., Part A: Polym. Chem*, 2004, **42**, 76.
159. G. Bouhadir, N. Legrand, B. Quiclet-Sire and S. Z. Zard, *Tetrahedron Lett.*, 1999, **40**, 277.
160. S. H. Thang, Y. K. Chong, R. T. A. Mayadunne, G. Moad and E. Rizzardo, *Tetrahedron Lett.*, 1999, **40**, 2435.

CHAPTER 4

NMP Under Homogenous Conditions in Bulk, Organic Solvents and Aqueous Systems

MILAN MARIC

Department of Chemical Engineering, McGill University, Montreal, QC Canada, H3A 0C5
Email: milan.maric@mcgill.ca

4.1 Introduction

Several excellent reviews about nitroxide mediated polymerization (NMP) have been written describing its development and implementation with various types of monomers[1–6] in various conditions.[1–8] The focus here is to hopefully summarize neatly the kinds of monomers and the types of nitroxide initiators used under strictly homogeneous conditions in bulk, organic solvent and water, and to highlight the novel aspects associated with each condition.

4.2 Bulk NMP

The simplest system to study is the bulk polymerization, which consists solely of the monomer and the initiating system. The earliest examples of bulk polymerizations consisted of styrenic monomers with 2,2,6,6-tetramethylpiperidine-*N*-oxyl (TEMPO) initiating systems and a free radical initiator like benzoyl peroxide.[9] Georges and co-workers then later fully

RSC Polymer Chemistry Series No. 19
Nitroxide Mediated Polymerization: From Fundamentals to Applications in Materials Science
Edited by Didier Gigmes

Published by the Royal Society of Chemistry, www.rsc.org

established the controlled bulk polymerization of styrene using TEMPO/benzoyl peroxide (BPO) as the initiating system to produce poly(styrene) resins with low dispersity and predictable linear growth of degree of polymerization with conversion.[10,11] The control of the polymerization is due to the equilibrium between dormant and active chains:

$$K = \frac{[P^\bullet][N^\bullet]}{[P-N]} \tag{4.1}$$

where K is the equilibrium constant between dormant and active chains, $[P^\bullet]$ is the concentration of actively growing macroradicals, $[N^\bullet]$ is the concentration of the nitroxide radical and $[P\text{–}N]$ is the concentration of the dormant, nitroxide capped chains. $[P^\bullet]$ is controlled by the relative low K of styrene (first reported to be 2.1×10^{-11} L mol^{-1} K^{-1} at 125 °C),[9] which retards irreversible termination reactions and thus allows the chain end to be active for a prolonged period of time during the polymerization. A more effective scheme for controlling the degree of polymerization is to form the adduct between styrene and TEMPO, which becomes a unimolecular initiator and by analogy to classical ionic or living polymerizations, the degree of polymerization can be determined by the ratio of monomer to unimolecular initiator (Figure 4.1).[12,13] For NMP in general,[9,14–18] and atom transfer radical polymerization (ATRP),[19–21] the persistent radical effect (PRE) is the key towards controlled polymerizations with linear evolution of degree of polymerization with monomer conversion and relatively narrow molecular weight distributions.

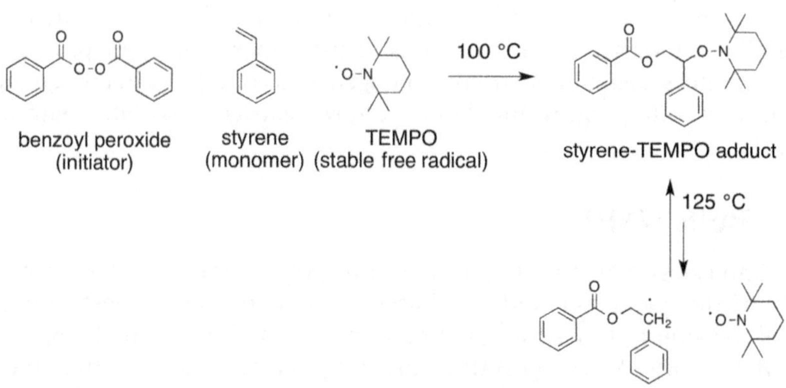

Figure 4.1 Typical steps involved in the TEMPO-mediated nitroxide mediated polymerization (NMP): (1) The TEMPO/monomer adduct is formed with a source of radicals from a conventional initiator like benzoyl peroxide (BPO). (2) With an increase in temperature, the equilibrium is shifted to favour addition of monomer, but the radical concentration is controlled by the persistent radical effect (PRE). (3) Propagation continues and the growing polymer is capped by TEMPO and the cycle continues.

$$R-Y \xrightarrow{k_d} R\cdot + Y\cdot$$

$$R\cdot + Y\cdot \xrightarrow{k_c} R-Y$$

$$R\cdot + R\cdot \xrightarrow{k_t} R_2$$

Figure 4.2 The combination of radicals involved in the persistent radical effect (PRE). $Y^\bullet$ is the persistent radical while $R^\bullet$ is the transient counter-radical.

These early modelling studies elucidated that the key species—the persistent or long-lived species (TEMPO; given as $Y^\bullet$) and the source of the transient, free radicals (from an initiator such as BPO, given as $R^\bullet$)—were participating species that formed at the same rate but one was much longer-lived than the other (see Figure 4.2). This build-up in the persistent species concentration (TEMPO), along with the subsequent decrease in the transient radical due to self-coupling is responsible for the control of the polymerization as the equilibrium reaction between the persistent radical and the capped species becomes favoured.[22,23]

The kinetics for homogeneous NMP are summarized below, which were derived in detail elsewhere.[18,22] The concentrations of the radical species found an unusual $t^{1/3}$ dependence time in the case of the persistent radical [eqn (4.2) and (4.3)]:

$$[Y^\bullet] = \left(\frac{3k_t k_d^2 [R-Y]_0^2}{k_c^2}\right)^{1/3} t^{1/3} \tag{4.2}$$

$$[R^\bullet] = \left(\frac{k_d [R-Y]_0}{3k_t k_c}\right)^{1/3} t^{-1/3} \tag{4.3}$$

where k_t is the irreversible termination rate constant, k_d is the rate constant for the dissociation of the nitroxide, k_c is the recombination rate constant and $[R-Y]_0$ is the initial concentration of the nitroxide (consisting of the persistent radical ($Y^\bullet$) and the transient radical ($R^\bullet$).

4.2.1 Homopolymerizations by Bulk NMP with TEMPO

Despite the seemingly limited avenue of TEMPO-based NMP, a wide variety of functional styrenic monomers were readily polymerized, leading to many functional materials. Some of these styrenics included 2-,[24–26] 3-[27] and 4-vinyl pyridine,[28,29] chloromethylstyrene,[30,31] *tert*-butoxystyrene,[32–34] *p*-acetoxystyrene,[35,36] *p*-(1-methylcyclohexyloxy)styrene,[37] 4-methylstyrene,[31,38,39] 4-methoxystyrene,[31] 2,5-bis[(4-butylbenzoyl)oxy]styrene,[40] 4-bromostyrene,[41,42] 4-fluorostyrene,[43] 4-(phenylethynyl) styrene,[44] 5-bis[(4-methoxyphenyl)oxycarbonyl]styrene[45,46] and dimethyl vinylbenzylphosponate.[47] Structures of these styrenic monomers are listed in Table 4.1.

Table 4.1 Common styrenic monomers polymerized in bulk using TEMPO.

Monomer	Ref.	Monomer	Ref.
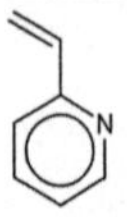 2-Vinyl pyridine	24–26	 4-Bromostyrene	41, 42
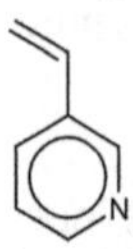 3-Vinyl pyridine	27	 4-Fluorostyrene	43
 4-Vinyl pyridine	28, 29	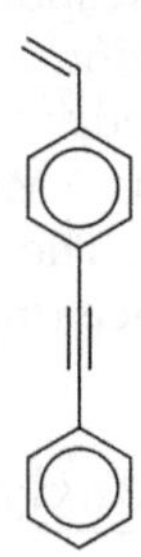 4-(Phenylethynyl) styrene	44
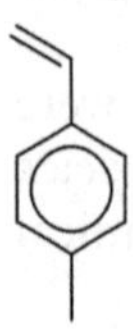 4-Methylstyrene	31, 38, 39	 4-Methoxystyrene	31
 Chloromethylstyrene	30, 31	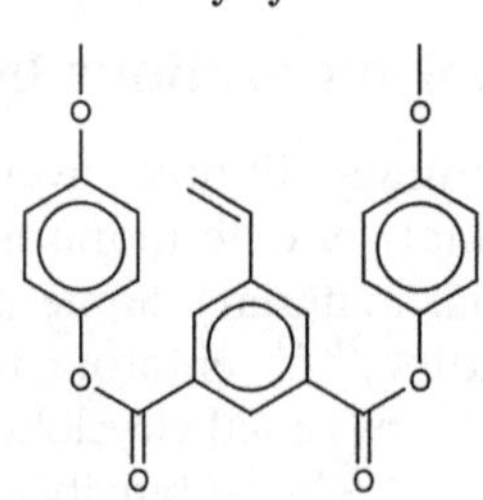 2,5-Bis-[(4-methoxyphenyl)-oxycarbonyl]styrene	40
 *tert*-Butoxystyrene	32–34		
 *p*-Acetoxystyrene	35, 36		

Table 4.1 (*Continued*)

Monomer	Ref.	Monomer	Ref.
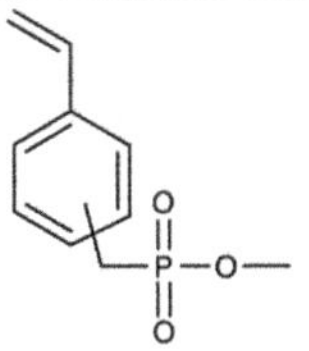 Dimethyl vinylbenzyl-phosponate	47	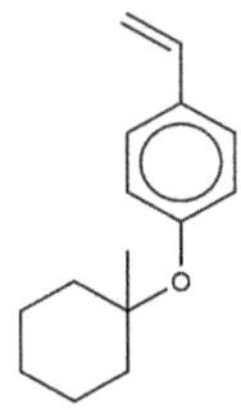 *p*-(1-Methylcyclo-hexyloxy)styrene	37

The polymers derived from these monomers were often subsequently used as macroinitiators for the polymerization of a second block to make segmented block copolymers or were derivatized for further applications, such as photoresists and membranes. The effect of the *para* substituent was not studied formally for NMP as it was for conventional radical polymerization[48] and ATRP,[48,49] where a Hammett parameter type of analysis was attempted to correlate the propagation rate constants with the electron donating/withdrawing *para* groups of styrenic monomers.

Acrylamide homopolymerizations were also attempted with TEMPO in bulk with various levels of success, with most focusing on *N,N*-dimethylacrylamide (DMAm).[50,51] The initial report by Li and Brittain using TEMPO in bulk at 120 °C or in toluene solvent at 84 °C with AIBN as the counter-radical resulted in poly(DMAm), which was rapidly polymerized with broad molecular weight distributions and poor ability to re-initiate a batch of a second monomer.[50] Later, Fischer *et al.* showed using benzoyl peroxide instead of AIBN permitted much better control of the polymerization with much slower polymerizations in bulk at 138 °C. The conversions were still however quite low for DMAm homopolymerizations (<10%) but dispersity $Đ = 1.20–146$ using various TEMPO : BPO ratios.[51]

Given the various homopolymerizations demonstrated using TEMPO in bulk, statistical copolymerizations started to be explored. In contrast to conventional radical polymerization, better chain-to-chain compositional heterogeneity was expected for the controlled polymerizations, leading to more carefully tailored polymers that would be more effective in a desired application.[1] For example, better compositional control would lead to more effective macromolecular surfactants for pigment dispersions.[52] Most of the copolymerization systems have been summarized previously. It should be noted here that I am omitting copolymerization where a small amount of co-monomer such as a styrenic (<10 mol% in the initial composition) is used to effectively control the polymerization of a methacrylate-rich system; this specific type of copolymerization will be addressed in the methacrylate "homopolymerizations" when using SG1 alkoxyamines.[53] Specifically, for TEMPO systems in bulk, the various pairs (typically using styrene as the co-monomer in each pair) included: styrene/acrylonitrile,[9,54–57] styrene/butyl methacrylate,[58,59] styrene/methyl methacrylate (MMA),[31,60] styrene/ethyl methacrylate (EMA),[59] methyl acrylate (MA),[59] ethyl acrylate (EA),[59] butyl acrylate (BA),[59] 2-hydroxyethyl acrylate (HEA),[59] and vinyl acetate (VAc)],[59] *N*-acryloyl morpholine (AMo),[60] 2-ethoxyethyl acrylate (EOEA),[60] isobornyl acrylate (iBoA),[60] vinylferrocene (VFc),[61] vinylcarbazole (VCz),[62,63] *N*-cyclohexylmaleimide,[64] *p*-chlorostyrene,[65] and *n*-vinyl pyrrolidone.[66]

4.2.1.1 *Complex Microstructures using TEMPO in Bulk*

During this initial period, rapid activity ensued to test how effectively TEMPO-mediated systems could mimic truly living polymerizations by forming block copolymers. Several strategies were adopted to show that the chains could be re-activated to add a second block consisting of another monomer. One method consisted of forming a linear chain of one more monomer, with the basis being that each chain was capped by the TEMPO group. A second batch of monomer would be added to chain-extend the macroinitiator by the same nitroxide mediated mechanism. This approach has been described extensively.[67–83]

An alternative strategy consisted of using a functional TEMPO mediator such as the hydroxyl functional TEMPO 4-hydroxy TEMPO (TEMPO-OH) where the hydroxyl group would not interfere with the polymerization. If the chain-end fidelity remained intact, the hydroxyl group could then be used to initiate a second batch of monomer by a different mechanism. This approach and its variants was initially explored by Yoshida and co-workers who quenched a living poly(styryl) lithium anion with sodium 4-oxy TEMPO to yield a poly(styrene) with a terminal TEMPO group that could be used as a macroinitiator for bulk NMP.[84] Similar transformations to a functional TEMPO were followed to produce diblock copolymers. For example, Yoshida and Sugita described the synthesis of sodium 4-oxy TEMPO that was used to couple to chains of living poly(tetrahydrofuran) [poly(THF)] made by cationic polymerization (Figure 4.3).[85,86] The poly(THF)-TEMPO polymer (12.6 kg mol^{-1}, $Đ = 1.27$) was subsequently used as a macroinitiator for the controlled bulk polymerization of styrene (ST) with BPO (95 °C for 3.5 h, 125 °C for 64 h) to yield poly(THF)-*b*-poly(ST) block copolymers with low *Đ* (1.27–1.40). ESR confirmed TEMPO chain end fidelity. Further, controlled nature of the ST polymerization was suggested by linear kinetic plots and shifts in the GPC peaks of the chain-extended species relative to the macroinitiator. A similar scheme was adopted by Yoshida and Osagawa where 4-hydroxy TEMPO was reacted with triethylaluminum to make a diethylaluminum-oxy TEMPO to initiate the ring-opening polymerization of caprolactone (CL).[86] The PCL-TEMPO ($M_n = 3250$ g mol^{-1}, $Đ = 1.28$) then served as a macroinitiator for the polymerization of a ST block to give a poly(CL)-*b*-poly(ST) copolymer in good yield (75%) with low dispersity ($M_n = 23\,600$ g mol^{-1}, $Đ = 1.37$) and a linear growth in the styrenic chain with polymerization time.

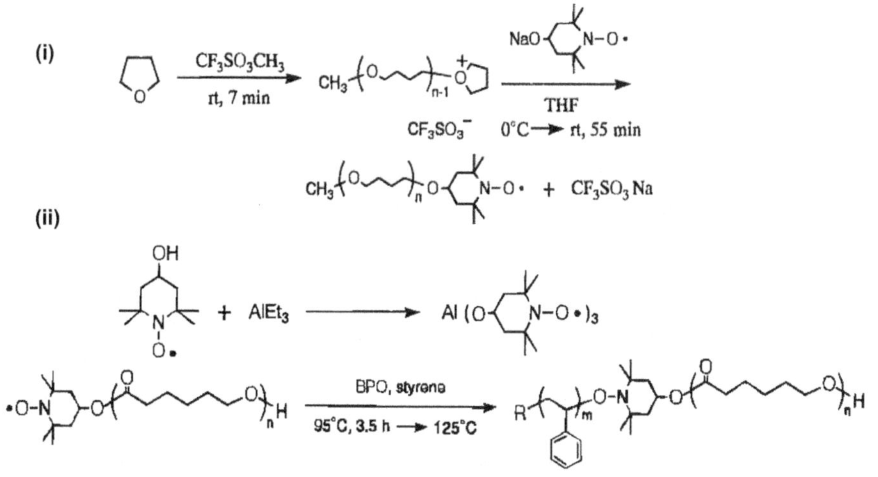

Figure 4.3 Some strategies for functionalizing TEMPO to convert it into a macroinitiator with NMP in bulk. In (i), tetrahydrofuran (THF) is polymerized by a ring-opening polymerization followed by capping with sodium methoxide capped TEMPO.[85] The resulting poly(THF)-TEMPO is used a macroinitiator for NMP. In (ii), a 4-hydroxy TEMPO is used as an initiator for ring-opening polymerization for caprolactone (CL) and then the PCL-TEMPO is used as a macroinitiator for NMP.[86]

More sophisticated architectures such as triblock copolymers were sought such as the coupling of TEMPO-OH with α, ω diisocyanate terminated poly(ethylene adipate) (PEA-$(NCO)_2$).[87] The resulting difunctional macroinitiator cleanly yielded the ABA triblock copolymer with linear M_n growth with conversion up to 60% although the molecular weight distribution was relatively broad ($Đ = 1.38$–1.54), likely due to the use of PEA made by a condensation polymerization. Another ABA type block copolymer was accessed by first anionically polymerizing methylphenylsilane to give an α,ω-dihalo(polymethylphenysilane), which was subsequently reacted with TEMPO-OH to provide a difunctional macroinitiator for bulk polymerization of styrene at 130 °C.[88] Aliphatic polyesters that can be made by ring-opening polymerization, such as PCL, were also coupled with TEMPO to make macroinitiators for subsequent TEMPO-mediated polymerization with functional styrenics such as 4-vinylpyridine (4-VP).[89,90] Coupling of PEOs either through end-functional PEOs by coupling tosylate terminated PEO with TEMPO-OH[91] or by using the method originally shown by Yoshida and co-workers where EO was anionically polymerized and then quenched with NaO-TEMPO that was subsequently a macroinitiator for TEMPO-mediated 4-VP polymerization.[92,93] The resulting PCL-P4VP and PEO-PVP block copolymers exhibited interesting self-assembly behaviour in aqueous media. Finally, lactide ring-opening polymerization was done using a TEMPO-OH with $AlEt_3$ to make a poly(lactide) macroinitiator for the controlled polymerization of styrene and 4-*tert*-butylstyrene di- and linear tri-block copolymers.[94] Many of the aforementioned macroinitiators were aliphatic polyesters, like poly(lactide), poly(adipates) and poly(caprolactones), which were likely envisioned to be used where degradability with mild conditions for biomaterials was desirable. However, macroinitiators based on poly(butadiene) (PB)[95] and poly(dimethylsiloxane) (PDMS)[96,97] were also used in bulk with styrene to form block copolymers. Other TEMPO-mediated polymerizations using the modification of 4-hydroxy TEMPO to a macroinitiating NMP where TEMPO-OH mediator was converted into a TEMPO with a fluorescent pyrene functional group (PYNOR) was also examined.[98] The PYNOR mediator was dissolved in styrene and the bulk polymerization ensued, with relatively low $Đ$ (1.14–1.42) and linear M_n *versus* conversion plots up to about 60% conversion before plateauing was observed.[98] Finally, a vinyl functionalized TEMPO was reported by Yoshida and Tanaka that was made as a block copolymer with styrene with the resulting TEMPO pendant groups available as initiating sites for graft copolymers or for oxidation/reduction reactions.[99] The latter application has been reported for methacrylate functional TEMPOs geared for the use of block copolymers for novel batteries.[100]

The use of multiple initiating sites for different polymerizations such as that described above has been widely discussed as orthogonal strategies. Not only have such strategies involved ionic or condensation polymerization modes combined with TEMPO-mediated polymerizations like the examples cited above, but multifunctional initiators have been designed to combine TEMPO initiators with several others.[101] Some strategies have deployed TEMPO mediated polymerizations in conjunction with other controlled radical polymerization methods, such as ATRP[102–104] and RAFT,[105,106] to polymerize one type of monomer suitable for one controller (typically an acrylate or methacrylate for the ATRP ligand or RAFT agent) with one more suited for the other (a styrenic for the TEMPO) as illustrated in Figure 4.4.

Besides block copolymers, graft copolymers were rapidly developed in the mid-1990s by NMP. Here, the initially reported approach relied on having the appropriate polymer terminated with a polymerizable vinyl group, which was then copolymerized with a styrenic to form the graft copolymer.[107] Hawker and co-workers showed that a TEMPO-derived unimolecular initiator could be used to make graft copolymers with styrene and methacrylate-terminated poly(ethylene glycol), poly(D,L lactide), poly(caprolactone) and acrylate-terminated poly(ethylene) (the macromonomomer molecular weights ranged from 800–8900 g mol^{-1}.[107] The graft feed ratios were 0.10–0.30 and the resulting copolymers were produced in high yield >70% and low $Đ = 1.15–1.40$. Interestingly, the authors noted that the relatively high temperatures employed (~125 °C) helped solubilize some macromonomers in styrene that would not be soluble at lower temperatures. A similar strategy was employed in xylene solution by Wang *et al.*, who used a PCL-functionalized TEMPO to control a copolymerization of styrene with a poly(ethylene glycol)-functionalized with a styryl end group.[108] Despite this

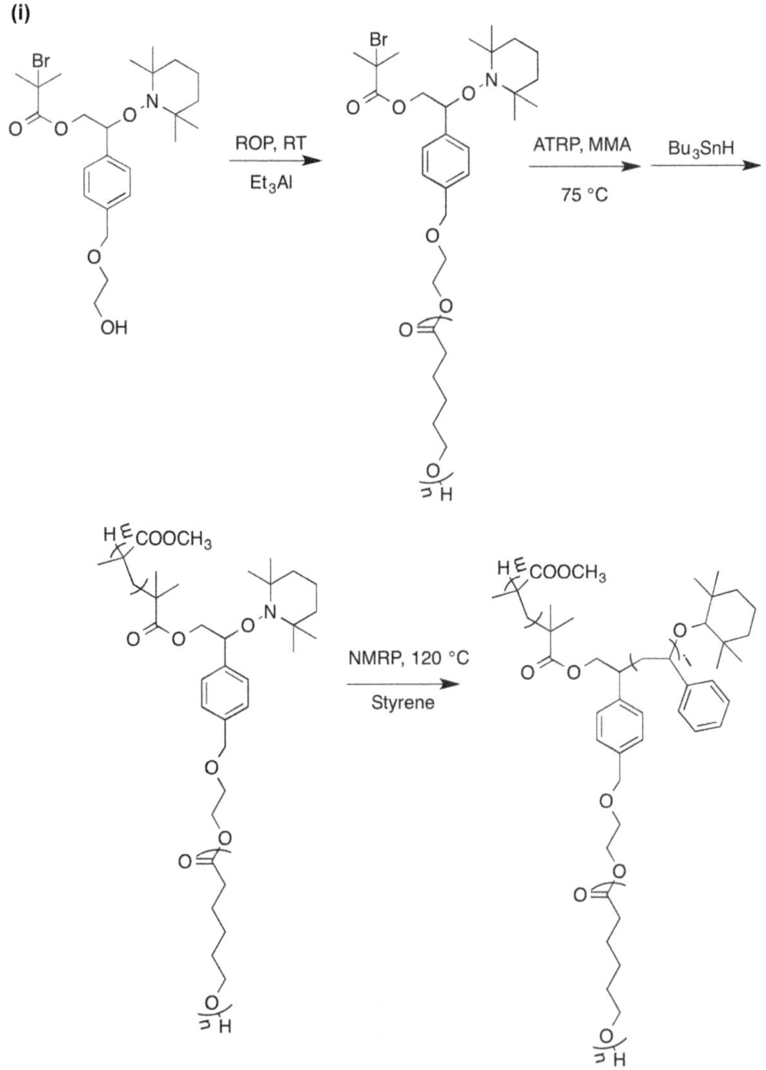

Figure 4.4 Examples of combining TEMPO with ATRP and RAFT methodologies to make copolymers of various microstructures. In (i), a trifunctional initiator is used to perform ring-opening polymerization of caprolactone, followed by ATRP and NMP.[102] In (ii), a similar multifunctional initiator that has TEMPO and a RAFT chain transfer agent is used to make various block copolymers.[106]

early report, graft and comb copolymers by NMP using TEMPO did not proliferate extensively in bulk conditions, in part due to the high viscosity of the macromonomers. Later studies, such as that by Ryan *et al.*, who copolymerized styrene with a poly(styrene) macromonomer using TIPNO and SG1 type nitroxides, have become more prevalent.[109] Holdcroft and co-workers

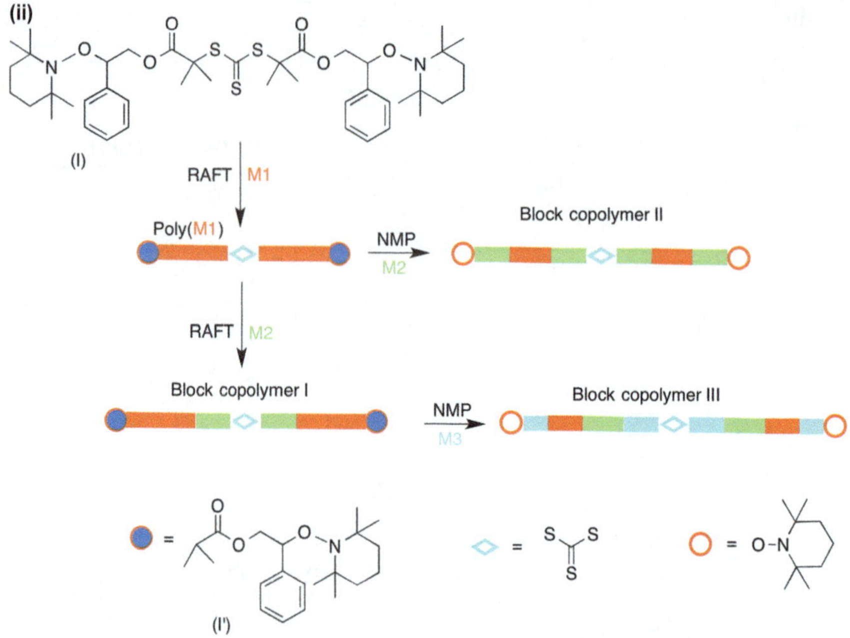

Figure 4.4 (Continued)

functionalized TEMPO with boronic acid groups that were coupled to bromide-terminated poly(3-hexylthiophene) (P3HT) chains *via* Suzuki coupling.[110,111] The P3HT-TEMPO macroinitiator was then used to initiate a styrene/chloromethylstyrene copolymerization *via* NMP in bulk. The chloromethyl groups were then converted to azide *via* reaction with sodium azide and then subsequently coupled to fullerenes so that the block copolymer had fullerene grafts from one segment. The resulting donor/acceptor copolymers were tested for photovoltaic properties (Figure 4.5).

Graft copolymer formation using TEMPO has been alternatively derived from reacting linear polymers, such as poly(propylene),[112,113] poly(ethylene)[114] or poly(styrene-*co*-butadiene) rubber,[115] with a typical free radical initiator like a peroxide followed by TEMPO. The resulting polymers had graft sites with TEMPO to initiate the polymerization of controlled graft lengths along the chain, typically poly(styrene).[112–116] Instead of creating radicals on the chain and then capping with TEMPO, monomers functionalized with TEMPO groups were copolymerized with a styrenic using conventional[91] or controlled radical polymerization.[117,118] With this methodology, the initial polymerization is active at much lower temperatures than that required for TEMPO-mediated polymerization. The subsequent grafting is accomplished by adding fresh monomer (styrene) at higher temperatures ~120–125 °C. Finally, instead of using peroxides to initiate a polymerization, ultrasonic scission has been employed to create radicals and then cap with a TEMPO moiety. The resulting TEMPO-capped polymer was

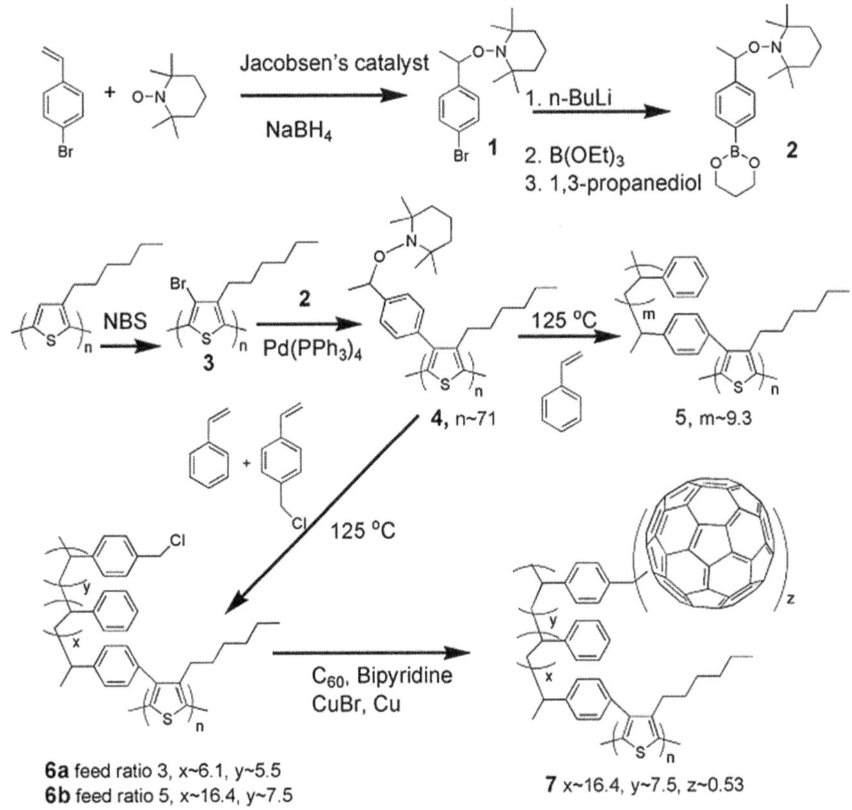

Figure 4.5 Reaction of TEMPO with P3HT to form a P3HT-TEMPO macroinitiator, which was used to initiate the NMP of a styrene/chloromethylstyrene mixture.[110]

used to initiate a controlled polymerization to produce a block copolymer (Figure 4.6).[119]

Besides linear and graft copolymers, TEMPO-mediated polymerizations were also used to access star and dendritic (co)polymers. Star-architectures have long been considered fascinating molecules to study due to their low solution viscosity compared to their linear counterparts of identical molecular weight.[120] Two approaches are commonly used to make such architectures by CRP: *"core-first"* and *"arm-first"*. In the former case, chains are grown from the centre of the molecule (typically a multi-functional initiator). This allows for the arms to be identical in length with a constant number of arms. In contrast, the arm-first approach used pre-formed linear polymers that were attached to a multi-functional coupling agent. The first route for star polymers by NMP, specifically using TEMPO-type initiators in bulk, was reported by Hawker.[121] Here, a core functionalized with 3 TEMPO groups was used to initiate the polymerization of deuterated styrene at 130 °C for 72 h, resulting in a three-arm star poly(d-styrene) with various arm lengths

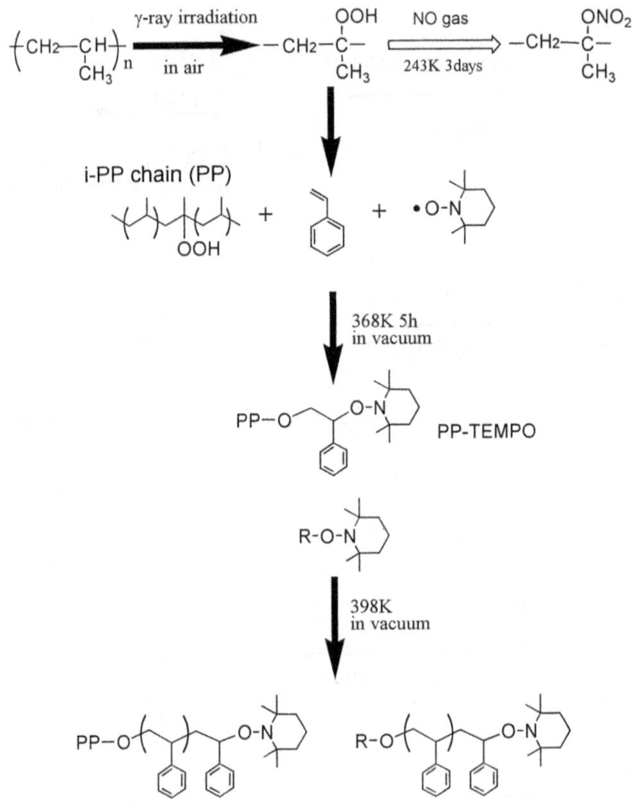

Figure 4.6 Graft copolymers were synthesized by first grafting poly(propylene) with TEMPO units, which were then used to initiate styrene NMP from the TEMPO sites.[112]

that could be adjusted by the ratio of monomer to initiator (Figure 4.7). The cleavage of the linear arms matched the theoretical expectation and confirmed the controlled nature and the architecture of the molecule.

Chessa *et al.* synthesized 3- and 6-arm star poly(styrene)s (PS) in bulk with relatively low total molecular weight to be able to investigate the structure by ^{1}H NMR easily (<15 kg mol^{-1}) using trifunctional or hexafunctional mesitylene cores.[122] The $Đ$ was <1.3 in all cases. Miura and Yoshida used a core functionalized with 6 TEMPO units to initiate the polymerization of S with relatively long PS arms ($M_n = 34.0 - 138$ kg mol^{-1}, $Đ = 1.29$–1.59)[123] in addition to 3-armed PS stars with $M_n = 29.5 - 43.7$ kg mol^{-1}, $Đ = 1.19$–1.59).[124] In contrast, Tsoukatos *et al.* prepared 4–644-arm PSs with $Đ = 1.15$–156 with varying star M_ns of 150–6250 g mol^{-1} using an arm first approach.[125] Miura and Dote described the use of a dodecafunctional macroinitiator made from the reaction of benzyl alcohol with 4 TEMPO-based alkoxyamines to synthesize 12-arm star PS at 120 °C with low

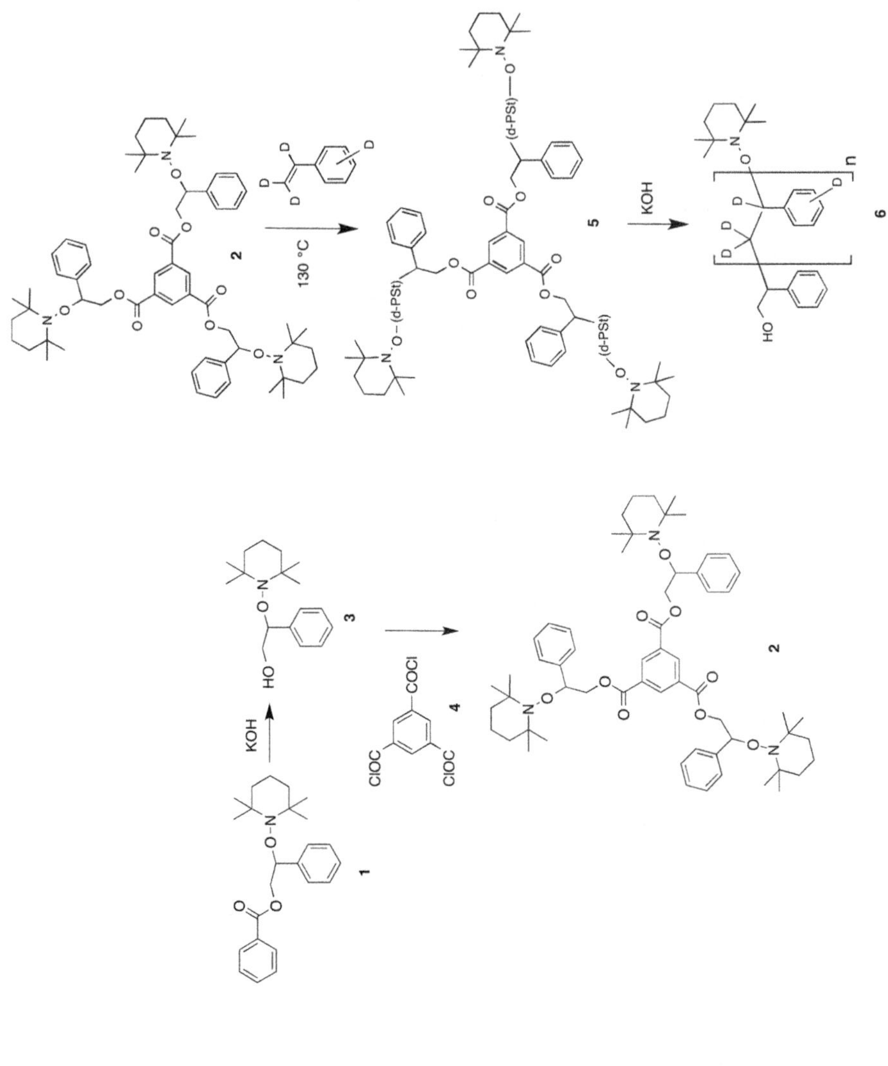

Figure 4.7 Three-arm star poly(styrene) synthesize using a core functionalized with three TEMPO groups.[121]

$Đ = 1.06–1.26$.[126] The chain-end fidelity remained sufficiently intact to make 12-arm star block copolymers with 4VP as the second block with $Đ = 1.18–1.19$. Similarly, a 16-arm star and star diblock copolymer and an AB8 9-miktoarm star copolymer were prepared from dendritic multi-TEMPO-functionalized macroinitiators.[127] 4-\srm star block copolymers were also prepared using a $Ti(OPr)_4$ core functionalized with TEMPO-OH, which subsequently initiated first the ring-opening polymerization of CL, which was then subsequently chain-extended by TEMPO-mediated polymerization of styrene at 120 °C to provide the corresponding 4-arm star block polymers with M_ns = 17.2–85.9 kg mol^{-1} and $Đ = 1.4–1.7$.[128] Dendritic star-burst PS was prepared from tris-(1,3,5)benzyloxy unimers[129] and tri-azofunctionalized unimers[130] with M_n as high as 126 kg mol^{-1}, although $Đ$ grew to 1.64.

4.2.2 Bulk NMP with TEMPO-derived Initiators

The structure of the TEMPO unit was subjected to intense scrutiny with the aim of altering the equilibrium constant so as to be capable of polymerizing acrylates or styrenics at much lower temperatures. Stüder's group[131–133] and others[134–137] have examined intermolecular hydrogen bonding in hydroxyl and silyloxy-substituted 2,2,6,6-tetramethylpiperidin-1-oxyl (TEMPO) derivatives, which led to faster C–O bond homolysis. In particular, the siloxy-substituted species was very effective for controlling styrene homopolymers at temperatures as low as 90 °C and was also shown to be capable of controlling *n*-butyl acrylate.[131] Using the TEMPO motif later, Stüder and co-workers showed that sterically hindered cyclic nitroxides enabled controlled *n*-BA homopolymerizations as low as 60 °C ($D = 1.11–1.23$) and styrene homopolymerizations as low as 50 °C ($Đ = 1.08–1.11$ for this particular case) (Figure 4.8)[133]

4.2.3 Bulk NMP with TIPNO

The limitations in using TEMPO became obvious to its practitioners by the late 1990s; excessively high temperatures restricted easy adoption to dispersed systems, the prominent role of thermal initiation of styrenics made chain end fidelity sometimes difficult to control and the ability to only control styrenic monomers (and in some cases copolymerization with other acrylates and methacrylates to a limit) were seen as barriers towards its wider adoption.[1–4,138] Further, the excessively long polymerization times characteristic of TEMPO mediated systems made industrial adaption unlikely.[4] Many groups thus attempted to examine the alkoxyamine structure to overcome these barriers.

The development of alkoxyamine unimolecular initiators like 2,2,5-trimethyl-4-phenyl-3-azahexane-3-oxyl (TIPNO)[139,140] and *N-tert*-butyl-*N*-[1-diethylphosphono-(2,2-dimethylpropyl)] nitroxide (DEPN or SG1)[141] in the late 1990s signalled a dramatic shift in applications of NMP towards functional materials (Figure 4.9). The TIPNO system was shown to effectively

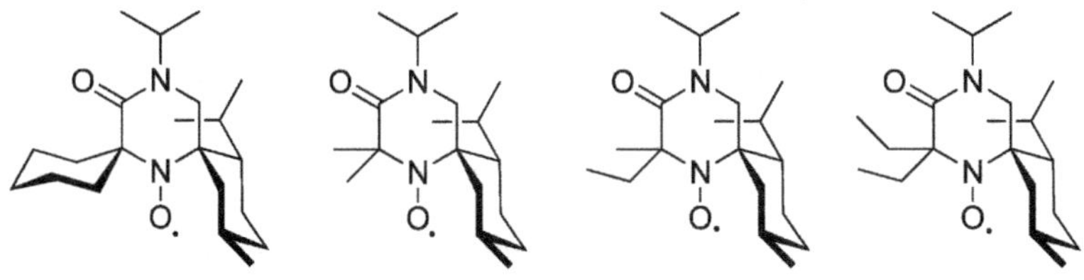

8 **9** **10** **11**

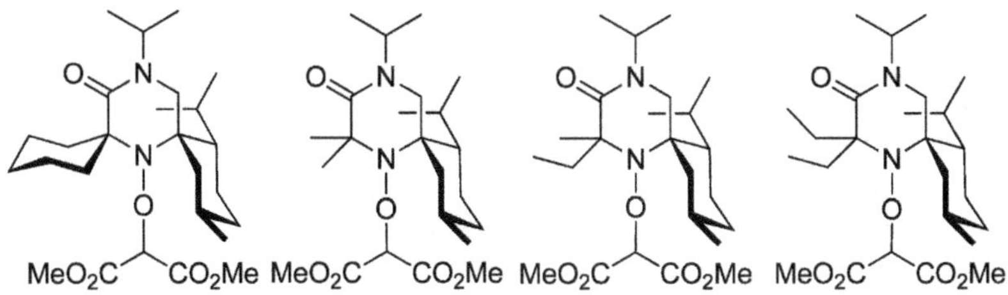

Mal-8 (46%) **Mal-9** (40%) **Mal-10** (43%) **Mal-11** (not stable)

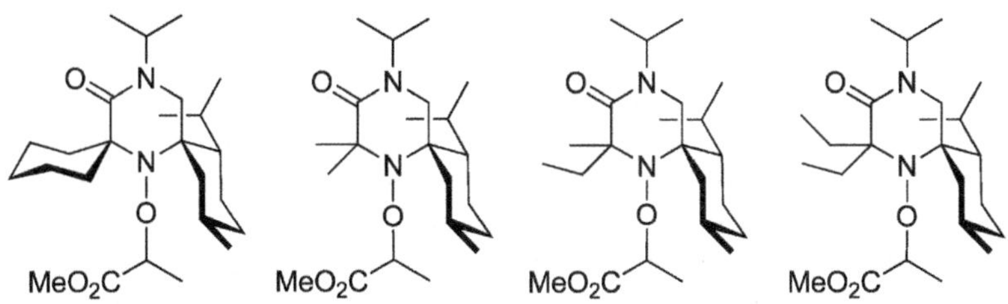

Pro-8 (93%) **Pro-9** (92%) **Pro-10** (90%) **Pro-11** (94%)

Figure 4.8 Sterically hindered TEMPO-based nitroxides that were used by the Stüder group for low temperature butyl acrylate (60 °C) and styrene (50 °C) homopolymerizations.[133]

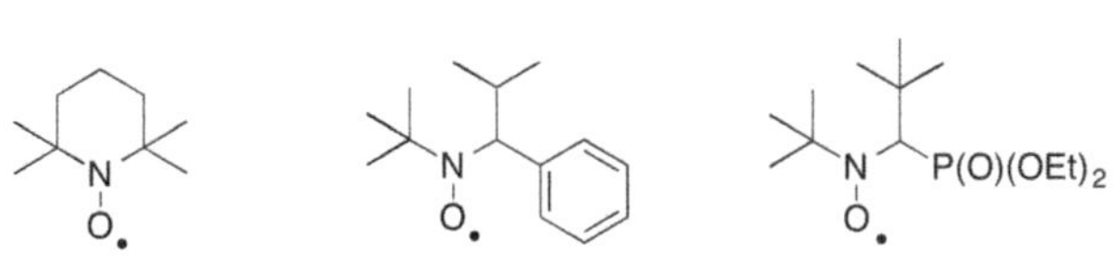

TEMPO TIPNO SG1

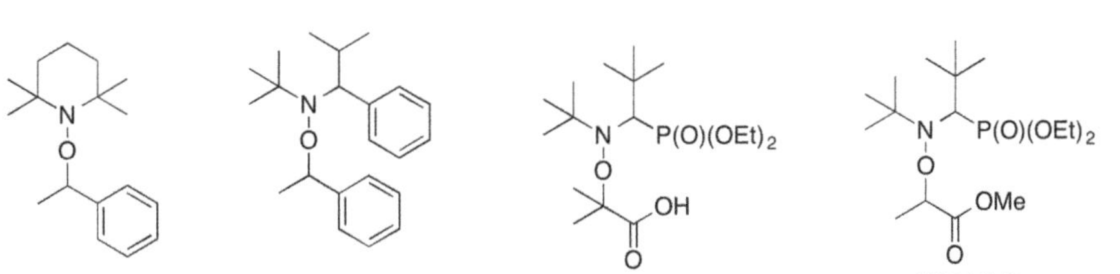

styryl-TEMPO styryl-TIPNO BlocBuilder MONAMS

Figure 4.9 Structures of TIPNO and the SG1-based alkoxyamines MONAMS and BlocBuilder.

polymerize acrylates at lower temperatures and even acrylamides could be polymerized with low $Đ$ and to relatively high degrees of polymerization.[140] TIPNO was also shown to be possible at much lower temperatures, thereby making emulsion polymerizations potentially viable without the use of pressurized reactors.

The simplicity of bulk homopolymerization using TIPNO and its derivatives was first undertaken by Benoit *et al.* using 1,3-dienes, such as isoprene, as a model monomer, since it was poorly controlled by TEMPO-mediated systems.[140] The polymerization was well-controlled using TIPNO (with any added free nitroxide, which would slow the polymerization sufficiently and allow better control, as it does for many acrylates), with $Đ$ as low as 1.07–1.14 and a linear relationship between M_n and conversion.[140] The bulk polymerization of isoprene using TIPNO was then expanded to include statistical copolymerizations with styrenics such as styrene, *p*-acetoxystyrene and chloromethylstyrene with acrylates such as *t*-butyl acrylate (feed compositions of *t*BA = 10–90%, M_n = 15.5–19.5 kg mol^{-1}, $Đ$ = 1.15–1.25), acrylic acid (10–20 mol% AA, M_n = 15.0–18.5 kg mol^{-1}, $Đ$ = 1.14–1.16, succinimidyl acrylate (ASA) (10–20 mol% of the acrylate, M_n = kg mol^{-1}, $Đ$ = 1.19–1.22, hydroxyethyl methacrylate (HEMA) (10–20 mol% HEMA, M_n = 14–17.0 kg mol^{-1}, $Đ$ = 1.18–1.19, MMA (20–80 mol% MMA, M_n = 15.5–18.0 kg mol^{-1}, $Đ$ = 1.12–1.26). Another interesting statistical copolymerization in bulk with TIPNO was with styrene and benzylcyclobutene, where the latter was used to induce cross-linked nanoparticles.[142] Further, block copolymers of P*t*BA-*block*-PI were synthesized in bulk at 120 °C using TIPNO with various compositions with excellent control of $Đ$ and M_ns of the block copolymer ranging from 9.7 kg mol^{-1} to 45.5 kg mol^{-1}. Interestingly, there was some dependence of sequence addition of the blocks with these initially reported TIPNO systems, analogous to some living ionic polymerizations. For PI-*b*-PS, if PI-TIPNO was the macroinitiator, the addition of styrene proceeded smoothly with homo-PI impurities. However, if PI was used as the macroinitiator for *t*-BA chain extension, homoPI was found as an impurity. Just as the case for TEMPO, TIPNO-based homopolymerizations in bulk proliferated enormously to include functional monomers capable of click reactions, such as benzyl substituted *N*-vinyl triazoles[143] and *para*-acetylene functional styrenes.[144] In the first case, *N*-benzyl ethyl vinyl triazole (VTA) was copolymerized with styrene with good control at low vinyl triazole feed concentrations (10 mol%, M_n = 42 kg mol^{-1}, $Đ$ = 1.17 after 12 h at 120 °C) and statistical incorporation of VTA but loss of control at 20 mol% VTA in the feed. Monomers such as acrylates, acrylamides and to some limited extent, methacrylates have been chronicled for NMP with TIPNO in bulk. Common acrylic monomers that were polymerized in bulk include *tert*-butyl acrylate,[140,145–150] *n*-butyl acrylate,[149–154] methyl acrylate,[155,156] and olefinic monomers, such as isoprene,[145,150,152] and norborene derivatives.[155] In almost all cases, to enhance the control of acrylate homopolymerization, a small excess of free nitroxide of ~5 mol% relative to unimolecular TIPNO initiator was added and ensured linear M_n *versus* conversion plots and narrow molecular weight distributions. Besides the ability to control acrylates and 1,3-dienes effectively, TIPNO mediators were also effective for bulk polymerizations of acrylamides such as *N,N*-dimethylacrylamide.[149,150,152]

4.2.3.1 Complex Microstructures with TIPNO in bulk

The first reports successfully showed how TIPNO initiators controlled the polymerization of isoprene to form block copolymers with styrene and *t*BA, respectively. An interesting pathway reported early in the application of TIPNO initiators towards "single-shot" block copolymers was reported by Benoit *et al.* using TIPNO in bulk *via* non-stoichiometric initial compositions of maleic anhydride and styrene (with styrene in excess) resulting in poly(MA-*ran*-S)-*b*-poly(S) block copolymers (Figure 4.10).[157] Such block copolymers were not successfully accessed *via* TEMPO initiators. The tendency of MA to alternate with styrene and not homopolymerize allowed such a pathway to exist, although the elevated temperatures required using the TIPNO system (120 °C) did not ensure strictly alternating MA/S sequences in the first block.

More conventionally, sequential linear block copolymerization where one of the blocks consisted of a homo-acrylic monomer was a frequently reported route. Just some of the examples of sequential block copolymers where the macroinitiator consisted of the TIPNO-mediated polymerization followed by use of the TIPNO-terminated species to re-initiate a second block included the

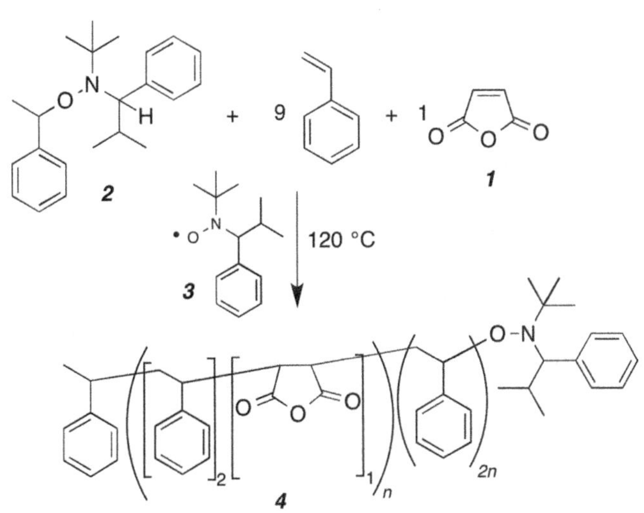

Figure 4.10 The one-shot block copolymerization of styrene (ST) with maleic anhydride (MA) using TIPNO. A non-stoichiometric mixture of ST : MA (9 : 1) was used. The MA was preferentially consumed initially, resulting in a ST/MA sequence followed a pure poly(styrene) sequence after MA was depleted. This is resulted in a poly(ST-*ran*-MA)-*b*-poly(ST) one-shot block copolymer.[157]

following architectures: linear AB diblock copolymers such as poly(*n*-butyl acrylate)-*b*-poly(styrene),[146,158] poly(*t*BA)-*b*-poly(4-acetoxystyrene),[159] poly(*t*BA)-*b*-poly(styrene),[148,159] poly(styrene)-*b*-poly(methyl acrylate),[160,161] poly(diethyl-4-vinyl benzyl phosphonate) with poly(*N,N*-dimethylacrylamide) and poly(styrene),[162] poly(2-vinyl-4,4-dimethyl-5-oxazolone) with poly(styrene) and poly(*n*-butyl acrylate),[163] and poly(styrene)-*b*-poly(isoprene).[133,161] Linear ABA triblocks emanating from suitable difunctional TIPNO initiators were polymerized in bulk also. Some copolymers synthesized included PDMA-PnBA-PDMA,[149] PS-PnBA-PS,[150,164] PS-*b*-PtBA-PS,[150] PI-P*t*BA-PI,[150] PnBA-P*t*BA-PnBA,[150] PDMAm-PI-PDMAm,[150] and PtBA-*b*-PS-P*t*BA.[149,150,165]

Further, functionalization of TIPNO-species was done in much the same way as TEMPO mediators were to combine different CRP strategies or different polymerization mechanisms, such as ROP and ROMP. In bulk, Hadziioannou and co-workers used a functionalized TIPNO macroinitiator based on poly(2,5-dioctyl-1,4-phenylenevinylene) (PPV-TIPNO) to initiate the NMP of a styrene/chloromethylstyrene (CMS) mixture.[151,166] The CMS was then attached to an electron accepting molecule such as C_{60} to form a donor/acceptor block copolymer.[166] Similarly, the TIPNO-terminated PPV ($M_n = 3.9$ kg mol^{-1}) also initiated the NMP of butyl acrylate in bulk ($M_n = 9.7$ kg mol^{-1} after 7 h polymerization but *Đ* was not reported).[151]

Another interesting functionalization strategy for TIPNO initiators was to add terpyridine groups to TIPNO with the resulting unimolecular initiator possessing a terpyridine group at one chain end and TIPNO at the other.[167–172] The inclusion of the terpyridine group enabled the polymer–metallic complexes to form and to induce supramolecular assemblies (the researchers coined this process the "LEGO toolbox"). Telechelic incorporation of terpyrdine groups in a polymer chain was done by reacting *N*-(4′-terpyridinyl)pent-5-oxy)maleimide with a terpyridine-terminated poly(styrene).[168] Other monomers used with the terpyridine-functionalized TIPNO were butyl acrylate, dimethylacrylamide, 4-vinylpyridine, 2-vinylpyridine and isoprene[167] and *tert*-butyl acrylate, and methyl acrylate.[170] Similar strategies to functionalize the counter-radical of TIPNO involved an alkyne and azido-functionalized TIPNO for click reactions,[173] a P3HT-functionalized TIPNO *via* a Grignard reagent route to make P3HT-poly(styrene) block copolymers[174] and various alcohol and carboxyl acid functionalized TIPNOs.[175]

Efforts to cleanly synthesize copolymers with grafts, stars or dendritic structures mimicked the strategies to make such microstructures with TEMPO in bulk. However, with the ability to control the polymerization of acrylates and *N,N*-substituted acrylamides, water-soluble polymers (after appropriate de-protection such as the case with *tert*-butyl acrylate) could be much more readily accessed. Early attempts with TIPNO-type initiators to make dendritic polymers involved using TIPNO with a copolymerization of styrene and a bismaleimide (99 : 1 ratio of styrene : bismaleimide) at 120 °C for 8 h to rapidly make a highly branched species with multiple alkoxyamine termini (Figure 4.11).[176] The bismaleimide was shown to be rapidly consumed compared to styrene and further polymerization resulted in linear

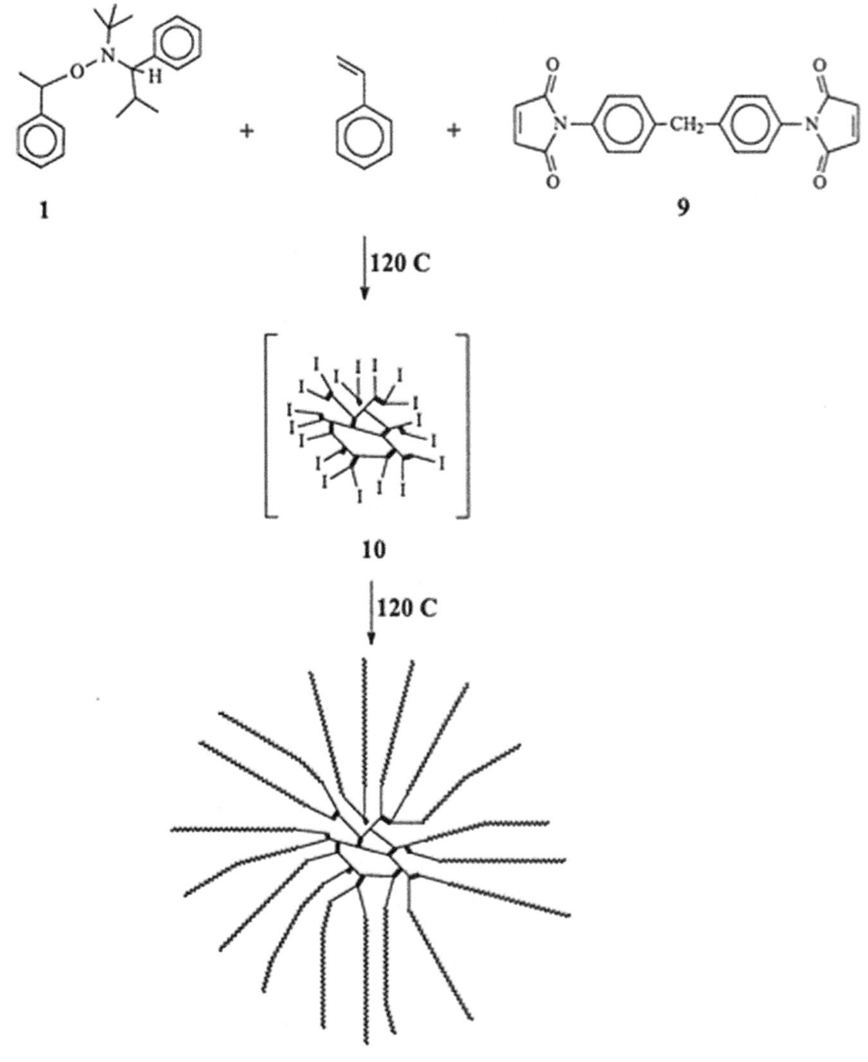

Figure 4.11 Dendritic poly(styrene) made by styrene/bismaleimide copolymerization from which multiple TIPNO sites would continue the polymerization of styrene.[176]

poly(styrene) branches emanating from the multiple chain ends. Confirmation of the controlled nature of the homopoly(styrene) chains was accomplished by cleaving the bismaleimide cross-links, and the resulting linear chains had very low $Đ = 1.1–1.2$.[176] Other cases in bulk have followed, such as making a TIPNO inimer with a TIPNO iniferter to make branched poly(styrene)s and poly(*n*-butyl acrylate)s.[177,178] Grafts have been prepared similarly.[179]

4.2.4 Bulk NMP with SG1

With the SG1-type initiators, the first reports centred on *n*-butyl acrylate (BA) homopolymerizations and block copolymers with styrene.[141] Consequently, since those initial reports, bulk NMP with acrylates was reported widely with SG1-type initiators. Acrylate monomers besides *n*-BuA[180,181] that were polymerized in bulk with the MONAMS form of SG1 include 2-hydroxyethyl acrylate (HEA)[182] and 2-dimethylamino ethyl acrylate (DMAEA),[183] while BlocBuilder initiated systems include oligoethylene glycol acrylate (OEGA),[184] *tert*-butyl acrylate[185] and hydroxyethyl acrylate,[186,187] hydroxypropyl acrylate,[187] and 3-acryloxylpropyltriethoxysilane.[188] It should be noted that in the latter case a styryl-SG1 unimolecular initiator was used. Statistical copolymerizations with MONAMS in bulk have been carried out using such monomer pairs as styrene/methyl acrylate.[189] Acrylamides such as *N,N*-dimethylacrylamide were also controlled well for the first time bulk using SG1-type (styryl-SG1 initiator plus excess free SG1) initiators.[190]

Like TIPNO, a small excess of free SG1 nitroxide improved the polymerization control markedly.[191–193] Initially, commercially available unimolecular initiators from Arkema were MONAMS, which was eventually replaced with BlocBuilder. BlocBuilder possessed several desirable advantages compared to MONAMS. It had a higher k_d,[194] and a carboxylic acid end group to solubilize it in water (for applications towards dispersed aqueous systems)[195] to modify/derivatize the end group,[196] and to use it as reactive group in blend compatibilization.[197]

With the SG1-type initiators, control of methacrylate polymerizations was accomplished by Charleux and co-workers, who used a small concentration of co-monomer such as 4.4–8.8 mol% styrene[53,198] or about 2.5 mol% acrylonitrile[199] in bulk at 80–100 °C to produce MMA-rich copolymers that were effectively poly(MMA) in terms of physical properties. Prior to this finding, even using BlocBuilder-type initiators led to uncontrolled homopolymerizations of MMA; as was the case for poly(*n*-BA) SG1-terminated macroinitiators (still block copolymers with broad molecular weight were obtained and indicated that self-assembly was possible).[200] Despite the simplicity of using such a copolymerization method, researchers sought out structural changes to the nitroxide to make methacrylate homopolymerizations by NMP feasible. Guillaneuf *et al.* developed 2.2-diphenyl-3-phenylimino-2,3 dihydroindol-1-yloxyl nitroxide (DPAIO), which was effective at MMA homopolymerization up to about 60% conversion, with a linear increase in M_n *versus* conversion and $Đ$ close to 1.4. However, using the resulting PMMA as a macroinitiator for *n*-BA resulted in a dead polymerization after only a few cycles.[201] It should also be noted interestingly that DPAIO nitroxides have been used for NMP at elevated temperatures, as high as 200 °C, for the controlled polymerization of styrene and butyl acrylate, resulting in polymers with $Đ \approx 1.2$, which makes such materials interesting for coatings, inks and varnishes.[202] Later, Grubbs and co-workers developed *N*-phenylalkoxyamines that were refined compared to the structure of DPAIO,

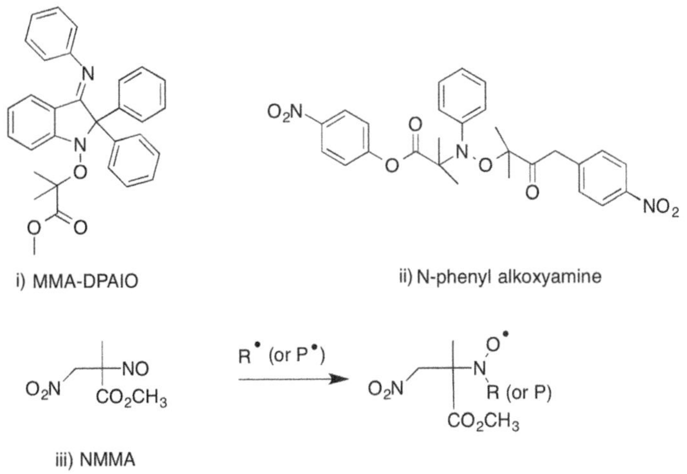

Figure 4.12 Nitroxides used for methacrylate homopolymerizations: (i) DPAIO,[201] (ii) *N*-phenylalkoxyamines,[203–204] and (iii) *in situ* hindered alkoxyamines such as methyl-2-methyl-3-nitro-2-nitrosopropionate (NMMA).[206]

and not only were able to homopolymerize MMA effectively, but chain extension to MMA or styrene was accomplished as well.[203,204] Further emphasis on methacrylate homopolymerizations *via* NMP was recently reported by Fukuyama *et al.*[205] and Detrembleur *et al.*,[206] who examined hindered alkoxyamines that could homopolymerize MMA at moderate temperatures of 40–55 °C. The latter approach involved an *in situ* NMP approach[207,208] in which the mediating species is formed during the polymerization. Several methacrylates, such as MMA, BzMA and trifluoroethyl methacrylate (TFEMA), were controlled well by this approach and chain extension experiments from a PMMA macroinitiator resulted in molecular weight growth with moderate increases in *Đ* from 1.22 to 1.38.[206] Figure 4.12 summarizes the initiating systems used to for methacrylate homopolymerization by NMP.

4.2.4.1 Complex Microstructures with SG1 in Bulk

Again, many of the same strategies used to make polymers with controlled architectures were employed using SG1-type initiators in bulk. Some have focused on functionalizing the initiator. For example, a dual initiator was developed by esterification of diols with a carboxylic acid terminated SG1 initiator, BlocBuilder.[209] Bertin and co-workers have used intermolecular radical addition (IRA) to make multi-functional macroinitiators (Figure 4.13).[210–219] For example, poly(caprolactone) diol was reacted with BlocBuilder and acryloyl or methacryolyl chloride which was subsequently used to initiate bulk polymerization of styrene, acrylonitrile and methyl methacrylate/styrene mixtures.[219] A similar procedure was done with poly(ethylene glycol) diol, which used IRA followed by styrene bulk

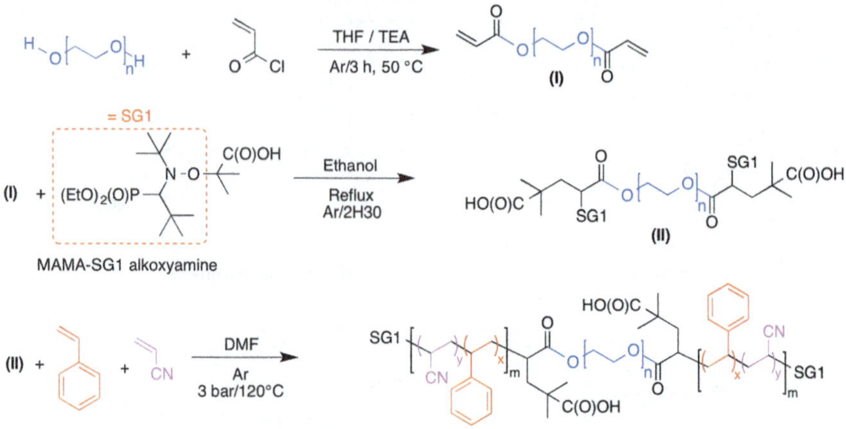

Figure 4.13 A representative example of intermolecular radical addition (IRA) that occurred between an olefin (in this case a telechelic poly(ethylene glycol) with vinyl end groups) and the BlocBuilder nitroxide followed by NMP of a styrene/acrylonitrile mixture.[218]

polymerization[220] to serve as templates for mesoporous silica with tunable pore sizes.[221–224] Some interesting bioconjugation has also been performed by using the so-called succinimidyl ester terminated NHS-BlocBuilder, which was derived from the esterification of BlocBuilder with *N*-hydroxysuccinimide.[225] The succinimidyl ester was used to couple with primary amines for biomedical materials such as those associated with peptides and proteins,[226–228] with other polymers such as poly(ethylene glycol)[227] and with small molecules to convert to a ring-opening polymerization caprolactone.[228]

Transformations from SG1 groups to other types of controlled polymerization processes have been reported. For example, cobalt-mediated radical polymerization (CMRP) was used to polymerize a poly(vinyl acetate)-poly(acrylonitrile) diblock copolymer terminated with a $Co(acac)_2$ group, which was exchanged with SG1.[229] The resulting PVAc-PAN-SG1 macroinitator was used to polymerize a third segment of either styrene, 4-vinylpyridine, or butyl acrylate. Another transformation involving RAFT agents was reported where the thioester RAFT groups were exchanged for SG1 groups.[230] Finally, the carboxylic terminated polystyrene with the complementary SG1 group was transformed by three different routes.[231] In one, the SG1 groups were converted to alcohols, thereby creating α,ω functional poly(styrene). In the second, a bromide was added to permit ATRP and finally, azido groups were derived (Figure 4.14).

In summary, bulk polymerizations with TEMPO-mediated systems have been remarkably effective at homopolymerizing styrenic monomers and at some statistical copolymerizations, and have allowed access to more complex architectures like block copolymers, graft copolymers and star/dendrimers. Characteristic of all the TEMPO systems is that high

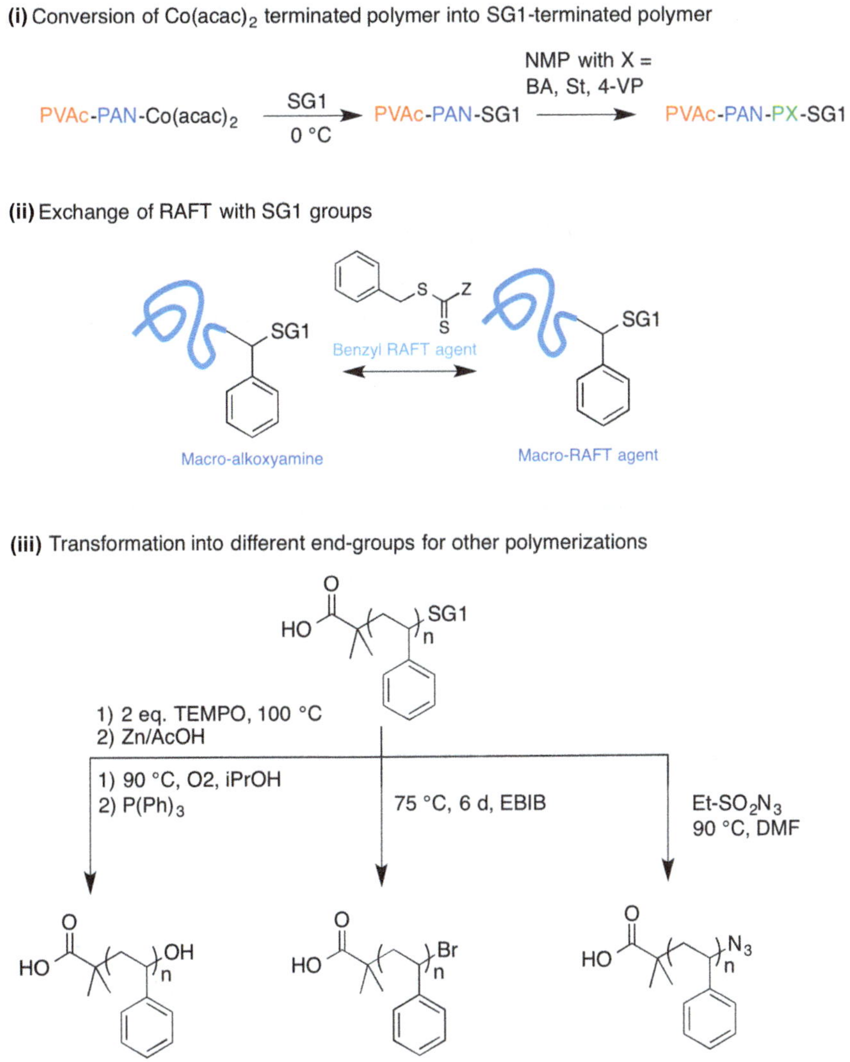

Figure 4.14 Examples of transformation of SG1 with other controlled polymerization strategies, such as cobalt-mediated radical polymerization (CMRP),[229] RAFT[230] and ATRP.[231]

temperatures (120–130 °C) were necessary to push the equilibrium sufficiently high to produce a sufficient concentration of radicals. This was aided by the contribution of thermal autopolymerization of styrenics, which allow the polymerization rate to be enhanced at higher temperatures due to the additional source of radical species.[134,135] Bulk NMP is also important with regards to scale-up potential as the gel or Trommsdorff effect is largely absent,[232,233] which is not the case for conventional radical polymerization. TIPNO and SG1-based initiators have greatly expanded the range of monomers that are polymerizable in bulk conditions and even though lower temperatures are accessible, the same limitations are faced (high viscosity, solubility of monomers) as with TEMPO initiators. Thus, NMP in solvents becomes a necessity to access many materials, particularly those with solid monomers (for stimuli-responsive and organic electronics, for example).

4.3 NMP in Organic Solvents

The possibility of excessive exotherms in bulk was noted for NMP in addition to some monomers being solids at ambient conditions. Consequently, solvent was necessary to affect controlled polymerizations. For example, acrylic acid could not be polymerized in bulk by NMP; dioxane solvent at ~30 wt% initial monomer composition was required to have effective control of the molecular weight and the molecular weight distribution, and to give the ability to chain-extend cleanly with a second monomer.[234–236] In other cases, solid monomers needed to be dissolved in solvents prior to polymerization. For example, NIPAM, an important monomer for making thermoresponsive polymers, requires solvent for homogeneous polymerization.[237–243] Furthermore, many monomers used for organic electronic applications are solids at room temperature and require solvent for dissolution.

4.3.1 NMP with TEMPO in Organic Solvents

Perhaps the most convenient approach to summarize the NMP of various monomers with TEMPO in organic solvents is to broadly describe such polymerizations as a function of monomer type: namely styrenics, acrylates and methacrylates.

4.3.1.1 Styrenic Polymerizations with TEMPO in Organic Solvents

N-vinyl carbazole was copolymerized with styrene at various initial compositions; it was necessary to dissolve the mixture in dioxane to continue the polymerization.[62,63] Liquid crystal monomers such as 2,5 bis[4-butylbenzoyl]oxystyrene (BBOS) were polymerized by TEMPO in solution as well.[244–251] The main application for using styrenic monomers with TEMPO in solution was for highly conjugated monomers directed towards organic light emitting diodes (OLEDs) and photovoltaic materials.[252–256] NMP was considered advantageous for such materials compared to RAFT or ATRP. For example, since NMP does not use metallic species like ATRP requires, any possible contamination that might affect electrical properties was avoided. Further, RAFT processes were not favoured necessarily due to the thioester chain transfer agents, which often left polymers discoloured (usually yellow). However, this can be overcome by easy cleavage and a review of RAFT polymerizations for OLEDs and PVs shows the versatility in applying RAFT towards such materials.[255]

4.3.1.2 Acrylic Polymerizations with TEMPO in Organic Solvent

Interestingly, despite the strong connection between TEMPO controlling agents and styrenic monomers, one of the first attempts to use TEMPO as a mediator in organic solvent was for methyl acrylate.[256] Later, *tert*-butyl acrylate (TBA) was homopolymerized with TEMPO in solution.[257] Homopolymerization of various alkyl acrylates with TEMPO in solvent was reported and the general conclusion unsurprisingly was that it was ineffective, although some reducing agents have helped at elevated temperatures, such as camphorsulfonic acid and ascorbic acid,[258] in addition to using initiators that more slowly decompose at elevated temperatures.[259,260]

4.3.1.3 Methacrylic Polymerizations with TEMPO in Organic Solvent

A few methacrylates were copolymerized with styrenics using TEMPO at moderate initial compositions; solvent was used as it was thought to help repress some of the kinetic problems associated with methacrylate polymerizations by TEMPO, namely the excessive exotherms caused by radical buildup.[259,260] Araujo *et al.* described the formation of PS-*b*-PMMA diblock copolymers with TEMPO; however, the addition of the MMA block was limited to a few monomer units.[261] In another case for styrene/BMA/maleimide terpolymerizations in anisole solution at 130 °C, Butz *et al.* reported high $Đ \approx 1.5$–1.6 when including BMA at 45–50 mol% in the initial monomer mixture.[262] The inadequacy of TEMPO to control methacrylate polymerizations was later explicitly shown by Ananchenko *et al.* whereby disproportionation of TEMPO into an alkene and hydroxylamine made controlled radical polymerization of methacrylate infeasible.[263,264] Despite the inadequacies of TEMPO to control methacrylates in bulk or in solution, many have continued using TEMPO in solution polymerizations, particularly in copolymerization systems. Araujo and Pinto recently reported styrene/MMA copolymerizations in toluene in tubular reactors with $M_n = 30$–60 kg mol^{-1} and $Đ = 1.3$–1.6 that were with other monomers besides methacrylates.[261] They found by manipulating residence times that linear M_n with conversion relationships were observed up to 70 mol% conversion.

Johnson *et al.* reported the copolymerization of 1-alkylvinazenes with styrene across a wide feed composition in DMF using TEMPO.[265] Compared to the conventional radical polymerization of the vinazene with styrene where the copolymer was much richer in vinazene compared to the initial monomer composition, the copolymers using TEMPO were more statistical in composition. The authors suggested that the highly reactive vinazyl radical was inhibited more strongly compared to the styryl radical. Grafting reactions have also continued in solution using TEMPO. For example, TEMPO was grafted onto poly(butadiene), which was then used to initiate the controlled polymerization of styrene.[266] Further, novel dendritic monomers have been polymerized in *o*-xylene using TEMPO as the mediator at 120 °C.[267] Also, polymerizations in ionic liquids has been attempted using TEMPO,[268–270] but polymerizations had to be done at high temperatures, which resulted in broad molecular weight distributions and relatively poor control of the process. Controlled radical polymerizations in ionic liquids that have been done successfully were organotellurium mediated polymerization (TERP),[271] organostilbine mediated polymerization,[272] RAFT[273,274] or ATRP,[275–279] and only recently have polymerizations been done with SG1-type initiators with adequate results.[280] Just as was shown in the section regarding bulk polymerizations with TEMPO, a myriad of microstructures can still be accessed, despite the limitations of the monomers that can be used with TEMPO.

4.3.2 NMP with TIPNO in Organic Solvents

The majority of cases with the TIPNO type of alkoxyamine involve the wider range of monomers that it is capable of polymerizing in a controlled fashion. Of course, it has been used for various functional styrenics.[281–283] In contrast to TEMPO-mediated systems, acrylamides such as *N*-isopropylacrylamide (NIPAM)[237,238,241] and various acrylates[283–296] have been polymerized effectively when using TIPNO-related initiators in solution. Many monomers used for organic electronics are bulky and are typically solids at room temperature; they thus require solvent to provide a homogeneous solution that will not be excessively viscous during polymerization.[285–296]

4.3.2.1 *Complex Microstructures* via *NMP with TIPNO in Organic Solvents*

Block copolymers have been polymerized in solution with TIPNO and of course display a wider variety of monomers compared to TEMPO-mediated systems. For example, Wegrzyn synthesized a TIPNO-terminated PEO macroinitiator *via* a halogen/nitroxide exchange reaction of a PEO-bromo ester with TIPNO.[297] This was subsequently used to initiate isoprene polymerization in *m*-xylene solution (from bulk to 80 wt% solvent) at 125 °C to make PEO-*b*-PI block copolymers (43–70% PI content) with low $Đ$ (1.05–1.18) but conversions were kept low, typically from 16–39%. The resulting block copolymers were micellized in aqueous media and cross-linked, retaining their structural integrity after removal of water. Lipo-glycopolymer amphiphiles were also synthesized with a styryl-TIPNO and TIPNO-functionalized with a lipophilic *N,N*-di(octadecyl)amine group that was used to initiate *t*BA/*N,N* dimethylacrylamide mixtures, *N,N*-di(octadecyl)acrylamide and

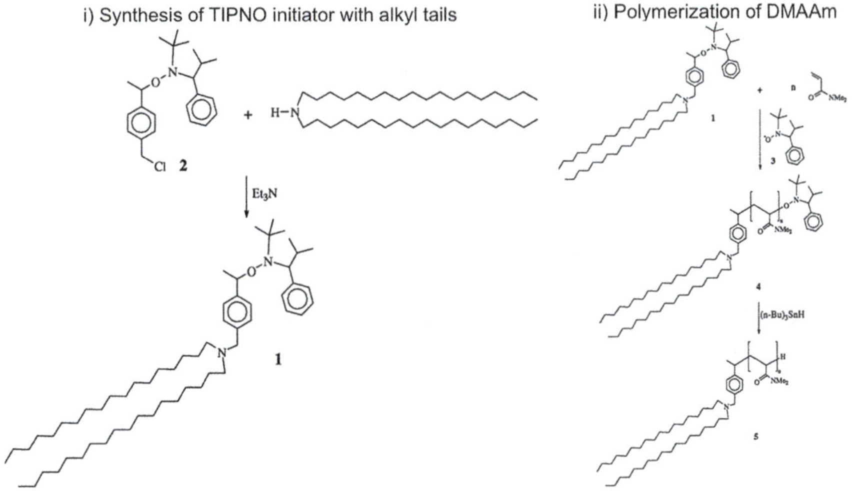

Figure 4.15 Using TIPNO-based NMP to access biomimetic materials resembling glycocalix in cell membranes.[298]

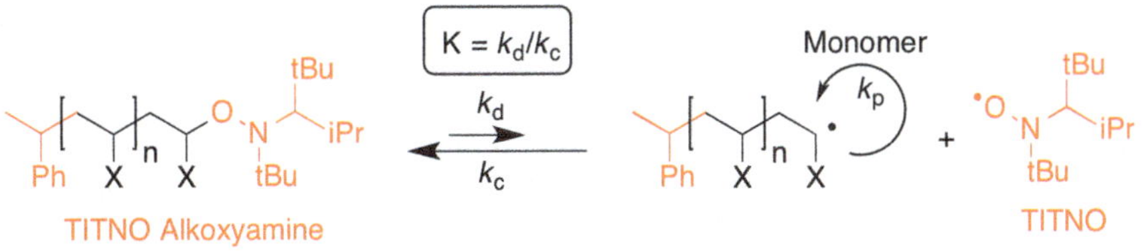

Figure 4.16 Structure of the TITNO nitroxide for low temperature NMP.[300]

1,2,5,6-di(isopropylidene)-D-glucose-2-propenoate in DMF solution at 105 °C.[298] Excellent control was achieved in terms of controlled character and low $Đ$ (1.10–1.16 for *t*BA/DMAm mixtures; 1.10–1.19 for *N,N*-di(octadecyl)-acrylamide/1,2,5,6-di(isopropylidene)-D-glucose-2-propenoate mixtures). The resulting copolymers were used as biomimetic models of the glycocalix in cell membranes (Figure 4.15).

Dendritic polymers were also polymerized using TIPNO initiators in solution. Using 6- and 12-armed initiators, styrene, methyl acrylate, isoprene and butyl acrylate were polymerized with the styrene and the methyl acrylate being controlled very well ($Đ = 1.10$–1.18).[299] As is the case earlier, researchers have continually tried to effect controlled NMP at lower temperatures (<100 °C) to be feasible ultimately in dispersed phase polymerizations like emulsion or mini-emulsion polymerization or for temperature-sensitive materials and applications. A variant of the TIPNO structure titled TITNO (see Figure 4.16) was recently reported, which was effective for polymerizing styrene at temperatures as low as 70 °C and *n*BA at temperatures as low as 90 °C.[300]

4.3.3 NMP with SG1 in Organic Solvents

Besides controlling exotherms, using solvents can introduce side reactions such as chain transfer to monomer, polymer or solvent.[301,302] For NMP in solvent, acrylate monomers such as *t*BA[185] and acrylamides such as *tert*-butyl acrylamide,[303] *N*-(2-morpholin-4-ylethyl)acrylamide,[303] and NIPAM[304] displayed chain transfer to solvent that was not insignificant, in some cases being as high as 4 mol%, depending on the solvent.[185] Despite the possibilities of side reactions such as chain transfer, NMP with SG1-based initiators in solvent has been effective in producing block copolymers that self-assembled into free-standing structures[200] and block copolymers with poly(ionic liquids).[305]

4.3.3.1 Methacrylate Polymerizations with SG1 in Organic Solvent

Perhaps the most prominent application of NMP in organic solvent using the SG1 type of initiators is for controllers of methacrylates using the addition of a small amount of controlling co-monomer, like styrene.[51] Initially, this was illustrated for MMA polymerizations with a fairly low temperature (90 °C), additional free nitroxide (~10 mol% SG1/BlocBuilder) and about 8 mol% styrene in the feed. Rapidly, other methacrylates began to be synthesized with this procedure using BlocBuilder. For the following using styrene as the co-monomer, methacrylic acid,[306] ethyl methacrylate in dioxane,[307] butyl methacrylate in dioxane,[307] glycidyl methacrylate in dioxane,[308] *t*-butyl methacrylate (*t*BMA) in dioxane,[309] 5-methacryloyloxy-2,6-norboranecarbolactone (NLAM) in dioxane[310] and 2-(2′,3′,4′,6′-tetra-*O*-acetyl-β-D-galactosyloxy)ethyl methacrylate (AcGalEMA) in dioxane.[311] Using VBK as the controller in DMF solutions, oligo(ethyleneglycol) methacrylate,[312] benzyl methacrylate,[313] dimethylaminoethyl methacrylate (DMAEMA),[314,315] 2-*N* morpholinoethyl methacrylate (MEMA),[316] and 2-methacryloyloxyethylpyrrolidone (MAEPYR)[317] have resulted in polymerizations with relatively narrow molecular weight distributions ($Đ = 1.2–1.4$) and linear M_n *versus* conversion up to relatively high conversions ~0.6. Pentafluorostyrene (PFS) was also used as a controller for DMAEMA and OEGMA in toluene solutions,[307] GMA in dioxane[308] and for NLAM in dioxane solutions.[309] In the latter case, NLAM was also controlled with BlocBuilder/SG1 in dioxane with the following styrenic monomers: *p*-acetoxystyrene and 2-vinyl naphthalene.[309] It should be noted that the concentration of the controlling co-monomer required varies, depending on its chemical structure and that of the methacrylate. For VBK, as little as 1 mol% was effective in controlling MMA polymerizations at 80 °C[318]—the reason for this is not obvious (and not necessarily related to the K or propagation rate constant k_p of the VBK/nitroxide pair), as the reactivity ratios for the monomers used play a critical role.[319] Since VBK prefers to add to itself much more than to MMA, the chain ends are overwhelmingly terminated by VBK units, which makes control much more effective at low concentrations, compared to styrene. Finally, trioctylammonium *p*-styrene sulfonate (SSTOA) was used to control GMA polymerizations with NHS-BlocBuilder at 90 °C in toluene solution.[320]

4.3.3.2 Styrenic and Acrylic Polymerizations with SG1 in Organic Solvent

Styrenic and acrylic monomers in solvent with SG1-type initiators have been applied in some cases where the monomer is a solid, such as a glucosylated monomer based on PFS, which was polymerized for a biocompatible coating for medical imaging materials[321] and also used in click polymerizations.[322,323] Other functional styrenic monomers polymerized in solution include SSTOA with NHS-BlocBuilder,[320] and pentafluorophenyl 4-vinylbenzene sulfonate with BlocBuilder alone in THF at 120 °C.[324]

Acrylic monomers have been polymerized in solution with SG1-type nitroxides; most commonly butyl acrylates such as *t*-butyl acrylate[325] and acrylic acid[209–211] and associated statistical copolymers with styrene.[326–328] Other acrylates polymerized in solution with SG1-type initiators include acryloylcarbazole (VAK) in DMF.[329] Many statistical copolymers, often with styrenics, with evidence of controlled behaviour (linear M_n *versus* conversion relationship, $Đ \approx 1.2–1.4$) have been described for SG1-mediated solution polymerization such as octadecyl acrylate, resulting in semi-crystalline polymers.[330] The synthesis of star or dendritic polymers by NMP with SG1 is rarer; the flexibility of such a system was highlighted recently where not only linear block copolymers were made, but palm tree, comb and pom-pom microstructures were also derived.[267] All of these microstructures were derived from a relatively small set of starting materials, such as styrene and oligoether functional monomers like OEGMA (Figure 4.17).

4.4 NMP in Aqueous Solutions

Relatively few examples exist of homogeneous NMP in water. Although many cases have derived water-soluble polymers after appropriate de-protection of water-insoluble monomers polymerized in bulk or organic solvent, very few examples exist of homogeneous NMP in aqueous media. With TEMPO, due to the monomer restrictions, only styrenic monomers such as styrenesulfonic acid sodium salt (SSNa) were attempted (in some cases it was polymerized in an ethylene glycol/water solution).[331,332] Purely aqueous solvent systems were reported later for TEMPO-based initiators, which were used at 130 °C to initiate the polymerization of SSNa in water with a co-initiator ($Na_2S_2O_5$).[333] The resulting PSSNa homopolymers were characterized by narrow molecular weight distributions ($Đ = 1.1–1.4$) up to relatively high conversions near 90% to molecular weights approaching 10 kg mol^{-1} after relatively long polymerization times (20 h). Subsequent re-initiation with sodium acrylate in water to produce block copolymers indicated monomodal shifts in the GPC chromatograms. More recently, an intriguing

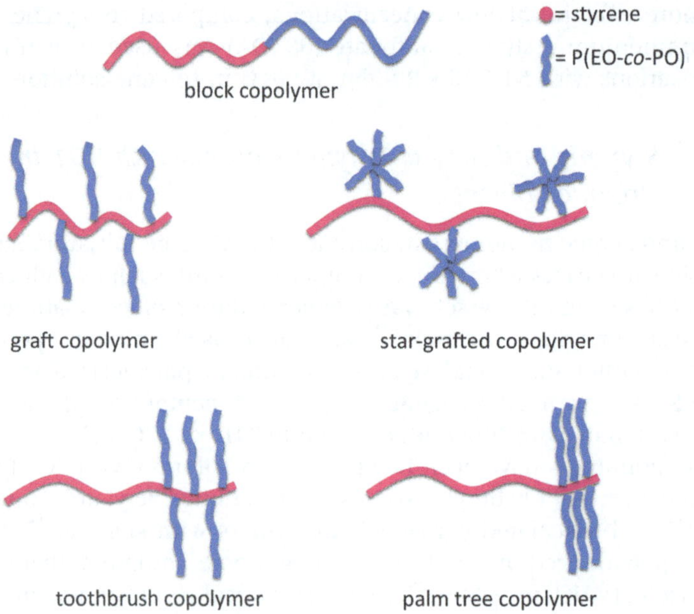

Figure 4.17 Various microstructures accessed by NMP (using either TEMPO or BlocBuilder) using a relatively simple set of monomers.[267]

room temperature photopolymerization of sodium methacrylate in water using 4-hydroxy TEMPO as the mediator and 2,2′-azobis[2-methyl-*N*-[1,1-bis(hydroxymethyl)-2-hydroxyethyl]-propionamide] as a radical source was reported.[334] The resulting polymers had relatively high $Đ \approx 2$ but the mechanism was deemed living (Figure 4.18).

Later, a TIPNO-based initiator that was modified to become water-soluble in basic form was subsequently used to initiate SSNa polymerization at 100 °C effectively.[335] Concurrently, an initiating system of water-soluble azo initiators with water-soluble TEMPO-derived nitroxides was reported to provide high molecular weight PSSNa homopolymers (M_n up to 106 kg mol^{-1}) with low $Đ$ (1.11–1.32).[336] The subsequent reinitiation of a PSSNa ($M_n = 26.6$ kg mol^{-1}, $Đ = 1.11$) with a fresh batch of SSNa in water resulted in a monomodal chain-extended species with $M_n = 100$ kg mol^{-1} and $Đ = 1.28$, suggesting excellent chain end fidelity. This is an important feature as it permitted the polymerization to be done without pressurized vessels.

With BlocBuilder-type initiators, SSNa, *N,N*-dimethylacrylamide (DMAAm) and 2-acryloxyethyl benzyldimethylammonium chloride (ABC) were polymerized cleanly with excess SG1.[337] It should be noted that without added nitroxide, the $Đ$s were high: 1.69–1.93 for SSNa but at 5.2 mol% SG1 relative to BlocBuilder, $Đ$s were reduced to 1.27–1.31 at 85 °C in 25 wt% solutions. Similarly good control was achieved for DMAAm and ABC and chain extensions revealed that the SG1 end groups were sufficiently retained (Figure 4.19).

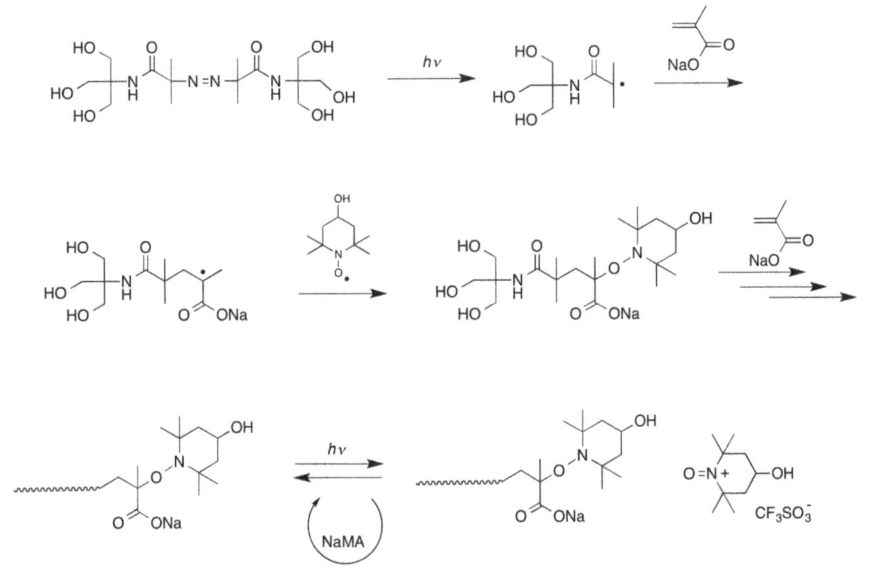

Figure 4.18 Ambient aqueous-based photopolymerization of sodium methacrylate with 4-hydroxy TEMPO.[334]

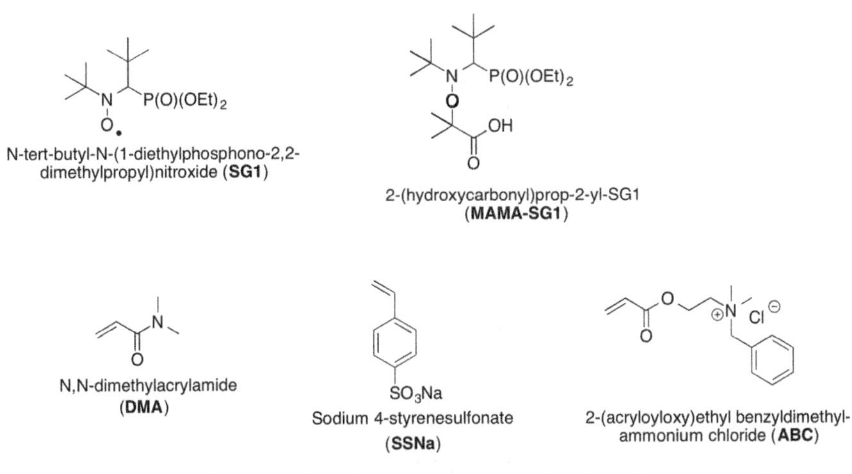

Figure 4.19 BlocBuilder-mediated aqueous polymerization of styrene sulfonic acid sodium salt (SSNa), *N,N*-dimethylacrylamide (DMAm) and 2-acryloxyethyl benzyldimethylammonium chloride (ABC).[337]

Using a bicomponent initiating system, Grassl *et al.* polymerized acrylamide in water using a dual initiator system of SG1 and water-soluble initiator VA-067.[338,339] They reported controlled polymerizations with low $Đ = 1.2–1.3$. Chain extension experiments to confirm livingness were not done, however. More recent studies have focused on using NMP in aqueous systems as macroinitiator/surfactants for mini-emulsion polymerization[340–342] or for one-pot polymerization induced self-assembly (PISA) in water.[343] Brusseau *et al.* synthesized poly(methacrylic acid-*ran*-styrene sulfonic acid) *via* BlocBuilder mediated NMP in DMSO at 76 °C[341] before conducting the same copolymerization in water at 76 °C in acidic conditions (pH = 3.5, which was unexpected for NMP of BlocBuilder). This resulting copolymer was then used as a water-soluble initiator/surfactant for subsequent emulsion copolymerization of MMA/S.[341] More recently, diethylaminoethyl methacrylate (DEAEMA), using a small percentage of acrylonitrile or styrene in the feed as a controlling co-monomer, was directly polymerization in aqueous media at pH = 6 with the succinimidyl ester form of BlocBuilder, commonly termed NHS-BlocBuilder, with no added SG1.[343] The polymerization was indicative of a controlled polymerization; linear M_n *versus* conversion and low $Đ$ (~1.3–1.4) were observed. Using these conditions, a one-pot PISA process was described using the poly(DEAEMA-*ran*-S) as the first block followed by the addition of MMA/S mixtures.[344] This recent progress hints at the potential still to be explored for NMP in aqueous systems. In particular, the recent reports for tertiary amine (meth)acrylates are particularly intriguing as they are capable of forming amidine salts, which are CO_2-responsive and can thus be applied as re-dispersible latexes, surfactants and hydrogels.[345–347]

4.5 Conclusions and Outlook

NMP in bulk or homogeneous solution polymerization in organic solvent or water has made enormous progress. Despite what superficially seems to be a gross limitation in that it can only polymerize styrenics in a controlled manner, TEMPO-based systems are highly versatile. Besides polymerization of various functional styrenics, TEMPO mediators are effective in making polymers with controlled microstructures such as grafts, block copolymers and in combining different types of polymerization mechanisms. The development of alkoxyamines such as TIPNO and the SG1 type of initiators markedly increased the versatility of NMP, permitting controlled homopolymerization of acrylates and acrylamides (and methacrylates with some manipulations). In particular, NMP in organic solvents has found distinct advantages in application for organic electronic materials. Finally, an emerging area for NMP is in its application in homogeneous aqueous media, which is yielding interesting results in areas such as polymerization induced self-assembly (PISA) and CO_2-switchable latexes.

References

1. J. Nicolas, Y. Guillaneuf, C. Lefay, D. Bertin, D. Gigmes and B. Charleux, *Prog. Polym. Sci.*, 2013, **38**, 63.
2. R. B. Grubbs, *Polym. Rev.*, 2011, **51**, 104.
3. C. J. Hawker, *Acc. Chem. Res.*, 1997, **30**, 373.
4. C. J. Hawker, A. W. Bosman and E. Harth, *Chem. Rev.*, 2001, **101**, 3661.

5. V. Sciannamea, R. Jérôme and C. Detrembleur, *Chem. Rev.*, 2008, **108**, 1104.
6. A. Studer and T. Schulte, *Chem. Rec.*, 2005, **1**, 27.
7. D. Bertin, D. Gigmes, S. R. Marque and P. Tordo, *Chem. Soc. Rev.*, 2011, **40**, 2189.
8. D. H. Solomon, *J. Polym. Sci., Part A: Polym. Chem.*, 2005, **43**, 5748.
9. T. Fukuda, T. Terauchi, A. Goto, K. Ohno, Y. Tsujii, T. Miyamoto and B. Yamada, *Macromolecules*, 1996, **29**, 6393.
10. M. K. Georges, R. P. N. Veregin, P. M. Kazmaier and G. K. Hamer, *Macromolecules*, 1993, **26**, 2987.
11. R. P. N. Veregin, M. K. Georges, P. M. Kazmaier and G. K. Hamer, *Macromolecules*, 1993, **26**, 5316.
12. C. J. Hawker, *J. Am. Chem. Soc.*, 1994, **116**, 11185.
13. C. J. Hawker and J. L. Hedrick, *Macromolecules*, 1996, **28**, 2993.
14. H. Fischer, *Macromolecules*, 1997, **30**, 5666.
15. H. Fischer, *J. Polym. Sci., Part A: Polym. Chem.*, 1999, **37**, 1885.
16. K. Ohno, Y. Tsujii and T. Fukuda, *Macromolecules*, 1997, **30**, 2503.
17. S. A. F. Bon, G. Chambard and A. L. German, *Macromolecules*, 1999, **32**, 8269.
18. A. Goto and T. Fukuda, *Macromolecules*, 1997, **30**, 4272.
19. V. Coessens, T. Pintauer and K. Matyjaszewski, *Prog. Polym. Sci.*, 2001, **26**, 337.
20. K. A. Davis, K. Matyjaszewski and H. Hocker, *Adv. Polym. Sci.*, 2002, **159**(III–XII), 1.
21. W. Braunecker and K. Matyjaszewski, *Prog. Polym. Sci.*, 2007, **32**, 93.
22. H. Fischer, *Chem. Rev.*, 2001, **101**, 3581.
23. A. Studer, *Chem. – Eur. J.*, 2001, 7, 1159.
24. J. Lokaj and P. Holler, *J. Appl. Polym. Sci.*, 2001, **80**, 2024.
25. I. Chalari, S. Pispas and N. Hadjichristidis, *J. Polym. Sci., Part A: Polym. Chem.*, 2001, **39**, 2889.
26. S. Wohlrab and D. Kuckling, *J. Polym. Sci., Part A: Polym. Chem.*, 2001, **39**, 3797.
27. X. Z. Ding, A. Fischer, A. Brembilla and P. Lochon, *J. Polym. Sci., Part A: Polym. Chem.*, 2000, **38**, 3067.
28. J. Bohrisch, U. Wendler and W. Jaeger, *Macromol. Rapid Commun.*, 1997, **18**, 975.
29. A. Fischer, A. Brembilla and P. Lochon, *Macromolecules*, 1999, **32**, 6069.
30. M. Baumann and G. Schmidt-Naake, *Macromol. Chem. Phys.*, 2000, **201**, 2751.
31. W. Devonport, L. Michalak, E. Malmstrom, M. Mate, B. Kurdi, C. J. Hawker, G. G. Barclay and R. Sinta, *Macromolecules*, 1997, **30**, 1929.
32. P. Lacroix-Desmazes, T. Delair, C. Pichot and B. Boutevin, *J. Polym. Sci., Part A: Polym. Chem.*, 2000, **38**, 3845.
33. K. Ohno, M. Ejaz, T. Fukuda, T. Miyamoto and T. Shimizu, *Macromol. Chem. Phys.*, 1998, **199**, 291.
34. E. Yoshida and Y. Tchiguchi, *Polym. J.*, 1999, **31**, 429.

35. G. G. Barclay, C. J. Hawker, H. Ito, A. Orellana, P. R. L. Malenfant and R. F. Sinta, *Macromolecules*, 1998, **31**, 1024.
36. J. D. Dang, J. D. Virdo, B. H. Lessard, E. Bultz, A. S. Paton and T. P. Bender, *Macromolecules*, 2012, **45**, 7791.
37. H. Otsuka, T. Sato and T. Endo, *Macromol. Rapid Commun.*, 2000, **21**, 48–52.
38. M. K. Gray, H. Zhou, S. T. Nguyen and J. M. Torkelson, *Macromolecules*, 2004, **37**, 5586.
39. S. Pragliola, C. K. Ober, P. T. Mather and H. G. Jeon, *Macromol. Chem. Phys.*, 1999, **200**, 2338.
40. M. K. Gray, M. I. Kinsinger and J. M. Torkelson, *Macromolecules*, 2002, **35**, 8261.
41. E. Yoshida, *J. Polym. Sci., Part A: Polym. Chem.*, 1996, **34**, 2937.
42. L. B. Sessions, B. R. Cohen and R. B. Grubbs, *Macromolecules*, 2007, **40**, 1926.
43. K. Daimon, P. M. Kazmaier, M. K. Georges, G. K. Hamer and R. P. N. Veregin, *ACS Polym. Prepr.*, 1997, **38**, 653.
44. L. B. Sessions, L. A. Milnea, K. D. Ericson, D. S. Glueck and R. B. Grubbs, *Macromolecules*, 2003, **38**, 2116.
45. L.-C. Gao, X. H. Fan, Z.-H. Shen, X. Chen and Q.-F. Zhou, *J. Polym. Sci., Part A: Polym. Chem.*, 2009, **47**, 319.
46. B. Boutevin, Y. Hervaud, A. Boulahna and M. El. Asri, *Macromolecules*, 2002, **35**, 6511.
47. M. L. Coote and T. P. Davis, *Macromolecules*, 1999, **32**, 4290.
48. Y. Kotani, M. Kamigaito and M. Sawamoto, *Macromolecules*, 2000, **33**, 6746.
49. J. Qiu and K. Matyjaszewski, *Macromolecules*, 1997, **30**, 5643.
50. D. Li and W. J. Brittain, *Macromolecules*, 1998, **31**, 3852.
51. A. Fischer, A. Brembilla and P. Lochon, *Eur. Polym. J.*, 2001, **37**, 33.
52. C. Lefay, B. Charleux, M. Save, C. Chassenieux, O. Guerret and S. Magnet, *Polymer*, 2006, **47**, 1935.
53. B. Charleux, J. Nicolas and O. Guerret, *Macromolecules*, 2005, **38**, 5485.
54. M. Baumann and G. Schmidt-Naake, *Macromol. Chem. Phys.*, 2001, **202**, 2727.
55. M. Baumann, A. I. Roland, G. Schmidt-Naake and H. Fischer, *Macromol. Mater. Eng.*, 2000, **280/281**, 1.
56. M. Baumert and R. Mulhaupt, *Macromol. Rapid Commun.*, 1997, **18**, 787.
57. J. Lokaj, L. Brozova, P. Holler and Z. Pientka, *Collect. Czech. Chem. Commun.*, 2002, **67**, 26.
58. R. Cuervo-Rodriguez, C. Fernandez-Monreal and E. L. Madruga, *J. Polym. Sci., Part A: Polym. Chem.*, 2002, **40**, 2750.
59. H. Jianying, C. Jiayan, Z. Jiaming, C. Yihong, D. Lizong and Z. Yousi, *J. Appl. Polym. Sci.*, 2006, **100**, 3531.
60. M. Appelt and G. Schmidt-Naake, *Macromol. Chem. Phys.*, 2004, **205**, 637.

61. M. Baumert, J. Fröhlich, M. Stieger, H. Frey, R. Mülhaupt and H. Plenio, *Macromol. Rapid Commun.*, 1999, **20**, 203.
62. H. Baethge, S. Butz and G. Schmidt-Naake, *Macromol. Rapid Commun.*, 1997, **18**, 911.
63. H. Baethge, S. Butz, C.-H. Han and G. Schmidt-Naake, *Angew. Makromol. Chem.*, 1999, **267**, 52.
64. G. Schmidt-Naake and S. Butz, *Macromol. Rapid Commun.*, 1996, **17**, 661.
65. S. Butz, J. Hillermann, G. Schmidt-Naake, J. Kressler, R. Thomann, B. Heck and B. Stühn, *Acta Polym.*, 1998, **49**, 693.
66. N. Arsalani, H. Fattahi and A. A. Entezami, *Iran. Polym. J.*, 2006, **15**, 997.
67. J. Kim, M. M. Mok, R. W. Sandoval, D. J. Woo and J. M. Torkelson, *Macromolecules*, 2006, **39**, 6152.
68. K. Ohno, M. Ejaz, T. Fukuda, T. Miyamoto and Y. Shimizu, *Macromol. Chem. Phys.*, 1998, **199**, 291.
69. P. Lacroix-Desmazes, P. Andre, J. M. Desimone, A. V. Ruzette and B. Boutevin, *J. Polym. Sci., Part A: Polym. Chem.*, 2004, **42**, 3537.
70. L. B. Sessions, L. A. Miinea, K. D. Ericson, D. S. Gluek and R. B. Grubbs, *Macromolecules*, 2005, **38**, 2116.
71. B. Boutevin, Y. Hervaud, A. Boulahna and M. El Asri, *Macromolecules*, 2002, **35**, 6511.
72. K. Ohno, Y. Izu, Y. Tsujii, T. Fukuda and H. Kitano, *Eur. Polym. J.*, 2004, **40**, 81.
73. M. Bouix, J. Gouzi, B. Charleux, J.-P. Vairon and P. Guinot, *Macromol. Rapid Commun.*, 1998, **19**, 209.
74. A. Narumi, T. Matsuda, H. Kaga, T. Satoh and T. Kakuchi, *Polym. J.*, 2001, **33**, 939.
75. Y. Chen and G. Wulff, *Macromol. Chem. Phys.*, 2001, **202**, 3273.
76. C. A. Barbosa and A. S. Gomes, *Polym. Bull.*, 1998, **41**, 15.
77. Z. Yousi, L. Jian, Z. Rongchua, Y. Jianliang, D. Lizong and Z. Lansun, *Macromolecules*, 2000, **33**, 4745.
78. M. Baumert, J. Frohlich, M. Steiger, H. Frey, R. Mulhaupt and H. Plennio, *Macromol. Rapid Commun.*, 1999, **20**, 203.
79. H. Okamura, Y. Takatori, M. Tsunooka and M. Shirai, *Polymer*, 2002, **43**, 3155.
80. M. Mariani, M. Lelli, K. Sparnacci and M. Laus, *J. Polym. Sci., Part A: Polym. Chem.*, 1999, **37**, 1237.
81. S. Meuer and R. Zentel, *Macromol. Chem. Phys.*, 2008, **209**, 158.
82. W. Lin, Y. Zheng, J. Zhang and X. Wan, *Macromolecules*, 2011, **44**, 5146.
83. L. I. Gabaston, S. A. Furlong, R. A. Jackson and S. P. Armes, *Polymer*, 1999, **40**, 4505.
84. E. Yoshida, T. Ishizone, A. Hirao, S. Nakahama, T. Takata and T. Endo, *Macromolecules*, 1994, **27**, 3119.
85. E. Yoshida and A. Sugita, *Macromolecules*, 1996, **29**, 6422.
86. E. Yoshida and Y. Osagawa, *Macromolecules*, 1998, **31**, 1446.

87. E. Yoshida and M. Nakamura, *Polym. J.*, 1998, **30**, 915.
88. N. A. A. Rossi, R. M. Anderson, R. G. Jones and S. J. Holder, *Polym. Int.*, 2004, **53**, 465.
89. Z. Li, G. Lu and J. Huang, *J. Appl. Polym. Sci.*, 2004, **94**, 2280.
90. S.-C. Chan, S.-W. Kuo, C.-H. Lu, H.-F. Lee and F.-C. Chang, *Polymer*, 2007, **48**, 5059.
91. G. Q. Lu, Z. F. Jia, W. Yi and J. L. Huang, *J. Polym. Sci., Part A: Polym. Chem.*, 2002, **40**, 4404.
92. X. Chen, B. Gao, J. Kops and W. Batsberg, *Polymer*, 1998, **39**, 911.
93. Y. B. Wang, S. Chen and J. L. Huang, *Macromolecules*, 1999, **32**, 2480–2483.
94. R. Jabbar, A. Graffe and M. Maric, *J. Appl. Polym. Sci.*, 2008, **109**, 3185.
95. S. Kobatake, H. J. Harwood, R. P. Quirk and D. B. Priddy, *Macromolecules*, 1998, **31**, 3735.
96. Y. Miura, K. Hirota, H. Moto and B. Yamada, *Macromolecules*, 1999, **32**, 8356.
97. A. M. Morgan, S. K. Pollack and K. Beshah, *Macromolecules*, 2002, **35**, 4238.
98. L. Búcsiová, M. Yin, S. Chmela and W. D. Habicher, *J. Macromol. Sci., Part A: Pure Appl. Chem.*, 2008, **45**, 761.
99. E. Yoshida and T. Tanaka, *Colloid Polym. Sci.*, 2008, **286**, 827.
100. T. Janoschka, M. D. Hager and U. S. Schubert, *Adv. Mater.*, 2012, **24**, 6397.
101. R. D. Puts and D. Y. Sogah, *Macromolecules*, 1997, **30**, 7050.
102. T. He, D. Li, X. Sheng and B. Zhao, *Macromolecules*, 2004, **37**, 3128.
103. Y. Shi, Z. Fu, B. Li and W. Yang, *J. Polym. Sci., Part A: Polym. Chem.*, 2006, **44**, 2468.
104. W. Tang, J. He and Y. Yang, *J. Macromol. Sci. A: Polym. Chem.*, 2006, **43**, 1553.
105. C. Thomas, H. Maldonado-Textle, A. Rockenbauer, L. Korecz, N. Nagy and R. Guerrero-Santos, *J. Polym. Sci., Part A: Polym. Chem.*, 2012, **50**, 2944.
106. C. Thomas, H. Maldonado-Textle, J. N. Cabello-Romero, J. Macossay, X. Zhang, N. Esturau-Escofet and R. Guerrero-Santos, *Polym. Chem.*, 2014, **5**, 3089.
107. C. J. Hawker, D. Mecerreyes, E. Eke, J. Duo, J. L. Hedrick, I. Barakat, P. Dubois, R. Jerome and W. Volksen, *Macromol. Chem. Phys.*, 1998, **198**, 155.
108. Y. Wang, G. Lu and J. Huang, *J. Polym. Sci., Part A: Polym. Chem.*, 2004, **42**, 2093.
109. J. Ryan, F. Aldabbagh, P. B. Zetterlund and B. Yamada, *React. Funct. Polym.*, 2008, **68**, 692.
110. X. Chen, B. Gholamkhass, X. Han, G. Vamvounis and S. Holdcroft, *Macromol. Rapid Commun.*, 2007, **28**, 1792.
111. B. Gholamkhass, T. J. Peckham and S. Holdcroft, *Polym. Chem.*, 2010, **1**, 708.

112. Y. Miwa, K. Yamamoto, M. Sakaguchi and S. Shimada, *Macromolecules*, 1999, **32**, 8234.
113. F. Sun, Z. Fu, Q. Deng and Z. Fan, *J. Appl. Polym. Sci.*, 2009, **112**, 275.
114. E.-S. Park, H.-J. Jin, I.-M. Lee, M.-N. Kim, H. S. Lee and J.-S. Yoon, *J. Appl. Polym. Sci.*, 2002, **83**, 1103.
115. M. Abbasian, H. Namazi and A. A. Entezami, *Polym. Adv. Technol.*, 2004, **15**, 606.
116. H. Zhang, Z. Guo and J. Huang, *J. Polym. Sci., Part A: Polym. Chem.*, 2002, **40**, 4398.
117. Y. Tao, J. He, Z. Wang, J. Pan, H. Jiang, S. Chen and Y. Yang, *Macromolecules*, 2001, **34**, 4742.
118. F. J. Hua, B. Liu, C. P. Hu and Y. L. Yang, *J. Polym. Sci., Part A: Polym. Chem.*, 2002, **40**, 1876.
119. H. Çatalgil-Giz and Y. Hepuzer, *J. Appl. Polym. Sci.*, 2000, 77, 1950.
120. N. Hadjichristidis, M. Pitsikalis, H. Iatrou, P. Driva, G. Sakellariou and M. Chatzichristidi, 6.03 – Polymers with Star-Related Structures: Synthesis, Properties, and Applications, in *Polymer Science: A Comprehensive Reference*, ed. K. Matyjaszewski and M. Möller, Elsevier, Amsterdam, 2012, pp. 29–111.
121. C. J. Hawker, *Angew. Chem., Int. Ed. Engl.*, 1995, **34**, 1456.
122. G. Chessa, A. Scrivanti, U. Matteoli and V. Castelvetro, *Polymer*, 2001, **52**, 9347.
123. Y. Miura and Y. Yoshida, *Polym. J.*, 2002, **34**, 748.
124. Y. Miura and Y. Yoshida, *Macromol. Chem. Phys.*, 2002, **203**, 879.
125. T. Tsoukatos, S. Pispas and N. Hadjichristidis, *J. Polym. Sci., Part A: Polym. Chem.*, 2001, **39**, 320.
126. Y. Miura and H. Dote, *J. Polym. Sci., Part A: Polym. Chem.*, 2005, **43**, 3689.
127. Y. Miura, H. Dote, H. Kubonishi, K. Fukuda and T. Saka, *J. Polym. Sci., Part A: Polym. Chem.*, 2007, **45**, 1159.
128. X. Li, T. Jeanmaire and A. Zerroukhi, *J. Appl. Polym. Sci.*, 2008, **107**, 3491.
129. D. Abdallah, M. A. A. Ghani, M. F. Cunningham, P. M. Kazmaier, B. Keoshkerian and E. Buncel, *Can. J. Chem.*, 2004, **82**, 1393.
130. M. A. A. Ghani, D. Abdallah, P. M. Kazmaier, B. Keoshkerian and E. Buncel, *Can. J. Chem.*, 2004, **82**, 1403.
131. C. A. Knoop and A. Studer, *J. Am. Chem. Soc.*, 2003, **125**, 16327.
132. A. Studer, K. Harms, C. Knoop, C. Müller and T. Schulte, *Macromolecules*, 2004, **37**, 127.
133. Y. Jing, A. Mardyukov, K. Bergander, C. G. Damiliuc and A. Studer, *Macromolecules*, 2014, **47**, 3595.
134. K. Matyjaszewski, S. Gaynor, D. Greszta, D. Mardare and T. J. Shigemoto, *Phys. Org. Chem.*, 1995, **8**, 306.
135. K. Matyjaszewski, S. Gaynor, D. Greszta, D. Mardare and T. J. Shigemoto, *Macromol. Symp.*, 1995, **98**, 73–89.

136. A. Goto, Y. Kwak, C. Yoshikawa, Y. Tsujii, Y. Sugiura and T. Fukuda, *Macromolecules*, 2002, **35**, 3520.
137. E. Harth, B. Van Horn and C. J. Hawker, *Chem. Commun.*, 2001, 823–824.
138. D. A. Shipp, *J. Macromol. Sci., Part C: Polym. Rev.*, 2005, **45**, 171.
139. D. Benoit, V. Chaplinski, R. Braslau and C. J. Hawker, *J. Am. Chem. Soc.*, 1999, **121**, 3904.
140. D. Benoit, E. Harth, P. Fox, R. M. Waymouth and C. J. Hawker, *Macromolecules*, 2000, **33**, 363.
141. D. Benoit, S. Grimaldi, S. Robin, J.-P. Finet, P. Tordo and Y. Gnanou, *J. Am. Chem. Soc.*, 2000, **122**, 5929.
142. E. Harth, B. Van Horn, V. Y. Lee, D. S. Germack, C. P. Gonzales, R. D. Miller and C. J. Hawker, *J. Am. Chem. Soc.*, 2002, **124**, 8653.
143. H. Nulwala, D. J. Burke, A. Khan, A. Serrano and C. J. Hawker, *Macromolecules*, 2010, **43**, 5474.
144. R. K. O'Reilly, M. J. Joralemon, C. J. Hawker and K. L. Wooley, *Chem. – Eur. J.*, 2006, **12**, 6776.
145. M. Rodlert, E. Harth, I. Rees and C. J. Hawker, *J. Polym. Sci., Part A Polym. Chem.*, 2000, **38**, 4749.
146. R. K. O'Reilly, M. J. Joralemon, C. J. Hawker and K. L. Wooley, *Chem. – Eur. J.*, 2006, **12**, 6776.
147. C. Cheng, K. Qi, E. Khoshdel and K. L. Wooley, *J. Am. Chem. Soc.*, 2006, **128**, 6808.
148. A. Tsimelzon, D. Deamer and R. Braslau, *Macromol. Rapid. Commun.*, 2005, **26**, 1872.
149. N. L. Hill and R. Braslau, *Macromolecules*, 2005, **38**, 9066.
150. J. Ruehl, A. Nilsen, S. Born, P. Thoniyot, L.-P. Xu, S. Chen and R. Braslau, *Polymer*, 2007, **48**, 2564.
151. U. Stalmach, B. de Boer, A. D. Post, P. F. van Hutten and G. Hadziioannou, *Angew. Chem., Int. Ed.*, 2001, **40**, 428.
152. B. G. G. Lohmeijer and U. S. Schubert, *J. Polym. Sci., Part A: Polym. Chem.*, 2005, **43**, 6331.
153. L. Marx and P. Hemery, *Polymer*, 2009, **50**, 2752, nBA.
154. L. Marx, Z. Eskandani and P. Hemery, *React. Funct. Polym.*, 2009, **69**, 306, nBA.
155. B. Gu, S. Liu, J. D. Leber and A. Sen, *Macromolecules*, 2004, **37**, 5142.
156. Y. Miura and K. Miyake, *J. Polym. Sci., Part A: Polym. Chem.*, 2005, **43**, 6153.
157. D. Benoit, C. J. Hawker, E. E. Huang, Z. Lin and T. P. Russell, *Macromolecules*, 2000, **33**, 1505.
158. R. Nicolay, L. Marx, P. Hemery and K. Matyjaszewski, *Macromolecules*, 2007, **40**, 6067.
159. N. S. Lee, Y. Li, C. M. Ruda and K. L. Wooley, *Chem. Commun.*, 2008, **42**, 5339.
160. K.-I. Fukukawa, L. Zhu, P. Gopalan, M. Ueda and S. Yang, *Macromolecules*, 2005, **38**, 263.

161. C. Ott, B. G. G. Lohmeigher, D. Wouters and U. S. Schubert, *Macromol. Chem. Phys.*, 2006, **207**, 1439.
162. A. Britze, K. Moosmann, E. Jahne, H. J. Adler and D. Kuckling, *Macromol. Rapid Commun.*, 2006, **27**, 1906.
163. D. C. Tully, M. J. Roberts, B. H. Geierstanger and R. B. Grubbs, *M*acromolecules, 2003, **36**, 4302.
164. M. Bothe and G. Schmidt-Naake, *Macromol. Chem. Phys.*, 2004, **205**, 208.
165. J. Ruehl, N. L. Hill, E. D. Walter, G. Millhauser and R. Braslau, *Macromolecules*, 2008, **41**, 1972.
166. K. Van De Wetering, C. Brochon, C. Ngov and G. Hadziioannou, *Macromolecules*, 2006, **39**, 4289.
167. B. G. G. Lohmeijer and U. S. Schubert, *J. Polym. Sci., Part A: Polym. Chem.*, 2003, **41**, 1413.
168. B. G. G. Lohmeijer and U. S. Schubert, *J. Polym. Sci., Part A: Polym. Chem.*, 2004, **42**, 4016.
169. B. G. G. Lohmeijer and U. S. Schubert, *J. Polym. Sci., Part A: Polym. Chem.*, 2005, **43**, 6331.
170. C. Ott, B. G. G. Lohmeijer and U. S. Schubert, *Macromol. Chem. Phys.*, 2006, **207**, 1439.
171. C. Ott, C. Ulbricht, R. Hoogenboom and U. S. Schubert, *Macromol. Rapid Commun.*, 2012, **33**, 556.
172. C. Ott, R. Hoogenboom, S. Hoeppener, D. Wouters, J.-F. Gohy and U. S. Schubert, *Soft Matter*, 2009, **5**, 84.
173. S. Fleischmann, H. Komber, D. Appelhans and B. I. Voit, *Macromol. Chem. Phys.*, 2007, **208**, 1050.
174. A. Britze, V. Möllmann, G. Grundmeier, H. Luftmann and D. Kuckling, *Macromol. Chem. Phys.*, 2011, **212**, 679.
175. J. Ruehl, C. Morimoto, D. J. Stevens and R. Braslau, *React. Funct. Polym.*, 2008, **68**, 1563.
176. E. Harth, A. Bosman, D. Benoit, B. Helms, J. M. J. Fréchet and C. J. Hawker, *Macromol. Symp.*, 2001, **174**, 85.
177. L. Marx, Z. Eskandani and P. Hémery, *React. Funct. Polym.*, 2009, **69**, 306.
178. P. Souchal, T. Roels, P. Hémery and L. Marx, *J. Polym. Sci., Part A: Polym. Chem.*, 2012, **50**, 4073.
179. U. Meyer, F. Svec, J. M. J. Frechet, C. J. Hawker and K. Irgum, *Macromolecules*, 2000, **33**, 7769.
180. C. Farcet, J. Belleney, B. Charleux and R. Pirri, *Macromolecules*, 2002, **35**, 4912.
181. C. Farcet, B. Charleux and R. Pirri, *Macromolecules*, 2001, **34**, 3823.
182. K. Bian and M. F. Cunningham, *Macromolecules*, 2005, **38**, 695.
183. K. Bian and M. F. Cunningham, *J. Polym. Sci., Part A: Polym. Chem.*, 2006, **44**, 414.
184. B. Lessard and M. Maric, *Macromolecules*, 2008, **41**, 7180.

185. B. Lessard, C. Tervo and M. Marić, *Macromol. React. Eng.*, 2009, **3**, 245.
186. T. M. Eggenhuisen, C. Renzi Becer, M. W. M. Fijten, R. Eckhardt, R. Hoogenboom and U. S. Schubert, *Macromolecules*, 2008, **41**, 5132.
187. R. Hoogenboom, D. Popescu, W. Steinjhauer, H. Keul and M. Moller, *Macromol. Rapid Commun.*, 2009, **30**, 2042.
188. C. G. Gamys, E. Beyou, E. Bourgeat-Lami, L. David and P. Alcouffe, *Macromol. Chem. Phys.*, 2012, **213**, 10.
189. K. Karaky, E. Péré, C. Pouchan, H. Garay, A. Khoukh, J. François, J. Desbrières and L. Billon, *New J. Chem.*, 2006, **30**, 698.
190. T. Diaz, A. Fischer, A. Jonquirères, A. Brembilla and P. Lochon, *Macromolecules*, 2003, **36**, 2235.
191. S. Grimaldi, J.-P. Finet, F. Le Moigne, A. Zeghdaoui, P. Tordo, D. Benoit, M. Fontanille and Y. Gnanou, *Macromolecules*, 2000, **33**, 1141.
192. A. Goto and T. Fukuda, *Prog. Polym. Sci.*, 2004, **29**, 329.
193. S. Robin, O. Guerret, J.-L. Couturier, R. Pirri and Y. Gnanou, *Macromolecules*, 2002, **35**, 3844.
194. F. Chauvin, P. E. Dufils, D. Gigmes, Y. Guillaneuf, S. R. Marque, P. Tordo and D. Bertin, *Macromolecules*, 2006, **39**, 5238.
195. J. Nicolas, B. Charleux, O. Guerret and S. Magnet, *Angew. Chem., Int. Ed.*, 2004, **43**, 6186.
196. D. Gigmes, J. Vinas, N. Chagneux, C. Lefay, T. N. T. Phan, T. Trimaille, P.-E. Dufils, Y. Guillaneuf, G. Carrot, F. Boue and D. Bertin, *ACS Symp. Ser.*, 2009, **1024**, 245.
197. B. Lessard, C. Aumand-Bourque, R. Chaudhury, D. Gomez, A. Haroon, N. Ibrahimian, S. Mackay, M.-C. Noel, R. Patel, S. Sitaram, S. Valla, B. White and M. Marić, *Int. Polym. Process.*, 2011, **26**, 197–204.
198. J. Nicolas, C. Dire, L. Mueller, J. Belleney, B. Charleux, S. R. A. Marque, D. Bertin, S. Magnet and L. Couvreur, *Macromolecules*, 2006, **39**, 8274.
199. J. Nicolas, S. Brusseau and B. Charleux, *J. Polym. Sci., Part A: Polym. Chem.*, 2010, **48**, 34.
200. A.-V. Ruzette, S. Tencé-Girault, L. Leibler, F. Chauvin, D. Bertin, O. Guerret and P. Gérard, *Macromolecules*, 2006, **39**, 5804.
201. Y. Guillaneuf, D. Gigmes, S. R. A. Marque, P. Astolfi, L. Greci, P. Tordo and D. Bertin, *Macromolecules*, 2007, **40**, 3108.
202. K. A. Payne, P. Nesvadba, J. Debling, M. F. Cunningham and R. A. Hutchinson, *ACS Macro Lett.*, 2015, **4**, 280.
203. A. C. Greene and R. B. Grubbs, *Macromolecules*, 2009, **42**, 4388.
204. A. C. Greene and R. B. Grubbs, *Macromolecules*, 2010, **43**, 10320.
205. T. Fukuyama, Y. Kajihara, I. Ryu and A. Studer, *Synthesis*, 2012, **44**, 2555–2559.
206. C. Detrembleur, C. Jérôme, J. De Winter, P. Gerbaux, J.-L. Clément, Y. Guillaneuf and D. Gigmes, *Polym. Chem.*, 2014, **5**, 335.
207. C. Detrembleur, P. Teyssie and R. Jérôme, *Macromolecules*, 2002, **35**, 1611–1621.

208. C. Detrembleur, M. Claes and R. Jérôme, *Advances in Controlled/Living Radical Polymerization*, ACS Symposium Series 854, 2009, ch. 35, pp. 496–518.
209. S. Magnet, O. Guerret and J.-L. Couturier, *Eur. Pat. Appl.*, EP 1526138 A1 20050427, 2005.
210. P.-E. Dufils, N. Chagneux, D. Gigmes, T. Trimaille, S. R. A. Marque, D. Bertin and P. Tordo, *Polymer*, 2007, **48**, 5219.
211. D. Gigmes, P.-E. Dufils, D. Glé, D. Bertin, C. Lefay and Y. Guillaneuf, *Polym. Chem.*, 2011, **2**, 1624.
212. C. Wetter, K. Jantos, K. Woithe and A. Studer, *Org. Lett.*, 2003, **5**, 2899.
213. C. Leroi, D. Bertin, P.-E. Dufils, D. Gigmes, S. Marque, S. P. Tordo, J.-L. Couturier, O. Guerret and M. A. Ciufolini, *Org. Lett.*, 2003, **5**, 4943.
214. S.-J. Mougnier, C. Brochon, E. Cloutet, S. Magnet, C. Navarro and G. Hadziioannou, *J. Polym. Sci., Part A: Polym. Chem.*, 2012, **50**, 2463.
215. J. Nicolas, B. Charleux and S. Magnet, *J. Polym. Sci., Part A: Polym. Chem.*, 2006, **44**, 4142.
216. J. Nicolas, B. Charleux, O. Guerret and S. Magnet, *Macromolecules*, 2005, **38**, 9963.
217. S. Magnet, O. Guerret and J.-L. Couturier, *Eur. Pat. Appl. EP 1 526 138*, 2005.
218. D. Quémener, G. Bonniol, T. N. T. Phan, D. Gigmes, D. Bertin and A. Deratani, *Macromolecules*, 2010, **43**, 5060.
219. B. Clement, T. Trimaille, O. Alluin, D. Gigmes, K. Mabrouk, F. Feron, P. Decherchi, T. Marqueste and D. Bertin, *Biomacromolecules*, 2009, **10**, 1436.
220. E. Bloch, T. Phan, D. Bertin, P. Llewellyn and V. Hornebecq, *Microporous Mesoporous Mater.*, 2008, **112**, 612.
221. E. Bloch, P. L. Llewellyn, T. Phan, D. Bertin and V. Hornebecq, *Chem. Mater.*, 2009, **21**, 48.
222. M. Girod, M. Mazarin, T. N. T. Phan, D. Gigmes and L. Charles, *J. Polym. Sci., Part A: Polym. Chem.*, 2009, **47**, 3380.
223. C. Barrere, M. Mazarin, R. Giordanengo, T. Phan, A. Thevand, S. Viel and L. Charles, *Anal. Chem.*, 2009, **81**, 8054.
224. F. A. Bonk, S. Caldarelli, T. Phan, D. Bertin, E. R. Deazevedo, G. L. Mantovani, T. J. Bonagamba, T. S. Plivelic and I. L. Torriani, *J. Polym. Sci., Part B: Polym. Phys.*, 2010, **48**, 55.
225. J. Vinas, N. Chagneux, D. Gigmes, T. Trimaille, A. Favier and D. Bertin, *Polymer*, 2008, **49**, 3639.
226. M. Chenal, C. Boursier, Y. Guillaneuf, M. Taverna, P. Couvreur and J. Nicolas, *Polym. Chem.*, 2011, **2**, 1523.
227. C. Zhang and M. Marić, *J. Polym. Sci., Part A: Polym. Chem.*, 2012, **50**, 4341.
228. S. Harrisson, J. Nicolas, A. Maksimenko, D. T. Bui, J. Mougin and P. Couvreur, *Angew. Chem., Int. Ed.*, 2013, **52**, 1678.

229. C. Detrembleur, A. Debuigne, C. Jerôme, T. N. T. Phan, D. Bertin and D. Gigmes, *Macromolecules*, 2009, **42**, 8604.
230. A. Favier, B. Luneau, J. Vinas, N. Lassaoui, D. Gigmes and D. Bertin, *Macromolecules*, 2009, **42**, 5953.
231. Y. Guillaneuf, P.-E. Dufils, L. Autissier, M. Rollet, D. Gigmes and D. Bertin, *Macromolecules*, 2010, **43**, 91.
232. M. D. Saban, M. K. Georges, R. P. N. Veregin, G. K. Hamer and P. M. Kazmaier, *Macromolecules*, 1995, **28**, 7032.
233. M. K. Georges, R. A. Kee, R. P. N. Veregin, G. K. Hamer and P. M. Kazmaier, *J. Phys. Org. Chem.*, 1995, **8**, 301.
234. L. Couvreur, C. Lefay, J. Belleney, B. Charleux, O. Guerret and S. Magnet, *Macromolecules*, 2003, **36**, 8260.
235. L. Couvreur, C. Lefay and B. Charleux, *Macromol. Chem. Phys.*, 2003, **204**, 2055.
236. C. Lefay, J. Belleney, B. Charleux, O. Guerret and S. Magnet, *Macromol. Rapid Commun.*, 2004, **25**, 1215.
237. E. Harth, A. Bosman, D. Benoit, B. Helms, J. M. J. Frechet and C. J. Hawker, *Macromol. Symp.*, 2001, **174**, 85.
238. A. W. Bosman, R. Vestberg, A. Heumann, J. M. J. Frechet and J. C. J. Hawker, *J. Am. Chem. Soc.*, 2003, **125**, 715.
239. W. H. Binder, D. Gloger, H. Weinstabl, G. Allmaier and E. Pittenauer, *Macromolecules*, 2007, **40**, 3097.
240. T. Schulte, K. O. Siegenthaler, H. Luftmann, M. Letzel and A. Studer, *Macromolecules*, 2005, **38**, 6833.
241. O. Gibbons, W. M. Carroll, F. Aldabbagh and B. Yamada, *J. Polym. Sci., Part A: Polym. Chem.*, 2006, **44**, 6410.
242. A. J. Convertine, N. Ayres, C. W. Scales, A. B. Lowe and C. L. McCormick, *Biomacromolecules*, 2004, **5**, 1177.
243. X. Yan, N. A. D. Burke and H. D. H. Stöver, *Macromolecules*, 2006, **39**, 2275.
244. C. Y. Li, K. K. Tenneti, D. Zhang, H. Zhang, X. Wan, E.-Q. Chen, Q.-F. Zhou, A.-O. Carlos, S. Igos and B. S. Hsiao, *Macromolecules*, 2004, **37**, 2854.
245. P. Gopalan, Y. Zhang, X. Li, U. Wiesner and C. K. Ober, *Macromolecules*, 2003, **36**, 3357.
246. S. Pragliola, C. K. Ober, P. T. Mather and H. G. Jeon, *Macromol. Chem. Phys.*, 1999, **200**, 2338.
247. C.-F. Li, Q.-K. Zhang, H.-J. Tian, C.-P. Chai, L.-C. Gao, Z. Shen, X.-H. Fan and Y.-J. Luo, *J. Polym. Sci., Part A: Polym. Chem.*, 2015, **52**, 1737.
248. X.-H. Wan, H.-L. Tu, Y.-F. Tu, D. Zhange, L. Sun, Q.-F. Zhou, Y.-P. Dong and M. Tang, *Chin. J. Polym. Sci.*, 1999, **17**, 189.
249. X.-H. Wan, Y.-F. Tu, D. Zhang and Q.-F. Zhou, *Polym Int.*, 2000, **49**, 243.
250. Y.-F. Tu, X.-F. Wan, D. Zhang, Q.-F. Zhou and C. Wu, *J. Am. Chem. Soc.*, 2000, **122**, 10201.
251. Y.-F. Tu, X. Wan, H. Zhang, X. Fan, X. Chen, Q.-F. Zhou and K. Chau, *Macromolecules*, 2003, **36**, 6565.

252. M. Behl, E. Hattemer, M. Brehmer and R. Zentel, *Macromol. Chem. Phys.*, 2002, **203**, 503–510.
253. J. D. Quinn and R. A. Register, *Polym. Adv. Technol.*, 2008, **19**, 556.
254. B. Ma, B. J. Kim, L. Deng, D. A. Poulsen, M. E. Thompson and J. M. J. Frechet, *Macromolecules*, 2007, **40**, 8156.
255. G. Moad, M. Chen, M. Haussler, A. Postma, E. Rizzardo and S. H. Thang, *Polym. Chem.*, 2011, **2**, 492.
256. A. Goto and T. Fukuda, *Macromolecules*, 1999, **32**, 618.
257. N. A. Listigovers, M. K. Georges, P. G. Odell and B. Keoshkerian, *Macromolecules*, 1996, **29**, 8992.
258. M. K. Georges, J. L. Lukkarila and A. R. Zhurkan, *Macromolecules*, 2004, **37**, 1297.
259. D. Chan-Seng, A. Debuigne and M. K. Georges, *Eur. Polym. J.*, 2009, **45**, 211.
260. C. Burguière, M. Dourges, B. Charleux and J. P. Vairon, *Macromolecules*, 1999, **32**, 3883.
261. C. L. Araujo and J. C. Pinto, *Macromol. Symp.*, 2013, **333**, 62.
262. S. Butz, H. Baethge and G. Schmidt-Naake, *Macromol. Chem. Phys.*, 2000, **201**, 2143.
263. G. S. Ananchenko and H. Fischer, *J. Polym. Sci., Part A: Polym. Chem.*, 2001, **39**, 3604.
264. G. S. Ananchenko, M. Souaille, H. Fischer, C. LeMercier and P. Tordo, *J. Polym. Sci., Part A: Polym. Chem.*, 2002, **40**, 3264.
265. D. M. Johnson, N. A. Clarke and P. G. Rasmussen, *J. Macromol. Sci., Part A: Pure Appl. Chem.*, 2001, **38**, 773.
266. J. Bonilla-Cruz, E. Saldívar-Guerra, J. R. Torres-Lubián, R. Guerrero-Santos, B. López-Carpy and G. Luna-Bárcenas, *Macromol. Chem. Phys.*, 2008, **209**, 2268.
267. L. Petton, E. P. C. Mes, H. Van Der Wal, S. Claessens, F. Van Damme, S. Verbrugghe and F. E. Du Prez, *Polym. Chem.*, 2013, **4**, 4697.
268. H. Zhang, K. Hong and J. W. Mays, *Polym. Bull.*, 2004, **52**, 9.
269. J. Ryan, F. Aldabbagh, P. B. Zetterlund and B. Yamada, *Macromol. Rapid Commun.*, 2004, **25**, 930.
270. H. Zhang, K. Hong, J. W. Mays and C. J. Hawker, *Polym. Prepr.*, 2004, **45**, 325.
271. S. Feng, W. Xu, K. Nakanishi and S. Yamago, *ACS Macro Lett.*, 2012, **1**, 146.
272. G. Cui, J. Li, Z. Liu and J. Dou, *Macromol. Chem. Phys.*, 2010, **211**, 1222.
273. S. Perrier, T. P. Davis, A. J. Carnichael and D. M. Haddleton, *Eur. Polym. J.*, 2003, **39**, 417–422.
274. G. Johnston-Hall, J. R. Harjani, P. J. Scammells and M. J. Monteiro, *Macromolecules*, 2009, **42**, 1604.
275. T. Biedron and P. Kubisa, *Macromol. Rapid Commun.*, 2001, **22**, 1237.
276. T. Biedron and P. Kubisa, *J. Polym. Sci., Part A: Polym. Chem.*, 2002, **40**, 2799.

277. N. V. Tsarevsky and K. Matyjaszewski, *J. Polym. Sci., Part A: Polym. Chem.*, 2006, **44**, 5098.
278. S. Maria, T. Biedron, R. Poli and P. Kubisa, *J. Appl. Sci.*, 2007, **105**, 278.
279. H. Zhang, Y. Zhang, W. Liu and H. Whang, *J. Appl. Polym. Sci.*, 2008, **110**, 244.
280. S. Brusseau, O. Boyron, C. Schikaneder, C. C. Santini and B. Charleux, *Macromolecules*, 2011, **44**, 215.
281. M. Messerschmidt, M. Millaruelo, H. Komber, L. Häussler, B. Voit, T. Krause, M. Yin and W. D. Habicher, *Macromolecules*, 2008, **41**, 2821.
282. L.-C. Gao, Q.-W. Pan, C. Wang, Y. Yi, X.-F. Chen, X.-H. Fan, Q.-F. Zhou, *J. Polym. Sci., Part A: Polym. Chem.*, **45**, 5949.
283. P. O. Shipman, C. Cui, P. Lupinska, R. A. Lalancette, J. B. Sheridan and F. Jäkle, *ACS Macro Lett.*, 2013, **2**, 1056.
284. A. S. Lang, A. Neubig, M. Sommer and M. Thalakkat, *Macromoledules*, 2010, **43**, 7001.
285. M.-A. Muth, M. Carrasco-Orozco and M. Thelakkat, *Adv. Funct. Mater.*, 2011, **21**, 4510.
286. E. Kaul, V. Senkovskyy, R. Tkachov, V. Bocharova, H. Komber, M. Stamm and A. Kiriy, *Macromolecules*, 2010, **43**, 77–81.
287. S. Maria, A. S. Susha, M. Sommer, D. V. Talapin, A. L. Rogach and M. Thelakkat, *Macromolecules*, 2008, **41**, 6081.
288. M. Sommer, S. Lindner and M. Thelakkat, *Adv. Funct. Mater.*, 2007, **17**, 1493.
289. S. M. Lindner, N. Kaufmann and M. Thelakkat, *Org. Electron.*, 2007, **8**, 69.
290. L. Deng, P. T. Furuta, S. Garon, J. Li, D. Kavulak, M. E. Thompson and J. M. J. Fréchet, *Chem. Mater.*, 2006, **18**, 386.
291. L. Boiteau, M. Moroni, A. Hilberer, M. Werts, B. de Boer and G. Hadziioannou, *Macromolecules*, 2002, **35**, 1543.
292. K. Tsuchiya, H. Kasuga, A. Kawakami, H. Taka, H. Kita and K. Ogino, *J. Polym. Sci., Part A: Polym. Chem.*, 2010, **48**, 1461.
293. D. G. Ramirez-Wong, C. A. Posada-Velez, E. Saldivar-Guerra, J. G. Luna-Barcenas, C. Ott and U. S. Schubert, *Macromol. Symp.*, 2009, **283/284**, 120.
294. R. Maeda, T. Hayakawa, M. Tokita, R. Kijuchi, J. Kouki, M.-A. Kakimoto and H. Urushibata, *React. Funct. Polym.*, 2009, **69**, 519.
295. S. B. Maria, A. S. Susha, M. Sommer, D. V. Talapin, A. L. Rogach and M. Thelakkat, *Macromolecules*, 2008, **41**, 6081.
296. R. H. Lohwasser, G. Gupta, P. Kohn, M. Sommer, A. S. Lang, T. Thurm-Albrecht and M. Thelakkat, *Macromolecules*, 2013, **46**, 4403.
297. J. K. Wegrzyn, T. Stephan, R. Lau and R. B. Grubbs, *J. Polym. Sci., Part A: Polym. Chem.*, 2005, **43**, 2977.
298. H. Gotz, E. Harth, S. M. Schiler, C. W. Frank, W. Knoll and C. J. Hawker, *J. Polym. Sci., Part A: Polym. Chem.*, 2002, **40**, 3379.
299. Y. Miura and M. Dairoku, *J. Polym. Sci., Part A: Polym. Chem.*, 2007, **45**, 4364–4376.

300. N. R. Cameron, O. Lagrille, P. A. Lovell and B. Thongnuanchan, *ACS Macro Lett.*, 2012, **1**, 1262.
301. N. M. Ahmad, B. Charleux, C. Farcet, C. J. Ferguson, S. G. Gaynor, B. S. Hawkett, F. Heatley, B. Klumperman, D. Konkolewicz, P. A. Lovell, K. Matyjaszewski and R. Venkatesh, *Macromol. Rapid Commun.*, 2009, **30**, 2002–2021.
302. N. M. Ahmad, F. Heatley and P. A. Lovell, *Macromolecules*, 1998, **31**, 2822.
303. C. Magee, Y. Sugihara, P. B. Zetterlund and F. Aldabbagh, *Polym. Chem.*, 2014, **5**, 2259.
304. Y. Sugihara, P. O'Connor, P. B. Zetterlund and F. Aldabbagh, *J. Polym. Sci., Part A: Polym. Chem.*, 2011, **49**, 1856.
305. S. Cheng, F. L. Beyer, B. D. Mather and T. E. Long, *Macromolecules*, 2011, **44**, 6509.
306. C. Dire, B. Charleux, S. Magnet and L. Couvreur, *Macromolecules*, 2007, **40**, 1897.
307. B. Lessard and M. Marić, *J. Polym. Sci., Part A: Polym. Chem.*, 2009, **47**, 2574.
308. A. Moayeri, B. H. Lessard and M. Marić, *Polym. Chem.*, 2011, **2**, 2084.
309. B. H. Lessard, C. Tervo, S. De Wahl, F. Clerveaux, Jr, K. K. Tang, S. Yasmine, S. Andjelić, A. D'Alessandro and M. Marić, *Macromolecules*, 2010, **43**, 868.
310. Z.-J. Wang and M. Marić, *Polymers*, 2014, **6**, 565.
311. S. R. S. Ting, E. H. Min, P. Escalé, M. Save, L. Billon and M. H. Stenzel, *Macromolecules*, 2009, **42**, 9422.
312. B. H. Lessard, E. J. Y. Ling and M. Marić, *Macromolecules*, 2012, **45**, 1879.
313. C. Zhang and M. Marić, *J. Polym. Sci., Part A: Polym. Chem.*, 2013, **51**, 4702.
314. B. Lessard and M. Marić, *J. Polym. Sci., Part A: Polym. Chem.*, 2011, **49**, 5270.
315. C. R. Becer, K. Kokado, C. Weber, A. Can, Y. Chujo and U. S. Schubert, *J. Polym. Sci., Part A: Polym. Chem.*, 2010, **48**, 1278.
316. B. H. Lessard, X. Savelyeva and M. Marić, *Polymer*, 2012, **53**, 5649.
317. X. Savelyeva and M. Marić, *J. Polym. Sci., Part A: Polym. Chem.*, 2014, **52**, 2011.
318. Z.-J. Wang and M. Marić, *J. Polym. Sci., Part A: Polym. Chem.*, 2013, **51**, 2970.
319. B. H. Lessard, E. J. Y. Ling, M. S. T. Morin and M. Maric, *J. Polym. Sci., Part A: Polym. Chem.*, 2011, **49**, 1033.
320. B. H. Lessard, Y. Guillaneuf, M. Mathew, K. Liang, J.-L. Clement, D. Gigmes, R. A. Hutchinson and M. Marić, *Macromolecules*, 2013, **46**, 805.
321. V. Consolante and M. Marić, *Macromol. React. Eng.*, 2011, **5**, 575.
322. K. Babiuch, R. Wyrwa, K. Wagner, T. Seemann, S. Hoeppener, C. R. Becer, R. Linke, M. Gottschaldt, J. Weisser, M. Schnabelrauch and U. S. Schubert, *Biomacromolecules*, 2011, **12**, 681.

323. C. Ott, R. Hoogenboom and U. S. Schubert, *Chem. Commun.*, 2008, 3516.
324. C. R. Becer, K. Babiuch, D. Pilz, S. Hornig, T. Heinze, M. Gottschaldt and U. S. Schubert, *Macromolecules*, 2009, **42**, 2387.
325. B. Lessard, A. Graffe and M. Marić, *Macromolecules*, 2007, **40**, 9284.
326. B. Lessard and M. Marić, *Macromolecules*, 2008, **41**, 3446.
327. L. Couvreur, B. Charleux, O. Guerret and S. Magnet, *Macromol. Chem. Phys.*, 2003, **204**, 2055.
328. E. Mignard, T. Leblanc, D. Bertin, O. Guerret and W. F. Reed, *Macromolecules*, 2004, **37**, 966.
329. F. Dumur, B. H. Lessard, L. Beouch, F. Goubard, M. Marić and D. Gigmes, *Org. Electron.*, 2015, **17**, 377.
330. K. Karaky, G. Clisson, G. Reiter and L. Billon, *Macromol. Chem. Phys.*, 2008, **209**, 715.
331. B. Keosherian, M. K. Georges and D. Boils-Boissier, *Macromolecules*, 1995, **28**, 6381.
332. M. Bouix, J. Gouzi, B. Charleux, J.-P. Vairon and P. Guinot, *Macromol. Rapid Commun.*, 1998, **19**, 209.
333. W. Huang, B. Charleux, R. Chiarelli, L. Marx, A. Rassat and J.-P. Vairon, *Macromol. Chem. Phys.*, 2002, **203**, 1715.
334. E. Yoshida, *ISRN Polym. Sci.*, 2012, 630478, 1–6.
335. R. Nicolay, L. Marx and P. He, *Macromolecules*, 2007, **40**, 6067.
336. M. A. Mannan, K. Fukuda and Y. Miura, *Polym. J.*, 2007, **39**, 500.
337. T. N. T. Phan and D. Bertin, *Macromolecules*, 2008, **41**, 1886.
338. B. Grassl, G. Clisson, A. Khoukh and L. Billon, *Eur. Polym. J.*, 2008, **44**, 50.
339. J. Rigolini, B. Grassl, L. Billon, S. Reynaud and O. F. X. Donard, *J. Polym. Sci., Part A: Polym. Chem.*, 2009, **47**, 6919.
340. C. Dire, J. Nicolas, S. Brusseau, B. Charleux, S. Magnet and L. Couvreur, *ACS Symp. Ser.*, 2009, **1024**, 303.
341. S. Brusseau, J. Belleney, S. Magnet, L. Couvreur and B. Charleux, *Polym. Chem.*, 2010, **1**, 720.
342. S. Brusseau, F. D'Agosto, S. Magnet, L. Couvreur, C. Chamignon and B. Charleux, *Macromolecules*, 2011, **44**, 5590.
343. A. Darabi, A. R. Shirin-Abadi, P. G. Jessop and M. F. Cunningham, *Macromolecules*, 2015, **48**, 72.
344. A. Darabi, P. G. Jessop and M. F. Cunningham, *Macromolecules*, 2015, **48**, 1952.
345. J. Pinaud, E. Kowal, M. F. Cunningham and P. G. Jessop, *ACS Macro Lett.*, 2012, **1**, 1103.
346. Q. Zhang, G. Yu, W.-J. Wang, B.-G. Li and S. Zhu, *Macromol. Rapid Commun.*, 2012, **33**, 916.
347. A. J. Morse, S. P. Armes, K. L. Thompson, D. Dupin, L. A. Fielding, P. Mills and R. Swart, *Langmuir*, 2013, **29**, 5446.

CHAPTER 5

Nitroxide-mediated Polymerization in Dispersed Media

GUILLAUME DELAITTRE[a,b]

[a] Institute of Toxicology and Genetics, Karlsruhe Institute of Technology (KIT), Hermann-von-Helmholtz-Platz 1, 76344 Eggenstein-Leopoldshafen, Germany; [b] Preparative Macromolecular Chemistry, Institut für Technische Chemie und Polymerchemie, Karlsruhe Institute of Technology (KIT), Engesserstr. 18, 76131 Karlsruhe, Germany
Email: guillaume.delaittre@kit.edu

5.1 Introduction

Before describing advances in the implementation of radical polymerization in the presence of nitroxides in dispersed media, it is certainly important to answer the question: why dispersed media? Processes in dispersed media exhibit very particular features in terms of mechanisms as well as final products. These features typically give access to various advantages but may also lead to complications. Perhaps it is worth mentioning that the majority of polymers produced *via* a radical mechanism in the industry originates from this type of processes.[1–4] Indeed, polymerizations in bulk, although popular in the academic field, are not suitable for large-scale synthesis as they require high-viscosity processing equipment, for instance. Solution polymerization then poses an issue regarding the great amounts of solvents needing to be removed and—hopefully—recycled. In aqueous and

RSC Polymer Chemistry Series No. 19
Nitroxide Mediated Polymerization: From Fundamentals to Applications in Materials Science
Edited by Didier Gigmes

Published by the Royal Society of Chemistry, www.rsc.org

CO_2-based dispersed media, a clear environmental—and hence economical—factor is addressed.

Although supercritical CO_2 gains interest (see Sections 5.3.2, 5.4, and 5.5), the vast majority of related reports concerns water as a dispersing phase. In aqueous dispersed media, water acts as a very efficient heat transporter. It absorbs the energy released during the exothermic polymerization and prevents autoacceleration and thermal runaway, which is crucial for large-scale reactors. Kinetics are usually faster and final conversion and degree of polymerization higher. These aspects lead us to consider that, if one day reversible-deactivation radical polymerization protocols (RDRP) are finally to break into the industrial arena—which has not been the case so far despite 20 years of intense academic development[5]—it will likely be *via* dispersed media processes.[6]

The final product of such a polymerization in aqueous dispersed media is a latex. It is basically a suspension of stable nano- or microparticles that can be further used as such, or in solid form after drying by freeze-drying at the laboratory scale, or more commonly by spray-drying or coagulation followed by transformation into pellets in an industrial setup. Using various methods, the particle morphology and surface composition can be controlled to yield charged particles, core–shell particles, or micro-/nanogels, for instance. Lattices find applications in elastomers, adhesives, paints, and to a lower extent in biotechnology.[7,8] It is however envisioned that the application of RDRP protocols, *e.g.*, nitroxide-mediated polymerization (NMP), in such processes could eventually give access to new features, and hence novel applications. In addition, it was found that compartmentalization of the polymerization in nanoreactors can improve control over molar mass distributions (MMDs) and afford a higher degree of end functionality, often referred to as livingness.[9]

Nevertheless, the translation of RDRP techniques from homogeneous to dispersed media has faced some hurdles. Indeed, RDRP recipes inherently consist of more components than classical free-radical polymerization. In homogeneous media, this is not an issue as there is one general polymerization locus. In dispersed media, there are several possible domains where various species can be localized. Particularly, localizing the controlling agents that must tame the propagating radicals where these radicals are was the main obstacle to implementing RDRP in aqueous emulsion polymerization. In addition, the success of an RDRP in dispersed media does not only rely on the control of the macromolecular chains but also on the colloidal stability of the final dispersion. While many examples have led to control of either of these, the challenge is to succeed in both aspects.

In the present chapter, an attempt is given at comprehensively compiling examples of NMP carried out in dispersed media. This encompasses various types of processes that differ in their initial and/or in their final state(s), depending on the solubility of initiator and monomer, employed additives, and pre-polymerization procedures. A summary of the general features for

Table 5.1 Typical features of polymerization carried out in dispersed media. These features are generally applicable to polymerizations in aqueous dispersed media and often, but not systematically, transferable to other solvents used as continuous phase. However, colloidal features may vary in the frame of NMP, as compared to conventional radical polymerization.

	Diameter		Solubility in the continuous phase	
Process	Final particles	Initial monomer droplets	Initiator	Monomer
Emulsion	30–600 nm	~2–20 μm	Yes	No
Precipitation	100–600 nm	–	Yes	Yes
Suspension	≥2 μm	~2–20 μm	No	No
Dispersion	≥2 μm	–	Yes	Yes
Microemulsion	20–60 nm	~20 nm	Yes	No
Miniemulsion	60–200 nm	~60 nm	Yes/No	No

each process is given in Table 5.1. Additional information will be given in the relevant subsections.

For general aspects pertaining to NMP and its mechanism, the reader is referred to the relevant chapters of the present book.

5.2 Nitroxide-mediated (Mini)Emulsion Polymerization

Emulsion and miniemulsion polymerizations—together with dispersion polymerization—have been the most studied techniques in RDRP in dispersed media. Before describing the advances of NMP adapted to these processes and especially to understand the challenges encountered, it is important to describe these methods in the frame of conventional free-radical polymerization. Simultaneously, this will give an overview of the particular mechanisms taking place in systems where the monomer is not soluble in the continuous phase as well as enable the introduction of the terminology specific to this area.

5.2.1 Brief Account of Free-radical Polymerization in Miniemulsion and Emulsion

In emulsion polymerization, the monomer is hydrophobic and is initially mainly present in the form of micrometric droplets dispersed in water and stabilized by surfactant molecules (Figure 5.1, left). The polymerization starts in the continuous aqueous phase through the use of water-soluble initiator. The initial emulsion is unstable. Polymerization yields an aqueous suspension of particles, ideally spherical and uniform in size, stabilized by the surfactant (Figure 5.1, right).

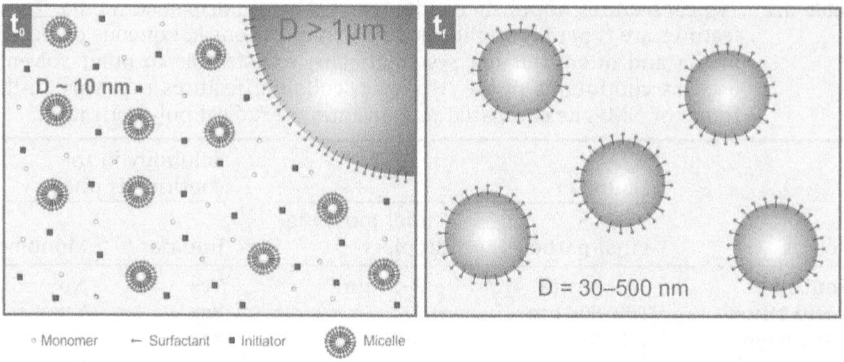

Figure 5.1 Schematic depiction of the initial (left) and final (right) states of an emulsion polymerization.

As mentioned above, a typical emulsion polymerization recipe consists of a few essential components:

- Water, which is the dispersing continuous medium;
- The monomer, which is typically hydrophobic but should still exhibit a minimal solubility in water as it is essential for the initiation step and the transport from the droplets to the particles (*vide infra*);
- The initiator, which is water-soluble and can be decomposed thermally or through redox activation;
- The surfactant, which is required for the stabilization of the polymeric particles and can be of various natures (charged or neutral, molecular or macromolecular).

In particular situations, a few additives are sometimes employed, *e.g.*, water-soluble monomers, transfer agents, buffers, and crosslinkers.[1]

Beyond the actual compositional recipe and depending on aspects such as the morphology of the final latex or the viscosity during the polymerization, three main types of processes can be implemented.[2] In a *batch* process, all ingredients are present in the reactor vessel from the beginning. In a *semi-batch* process, the reaction starts with a portion of the ingredients (obviously, water, initiator, and some monomer are present) and at a given time the rest fraction is added in order to reach the desired final recipe. This addition can be done in different ways: continuously or in several portions. Finally, albeit rare in academic laboratories, fully *continuous* processes can be applied, where the initial components of the emulsion flow through one or several reactors with controlled residence times.

Besides the description of the overall emulsion handling, an additional terminology is available to distinguish between various initial states of an emulsion polymerization. Indeed, when the process starts from only monomer droplets, one can employ the term *ab initio*. However, sometimes a small quantity of pre-formed latex is present from the start in what is named *seeded* emulsion polymerization. Most academic reports on emulsion polymerization are based on or aim at *batch ab initio* processes, as it is clearly a simpler approach. Nevertheless, we will see in the following that semi-batch and seeded approaches proved to be essential for the successful implementation of NMP in emulsion.

Kinetics and related mechanisms of emulsion polymerization are rather complex but it is nevertheless common to distinguish three periods named intervals,[1] which differ in the rate of polymerization and are related to phenomena such as the nucleation—the event of forming particles, *i.e.*, a segregated domain containing polymer—and the transport of monomer from the droplets to the particles. In *interval I*, which typically spans a few minutes, nucleation takes place. Following initiation in the aqueous phase, particles form and their number increases as well as the polymerization rate until all surfactant micelles disappear. In *interval II*, the particle number and polymerization rate are constant. The particle growth is supported *via* monomer supply by transport through the aqueous phase from the droplets. During this interval, the monomer concentration in the growing particles is also constant. In the final *interval III*, droplets have disappeared. Subsequently, the monomer concentration inside the particles decreases and so does the polymerization rate. During a seeded emulsion polymerization, *interval I* (*i.e.*, nucleation) is obviously absent.

The nucleation step is essential as it determines the number of particles, which in turn influences various features, such as the polymerization rate, the molar mass, and the latex stability, among others. The number of particles depends on the nature of the monomer(s), the structures and concentrations of the surfactant and initiator, the temperature, and the nature of the process (batch, semi-batch, continuous). Two main types of nucleation were proposed: (i) micellar or (ii) homogeneous. The former was introduced by Harkins[10] and takes place when the surfactant concentration is superior to its critical micellar concentration (CMC) and/or when the solubility of the monomer in the aqueous phase is very low. In this case, when the oligoradicals formed in the aqueous phase reach a critical degree of polymerization z (z-mers) and become surface-active, they enter a micelle to create a new particle or may alternatively enter an existing particle (Figure 5.2a). The homogeneous nucleation was proposed by Roe[11] and described by Fitch and Tsai.[12] Irrespective of the surfactant concentration, it can occur when the water-solubility of the monomer is significant. When water-borne oligoradicals are long enough (degree of polymerization j_{crit}), they form unstable colloidal objects called primary particles or nuclei, which then grow by monomer swelling and polymerization or flocculation with other nuclei. The specific characteristics of the stabilizer will have a strong influence on determining which of these events dominate. Homogeneous nucleation obviously takes place in systems where no surfactant is present and for which particle stabilization is maintained solely by the charged initiator fragments (Figure 5.2b).

The two aforementioned nucleation theories take into account only one event: either the entry of a radical in a micelle or the precipitation of this

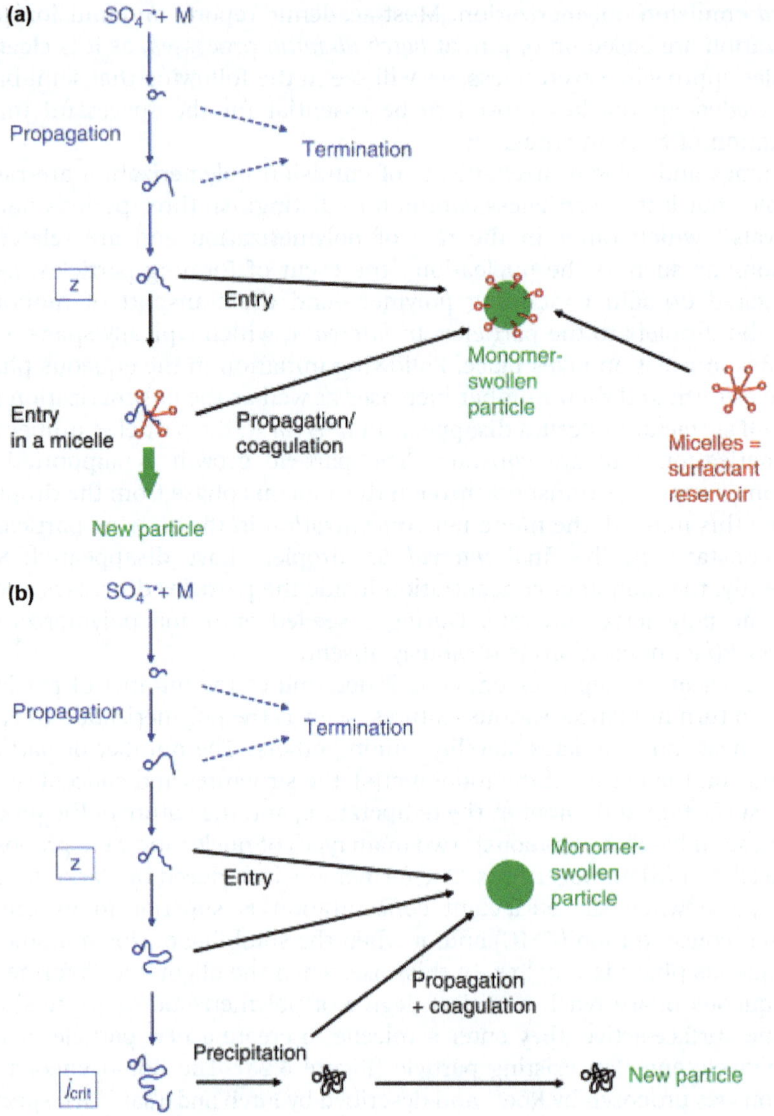

Figure 5.2 Schematic representation of (a) micellar nucleation and (b) homogeneous-coagulative nucleation, for an emulsion polymerization initiated by sulfate radicals (*i.e.*, persulfate initiator).
Reproduced from ref. 14 with permission from Elsevier.

oligoradical. In the 1980s, Gilbert proposed the theory of coagulative nucleation in two steps.[13] The first step consists of the formation of the nuclei by one of the aforementioned mechanisms (micellar or homogeneous nucleation). Being very small (2–3 nm), these nuclei are rather unstable and not prone to monomer swelling. Gilbert thus proposed that the growth of these precursors occurs more by mutual aggregation rather than propagation (Figure 5.2b).

Finally, another type of nucleation can occur when oligoradicals enter monomer droplets. This is highly undesirable in emulsion polymerization as this usually leads to the destabilization of the dispersion. Nevertheless, this phenomenon was exploited in techniques were the size of the monomer droplets is reduced in a controlled manner by high-shear treatment or by employing a very high surfactant concentration. The latter option leads to microemulsion polymerization, where only monomer-swollen micelles of an extremely low diameter (down to 5 nm) are generated. The initial dispersion is optically transparent and thermodynamically stable. Usually not all micelles are entered by a radical and those which are not are eventually cannibalized by those in which polymer is growing.[15] The final particle size typically ranges from 20 to 60 nm. Ultrasonication is employed in the frame of miniemulsion polymerization,[16] where monomer droplets of approximately 50 nm in diameter are created. Thanks to this reduced size, nucleation takes place only in droplets and, in the ideal scenario, the final latex is a copy of the initial dispersion (Figure 5.3). Miniemulsions are initially kinetically stable yet not thermodynamically. A simplified (sometimes simplistic) way of looking at miniemulsion polymerization is to consider the droplets as bulk nanoreactors.

5.2.2 Challenges to Overcome

While particle formation is already a complex process in conventional emulsion polymerization and has been only partially elucidated for specific recipes,[15] it is necessary to take into account the particularities arising from the slow growth of polymer chains in RDRP. Indeed, particles are classically formed from oligoradicals generated in the aqueous phase, which add onto monomer and propagate until they reach a critical size and enter a particle.

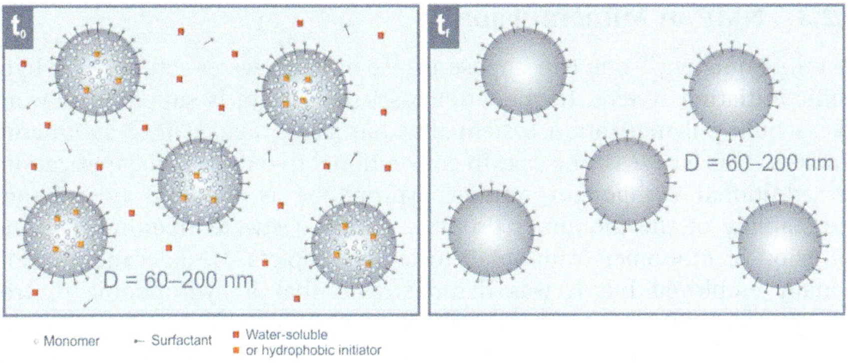

Figure 5.3 Schematic depiction of the initial (left) and final (right) states of a miniemulsion polymerization. In miniemulsion, the initiator can be either organosoluble or water-soluble.

At this stage, propagation becomes extremely fast as a result of the segregation effect—as the radical concentration is reduced, termination reactions are reduced as well—and the high monomer concentration in the particles. Therefore, this mechanism is generally irreversible and it is impossible for propagating radicals to return to the aqueous phase. In RDRP, in the early stage of the polymerization, short chains are created and reversibly terminated. Consequently, they may not possess a marked preference for either phase. This leads to possible redistribution in various loci of the emulsion, including the droplets, which can then be nucleated and eventually induce broadening of the particle size distribution (PSD) or even destabilization of the entire dispersion.

In addition, an inherent feature of NMP—at least in the early systems based on the TEMPO nitroxide (see Scheme 5.1)—is the requirement for temperatures typically above 100 °C, and more generally above 120 °C. At such temperatures, particularly in the case of styrene, spontaneous polymerization by thermal initiation can occur and is therefore a complication with respect to droplet nucleation.

Finally, as mentioned earlier, localization of the nitroxide in the particles during intervals II and III is essential for the control of the polymerization. This implies that the nitroxide must be soluble in the monomer and the polymer, while it must also be hydrophilic to some extent to enable its transfer from the droplets to the particles. Clearly, finding the balance between the hydro- and lipophilicity of the controlling agent is key to achieving a successful polymerization in emulsion.

Although the first attempts at applying NMP to dispersed media were made in emulsion,[17] the following part will concern nitroxide-mediated miniemulsion polymerization as it led to the first successful results, certainly due to its simplified mechanism for nucleation. In this area, the teams of Cunningham, Charleux, Georges, Okubo, and Zetterlund, respectively, have made a significant contribution.

5.2.3 NMP in Miniemulsion

In miniemulsion,[16] one can choose to use either a water-soluble or a lipophilic initiating system. In the latter case, the system is supposed to work like a bulk polymerization system that has been divided into submicron reactors. This is at least the case in conventional free-radical polymerization. An additional component termed hydrophobe is required to enhance the stability of the monomer droplets against Ostwald ripening, *i.e.*, the diffusion of monomer from small to larger droplets. Hexadecane is traditionally employed but it was demonstrated that a hydrophobic macroalkoxyamine could be an efficient substitute while also playing the role of the initiator and the controlling agent source.[18–21] The introduction of a few percent of a very hydrophobic monomer (*e.g.*, long alkyl chain (meth)acrylates) could potentially be an interesting alternative as it does not require a preliminary polymerization step.[6] The current section is divided according

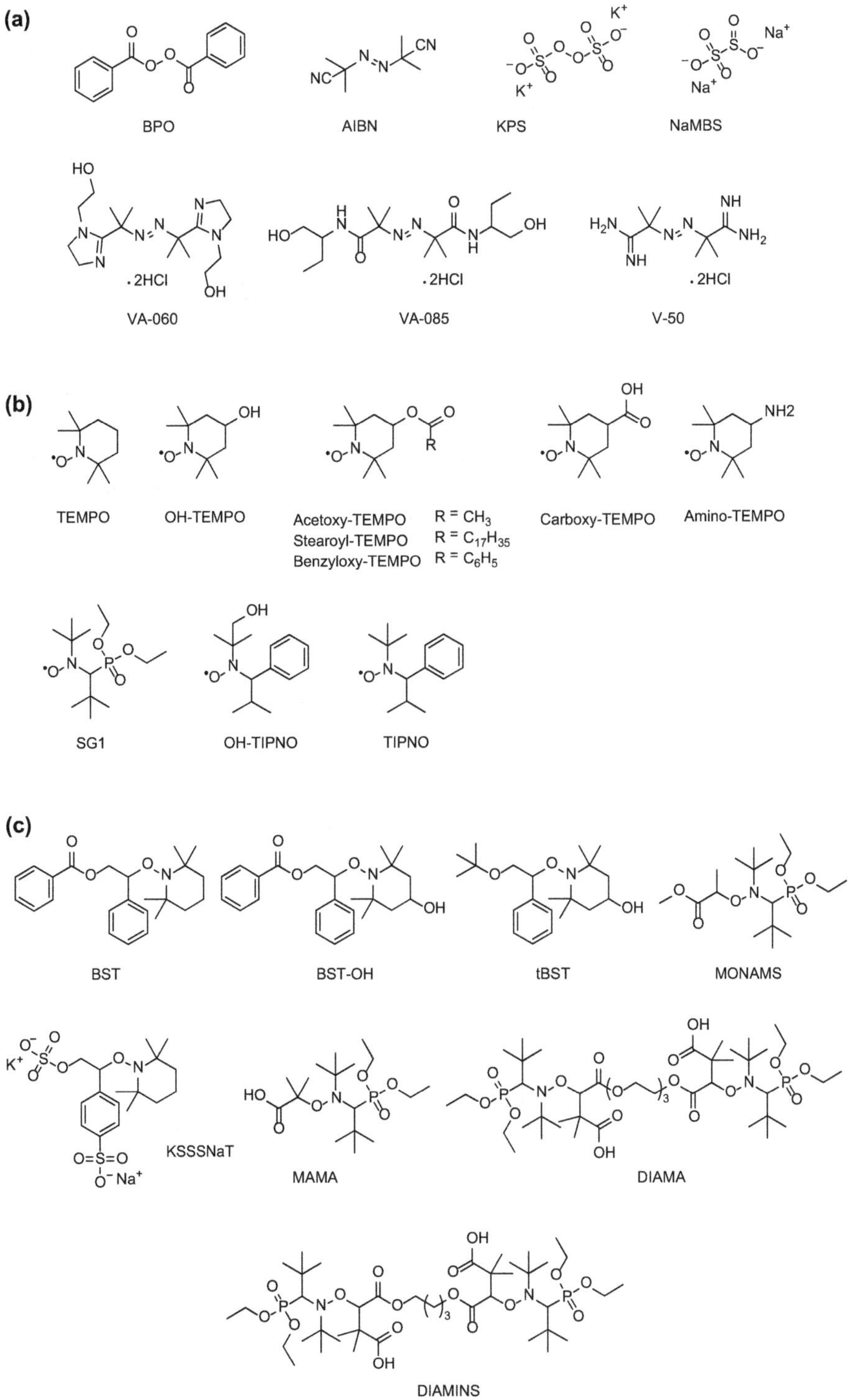

Scheme 5.1 Structure of the nitroxides (a) and alkoxyamines (b) employed in NMP in aqueous dispersed media.

to the solubility of the initiating system, due to the important influence of this physical property on the kinetics and the stability of the dispersion. In addition, as NMP can be initiated either by an alkoxyamine solely (monocomponent system) or a conventional radical initiator in conjunction with free nitroxide (bicomponent system), we will take this into account as well.

The various initiator, nitroxide, and alkoxyamine structures mentioned in the current chapter can be found in Scheme 5.1.

5.2.3.1 Oil-soluble Bicomponent Initiating System

Only a handful of studies have been reported on the employment of a lipophilic initiator in conjunction with a nitroxide for NMP in miniemulsion. These studies concern only styrene polymerization. The first account of a successful NMP in miniemulsion came from El-Aasser and co-workers, who initiated the polymerization with benzoyl peroxide (BPO) in the presence of TEMPO at 125 °C.[22] The system was stabilized by the anionic surfactant Dowfax 8390 (see Scheme 5.2), a mixture of mono- and di-hexadecyl disulfonated diphenyloxide disodium salt, and hexadecane for a theoretical solids content (*i.e.*, mass fraction of monomer over the entire reaction mixture) of 19 wt%. Interestingly, it was shown that in the absence of BPO and TEMPO, styrene was completely polymerized in just 6 h, evidencing the strong influence thermal initiation can have at high temperatures typical for NMP. In the presence of TEMPO, a clear induction period was observed. When both BPO and TEMPO were present, no induction took place and 90% conversion could be reached in approximately 12 h. Molar masses increased linearly with conversion, up to approximately 40 000 g mol^{-1}, and dispersity values remained below 1.7 ($Đ<1.5$ until 80% conversion). The particle size became larger in the presence of TEMPO. Actually, the number of particles

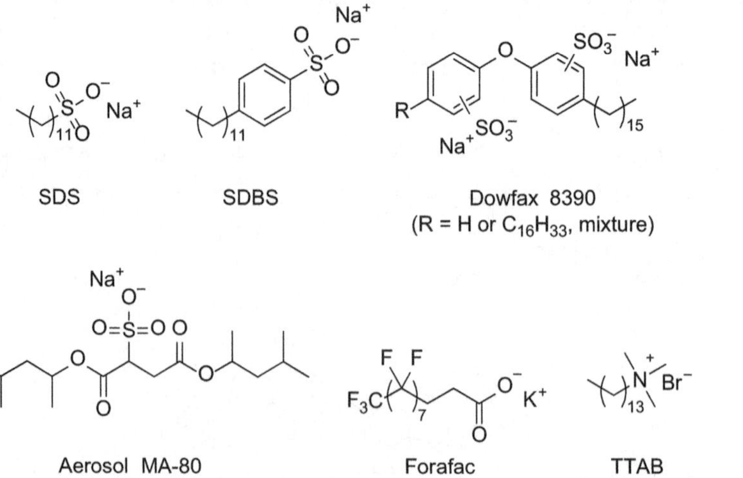

Scheme 5.2 Structure of the surfactants mentioned in the current book chapter.

remained rather constant throughout the polymerization, which was explained by the reduced influence of thermal initiation. In these experiments, camphorsulfonic acid (CSA) was added as it was earlier shown in bulk that it enhances the rate of polymerization.[23]

Cunningham and co-workers later showed that the influence of CSA on TEMPO-mediated miniemulsion polymerization at 135 °C was significant above a certain TEMPO/BPO molar ratio, *e.g.*, for a nitroxide/BPO molar ratio of 3.6.[24] This observation was valid for various BPO concentrations and for nitroxides having a very different water solubility, as demonstrated by the same results obtained with TEMPO and OH-TEMPO. Nevertheless, when CSA was introduced, the MMDs were systematically broader and the apparent initiating efficiency lower. The teams of Cunningham and Georges further examined the influence of the solubility of both nitroxide and initiator for a nitroxide/initiator molar ratio of 1.7 : 1.[25] Particularly, in the case of a hydrophobic radical initiator (BPO), the solubility of the nitroxide had little influence on the kinetics and the evolution of molar masses, which was in strong contrast with the use of a water-soluble initiator (see Section 5.2.3.2).

In parallel to TEMPO-based studies, Charleux and co-workers showed early that NMP of styrene could be carried out in miniemulsion at reduced temperatures, *i.e.*, 90 °C, employing SG1—an acyclic bulky nitroxide known for its lower temperature dissociation within corresponding alkoxyamines[26]—in combination with AIBN.[27] SDS was employed as the surfactant (see Scheme 5.2), with hexadecane and high molar mass polystyrene (330 kg mol^{-1}) acting as co-stabilizers. The best compromise between fast polymerization and good control was for a SG1/AIBN molar ratio of 1.2 : 1. Polymers exhibited low dispersity values but only approximately 60% conversion could be reached in 24 h. This could be explained by the large quantity of free nitroxide released by persistent radical effect (PRE),[28,29] which is not counterbalanced by any thermal initiation of styrene at these lower temperatures. At the same time, the system may be complicated by the instability of SG1 at 90 °C in the water phase, depending on the conditions and particularly in an acidic continuous phase. Recourse to buffers such as $NaHCO_3$ and K_2CO_3 was satisfactory in this respect.

Many years later, Zetterlund reported another miniemulsion system where AIBN and SG1 were employed together but this time in the presence of an *in situ* formed surfactant system (Figure 5.4).[30] In this method, oleic acid dissolved in the organic phase—together with hexadecane—is neutralized by contacting with the aqueous phase containing potassium hydroxide (2 equivalents) before magnetic stirring and ultrasonication. Assuming a complete conversion of oleic acid into potassium oleate, the surfactant content would be 120 wt% compared to styrene. This is largely above the concentration typically employed in miniemulsion but lower than for microemulsion (see Section 5.2.5). Nevertheless, this process led to a very small initial droplet size of about 25 nm, which is more akin to microemulsion and significantly lower than the droplet size generally witnessed in

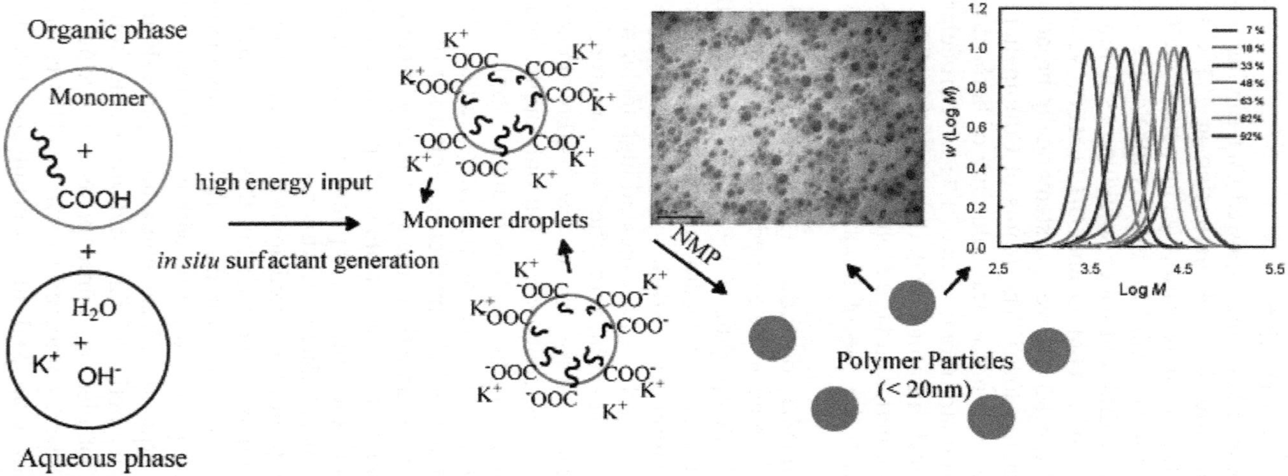

Figure 5.4 Illustration of the general principle of the miniemulsion protocol developed by Zetterlund and co-workers and based on the *in situ* formation of the surfactant.
Reproduced from ref. 30 with permission from Wiley.

miniemulsion (50–200 nm). A larger SG1/AIBN molar ratio of 1.95 : 1 was employed. In that case, conversion reached more than 90% in 24 h while molar masses increased linearly (final dispersion: $M_n = 27$ kg mol^{-1}, $Đ < 1.2$). The initial dispersion and the final latex were both translucent/transparent. Surprisingly, while the particle size distribution (PSD) remained narrow throughout the polymerization, the particle size steadily decreased until 40% conversion to then remain constant until the end, with a final number-average diameter d_n of 13 nm. The authors discarded a possible explanation based on the occurrence of secondary nucleation, which would have been at the origin of new particles and thus the reason for the expected particle growth not to happen. To date, no explanation has been found.

5.2.3.2 Water-soluble Bicomponent Initiating System

Another handful of reports disclosed the use of a water-soluble radical initiator in combination with nitroxide for the miniemulsion polymerization of styrene. Here, the polymerization obviously starts in the aqueous phase through creation of primary radicals that add onto monomer present at saturation in the aqueous phase, leading to the formation of oligoradicals that are hydrophobic or at least surface-active enough to enter particles and trigger the nucleation. In this way, some issues related to pure emulsion polymerization can be addressed. This was the declared goal of McLeod *et al.*, who replaced BPO by potassium persulfate (KPS) in the miniemulsion recipes previously reported by the teams of Cunningham and Georges, but without rate accelerant.[31] While the first experiment with a classical nitroxide/initiator molar ratio of 1.2 : 1 gave a stable latex, the polymerization did not follow a clear controlled behavior, *i.e.*, no clear shift of the MMD with conversion and rather high $Đ$. Increasing only the amount of nitroxide resulted in an unstable latex. Nevertheless, a concomitant increase of the concentrations in stabilizers (hexadecane and SDBS, see Scheme 5.2) and nitroxide proved to be rewarding. For a TEMPO/KPS 2.9 : 1, a final latex at 87% conversion and with an average diameter of 50 nm was obtained in 6 h ($M_n = 11\ 000$ g mol^{-1}, $Đ = 1.1$). The polymerization was considerably faster than in bulk, despite the absence of an accelerant, *e.g.*, CSA. In further experiments, the influence of the polymerization time and temperature on the living character of the chains was evaluated.[18] For this purpose, the polymer produced in miniemulsion was isolated and reused in the bulk polymerization of styrene at 135 °C. The higher the polymerization temperature or the conversion, the lower the degree of livingness (DOL).

Cunningham and co-workers, as they did for BPO, evaluated the influence of the nitroxide solubility (TEMPO *vs.* OH-TEMPO) on KPS-initiated emulsion polymerization.[25] As compared to a BPO-initiated polymerization and irrespective of the nature of the nitroxide, the polymerization rate was not strongly affected when KPS was used. However, in the particular case of TEMPO/KPS, the molar masses were significantly lower at any conversion, while still increasing linearly (Figure 5.5). Inspection of the evolution of the

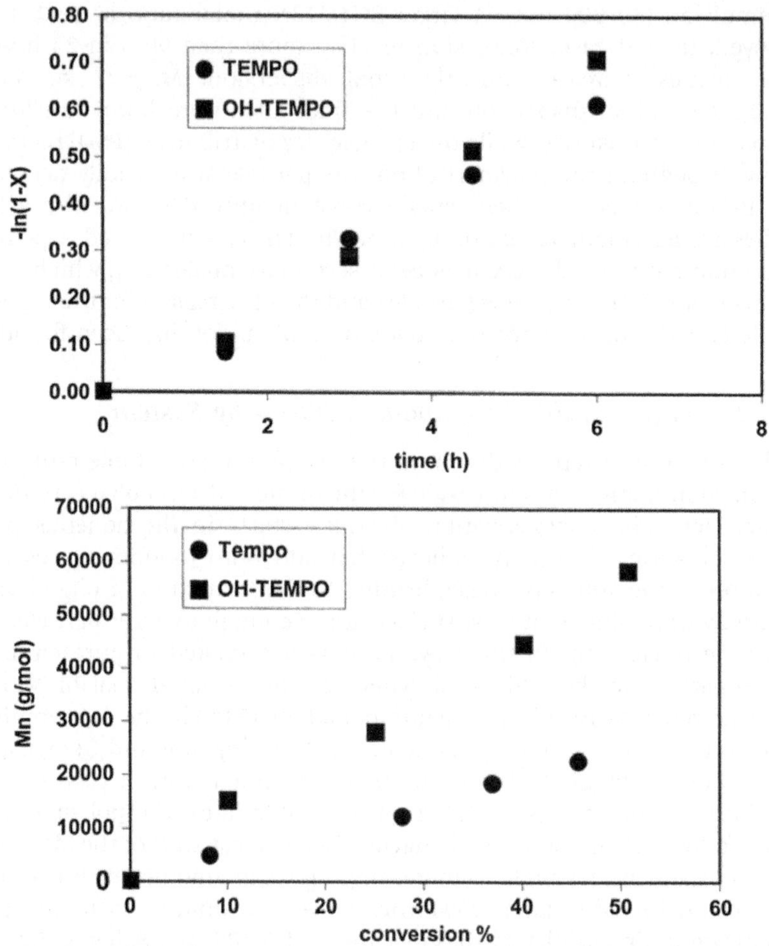

Figure 5.5 Influence of the solubility of the TEMPO derivative employed for the KPS-initiated miniemulsion polymerization of styrene at 135 °C. Reproduced from ref. 25 with permission from Wiley.

number of chains with conversion was revealing. While the number of chains remained relatively constant for TEMPO-mediated BPO-initiated polymerization, the KPS-initiated counterpart underwent an increase in chain number at low conversion. For OH-TEMPO, it increased constantly over an extended conversion range with both BPO and KPS. This phenomenon was explained by the loss of oligomers in the early stage of the polymerization, which became more pronounced for theoretically more hydrophilic systems (KPS and/or OH-TEMPO), and highlighted the importance of the aqueous phase kinetics as well as the partitioning of nitroxides. These aspects were further emphasized in a thorough study where the same team focused on the interplay between the absolute KPS concentration and its relation with the TEMPO/KPS ratio.[32] They found that with TEMPO/KPS at

4 : 1 mol/mol the initiator efficiency was around 100% and explained it by the reversible termination of the aqueous oligomeric radicals, which lowers the probability of irreversible bimolecular termination in the aqueous phase. For a large KPS concentration (9.25 mM) together with a reduced TEMPO/KPS molar ratio of 1.7 : 1, a compartmentalization effect could be detected as the kinetics were significantly faster than for any other conditions, along with a reduced particle size. This feature reflected an increase in particle number, which was explained by the possibility for homogeneous nucleation in these conditions, where insufficient nitroxide was present in the aqueous phase to prevent styrene oligomers from reaching a critical degree of polymerization and forming new particles.

SG1-mediated miniemulsion polymerization was also initiated with a water-soluble initiator.[33] KPS was employed together with sodium metabisulfite (NaMBS) as it was found that the former alone inexplicably led to poor control. A molar ratio of SG1/KPS/NaMBS of 1.2 : 1 : 1 was employed in conditions similar to those employed with of a hydrophobic initiator.[27] After a ubiquitous induction period of about 45 min, during which the essential event is the trapping of short propagating radicals by SG1 largely in excess, polymerizations were generally fast: for instance, at a theoretical solids content of approximately 10 wt%, near-to-quantitative conversion was attained in 7 h. Molar masses increased linearly with conversion. However, dispersity values were rather high for the lowest solids content, which was explained by the high partitioning tendency of SG1 leading to a reduced controlling agent concentration in the particles. When the quantity of styrene was increased to 25 wt%, the MMD became clearly narrower, at the expense of the fast kinetics. In addition, this study dealt with an important general aspect for SG1-mediated polymerizations in dispersed media: the instability of SG1 in acidic conditions. Since the initiating system (KPS alone or in combination with NaMBS) clearly induced acidity, buffers such as $NaHCO_3$ or K_2CO_3 were evaluated and found to be effective. In their absence, the polymerization was extremely fast and uncontrolled. For K_2CO_3, a complete inhibition was witnessed for the highest concentration ($pH > 7.0$). Too low a concentration resulted again in uncontrolled polymerizations, yet at an ideal 3 mM_{aq} concentration—leading to a pH of 4.8 after the induction period—control could be achieved. Successful chain extension after addition of a second load of styrene was also demonstrated.

5.2.3.3 Oil-soluble Monocomponent Initiating System

The largest body of literature on NMP in miniemulsion concerns the use of organosoluble monocomponent initiating systems, *i.e.*, hydrophobic alkoxyamines. Among these reports, it is possible to distinguish two types of alkoxyamines: low molar mass or macromolecular. In addition to reducing the number of components in the recipe, this method presents the critical advantage of ridding the system of an ambiguous parameter, that is, the initiator efficiency. It must be noted that in most of the studies described in the following attention has been more focused on the kinetics of the polymerization, rather than on the colloidal characteristics of the final latex. In general, broad monomodal distributions were obtained, which often does not allow the exclusion of homogeneous nucleation from the picture.

The macromolecular alkoxyamines typically employed are oligomeric styrene sequences capped with a nitroxide moiety, which are obtained through a previous NMP step in homogeneous conditions. In some cases, they are purified before their use in miniemulsion, while in some others the nonpurified polystyrene/styrene solution is directly emulsified, with potential addition of monomer. Except for the first report from the team of El-Aasser, where a PS-TEMPO of about 7000 g mol^{-1} was employed,[34] these oligoalkoxyamines are usually in the 1300–2700 g mol^{-1} range. No discussion on the potential influence of the molar mass was reported so far. Curiously, although these oligomers could also be simultaneously considered as hydrophobes, they were often used in conjunction with hexadecane, probably to decouple [hydrophobe] and targeted M_n.

El-Aasser and co-workers first described the use of PS-TEMPO ($M_n = 7500$ g mol^{-1}) to initiate the miniemulsion polymerization of styrene at 125 °C at 20 wt%.[34] The initial concentration of alkoxyamine had a clear influence on the molar masses, as expected. Nevertheless, rather high dispersity values were obtained in all cases (up to 1.8–1.9 at 75% conversion). Stables lattices with large PSDs were obtained. Later, the same team probed the effect of varying the surfactant (Dowfax 8390) concentration for polymerizations initiated with a shorter PS-TEMPO ($M_n = 1500$ g mol^{-1}).[35] Logically, the particle size diminished and the particle number increased, yet no meaningful difference in kinetics or macromolecular control was detected. This led the authors to conclude that the influence of the reversible termination by TEMPO surpassed the compartmentalization effect typical of emulsion systems. Different trends were published a few years later by Okubo and co-workers for polymerizations initiated at 125 °C with a similar PS-TEMPO ($M_n = 2100$ g mol^{-1}) and stabilized by either SDBS or poly(vinyl alcohol) (PVA).[36] Decreasing d_n from 200 nm to 70 nm resulted in a dramatic increase of the polymerization rate: at 70 nm, the polymerization proceeded 40 times faster than for systems above 170 nm, which behaved similarly to bulk polymerizations (Figure 5.6). However, loss of control was noticed for such fast kinetics. At 110 nm, low molar mass material remained in small quantity. This phenomenon was more pronounced at 70 nm, where in fact the molar mass was essentially independent of the conversion. The authors hypothesized that a low diameters TEMPO had a tendency to remain at the oil/water interface, thus being unavailable for growth control. These findings were supported by further experiments showing that the faster the polymerization the lower the DOL.[37] At 70 nm and 92% conversion, a large majority of chains were found to be dead in a chain extension experiment. This is in good agreement with theoretical predictions formulated by Cunningham for polymerizations at 135 °C.[38] In addition, a polymeric TEMPO-based nitroxide capping a polystyrene segment (overall $M_n = 3850$ g mol^{-1},

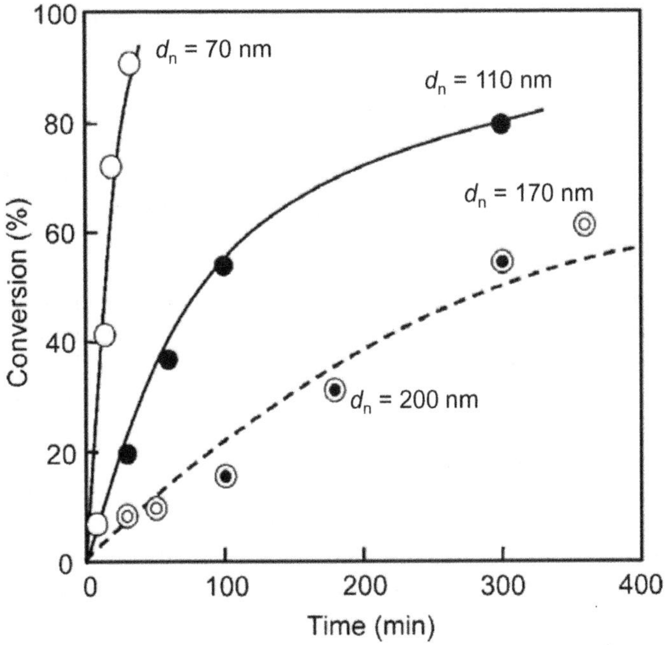

Figure 5.6 Influence of the particle size on the rate of miniemulsion polymerization of styrene initiated by PS-TEMPO at 125 °C in the conditions reported by Okubo and co-workers.
Adapted from ref. 36 with permission from Wiley.

with the initiating fragment M_n being 2400 g mol^{-1}) that cannot exit the organic phase was employed and showed no significant difference as compared to classic TEMPO, thereby demonstrating that nitroxide partitioning had no effect in these conditions. This is also in line with predictions stating that nitroxide partitioning influences kinetics only in the absence of thermal initiation, *i.e.*, at temperatures below 100 °C.[38] Furthermore, it was then postulated that for such diameters the influence of the SDBS concentration on the enhanced production of radicals cannot be discarded, as suggested earlier.[21] Indeed, Cunningham *et al.* reported that SDBS concentration had a strong influence on the kinetics, whereas no such observation could be made in similar experiments with Dowfax 8390.[39] However, later work of Okubo and co-workers ultimately revealed that the enhanced thermal initiation was not specific to SDBS as Dowfax 8390-stabilized miniemulsions displayed similar characteristics.[40] The origin of these discrepancies is rather unclear. In the same paper, Okubo showed that the initially witnessed loss of control found in particles of 70 nm and less could be resolved by employing a higher initial PS-TEMPO concentration.

One of the very first reports on the use of PS-TEMPO oligomers gave rather surprising results.[18] Starting from a 1500 g mol^{-1} PS-TEMPO solution, which was then emulsified, miniemulsion polymerization of styrene at 135 °C could be brought to quantitative conversion (99.5%) while retaining low dispersity values in the 1.15 range. Even more surprisingly for a TEMPO-mediated polymerization, addition of *n*-butyl acrylate (BA) to this final polystyrene latex and re-heating the miniemulsion led to successful production of a PS-*b*-PBA block copolymer, also at full conversion (99.4%) with a narrow MMD ($Đ = 1.18$). To date, such results remain marginal. However, in the light of a later report by Lin *et al.*,[19] the high polymerization rate—at least—could potentially be explained by the employed high surfactant concentration.

While Keoshkerian *et al.*[18] did not use a rate accelerant, Lin *et al.* probed the influence of such an additive with the aim of shortening the polymerization time.[19] While CSA was known to be efficient in this respect, it is actually not very practical since, being a solid, it must be added directly at the beginning of the polymerization, resulting in a dramatic drop in nitroxide concentration from the start and leading to more termination. On the contrary, ascorbic acid, being water-soluble, can easily be added in semi-batch mode, thus allowing more flexibility and access to better control of the nitroxide concentration. This method allowed NMP of styrene in miniemulsion initiated by PS-TEMPO to reach full conversion in less than 3 h, with a linear increase of M_n and low $Đ$ values throughout the process (Figure 5.7).[41] High DOLs could be maintained even at very high conversions (*e.g.*, 0.8 for conversion higher than 99%) while polymerization time was indicated to be the main contributing factor to loss of livingness. Indeed, when a latex was kept at polymerization temperature for extended periods of time after full conversion was reached, a dramatic decrease of the DOL was observed (0.4 after 6 h of reaction).[19] Motivated by the successful use of ascorbic acid and having in mind that the requirement for pressure reactors is a serious complication for industrial translation, the same team aimed at lowering the reaction temperature from 135 °C to 100 °C. Besides the addition of ascorbic acid,[20] introduction of a water-soluble radical initiator was also assessed.[41] The latter is used to counter-balance the strong decrease of thermal initiation at these lower temperatures in comparison to the usual experiments at 135 °C, as it is a decisive factor in the determination of polymerization rate in TEMPO-mediated polymerizations.[42,43] Two initiators were tested: VA-060 and VA-085 (see Scheme 5.1). The former, which has a lower 10-hour half-life temperature (60 *vs.* 85 °C for VA-085), helped increase the polymerization rate but yielded unsatisfactory control over the polymerization. A slower production of water-borne radicals with VA-085 enabled a complete styrene consumption in 12 h at 100 °C and even permitted 75% conversion to be achieved at 95 °C for the same polymerization time. Nonetheless, ascorbic acid was finally determined to be more efficient in terms of DOL.

In order to look at the combined influence of the droplet/particle diameter and the solubility of the nitroxide species, Zetterlund's team employed a method based on the *in situ* formation of the surfactant, which leads to a number-average diameter in the 10–15 nm range, and compared TEMPO,

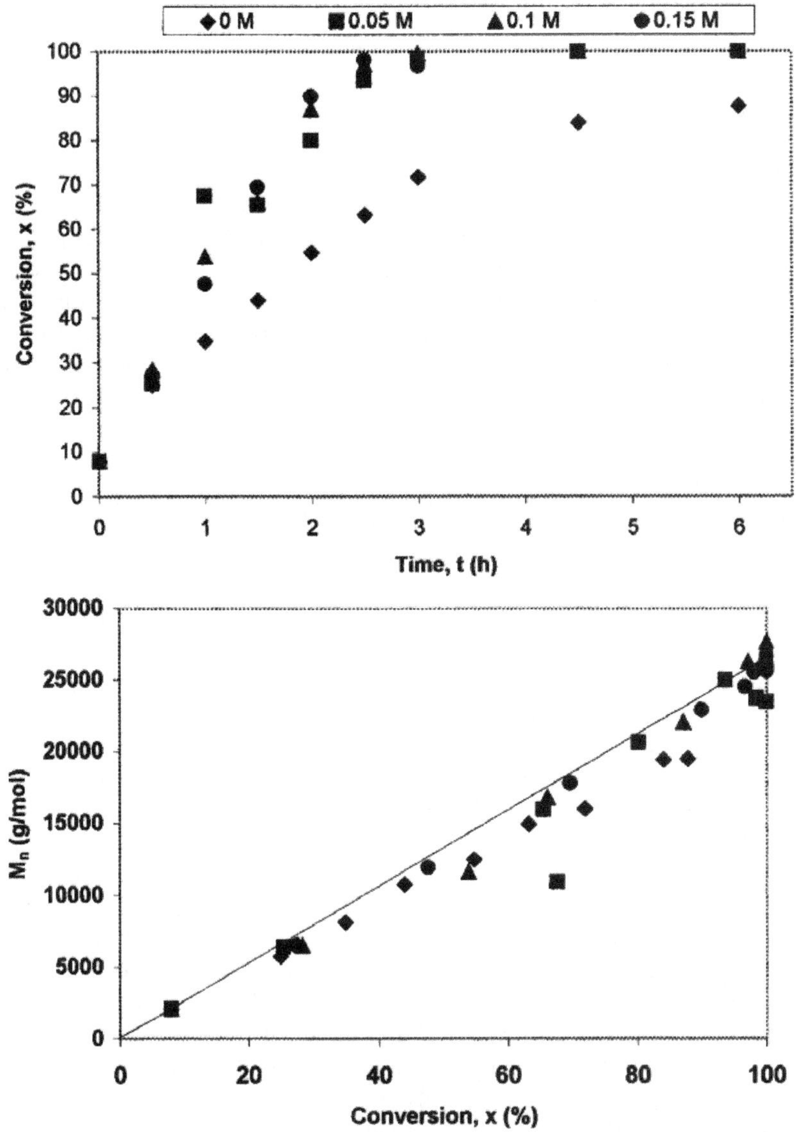

Figure 5.7 Influence of the concentration of ascorbic acid added in semi-batch mode on the kinetics (top) and molar masses (bottom) for the miniemulsion polymerization of styrene initiated by TEMPO-PS at 135 °C and 20 wt% theoretical solids content.
Reproduced from ref. 19 with permission from Wiley.

OH-TEMPO, and stearoyl-TEMPO (see Scheme 5.1) in this context.[44] As opposed to previous results obtained for larger particles[37] and predictions indicating an influence of the nitroxide partitioning only at low temperatures,[38] the solubility of the nitroxide had a clear influence on the outcome of the miniemulsion polymerization carried out in these conditions. For TEMPO-mediated miniemulsion polymerization, a 30-fold rate enhancement of the bulk kinetics was witnessed—which represents the fastest TEMPO-mediated styrene polymerization to date—along with reasonable control. Interestingly, molar masses and dispersity indices reached the same final values in bulk and miniemulsion but in the latter case were lower at intermediate conversions. Stearoyl-TEMPO gave kinetics and macromolecular characteristics similar to those obtained in bulk polymerization, whereas a drastically opposite behavior was observed when OH-TEMPO was employed. In the latter case, miniemulsion polymerization was even faster than for TEMPO, accompanied by a complete loss of control. It must be noted that in the present case, as opposed to experiments with bicomponent systems, there is no excess nitroxide here, which may enhance nitroxide exit from the particles. All in all, these results demonstrate that nitroxide exit can play an important role in the given conditions. In addition, the direct influence of oleic acid on partial nitroxide loss, hence kinetics, was alluded to. The authors therefore concluded that an interplay between particle size, nitroxide/initiator molar ratio, and oleic acid-induced nitroxide loss must have permitted the rate enhancement. Rate enhancement had actually been predicted for a lower particle size when nitroxide entry/exit operates, however together with a loss of control.[45] Differences may arise from broad particle distributions, where polymerization may proceed differently in the smaller and larger particles.

The *in situ* surfactant formation method was actually first assessed for SG1-mediated miniemulsion polymerization of styrene at 100 °C.[46] In that case, PS-SG1 ($M_n = 2370$ g mol^{-1}) was employed as initiator in the absence of a high-shear treatment step. Consequently, the droplet size was not as low as when the same system is ultrasonicated but produced droplets similar in size to classic miniemulsion. It must however be noted that, in spite of number-average monomodal populations, the volume-weighed distributions evidenced the presence of rather large particles (500–3500 nm),[47] in which the vast majority of the polymerization was certainly occurring. The polymerization exhibited a controlled character until approximately 80% conversion with $Đ < 1.2$. No comparison with bulk systems was drawn. In another report, SG1 was also involved in an oleic acid-based system in conjunction with ultrasonication, leading to an initial droplet size of approximately 15 nm.[48] As for TEMPO-mediated polymerization in these conditions,[44] the polymerization rate increased as compared to bulk polymerization, however to a lower extent (approximately one order of magnitude). The MMD remained narrow until at least 90% conversion but after 70% the molar masses became significantly higher than the theoretical ones, indicating potential bimolecular termination. EPR measurements revealed an important decrease in the half-time of SG1 in oleic acid at 90 °C (3 h *vs.* 15 h at 120 °C in other conditions, as previously reported). This could have potentially explained the increase in polymerization rate. However, solution polymerization in the presence of oleic acid in similar amounts displayed only a limited increase in polymerization rate. This study thus points at a favorable combination of oleic acid-based SG1 degradation and nitroxide exit leading to reduced deactivation, together with the classic effects of compartmentalization (segregation reducing irreversible termination and confined space favoring deactivation) to obtain fast polymerization with good control.

Besides all the aforementioned reports on "classical" styrene nitroxide-mediated miniemulsion polymerization, a few reports based on initiation with organosoluble macroalkoxyamines distinguish themselves either by the process employed or by the nature of the monomers. For instance, Cunningham and co-workers proposed a tubular reactor setup[49–51] as it may offer several interesting possibilities, such as manipulating the flow rate and developing controlled macromolecular microstructure by introduction of different monomers along the reactor line (see Figure 5.8). Also, pressure is more easily handled in such tubular reactors in comparison to stirred tank-type reactors. In their first report on this topic, the PS-TEMPO macroinitiator was synthesized by bulk polymerization of styrene in a batch reactor for a short time (up to 20% conversion) before emulsification of the PS-TEMPO/styrene raw mixture using a piston homogenizer.[49] Later, it was shown that this first step could also be carried out in a similar tubular reactor.[50] In that case, it was found that the polymerization proceeds slower than in a batch reactor, potentially due to significant thermal inertia during heating and cooling procedures inherent to batch reactors. Kinetics of miniemulsion polymerization were compared on the basis of residence time in the tubular reactor. For the TEMPO-mediated miniemulsion polymerization of styrene, kinetic trends in batch and tubular reactors were roughly identical, albeit slightly lower conversions were achieved for the flow reaction. This was initially attributed to possible mixing phenomena and later confirmed through a thorough residence time distribution study employing a dye tracer.[51] Macromolecular characteristics were conserved when going from a batch reactor setup to the flow reaction. Particularly, it was possible to carry out a rather clean chain extension, also in miniemulsion in the tubular reactor, after addition of a new load of styrene.[49] When BA was employed for chain extension, rather low polymerization rates were observed.[50] Therefore, ascorbic acid was introduced. This indeed resulted in faster kinetics and higher conversions but also in significantly broader MMDs. Nevertheless, so-formed PS-*b*-PBA diblock copolymers could partially reinitiate the polymerization of styrene to yield a mixture of diblock and triblock copolymers. Analogous to earlier work,[18] reducing reaction times (here residence times) permitted to slightly improve the situation.

Besides classical styrene miniemulsion polymerization, PS-based macroalkoxyamines have been used to initiate BA polymerization as well as the copolymerization of styrene and divinylbenzene (DVB). In the latter case, Okubo and co-workers observed that miniemulsion products could be rather different from those obtained in bulk for similar comonomer compositions, thereby warning against a simple assumption that miniemulsion can

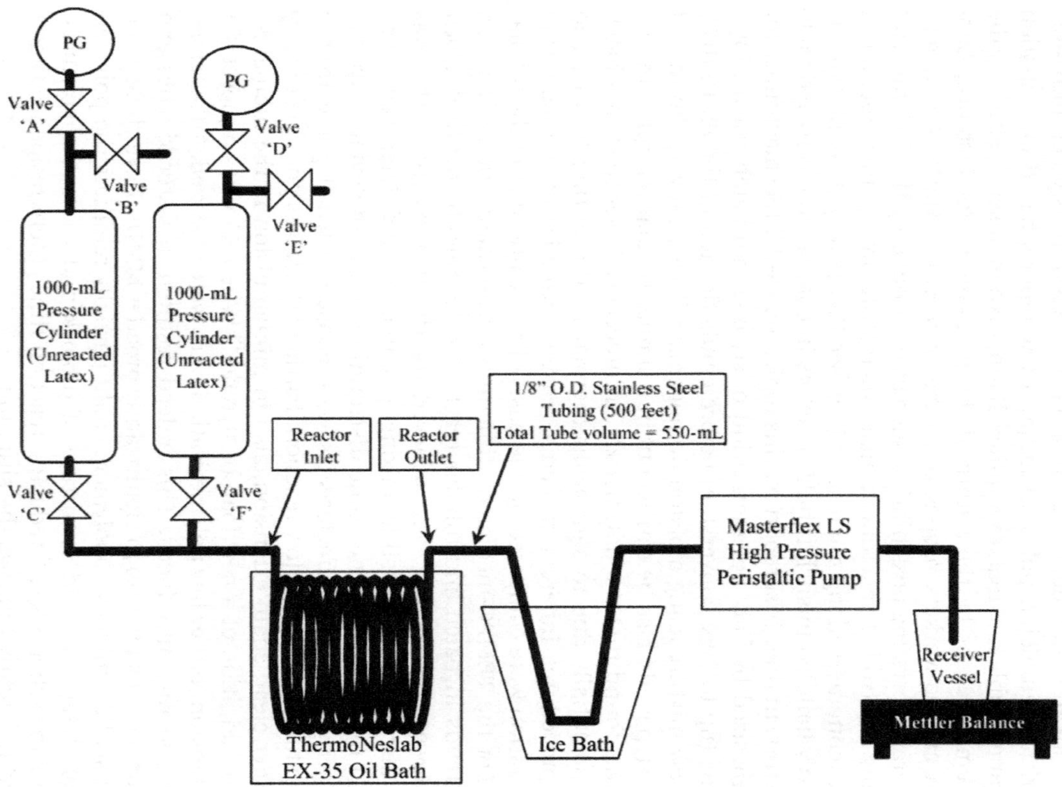

Figure 5.8 Example of a tubular reactor setup utilized by Cunningham and co-workers for the TEMPO-mediated miniemulsion polymerization of styrene at 135 °C. Reproduced from ref. 50 with permission from Wiley-VCH.

universally be considered as a bulk polymerization in nanoreactors.[52] Indeed, in addition to the rate enhancement previously observed, crosslinking reactions seemed to occur very differently. Both pendant conversions (*i.e.*, addition onto the second double bond of DVB units previously incorporated in polymer chains by addition on the first double bond) and crosslinking density increase with increasing styrene conversion, while the former is rather constant and the latter constantly diminishes in bulk experiments. While, as mentioned before, styrene polymerized faster in miniemulsion, DVB incorporation was interestingly slower in dispersed media and with decreasing particle size. This was attributed to the fact that DVB is more soluble in water than styrene, which leads to less DVB initially present on the locus of polymerization. In addition, a higher degree of intramolecular crosslinking in bulk was proposed as a possible explanation for the lower apparent crosslinking density in miniemulsion. This was corroborated by the fact that MMDs were not broader in bulk. It was rather the opposite, particularly at moderate conversions.[53] Further elements indicating an influence of the particle/water interface on the crosslinking process were disclosed. When a hydrophobe (tetradecane) was added to the recipe, a clear change in apparent pendant reactivity and MMDs occurred. Hydrophobes are known to promote the migration of PDVB to the interface of toluene droplets dispersed in water.[54] Finally, MMDs in miniemulsion for small particles (approximately 85 nm) exhibited two main peaks: one at low molar mass seemingly corresponding to non-crosslinked chains and another one at higher molar mass. The latter appeared to reach a maximal value at around 40% conversion. It was suggested that the limited space effect (first introduced by Tobita) was responsible for this observation: in a dispersed system, a given chain can only enter a crosslinking event with the chains present in the same particle but not with those located in the other particles, thereby limiting the extent of molar mass increase far away from the theoretical infinite value at the gel point in homogeneous crosslinking polymerization.[55,56] Beyond all these mechanistic considerations and the fact that understanding remains relatively poor, it is noteworthy that the present method is a practical way of obtaining non- or lightly crosslinked polymers with pendant double bonds.

Miniemulsion polymerization of BA initiated by PS-based macroalkoxyamines was rather early reported but only in a couple of instances. Tortosa *et al.* actually employed PS-TEMPO and PS-(OH-TEMPO) of significantly higher molar masses than in the aforementioned reports ($M_n = 12\,000$–$29\,000$ g mol^{-1}, $Đ = 1.15$–1.27) with the aim of producing PS-*b*-PBA diblock copolymers.[57] When these macroinitiators were used to initiate the miniemulsion polymerization of BA at 135 °C, in the presence of hexadecane and SDBS, OH-TEMPO-mediated polymerizations proceeded much faster than those with TEMPO. This, together with the presence of pure PBA, was attributed to the important partitioning of OH-TEMPO in the aqueous phase. MMDs were narrower when TEMPO was employed. Georges and co-workers also reported the successful NMP of BA in miniemulsion employing a styrene oligomer capped with a hydroxylated analogue of TIPNO ($M_n = 1150$ g mol^{-1}, $Đ = 1.15$), the nitroxide reported earlier by Braslau and Hawker and known to efficiently mediate acrylate polymerization.[58] After 3 h of polymerization at 135 °C in a miniemulsion stabilized by SDBS and co-stabilized by hexadecane, the conversion reached 86% with M_n and $Đ$ values of 12 000 g mol^{-1} and 1.27, respectively.[59] This implies that the experimental molar mass was twice lower than theoretically expected ($[BA]_0/[PS\text{-}(OH\text{-}TIPNO)]_0 = 200$). No comment on this aspect can be found. Comparatively, when Farcet *et al.* switched to a PBA-SG1 macroalkoxyamine ($M_n = 8800$ g mol^{-1}, $Đ = 1.76$) for the polymerization of BA in miniemulsion at 112 °C, an initiator efficiency close to unity was obtained at 80% conversion.[60]

Next to those based on organosoluble macroalkoxyamines, reports on the use of hydrophobic molecular alkoxyamines are significantly less abundant. Two compounds have been used: (i) BST, which is an alkoxyamine obtained by the addition of a BPO primary radical onto one styrene unit followed by direct trapping with TEMPO and (ii) MONAMS, which is a secondary alkoxyamine derived from methyl acrylate and SG1 (see Scheme 5.1). A hydrophobe was systematically employed in these cases (hexadecane, sometimes in conjunction with high molar mass PS). Georges and co-workers first used BST for the miniemulsion polymerization of BA at 135 °C with SDBS and hexadecane as stabilizing system.[59] After 4 to 6 h, only low conversions were observed and oligomers were obtained. The addition of ascorbic acid yielded faster initial polymerization rates and molar masses but again—probably due to the decrease in ascorbic acid concentration with time—kinetics became slower after a certain amount of time. Adding more ascorbic acid resulted in uncontrolled MMDs. The situation could be further improved by adding ascorbic acid continuously throughout the polymerization rather than entirely from the beginning.[61] In that case, in 6 h, half of the monomer was consumed to yield polymers with M_n and $Đ$ values of 18 500 g mol^{-1} and 1.35, respectively. Notably, the so-obtained PBA chains could well reinitiate the copolymerization of BA and styrene (equimolar ratio) in bulk conditions. Cunningham *et al.* also employed BST for polymerizing styrene in miniemulsion when they found that SDBS concentration could have a strong influence on the kinetics.[39] Together with the team of Georges, they also compared BST with a hydroxylated version (BST-OH, see Scheme 5.1), where the TEMPO fragment is simply replaced by OH-TEMPO.[38] Along with modelling, the comparison helped assess the role of thermal initiation (*vide supra*) and nitroxide partitioning on the kinetics and the controlled character of styrene miniemulsion polymerization. At high temperatures, similar results were obtained for both alkoxyamines and fitted the established mathematical model. Once thermal initiation was removed from this model, clear discrepancies arose.

Motivated by the development of a process in which polyacrylates are readily obtained with good control, the team of Charleux concentrated on MONAMS. A small amount of free SG1 (2.5 mol% with respect to MONAMS) was employed in order to control the polymerization from the beginning by artificially establishing the PRE. In 6 h at 115 °C, 80% conversion could be achieved while it took the same time at 120 °C to each 95%.[62] Molar masses increased linearly with conversion while *Đ* values remained lower than 1.4 until 80% conversion at 115 °C, however with broadening of MMDs occurring at 120 °C and high conversions. A latex with a 20 wt% theoretical solids content stabilized by a mixture of Forafac (see Scheme 5.2) and SDS exhibited good colloidal stability, however with poor control over the particle size (variation between experiments and ubiquitous broad distribution). It was then demonstrated that well-defined PBA-*b*-P(BA-*co*-S) diblock copolymers could be obtained by stopping the first BA polymerization at 80% conversion, injecting styrene into the non-purified latex (*i.e.*, containing residual BA), and re-heating at 115 °C. An overall conversion (BA and styrene) of 80% could be reached in approximately 8 additional hours. The molar masses matched the expected values and the MMD even became narrower. Liquid adsorption chromatography (LAC) analysis of the final copolymer evidenced the absence of pure PBA, hence complete re-initiation after addition of styrene. However, differential scanning calorimetry (DSC) measurements were unable to confirm a block copolymer structure as only a single transition was observed. Homopolymerization of BA could also be performed at lower temperature (112 °C) and it was shown that with the initial addition of a small amount of free SG1 the polymerization proceeded slower with a better control over MMDs.[63] In contrast to experiments at 115 °C, the initiator efficiency seemed lower, that is, the experimental molar masses exceeded the expected values. This issue was addressed in a further study where $[\text{MONAMS}]_0$, $[\text{SG1}]_0/[\text{MONAMS}]_0$, (the so-called *r* ratio) and temperature were varied.[60] While $[\text{MONAMS}]_0$ logically set the molar mass range, *r* had little influence on the number of chains (*i.e.*, M_n) yet significantly affected both the polymerization rate and dispersity. The higher *r* the slower the polymerization and the lower *Đ*. Interestingly, when the polymerization was initiated and carried out in bulk to low conversion before being pursued in miniemulsion (employment of PBA-SG1 as initiator), the initiator efficiency was markedly higher with all other values remaining mostly unchanged. Increasing temperature from 112 to 125 °C improved the initiator efficiency up to 100%. However, at higher temperatures a larger initial excess of free SG1 was required to counter the broadening of MMDs. For any $[\text{SG1}]_0$ this broadening was inevitable at conversion values over 80% and ascribed to intermolecular chain transfer to the polymer, leading to branched structures.

In RDRP, thanks to adequate copolymerization ratios and favorable kinetics (particularly similar $K.k_p$ products, as the low activation–deactivation equilibrium constant (K) of BA in SG1-mediated NMP compensates for its large k_p), it is possible to synthesize gradient copolymers of styrene and BA. This was demonstrated by their batch miniemulsion copolymerization initiated by MONAMS at 120 °C.[63] The fraction of styrene indeed constantly decreased with conversion while MMDs remained narrow and M_n increased linearly with conversion, suggesting a constant concentration of chains. LAC

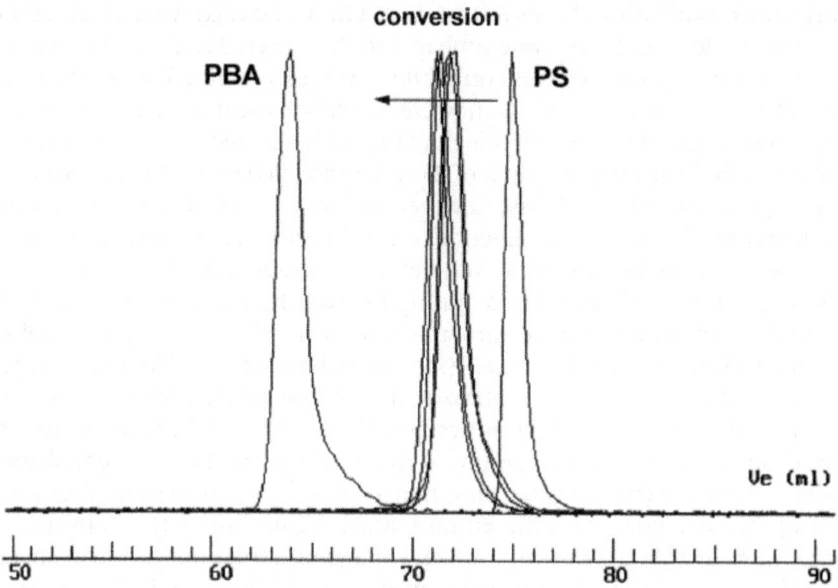

Figure 5.9 Liquid adsorption chromatograms of products obtained by miniemulsion copolymerization of BA and styrene initiated by MONAMS at 120 °C and comparison with PBA and PS homopolymers.
Adapted from ref. 63 with permission from Wiley.

assessed the pure copolymer-like nature of the chains and showed an enrichment of BA with conversion (Figure 5.9). This was finally corroborated by the decrease of the single glass transition temperature observed with conversion.

5.2.3.4 *Water-soluble Monocomponent Initiating System*

Only one example of water-soluble alkoxyamine initiator was reported in the frame of miniemulsion polymerization. Starting the polymerization in the aqueous phase brings the process closer to an emulsion mechanism. For this purpose, the methyl acrylate fragment of MONAMS was replaced by a methacrylic acid one in the new water-soluble alkoxyamine MAMA (commercial name Blocbuilder MA, see Scheme 5.1).[64] The carboxylic group imparts water-solubility above pH = 4 and the tertiary carbon connected to the nitroxide ensures a high dissociation rate. The latter feature permits a faster establishment of the PRE and often removes the need for added SG1. For the polymerization of BA at 112 °C, polymerizations proceeded with very high initiator efficiency, demonstrating adequate kinetics in the aqueous phase and efficient entry of the oligoradicals in the droplets. Styrene—at 120 °C—behaved rather differently: While all other living polymerization criteria were fulfilled, initiator efficiency was initially very low (approximately 30–35%) and slowly increased with conversion (50% at 60% conversion). In addition, before exhibiting classic first-order kinetics, an induction period of an hour during which very small amounts of monomer are consumed was observed. This deviation from the ideal behavior was explained by two phenomena: (i) the large dissociation rate of MAMA and the large equilibrium constant *K* for PS-SG1 lead to fast termination in the aqueous phase, thereby releasing an important quantity of free SG1 by PRE and (ii) due to this large concentration of SG1 in the aqueous phase combined with a slow propagation rate, reversible deactivation of short water-soluble oligomers may also happen in water. This would lead to a slow and incomplete entry in the droplets. To improve this situation, the trick was to introduce a small amount of a monomer exhibiting a greater water solubility, a larger polymerization rate constant, and a lower *K*. Methyl acrylate (MA) fulfils these requirements. With a MA/MAMA molar ratio as low as 2.5 : 1 (corresponding to $[MA]_0/[styrene]_0 = 0.008$), the induction period disappeared and $Đ$ values did not exceed 1.25, even at more than 90% conversion. With this ratio equal to 10 (*i.e.*, $[MA]_0/[styrene]_0 = 0.031$), initiator efficiency approached 100%.

A common concern in NMP in miniemulsion is the systematic development of broad PSD. The use of MAMA clearly improved this aspect. Compared to similar experiments conducted with MONAMS, the PSD was narrower and also drastically shifted to lower diameters, as demonstrated by capillary hydrodynamic fractionation (CHDF) (Figure 5.10). The additional charges present at the α chain ends originating from the MAMA initiating fragment were suspected to be responsible for this enhanced stability.

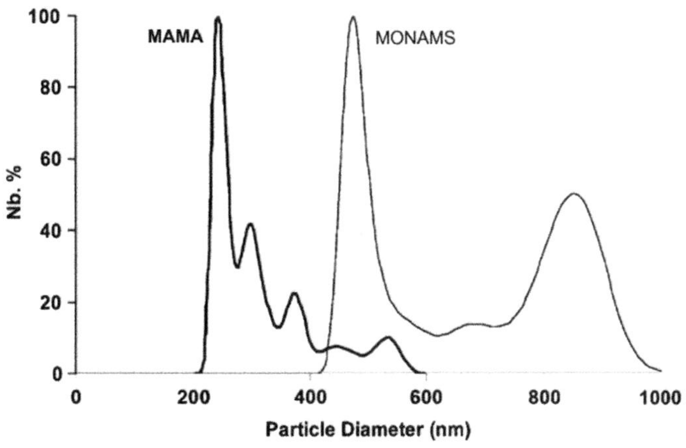

Figure 5.10 Number distributions of particle diameters obtained by CHDF for the nitroxide-mediated miniemulsion polymerization of BA initiated by MAMA (thick line) and MONAMS (thin line) in identical conditions. Adapted from ref. 64 with permission from the American Chemical Society.

5.2.4 NMP in Emulsion

Despite progress in nitroxide-mediated miniemulsion polymerization, the pursuit of a true emulsion process was never abandoned as it avoids the high-shearing step, which can be problematic for upscaling. In addition, as mentioned just above, the PSD remained broad in most cases.

On the other hand, many nitroxide-mediated emulsion polymerization attempts have failed in terms of colloidal stability, due to the complexity of the nucleation step. For instance, one issue of styrene NMP in emulsion is that the required elevated temperatures lead to creation of radicals in the droplets by thermal initiation and subsequently destabilization. It is also for this reason that initiation must occur in the aqueous phase. Nitroxide partitioning is also a great concern as a too high concentration of nitroxide in the aqueous phase can lead to a retardation of the nucleation, hence slow polymerizations. All these factors may also induce a poor control of the polymerization. Ideally, the initiation should only take place in the aqueous phase and lead to the rapid formation of oligomers reaching a critical length for exclusion from the aqueous phase, in combination with appropriate concentrations of nitroxide in both aqueous and organic phases.

5.2.4.1 *Initiation by a Water-soluble Radical Initiator in the Presence of Nitroxide*

Rather logically, the first attempts were based on classic emulsion radical polymerization where a nitroxide was simply added. The team of Claverie evaluated a series of TEMPO derivatives at 130 °C with KPS initiation at 9 wt% theoretical solids content: TEMPO, OH-TEMPO, carboxy-TEMPO, and amino-TEMPO (see Scheme 5.1).[65] The authors first commented on the stability of the emulsions with regards to surfactants. Since long polymerization times at elevated temperatures are often required for TEMPO-mediated styrene polymerization, degradation of surfactants can be an issue. Indeed, dodecanol could be detected after 21 h when SDS was employed. Nevertheless, stable lattices with amino-TEMPO could be obtained. However, SDBS, Aerosol-MA80 (see Scheme 5.2) and other derivatives did not lead to stable lattices despite their higher hydrolytic stability. Dodecanol was actually found to play a decisive role in the colloidal stability, as demonstrated by an experiment where it was deliberately introduced in the recipe, even when SDS was replaced by SDBS. As just mentioned, apart from amino-TEMPO, all polymerizations with other nitroxides resulted in almost no polymerization. The best results were obtained with amino-TEMPO and dodecanol: 70% conversion was achieved in 55 h. However, only low molar mass polymer was obtained in this way ($M_n = 6000$ g mol^{-1} and $Đ = 1.7$). Replacing KPS with another radical initiator, V-50 (see Scheme 5.1), only induced a slight decrease of the polymerization rate but had no influence on other parameters. In any case, all lattices exhibited excessively broad PSDs.

Cao *et al.* carried out a similar study, yet at relatively higher theoretical solids content (33 wt%) with nitroxides of various hydrophilicities: TEMPO, OH-TEMPO, acetoxy-TEMPO, and benzoyloxy-TEMPO (see Scheme 5.1).[66] The polymerization was initiated by KPS and carried out at 120 °C in the presence of SDS. With TEMPO and benzoyloxy-TEMPO, no signs of a controlled polymerization were observed. With the latter, coagulation happened during polymerization. With OH-TEMPO, an improvement was obtained. Despite only 1% conversion and the presence of a large amount of oligomers after 1 h, almost 40% conversion was reached in 12 h with $M_n = 30\,000$ g mol^{-1} and $Đ = 1.4$. A slight flocculation occurred in the course of the polymerization. Acetoxy-TEMPO yielded the best results, both in terms of stability (no flocculation) and polymerization control with $Đ = 1.3$ at 80% conversion in 12 h. A rather small number-average particle size was obtained for both OH-TEMPO and acetoxy-TEMPO after 12 h: 46 and 65 nm, respectively. Nevertheless, no detail was given regarding the PSD.

Later, Maehata *et al.* investigated the influence of a more hydrophobic TEMPO derivative for the emulsion polymerization of styrene at approximately 30 wt% theoretical solids content at 135 °C, initiated by KPS in the presence of SDBS.[67] With TEMPO, controlled polymerization occurred (50% conversion, $Đ = 1.19$) but large particles and 25% of coagulum were formed (Figure 5.11). When TEMPO was replaced by stearoyl-TEMPO, a narrow and monomodal distribution of small particles was achieved ($d_n = 45$ nm, $d_v = 51$ nm), however with absolutely no control of the polymerization. It is therefore implied that stearoyl-TEMPO remains in the monomer droplet and does not participate in the main polymerization step. It only acts as a trap for radicals

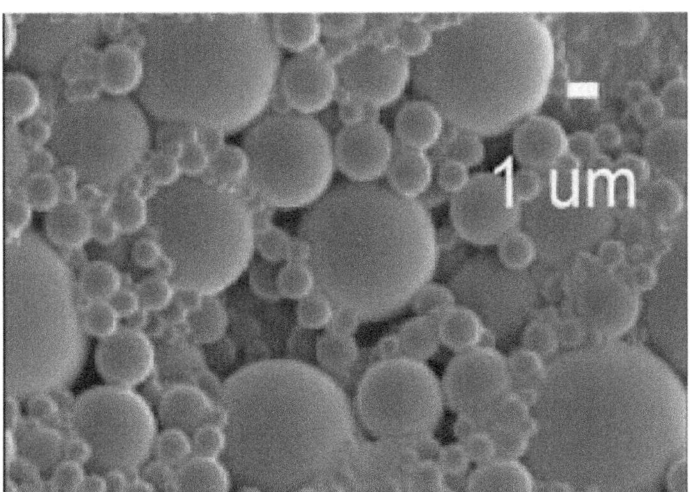

Figure 5.11 Scanning electron micrograph showing the classical type of large particles present in coagulum and suspected to originate from thermal initiation in the droplets.
Reproduced from ref. 67 with permission from Wiley.

created by thermal initiation in the droplets, which are suspected to be the cause of coagulation.[68] This led the authors to a clever trick: employing both TEMPO and stearoyl-TEMPO in the same emulsion polymerization. With a stearoyl-TEMPO/TEMPO molar ratio of 1.3 : 1, not only was the polymerization controlled but also small particles with a narrow distribution were obtained, perfectly marrying the best features of the polymerizations carried out with each of the nitroxides independently. Lowering the polymerization temperature to 120 °C while keeping all other conditions identical did not drastically change the kinetics but the control of both molar mass and particle size slightly deteriorated.

Two more recent reports on TEMPO-mediated emulsion polymerization initiated by KPS deal with the effect of additives, either sodium bicarbonate or methanol. Montezuma *et al.* examined the influence of sodium bicarbonate for low-temperature polymerizations.[69] It was demonstrated that quantitative conversion could be achieved in less than 90 min at 90 or 95 °C after an initial inhibition period, for which the duration depends on the amount of additive and the temperature. Molar masses increased linearly with conversion and *Đ* remained mainly below 1.3. Decreasing the temperature led to significantly higher molar masses, which was attributed to a lesser amount of chains created at lower temperature. Adding more sodium bicarbonate led to an increase in pH and narrower MMDs, which could be assumed to be linked to the relative instability of nitroxide in a more acidic environment. Stable lattices at 20 wt% solids content were obtained with a relatively narrow population of small particles (30 nm) accompanied by a limited amount of larger particles (approximately 300 nm).

Saldívar-Guerra and co-workers proposed to introduce methanol in the typical emulsion recipe to counter the so-called superswelling effect.[70] Colloidal instability in nitroxide-mediated emulsion polymerization, as well as other RDRP protocols carried out in emulsion, originates from the intrinsic features related to a relatively slower (and simultaneous) growth of most—ideally all—polymer chains as compared to conventional radical polymerization. This leads to the presence of a very large concentration of oligomers at the beginning of the polymerization. The issue arises from the fact that oligomers are efficient swelling agents, leading to at least 100-fold swelling factors,[71,72] *i.e.*, superswelling.[73] When superswelling is too important, micrometric particles can be formed for which buoyancy phenomena lead to the formation of an upper organic layer.[74] Various parameters—*e.g.*, droplet size, concentration of controlling agent and costabilizer, and molar mass of the latter—can affect swelling.[73] In a study involving a rather complex multi-step process, the addition of methanol (25 wt% with respect to styrene) helped going from a multimodal particle distribution to a monomodal one. In addition to the instability of the lattices manifested shortly after the polymerization and the use of an organic cosolvent (methanol), the fact that only 30–40% conversion was achieved at 5 wt% theoretical solids content makes this report of rather limited practical interest.

Besides TEMPO derivatives, SG1 has also been evaluated as an additive in classic emulsion polymerization recipes, allowing for lower polymerization temperatures. Following previous results obtained in miniemulsion,[27,33] Lansalot *et al.* employed a co-initiator couple (KPS/NaMBS) with $[SG1]_0/[KPS]_0 = 1.2$ at 90 °C and 20 wt% theoretical solids content.[27] Rather slow kinetics (40% conversion in 6 h) accompanied a linear increase of molar masses, however with large MMDs ($Đ = 2.2$). In addition, the obtained latex was seemingly unstable, as evidenced by the formation of 5–10 wt% coagulum with respect to the monomer content (*i.e.*, 12.5–25 wt% with respect to the formed polymer at 40% conversion). Simms *et al.* evaluated a two-step process with $[SG1]_0/[KPS]_0 = 1.7$ at 90 °C in the absence of surfactant.[75] After a first step where a potassium carbonate solution containing only 1.6 wt% styrene was heated for 3.5 h, an additional charge of styrene was introduced for a 10 wt% overall theoretical solids content. After the onset of the second step, the polymerization proceeded slightly faster than in the study of Lansalot *et al.* The molar masses increased with conversion yet not strictly following a linear trend—inspection of the data could reveal some decrease of the number of chains—and with rather broad MMDs (final, $Đ = 1.77$). Nevertheless, the polymer chains exhibited a good livingness, as demonstrated by successful chain extension with BA in bulk. Replacement of potassium carbonate by sodium bicarbonate required a doubled concentration, which led to a decrease in stability due to an electrolyte effect (only the charges offered by KPS initiating fragments act as stabilizers). Decreasing the amount of styrene in the first step did not provide better results.

While some studies on the use of a bicomponent system gave results that are reasonably satisfactory, the inherent issue of initiator efficiency persists with regards to obtaining predictable molar masses. Therefore, efforts were directed to employing alkoxyamines in emulsion polymerization, particularly after the reasonable success met in miniemulsion.

5.2.4.2 *Molecular Alkoxyamines*

Two of the earliest reports on the use of alkoxyamines in emulsion were based on organosoluble species. Obviously, in that case, a true *ab initio* process is not possible and seeded polymerization emulsion was performed. Actually, the very first reported attempt at applying NMP to an emulsion process by Bon *et al.* was based on a *tert*-butylperoxide styrene monoadduct trapped by TEMPO (see Scheme 5.1, tBST).[17] A polystyrene seed latex (90 nm) obtained by conventional free-radical polymerization was swollen with a styrene solution of tBST at a styrene/PS mass ratio of 3.1 : 1. At 125 °C the conversion reached 80% in 24 h. Full conversion was achieved in at most 36 h. Molar masses increased with time while dispersity values remained between 1.4 and 1.55, even at full conversion. The experimental number-average molar masses were notably lower than the theoretical values at any conversion, accounting for the creation of a rather large amount of additional chains, possibly through thermal initiation. The authors indicated the absence of secondary nucleation or coagulation. Szkurhan and Georges prepared a living PS seed by a nanoprecipitation technique.[76] A PS-TEMPO ($M_n = 3300$ g mol^{-1}; $Đ = 1.13$) was produced by bulk polymerization, dissolved in acetone, and added dropwise to a PVA aqueous solution (1 wt%) under stirring. After acetone evaporation, a monomodal seed of 470 nm average diameter was obtained and further swollen with styrene (up to 20 wt%). After heating at 135 °C for 6 h, typical conversions were between 60 and 70%, depending on the solids content. Final MMDs were narrow ($Đ = 1.11$–1.12, up to 80% conversion at 8 h) and the lattices were stable with a monomodal PSD (400–500 nm in average).

In the view of establishing a successful *ab initio* process, developing processes with water-soluble alkoxyamines in essential. In fact, Marestin *et al.* early reported such as attempt.[65] KSSSNaT, an alkoxyamine based on a KPS-based monoadduct of sodium styrene sulfonate trapped by TEMPO (see Scheme 5.1), was synthesized and used to initiate the batch *ab initio* polymerization of styrene at 130 °C. Even in the absence of additional surfactant, narrow MMDs could be obtained, yet substantial flocculation appeared for solids content as low as 2 wt%. With the addition of SDS or SDBS, stability is achieved, however with a bimodal distribution evenly split (in weight) between narrow and broad populations centered on 70 nm and 200 nm, respectively. Initial addition of 0.3 equivalents of amino-TEMPO reduced the final dispersity value from 1.7 to 1.2 at the expense of the polymerization rate.

Pleased with the progress obtained in miniemulsion using MAMA,[64] the team of Charleux sought to employ this water-soluble alkoxyamine in emulsion polymerization. First attempts at a batch *ab initio* process resulted in controlled MMDs but lattices were systematically destabilized around 50% conversion. A two-step process was therefore utilized in order to avoid nucleation in the droplets by purely suppressing their existence.[77] The first step consisted of the batch *ab initio* polymerization of a small amount of BA (0.7 wt%) initiated by MAMA in its ionized form, in the presence of a concentration of surfactant unusually large for an emulsion process and more akin to that of a microemulsion. After 8 h at 112 °C, a stable dispersion was produced with an average particle size between 150 and 230 nm, depending on the surfactant concentration, yet with a rather broad distribution, possibly due to limited superswelling. Importantly, well-defined PBA oligomers were obtained with $M_n = 1100$ g mol^{-1} and $Đ = 1.14$–1.20. The second step consisted of the swelling of this dispersion with either BA or styrene to reach a theoretical content of approximately 16 wt%. No additional surfactant was introduced, which therefore led to a concentration in the typical range for emulsion polymerization (2.2–5.1 wt% with respect to monomer). The kinetics of polymerization at 112 °C and 120 °C for BA and styrene, respectively, were rather fast (85 and 95% conversion in 8 h, respectively). In the case of BA, for an initially larger seed particle size, the polymerization proceeded faster. Characteristics of a controlled polymerization were observed: the molar masses increased linearly with conversion and were close to expected, while Đ remained below 1.3 and 1.4 until 90% conversion for styrene and BA, respectively. Only above 95% of BA converted into PBA was a sudden broadening of MMDs observed. Shelf-stable dispersions were obtained in all cases, nevertheless with broad PSDs. Gratifyingly, the number of particles seemed to remain constant over the entire course of the polymerization. Finally, a three-step process was developed for the synthesis of a PBA-*b*-P(BA-*co*-S) diblock copolymer, analogous to earlier work from the same team.[62] For this purpose, the BA emulsion polymerization mixture was stopped at 55% conversion in the first step and cooled down before styrene was added for an overall theoretical solids content of 26 wt%. After swelling, deoxygenation, and increase of the temperature to 112 °C, the polymerization proceeded with a controlled character again. For a reaction time of 14 h (second and third step, at polymerization temperature), an overall conversion of 80% was achieved with $M_n > 60\,000$ g mol^{-1} and $Đ < 1.4$, within stable particles (Figure 5.12).

When the same two-step protocol was employed with a difunctional water-soluble alkoxyamine DIAMA (see Scheme 5.1), synthesized by the radical 1,2 addition of two MAMA molecules onto one molecule of tri(ethylene glycol) acrylate, a great improvement of the colloidal characteristics occurred.[78] Allegedly due to the double charge borne by DIAMA, the initial PBA seed exhibited a reduced diameter in comparison to MAMA-initiated seeds (Figure 5.13). It was also possible to control the particle size of the seed by varying the amount of Dowfax 8390. Nevertheless, it was found—at least in the studied range—that this parameter had little influence on the kinetics and macromolecular characteristics during the seeded emulsion polymerization. Indeed, while seeds were made with various surfactant amounts, the Dowfax 8390 concentration was adjusted to the same value in all seeded polymerizations by addition of surfactant between the two steps. The only differences were found in the greater narrowness of the MMDs at high conversion for the small seed particles in the case of BA. The initiator efficiency was close to 1 in the case of BA while it was notably lower for styrene. Interestingly, the styrene-based lattices exhibited stable pH values above 8, while it steadily decreased with conversion for PBA lattices. Also, in one case, after 70% conversion of styrene, molar masses seemed to come closer to theoretical values and a tailing appeared in the SEC traces. This was ascribed to hydrolysis of some polymer chains in their middle, at one of the ester bonds of DIAMA. The most important difference with MAMA-based lattices, as mentioned above, could be found in the transition to monomodal PSDs. Particularly, despite similar final surfactant concentrations, particle sizes, and reciprocally particle numbers, could be controlled *via* the first step, *i.e.*, the seed preparation. The higher the surfactant concentration in the first step, the greater the particle number and the smaller the final particles. Not only does the difunctional nature of DIAMA allow for a better colloidal stability but also it enables the synthesis of triblock copolymers in two steps of propagation. Therefore, a PBA latex was brought to approximately 70% conversion by seeded emulsion polymerization, swollen with styrene in the

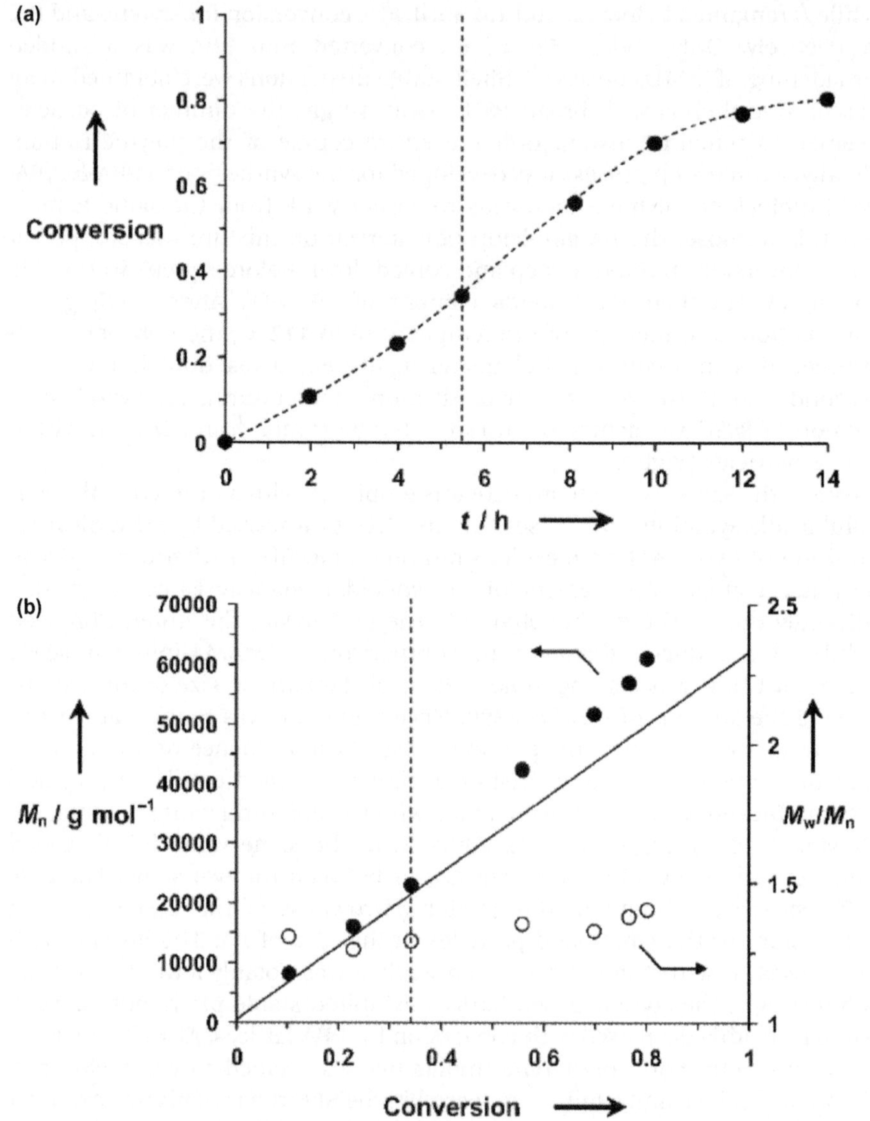

Figure 5.12 Synthesis of a PBA-*b*-P(BA-*co*-S) diblock copolymer by a three-step emulsion polymerization process initiated by MAMA at 112 °C and 26 wt% theoretical solids content.
Reproduced from ref. 77 with permission from Wiley.

presence of additional surfactant, and re-heated to 112 °C. Clear shifts of SEC traces were detected at least until 70%, after which a shoulder started to appear, potentially due to termination by combination. LAC characterization ascertained the absence of PBA homopolymer.

The two-step process described above with MAMA was further improved in terms of reaction time.[80] Indeed, while the seed was previously produced in

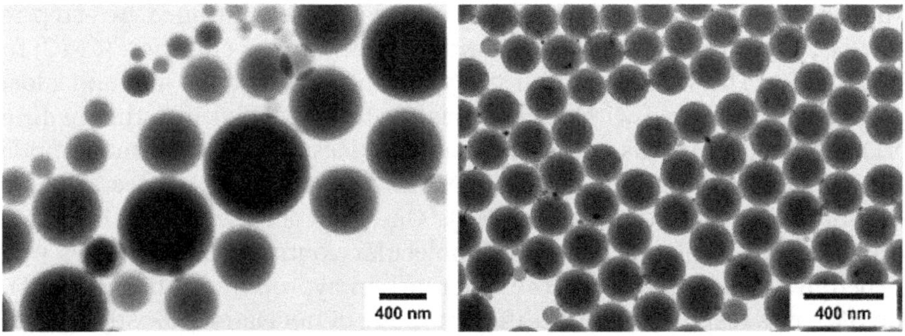

Figure 5.13 Transmission electron micrographs of the final PS latexes synthesized by the two-step SG1-mediated emulsion polymerization process, initiated with either MAMA (left) or with DIAMA (right) at equal surfactant concentration.
Reproduced from ref. 79 with permission from Elsevier.

8 h, it was shown that satisfying results could be obtained with a seed latex polymerized for only 2 h as there was only a small variation in terms of conversion and molar mass in between and a stable dispersion could already be produced in 2 h. Pleasingly, it was even found that a shorter seed polymerization time favored a faster seeded emulsion polymerization in the case of BA. This is explained by the accumulation of free SG1 with increasing conversion during the seed synthesis due to irreversible termination. For a seed polymerization time of 2 h, 90% conversion in the seeded emulsion polymerization could be achieved in 4 h, even with slightly superior macromolecular features (lower $Đ$ and higher initiator efficiency). In the case of styrene, no such effect on the kinetics was observed as the establishment of the PRE in this case induces a greater amount of released SG1, overcoming the amounts arising from the seed. Nonetheless, a slight improvement in initiator efficiency and MMD narrowness could be detected. A further improvement in overall reaction times could be reached by polymerizing the seed for only 30 min before starting the semi-continuous addition of BA for 3.5 h. Nearly full conversion was obtained in a total reaction time of 6.5 h for the two steps. Shifting the start of the addition to a later point in time only resulted in longer polymerization times without any noticeable change in MMDs. Stables lattices with a broad distribution similar to those obtained in the 2-step process were produced. Substituting MAMA with DIAMA yielded the same effect as in the 2-step process, *i.e.*, no change in MMDs but a dramatic narrowing of the PSD.

Besides the polymerization of styrenics and acrylates, alkoxyamines based on SG1 are able to mediate the polymerization of methacrylates at reduced temperature (75–90 °C), provided a small amount of a comonomer with a low K (and preferably a lower k_p) is added.[81–83] Therefore, Thomson *et al.* envisioned the possibility of applying the semi-batch emulsion polymerization process to *n*-butyl methacrylate (BMA) in the presence of 10 mol% of styrene at 90 °C, for theoretical solids content of approximately 9 wt%.[84] First experiments were carried out similarly to the aforementioned two-step recipe, *i.e.*, with surfactant above its critical micellar concentration (CMC) for the first step. The seed formation step was conducted over 20 min and a load of methacrylate mixed with styrene was continuously fed into the medium for 3 h. When MMA was employed, stable lattices with a broad monomodal PSD were obtained along a narrow MMD ($Đ = 1.33$; 65% conversion, approximately 80% initiator efficiency). On the contrary, when BMA was polymerized, despite a good macromolecular control, bimodal PSDs were formed. It was suggested that the difference in hydrophilicity between MMA and BMA was responsible for a different mode of nucleation (mostly micellar nucleation for MMA *vs.* a combination of micellar and homogeneous nucleation for BMA). It was therefore cleverly foreseen that the absence of surfactant would lead to monomodal PSDs. Indeed, the surfactant-free semi-batch emulsion polymerization yielded very narrow PSD centered on 230–250 nm. Unfortunately, this was at the expense of the initiator efficiency, which decreased from 70 to 55%. Inspired by earlier work from Nicolas *et al.* in miniemulsion,[64] the addition of methyl acrylate—more hydrophilic and fast propagating, thereby enhancing particle entry—permitted 65–70% initiator efficiency to be reached. Addition of free SG1 at the very start of the whole surfactant-free process resulted in higher initiating efficiencies but again bimodal PSDs arose, possibly due to an incomplete seed formation at the time the monomer feed was commenced.

The same team further evaluated the semi-batch methodology for PBA lattices with rather high solids content (up to 45 wt%). In that case, a difunctional alkoxyamine (DIAMINS) similar to DIAMA yet slightly less hydrophilic was employed.[85] Particular attention was paid on the conditions employed during the first stage (seed formation). Various factors, such as the nature of the buffer and of the base employed to neutralize DIAMINS, the concentration of the latter, and the ionic strength, were found to greatly impact the stability. Superswelling was identified to be responsible for destabilization of the latex in general. It was necessary to keep the particles in the first stage as small as possible to ensure a sufficient colloidal stability during the seeded emulsion polymerization. In parallel, when $[\text{DIAMINS}]_0$ increased, particle size also increased (see Figure 5.14), which in turn made it complicated to synthesize lattices with low molar masses. When small particles were synthesized, narrow MMDs could not be maintained until the end of the polymerization, whereas this was possible to a better extent with larger particles. Nevertheless, stable dispersions with 45 wt% solids content were obtained with narrow MMDs up to 60–70% conversion, after which broadening was observed.

In a study dedicated to the synthesis of P(MMA-*co*-BA)-*b*-PBA-*b*-P(MMA-*co*-BA) triblock block copolymers and their thorough characterization by two-dimensional (2D) chromatography, Moyses *et al.* also employed a multi-step protocol with DIAMINS.[86] In this particular case, a conventional PMMA seed was first synthesized, which was followed by seeded emulsion polymerization of BA in three stages (hourly sequential addition of 15, 42.5, and

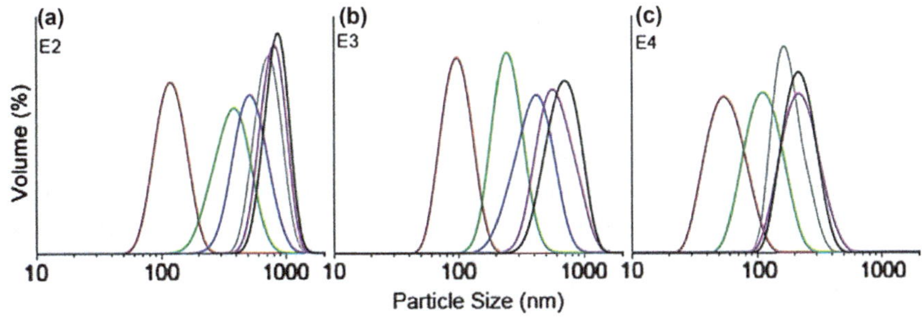

Figure 5.14 Volume-average particle size distribution obtained by dynamic light scattering (DLS) for the semi-batch emulsion polymerization of BA at 120 °C and 40 wt% theoretical solids for various targeted molar masses. (a) 70 kg mol^{-1}, (b) 87 kg mol^{-1}, (c) 131 kg mol^{-1}.
Reproduced from ref. 85 with permission from the Royal Society of Chemistry.

42.5% of the total BA content for a total theoretical solids content of approximately 40 wt%). At the end of this first stage, 83% BA conversion was obtained and the latex was swollen with MMA (overall mass ratio BA/MMA = 2.65 : 1). 2D chromatography with gradient polymer elution chromatography in the first dimension (hexane to tetrahydrofuran) and conventional size-exclusion chromatography in the second one enabled thorough analysis of the products formed in these conditions. Although no real control of the polymerization *per se* was reported ($Đ > 2$ for the first block and $Đ > 4$ for the triblock copolymers), results from 2D chromatography showed that upon appropriately chosen conditions block copolymers can be obtained.

5.2.4.3 *Water-soluble Macromolecular Alkoxyamines (NMPISA)*

Following the breakthrough achieved by using water-soluble alkoxyamines such as MAMA and more importantly DIAMA,[79] the team of Charleux sought to investigate the ability of water-soluble macroalkoxyamines as initiators in batch *ab initio* emulsion polymerization. Notably inspired by the work of Gilbert and co-workers on a reversible addition-fragmentation transfer polymerization (RAFTP) variant using starved-feed conditions,[87] it was envisioned that not only could macromolecular water-soluble chains capped with SG1 mediate the polymerization but they would also be able to stabilize the formed nanoparticles. Hence, first trials were directly performed in surfactant-free conditions. These experiments were rendered possible by the then-recent discovery of conditions to produce well-defined poly(acrylic acid) (PAA) by SG1-mediated polymerization.[88] It was indeed shown that PAA with degrees of polymerization up to 40–50 could be obtained with high DOL (>90%).[89] Therefore, a PAA-SG1 ($M_n = 1900$ g mol^{-1}; $Đ = 1.17$) was produced and directly dissolved in basic water (1 eq. NaOH with respect to acrylic acid units, 36 mM K_2CO_3) before styrene or BA was added.[90] The theoretical solids content was 20 wt% and the mass ratio of PAA to monomer was 5 wt%, which is in the range of surfactant concentration typically employed in more conventional recipes. The polymerization proceeded rather fast with essentially full conversion achieved in 6–7 h and 8 h for BA and styrene, respectively. M_n increased linearly with conversion and $Đ$ did not exceed 1.5 until at least 80% conversion. In the case of BA, a dramatic broadening of MMD was observed above this value, which was ascribed to potential transfer to the polymer and irreversible terminations.[91] All lattices were stable without noticeable fouling. Importantly, the obtained PSD were very narrow with exceptionally low particle size: 65 and 90 nm for PS and PBA, respectively, as determined by dynamic light scattering (DLS) at pH 7. The homogeneity of the particle size distribution was confirmed by transmission electron microscopy (TEM) in the case of PS lattices (Figure 5.15).

When the pH of the final latex was decreased, the average particle size also decreased as a consequence of the responsive character of the PAA shell. Below pH 4, destabilization was observed, which can be seen as a practical way to recover the polymer. Various parameters were further investigated, such as the length and the concentration of PAA-SG1, the nature of the added salt, and the initial presence of SG1.[92] It should be noted that initiator efficiencies in this system were relatively low and depended more on the type of polymerized monomer than on other parameters. They were typically in the 25–50% and 40–60% ranges for styrene and BA, respectively.

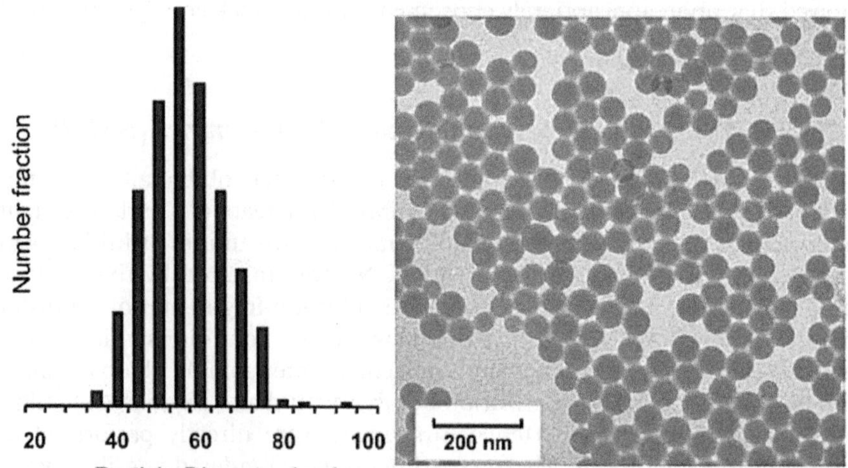

Figure 5.15 (Left) Particle size distribution obtained by manual count from TEM images. (Right) Typical TEM image obtained for a polystyrene latex synthesized by surfactant-free polymerization initiated by PAA-SG1 at 120 °C.
Reproduced from ref. 90 with permission from the Royal Society of Chemistry.

Interestingly, varying the length of PAA-SG1 enabled tuning of the particle size. The longer PAA-SG1, the larger the particles. For a PAA-SG1 of a given degree of polymerization, decreasing its concentration had the same effect.[93] With this method, it was possible to obtain stable lattices with solids contents as high as 39 wt% with an invariable level of quality in terms of stability.[94] The calculation of the number of chains per particle and the area occupied per chain at the surface of the particles indicated values close to those obtained by self-assembly of amphiphilic block copolymer into crew-cut micelles, while the self-assembly process should have been more likely proceeding with star-like aggregates. Morphological rearrangements observed in further studies (*vide infra*) may explain this phenomenon. These experiments represent the first polymerization-induced self-assembly (PISA) example ever carried out in *ab initio* conditions, Gilbert's work being the true pioneer employing seeded conditions.[87] This term was first employed by Hashimoto in solid state,[95] employed by Pan and co-workers for dispersion polymerization in the frame of the RAFTP variant,[96] and is now commonly referred to as PISA.[97] PISA particularly enables the synthesis of the entire range of morphologies attainable by classic self-assembly of amphiphilic block copolymers.[98] Notably, the first display of such possibilities was based on the PAA-SG1-initiated polymerization of 4-vinylpyridine (4VP) in aqueous emulsion.[99] Simply substituting BA or S by 4VP without changing any parameter in the previously reported procedure serendipitously led to the synthesis of various aggregates ranging from spheres to worm-like structures and (multicompartment) vesicles (Figure 5.16).

It was believed that the poor initiating efficiency observed in nitroxide-mediated PISA (NMPISA) with PAA-SG1 was related to the kinetic characteristics of the macroalkoxyamine, particularly a slow initiation ability due to the secondary nature of the formed carbon radical upon dissociation. Introduction of the PAA-SG1 at high temperature or addition of methyl acrylate at the start of the polymerization did not bring any improvement.[100] However, the simultaneous advances operated in the control of methacrylic esters by NMP *via* copolymerization,[81,83] as discussed above, offered an adequate solution. A poly(methacrylic acid-*co*-styrene)-SG1 (PMAA-SG1) was synthesized and used as a substitute for PAA-SG1.[101] This also enabled the polymerization to be carried out at lower temperatures, avoiding the requirement for pressurized vessels. PMAA-SG1 with degrees of polymerization ranging between 17 and 40 were produced. They contained only 2–4 styrene units (including the terminal one) and were therefore highly water-soluble, in analogy to PAA-SG1. A first ambitious attempt was made at polymerizing pure MMA in emulsion at only 40 °C and 20 wt%. The polymerization was initially very fast for such a temperature (50% conversion reached in approximately 5 h), reached a plateau at about 24 h and 70% conversion, but was not controlled. Nevertheless, stable lattices were obtained with narrow PSD. Consequently, styrene was also employed as a comonomer in the emulsion step. Although the polymerization was controlled at 60 °C with only 3 mol% of styrene, the best compromise in terms of control, kinetics,

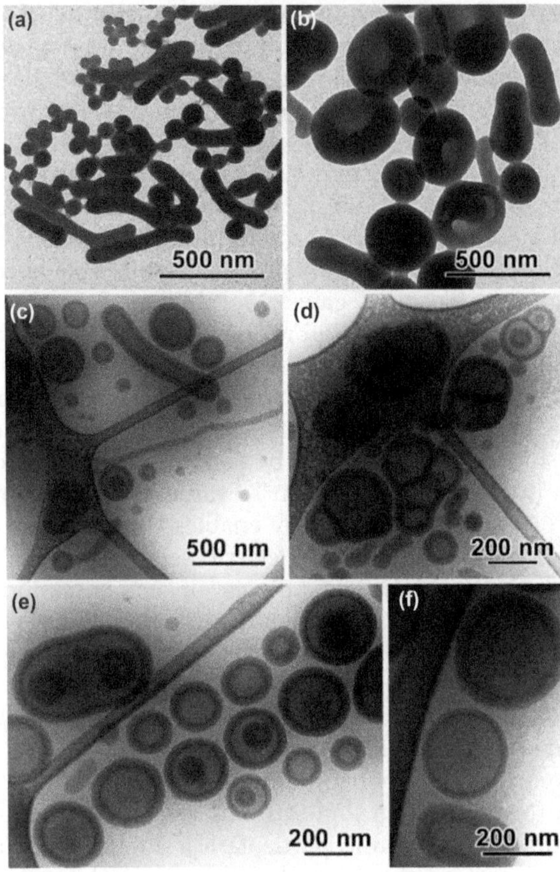

Figure 5.16 TEM (a and b) and cryo-TEM (c, d, e, and f) images of the final latex obtained by PAA-SG1-initiated PISA of 4VP.
Reproduced from ref. 99 with permission from the Royal Society of Chemistry.

and final conversion was found at 90 °C and 9 mol% of styrene. A final conversion of 90% was reached in less than 3 h with a linear increase of molar mass with conversion and narrow MMDs ($Đ = 1.2$). In contract to results obtained previously with PAA-SG1, high initiator efficiencies were achieved. The lattices exhibited slightly broader PSDs than in the case of PAA-SG1 with styrene and BA. Similar results were obtained when the macroinitiator was a terpolymer of MAA and poly(oligoethylene glycol methyl ether methacrylate) (MeOEGMA), and styrene for the preparation of PEGylated lattices.[102]

Since the presence of styrene units in the macroinitiator, hence in the particle shell, was of concern for the colloidal stability, this comonomer was later replaced by sodium styrene sulfonate (NaSS).[103] Copolymerization of MMA with a small amount of styrene in PISA conditions at 90 °C was

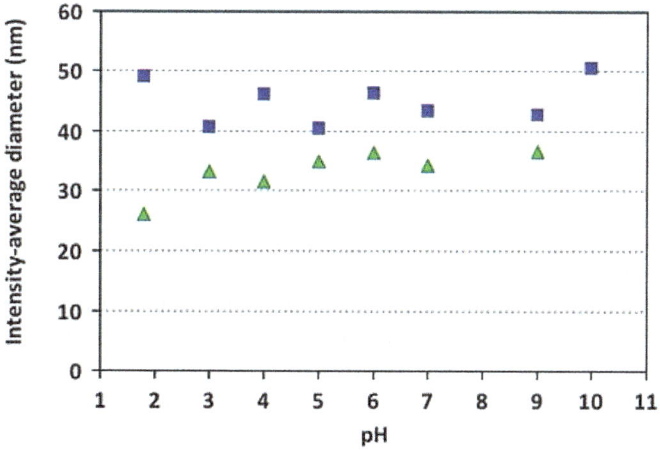

Figure 5.17 Particle size *vs.* pH for lattices of PMMA obtained by PISA initiated by P(MAA-*co*-SSNa)-SG1 at two different concentrations.
Reproduced from ref. 103 with permission from the American Chemical Society.

extremely fast, with maximal conversions reached in 1.5 h, and controlled. Interestingly, lattices were exceptionally stable over almost the entire pH range, thanks to the presence of SSNa units in the corona (Figure 5.17). By appropriate choice of the macroinitiator concentration and depending on the conversion, various arbitrary morphologies were generated with P(MAA-*co*-SSNa) as the stabilizing shell.[104] It was also demonstrated that a one-pot process could be applied in order to create the stabilizer/macro-initiator *in situ* and without further purification, before addition of the core-forming monomers and PISA.[105] The only requirement was to limit the reaction time for the first step as it was carried out in acidic conditions and could impair the stability of SG1.

With respect to the instability of SG1 in acidic media, the recent reports of Darabi *et al.* arouse interest. Indeed, they employed SG1-capped macroalk-oxyamines obtained by copolymerization of either 2-(dimethylamino)ethyl methacrylate (DMAEMA) or 2-(diethylamino)ethyl methacrylate (DEAEMA) with a small amount of styrene for the emulsion copolymerization of MMA in PISA conditions at 90 °C.[106] In order to provide sufficient stability to the particles, the macroinitiators needed to be employed in their protonated form, hence hydrochloric acid was employed for neutralization of the tertiary amine groups. However, care was taken to adjust the pH to a reasonable value of 6 and the polymerization appeared to follow a controlled character and the obtained final MMDs were narrow (*e.g.*, $Đ = 1.18$ at 70% conversion for P(DMAEMA-*co*-S)-SG1 initiation). A one-pot synthesis similar to that of P(MAA-*co*-SSNa)-SG1[105] was applied to generate *in situ* P(DEAEMA-*co*-S)-SG1 in water before direct addition of MMA and styrene to perform PISA.[107] In this particular case, MAMA could not be employed as an initiator for the solution polymerization due to solubility issues at such a low pH. A bicomponent system (VA-044 and SG1) was therefore employed. It was also found in this case that a short polymerization time (15 min) in the first step was required to preserve a high DOL.

Bourgeat-Lami and co-workers focused on PEGylated latex particles with a special interest in the ability of PEG-based materials to adsorb onto inorganic surfaces such as clay and silica. A first report was dedicated to the synthesis of PBMA-core nanoobjects by PISA using macroalkoxyamines based on two distinct MeOEGMA macromonomers ($M_n = 300$ or 950 g mol^{-1}) and a small amount of styrene.[108] It was found that the lowest molar mass macromonomer was not suitable for the further step of nitroxide-mediate NMPISA as its polymer exhibits a lower critical solution temperature (LCST) in water of approximately 50 °C, well below the usual polymerization temperatures in NMP. NMPISA was thus performed at 85 °C with two distinct alkoxyamines, one purely (at the exception of a small styrene content) based on the 950 g mol^{-1} macromonomer and the other being a 50 : 50 mol/mol mixture of the two MeOEGMA derivatives. Both nevertheless possessed the same molar mass (same amount of ethylene glycol units) and were able to mediate PISA of BMA with 8 mol% of styrene. Since the reaction time was short, a pH value of around 4 was considered acceptable. This turned out to be true as the polymerizations were well-controlled. Despite their identical molar masses, the OEGMA950-based macroalkoxyamine was a better stabilizer, leading to a reduced particle size. Since the macroinitiator was obtained by MAMA-initiated NMP, all chains bore a carboxylic acid at their α end. For this reason, the influence of pH was examined. Increasing the pH to 6.0 and 6.7 by NaOH addition logically slowed down the polymerization, as SG1 became more stable. The MMDs were not affected. However, the particle morphology underwent a dramatic transition to fibers and vesicles, as seen before for different systems in other conditions. Since such a morphological transition could not have not plausibly been triggered by the neutralization of a single carboxyl unit at the end of the chains, it was suspected that the change in ionic strength originating from the addition of NaOH was responsible. As a conclusive series of experiments, it was found that with increasing amount of NaCl at a given pH, various particle morphologies were generated. This was linked to a lowering of the LCST of the PMeOEGMA chains due to a salting out effect. This type of macroinitiator was further used to carry out NMPISA of BMA in the presence of various silica nanoparticles.[109] In a preliminary phase, the authors demonstrated the actual adsorption behavior of the macroinitiators onto the inorganic material. In the presence of silica nanoparticles, the control of the polymerization decreased but formation of diblock copolymers could be ascertained by the clear shift in SEC traces with conversion. Pleasingly, TEM images revealed the presence of organic/inorganic supracolloidal aggregates. Observed morphologies ranged from dumbbell- to raspberry- and daisy-like shapes and could be tuned by changing the size of the silica nanoparticles or the macroinitiator concentration, independently (Figure 5.18).

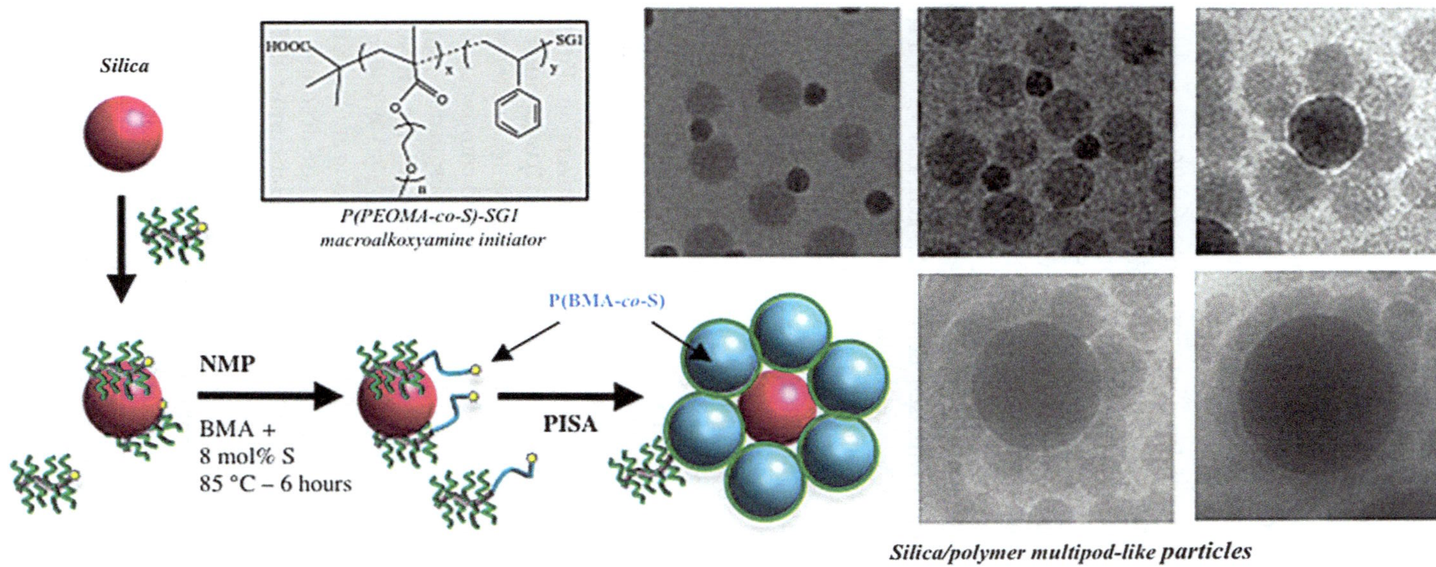

Figure 5.18 Overall strategy for the synthesis of hybrid polymer latex/silica particle supracolloidal assemblies and typical examples of obtained morphologies.
Reproduced from ref. 109 with permission from the American Chemical Society.

5.2.5 NMP in Microemulsion

Microemulsions are formed spontaneously upon mixing of an oil in water with a very high concentration of surfactant, without the need for high shearing. They are optically clear, macroscopically homogeneous, and thermodynamically stable.[110] Only a handful of reports can be found on NMP applied to a microemulsion process.

As mentioned earlier in the frame of the multi-step emulsion process developed by Nicolas *et al.*,[77,78,80] the first stage carried out in the presence of high concentrations of Dowfax 8390 could qualify as a microemulsion polymerization. In that case, only short oligomeric sizes were targeted and polymerizations were performed up to a maximum of 60% conversion to maintain a living character for the next step. Starting from a transparent clear solution, cloudy dispersions were obtained with particle sizes ranging from 25 to 290 nm, inversely proportional to the Dowfax 8390 concentration.[111]

Okubo and co-workers investigated the polymerization of styrene in microemulsion systems formed under *n*-tetradecyltrimethylammonium bromide stabilization (TTAB, see Scheme 5.2) at 125 °C.[112] TTAB/styrene weight ratios of 2.5 to 6.7 were examined. Higher concentrations of TTAB systematically resulted in smaller particles, as expected. Both TEMPO and SG1 were employed. In the case of TEMPO, organosoluble (BPO and AIBN) and water-soluble (KPS) initiations were assessed with very similar results under all aspects of kinetics, macromolecular control, and colloidal features. Polymerizations were slow and final particle sizes were in the range of 100–150 nm. When SG1 was employed together with AIBN, polymerizations became substantially faster (full conversion in 4–6 h *vs.* 7–48% in 24 h for similar experiments with TEMPO) and the particle size was drastically reduced to 25–35 nm. The macromolecular control could be qualified of average for M_n reaching 31 000 g mol^{-1} at full conversion. Further investigations were performed at lower temperature (100 °C) with either TIPNO (see Scheme 5.1) or SG1, the former being significantly less water-soluble than the latter.[113] AIBN was employed as an initiator and comparisons were made with bulk systems. The polymerizations in microemulsion (approximately 6 wt% styrene) were slower than in bulk, particularly for TIPNO with 20% conversion barely reached in more than 35 h (Figure 5.19). Given that SG1 is more hydrophilic than TEMPO, faster polymerizations are logically expected. TIPNO was employed as a point of comparison to decouple activation–deactivation features from water-solubility in comparing different nitroxides, as activation and deactivation rate constants of styrene-type SG1 and TIPNO alkoxyamines are in the same order of magnitude at 120 °C.[114,115] Since TIPNO is more hydrophobic than SG1, a stronger confined space effect is expected (see Section 5.2.6), leading to a reduction in the rate of polymerization. MMDs were narrower at low conversions but increased steadily in the case of SG1, which was ascribed to a heterogeneous distribution among the particles of oligoalkoxyamines formed in the early stages. In addition, an experiment was carried out where a microemulsion was

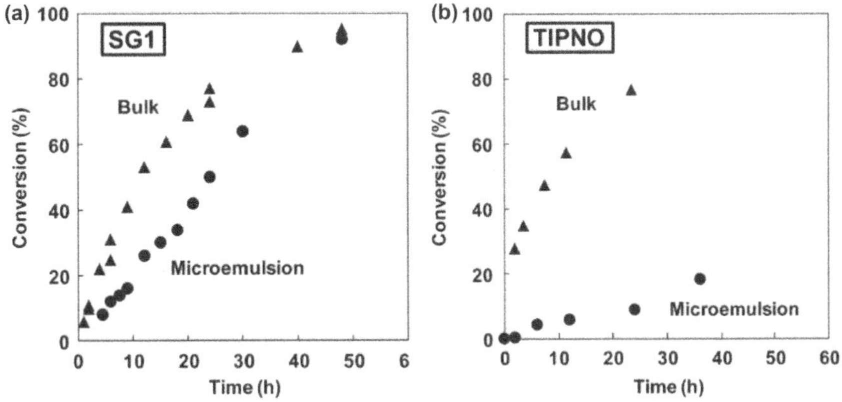

Figure 5.19 Conversion *vs.* time for nitroxide-mediated polymerization of styrene at 100 °C with $[\text{nitroxide}]_0/[\text{AIBN}]_0 = 1.8$ for (a) SG1 and (b) TIPNO in bulk and microemulsion.
Adapted from ref. 113 with permission from the American Chemical Society.

prepared by mixing two microemulsions: one with no TEMPO and the other with the double amount of TEMPO, thus resulting in an overall TEMPO concentration identical as previously. This showed that this nitroxide was able to diffuse homogeneously on a time frame of 15 min or less in the employed conditions. Polymerizations of BA initiated by AIBN in the presence of SG1 in similar conditions were found to proceed in a controlled way, while carried out in bulk it led to extremely fast and completely uncontrolled polymerization, which was anew ascribed to confined space effects.[116]

Quite differently, Thomson *et al.* employed a monocomponent water-soluble initiating system (MAMA) for the microemulsion polymerization of MMA in the presence of styrene at 90 °C.[117] A one-pot two-step process was applied, similarly to that employed by Charleux and co-workers earlier in emulsion, where a low amount of monomer was polymerized in the presence of a high concentration of surfactant. While controlled polymerization of MMA and other methacrylates usually required only approximately 9 mol% of styrene, it was found that a greater amount (30 mol%) was preferable in the first stage of the polymerization in order to gain satisfactory results in the second stage, *i.e.*, when the main part of the monomer is added. With such a high styrene/MMA ratio, the presence of a majority of chains with a styrene–styrene-SG1 sequence is very likely. We therefore could hypothesize that the situation is close to what would happen if a PS-SG1 macroinitiator was formed. The advantage in the present case lies in the faster kinetics of a system containing MMA as the main monomer. Since the conversion in the first stage reached a plateau in one hour, irrespective of the initial amount of styrene, it was important to limit its duration—typically to 20 min—to avoid excessive terminations and therefore maintain narrow MMDs in the second stage. Remarkably, translucent lattices up to 40 wt% solids content were achieved with narrow PSDs centered on 25–40 nm. Polymers with M_n as high as 15 0000 g mol^{-1} and $Đ$ values as low as 1.4 could be obtained. Block copolymers with BMA were also synthesized by simply adding a load of BMA and styrene to a running microemulsion polymerization at more than 85% conversion. When the first block was made by first polymerizing BMA and 8 mol% styrene before addition of MMA and styrene, bimodal PSDs were obtained, thereby evidencing that this procedure may not be applicable to any monomer, depending on its hydrophilicity.

More recently, the team of Cunningham managed to decouple molar mass from particle size in the frame of MAMA-initiated microemulsion polymerization of BA (20 wt%) at 120 °C and at relatively low Dowfax 8390 concentrations for a microemulsion process (surfactant/BA 0.2 : 1 to 0.5 : 1 w/w).[118] Indeed, while varying the monomer-to-initiator ratio is usually sufficient in homogeneous media, doing so is not trivial in dispersed media due to the difference in kinetics for different particles sizes as well as between the aqueous and organic phases, which results in various chain concentrations in the particles. Several factors may influence these kinetics. Here, manipulation of the surfactant-to-monomer ratio, the MAMA/buffer ratio, and the monomer content in the first stage, as well as the addition of surfactant in semi-batch mode proved successful for arbitrarily obtain particles sizes in the range of 40–70 nm for any targeted molar masses in the 20 000–80 000 g mol^{-1} range (Figure 5.20).

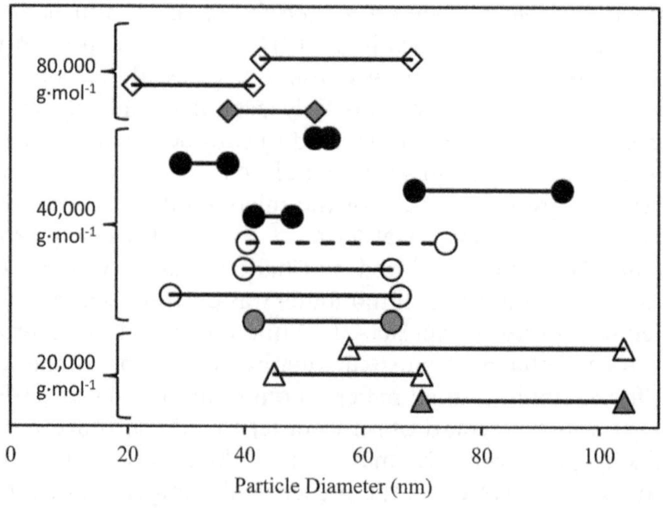

Figure 5.20 Range of particle sizes achieved by SG1-mediated microemulsion polymerization of BA at targeted M_n of 20 000 g mol^{-1} (▲), 40 000 g mol^{-1} (●), and 80 000 g mol^{-1} (◆) by varying the surfactant-to-monomer ratio (w/w) (gray), by varying the monomer content in the seed stage (empty solid line), by semi-batch addition of surfactant (empty dotted line), and by varying the MAMA to Na_2CO_3 molar ratio (black).
Reproduced from ref. 118 with permission from the Royal Society of Chemistry.

5.2.6 Compartmentalization Effects

As mentioned earlier, it is not possible to consider polymerization kinetics in dispersed media as simply as in homogeneous media due to the compartmentalization of propagating radicals, which has an influence on the termination rate. Overall, the concentration of radicals in the entire organic phase of a latex is largely superior to that of the corresponding homogeneous system as terminations are reduced. Although simple to understand from this simple picture, the underlying mechanism is obscured by various events. When particle size decreases below a certain value, termination occurs instantaneously upon entry of a surface-active radical in a particle already containing another radical. This is often called the Smith–Ewart theory (although it is only part of it) or sometimes the zero-one approach, *i.e.*, small particles contain at most one radical. Local termination events are then enhanced by the small volume and the high local concentration: a confined space effect is exerted on this particular reaction. The rate of termination for the entire latex is therefore determined by the rate of entry of radicals and the number of particles (the greater the number of particles, the lesser the probability for one particle to be entered by a radical). Since entry is typically slow and the number of particles is large, the lifetime of radicals is substantially increased as well as their overall concentration, which results in typically faster overall kinetics and higher molar masses in emulsion polymerization in comparison to homogeneous systems.

However, compartmentalization effects in NMP are generally more complex as not only radicals but also nitroxides can be compartmentalized. Intrinsically, radicals in NMP are additionally involved in reversible terminations. In the case of initiation by an alkoxyamine, oligoradicals grow and, once they have reached a given length, enter particles. Past the nucleation stage, no entry is supposed to occur, as there is no radical-generating species in the aqueous phase. Radicals are thus entirely confined in the particles and can either reversibly react with a nitroxide or irreversibly with another radical, which will be less likely for small particles. However, the kinetics are not only influenced in this case by irreversible termination but also by the activation–deactivation equilibrium and the local concentration of nitroxide, hence its partitioning ability. Depending on this feature, two extreme cases were considered: complete compartmentalization or easy exit/re-entry of the nitroxide. For both cases, theory shows that above a given particle size, no influence on polymerization rate and irreversible termination should occur. Below this size, the situation may become drastically different. For a very hydrophobic nitroxide, a confined space effect on reversible termination should lead to decreasing polymerization rates and higher DOLs. On the contrary, possible exit of the nitroxide combined with a small particle size, which by a segregation effect reduces termination events and therefore the amount of released nitroxide by PRE, should lead to faster polymerizations. However, opposite trends have been observed in some occurrences. Overall, compartmentalization effects remain complex to predict as experience shows a significant system dependence. They were summarized and discussed in details by Zetterlund in the frame of RDRP, including NMP.[9] Recent simulation data shed some light on the influence of the nitroxide ability to exit and re-enter particles in relation to these effects.[45]

5.3 Nitroxide-mediated Dispersion Polymerization

Dispersion polymerization[119,120] was developed in the 1960s by the company ICI.[121] In its most common form, this process operates in an organic solvent and yields rather large particles (200 nm–10 microns). The polymerization starts from a homogeneous mixture, *i.e.*, all components are soluble in the continuous medium (Figure 5.21, left). However, the targeted polymer is insoluble and is therefore excluded from the dispersing phase as it grows. In the presence of a stabilizer, particles are formed (Figure 5.21, right).

Dispersion polymerization is considered simpler than (mini)emulsion as the monomer does not form a separate phase and there are consequently few mechanistic reports. Nevertheless, Cawse and El-Aasser proposed a five-step scenario:

- The initiator decomposes in a solution containing the monomer and the soluble stabilizer (often poly(*N*-vinylpyrrolidone), PVP);
- Oligoradicals are formed and start precipitating when they reach a critical degree of polymerization;
- Homogeneous nucleation through nuclei aggregation occurs;
- The monomer swells the particles, which are stabilized by the soluble polymer by adsorption; and
- Finally, a stable latex is obtained.

Among the few reports on nitroxide-mediated dispersion polymerizations, it is possible to make a distinction based on the employed solvent. The first group of reports is based on the use of common solvents such as alkanes,

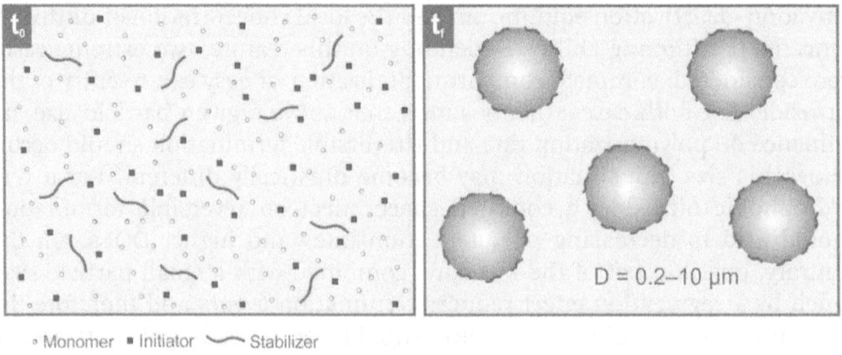

Figure 5.21 Schematic depiction of the initial (left) and final (right) states of a classic dispersion polymerization.

alcohols, water, or hydro-alcoholic mixtures, while the second group employs supercritical carbon dioxide ($scCO_2$) as a continuous medium.

5.3.1 Dispersion Polymerization in Common Solvents

Mülhaupt and co-workers were the first to attempt nitroxide-mediated dispersion polymerization in 1997.[122] A bicomponent initiating system with $[TEMPO]_0/[BPO]_0 = 1.6$ was employed to polymerize styrene at 135 °C in decane in the presence of Kraton G1701 [PS-*b*-poly(ethylene-*alt*-propene)] as a stabilizer. Three dispersion polymerizations with various styrene/TEMPO ratio were carried out and compared to bulk polymerization with otherwise unchanged conditions. Dispersion polymerizations proceeded extremely slower and in the case of the lowest TEMPO/styrene, it was possible to notice an acceleration of the polymerization concomitant with the apparition of turbidity, *i.e.*, particle formation. It was hypothesized that deviation from first-order kinetics in dispersion arose from simultaneous polymerization in both phases accompanied by a difference in solubility of TEMPO in either of them. Nevertheless, number-average molar masses increased with conversion, yet exhibiting significantly broader MMDs than in bulk. PS particles with extremely broad PSDs particles ranging from 50 nm to 10 μm were generated. This was attributed to potential additional particle formation by precipitation when cooling the system as high temperatures required for TEMPO-mediated NMP and the presence of styrene in the continuous medium may contribute to partial solubility of PS chains in the latter during the polymerization.

Dispersion polymerization of styrene is also commonly carried out in alcohols. Alcohols are significantly poorer solvents for PS and it was thus envisioned that the phenomenon assumed by Mülhaupt and co-workers could not occur. Therefore, Gabaston *et al.* carried out investigations by varying the nature of the alcohol.[123] Obviously, high boiling point alcohols are preferable for TEMPO-mediated NMP. *n*-Butanol, *n*-pentanol, *n*-octanol, and ethylene glycol were evaluated, as well as aqueous mixtures, in the presence of PVP. Only with ethylene glycol and water-ethylene glycol mixtures would the polymerization occur *via* a true dispersion polymerization mechanism, as the other alcohols were still able to solubilize low molar mass PS at elevated temperatures. Particle formation only occurred upon cooling in the latter case. In the case of pure ethylene glycol, the polymerization at 115–120 °C was well controlled over 20 000 g mol^{-1}. Employing hydroalcoholic mixtures, lower molar masses were obtained within stable particles in the range of 700 to 1500 nm. PSDs were generally broader than in the case of dispersion polymerization carried out without TEMPO, yet no explanation was provided.

Choe and co-workers showed that the addition of CSA in the TEMPO-mediated dispersion polymerization of styrene in tripropylene glycol in similar conditions to those employed by Gabaston *et al.* not only accelerated the BPO-initiated polymerization but also greatly reduced the breadth of the

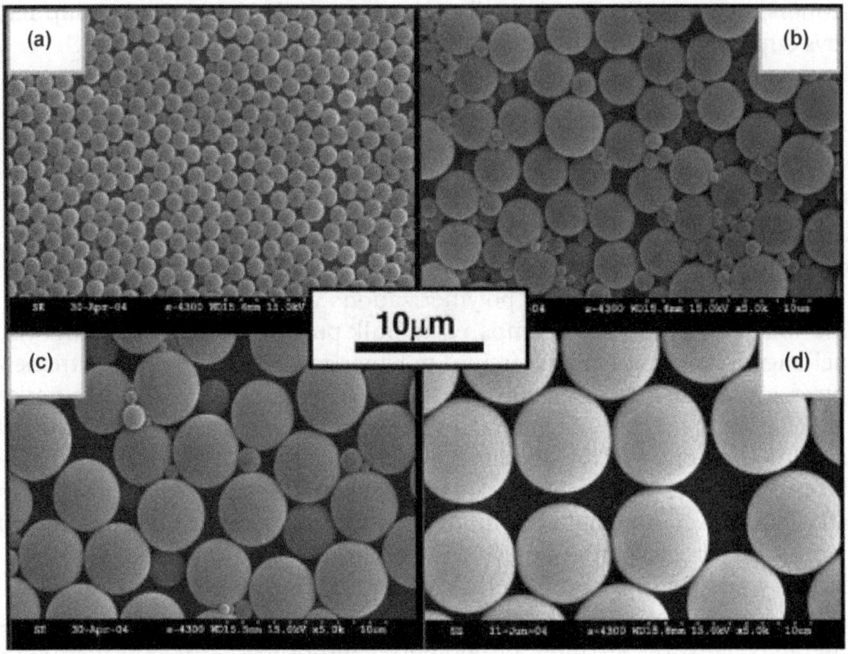

Figure 5.22 Scanning electron micrographs of PS beads prepared by dispersion polymerization of styrene in tripropylene glycol under various conditions at 125 °C. (a) No TEMPO, no CSA; (b) [TEMPO]/[BPO] = 1, no CSA; (c) [TEMPO]/[BPO] = 1, [CSA]/[BPO] = 0.5; and (d) [TEMPO]/[BPO] = 1, [CSA]/[TEMPO] = 1.
Reproduced from ref. 124 with permission of Wiley.

PSD (Figure 5.22).[124] This was however achieved at the expense of the macromolecular control: molar masses increased linearly but were substantially higher than theoretical values, while MMDs became slightly broader.

Yoshida also investigated TEMPO-mediated dispersion polymerization of MMA in 1:3 vol/vol methanol/water mixture proceeding *via* a yet-to-be-elucidated mechanism. Up to 50% conversion, apparent characteristics of a controlled polymerization were observed and relatively homogeneous microparticles (approximately 6 μm) were generated.[125]

The dispersion polymerization is the process of choice for the synthesis of microgels. Typically, the polymerization of monomers for which the polymer is stimulus-sensitive is carried out in conditions that induce phase separation and in the presence of a crosslinker. Charleux and co-workers investigated the PAA-SG1-initiated polymerization of *N,N*-diethylacrylamide (DEAAm)—a monomer that has been shown to polymerize in a controlled way in the presence of SG1[126]—in water up to 40 wt% monomer concentration and at 112–120 °C,[127] well above the reported LCST of PDEAAm in water. Polymerizations in the absence of free SG1 were extremely fast with 90% conversion reached in 2 and 3 h at 120 and 112 °C, respectively.

Addition of 5–10 mol% SG1 (with respect to PAA-SG1) effectively slowed down the polymerization. In any case, linear first-order kinetics were observed for the main part of the polymerization. The macroalkoxyamine concentration exerted very limited influence on the kinetics in the studied range, in contrast to the accelerated polymerization observed when varying the monomer concentration from 20 to 30 and 40 wt%. Accurate SEC analysis of such block copolymers consisting of a PAA block and a polyacrylamide block was not trivial but still allowed to evidence a clear shift of the MMDs towards lower elution volume. This was particularly clear when MMDs were converted to hydrodynamic volume distributions. Direct analysis by DLS of the dispersions directly taken from the reactor without cooling below the phase transition temperature revealed reasonably homogeneous PSDs with diameters ranging from 30 to 110 nm, which could be correlated to the macroalkoxyamine concentration. The process of self-assembly was assumed to be following the mechanism of PISA (Figure 5.23). Upon cooling, the particles progressively swelled until below 32–34 °C where they disrupted completely into free diblock double hydrophilic block copolymer chains. Addition of styrene during the course of the dispersion polymerization enabled the generation of particles with a hydrophobic core, a thermosensitive shell, and a pH-sensitive corona. Alternatively, a crosslinker (*N*,*N*-methylenebisacrylamide) (MBAAm) could be employed to stabilize the particles and yield nanogels.[128] Thermosensitive nanogels could be made at 20 wt%, a substantially higher concentration that those used in classical nanogel preparation protocols. For a batch dispersion crosslinking polymerization to proceed without macroscopic gelation, MBAAm concentration had to remain below 3 mol% (with respect to all monomers). However, it was possible to increase the crosslinker concentration by employing a semi-batch process, so that the nanoparticle formation was decoupled from the crosslinking events. Interestingly, the nanogel dispersion was a fluid milky latex at elevated temperatures while it turned into a physical gel at room temperature. Apart from allowing significantly more concentrated conditions, this method addresses various challenges related to the synthesis of nano-/microgels such as the presence of surfactant (absent here) which usually influences the phase transition or the difficulties to reach small diameters and to generate structured particles.[129]

Hadziioannou and co-workers employed a similar approach in a different context with the aim of producing particles with an ionizable core.[130] SG1-capped macroinitiators prepared from long alkyl chain acrylates were employed for the dispersion polymerization of various monomers in toluene or octane. The authors paid particular attention to the colloidal features of the obtained dispersions and did not report on the macromolecular characteristics. Poly(2-ethylhexyl acrylate)-SG1 (P2EHA-SG1) was employed in toluene to initiate the polymerization of 4VP, AA, and MAA. Since P2EHA was not soluble in octane, poly(dodecyl acrylate)-SG1 (PDoA-SG1) was utilized in lieu to produce block copolymer aggregates with a second block based on 2-aminostyrene, 4VP, styrene, AA, MAA, 4-vinylbenzoic acid, MMA, and

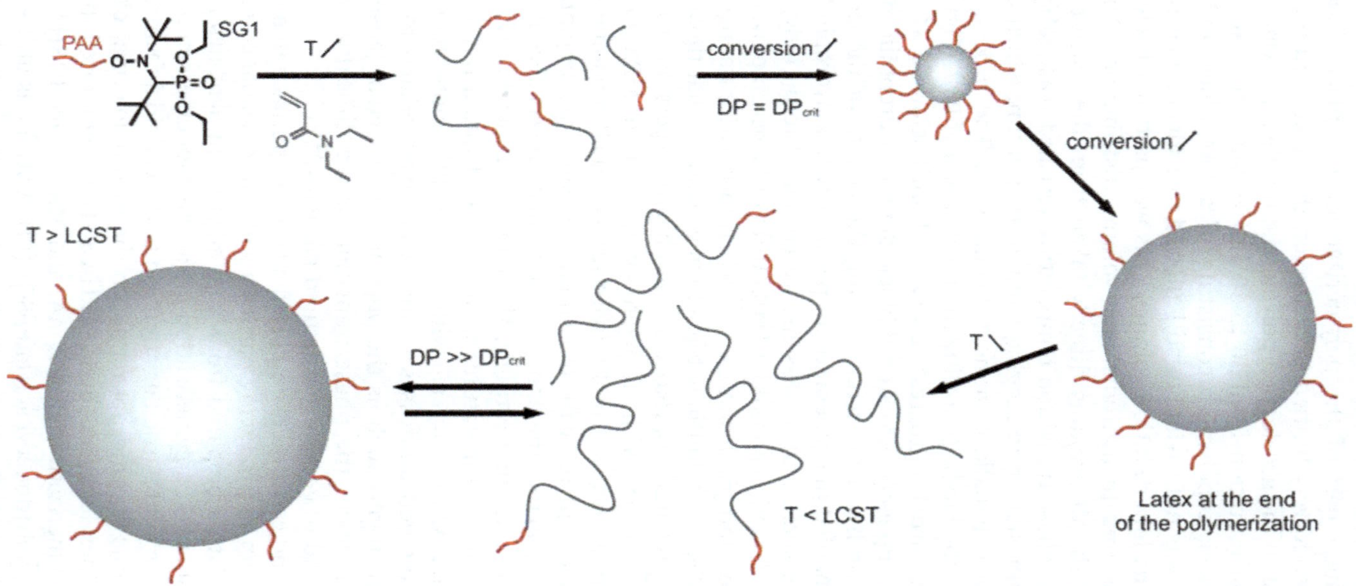

Figure 5.23 Proposed mechanism for PAA-SG1-initiated DEAAm dispersion polymerization and solubilization–micellization behavior of PAA-*b*-PDEAAm copolymers obtained thereof.
Reproduced from ref. 127 with permission of the Royal Society of Chemistry.

tert-butyl acrylate (tBA). In the former case, well-defined P2EHA-stabilized particles were obtained only with 4VP, while PDoA-SG1 gave rise to stable particles only with 4VP and MMA. Nevertheless, it was possible to introduce a negatively ionizable monomer such as AA and MAA by delayed addition in the MMA polymerization. 4VP could also be copolymerized with DMAEMA in toluene. In analogy to previous results obtained with PAA-SG1 in emulsion or dispersion, increasing the concentration of P2EHA-SG1 in toluene gave rise to smaller particles. However, as opposed to these same results, it was here found that the macroinitiator length did not induce a change in particle size. Other studies' parameters concern the presence of an additional molecular initiating species (MAMA) or of a crosslinker (ethylene glycol dimethacrylate). In addition disperse red-labelled particles could be obtained by copolymerization of the corresponding methacrylate derivative with 4VP.

5.3.2 Dispersion Polymerization in Supercritical Carbon Dioxide ($scCO_2$)

$scCO_2$ is the state of carbon dioxide when temperature and pressure are simultaneously above approximately 30 °C and 7.4 MPa, respectively. It is an inexpensive solvent that is also considered environmentally friendly. Non-polar as well as most polar low molar mass compounds are readily soluble in $scCO_2$. However, most polymers of high molar mass do not dissolve in scCO2, except silicon-based polymers and fluoropolymers. This makes it an adequate solvent for dispersion polymerization, as initially introduced by DeSimone *et al.*[131] More specific details can be found in the feature article of Zetterlund, Aldabbagh, and Okubo.[132] Actually, an overwhelming majority of the work reported on dispersed polymerization in $scCO_2$ originates from their teams.

In 2005, they reported the first NMP of styrene carried out in $scCO_2$.[133] This first attempt can actually be related to a PISA process as the polymerization was initiated, in the presence of SG1, using a soluble macroinitiator acting as well as a stabilizer. Nevertheless, compared to classical PISA and due to the bicomponent nature of the initiating system, continuous production of radicals does not allow for the formation of nanoparticles through the exact same process. Polymerizations were carried out at 110 °C under approximately 42 MPa. Since the macroinitiator was actually a multiblock made of polydimethylsiloxane (PDMS) segments connected to each other by diazo moieties, the polymerization led to multimodal distributions. Monomodality of MMDs—nevertheless remaining rather broad—was achieved by using a mixture of initiators consisting of this macroinitiator and AIBN designed in such a way that 95% of the chains should originate from AIBN fragments and thus mostly lead to PS homopolymers. Nonetheless, this amount of macroinitiator was sufficient to ensure efficient stabilization. Chain extensions in bulk evidenced the relative living character of the polymerization in dispersion. A PDMS-*b*-PS-SG1 macroalkoxyamine based on the azo macroinitiator was also evaluated. MMDs were not narrower than before but a better agreement between experimental and theoretical molar masses could be achieved. Although it is possible that CO_2 removal at the end of the polymerization affected the particle morphology, ill-defined aggregates were systematically recovered. It was also hypothesized that the presence of azo groups in the middle of the macroinitiator (also in the macroalkoxyamine), which leads to growth from both sides of the PDMS block, could be responsible for the poor stabilization. Therefore, an advanced macroalkoxyamine, still based on PDMS and SG1, was synthesized by a combination of atom-transfer radical polymerization and atom-transfer nitroxide radical coupling.[134] Well-defined PDMS-*b*-PS-SG1 diblock copolymers were thus obtained with a limited styrene fraction to ensure complete solubility in $scCO_2$. This macroinitiator was then employed for the dispersion polymerization of styrene. A 100% excess SG1 was introduced to obtain well-controlled MMDs. Pleasingly, better-defined nanoparticles were generated through this new recipe (132 nm). Similar results were also acquired with a rather different approach: the polymerization of styrene at 40 wt% was initiated by AIBN in the presence of SG1 and a non-reactive stabilizer PDMS-*b*-PMMA.[135] Again, a large excess of SG1 was required to achieve excellent control.

Saldívar-Guerra and co-workers synthesized similar PS-*b*-PDMS-*b*-PS stabilizers by polymerization of styrene initiated by a TEMPO-PDMS-TEMPO difunctional macroalkoxyamine.[136] This product was actually ill-defined and contained PS and PDMS homopolymer as well as potentially PDMS-*b*-PS diblock copolymer. It was then involved in the dispersion polymerization of styrene at 130 °C in scCO2 with a TEMPO/AIBN bicomponent system (1.6 : 1 mol/mol). Very low molar masses were reached in 4 hours of polymerization and no detail was provided regarding colloidal aspects. In the same study, fluorinated surfactants based on pentafluorostyrene and styrene blocks obtained by TIPNO-mediated NMP were reported. The control during the dispersion polymerization step seemed improved but again molar masses were very low.

The final example of nitroxide-mediated dispersion polymerization in $scCO_2$ is certainly the most successful with respect to the control of the polymerization and the colloidal features. Detrembleur and co-workers employed a PISA-like approach to polymerize both BA and MMA.[137] A perfluorinated SG1-capped macroalkoxyamine was first synthesized in $scCO_2$ by MAMA-initiated homogeneous polymerization of heptadecafluorodecyl acrylate (HDFDA). A well-defined oligomer (3800 g mol^{-1}; $Đ = 1.09$) was recovered at 90% conversion. This macroinitiator was then used to initiate the polymerization and simultaneously stabilize the particles. An asymmetrical PHDFDA-*b*-PBA block copolymer with a PBA block of 45 000 g mol^{-1} and an overall dispersity value of 1.45 was generated in 24 h at 100 °C in the absence of free SG1. A high molar mass PHDFDA-SG1 macroinitiator (48 000 g mol^{-1}) was then used in conjunction with MAMA (PHDFDA-SG1/MAMA 5 : 95 mol/mol) to initiate the polymerization of MMA at 70 °C. A small amount of styrene was added to allow better control of the

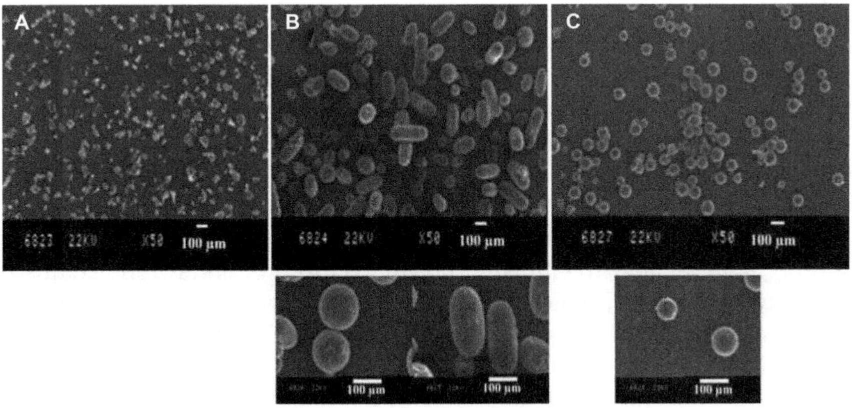

Figure 5.24 Scanning electron micrographs of PMMA particles obtained by MAMA-*co*-initiated dispersion polymerization in $scCO_2$ at 70 °C in the presence of 5 wt% PHDFDA-SG1 at various $[MMA]_0/[MAMA]_0$. A: $[MMA]_0/[MAMA]_0 = 200$, 89% conversion, $M_n = 18\,500$ g mol^{-1}, $Đ = 1.29$; B: $[MMA]_0/[MAMA]_0 = 550$, $M_n = 57\,000$ g mol^{-1}, $Đ = 1.26$; C: $[MMA]_0/[MAMA]_0 = 1000$, $M_n = 94\,000$ g mol^{-1}, $Đ = 1.31$.
Reproduced from ref. 137 with permission from the Royal Society of Chemistry.

polymerization. Various monomer-to-initiator ratios enabled the successful targeting of different molar masses (18 500–94 000 g mol^{-1}) with satisfactory control ($Đ = 1.29$–1.31). As previously witnessed in distinct techniques, varying the reactive macroinitiator concentration triggered a particle size change (Figure 5.24). Interestingly, although this was not particularly highlighted, non-spherical elongated particle shapes were observed at intermediate molar mass values.

5.4 Nitroxide-mediated Precipitation Polymerization

The precipitation process can very simply be regarded as a dispersion polymerization carried out in the absence of a surfactant. This usually leads to a larger particle size and often to a macroscopic solid phase.

This is precisely what Gabaston *et al.* reported in the frame of their study on the dispersion polymerization of styrene.[123] When PVP was omitted, a macroscopic precipitate was recovered. SEC analysis of this polystyrene revealed a rather low molar mass with a relatively narrow MMD ($Đ = 1.15$).

All further reports of nitroxide-mediated precipitation polymerization were based on $scCO_2$ as the continuous phase. Most of them concerned styrene as a monomer and came from the team of Aldabbagh, often in collaboration with Okubo and Zetterlund. Nevertheless, Odell and Hamer first published on the TEMPO-mediated polymerization of styrene at 125 °C at 30 and 50 wt%.[138] High conversion (approximately 85%) was achieved but in some cases low molar masses were obtained, for which the solubility in $scCO_2$ is rather unquestionable, making this process in these instances more akin to solution polymerization. In one instance, a molar mass of approximately 40 000 g mol^{-1} was reached. In this case, the occurrence of a precipitation protocol is likely but no detail on conversions was given and it is therefore not possible to know at which stage precipitation actually took place. With lessons learned from their previous studies on dispersion polymerization in $scCO_2$,[135] McHale *et al.* carried out polymerization of styrene at 110 °C in the presence of a large excess of nitroxide.[139] Indeed, it was found that the critical degree of polymerization required for precipitation (j_{crit}, in analogy to emulsion polymerization) of polystyrene chains in $scCO_2$/styrene was 28, which corresponds to a conversion of 7.5% in the present conditions if control occurs. Since the solubility of SG1 is rather limited, it is possible that the level of control in the early stage is poor, which would have consequences at a later stage as well. At low monomer concentration (20 wt%), polymerizations were judged too slow (27% in 48 h) and control was not optimal. Increasing monomer content to 40 and 60 wt% improved the situation in terms of kinetics (45 and 50% conversion in 24 h, respectively) and control ($Đ = 1.2$). Experimental and theoretical number-average molar masses were in close agreement. Decreasing the SG1 excess from 100 to 30 and 10 mol% (taking into account the initiating efficiency of AIBN) resulted in faster kinetics but resulted in the apparition of shoulders at high molar mass in the SEC traces. Polymers were usually recovered in the form of a white powder. For an experiment at 40 wt% monomer and 100% excess nitroxide, it was found *via* a chain extension experiment that the polymer obtained at 75% conversion still contained about two thirds of living chains. Continuous efforts revealed that a high monomer concentration was preferable. It particularly appeared that a high j_{crit} is favorable for the control of MMDs.[140] This means that the polymerization occurs for an extended period of time in solution while the solubility of the nitroxide is simultaneously higher due to a lower $scCO_2$ concentration. This leads to an increase in livingness in the early stage of the polymerization. Precipitation polymerization in these conditions actually provided a better control than for SG1- or TIPNO-mediated polymerization in toluene solution (all other conditions being equal). This is ascribed to a higher nitroxide concentration at the locus of polymerization (particles). At higher $scCO_2$ content (or lower monomer concentration), the termination rate increases as a result of poor solubility in the continuous medium. Nitroxide partitioning also plays a particular role in such conditions.[141] Differences in control between TIPNO and SG1 were significantly larger at low monomer loads, to the disadvantage of the former. Finally, calculated and experimentally validated j_{crit} values for styrene and BA were shown to depend not only on the monomer load (*vide supra*) but also on the pressure and the monomer-to-initiator ratio.[142] These results were rightly claimed to be also applicable to dispersion polymerization in $scCO_2$. The aspect of nitroxide partitioning in scCO2 was further addressed with the use of TIPNO alkoxyamines being fluorinated either on the nitroxide side or on the initiating fragment part in comparison to non-fluorinated TIPNO and SG1 alkoxyamines.[143] TIPNO revealed to be more efficient than SG1 in controlling the polymerization of styrene (50 wt%) at 110 °C. Fluorination has an impact on the breadth of MMDs but not significantly on molar masses.

To be complete in the area of nitroxide-mediated precipitation polymerization in $scCO_2$, it is worth mentioning that during their study on dispersion polymerization, Grignard *et al.* also carried out an experiment in the absence of PHDFDA-SG1 (under MAMA initiation).[137] While PMMA was collected as a "single chunk", nearly quantitative conversion was achieved with satisfactory values of $M_n = 52\,000$ g mol^{-1} and $Đ = 1.3$.

5.5 Nitroxide-mediated Suspension Polymerization

In suspension polymerization, a monomeric hydrophobic phase is dispersed with the help of a stabilizer, typically PVA as in dispersion polymerization. The initial state may resemble that of emulsion polymerization. However, as opposed to this process, the initiator is organosoluble and leads to initiation inside micrometric droplets. Schmidt-Naake and co-workers reported the first attempt at applying NMP to the aqueous suspension polymerization of styrene (30 wt%) at elevated temperatures.[144] PS-TEMPO ($M_n = 28\,500$ g mol^{-1} and $Đ = 1.13$) was employed as the initiator. At 122 °C, approximately 80% conversion could be reached in 24 h. High molar masses up to 150 000 g mol^{-1} were reached with reasonably narrow MMDs. This method produced stable microparticles for which the morphology could be altered at too high a stirring rate. Braun and co-workers utilized di-*tert*-butyl nitroxide (DBNO) as well as two structures similar to DBNO and TEMPO, respectively, for the suspension polymerization of vinyl chloride (70 wt%).[145] The latter has a very low boiling point (approximately −13 °C). Therefore, pressures that are rather unusual for aqueous NMP were applied (12–18 bars). Polymerizations were carried out between 70 and 90 °C and were found to be significantly slowed down by the presence of nitroxides. MMDs were rather broad for NMP—which was ascribed to transfer to the monomer reactions—yet markedly narrower than in the absence of nitroxides. Random copolymerization with vinyl acetate as well as blocking experiments using a PVC macroinitiator with various vinyl monomers, *e.g.*, styrene, maleic anhydride, and butyl methacrylate, were successful, with however rather large MMDs.

In relation to their aforementioned study on the crosslinking polymerization of styrene and DVB in miniemulsion leading to nanogels, Okubo and co-workers employed a suspension process to generate microgels.[146] Under BPO initiation and PVA stabilization, monomers (9 wt% overall, 1 mol% DVB) were polymerized at 125 °C in the presence of TEMPO (1.86 eq. with respect to BPO). While gelation starts typically at 7–16% conversion for such a system in the absence of TEMPO, nitroxide-mediated polymerization induced a shift to 47–72%. Importantly, determination of the mechanical properties for single microgels was very different between conventional and controlled polymerizations. While compressive strength, deformation at break, and breaking energy remained identical for any conversion for microgel particles obtained by conventional polymerization, they all increased with conversion for the NMP sample. This was an indication that crosslinking is occurring in a more homogeneous fashion in RDRP.

Finally, Aldabbagh and co-workers developed a method to produce poly(*N*-isopropylacrylamide) (PNiPAAm) by AIBN-initiated polymerization at 120 °C in the presence of TEMPO in $scCO_2$, having in mind the economy of all solvents necessary for usual purification (precipitation, dialysis). The process is referred to as an inverse suspension as in this case the continuous medium is hydrophobic, while the monomer and polymer are hydrophilic.[147] Polymerizations were controlled at 20 and 30 MPa, but not at 10 MPa. Monomer concentration had however no significant influence in these conditions. Until 70% conversion, the polymerization appeared well-controlled, before important coagulation occurred and was accompanied by a dramatic broadening of the MMD. It is interesting to note that residual monomer could be removed by consecutive $scCO_2$ purging cycles. The same team further demonstrated the possibility of employing this method to synthesize PNiPAAm-based block copolymers starting from SG1-capped macroinitiators [PS, PtBA, and poly(*N*,*N*-dimethylacrylamide)], which were themselves prepared the same way.[148]

5.6 Conclusion

Since the inception of nitroxide-mediated polymerization in dispersed media in the late 1990s, significant progress has been made both in terms of macromolecular control and colloidal features. While the former has been achieved in numerous occurrences, it was essential to take into account mechanistic aspects in order to gain control over particle stabilization. Nevertheless, marrying both aspects is still not trivial in many cases. Clearly introducing a nitroxide into a conventional polymerization mixture requires thoughtful considerations.

Likewise, appropriate choice of the process depending on the requirements for the final product is important. For instance, while emulsion polymerization is popular, a simpler method such as precipitation polymerization seems to be the most convenient—economically and environmentally—way to produce well-defined PS in a dispersed phase when control of particle size or morphology is not necessary. Besides, it is interesting to realize that extremely narrow MMDs are always targeted while many applications would probably not require such a high level of definition.

Nevertheless, polymerizations in dispersed media present particularities that make them a good complement to homogeneous systems. Particularly, those based on small particles—for which compartmentalization effects operate—may offer solutions to polymerization systems that are difficult to control or to reach very high molar masses in other cases. An advanced method like PISA also opens a myriad of possibilities in the field of materials science.

In spite of some attempts in this direction, considerably more efforts are needed for the establishment of protocols in industrially relevant conditions, *e.g.*, lower temperatures, shorter reaction times, and high concentrations. This could potentially help NMP in general take a leap forward in the commercial arena.

References

1. R. G. Gilbert, *Emulsion Polymerization: A Mechanistic Approch*, Academic Press, London, 1995.
2. *Emulsion Polymerization and Emulsion Polymers*, ed. P. Lovell and M. S. El-Aasser, John Wiley & Sons, Chichester, 1997.
3. *Chemistry and Technology of Emulsion Polymerisation*, ed. A. van Herk, John Wiley & Sons Ltd, 2005.
4. *Colloidal Polymers: Synthesis and Characterization*, ed. A. Elaissari, Surfactant Science Series, 2003, vol. 115.
5. M. Destarac, *Macromol. React. Eng.*, 2010, **4**, 165–179.
6. M. J. Monteiro and M. F. Cunningham, *Macromolecules*, 2012, **45**, 4939–4957.
7. *Polymer Dispersions and Their Industrial Applications*, ed. D. Urban and K. Takamura, Wiley-VCH Verlag GmbH & Co. KGaA, 2005.
8. *Colloidal Nanoparticles in Biotechnology*, ed. A. Elaissari, John Wiley & Sons, Inc., 2008.
9. P. B. Zetterlund, *Polym. Chem.*, 2011, **2**, 534–549.
10. W. D. Harkins, *J. Am. Chem. Soc.*, 1947, **69**, 1428–1444.
11. C. P. Roe, *Ind. Eng. Chem. Res.*, 1968, **60**, 20–33.
12. R. M. Fitch and C. H. Tsai, *Polym. Colloids, Proc. Symp.*, 1971, 73–102.
13. P. J. Feeney, D. H. Napper and R. G. Gilbert, *Macromolecules*, 1984, **17**, 2520–2529.
14. B. Charleux, M. Cunningham and J. R. Leiza, in *Polymer Science: A Comprehensive Reference*, ed. K. M. Möller, Elsevier, Amsterdam, 2012, pp. 463–499.
15. M. Save, Y. Guillaneuf and R. G. Gilbert, *Aust. J. Chem.*, 2006, **59**, 693–711.
16. J. M. Asua, *Prog. Polym. Sci.*, 2002, **27**, 1283–1346.
17. S. A. F. Bon, M. Bosveld, B. Klumperman and A. L. German, *Macromolecules*, 1997, **30**, 324–326.
18. B. Keoshkerian, P. J. MacLeod and M. K. Georges, *Macromolecules*, 2001, **34**, 3594–3599.
19. M. Lin, M. F. Cunningham and B. Keoshkerian, *Macromol. Symp.*, 2004, **206**, 263–274.
20. M. F. Cunningham, D. C. T. Ng, S. G. Milton and B. Keoshkerian, *J. Polym. Sci., Part A: Polym. Chem.*, 2006, **44**, 232–242.
21. M. Lin, J. C. C. Hsu and M. F. Cunningham, *J. Polym. Sci., Part A: Polym. Chem.*, 2006, **44**, 5974–5986.

22. T. Prodpran, V. L. Dimonie, E. D. Sudol and M. S. El-Aasser, *Macromol. Symp.*, 2000, **155**, 1–14.
23. M. K. Georges, R. P. N. Veregin, P. M. Kazmaier, G. K. Hamer and M. Saban, *Macromolecules*, 1994, **27**, 7228–7229.
24. M. F. Cunningham, K. Tortosa, M. Lin, B. Keoshkerian and M. K. Georges, *J. Polym. Sci., Part A: Polym. Chem.*, 2002, **40**, 2828–2841.
25. M. F. Cunningham, K. Tortosa, J. W. Ma, K. B. McAuley, B. Keoshkerian and M. K. Georges, *Macromol. Symp.*, 2002, **182**, 273–282.
26. D. Benoit, S. Grimaldi, S. Robin, J.-P. Finet, P. Tordo and Y. Gnanou, *J. Am. Chem. Soc.*, 2000, **122**, 5929–5939.
27. M. Lansalot, C. Farcet, B. Charleux, J. P. Vairon, R. Pirri and P. Tordo, in *Controlled/Living Radical Polymerization*, American Chemical Society, 2000, ch. 10, vol. 768, pp. 138–151.
28. H. Fischer, *Macromolecules*, 1997, **30**, 5666–5672.
29. H. Fischer, *Chem. Rev.*, 2001, **101**, 3581–3610.
30. Y. Guo and P. B. Zetterlund, *Macromol. Rapid Commun.*, 2011, **32**, 1669–1675.
31. P. J. MacLeod, R. Barber, P. G. Odell, B. Keoshkerian and M. K. Georges, *Macromol. Symp.*, 2000, **155**, 31–38.
32. M. F. Cunningham, M. Xie, K. B. McAuley, B. Keoshkerian and M. K. Georges, *Macromolecules*, 2002, **35**, 59–66.
33. C. Farcet, M. Lansalot, B. Charleux, R. Pirri and J. P. Vairon, *Macromolecules*, 2000, **33**, 8559–8570.
34. G. Pan, E. D. Sudol, V. L. Dimonie and M. S. El-Aasser, *Macromolecules*, 2001, **34**, 481–488.
35. G. Pan, E. D. Sudol, V. L. Dimonie and M. S. El-Aasser, *Macromolecules*, 2002, **35**, 6915–6919.
36. T. Nakamura, P. B. Zetterlund and M. Okubo, *Macromol. Rapid Commun.*, 2006, **27**, 2014–2018.
37. P. B. Zetterlund, T. Nakamura and M. Okubo, *Macromolecules*, 2007, **40**, 8663–8672.
38. J. W. Ma, J. A. Smith, K. B. McAuley, M. F. Cunningham, B. Keoshkerian and M. K. Georges, *Chem. Eng. Sci.*, 2003, **58**, 1163–1176.
39. M. Cunningham, M. Lin and B. Keoshkerian, *JCT Res.*, 2004, **1**, 33–39.
40. M. N. Alam, P. B. Zetterlund and M. Okubo, *Polymer*, 2008, **49**, 3428–3435.
41. M. Cunningham, M. Lin, C. Buragina, S. Milton, D. Ng, C. C. Hsu and B. Keoshkerian, *Polymer*, 2005, **46**, 1025–1032.
42. T. Fukuda, T. Terauchi, A. Goto, K. Ohno, Y. Tsujii, T. Miyamoto, S. Kobatake and B. Yamada, *Macromolecules*, 1996, **29**, 6393–6398.
43. D. Greszta and K. Matyjaszewski, *Macromolecules*, 1996, **29**, 7661–7670.
44. Y. Guo, M. E. Tysoe and P. B. Zetterlund, *Polym. Chem.*, 2013, **4**, 3256–3264.
45. Y. Sugihara and P. B. Zetterlund, *ACS Macro Lett.*, 2012, **1**, 692–696.
46. Y. Guo, J. Liu and P. B. Zetterlund, *Macromolecules*, 2010, **43**, 5914–5916.

47. Y. Guo and P. B. Zetterlund, *Polymer*, 2011, **52**, 4199–4207.
48. Y. Guo and P. B. Zetterlund, *ACS Macro Lett.*, 2012, **1**, 748–752.
49. T. E. Enright, M. F. Cunningham and B. Keoshkerian, *Macromol. Rapid Commun.*, 2005, **26**, 221–225.
50. T. E. Enright, M. F. Cunningham and B. Keoshkerian, *Macromol. React. Eng.*, 2010, **4**, 186–196.
51. T. E. Enright and M. F. Cunningham, *Macromol. React. Eng.*, 2011, **5**, 431–442.
52. P. B. Zetterlund, M. N. Alam, H. Minami and M. Okubo, *Macromol. Rapid Commun.*, 2005, **26**, 955–960.
53. M. Nur Alam, P. B. Zetterlund and M. Okubo, *Macromol. Chem. Phys.*, 2006, **207**, 1732–1741.
54. Y. Konishi, M. Okubo and H. Minami, *Colloid Polym. Sci.*, 2003, **281**, 123–129.
55. H. Tobita and Y. Yoshihara, *J. Polym. Sci., Part B: Polym. Phys.*, 1996, **34**, 1415–1422.
56. H. Tobita, M. Kumagai and N. Aoyagi, *Polymer*, 2000, **41**, 481–487.
57. K. Tortosa, J.-A. Smith and M. F. Cunningham, *Macromol. Rapid Commun.*, 2001, **22**, 957–961.
58. D. Benoit, V. Chaplinski, R. Braslau and C. J. Hawker, *J. Am. Chem. Soc.*, 1999, **121**, 3904–3920.
59. B. Keoshkerian, A. R. Szkurhan and M. K. Georges, *Macromolecules*, 2001, **34**, 6531–6532.
60. C. Farcet, J. Nicolas and B. Charleux, *J. Polym. Sci., Part A: Polym. Chem.*, 2002, **40**, 4410–4420.
61. M. K. Georges, J. L. Lukkarila and A. R. Szkurhan, *Macromolecules*, 2004, **37**, 1297–1303.
62. C. Farcet, B. Charleux and R. Pirri, *Macromolecules*, 2001, **34**, 3823–3826.
63. C. Farcet, B. Charleux and R. Pirri, *Macromol. Symp.*, 2002, **182**, 249–260.
64. J. Nicolas, B. Charleux, O. Guerret and S. Magnet, *Macromolecules*, 2004, **37**, 4453–4463.
65. C. Marestin, C. Noël, A. Guyot and J. Claverie, *Macromolecules*, 1998, **31**, 4041–4044.
66. J. Cao, J. He, C. Li and Y. Yang, *Polym. J.*, 2001, **33**, 75–80.
67. H. Maehata, X. Liu, M. Cunningham and B. Keoshkerian, *Macromol. Rapid Commun.*, 2008, **29**, 479–484.
68. J. Pohn, C. Buragina, M. K. Georges, B. Keoshkerian and M. F. Cunningham, *Macromol. Theory Simul.*, 2008, **17**, 73–85.
69. G. G. C. Montezuma, S. Contant and L. M. F. Lona, *Macromol. React. Eng.*, 2012, **6**, 516–522.
70. R. González-Blanco, E. Saldívar-Guerra, J. Herrera-Ordóñez and A. Cano-Valdez, *Macromol. Symp.*, 2013, **325–326**, 89–95.
71. J. Ugelstad, P. C. Mórk, K. H. Kaggerud, T. Ellingsen and A. Berge, *Adv. Colloid Interface Sci.*, 1980, **13**, 101–140.

72. J. Ugelstad, K. H. Kaggerud, F. K. Hansen and A. Berge, *Makromol. Chem.*, 1979, **180**, 737–744.
73. Y. Luo, J. Tsavalas and F. J. Schork, *Macromolecules*, 2001, **34**, 5501–5507.
74. P. B. Zetterlund, Y. Kagawa and M. Okubo, *Chem. Rev.*, 2008, **108**, 3747–3794.
75. R. W. Simms, M. D. Hoidas and M. F. Cunningham, *Macromolecules*, 2008, **41**, 1076–1079.
76. A. R. Szkurhan and M. K. Georges, *Macromolecules*, 2004, **37**, 4776–4782.
77. J. Nicolas, B. Charleux, O. Guerret and S. Magnet, *Angew. Chem., Int. Ed.*, 2004, **43**, 6186–6189.
78. J. Nicolas, B. Charleux, O. Guerret and S. Magnet, *Macromolecules*, 2005, **38**, 9963–9973.
79. B. Charleux and J. Nicolas, *Polymer*, 2007, **48**, 5813–5833.
80. J. Nicolas, B. Charleux and S. Magnet, *J. Polym. Sci., Part A: Polym. Chem.*, 2006, **44**, 4142–4153.
81. B. Charleux, J. Nicolas and O. Guerret, *Macromolecules*, 2005, **38**, 5485–5492.
82. J. Nicolas, S. Brusseau and B. Charleux, *J. Polym. Sci., Part A: Polym. Chem.*, 2010, **48**, 34–47.
83. E. Guégain, Y. Guillaneuf and J. Nicolas, *Macromol. Rapid Commun.*, 2015, **26**, 1227–1247.
84. M. E. Thomson, A.-M. Manley, J. S. Ness, S. C. Schmidt and M. F. Cunningham, *Macromolecules*, 2010, **43**, 7958–7963.
85. M. E. Thomson, J. S. Ness, S. C. Schmidt, N. Macy, T. F. L. McKenna and M. F. Cunningham, *Polym. Chem.*, 2013, **4**, 1803–1814.
86. S. Moyses, J. Ness and G. Papakonstantopoulos, *J. Sep. Sci.*, 2010, **33**, 3511–3518.
87. C. J. Ferguson, R. J. Hughes, B. T. T. Pham, B. S. Hawkett, R. G. Gilbert, A. K. Serelis and C. H. Such, *Macromolecules*, 2002, **35**, 9243–9245.
88. L. Couvreur, C. Lefay, J. Belleney, B. Charleux, O. Guerret and S. Magnet, *Macromolecules*, 2003, **36**, 8260–8267.
89. C. Lefay, J. Belleney, B. Charleux, O. Guerret and S. Magnet, *Macromol. Rapid Commun.*, 2004, **25**, 1215–1220.
90. G. Delaittre, J. Nicolas, C. Lefay, M. Save and B. Charleux, *Chem. Commun.*, 2005, 614–616.
91. C. Farcet, J. Belleney, B. Charleux and R. Pirri, *Macromolecules*, 2002, **35**, 4912–4918.
92. G. Delaittre, J. Nicolas, C. Lefay, M. Save and B. Charleux, *Soft Matter*, 2006, **2**, 223–231.
93. B. Charleux, F. D'Agosto and G. Delaittre, *Adv. Polym. Sci.*, 2010, **233**, 125–183.
94. G. Delaittre and B. Charleux, *Macromolecules*, 2008, **41**, 2361–2367.
95. H. Tanaka, K. Yamauchi, H. Hasegawa, N. Miyamoto, S. Koizumi and T. Hashimoto, *Phys. B*, 2006, **385–386**(Part 1), 742–744.

96. W.-M. Wan, C.-Y. Hong and C.-Y. Pan, *Chem. Commun.*, 2009, 5883–5885.
97. B. Charleux, G. Delaittre, J. Rieger and F. D'Agosto, *Macromolecules*, 2012, **45**, 6753–6765.
98. L. Zhang and A. Eisenberg, *J. Am. Chem. Soc.*, 1996, **118**, 3168–3181.
99. G. Delaittre, C. Dire, J. Rieger, J.-L. Putaux and B. Charleux, *Chem. Commun.*, 2009, 2887–2889.
100. G. Delaittre, *Polymérisation radicalaire contrôlée en émulsion et dispersion sans tensioactif*, Edilivre, 2010.
101. C. Dire, S. Magnet, L. Couvreur and B. Charleux, *Macromolecules*, 2009, **42**, 95–103.
102. C. Dire, J. Nicolas, S. Brusseau, B. Charleux, S. Magnet and L. Couvreur, in *Controlled/Living Radical Polymerization: Progress in RAFT, DT, NMP & OMRP*, American Chemical Society, 2009, ch. 20, vol. 1024, pp. 303–318.
103. S. Brusseau, J. Belleney, S. Magnet, L. Couvreur and B. Charleux, *Polym. Chem.*, 2010, **1**, 720–729.
104. E. Groison, S. Brusseau, F. D'Agosto, S. Magnet, R. Inoubli, L. Couvreur and B. Charleux, *ACS Macro Lett.*, 2012, **1**, 47–51.
105. S. Brusseau, F. D'Agosto, S. Magnet, L. Couvreur, C. Chamignon and B. Charleux, *Macromolecules*, 2011, **44**, 5590–5598.
106. A. Darabi, A. R. Shirin-Abadi, J. Pinaud, P. G. Jessop and M. F. Cunningham, *Polym. Chem.*, 2014, **5**, 6163–6170.
107. A. Darabi, P. G. Jessop and M. F. Cunningham, *Macromolecules*, 2015, **48**, 1952–1958.
108. X. G. Qiao, M. Lansalot, E. Bourgeat-Lami and B. Charleux, *Macromolecules*, 2013, **46**, 4285–4295.
109. X. G. Qiao, P. Y. Dugas, B. Charleux, M. Lansalot and E. Bourgeat-Lami, *Macromolecules*, 2015, **48**, 545–556.
110. C. Yong and L. Gan, in *Polymer Particles*, ed. M. Okubo, Springer Berlin Heidelberg, 2005, ch. 3, vol. 175, pp. 257–298.
111. J. Nicolas, PhD, Université Pierre et Marie Curie, Paris, 2005.
112. J. Wakamatsu, M. Kawasaki, P. B. Zetterlund and M. Okubo, *Macromol. Rapid Commun.*, 2007, **28**, 2346–2353.
113. P. B. Zetterlund, J. Wakamatsu and M. Okubo, *Macromolecules*, 2009, **42**, 6944–6952.
114. S. Marque, C. Le Mercier, P. Tordo and H. Fischer, *Macromolecules*, 2000, **33**, 4403–4410.
115. J. Sobek, R. Martschke and H. Fischer, *J. Am. Chem. Soc.*, 2001, **123**, 2849–2857.
116. S. Tomoeda, Y. Kitayama, J. Wakamatsu, H. Minami, P. B. Zetterlund and M. Okubo, *Macromolecules*, 2011, **44**, 5599–5604.
117. M. E. Thomson, J. S. Ness, S. C. Schmidt and M. F. Cunningham, *Macromolecules*, 2011, **44**, 1460–1470.
118. W. S. J. Li and M. F. Cunningham, *Polym. Chem.*, 2014, **5**, 3804–3816.
119. J. L. Cawse, in *Emulsion Polymerization and Emulsion Polymers*, ed. P. A. Lovell and E.-A. M. S., John Wiley & Sons: Chichester, 1997.

120. P. Lacroix-Desmazes, in *Les latex synthétiques - Elaboration, propriétés, applications*, ed. J.-C. Daniel and C. Pichot, Tec & Doc Lavoisier, 2006.
121. Application: *GB Pat.*, UK 893429, 1962.
122. M. Hölderle, M. Baumert and R. Mülhaupt, *Macromolecules*, 1997, **30**, 3420–3422.
123. L. I. Gabaston, R. A. Jackson and S. P. Armes, *Macromolecules*, 1998, **31**, 2883–2888.
124. S. Oh, K. Kim, B. H. Lee, S. E. Shim and S. Choe, *J. Polym. Sci., Part A: Polym. Chem.*, 2006, **44**, 62–68.
125. E. Yoshida, *Colloid Polym. Sci.*, 2011, **289**, 1625–1630.
126. G. Delaittre, J. Rieger and B. Charleux, *Macromolecules*, 2011, **44**, 462–470.
127. G. Delaittre, M. Save, M. Gaborieau, P. Castignolles, J. Rieger and B. Charleux, *Polym. Chem.*, 2012, **3**, 1526–1538.
128. G. Delaittre, M. Save and B. Charleux, *Macromol. Rapid Commun.*, 2007, **28**, 1528–1533.
129. R. Pelton, *Macromol. Symp.*, 2004, **207**, 57–66.
130. A. Charbonnier, C. Brochon, E. Cloutet, C. Navarro and G. Hadziioannou, *J. Polym. Sci., Part A: Polym. Chem.*, 2013, **51**, 4608–4617.
131. J. M. DeSimone, E. E. Maury, Y. Z. Menceloglu, J. B. McClain, T. J. Romack and J. R. Combes, *Science*, 1994, **265**, 356–359.
132. P. B. Zetterlund, F. Aldabbagh and M. Okubo, *J. Polym. Sci., Part A: Polym. Chem.*, 2009, **47**, 3711–3728.
133. J. Ryan, F. Aldabbagh, P. B. Zetterlund and M. Okubo, *Polymer*, 2005, **46**, 9769–9777.
134. R. McHale, F. Aldabbagh, P. B. Zetterlund and M. Okubo, *Macromol. Rapid Commun.*, 2006, **27**, 1465–1471.
135. R. McHale, F. Aldabbagh, P. B. Zetterlund, H. Minami and M. Okubo, *Macromolecules*, 2006, **39**, 6853–6860.
136. D. G. Ramírez-Wong, C. A. Posada-Vélez, E. Saldívar-Guerra, J. G. Luna-Bárcenas, C. Ott and U. S. Schubert, *Macromol. Symp.*, 2009, **283–284**, 120–129.
137. B. Grignard, T. Phan, D. Bertin, D. Gigmes, C. Jerome and C. Detrembleur, *Polym. Chem.*, 2010, **1**, 837–840.
138. G. K. H. P. G. Odell, *Polym. Mater. Sci. Eng.*, 1996, **74**, 404–405.
139. R. McHale, F. Aldabbagh, P. B. Zetterlund and M. Okubo, *Macromol. Chem. Phys.*, 2007, **208**, 1813–1822.
140. F. Aldabbagh, P. B. Zetterlund and M. Okubo, *Macromolecules*, 2008, **41**, 2732–2734.
141. F. Aldabbagh, P. B. Zetterlund and M. Okubo, *Eur. Polym. J.*, 2008, **44**, 4037–4046.
142. P. O'Connor, P. B. Zetterlund and F. Aldabbagh, *Macromolecules*, 2010, **43**, 914–919.
143. C. Magee, A. Earla, J. Petraitis, C. Higa, R. Braslau, P. B. Zetterlund and F. Aldabbagh, *Polym. Chem.*, 2014, **5**, 5725–5733.

144. G. Schmidt-Naake, M. Drache and C. Taube, *Angew. Makromol. Chem.*, 1999, **265**, 62–68.
145. T. Wannemacher, D. Braun and R. Pfaendner, *Macromol. Symp.*, 2003, **202**, 11–24.
146. T. Tanaka, T. Suzuki, Y. Saka, P. B. Zetterlund and M. Okubo, *Polymer*, 2007, **48**, 3836–3843.
147. P. O'Connor, P. B. Zetterlund and F. Aldabbagh, *J. Polym. Sci., Part A: Polym. Chem.*, 2011, **49**, 1719–1723.
148. P. O'Connor, R. Yang, W. M. Carroll, Y. Rochev and F. Aldabbagh, *Eur. Polym. J.*, 2012, **48**, 1279–1288.

CHAPTER 6

Recent Developments in Nitroxide Mediated Polymerization

THOMAS JUNKERS,*[a] CHRISTOPHER BARNER-KOWOLLIK[b,c] AND JACQUES LALEVÉE[d]

[a] Polymer Reaction Design Group, Institute for Materials Research (IMO), Hasselt University, Agoralaan D, 3590 Diepenbeek, Belgium; [b] Preparative Macromolecular Chemistry, Institut für Technische Chemie und Polymerchemie, Karlsruhe Institute of Technology (KIT), Engesserstr. 18, 76128 Karlsruhe, Germany; [c] Institut für Biologische Grenzflächen, Karlsruhe Institute of Technology (KIT), Hermann-von-Helmholtz-Platz 1, 76344 Eggenstein-Leopoldshafen, Germany; [d] Institut de Science des Matériaux de Mulhouse IS2M, UMR CNRS 7361, UHA, 15, rue Jean Starcky, 68057 Mulhouse Cedex, France
*Email: Thomas.junkers@uhasselt.be

6.1 Introduction

The past two decades have been characterized by an unprecedented step-change in our abilities to generate precision engineered macromolecular architectures based on radical free polymerization processes. This change was brought about by the introduction of three main controlling protocols, namely the reversible addition fragmentation chain transfer (RAFT) process,[1] atom transfer radical polymerization (ATRP)[2] and—the focus of the current chapter—nitroxide mediated polymerization (NMP).[3,4] The latter

RSC Polymer Chemistry Series No. 19
Nitroxide Mediated Polymerization: From Fundamentals to Applications in Materials Science
Edited by Didier Gigmes

Published by the Royal Society of Chemistry, www.rsc.org

technique was initially devised by Rizzardo and Solomon at the Commonwealth Scientific and Industrial Research Organisation (CSIRO) in the late 1980s[5,6] and has since developed into a powerful macromolecular construction tool. As NMP has recently been reviewed in-depth,[3,4] and is the subject of discussion in the other chapters, we herein wish to concentrate on some of the later key developments made in the areas of enhanced spin capturing polymerization (ESCP; a methodology for generating macroalkoxyamines),[7] related to *in situ* NMP[8] as well as light-induced NMP processes. The first noted technique employs nitrones as additives to conventional free radical polymerization, which immediately generates a nitroxide radical that subsequently terminates with a growing macroradical. The process yields alkoxyamines flanked by a polymer chain on either side, allowing for the facile construction of ABA'-type block copolymer structures under NMP conditions. In addition, functional nitrone variants can be designed, providing access to star polymer structures. Further, nitrones can be used as radical trapping agents for macromolecular design. Thus, the ESCP technique essentially provides a simple access route to macroalkoxyamines of finely tunable molecular weight. ESCP, however, is not to be confused with *in situ* NMP, in which NMP-typical control agents are directly obtained from spin trap agents rather than midchain functionalized polymers in a one-pot approach and thus allows for a more direct access to NMP-polymers without the requirement of a preceding alkoxyamine synthesis.

The use of light to induce chemical processes offers significant advantages over thermally induced polymerizations, as the important features of spatial and temporal control allow applications in precision surface design and in fields where high reaction temperatures can be detrimental (*e.g.* in the presence of biological molecules). While both photo-induced ATRP as well as RAFT have been implemented successfully,[9–12] the situation with regard to photo-induced NMP (NMP2) is far more challenging, as the initial alkoxyamine—stable up to high temperatures—has to be cleaved along the carbon–oxygen bond *via* an efficient energy transfer. Designing suitable chromophores is no simple task, yet some have been identified—taking inspiration from classical photoiniferter species—and shown to work effectively, allowing for the generation of block copolymers and further well-controlled structures. Furthermore, NMP2 can be employed for the spatially resolved gradient functionalization of surfaces. In summary, the current chapter aims at providing an overview of the latest developments in the above three fields, with work from our own groups critically embedded in the context of the existing literature. While we do not aim at providing an exhaustive review of these three topics, we hope to succeed in highlighting the most important developments in these areas.

6.2 Nitrone Spin Trapping Methods

6.2.1 Spin Trapping in Radical Polymerization

Spin traps are agents that are commonly used in electron spin resonance spectroscopy (ESR) as a tool to convert transient radicals into a more or less stable form in order to be able to study short-lived radicals in higher detail. In medicinal chemistry, spin traps also find use as therapeutics and as a tool to quantify free-radical concentration profiles in living organisms.

In polymerizations, spin traps act as radical scavengers and hence mostly inhibit chain growth. As a matter of fact, nitroxides as commonly employed in NMP can be seen as such radical spin trap scavengers and indeed the very well-known TEMPO nitroxide finds use in ESR studies to quantify radical initiation, to name only one prominent example. Absence of polymerization in the presence of TEMPO is also often used as proof of the existence of a radical polymerization pathway.[13] A large variety of spin trap agents exist, whereby the class of nitrones takes a very prominent position. Nitrones react rapidly with a broad range of radicals to form nitroxides, which—if not terminated by transient radicals—are stable. It thus comes not as a surprise that nitrones have found application related to NMP processes. In fact, many available NMP nitroxides are synthesized *via* nitrone intermediates.

6.2.2 Enhanced Spin Capturing Polymerization

When nitrones are added to a radical polymerization mixture, a process coined enhanced spin capturing polymerization (ESCP) is established.[14,15] In ESCP, (macro)radical spin trapping is in competition with conventional macroradical chain growth, leading to an effective chain length control of the polymerization. The more nitrone that is added to the reaction mixture, the faster the macroradical chains will be trapped and hence the shorter the residual polymer chains become. Spin trapping, is, however, not the final reaction taking place, as the trapped species constitutes a macronitroxide, which by itself is able to trap a further transient radical in a radical recombination event (see Scheme 6.1). Thus, the polymer product consists of a coupling product of two polymers, which is bridged by an alkoxyamine functionality. A midchain-functional polymer is thus obtained, which features the additional benefit of being able to reinitiate in an NMP process.

The mechanism of ESCP is thus based on the ability of the nitrones to limit the lifetime of a propagating radical combined with enforced termination. As termination is enhanced, the number of radicals in ESCP is significantly below the steady state concentration of analogous free-radical polymerizations. As a consequence, the reactions are rate retarded. It is important to note that ESCP is a controlled polymerization pathway, yet polymerizations are strictly non-living—even though a "living polymer" is obtained as the product. A reinitiation within an ESCP process is impossible. However, when reaction temperatures are increased above the temperature where alkoxyamine chain scission may occur, classical NMP reactions can be used to continue polymerization.

In an ideal ESCP process, one would expect the alkoxyamine function to be situated precisely in the middle of the polymer chains, a fact that can be demonstrated *via* radical quenching experiments.[16] For a quenching experiment, the polymer is dissolved in a suitable solvent in the presence of a

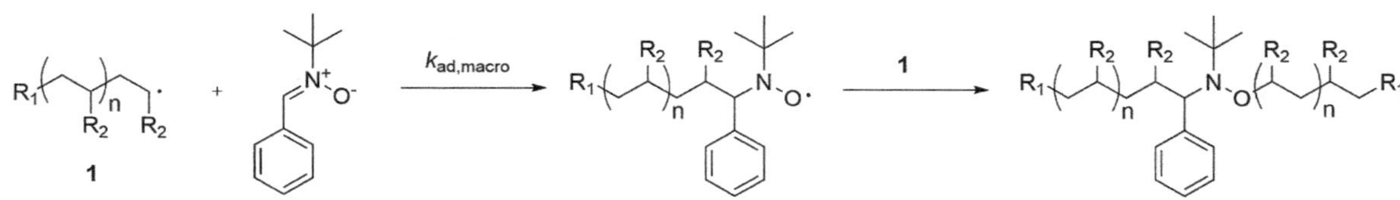

Scheme 6.1 Reaction mechanism of the spin trapping during an ESCP, employing *N*-tert-butyl-α-phenylnitrone (PBN) as the spin trap agent.

radical scavenger. When the reaction mixture is heated well above the alkoxyamine bond dissociation temperature, the scavenger can react with the transient radicals, prohibiting recombination of the macronitroxide. As could be shown, the polymer chains break in half during quenching as one would expect when the alkoxyamine functionality was located in midchain position. After quenching, the number-average molecular weight is reduced by a factor of two. Concomitantly, an increase in dispersity from approximately 1.5 to 2 is observed. This is in line with the ESCP product being a pure radical combination product, whereas the quenched distribution is a product as expected in a polymerization dominated by 100% disproportionation events. This experimental result can also be followed by kinetic modeling, which underpins the identified mechanism.[16] While the midchain location and its thermal lability can be synthetically employed, as will be discussed below, this of course also implies that the polymers are inherently thermally instable and may change in molecular weight distribution, *e.g.* during extruding or thermal insult. While the thermal instability may be of concern for certain applications, it must, however, be noted that the alkoxyamine structure can be tuned (by selection of suitable nitrones) in order to increase the bond dissociation temperature (a task not often undertaken in classical NMP where usually a lower dissociation temperature is seen as favorable), thus giving in principle access to more stable ESCP products. A good example in this respect is the ESCP of ethylene, where a very stable alkoxyamine is formed by reaction with the primary polyethylene macroradicals.

Yet, non-idealities can occur in ESCP—like in any other polymerization technique. For the maximum formation of mid-chain functionalized polymers *via* ESCP, the choice of the radical initiator (or more precisely the attacking initiator radical fragment) is crucial (see also the following section on *in situ* NMP). The selection of an appropriate attacking radical is critical due to the existence of competition reactions between the addition of small radical fragments from the initiator and the addition of propagating macroradicals to the nitrones to form either small nitroxides or macronitroxide species, respectively. To minimize the small nitroxide formation reaction that is caused by the initiator fragment, oxygen-centered radicals (*e.g.* derived from peroxide initiator decomposition) should be avoided and tertiary carbon-centered radicals (*e.g.* derived from azoinitiators) should be preferably used.[16] This is simply due to the higher affinity of nitrones towards reacting with oxygen-centered radicals, followed by primary radicals, acrylates, styrenic and tertiary carbon-centered radicals.[14,17–19] Nevertheless, the initiator selection criterion is only important if a mid-chain functionalized polymer is required. An improper choice of the attacking radical (with respect to its ability to affect the formation of mid-chain functional alkoxyamines) neither limits the ability of ESCP as a molecular weight control tool nor disproves its overall mechanism: A non-reversible enhanced (double) spin capturing reaction takes place regardless whether the radicals are originating from the initiator or propagating chains.

ESCP is with regard to its polymerization outcome closer to chain-transfer polymerization (CTP) than to any other controlled polymerization methodology. As mentioned above, the control over molecular weight in ESCP is governed by the amount (*i.e.* concentration) of nitrone employed, which is also similar to CTP where the amount of chain transfer reagent (CTA) employed predetermines the molecular weight of the generated polymers. Both spin capping as well as chain transfer can be regarded as chain-terminating events. It comes thus at no surprise that the kinetic equation derived to describe the chain length control in ESCP is very similar to the classical Mayo equation describing CTPs.

$$DP_n^{-1} = DP_\infty^{-1} + 0.5C_{SC}\frac{[\text{Nitrone}]}{[\text{Monomer}]} \tag{6.1}$$

Analogous to the transfer constant in the Mayo equation, a spin capturing constant, C_{SC}, can be defined for ESCP, representing the ratio of $k_{ad,macro}/k_p$ whereby $k_{ad,macro}$ and k_p are the rate coefficients for the addition of macroradicals to the nitrones and propagation, respectively. Thus, from measuring the average molecular weight of polymers synthesized in the presence of various nitrone concentrations, the rate coefficient for the spin capturing of the macroradicals, $k_{ad,macro}$, can be determined (Figure 6.1).

Besides its practical value in radical coupling (see below), the value of C_{SC} has important implications for the ESCP process. For a C_{SC} value close to unity (which leads to an equal consumption of nitrones and monomer during the course of the reaction), a constant M_n can be observed throughout the entire conversion regime. The same also applies to CTP in cases when the chain transfer constant, C_T, is unity. Despite some of the described similarities, it must be emphasized that both ESCP and CTP are still distinguishable from one another because: (i) the main mode of chain termination is intrinsically different and (ii) no re-initiating species is present in ESCP. Thus, ESCP polymers bear the initiator-derived moieties as end groups on the polymer whereas in CTP, the transferred functionality and the re-initiating fragment are primarily found. ESCP thus leads to homotelechelic materials while CTP yields heterotelechelic chains.

One key factor that we have identified to be crucial in determining the C_{SC} or $k_{ad,macro}$ values is the nitrones structure. Investigations into the structure–reactivity correlations of different nitrones revealed that the type of *N*-alkyl groups or the attachment of different functionalities on the aromatic ring of phenylic-typed nitrones can have a significant effect on $k_{ad,macro}$. More bulky *N*-alkyl substituents such as *N*-tert butyl result in lower reactivity compared to the less sterically hindered *N*-methyl group. However, the use of the *N*-tert butyl variants is largely preferred because of the better performance of such compounds in later NMP reactions in which methyl moieties would promote disproportionation and hence reduce the livingness of the reaction. Comparison of the determined $k_{ad,macro}$ values

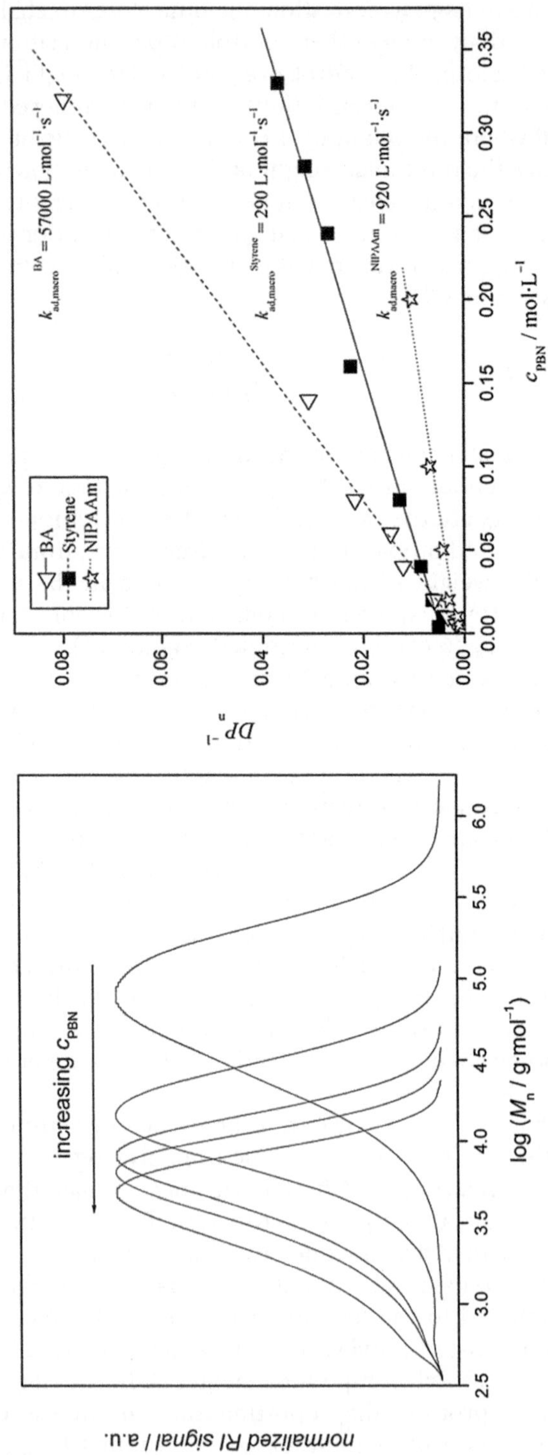

Figure 6.1 Molecular weight distributions obtained for ESCP of polystyrene at different nitrone concentrations (left) and determination of $k_{ad,macro}$ for different monomer/nitrone pair systems utilizing the ESCP equation (right). Reproduced with permission from ref. 14 and 15.

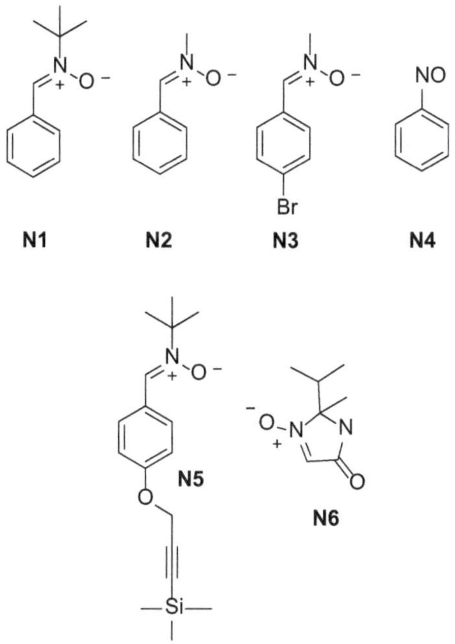

Scheme 6.2 Structures of various nitrones employed in ESCP.

Table 6.1 $k_{ad,macro}$ for various nitrones (see Scheme 6.2) in styrene polymerization at 60 °C.

	C_{SC}	$k_{ad,macro}$ (L mol^{-1} s^{-1})
N1	1.7	580
N2	8.2	2800
N3	10.0	3400
N4	7.6	2600
N5	0.8	270
N6	12.2	8000

for nitrones with various substituents on the aromatic structure clearly indicates that the attachment of electron-withdrawing groups enhances the nitrone reactivity, while electron donating groups impose the opposite.[17,20] Structures of nitrones studied so far with regard to $k_{ad,macro}$ in styrene polymerization are depicted in Scheme 6.2 with kinetic data being listed in Table 6.1.[15,18,19,21]

Among the various nitrones studied in ESCP, a special role is clearly taken by *N*-tert-butyl-α-phenylnitrone, PBN. PBN features a C_{SC} of close to unity in many systems, giving rise to stable ESCP conditions. At the same time, PBN is readily commercially available and non-demanding in its synthesis. Further, the alkoxyamine that evolves from reactions with macroradicals resembles the TIPNO-type NMP system. Later reinitiations making use of NMP are thus versatile and widely applicable to a range of monomers. As an

Table 6.2 $k_{ad,macro}$ for PBN (N1) for different monomers at 60 °C.

	STY	BA	NIPAAm	MMA	Ethylene
C_{SC}	1.7	1.6	0.04	$\sim 0.3^a$	77
$k_{ad,macro}$ (L mol^{-1} s^{-1})	580	57 000	920	—	$\sim 10^{5\,b}$

[a]ESCP of MMA was carried out with low amounts of comonomer styrene to facilitate the spin trapping process, hence no clearly defined C_{SC} or $k_{ad,macro}$ can be obtained.
[b]Ethylene k_p is not known to full accuracy, hence no $k_{ad,macro}$ can be derived.

additional benefit, the aromatic ring of PBN can be used to introduce secondary functionalities *via* aromatic substitution—a feature that allows for a facile post-polymerization modification and hence the construction of complex macromolecular architectures. The specific $k_{ad,macro}$ of PBN for macroradical addition in different polymerizations is collated in Table 6.2.[15,18,19]

6.2.3 Nitrone-mediated Radical Coupling

Radical coupling reactions are a very useful tool in polymer synthesis.[22] Based on atom transfer radical polymerization, first macroradical coupling methods were developed.[23,24] In these, ATRP-derived polymers are reacted in the presence of copper/ligand but in the absence of any monomer. In such system, the ATRP-typical reversible deactivation equilibrium is established. Since no monomer is available for chain growth, termination will eventually predominate and termination product is obtained. In case termination is governed by combination, a doubling in molecular weight is observed. This method—coined atom transfer radical coupling (ATRC)—is synthetically interesting, yet suffers from the drawback that many macroradicals undergo termination—at least partially—by disproportionation. Thus, several strategies were developed to counteract disproportionation. For example, by employing macronitroxides, radical combination can be enforced similar to the ESCP process.[25,26] Alternatively, non-propagating comonomers can be used, such as isoprene—a strategy employed in the field of cobalt-mediated radical polymerization.[27,28] A third pathway is to employ spin trap agents.[22,29,30] Various spin traps can be used, with nitrones and nitroso compounds being the most prominent classes. For nitrones, the method is called nitrone-mediated radical coupling (NMRC). The reactions of nitroso-compounds and nitrones are virtually identical.

The mechanism of NMRC follows the same general pathway as given in Scheme 6.1 for ESCP, with the only difference that the macroradical source in NMRC is a pre-made polymer activated *via* reversible deactivation, *e.g.* from ATRP polymer by the action of CuBr/ligand-initiated equilibrium. NMRC yields virtually identical products as ESCP with the only difference being that the dispersity of the product is given by the precursor polymer, and not by the statistical chain growth in ESCP. Narrowly distributed mid-chain functionalized polymers are thus accessible. Importantly, NMRC often proceeds with a higher rate compared to ESCP and final products can be obtained within minutes rather than hours. An important factor in NMRC is the size of the macroradical addition rate. If too low, either conventional termination (ATRC) will occur, or the polymers will again be deactivated towards their dormant state. In this respect, ESCP is an easy tool to determine $k_{ad,macro}$ (see above) and ESCP is often carried out on a certain nitrone/monomer combination before NMRC is carried out.

In NMRC, quenching of the distribution leads to a return of the molecular weight distribution to its pre-spin capturing state. This is visualized in Figure 6.2 for the NMCR of ATRP-made polystyrene and two poly(isobornyl acrylate) samples. After NMRC, a doubling in molecular weight is observed, while after quenching almost the original distribution is recovered, demonstrating that all chains were successfully bridged by alkoxyamine functions.

NMRC (and its related techniques) can be carried out on a variety of substrates and materials. A comprehensive review on these reactions is found in ref. 22 and 30.

6.2.4 Block Copolymer Formation Based on ESCP and NMRC Polymers

As mentioned, ESCP and NMRC polymers have the ability to re-initiate polymerization in classical NMP reactions. Especially with the nitrone PBN (and its derivatives), efficient NMP reactions are accessible. As the alkoxyamine functionality is in both cases embedded in the polymer rather than being located at the chain end, symmetrical triblock copolymers are obtained when performing NMP.[16,31] Naturally, triblock copolymers of ESCP-derived material will have larger dispersities when compared to NMRC starting material, yet also here reactivity of both kinds of polymers is largely identical. Figure 6.3 shows the molecular weight distributions with increasing monomer conversion in the NMP chain extension of ESCP-derived polystyrene. Successful chain extension can be carried out with midchain functionalized materials with molecular weights of up to 20 000 g mol^{-1}, however, with increasing chain length slight deviations from ideality are observed due to low amounts of conventionally terminated polymers being present in the reaction mixture.

Overall, any polymer that is accessible for polymerization with TIPNO-type alkoxyamines can be inserted into an ESCP product, allowing for the synthesis of amphiphilic triblock copolymers. It should be noted however that the insertion process does not necessarily need to proceed *via* a controlled polymerization pathway. Similar to polymer activation followed by uncontrolled radical polymerization and concomitant ESCP-like spin capturing,[32] also an uncontrolled polymerization can be used for block insertions after bond dissociations of the alkoxyamine. For example, when ESCP-made polystyrene is activated at high temperatures to form macronitroxides and reactive macroradicals in the presence of ethylene, a controlled

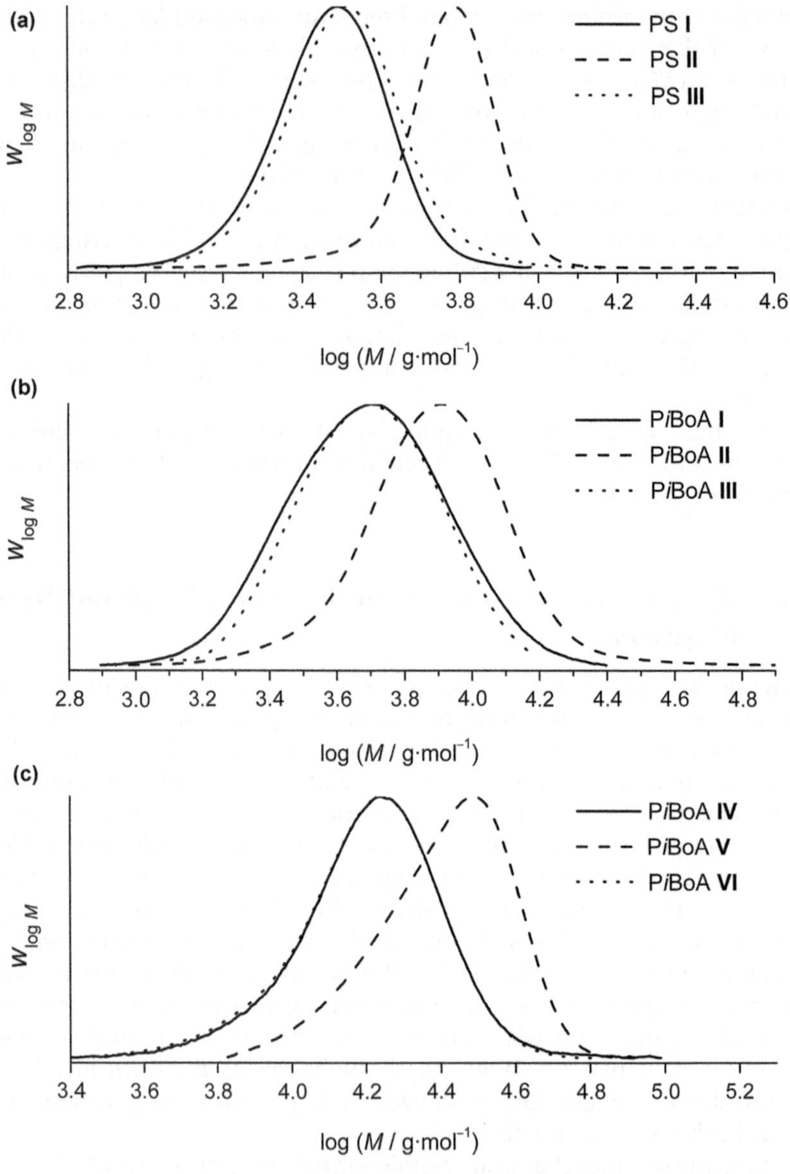

Figure 6.2 Molecular weight distributions of polymer before nitrone coupling (I, IV), after coupling (II, V) and after quenching of the alkoxyamine bond (III, VI) for polystyrene (a) and poly(isobornyl acrylate) (b, c). Reproduced with permission from ref. 29.

polymerization will occur as ethylene terminated macroradicals are irreversibly deactivated by reaction with the macronitroxide. Despite the uncontrolled nature of ethylene polymerization in this example, pristine triblock copolymers are obtained with reasonable dispersity.

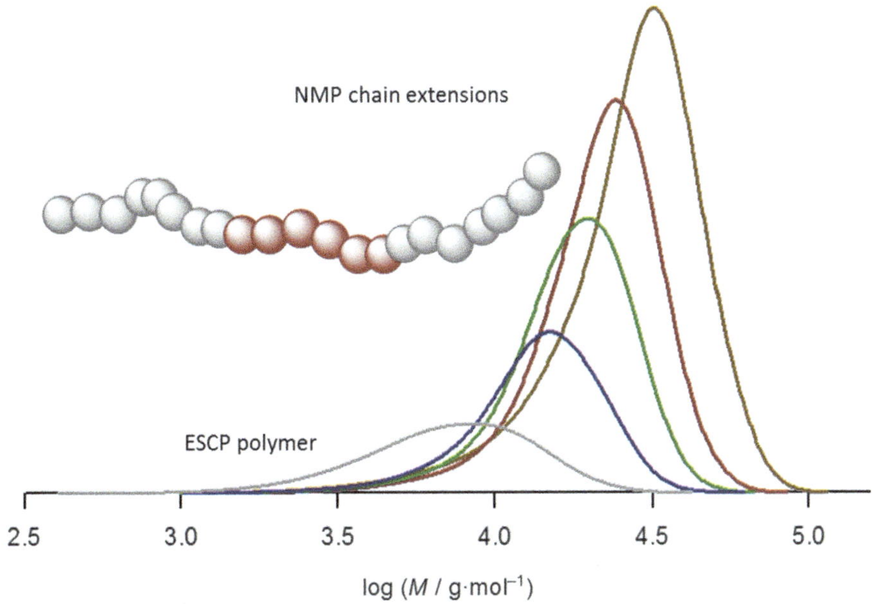

Figure 6.3 Evolution of molecular weight distributions in NMP chain extensions of polystyrene derived from ESCP.
Reproduced with permission from ref. 31.

6.2.5 Complex Macromolecular Architectures Based on ESCP/NMRC

As shown above, triblock copolymers and—if ATRP-made block copolymers are used—also symmetrical multiblock copolymers can be accessed with ease by tandem ESCP/NMP or NMRC/NMP. A further way to construct complex macromolecular architectures is to employ functional nitrones. In most cases, *para*-substituted PBN is used to introduce a secondary functionality in the midchain position of a chain, which can for example be an alkyne for copper-assisted azide alkyne cycloaddition[33] or an *N*-hydroxysuccinimide ester for bioconjugation.[34] Figure 6.4 and Scheme 6.3 depict an example for such a reaction based on CuAAC chemistry.[35] In this case, polystyrene or poly(isobornyl acrylate)s were reacted *via* ESCP or with corresponding precursors *via* NMRC to form the midchain functionalized coupling products. Since the alkoxyamine is not a carrier for an alkyne group, a click reaction can performed in this location, giving access to miktoarmed star polymers. The high efficiency of both the spin trapping and the click reactions is visualized in the 2-D chromatogram given in Figure 6.4.

In a related approach, nitrone spin capturing with virtually the same PBN-derivative was carried out, starting from a diblock copolymer [poly(vinyl acetate)-*b*-poly(acetonitrile)] that was endcapped by a labile cobalt complex (the diblock copolymers were obtained from cobalt-mediated radical polymerization, CMRC), see Scheme 6.4.[27] In addition, these CMRC derived

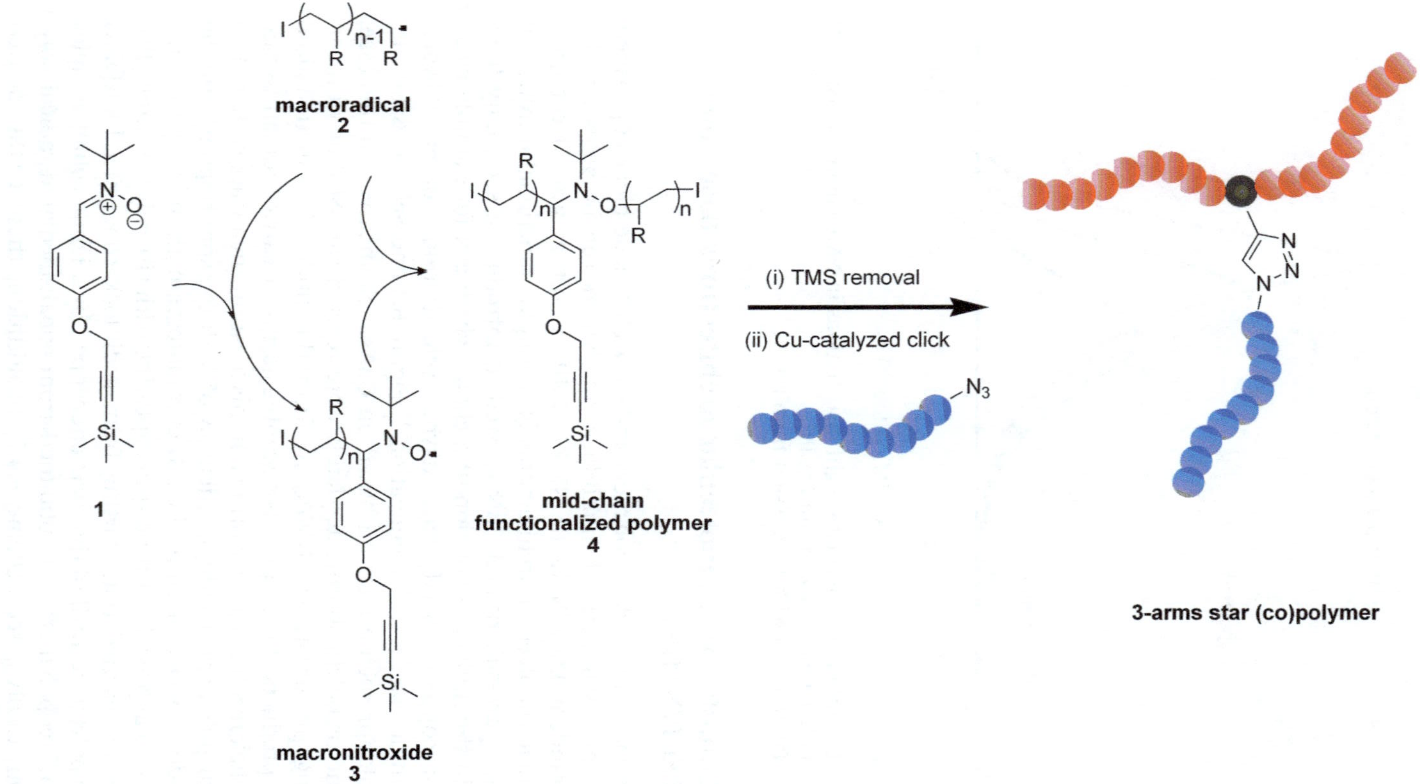

Scheme 6.3 Spin capturing of macroradicals with an alkyne-derivative of PBN and the subsequent Cu-catalyzed click reaction forming 3-armed star (co)polymers. I denotes the polymer end group resulting from the employed initiator and R represents the monomer side group (COOR for iBoA or phenyl for styrene).
Reproduced with permission from ref. 33.

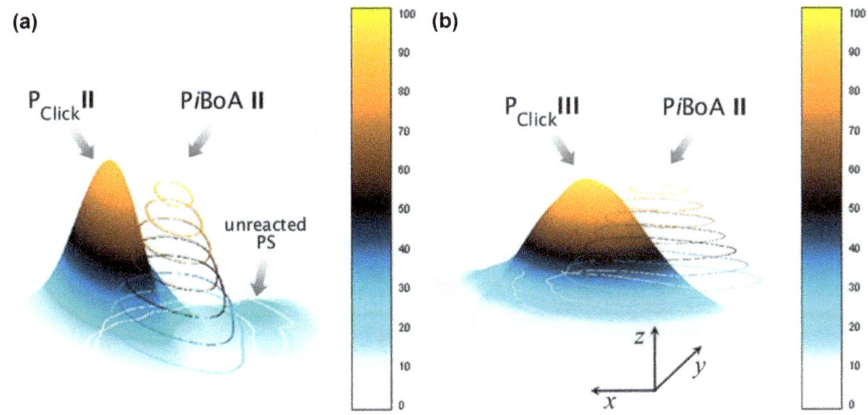

Figure 6.4 LACCC-SEC chromatograms in 3D view of the overlaps between NMRC-derived poly(isobornyl acrylate) and their click products with either azide-functional polystyrene or poly(isobornyl acrylate). The x, y and z axis corresponds to SEC, LACCC and concentration dimensions, respectively.
Reproduced with permission from ref. 33.

polymers can be efficiently trapped by the nitrone, leading to symmetrical alkyne-functionalized triblock copolymers. By reaction with a homotelechelic poly(ethylene glycol) azide, H-shape polymers can be obtained with ease, underpinning the rapidly developing complexity of macromolecules that can be accessed by combinations of ESCP/NMRC and click chemistry.[36]

Another way to use NMRC reactions for the design of complex macromolecules is to make explicit use of the fact that any nitrone always reacts with two radicals. Hence, the product features not only a doubling in molecular weight, but also a doubling in available chain-end functionality. Figure 6.5 depicts how this feature of NMRC can be used to construct dendrimers. Instead of polymeric precursors, only ATRP-initiators are used in this reaction to allow for monodisperse reaction products.[37] Starting from a tris-nitrone core molecule, a doubling in available functionalities is reached when carrying out the NMRC. By alternating NMRC and CuAAC reactions, a doubling in outer sphere functionalities is reached in each successive dendrimer generation.

Scheme 6.5 depicts the last example given herein for complex macromolecular architectures based on nitrone chemistry. Starting from polymers made by a bifunctional ATRP agent, bromo-homotelechelic polymer is obtained. When subjected to NMRC, a multi-segmented polymers are obtained where alkoxyamine functionalities are found along the backbone of the polymers with distinct equal spacing between each alkoxyamine.[38] These materials could be used in NMP reactions to obtain multi-segmented block copolymers and—as shown in the current example—to introduce secondary functionalities at each junction point of the material. With the help of the alkyne functionality that was introduced by each nitrone, additional

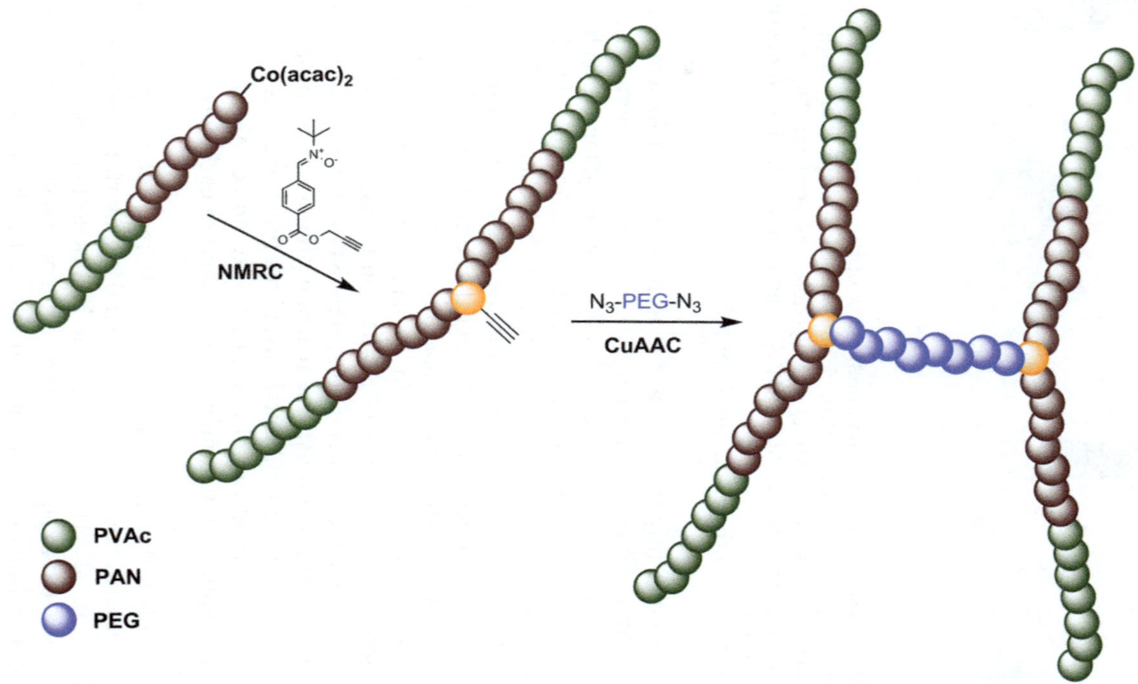

Scheme 6.4 Schematic representation of the synthesis of H-shaped polymers *via* a combination of CMRP and NMRC followed by CuAAC conjugation.
Reproduced with permission from ref. 36.

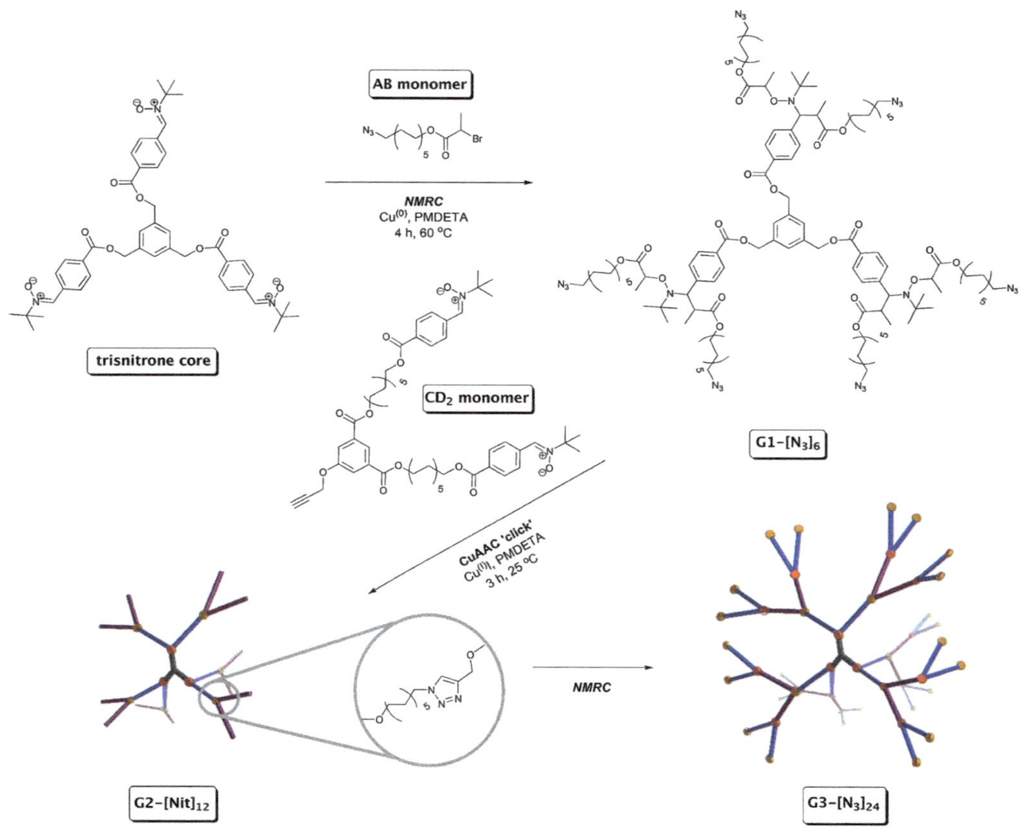

Figure 6.5 Synthesis of third-generation dendrimers from a combination of NMRC and CuAAC reactions. Reproduced with permission from ref. 37.

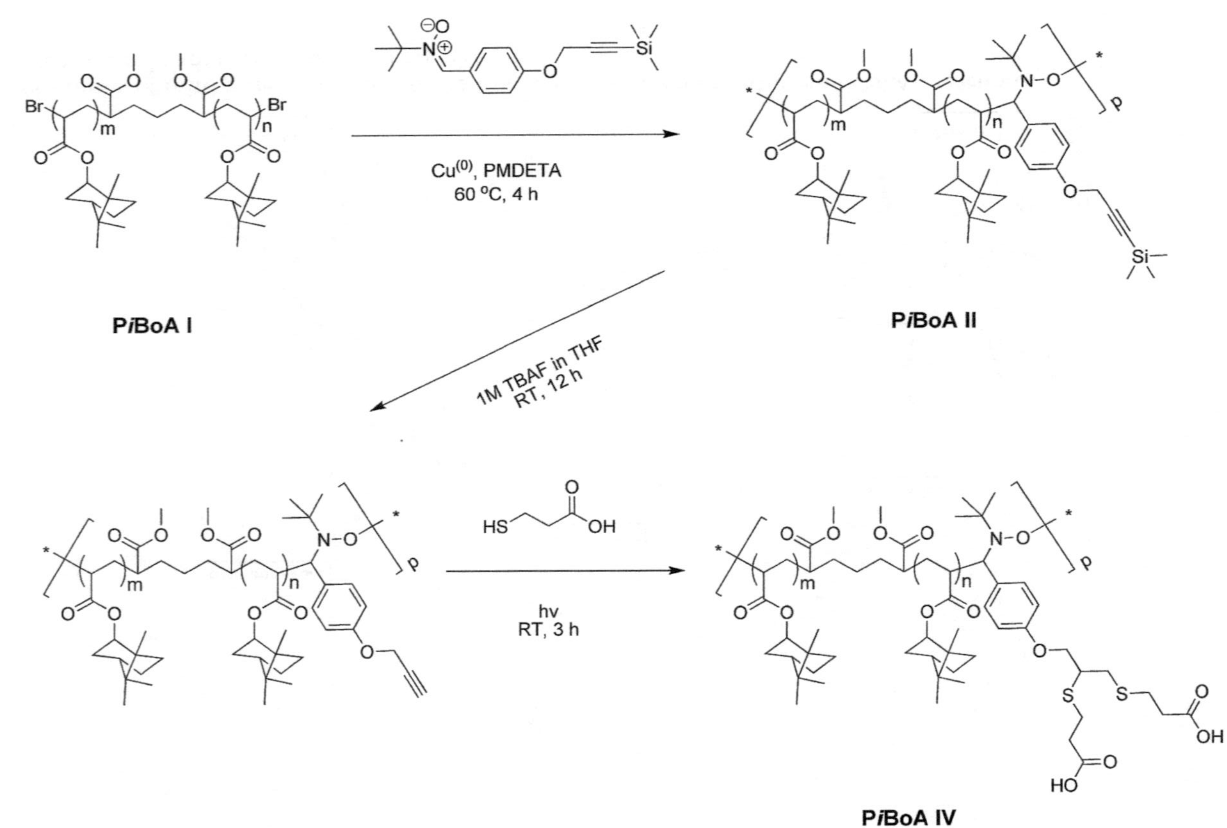

Scheme 6.5 Overall representation of the NMRC of α,ω-bromo functional poly(isobornyl acrylate) (P*i*BoA I) with a PBN-derived nitrone to form high molecular weight step growth polymers (P*i*BoA II) and its post-functionalization with 3-mercaptopropionic acid *via* UV-induced thiol-yne addition.
Reproduced with permission from ref. 38.

chemical moieties can be introduced, be it *via* CuAAC click reactions or *via* thiol-yne reactions.[39]

The above examples given for complex architectures give only an overview over the most important possibilities offered by ESCP and NMRC for the design of high-precision macromolecules, in fact many more studies exist in which nitrones have been used in the realm of polymer science.[40–44] Yet, any combination of the above methods can be in principle employed, leaving virtually endless possibilities in how polymer segments can be assembled. The variety of types of macroradicals susceptible to nitrone spin capturing is likewise very large as almost any transient radical can form (macro)nitroxides upon reaction with a nitrone. A significant broadening of applications of nitrone chemistry (and related spin traps) is envisaged and—without doubt—further potential of the reactions are yet to be explored.

6.3 *In Situ* Nitroxide Mediated Polymerization

While ESCP and NMRC are relatively new developments in the field of controlled polymerization, earlier uses of nitrones had led to the development of another type of reaction, the so-called *in situ* NMP reactions. *In situ* NMP generally denotes NMP systems which are controlled classically *via* the interchange of nitroxide and dormant alkoxyamine species, but in which the controlling nitroxide agent is formed *in situ* during the reaction itself. Thus, any precursor compound that is able to form a nitroxide in a more or less quantitative pathway can be used in a pre-reaction to form the active NMP agents, which can subsequently be followed by classical NMP. Nitrones play thereby a special role for the same reasons as in ESCP; nitrones are easily available and are mostly non-toxic (in contrast to many other precursors such as nitroso compounds). The various pathways used in *in situ* NMP are given in Scheme 6.6.

The given reaction pathways are at least partially similar to classical routes used in nitroxide and alkoxyamine synthesis and thus find broad application in preparation of typical NMP agents. The attractiveness of *in situ* NMP stems from the fact that these extra synthesis steps (and purifications) can be avoided and that NMP reaction mixtures can be formulated from simple and inexpensive sources. Thus, any reaction in which a precursor is reacted to a nitroxide/alkoxyamine followed directly in a one-pot approach by NMP can be described as an *in situ* NMP method. A detailed overview over all reported *in situ* NMP systems can be found in an excellent review by Detrembleur and colleagues.[8]

Within the scope of this overview, focus is put on the use of nitrones in *in situ* NMP. Not only are nitrone systems best understood, they can also be used to demonstrate the difference between *in situ* NMP and ESCP efficiently. The general features thereby discussed are, however, also transferable to the other precursor routes. It is noteworthy to note that many polymerizations reported as *in situ* NMP initially are probably today better understood on the basis of the ESPC mechanism and should be consequently referred to as such.

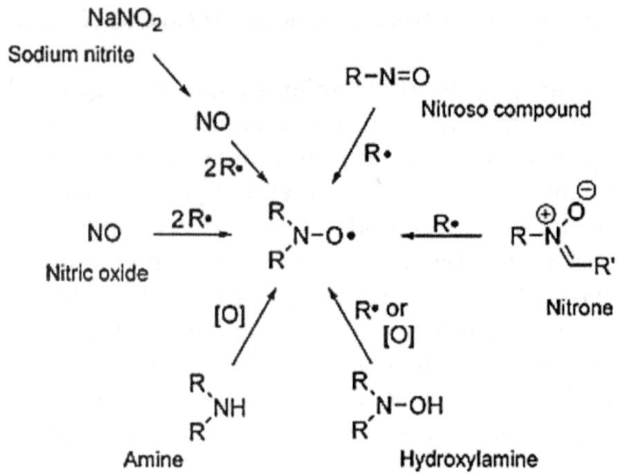

Scheme 6.6 Reaction pathways towards the formation of nitroxides by reaction with transient radicals or oxidation agents.
Reproduced with permission from ref. 8.

As described in the context of ESCP, nitrones react efficiently with almost any transient radical into nitroxides, and after termination with a second radical, into an alkoxyamine. A polymerization that is thus started by a conventional initiator in the presence of sufficient amounts of nitrone at a reaction temperature that allows the NMP equilibrium to be established can thus be regarded as an *in situ* NMP rather than ESCP (note that ESCP is typically performed well below the dissociation temperature of the involved alkoxyamines). Early reports on *in situ* NMP in fact describe reactions at lower temperatures and indeed relatively broad molecular weight distributions were observed, well in-line with the ESCP mechanism.[45–48] Soon, Detrembleur and coworkers identified that in order to achieve a high degree of livingness during the polymerization, the reaction should be carried out in two stages.[49] In addition, they tested various radical initiators as sources of transient radicals and found the choice of initiator crucial for the outcome of the reaction. While azo-initiators lead to broad molecular weight distributions in styrene polymerization, optimum outcomes were observed when the peroxide initiator benzoyl peroxide (BPO) was employed.[49,50] This is well in-line with the observation that AIBN was optimal for ESCP; since chain propagation is in competition with radical addition to the spin trap, a better *in situ* NMP process will be obtained when spin trapping is favored over chain initiation, as is the case for BPO/styrene.

Scheme 6.7 depicts the difference between the one-stage and the two-stage procedure for a 5,5-dimethyl-1-pyrroline-N–Oxide (DMPO)-mediated *in situ* NMP. While the one-stage reaction pathway is identical to the ESCP mechanism, the two-step reaction makes a distinct differentiation between the nitroxide and alkoxyamine formation and polymerization. In the pre-reaction, the reaction mixture is given time to convert the nitrone into the required NMP species before significant polymerization can take place.

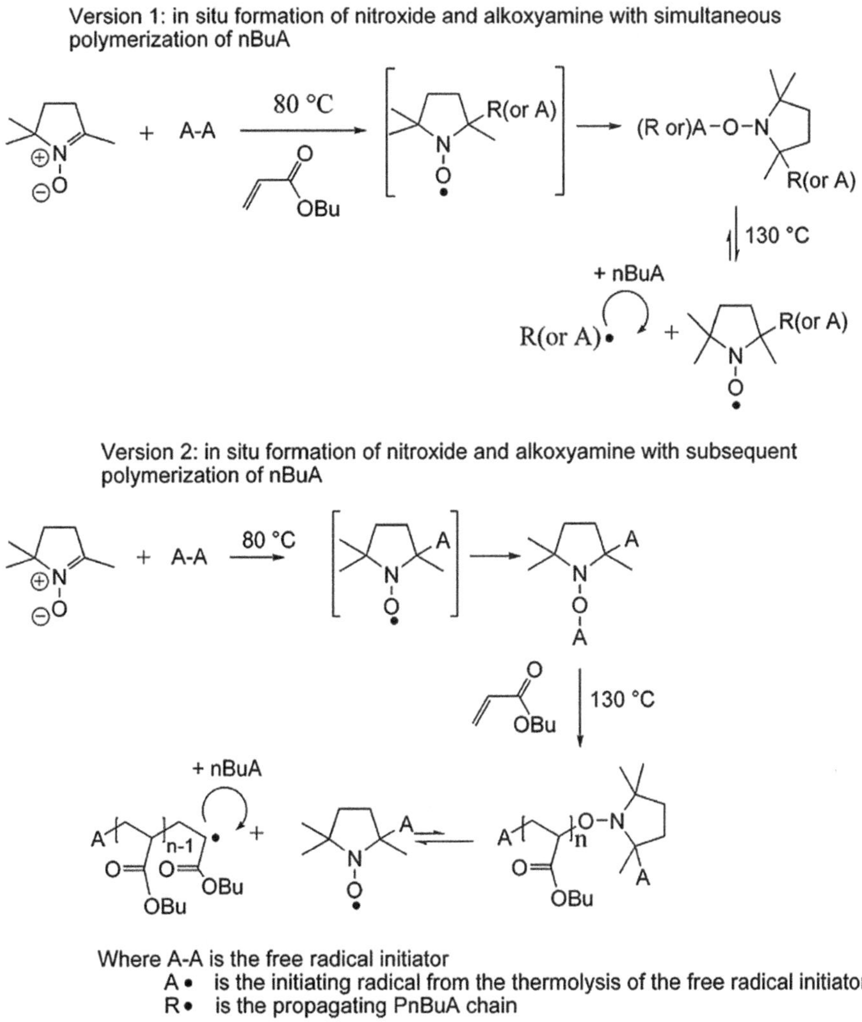

Scheme 6.7 Difference between *in situ* formation of a nitroxide with concomitant polymerization (compare ESCP process) and *in situ* nitroxide formation in a two-step pre-reaction procedure.
Reproduced with permission from ref. 8.

The polymerization itself occurs subsequently, after the nitrone is converted and the temperature is increased to NMP-characteristic temperatures. Here, the pre-reaction resembles ESCP, but due to the choice of initiator (see above), polymerization is largely avoided, leading in the following NMP to mostly end-functional macroalkoxyamines rather than midchain-functional chains. Even if some ESCP-like behavior should occur (no significant formation of polymer was detected in the pre-reaction stage), block lengths are insignificant compared to the NMP chain growth, and results much more similar to conventional NMP are obtained. In fact, in such two-step (but one-pot) *in situ* NMP, high livingness of polymerization was observed with concomitant low polydispersities in the range of 1.25 for example for polystyrene.[49] The detailed reaction pathways occurring during the pre-reaction stage of this *in situ* NMP are given in Scheme 6.8. Also for other

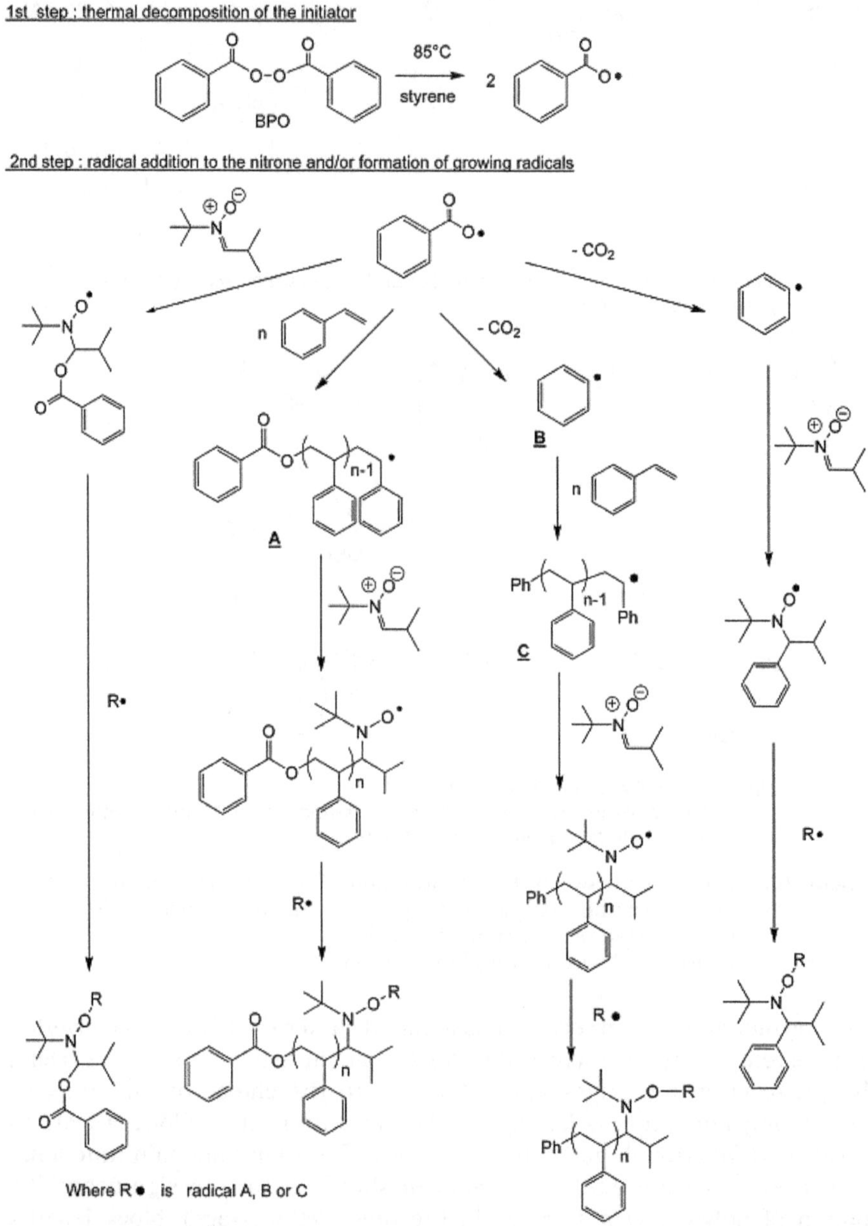

Scheme 6.8 Detailed reaction pathways in the nitrone/BPO *in situ* NMP process. Reproduced with permission from ref. 8.

monomers and copolymer mixtures, low dispersities and good livingness were only observed in two-step procedures, which can thus be seen as a general trend.[51–53] Pre-reactions are consequently strongly advised for any *in situ* NMP to proceed.

In conclusion, it may be stated that *in situ* NMP features rather complex reaction pathways and always leads to a certain mixture of products, which are in all important details identical to their classical NMP counterparts, yet feature a variety of polymer end groups. This may be seen as a disadvantage from a purist point of view, but is, however, of low relevance for many applications since chains are inherently of living character and thus open the entire range of reactions as known from the realm of NMP. Access to block copolymers or more complex polymeric structures is granted without penalty.

Moreover, the complexity of the pre-reaction can also be used to advantage over typical NMP. Since a variety of nitroxide species (with very different steric demands and electronic structures) are formed, NMPs become accessible that would be impossible to perform with just a single alkoxyamine control agent. A good example for such reaction is the low-temperature *in situ* NMP of MMA, which has recently been reported (see Figure 6.6).[54] In this study, methyl-2-methyl-3-nitro-2-nitrosopropionate (NMMA) was used in conjunction with the radical initiator V-70, a low-temperature azo-initiator (10 h half-life time at 30 °C). During the prereaction of NMMA with V-70 in conjunction with MMA, a very complex mixture of radical species and nitroxide species is formed, with each pair yielding very different decomposition behaviors. Due to the often high steric hindrance of the alkoxyamine adducts, decomposition of the alkoxyamines can proceed at much lower temperatures than usually required for NMP, while at the same time only a little disproportionation occurs, also in contrast to most MMA NMP systems. In consequence, *in situ* NMP provides in this case a polymerization that would be practically inaccessible by conventional NMP, yielding polymers with good molecular weight control and overall low dispersities (<1.5).

6.4 Light-sensitive Alkoxyamines: Towards Nitroxide Mediated Photopolymerization (NMP2)

Photopolymers represent a rapidly emerging field in material science. Photochemical processes are characterized by important advantages, *e.g.* thin films can be polymerized very rapidly (less than 1 second) by UV irradiation. For coatings, inks, photoresists, dual-cure systems and others, photopolymerization reactions are today recognized as particularly interesting and powerful.[55,56] Photochemistry is also an ecological alternative to thermal processes since less or even no volatile organic compounds (VOC) are usually used with such technology. Another interesting feature of photopolymerization processes is the possible modification or patterning of surfaces, exhibiting excellent spatial control. Here, we denote

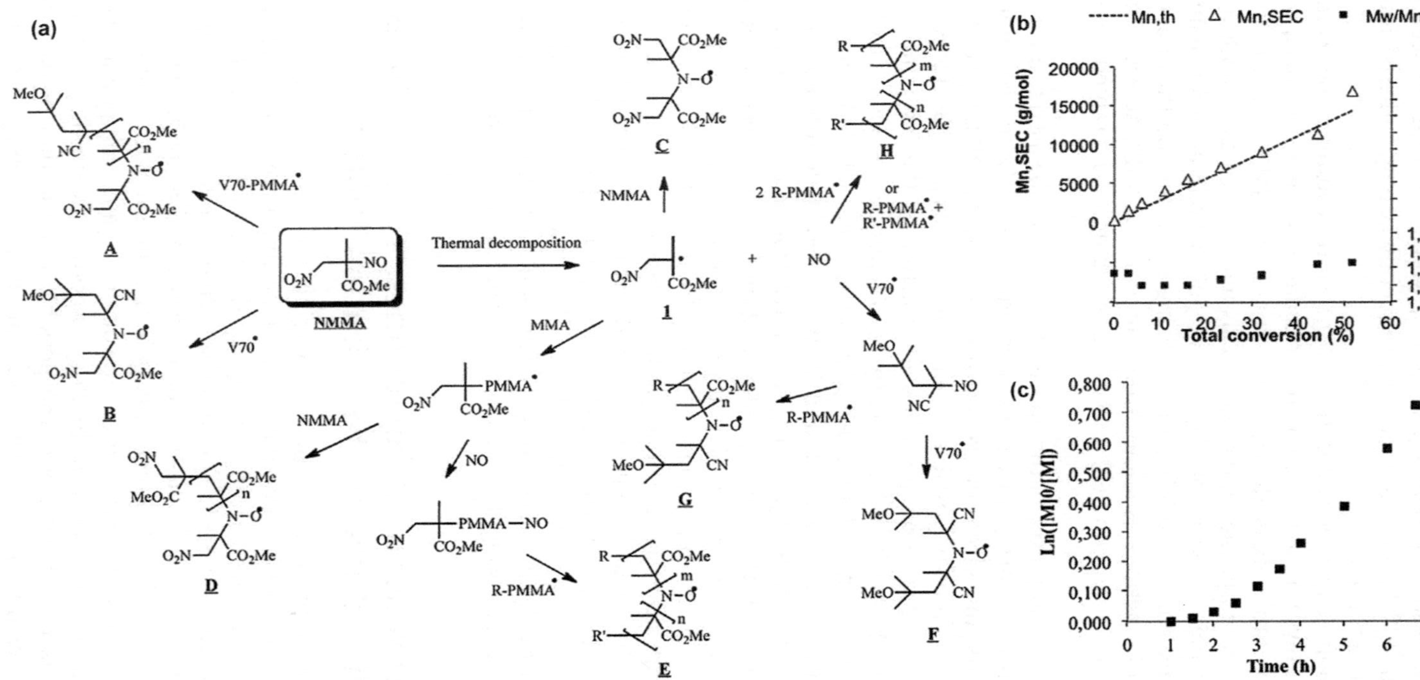

Figure 6.6 (a) Reaction pathways in the low-temperature MMA *in situ* NMP initiated by V70/NMMA. (b) Evolution of average molecular weight and dispersity and (c) first order kinetic plot for the same polymerization. Reproduced with permission from ref. 54.

photopolymerization from surfaces as a grafting-from approach, excluding the multitude of light-induced grafting-to processes.

However, a strong impediment to the development of photopolymers is related to the lack of control techniques for the polymerization reactions (even though recently more and more processes are being developed in this area).[9,10,12] Thus, access to complex architectures is often hindered in photopolymerizations and the search for new radical photoinitiating systems allowing for reversible deactivation process remains of great interest. In the following, light-sensitive alkoxyamines will be presented as photoinitiating systems for a controlled radical photo-polymerization (CRP2), termed nitroxide mediated photo-polymerization (NMP2). During NMP2, the photolysis of an alkoxyamine bearing a chromophore leads to a reversible equilibrium between the alkoxyamine and the generated nitroxide and alkyl radicals, allowing a controlled radical photopolymerization (Scheme 6.9).[57] The fundamental aim is to combine the advantages of the controlled radical polymerization (CRP) and photoassisted processes. For example, the new mode of NMP2 can be used for spatially controlled surface modifications, polymer grafting or lithography applications, to name a few important applications. Some recent examples will be described in the following.

The key idea of NMP2 is to replace the dithiocarbamyl radical mediator in the photoiniferter process (see below)—this latter structure being characterized by important drawbacks (see below)[58]—by a nitroxyl radical that is a fully persistent radical.[59] The side reactions observed for dithiocarbamyls are suppressed and significant control improvement can be expected. *Via* this approach, light-sensitive alkoxyamines are proposed as new photoiniferters. Typically, the chromophore group must be located on the nitroxide moiety to ensure a reversible cleavage of the macroalkoxyamine during the photopolymerization process. Significant improvement can be expected for the new approach: (i) better control for the photopolymerization process, (ii) increase of dormant species concentration essential for grafting, multilayers polymerization process and the re-initiation reaction and (iii) easier access to complex architectures (block copolymers, *etc.*).

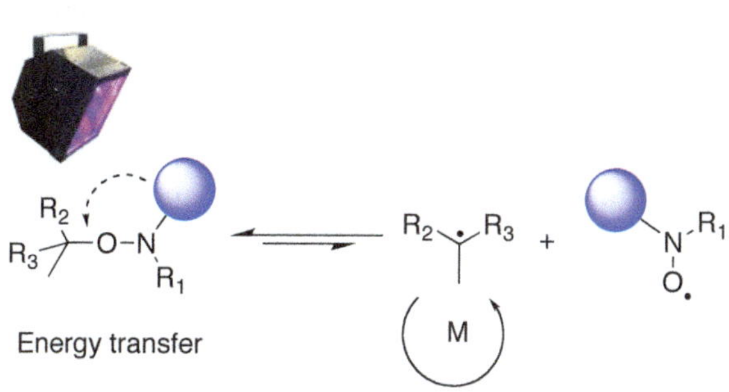

Scheme 6.9 Cleavage of light-sensitive alkoxyamines and nitroxide mediated photopolymerization (NMP2) mechanism.
Reproduced with permission from ref. 57.

The NMP2 approach can also be a highly original answer to the actual challenge to obtaining—in a reproducible fashion—well-defined topographical patterns ranging from micro- to sub-micro levels with a uniform chemistry. This approach can be relevant for different fields, such as microfluidics, superhydrophobic surfaces, biology, optics or photonics. Indeed, photopolymers and in particular surface patterning have become an increasingly important part of modern science and technology, such as in the areas of micro-electronics, information processing and storage, nano/microfluidic devices and biosensing. A straightforward methodology based on light-sensitive alkoxyamines to prepare such (micro)patterns with the advantage of the possibility of reinitiating the polymerization of a second monomer can be very interesting. In that case, multi-layered covalently linked micropatterns could be prepared.

In a first part, the chemical mechanisms associated with these new light-sensitive alkoxyamines will be presented to find the optimized compounds able to control the photopolymerization both spatially and structurally (*i.e.* the chain length and the end chain functionality). In a second part, some examples for the use of the NMP2 for controlled radical photopolymerization as well as surface modifications and (micro)patterning will be given.

6.4.1 Light-sensitive Alkoxyamines

Quite a few studies have been devoted to the photo-dissociation properties of alkoxyamines, which would allow the possibility of NMP2 (Scheme 6.9). Initially Scaiano *et al.* studied the cleavage of different TEMPO-based alkoxyamines (where TEMPO is 2,2,6,6-tetramethylpiperidinyloxy-4-yl radical) whose homolysis was photosensitized by a xanthone (Xa) triplet state or a pyrene (Py) excited singlet state through an energy transfer process [r1–r2, shown in eqn (6.3) and (6.4)].[60] A photosensitizer (Xa, Py) was necessary as these alkoxyamines exhibit very poor light absorption properties for $\lambda > 300$ nm. A possible quenching of the excited states of Xa or Py by the alkoxyamine (**A** or **B** in Scheme 6.10) was observed, yet the yield of alkyl radicals remains very low and in the presence of xanthone suggests a low efficiency for the bimolecular sensitization reaction.

$$\text{Xa (or Py)} \rightarrow {}^*\text{Xa (or }{}^*\text{Py)} \quad (h\nu) \tag{6.2}$$

$$^*\text{Xa (or }{}^*\text{Py)} + \text{alkoxyamine} \rightarrow \text{Xa (or Py)} + {}^*\text{alkoxyamine} \quad \text{r1} \tag{6.3}$$

$$^*\text{Alkoxyamine} \rightarrow \text{nitroxide} + \text{alkyl radical} \quad \text{r2} \tag{6.4}$$

To enhance the efficiency of the energy transfer between the chromophore and the C–O bond, Neckers *et al.* designed new alkoxyamines bearing different chromophores covalently linked to the TEMPO moiety (see *e.g.* the alkoxyamine **C–D** in Scheme 6.10).[61] For these alkoxyamines bearing a xanthone or a benzophenone moiety, a rapid quenching of the excited state

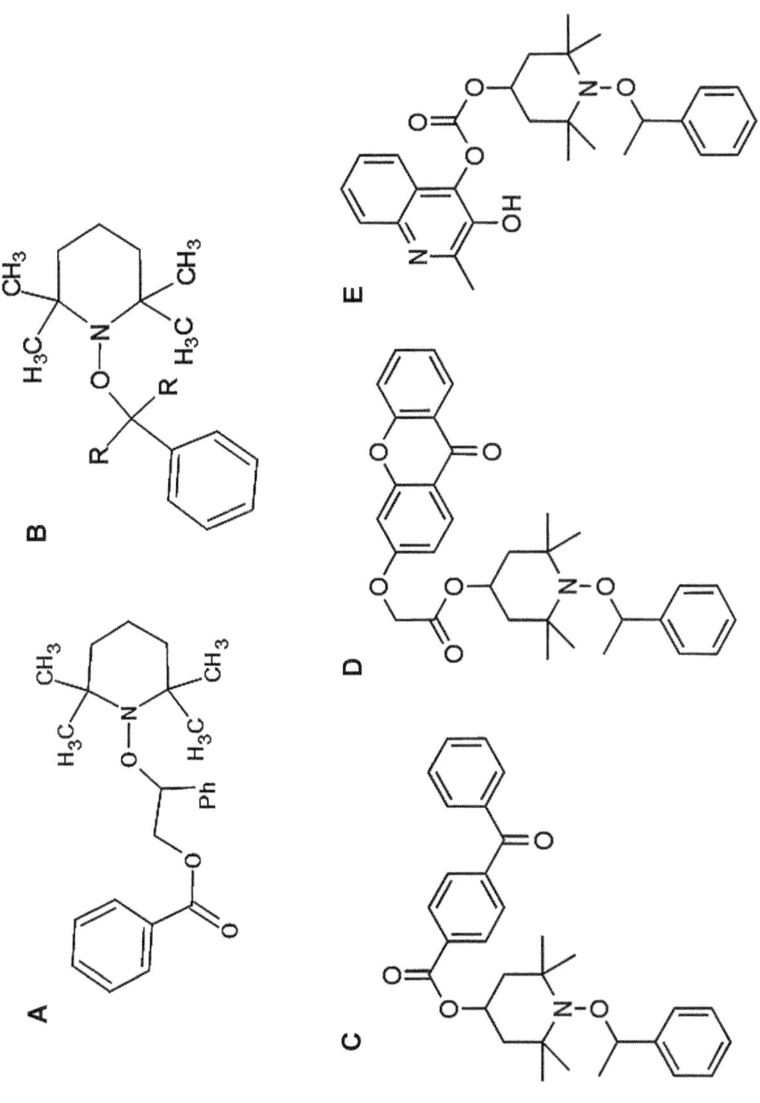

Scheme 6.10 Structures of NMP2 agents used in first trials as light-sensitive alkoxyamines.

during the laser flash photolysis (LFP) experiments was observed but the detection of the alkyl radical was not easy. The photopolymerization of methyl methacrylate in the presence of **D** has been investigated and exhibits one of the criteria of a CRP2 process, *i.e.* M_n is linked to the initial amount of **D**. However, the final polymer was not photoreactive and reactivation of the dormant species did not work, suggesting a non-living character.

More recently, Scaiano *et al.* used a quinoline as chromophore (alkoxyamine **E**, Scheme 6.10).[62] The photolysis upon UV light showed an increase of the nitroxide concentration in ESR experiments but their first attempt to photopolymerize styrene did not show any living character.

Yoshida also investigated the photopolymerization of various monomer in the presence of nitroxides or their corresponding alkoxyamines using a photoacid generator (bis(alkylphenyl)iodonium hexafluorophosphate) to produce radicals:[63]

$$\begin{aligned}&\text{Alkoxyamine} + \text{diaryliodonium}^{+}\\ &\quad\rightarrow \text{macroradical} + \text{oxopyrrolidiumn}^{+}\ {}^{*}\text{diaryliod}\quad (h\nu)\end{aligned} \tag{6.5}$$

However, this mechanism remains somewhat unclear since the role of the iodonium salt is not fully elucidated.

Since the cleavage of the C–O bond in these alkoxyamines is due to an energy transfer from the chromophore ("antenna") to the weak C–O bond, a more recent approach was to link the chromophore directly to the aminoxyl function or to introduce a minimal spacer (only one carbon atom) to facilitate the homolysis of the alkoxyamine.[57,64] An increase of the energy transfer efficiency can be expected leading to a higher yield in nitroxides. The alkoxyamines of this second generation (*i.e.* with a closer proximity between the chromophore and the C–O bond) are gathered in Scheme 6.11.[57,64] In this latter scheme, **F**, **G** and **I** can be used for NMP2 as the chromophore is located on the aminoxy moiety. **H**, **J** and **K** were only used as model systems to investigate the structure–reactivity relationships for the cleavage of alkoxyamines.

For alkoxyamines, severe limitations for their cleavage processes under light irradiation can be expected, *i.e.* as both the N–O and C–O bonds are relatively weak, the selectivity of the cleavage process (N–O *vs.* C–O) must be carefully investigated. Indeed, for NMP2, the cleavage must be highly selective to generate only nitroxides (C–O cleavage). A competitive N–O cleavage does not lead to persistent radicals (only aminyl and alkoxyl radicals); this will be associated with a loss of control during the light-assisted photopolymerization process and the formation of dead polymer chains. With this unwanted reaction (N–O cleavage), the yield in dormant species decreases during the polymerization and the re-initiation step is no longer accessible. Ideally, the C–O dissociation quantum yield must be 1.0 (each photon absorbed leads to a nitroxide formation). However, other specific points also need to be fulfilled for an optimal NMP2 mediator. Particularly, the behavior of the macroalkoxyamine (generated during the

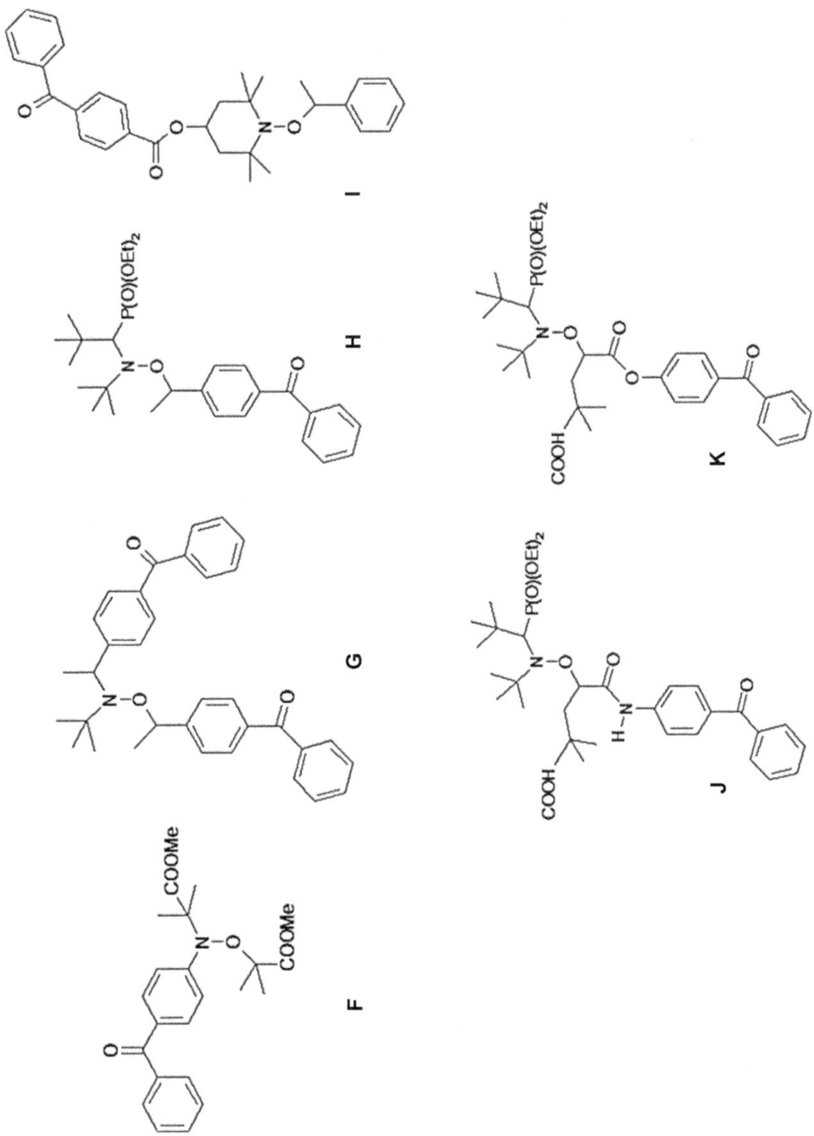

Scheme 6.11 Recently proposed light-sensitive alkoxyamines.

photopolymerization process) must be examined. Typically, the chromophore group must be located on the nitroxide moiety to ensure a reversible cleavage of the macroalkoxyamine during the photopolymerization process.

The observation of the nitroxides generated under light irradiation required specific selected conditions to obtain a rapid conversion of the alkyl radicals into unreactive species and to avoid their recombination with the nitroxides. Different trapping agents, such as other nitroxides, hydrogen donors or oxygen were already used to prevent this back-reaction.[65] For easy handling, the use of O_2 has generally been chosen.[64a,c] In that case, the carbon-centered radicals (resulting from the C–O bond dissociation) are trapped by oxygen yield peroxyl radicals, thereby preventing the recombination between a nitroxide and a carbon-centered radical. As a consequence, the intensity increase of the ESR signal was only related to the formation of the nitroxides. Rate coefficients for the dissociation process (k_d) are accessible by this approach (Figure 6.7). Nitroxides were generated upon UV irradiation for **F–I** (fast kinetic for **F–H** but slower for **I**; Figure 6.7B) but not for **J–K**. However, for all these alkoxyamines, the bond dissociation energy (C–O) was lower than the excited state energy level showing that this parameter is not the main driving factor for the C–O cleavage.[64d] The closer proximity for **F–H** between the benzophenone chromophore and the C–O bond compared to **I–K** associated with the higher efficiency of **F–H** for nitroxides generation has suggested that the distance plays a key role in the cleavage process.[64d] For **F**, quite stable aromatic aminyl radicals can be generated through N–O cleavage process and consequently a rather poor selectivity was observed (C–O and N–O cleavage). The best compromise (high cleavage efficiency and good selectivity) was found for **G**, *i.e.* with the introduction of a carbon atom between the chromophore and the aminoxy group, the formation of aliphatic aminyl radicals is unfavorable leading to good C–O cleavage selectivity. The decomposition of the alkoxyamines was also proven by various techniques (laser flash photolysis, molecular modelling and photoproducts analysis).[57,64]

6.4.2 Towards Nitroxide Mediated Photopolymerization: NMP2

In the following section, examples of NMP2 in solution will be discussed. Finally, to go a step further, the macroalkoyamine formed in the polymerization of a first monomer layer will correspond to a macrophotoinitiator for a second layer, which leads to covalently bonded multilayer devices. Examples of the use of NMP2 for surface modifications and patterning will be given. Indeed—since controlled/living graft polymerization proceeds only with UV irradiation and on the UV irradiated regions—controlled grafting can be realized and applied to various fields, such as regionally precise modification, photo-patternable grafting, grafting with controlled densities at spatial locations, and microfluidic devices.

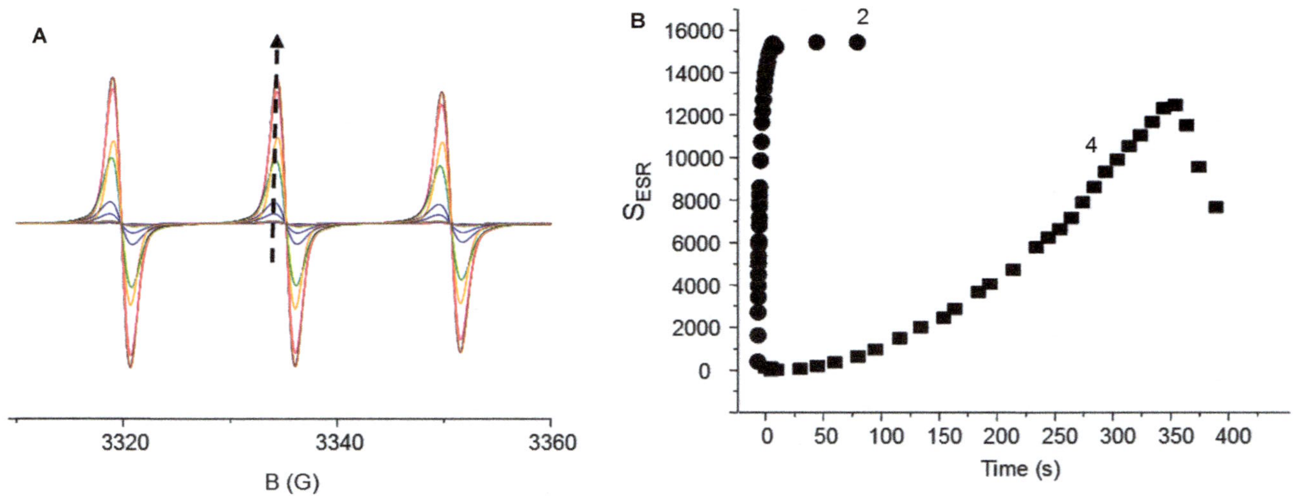

Figure 6.7 (A) Increase of the ESR spectrum of the nitroxide under light irradiation of **I** in *tert*-butylbenzene in aerated conditions. (B) Double integration of the ESR signal ascribed to the nitroxide *vs.* the irradiation time in **G** and **I** for similar light absorptions; light intensity = 44 mW cm^{-2}.
Reproduced with permission from ref. 64*d*.

6.4.2.1 NMP2 in Solution

Controlled/living radical polymerization (CRP) is among the most rapidly developing areas of chemistry and polymer science but its application in photopolymerization technology is still developing. Otsu pioneered this technique in 1982 using alkyl dithiocarbamate compounds.[58,66] Alkyl dithiocarbamates act as iniferters in radical photopolymerization; the latter process involves initiation, chain transfer and termination (Scheme 6.12). Decomposition of dithiocarbamates generates carbon radicals initiating polymerization of vinyl monomers and less-reactive dithiocarbamyl (DC) radicals acting as reversible terminators. This is, however, an idealized mechanistic picture of iniferter polymerization. Indeed, different studies have revealed a more complex behavior of the terminating groups: DC can dimerize to form dithiurame, initiate polymerization by adding to alkenes, or can eliminate CS_2 by side reactions resulting in complete loss of "livingness".[67]

Nevertheless, it was recently demonstrated that photoiniferter dithiocarbamate chemistry can be used to construct complex macromolecular architectures such as block and graft copolymers, stars, hyperbranched and other polymeric materials as well as for modification or patterning of surfaces (see also below).[58,66,68–76]

Apart from the above photoiniferter system, few efficient living photopolymerization techniques are currently available. Very recently, photoinduced ATRP and/or the photoredox catalysis approach has also emerged as another very elegant approach.[9,10,12,77] This approach usually required the presence of iridium or copper catalysts albeit a metal free approach has been given very recently.[78] In this section, the use of light-sensitive alkoxyamines as a photoiniferter for the polymerization of monofunctional monomers in solution is presented as a metal-free approach. Indeed, some of the reported light-sensitive alkoxyamines (**F** and **G**) were used to promote the first polymer synthesis using a NMP2 process.[57] Using compound **F** as a photoiniferter for the polymerization of *n*-butyl acrylate, a linear increase in ln $[M]_0/[M]$ was observed, and more interestingly, 80% of conversion was obtained in 500 s. A linear growth of the polymer chain combined with a partial

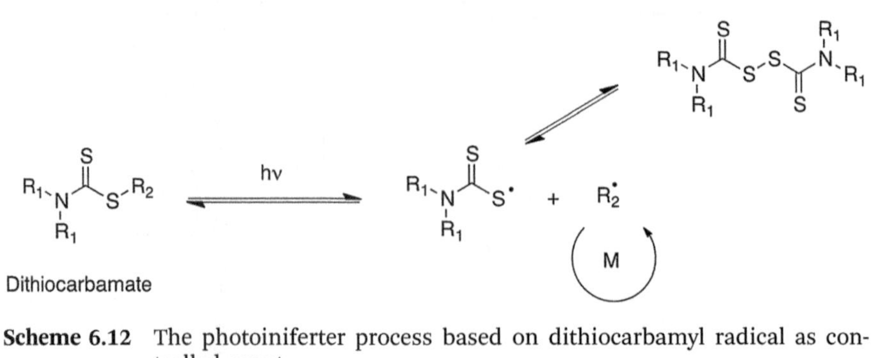

Scheme 6.12 The photoiniferter process based on dithiocarbamyl radical as controlled agent.

living character has been evidenced (Figure 6.8).[57] Interestingly, light re-initiation experiments were also performed in agreement with the proposed NMP2 reaction (Figure 6.9), *i.e.* a very clear shift of the molecular weight distributions before and after re-initiation (the two blocks being prepared by NMP2), showing good re-initiation even if a long tail toward lower molar masses is observed in agreement with a partial loss of control (dead chain due to N–O cleavage).

As **G** exhibits a better selectivity towards the C–O cleavage process compared to **F**, better control was found, *i.e.* narrower molecular weight distributions were obtained albeit they remain quite large compared to thermal NMP.[79] The design of molecules exhibiting excellent bond cleavage selectivity without side reactions (*e.g.* hydrogen abstraction from the benzophenone triplet state, photodegradation of the nitroxides) is still required to improve the control process.

Remarkably, for **G**, thermally driven polymerizations were also performed *via* the cleavage of the alkoxyamine functionality and its use for block copolymer syntheses investigated. Excellent control of the polymerization for **G** was found in thermal mode (monomer = styrene at 120 °C). This dual system (**G** being both a NMP and NMP2 mediator) was applied to the post-modification of polymer coatings for application in patterning and photografting (the first layer prepared by thermal NMP and the post-modification by light activation through an NMP2 process).[79]

6.4.2.2 *NMP2 as an Elegant Tool for Surface Modifications and/or (Micro)Patterning*

Different grafting-from polymerization techniques have been developed over the years. Free radical precursors immobilized or formed on surface regions can allow radical polymerization. However, these techniques cannot fully provide precision graft architectures due to the high reactivity of free radicals. Therefore, the regiospecificity of graft-polymerized surfaces and precision graft-chain architectures, including their surface graft-chain density, spatio segmental density, segmental or graded block chains, molecular shape or configuration, can only be controlled to some extent.

The combination of living radical photopolymerization and lithographic techniques such as the use of a mask can allow this regiospecificity.[70]

Using dithiocarbamate compounds as photoiniferters, the potential of photoinduced modification or patterning of surfaces has been already successfully illustrated by different groups.[68–70,80–83] Specifically, a photolithographic method with three distinct features has been proposed[80,81]: (i) liquid photoresists in contact with the photomask, (ii) readily removed sacrificial materials, and (iii) living radical processes enable multiple polymeric chemistries and mechanical properties while simultaneously allowing for the facile fabrication of 3D geometries and surface chemistry control. Moreover, CRP2 enables initiation and covalent adhesion of new polymer

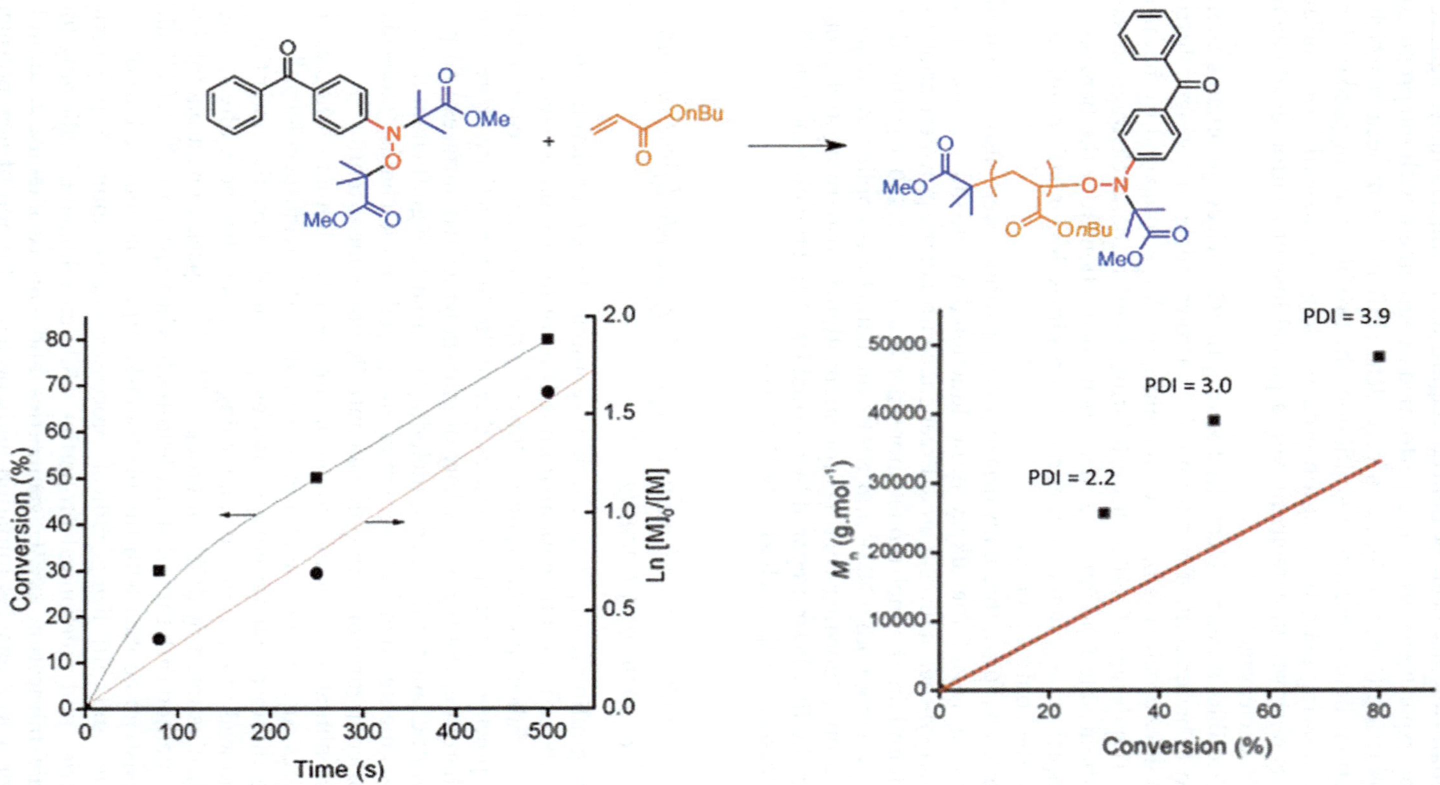

Figure 6.8 Kinetic plot and M_n and PDI *versus* conversion for the photopolymerization of *n*-butyl acrylate with the alkoxyamine **F** (1% w/w targeted $M_n = 41\,300$ g mol^{-1} at 100% conversion); for more detail see also ref. 57.

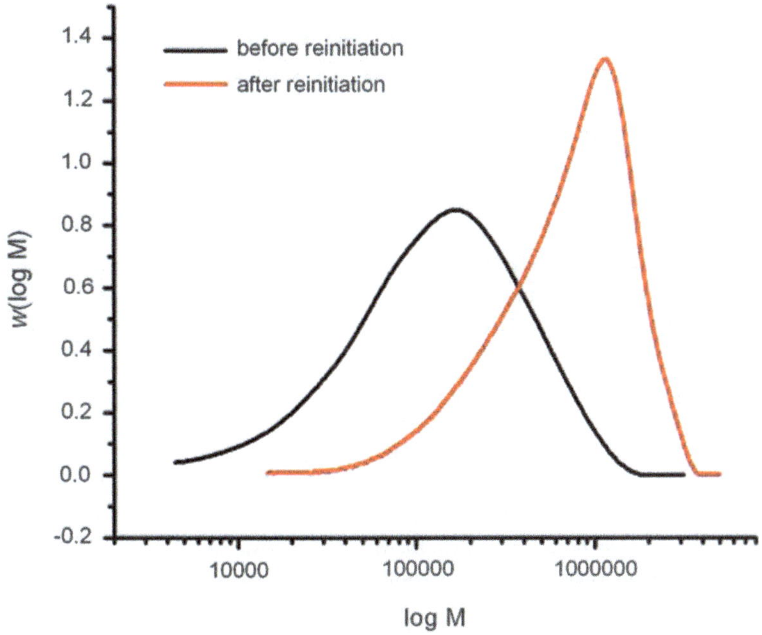

Figure 6.9 Molecular weight distribution before and after re-initiation by light; **F** is used as light-sensitive alkoxyamine; monomer = *n*-butyl acrylate; reproduced with permission from ref. 57.

chains or films to the surfaces of previously photopolymerized films. The advantage of this novel technique for fabricating patterned biomedical polymer surfaces that include the ability to make patterns on polymer films of any type of monomer that can be polymerized by UV free-radical polymerization has been given.[82]

Even if the use of dithiocarbamate compounds is very encouraging for such surface modifications, the technique suffers from severe limitations.[67] The main restraint is related to the generation of dithiocarbamyl radicals as controlling agents. Indeed, these radicals are also able to either dimerize and/or initiate extra chains leading to a dramatic loss of control and livingness (see above).[76] For example, these systems could suffer from some reproducibility concerns, poor quality in terms of topology or composition of the surface pattern and difficulty to access covalent bonded polymer layers. In order to fulfill the growing need in the elaboration of photoinduced modification or (micro/nano)patterned surface, the development of NMP2 as alternative to dithiocabarmates has been proposed recently.

The NMP2 technology has been used to prepare such patterned surfaces. Two sequential polymerizations have been performed, yet the lateral resolution as well as the potential for complex patterning have not been fully explored. The NMP2 process gives access to spatially controlled reactions. An example has been performed using a first layer prepared using the NMP2 process and an Ebecryl 605 matrix.[64a] The macroalkoyamine formed in the polymerization of the first layer will correspond to a macrophotoinitiator for the second layer of the same monomer, which leads to covalently bonded multilayer devices (see Figure 6.10). *Via* such an approach, it was possible to obtain thickness above 100 nanometers and a linear growth of the thickness *versus* irradiation time. Different monomers as well as irradiation conditions were investigated.[84]

Spatial control can be also obtained by re-initiation of a second layer only in the light irradiated area (use of a mask; Figure 6.11A).[79] Remarkably, a quite fast re-initiation was found, *i.e.* for 400 s of irradiation of a second acrylate monomer a final thickness of ~100 μm was obtained.

Surface re-initiation experiments can also be used to drastically change the surface properties.[79] The light-sensitive alkoxyamine was used for the polymerization of trimethylolpropane triacrylate (TMPTA) leading to a first surface exhibiting a quite hydrophilic character (Figure 6.11B). A second monomer (2,2,3,3,3-pentafluoropropylacrylate) can be coated on the top of the synthesized polymer allowing a re-initiation of the polymerization to generate a second layer upon light irradiation. A fast and efficient re-initiation reaction was noted as a tack-free surface was obtained after only

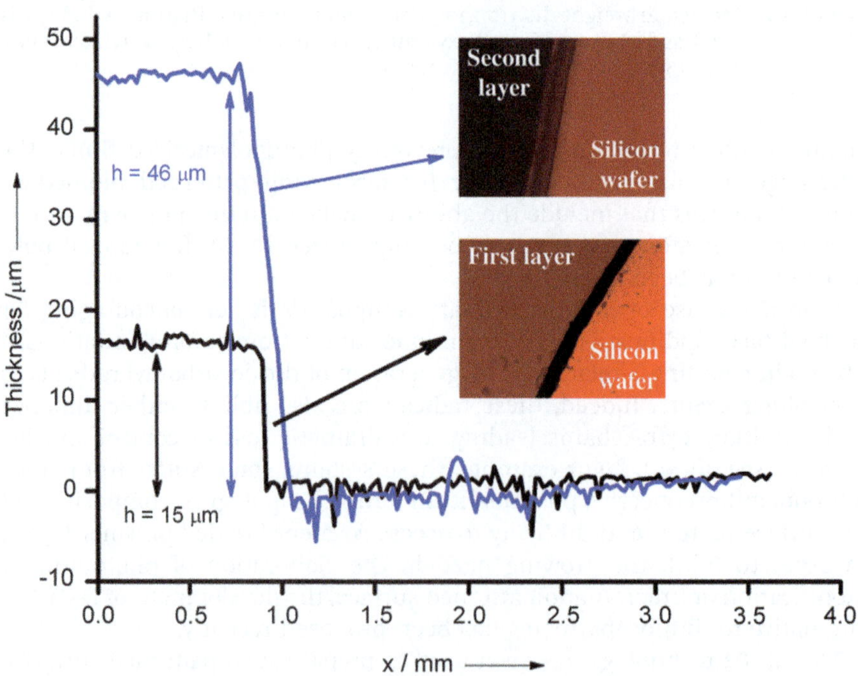

Figure 6.10 Surface re-initiation experiment using **G** in an Ebecryl 605 matrix: thickness before and after re-initiation determined by an altimeter; irradiation with a UV lamp.
Reproduced with permission from ref. 64*a*.

Figure 6.11 (A) Profilometric profile of the second layer; irradiation through a mask for the photopolymerization of the second layer upon a Xe–Hg lamp irradiation ($\lambda > 340$ nm); (B) Contact angle for the polymer obtained for the polymerization of TMPTA with the alkoxyamines **G** thermal polymerization at 120°; (C) Contact angle after re-initiation upon a Xe–Hg lamp irradiation with 2,2,3,3,3-pentafluoropropylacrylate.
Adapted from ref. 79.

5 minutes of irradiation. The second surface was hydrophobic thanks to the fluoro groups of this second layer (Figure 6.11C). The successful re-initation was also supported by XPS analysis of the two layers.

Thanks to the presence of dormant species (macroalkoxyamines) in the polymers, NMP2 was also used to bind two polymers together upon UV irradiation. The linkage was ascribed to the NMP re-initiation leading to carbon centered radicals that can be combined resulting in covalent bonds between the two polymers.[79] Further, the use of light-sensitive alkoxyamines in self-healing materials has been investigated recently *i.e.* upon light irradiation the network is progressively reformed after irradiation *via* the recombination of radical species. This self-healing behavior was evidenced by AFM.[85]

References

1. J. Chiefari, Y. K. Chong, F. Ercole, J. Krstina, J. Jeffery, T. P. T. Le, R. T. A. Mayadunne, G. F. Meijs, C. L. Moad, G. Moad, E. Rizzardo and S. H. Thang, *Macromolecules*, 1998, **31**, 5559–5562.
2. K. Matyjaszewski and J. Xia, *Chem. Rev.*, 2001, **101**, 2921–2990.
3. C. J. Hawker, A. W. Bosman and E. Harth, *Chem. Rev.*, 2001, **301**, 3661–3688.
4. J. Nicolas, Y. Guillaneuf, C. Lefay, D. Bertin, D. Gigmes and B. Charleux, *Prog. Polym. Sci.*, 2013, **38**, 63–235.
5. D. H. Solomon, E. Rizzardo and P. Cacioli, US 4581429, CSIRO, 1986.
6. D. H. Solomon, *J. Polym. Sci., Polym. Chem.*, 2005, **43**, 5748–5764.
7. E. H. H. Wong, T. Junkers and C. Barner-Kowollik, *Polym. Chem.*, 2011, **2**, 1008–1017.
8. V. Sciannamea, R. Jerone and C. Detrembleur, *Chem. Rev.*, 2008, **108**, 1104–1126.
9. T. G. Ribelli, D. Konkolewicz, S. Bernhard and K. Matyjaszewski, *J. Am. Chem. Soc.*, 2014, **136**, 13303–13312.
10. A. Anastasaki, V. Nikolaou, Q. Zhang, J. Burns, S. Samanta, C. Waldron, H. Haddleton, R. McHale, D. Fox, V. Percec, P. Wilson and D. Haddleton, *J. Am. Chem. Soc.*, 2014, **136**, 1141–1149.
11. S. Muthukrishnan, E. H. Pan, M. H. Stenzel, C. Barner-Kowollik, T. P. Davis, D. Lewis and L. Barner, *Macromolecules*, 2007, **40**, 2978.
12. J. Xu, K. Jung and C. Boyer, *Macromolecules*, 2014, **47**, 4217–4229.
13. I. Cosemans, L. Hontis, D. Van Den Berghe, A. Palmaerts, J. Wouters, W. Maes, T. J. Cleij, L. Lutsen, T. Junkers and D. J. M. Vanderzande, *Macromolecules*, 2011, **44**, 7610–7616.
14. E. H. H. Wong, T. Junkers and C. Barner-Kowollik, *J. Polym. Sci., Polym. Chem.*, 2008, **46**, 7273–7279.
15. E. H. H. Wong, M. H. Stenzel, T. Junkers and C. Barner-Kowollik, *J. Polym. Sci., Polym. Chem.*, 2009, **47**, 1098–1107.
16. T. Junkers, E. H. H. Wong, M. H. Stenzel and C. Barner-Kowollik, *Macromolecules*, 2009, **42**, 5027–5035.

17. T. J. Kemp, *Prog. React. Kinet. Mech.*, 1999, **24**, 287–358.
18. C. Dommanget, C. Boisson, B. Charleux, F. D'Agosto, V. Monteil, F. Boisson, T. Junkers, C. Barner-Kowollik, Y. Guillaneuf and D. Gigmes, *Macromolecules*, 2013, **46**, 29–36.
19. L. Zang, E. H. H. Wong, C. Barner-Kowollik and T. Junkers, *Polymer*, 2010, **51**, 3821–3825.
20. (a) V. Sciannamea, A. Guerrero-Sanchez, U. S. Schubert, J.-M. Catala, R. Jerome and C. Detrembleur, *Polymer*, 2005, **46**, 9632–9641; (b) V. Sciannamea, M. Bernard, J.-M. Catala, R. Jerome and C. Detrembleur, *J. Polym. Sci., Polym. Chem.*, 2008, **46**, 7273–7279; (c) V. Sciannamea, J.-M. Catala, R. Jerome and C. Detrembleur, *J. Polym. Sci., Polym. Chem.*, 2007, **45**, 1219–1235; (d) V. Sciannamea, J.-M. Catala, R. Jerome, C. Jerome and C. Detrembleur, *J. Polym. Sci., Polym. Chem.*, 2009, **47**, 1085–1097.
21. K. Ranieri, M. Conradi, P.-Y. Chavant, V. Blandin, C. Barner-Kowollik and T. Junkers, *Aust. J. Chem.*, 2012, **65**, 1110–1116.
22. A. Debuigne, M. Hurtgen, C. Detrembleur, C. Jérôme, C. Barner-Kowollik and T. Junkers, *Prog. Polym. Sci.*, 2012, **37**, 1004–1030.
23. C. Yoshikawa, A. Goto and T. Fukuda, *e-Polym.*, 2002, (No. 13).
24. T. Sarbu, K.-Y. Lin, J. Ell, D. J. Siegwart, J. Spanswick and K. Matyjaszewski, *Macromolecules*, 2004, **37**, 3120–3127.
25. Q. Fu, Z. Zhang, W. Lin and J. Huang, *Macromolecules*, 2009, **42**, 4381–4383.
26. J. Kulis, C. A. Bell, A. S. Micallef, Z. Jia and M. J. Monteiro, *Macromol. Rapid Commun.*, 2009, **42**, 8218–8227.
27. A. Debuigne, R. Poli, C. Jérôme, R. Jérôme and C. Detrembleur, *Prog. Polym. Sci.*, 2009, **34**, 211.
28. A. Debuigne, C. Jerome and C. Detrembleur, *Angew. Chem., Int Ed.*, 2009, **48**, 1422–1424.
29. E. H. H. Wong, C. Boyer, M. H. Stenzel, C. Barner-Kowollik and T. Junkers, *Chem. Commun.*, 2010, **46**, 1959–1961.
30. G. Wang and J. Huang, *Polym. Chem.*, 2014, **5**, 277–308.
31. T. Junkers, L. Zang, E. H. H. Wong, N. Dingenouts and C. Barner-Kowollik, *J. Polym. Sci. Polym. Chem.*, 2011, **49**, 4841–4850.
32. A. Debuigne, C. Detrembleur, C. Jerome and T. Junkers, *Macromolecules*, 2013, **46**, 8922–8931.
33. E. H. H. Wong, M. H. Stenzel, T. Junkers and C. Barner-Kowollik, *Macromolecules*, 2010, **43**, 3785–3793.
34. M. Peters, Master Thesis, Hasselt University, 2013.
35. H. C. Kolb, M. G. Finn and K. B. Sharpless, *Angew. Chem., Int Ed.*, 2001, **40**, 2004.
36. C. Detrembleur, A. Debuigne, O. Altintas, M. Conradi, E. H. H. Wong, C. Jérôme, C. Barner-Kowollik and T. Junkers, *Polym. Chem.*, 2012, **3**, 135–147.
37. E. H. H. Wong, O. Altintas, M. H. Stenzel, C. Barner-Kowollik and T. Junkers, *Chem. Commun.*, 2011, **47**, 5491–5493.

38. E. H. H. Wong, M. H. Stenzel, T. Junkers and C. Barner-Kowollik, *J. Polym. Sci. Polym. Chem.*, 2011, **49**, 2118–2126.
39. R. Hoogenboom, *Angew. Chem., Int. Ed.*, 2012, **49**, 3415–3417.
40. K. Ranieri, J. Vandenbergh, C. Barner-Kowollik and T. Junkers, *Macromol. Chem. Phys.*, 2014, **215**, 1991–2000.
41. L. Barner, A. S. Quick, A. P. Vogt, V. Winkler, T. Junkers and C. Barner-Kowollik, *Polym. Chem.*, 2012, **3**, 2266–2276.
42. M. Heinenberg and H. Ritter, *Macromol. Chem. Phys.*, 1999, **200**, 1792–1805.
43. L. Vretik and H. Ritter, *Macromolecules*, 2003, **36**, 6340–6345.
44. R. Husmann, S. Wertz, C. G. Daniliuc, S. W. Schäfer, C. B. McArdle and A. Studer, *Macromolecules*, 2014, **47**, 993–1000.
45. D. F. Grishin, L. L. Semenycheva and E. V. Kolyakina, *Polym. Sci. Ser. A*, 1999, **41**, 401.
46. (a) M. V. Pavlovskaya, E. V. Kolyakina, V. V. Polyanskova, L. L. Semenycheva and D. F. Grishin, *Russ. J. Appl. Chem.*, 2002, **75**, 1868; (b) D. F. Grishin, L. L. Semenycheva, S. A. Bulgakova, L. M. Mazanova and E. V. Kolyakina, *Polym. Sci., Ser. A*, 2003, **45**, 305.
47. E. V. Kolyakina, D. F. Grishin, L. L. Semenycheva and E. V. Sazonova, *Polym. Sci. Ser. B*, 2004, **46**, 10.
48. K. Matyjaszewski, S. Gaynor, D. Greszta, D. Mardare and T. Shigemoto, *Macromol. Symp.*, 1995, **98**, 73.
49. C. Detrembleur, V. Sciannamea, C. Koulic, M. Claes, M. Hoebeke and R. Jerome, *Macromolecules*, 2002, **35**, 7214.
50. M. Y. Zaremski, A. P. Orlova, E. S. Garina, A. V. Olenin, M. B. Lachinov and V. B. Golubev, *Polym. Sci. Ser. A*, 2003, **45**, 502.
51. K. Asada and S. Lee, *Jp. Pat. Specification* No. JP 2000344809, 2000.
52. V. Sciannamea, C. Guerrero-Sanchez, U. S. Schubert, J. M. Catala, R. Jerome and C. Detrembleur, *Polymer*, 2005, **46**, 9632.
53. V. Sciannamea, J. M. Catala, R. Jerome and C. Detrembleur, *J. Polym. Sci., Part A: Polym. Chem.*, 2007, **45**, 1219.
54. C. Detrembleur, C. Jerome, J. De Winter, P. Gerbaux, J. L. Clement, Y. Guillaneuf and D. Gigmes, *Polym. Chem.*, 2014, **5**, 335–340.
55. (a) K. D. Belfied and J. V. Crivello, *Photoinitiated Polymerization*, ACS Symposium series 847, Washington DC, 2003; (b) K. Dietliker, *A Compilation of Photoinitiators Commercially Available for UV Today*, Sita Technology Ltd, Edinburgh, London, 2002; (c) S. Davidson, *Exploring the Science, Technology and Application of UV and EB Curing*, Sita Technology Ltd, London, 1999.
56. (a) J. P. Fouassier and J. Lalevée, *Photoinitiators for Polymer Synthesis-Scope, Reactivity, and Efficiency*, Wiley-VCH Verlag GmbH & Co. KgaA, Weinheim, 2012; (b) J. P. Fouassier, *Photochemistry and UV Curing: New Trends*, Research Signpost, Trivandrum, India, 2006; (c) J. P. Fouassier, *Photoinitiation, Photopolymerization, and Photocuring: Fundamental and Applications*, Hanser Publishers, New York, 1995.
57. Y. Guillaneuf, D. Bertin, D. Gigmes, D. L. Versace, J. Lalevee and J. P. Fouassier, *Macromolecules*, 2010, **43**, 2204.

58. T. Otsu, *J. Polym. Sci., Part A: Polym. Chem.*, 2000, **38**, 2121.
59. H. Fischer, *J. Am. Chem. Soc.*, 1986, **108**, 3925.
60. J. C. Scaiano, T. J. Connolly, N. Mohtat and C. N. Pliva, *Can. J. Chem.*, 1997, 75, 92.
61. S. Hu, J. H. Malpert, X. Yang and D. C. Neckers, *Polymer*, 2000, **41**, 445.
62. A. Goto, J. C. Scaiano and L. Maretti, *Photochem. Photobiol. Sci.*, 2007, **6**, 833.
63. (a) E. Yoshida, *Colloid Polym. Sci.*, 2010, **288**, 7–13; (b) E. Yoshida, *Colloid Polym. Sci.*, 2010, **288**, 73–78; (c) E. Yoshida, *Colloid Polym. Sci. Colloid. Polym. Sci.*, 2009, **287**, 1417–1424; (d) E. Yoshida, *Colloid Polym. Sci.*, 2009, **287**, 767–772; (e) E. Yoshida, *Colloid Polym. Sci.*, 2008, **286**, 1663–1666.
64. (a) Y. Guillaneuf, D. L. Versace, D. Bertin, J. Lalevee, D. Gigmes and J. P. Fouassier, *Macromol. Rapid Commun.*, 2010, **31**, 1909; (b) D. L. Versace, J. Lalevee, J. P. Fouassier, D. Gigmes, Y. Guillaneuf and D. Bertin, *J. Polym. Sci., Part A: Polym. Chem.*, 2010, **48**, 2910; (c) D. L. Versace, J. Lalevee, J. P. Fouassier, Y. Guillaneuf, D. Bertin and D. Gigmes, *Macromol. Rapid Commun.*, 2010, **31**, 1383; (d) D. L. Versace, Y. Guillaneuf, D. Bertin, J. P. Fouassier, J. Lalevee and D. Gigmes, *Org. Biomol. Chem.*, 2011, **9**, 2892–2898.
65. (a) S. Marque, C. Le Mercier, P. Tordo and H. Fischer, *Macromolecules*, 2000, **33**, 4403; (b) D. Bertin, D. Gigmes, S. Marque and P. Tordo, *e-Polym.*, 2003, **2**, 1.
66. T. Otsu and M. Yoshida, *Makromol. Chem., Rapid Commun.*, 1982, **3**, 127.
67. (a) J. D. Manga, A. Polton, M. Tardi and P. Sigwalt, *Polym. Int.*, 1998, **45**, 14; (b) J. D. Manga, M. Tardi, A. Polton and P. Sigwalt, *Polym. Int.*, 1998, **45**, 243.
68. Y. Nakayama, M. Miyamura, Y. Hirano, K. Goto and T. Matsuda, *Biomaterials*, 1999, **20**, 963.
69. Y. Nakayama and T. Matsuda, *Langmuir*, 1999, **15**, 5560.
70. Y. Nakayama and T. Matsuda, *Macromolecules*, 1996, **29**, 8622.
71. T. Matsuda, *Surface-Initiated Polymerization I*, 2006, vol. 197, p. 67.
72. H. J. Lee, Y. Nakayama and T. Matsuda, *Macromolecules*, 1999, **32**, 6989.
73. B. Sellergren, B. Ruckert and A. J. Hall, *Advanced Materials*, 2002, **14**, 1204.
74. K. Ishizu, H. Katsuhara, S. Kawauchi and M. Furo, *J. Appl. Polym. Sci.*, 2005, **95**, 413.
75. B. de Boer, H. K. Simon, M. P. L. Werts, E. W. van der Vegte and G. Hadziioannou, *Macromolecules*, 2000, **33**, 349.
76. S. B. Rahane, A. T. Metters and S. M. Kilbey, *Macromolecules*, 2006, **39**, 8987.
77. (a) B. P. Fors and C. J. Hawker, *Angew. Chem., Int. Ed.*, 2012, **51**, 8850–8853; (b) D. Konkolewicz, K. Schröder, J. Buback, S. Bernhard and K. Matyjaszewski, *ACS Macro Lett.*, 2012, **1**, 1219–23; (c) M. Ciftci, M. A. Tasdelen, W. Li, K. Matyjaszewski and Y. Yagci, *Macromolecules*, 2013, **46**, 9537–43; (d) B. Wenn, M. Conradi, A. Demetrio Carreiras,

D. Haddleton and T. Junkers, *Polym. Chem.*, 2014, **5**, 3053–3060; (e) Y.-M. Chuang, A. Ethirajan and T. Junkers, *ACS Macro Lett.*, 2014, 732–737; (f) A. Anastasaki, V. Nikolaou, G. S. Pappas, Q. Zhang, C. Wan, P. Wilson, T. P. Davis, M. R. Whittaker and D. M. Haddleton, *Chem. Sci.*, 2014, DOI: 10.1039/C1034SC01374C; (g) J. Xu, K. Jung, A. Atme, S. Shanmugam and C. Boyer, *J. Am. Chem. Soc.*, 2014, **136**, 5508–5519; (h) N. J. Treat, B. P. Fors, J. W. Kramer, M. Christianson, C.-Y. Chiu, J. R. Alaniz and C. J. Hawker, *ACS Macro Lett.*, 2014, **3**, 580–584; (i) M. A. Tasdelen, M. Ciftci and Y. Yagci, *Macromol. Chem. Phys.*, 2012, **213**, 1391–1396; (j) X. Zhang, C. Zhao, Y. Ma, H. Chen and W. Yang, *Macromol. Chem. Phys.*, 2013, **214**, 2624–2631; (k) O. S. Taskin, G. Yilmaz, M. A. Tasdelen and Y. Yagci, *Polym. Int.*, 2014, **63**, 902–907; (l) J. Mosnáček and M. Ilčíková, *Macromolecules*, 2012, **45**, 5859–5865; (m) T. Zhang, T. Chen, I. Amin and R. Jordan, *Polym. Chem.*, 2014, **5**, 4790–4796.

78. (a) N. J. Treat, H. Sprafke, J. W. Kramer, P. G. Clark, B. E. Barton, J. Read de Alaniz, B. P. Fors and C. J. Hawker, *J. Am. Chem. Soc.*, 2014, **136**, 16096–16101; (b) A. Ohtsuki, A. Goto and H. Kaji, *Macromolecules*, 2013, **46**, 96–102.
79. J. Morris, S. Telitel, K. E. Fairfull-Smith, S. E. Bottle, J. Lalevée, J. L. Clément, Y Guillaneuf and D. Gigmes, *Polym. Chem.*, 2015, DOI: 10.1039/C4PY01270D.
80. J. B. Hutchison, K. T. Haraldsson, B. T. Good, R. P. Sebra, N. Luo, K. S. Anseth and C. N. Bowman, *Lab Chip*, 2004, **4**, 658.
81. K. T. Haraldsson, J. B. Hutchison, R. P. Sebra, B. T. Good, K. S. Anseth and C. N. Bowman, *Sens. Actuators, B*, 2006, **113**, 454.
82. J. H. Ward, R. Bashir and N. A. Peppas, *J. Biomed. Mater. Res.*, 2001, **56**, 351.
83. N. A. Peppas and J. H. Ward, *Adv. Drug Delivery Rev.*, 2004, **56**, 1587.
84. S. Telitel, S. Telitel, J. Bosson, A. Spangenberg, J. Lalevée, F. Morlet-Savary, J. L. Clément, Y. Guillaneuf, D. Gigmes and O. Soppera, *Adv. Mater Interfaces*,, DOI: 10.1002/admi.201400067.
85. S. Telitel, Y. Amamoto, J. Poly, F. Morlet-Savary, O. Soppera, J. Lalevee and K. Matyjaszewski, *Polym. Chem.*, 2014, **5**, 921–930.

CHAPTER 7

NMP of Methacrylic Esters: How to Circumvent a Long-time Obstacle

JULIEN NICOLAS,[a] ELISE GUÉGAIN[a] AND YOHANN GUILLANEUF*[b]

[a] Institut Galien Paris-Sud, UMR CNRS 8612, Univ Paris-Sud, Faculté de Pharmacie, 5 rue Jean-Baptiste Clément, F-92296 Châtenay-Malabry cedex, France; [b] Aix-Marseille Université, CNRS, Institut de Chimie Radicalaire ICR UMR 7273, Av. Escadrille Normandie-Niemen, Case 542, 13397 Marseille, France
*Email: yohann.guillaneuf@univ-amu.fr

7.1 Introduction

Among the various commonly available vinyl monomers, methyl methacrylate (MMA) and its derivatives are very attractive thanks to the interesting properties of the corresponding polymer (*e.g.*, good impact strength, transparency) and aspects (*e.g.*, brightness).[1] Poly(methyl methacrylate) (PMMA) is a versatile material and has been used in a wide range of applications, such as rear-lights and instrument clusters for vehicles, lenses for glasses, resistant panels for building windows, skylights, signs and displays, LCD screens, and furniture. Another interesting property is the exothermic mixing of PMMA with a wide range of commercially relevant plastics (*e.g.*, poly(vinyl chloride), PVC; poly(styrene-*co*-acrylonitrile), PSAN; poly(vinylidene

RSC Polymer Chemistry Series No. 19
Nitroxide Mediated Polymerization: From Fundamentals to Applications in Materials Science
Edited by Didier Gigmes

Published by the Royal Society of Chemistry, www.rsc.org

fluoride), PVDF; epoxy matrixes). The compatibility of PMMA with these brittle polymers makes it possible to incorporate pre-made rubber particles decorated with PMMA chains to be used as impact modifiers, leading to the so-called rubber-modified thermoplastics.[2]

A considerable effort has therefore been devoted to improving the controlled polymerization of methacrylate monomers. Although it is possible to perform living anionic polymerization of methacrylates derivatives,[3] the recent advances in reversible deactivation radical polymerization (RDRP) led to robust and simple methods to prepare well-defined polymer architectures with high complexity and degree of functionalization.[4] Unfortunately, MMA is one of the most difficult monomers to control, because of the high stabilization of the corresponding radical. Among the RDRP techniques, nitroxide-mediated polymerization[5] (NMP) was early criticized due to the limited number of monomers that could be controlled (*e.g.* styrene and its derivatives).[6,7] A major breakthrough was however witnessed with the development of acyclic nitroxides such as 2,2,5-trimethyl-4-phenyl-3-azahexane-3-oxyl (TIPNO)[8] and *N*-*tert*-butyl-*N*-[1-diethylphosphono-(2,2-dimethylpropyl)] nitroxide (SG1).[9] These nitroxides extended the range of monomers to be successfully controlled to acrylates, acrylamides and dienes, and also drastically reduced the polymerization temperature (typically down to 90 °C).[5] Due to the great deal of work devoted to NMP in the past few years, this technique has significantly gained in efficiency and robustness. This was illustrated by the various fields of application where NMP-derived materials have been successfully employed, including additives,[10] nanoporous materials,[11] bioconjugates,[12] micro/optoelectronic devices[13] and electrolytes for lithium-metal batteries.[14] However, among the different aspects of NMP that would benefit from some improvements certainly is the control of methacrylic esters.

This chapter aims at presenting the different strategies directed toward the control of MMA and its derivatives by NMP. Some examples of NMP-derived materials comprising controlled or uncontrolled methacrylate segments will also be given, showing that a high quality of control is not systematically mandatory and mainly depends on the envisioned application.

7.2 Reasons for the Failure of the NMP of MMA

The first attempts to perform the NMP of methacrylate derivatives were reported in 1998 by the CSIRO group.[15] They reported the bulk polymerization of MMA at 90 °C mediated by several nitroxides (Figure 7.1). Whatever the nitroxide, the reaction stopped after about 1 h and the analysis of the purified polymer showed a double bond at the chain end. The conversions were in the range 10–40% with experimental M_ns close to the theoretical values, showing an efficient initiation combined with no side-reaction leading to new polymer chains. Using numerical simulations, a high

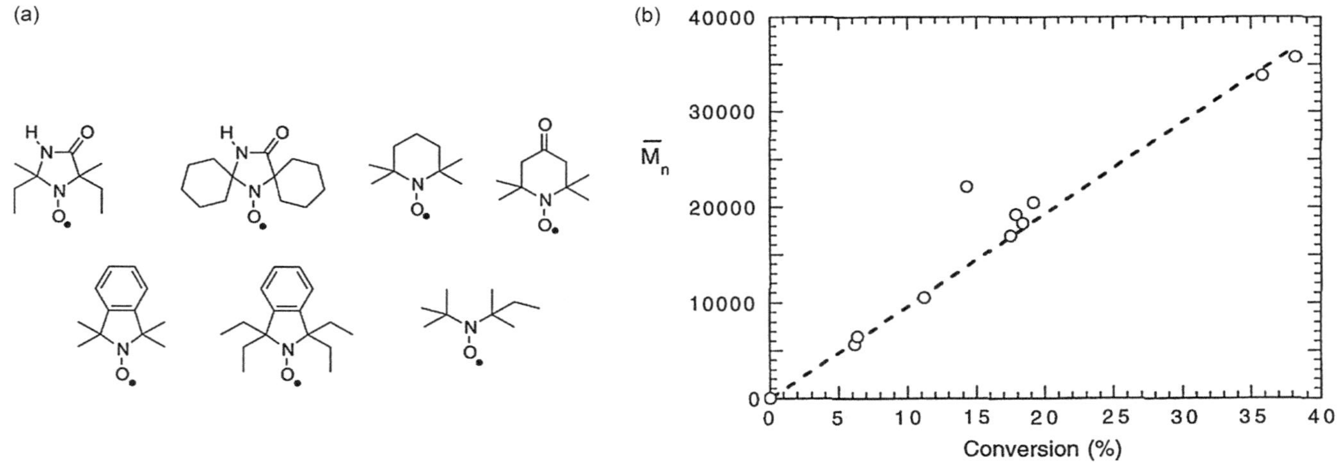

Figure 7.1 (a) Nitroxides used for the bulk polymerization of MMA at 90 °C initiated with azobis(2,4-dimethyl-2-pentanenitrile). (b) Evolution of M_n *versus* MMA conversion plot for the different polymerizations. The dashed line indicates the theoretical M_n if the initiator efficiency is 90%.
Reprinted with permission from ref. 15. Copyright (1998) American Chemical Society.

equilibrium constant $K = k_d/k_c$ for the MMA-nitroxide was determined, the value being higher than that of styrene, but they ascribed the loss of control to the disproportionation reaction.

The exact nature of the chain end was later extensively studied by Charleux and co-workers for the *n*-butyl methacrylate (*n*BMA)/TEMPO system.[16] The ^{1}H NMR analysis and a detailed MALDI-TOF study showed that the resulting P(*n*BMA) had the initiator attached at one end. The other terminal functionality of the polymer was not a TEMPO-based alkoxyamine but a methylene unsaturation exclusively. Particularly, no saturated polymer, which would also be formed by conventional disproportionation reaction between two propagating radicals, could be detected. From this result, it was concluded that the main chain-breaking event was the β-hydrogen transfer from a propagating radical to TEMPO (also called disproportionation reaction).

The formation of the unsaturated macromonomer and hydroxylamine could be in fact due to either a usual radical disproportionation reaction (rate constant k_{cD}), which competes with the cross-coupling (rate constant k_c), or a non-radical alkene elimination (rate constant k_{dD}) that parallels the cleavage (rate constant k_d) (Figure 7.2).

The theoretical influence of the disproportionation on the NMP process was studied by Fischer and co-workers.[18] They showed that the overall effects of the two modes of disproportionation are nearly macroscopically undistinguishable (Figure 7.3a). In the absence of other initiations, in both cases, the monomer conversion ceases at an approximate time of $2(k_d f_D)^{-1}$, where k_d is the rate constant for the homolytic dissociation and f_D is the fraction of disproportionation that could be defined by either $f_D = k_{cD}/(k_{cD} + k_c)$ or $f_D = k_{dD}/k_d$, where k_c is the rate constant for the recombination, k_{cD} is the rate constant for the disproportionation between (macro)radicals and nitroxide and k_{dD} is the rate constant for the disproportionation by non-radical elimination.

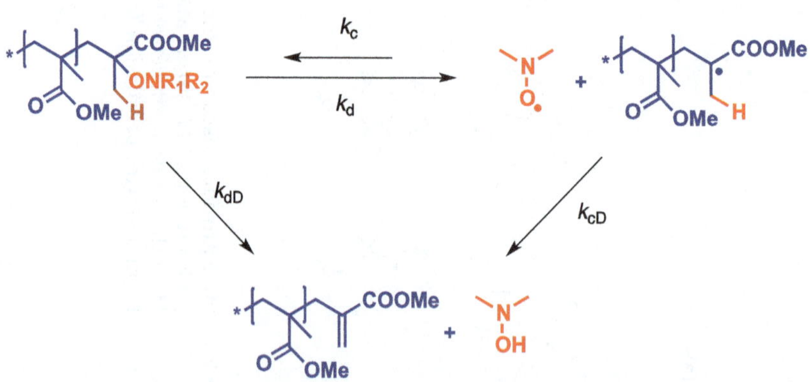

Figure 7.2 The two mechanisms of the disproportionation reaction. Reprinted with permission from ref 17. Copyright (2015) Wiley.

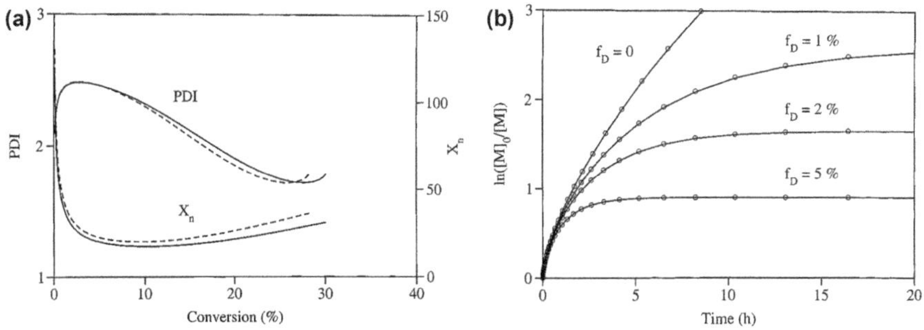

Figure 7.3 (a) Number-average degree of polymerization and dispersity, computed including 20% of hydroxylamine and macromonomer formation during the cross-termination (solid line) or during alkoxyamine decomposition (broken line). (b) $\ln([M]_0/[M])$ *vs.* time for different fractions of disproportionation. Parameters for both figures: $k_d = 4.5 \times 10^{-3}$ s^{-1}, $k_t = 10^8$ M^{-1} s^{-1}, $k_c = 2.2 \times 10^7$ M^{-1} s^{-1}, $k_p = 5000$ M^{-1} s^{-1}, $[I]_0 = 10^{-1}$ M, and $[M]_0 = 10$ M. Reprinted with permission from ref. 18. Copyright (2001) American Chemical Society.

Even for relatively small values of f_D, the end of conversion (c_{max}) may occur long time before complete monomer conversion (eqn 7.1), and it is accompanied by a final increase in the dispersity.[18]

$$C_{max} = 1 - \exp\left(-\frac{3}{2}k_p\left(\frac{[R-Y]_0}{3f_D^2 k_d(k_c + k_{cD})k_t}\right)^{1/3} \times F_\infty\right) \tag{7.1}$$

with $F_\infty = 2^{2/3}[\Gamma(2/3)]2/3\Gamma(1/3) \approx 1.08$

For example, using classic values of k_p, k_t, k_d and k_c, the maximum conversion for $f_D = 1\%$ is 93%, for $f_D = 2\%$ is 81% and for $f_D = 5\%$ is 59% (Figure 7.3b).

This disproportionation reaction during the homolysis of alkoxyamines has been already studied experimentally in a model system. The first studies of Priddy[19] and Fukuda[20] on the decomposition of the model styryl-TEMPO alkoxyamine showed that disproportionation occurred but at a very low extent with a probability factor that this reaction occurred close to 0.4%. In particular, they estimated the activation energy of this probability factor to be 29 kJ mol^{-1}. Fischer and co-workers thus extended this study to other alkoxyamines based on TEMPO (*e.g.*, cumyl, styryl, 1-*tert*-butoxy carbonyl-1methyl-ethyl, 1-*tert*-butoxy carbonyl-ethyl).[21] They first demonstrated that the alkenes and TEMPOH were formed by bulk disproportionation between a carbon-centered radical and free TEMPO. Only the alkoxyamine with the 1-*tert*-butoxy carbonyl-ethyl moiety also gave these products *via* a substantial non-radical elimination reaction. They subsequently managed to estimate the fraction of disproportionation (Table 7.1) using a detailed ^{1}H NMR analysis of the alkoxyamine decomposition in the presence or absence of styrene/MMA.

Table 7.1 Fraction of disproportionation f_D determined by NMR for various alkoxyamines.

Alkyl fragment	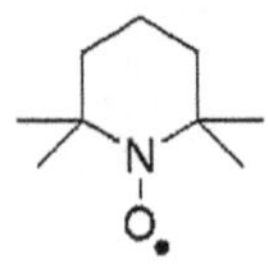 21	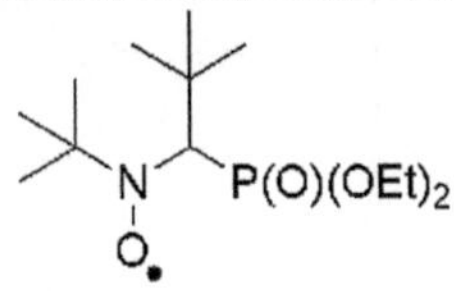 22
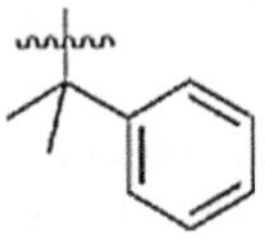	0.6%	—
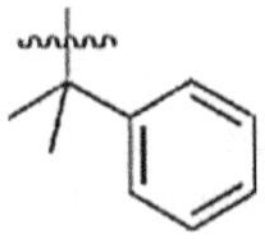 +MMA	20%	—
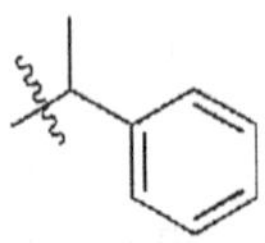	0.8%	0%
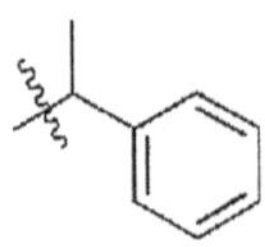 +MMA	20%	—
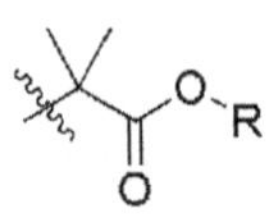	2.2%	0%
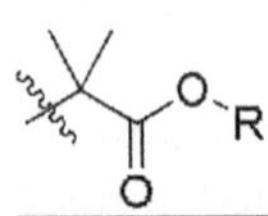 +MMA	20%	0%

In the case of model TEMPO-based alkoxyamines, they found that the styryl and cumyl alkyl fragment led to similar fractions of disproportionation (0.6%), whereas the 1-*tert*-butoxy carbonyl-1methyl-ethyl moiety, mimicking the methacrylate functionality, led to 2.2%. In the presence of MMA, which forms PMMA macroradicals, all the experiments led to a fraction of disproportionation at least equal to 20%. This showed that steric substituent factors are important and that disproportionation is favoured if the reversible coupling leads to a sterically strained bond. In a second study, Fischer and co-workers extended this analysis to SG1-based alkoxyamines.[22] In contrast to TEMPO, there was little or no disproportionation of SG1 with both 1-ethoxy carbonyl-ethyl and styryl alkyl moieties. For these alkyl moieties, in the absence of alkene and hydroxylamine products, they estimated an upper limit of f_D of 0.2% for the SG1 nitroxide. It was also shown that the decomposition of the SG1-based alkoxyamine in the presence of MMA did not lead to significant alkene formation, demonstrating the very low fraction of disproportionation (Table 7.1).

More recently, Marque and Bagryanskaya[23] developed a method based on the ^{1}H NMR analysis of the thermal decomposition of alkoxyamine in the presence and absence of a radical scavenger (*i.e.*, thiophenol) to discriminate the two different disproportionation modes for various alkoxyamines (Figure 7.4a). The experiment performed with the TEMPO-based alkoxyamine clearly demonstrated that the disproportionation is an intermolecular process (Figure 7.4b and c) with a $f_D = 3.5\%$ that is a k_{cD} value of 2.0×10^7 M^{-1} s^{-1}.

In a second study, the H-atom transfer reaction was studied for a series of imidazoline, imidazolidine, and pyrrolidine-based alkoxyamines containing the isobutyrate-2-yl moiety (Figure 7.5a).[24] Alkoxyamines based on cyclic nitroxides have recently been shown to be good model compounds for studying the influence of various substituents on the nitroxide structure. Depending of the nitroxide structure (Figure 7.5a), both intra and inter or inter only H-transfer reactions were observed and the various rate constant were determined.[24] Whatever the nitroxide structure, the fraction of disproportionation was higher than the one of TEMPO and thus these nitroxides cannot control the NMP of methacrylate derivatives.

Concerning the intramolecular mechanism, the reaction requires a cyclic transition state with good overlapping of the nitrogen orbital n_σ and the σ^* antibonding orbital of a C–H bond (Figure 7.5b). This reaction is similar to a Cope elimination, as already mentioned by Ananchenko and Fischer.[21] This explains why the steric demand of the substituents at positions 2 and 5 of the nitroxide is the main parameter to explain the occurrence of such reaction. For example the less sterically hindered compound (4 methyl group in position 2 and 5) led to intramolecular H-transfer whereas this reaction was not observed for the highly sterically hindered compound with one cyclohexyl and two ethyl groups (Figure 7.5b). Temperature dependence study of this reaction showed an Arrhenius type behaviour for k_{dD} with a frequency factor A close to 10^{15} s^{-1} and high activation energy ($E_a \approx 140$ kJ mol^{-1}). This high value for E_a denotes a strong dependence on the temperature and a late transition state, meaning a non-favoured reaction. Recently, Marque and co-workers[25] studied the intramolecular proton transfer (IPT) during the thermal dissociation of alkoxyamines and they concluded that the energetic parameters for such a reaction were not dependent on the nature of the nitroxide structure as well as the geometry of the transition state. Thus the occurrence of such side reaction is dependent of the activation energy of the homolytic dissociation of the alkoxyamine moiety, *i.e.*, occurrence of IPT is highly probable for C–ON bond homolysis with $E_a > 125$ kJ mol^{-1} and less probable for C–ON bond homolysis with $E_a < 115$ kJ mol^{-1}.[25]

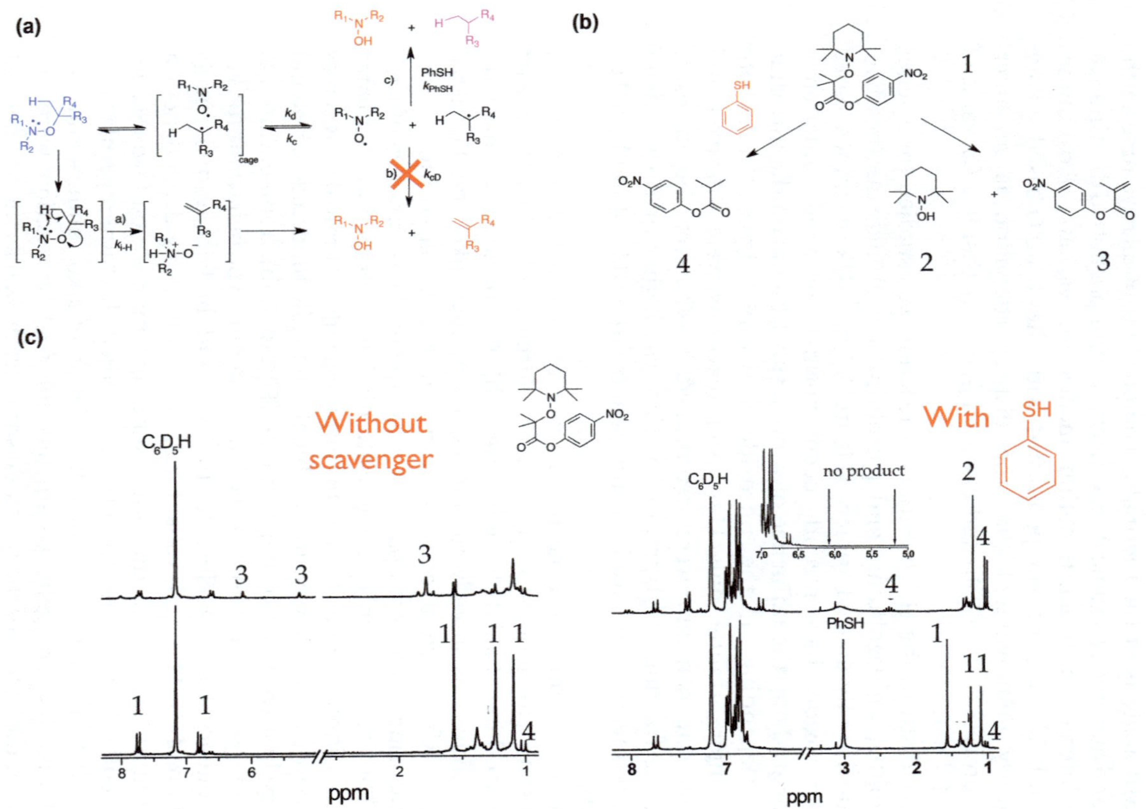

Figure 7.4 (a) Intra-/inter-hydrogen transfer reaction during thermal decomposition of alkoxyamines in the presence of scavenger. (b) Decomposition pathway of the 4-nitrophenyl 2-(2,2,6,6-tetramethylpiperidine-1-yloxy)-2-methylpropionate alkoxyamine. (c) Comparison of ^{1}H NMR spectra before (lower spectra) and after (upper spectra) thermolysis of 0.02 M solution of alkoxyamine 1 in benzene-d_6 in the presence of 0.24 M radical scavenger PhSH (right) and in its absence (left) at 351 K. Reprinted with permission from ref. 23. Copyright (2008) Wiley.

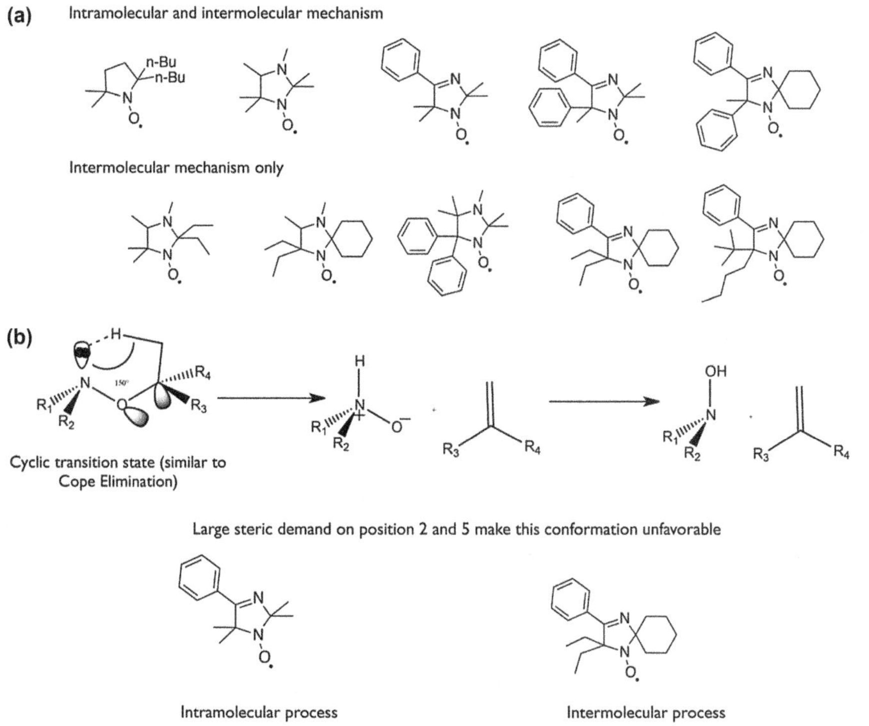

Figure 7.5 (a) Structure of the imizadoline, imizadolidine and pyrrolidine nitroxides whose disproportionation ratios were determined. (b) Conformation of the transition state for the intramolecular H-transfer.
Reprinted with permission from ref. 17. Copyright (2015) Wiley.

Unlike the intramolecular mechanism, the intermolecular H-atom transfer reaction requires a linear transition state between the aminoxyl group and the hydrogen. This led to a reaction occurring with a very low activation barrier (<15 kJ mol^{-1}), and rather high k_{cD} values. In a similar manner to the recombination rate constant k_c value, the k_{cD} value decreased from 10^8 to 8×10^6 M^{-1} s^{-1} with the increase of the bulkiness of the groups at positions 2 and 5 of the nitroxide. A thorough analysis of the k_{cD} values also showed ring conformation effects (imidazolidine, imidazoline) balancing the steric effect of the groups flanking the aminoxyl moiety.[24]

The work of Fischer *et al.*[22] showing that the SG1 nitroxide does not induce disproportionation in the presence of MMA led Benoit[26] to perform the polymerization of MMA at 120 °C using a bi-component system AIBN/SG1. No control was observed, which can be accounted for by the too high equilibrium constant *K*. From the phase diagram developed by Fischer[27,28] and the values of kinetic constants available in the literature ($k_{d,120\,°C} = 2.4$ s^{-1}, $k_c = 5.0 \times 10^5$ L mol^{-1} s^{-1})[29], Guillaneuf and co-workers[30] performed the polymerization of MMA at 45 °C using the highly labile BlocBuilder alkoxyamine[31,32] as initiator. They obtained a partial livingness at low conversion (<15%) and no disproportionation side products. To explain these results, they did some kinetic modelling using the PREDICI software. Good agreement for both the kinetic and the molar mass evolutions can be obtained with an estimated k_d value of 10^{-2} s^{-1} and an estimated k_c value of 1.4×10^4 L mol^{-1} s^{-1}. The kinetic rate constants determined by fitting thus give an equilibrium constant K of 7×10^{-7} M at 45 °C. This value is two orders of magnitude higher than the one expected by Fischer[29] and is also higher than the upper limit K_{Max}[29] [$K_{Max} = (k_p \times [I]_0)/(2 \times \ln(10) \times k_t) = 1.0 \times 10^{-7}$ M] that allowed for a controlled and living polymerization. With such a value, the nitroxide cannot recombine efficiently with the macroradicals to avoid bimolecular terminations. Secondly, this affords a high concentration of SG1 in the polymerization medium, which slows down the decomposition of the starting BlocBuilder alkoxyamine without exerting any effect on the control and livingness of the polymerization. To confirm these values that significantly differed from the literature, the authors[30] determined experimentally the recombination rate constant k_c using the RNR-PLP-SEC method.[33] The results showed a recombination rate constant for the PMMA-SG1 system of about 6.0×10^4 L mol^{-1} s^{-1} at 20 °C and 2.0×10^4 L mol^{-1} s^{-1} at 35 °C, in good agreement with the value predicted by the simulations (1.4×10^4 L mol^{-1} s^{-1} at 45 °C). Secondly, the PMMA-SG1 system seemed to exhibit a non-Arrhenius behaviour, as already reported by Fischer[34] for the model methyl carbonyl-1-methyl-ethyl radical. The apparent negative activation energy (*i.e.*, the recombination rate constant decreased when the temperature increased) was explained by Fischer[34] as the result of a barrierless single step reaction with a low exothermicity, in which the free activation energy was dominated by a large negative entropy.

Concerning the k_d value, the activation energy that was estimated by theoretical modelling led to 99.5 kJ mol^{-1}; that is a 35 + fold increase compared to the model methyl 1-carbonyl-1-methyl-ethyl moiety[35] ($E_a = 108.9$ kJ mol^{-1}). The penultimate effect, which was usually neglected, is very important when a bulky substituent (a tertiary alkyl group here) is present and can drastically change the k_d value. The decomposition of alkoxyamines was nicely described by structure–reactivity relationships linking the k_d value to polar, stabilization and steric parameters.[36] In the case of MMA and PMMA macroradical, an increase of the steric constant implied a 35-fold increase in the value of k_d at 120 °C, in good agreement with the modelling results. Matyjaszewski[37] observed the same trends in ATRP but to a lesser extent. When the penultimate unit was a methacrylate moiety, the activation rate constant k_{act} was increased (by a factor of 8 between model and PMMA macroradicals). Later on, the effect of a bulky substituent (*i.e.*, tertiary alkyl group) in the penultimate position on SG1-based alkoxyamines was experimentally characterized[38] on alkoxyamines that are 1,2 adducts of the BlocBuilder alkoxyamine onto activated olefins. The increase of the measured k_d values (18 times increase between the styryl-SG1 and the 1,2 adduct, see Figure 7.6) are in good agreement with the 35 times increase that is expected in the case of PMMA alkyl moiety.

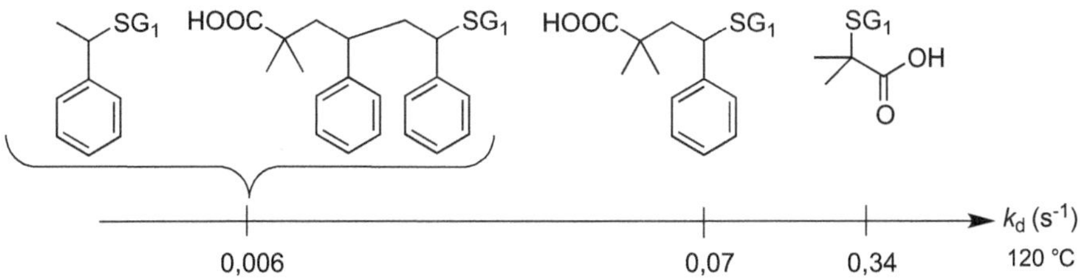

Figure 7.6 Influence of the penultimate unit on the k_d value of SG1-based alkoxyamines.
Reprinted with permission from ref. 17. Copyright (2015) Wiley.

With these unexpected k_d and k_c parameters, the homopolymerization of MMA mediated by the SG1 nitroxide appeared impossible. The authors[30] performed a theoretical study to investigate the influence of either the presence of excess SG1 or a higher initiating alkoxyamine concentration over the control and the livingness. From this study and the control of the MMA derivatives by the SG1 using the copolymerization approach, McHale *et al.*[39] thoroughly studied the homopolymerization of MMA in the presence of a large excess of free SG1 (from 38% to 315%) along with a conventional radical initiator (*i.e.*, bicomponent initiating system) and concluded that, in such conditions, intermolecular H-transfer from PMMA propagating radicals to the nitroxide played an important role at 110 °C and 45 °C. It has to be noted that the use of a bicomponent initiating system cannot fully represent a polymerization initiated by an alkoxyamine where the nitroxide and the alkyl radical are produced at the same rate. Charleux and co-workers[40] also investigated the homopolymerization of MMA at 70 °C initiated by the BlocBuilder alkoxyamine in the presence of excess SG1. With a detailed analysis of the polymer chain end by ^{1}H NMR and MALDI-TOF, they managed to determine the rate constant of β-hydrogen transfer, $k_{\beta Htr}$, which was estimated to be 1.69×10^3 L mol^{-1} s^{-1} at 70 °C. This value corresponds to a fraction of disproportionation of 7.8% and is also three orders of magnitude lower than the value of 1.4×10^6 L mol^{-1} s^{-1} found for the *n*-butyl methacrylate/TEMPO system at 130 °C,[16] confirming a lower propensity of the SG1 nitroxide for hydrogen abstraction in comparison with TEMPO.

Marque and co-workers[41] used their method based on the ^{1}H NMR analysis of the thermal decomposition of alkoxyamine in the presence and absence of radical scavenger to investigate the decomposition of SG1-based alkoxyamines bearing a tertiary alkyl moiety (*i.e.*, alkyl carbonyl-1-methylethyl derivatives). In the absence of a scavenger and in "highly degassed solution" (10^{-5}–10^{-6} mbar), the decomposition of SG1-based alkoxyamines proceeded with intramolecular H-transfer. The analysis of the experimental results led to a k_{cD} equal to 1.7×10^3 L mol^{-1} s^{-1} at 75 °C, which was in good agreement with the value reported by Charleux *et al.*[40] for PMMA macroradicals and SG1. The fraction of disproportionation for the model alkoxyamine was thus close to 0.1%. The authors also investigated the role of residual oxygen by studying the decomposition of the same alkoxyamines in solution degassed under low vacuum (10^{-3} mbar), which corresponds to degassing performed by conventional oil pump or by bubbling by gases such as nitrogen or argon. It has been found that the presence of residual oxygen can affect the amount of alkene detected in reaction products (25% with low vacuum instead of 90% at high vacuum, Figure 7.7b). At the same time, the starting materials completely decomposed and the formation of a new phosphorus species has been identified as an ester of the SG1 hydroxylamine (Figure 7.7a).

With all these data, we can conclude that the penultimate effect, which increases the decomposition rate constant and decrease the recombination rate constant, led to an uncontrolled polymerization since the K_{max} threshold is not fulfilled. In addition, this effect led to an increase of f_D from 0.1% to 7.8%, above the f_{Dmax} value of 3% that impedes the possibility to control the polymerization.

7.3 Development of Dedicated Nitroxides

As already mentioned above, the seminal works on the NMP of MMA were performed by Rizzardo and co-workers[15] with various nitroxides and in particular using TEMPO derivatives. Later on, the second generation acyclic nitroxides, such as SG1 and TIPNO, that led to improved performance[5] in NMP were also tested, but with limited success. The copolymerization approach (see below) enabled good control to be obtained with a nitroxide (*i.e.*, SG1) that was already designed and used in NMP,[42] but the homopolymerization of methacrylate derivatives was still a challenge.

It has to be noted that some research was focused on the use of additives to decrease the occurrence of β-hydrogen transfer for the TEMPO-mediated polymerization of MMA.[43–45] Strong acids, such as camphorsulfonic acid[45] (CSA) or sulfuric acid[43,44] (H_2SO_4), are claimed to protonate the weakly basic nitroxide to form a radical cation, which further reacts with another nitroxide to form the corresponding hydroxylamine and an oxoammonium derivative. For instance, using a bicomponent system (AIBN/TEMPO) at 70 °C and 5.8 mol% of H_2SO_4, >90% monomer conversion were obtained after 16 h with low dispersites (~1.1–1.3).[44] The livingness of the polymer was surprisingly confirmed by conducting the polymerization of S from a PMMA macroinitiator.[44]

To control the homopolymerization of methacrylate derivatives, other nitroxides than those previously used in NMP should be employed. First, these nitroxides must not be too reactive to avoid the disproportionation side reaction. Second, according to kinetic simulations,[46] the homolysis of the corresponding (macro)alkoxyamine must be slower than in the case of the SG1 analogue (by a factor close to 50) and the recombination of the nitroxide with macroradicals faster (by a factor close to 100).

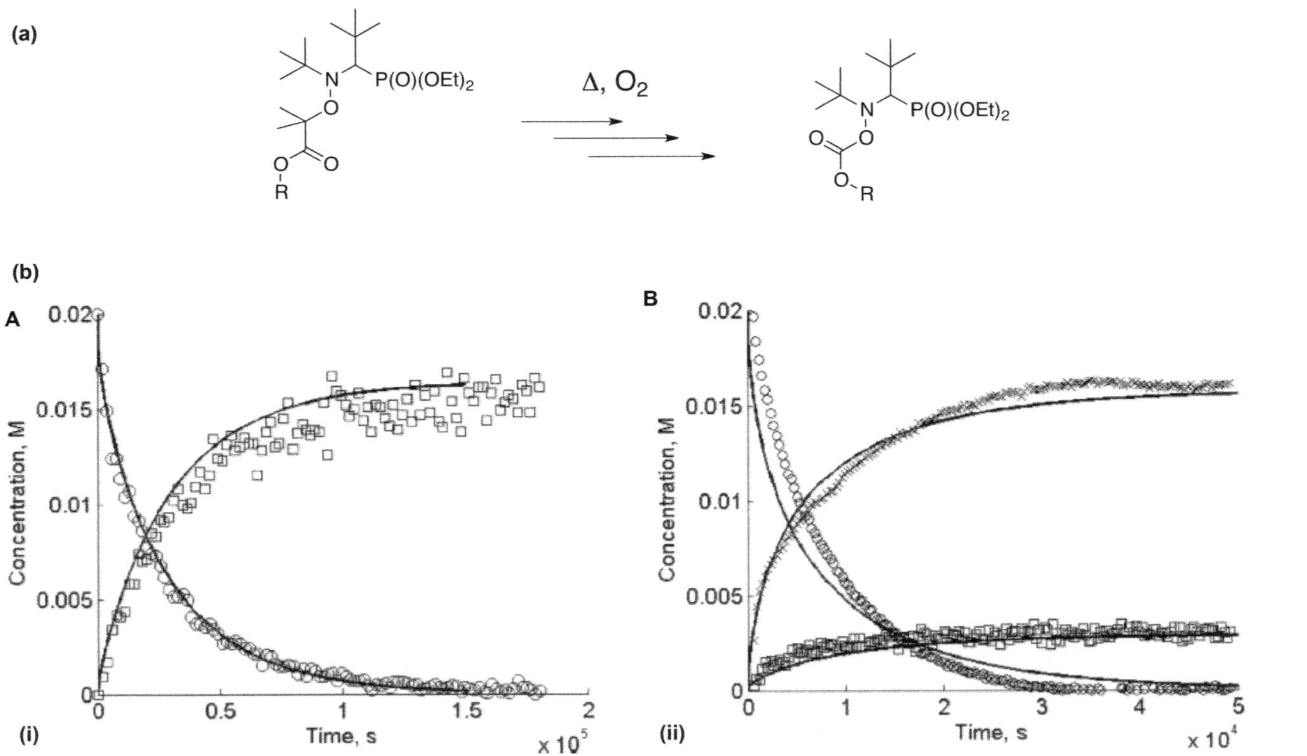

Figure 7.7 (a) Mechanism of decomposition that occurs during the thermolysis of alkyl carbonyl-1-methyl-ethyl SG1-based alkoxyamines with low-vacuum degassing. (b) Experimental (symbols) and calculated (full line) kinetics for decomposition of the methyl carbonyl-1-methy-ethyl-SG1 alkoxyamine in the absence of a scavenger for solution degassed under high vacuum (I) and for poorly degassed solution (II): (○) alkoxyamine, (□) alkene, (+) oxygen product.
Reprinted with permission from ref. 41. Copyright (2013) Wiley.

Guillaneuf and co-workers[46] focused their attention on the indolynoxyl radical 2,2-diphenyl-3-phenylimino-2,3-dihydroindol-1-yloxyl nitroxide (DPAIO, Figure 7.8). They thought that the radical delocalization on the phenyl ring may avoid the disproportionation side reaction. This nitroxides, which was previously synthesized by Greci,[47] is stable up to 200 °C. After the determination of k_d and k_c values ($k_c = 1.4 \times 10^6$ L mol^{-1} s^{-1} and $E_a = 125.6$ kJ mol^{-1} for the dissociation) the authors used the phase diagram approach[27,28] to find the optimized polymerization temperature. The bulk MMA polymerization was then performed at 100 °C (Figure 7.8) and showed a good control of the polymerization even for high targeted M_n (80 000 g mol^{-1}).[46] This result showed that the disproportionation side reaction did not occur with the DPAIO nitroxide. The ^{1}H NMR analysis of the model DPAIO based alkoxyamines also confirmed this result.[23,48]

To perform the polymerization of MMA derivatives, the choice of the initiating alkoxyamine is crucial. If the initiating alkoxyamine decomposes too slowly with respect to the macroalkoxyamine or to the polymerization time, no linear evolution of the M_n *vs.* conversion can be obtained.[28,31] The penultimate effect that occurred during the MMA polymerization implied to use an initial alkoxyamine with a very high dissociation rate constant and at least even higher than the MMA-nitroxide model alkoxyamine. By analogy with SG1 analogues alkoxyamines,[49,50] the acceleration of the dissociation rate constant can be achieved using a released alkyl moiety with a long range polar effect. With the phenyl carbonyl 1-methyl-ethyl moiety, the evolution of the molar masses *vs.* conversion both showed a regular increase of the M_n characteristic of a controlled polymerization. Nevertheless, at the beginning of the polymerization (conversion <30%), the experimental M_ns were slightly higher than the theoretical ones, which was due to a relatively slow initiation (Figure 7.8). The slow decomposition of the starting alkoxyamine caused relatively high $Đ$ (close to 1.4 for a targeted M_n of 40 000 g mol^{-1} at 60% conversion). The presence of the nitro group in the *para* position of the phenyl ring should increase k_{d1}, due to a long-range polar effect. With the nitro-based alkoxyamine, the polymerization was well controlled even at low conversion (Figure 7.8).

The livingness of the polymerization was also investigated. When the polymer was heated in *tert*-butyl benzene for 2 h, the ESR spectrum of the DPAIO was observed, confirming the presence of living chains and thus partial livingness. Gigmes and co-workers[51] examined by ESR the decomposition of various DPAIO based alkoxyamines and they concluded that when the alkyl moiety of the DPAIO-based alkoxyamines was a primary alkyl radical, only N–OC bond dissociation occurred. With a stabilized secondary radical (as the terminal *n*-butyl acrylate radical), both the N–OC bond and NO–C bond dissociation are in competition (Figure 7.9). For the stabilized tertiary radical such as the PMMA macroradical, the desired NO–C bond was the major breaking event that allowed the controlled polymerization of MMA derivatives. This phenomenon was also confirmed by ^{1}H NMR analyses.[23]

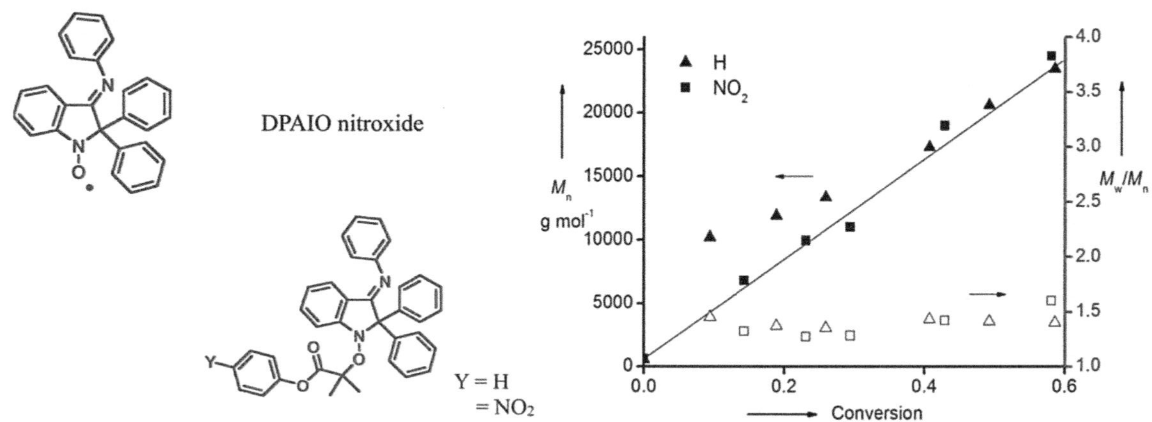

Figure 7.8 Evolution of number-average molar mass (M_n full symbols) and dispersity ($Đ$ empty symbols) *vs.* conversion for bulk MMA polymerization at 100 °C initiated with DPAIO-based alkoxyamines with a targeted $M_n = 40\,000$ g mol^{-1} ($[\text{alkoxyamine}]_0 = 2.5 \times 10^{-2}$ mol L^{-1}). The solid line corresponds to the theoretical M_n. Reprinted with permission from ref. 46. Copyright (2007) American Chemical Society.

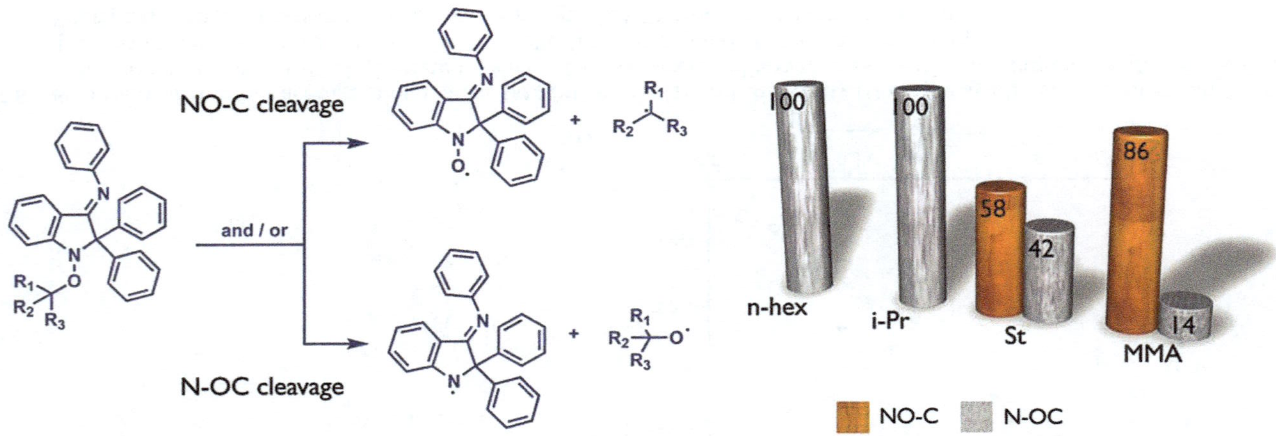

Figure 7.9 The two modes of decomposition of alkoxyamines based on the 2,2-diphenyl-3-phenylimino-2,3-dihydroindol-1-yloxyl (DPAIO) nitroxide.
Reprinted with permission from ref. 17. Copyright (2015) Wiley.

The breakthrough in the NMP of methacrylate derivatives introduced by the DPAIO nitroxide nevertheless has some limitations that prevent its further use to control the polymerization of others classes of monomers. For instance, in the presence of either styrene or acrylates, which afford less hindered and stabilized propagating radical chains than methacrylates, the required temperature (>150 °C) to dissociate the corresponding DPAIO macroalkoxyamine is too high to control the polymerization of such monomers. In addition, when the steric hindrance and the stabilization of the released alkyl moiety decreased, DPAIO alkoxyamines could suffer from undesired CO–N homolytic dissociation. Recently Guillaneuf, Greci and Gigmes[52] investigated the influence of the indolinic nitroxide structure on both the C–ON bond dissociation energy and the selectivity of the homolytic dissociation (C–ON and/or CO–N). The ideal candidate should exhibit a lower C–ON bond dissociation energy compared to the DPAIO-based alkoxyamine and undergo 100% of C–ON bond dissociation for either methacrylate and styrene based-alkoxyamines. Various alkoxyamines were synthesized, by changing essentially one of the phenyl rings attached to the carbon at the alpha position to the aminoxyl function by several alkyl fragments (Figure 7.10).

Using the ESR spin-trapping technique as well as HPLC, the different nitroxides can be discriminated according to the ability of their corresponding alkoxyamines to have a selective C–ON *vs.* CO–N bond homolysis. The steric hindrance introduced by either the isopropyl group or the neopentyl group led to a quasi-exclusive C–ON bond dissociation for the methacrylate derivatives and roughly 80% of desired C–ON bond dissociation for styrene. With a lower steric hindrance, the homolysis is less selective. Such higher steric hindrance is also responsible for an increase of the rate of dissociation ($E_a = 119$ kJ mol^{-1} compared to $E_a = 124.1$ kJ mol^{-1} for the isopropyl and phenyl moiety, respectively) that could allow the polymerization to be performed at lower temperature. The MMA polymerization was indeed carried out in bulk at 85 °C using the isopropyl-based DPAIO derivative. The polymerization was well controlled with $Đ$ close to 1.2 at 50% conversion (Figure 7.10). The improved selectivity during the homolysis of styryl-based alkoxyamines led the authors to test the isopropyl-based DPAIO derivative for the bulk polymerization of styrene (Figure 7.10).[52] The plot of the conversion index [ln([M]$_0$/[M])] is linear and the molar mass distribution (MMD) is monomodal and is shifted with the monomer conversion, contrary to the same polymerization initiated with the DPAIO-based alkoxyamine. Even if the dispersity was high (close to 2) and the experimental M_ns are above the theoretical ones, these experiments show that it is now possible to use the same nitroxide for both controlling the polymerization of methacrylate derivatives and at the same time obtaining decent control of styrene derivatives.

A PMMA-*b*-PS diblock copolymer was prepared using the isopropyl-based DPAIO derivative. It appeared then that there was no residual homo-PS in the copolymer sample at the end of the polymerization, which proved the

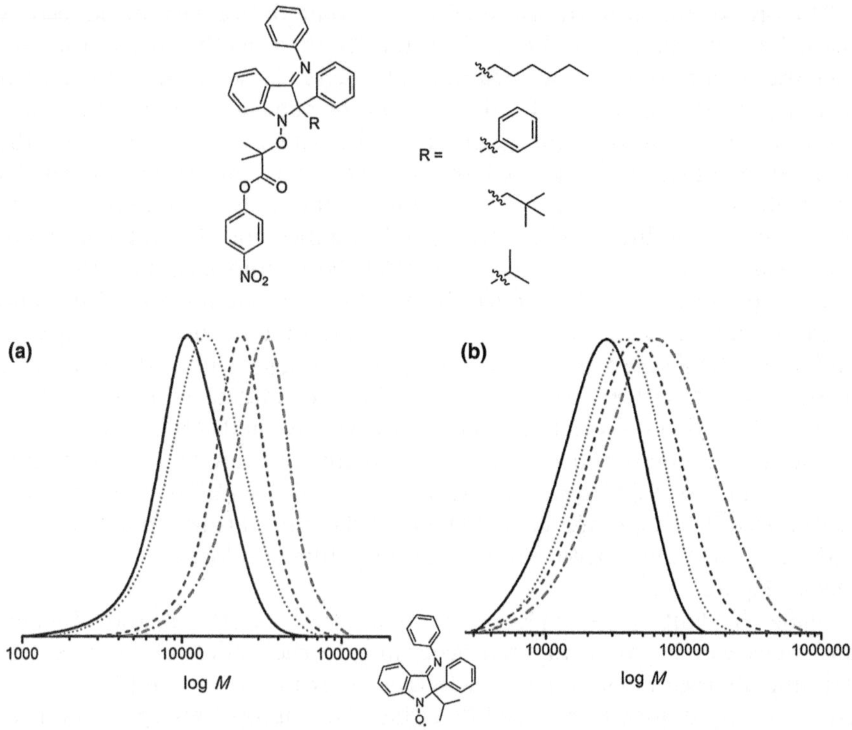

Figure 7.10 (a) Molar mass distribution obtained from the bulk MMA polymerization at 85 °C initiated with 2-phenyl-2-isopropyl-3-phenyimino-indoline-1-yloxy-*O*-(*p*-nitrophenyl)-methylpropanoate alkoxyamine (black solid line 20% conversion; green dotted line 33% conversion; blue dashed line 38% conversion; red dotted dashed line 53% conversion). (b) Molar mass distribution obtained from the bulk styrene polymerization at 120 °C initiated with 2-phenyl-2-isopropyl-3-phenyimino-indoline-1-yloxy-*O*-(*p*-nitrophenyl)-methylpropanoate alkoxyamine (black solid line 18% conversion; green dotted line 24% conversion; blue dashed line 34% conversion; red dotted dashed line 58% conversion).

absence of undesired CO–N cleavage of the macroinitiator that could initiate new PS chains. All these results showed that the development of a "universal" nitroxide that could control the polymerization of a wide range of monomer is still possible. Future research is still needed to modulate the reactivity of indolinic nitroxides to improve the polymerization of monosubstituted monomers.

As an alternative to DPAIO, Grubbs and co-workers[53,54] prepared *N*-phenyl alkoxyamines, using their methodology based on the addition of C-centered radicals obtained from alkyl halides precursors onto nitroso compounds (Figure 7.11). This approach has the advantage to prepare alkoxyamines in

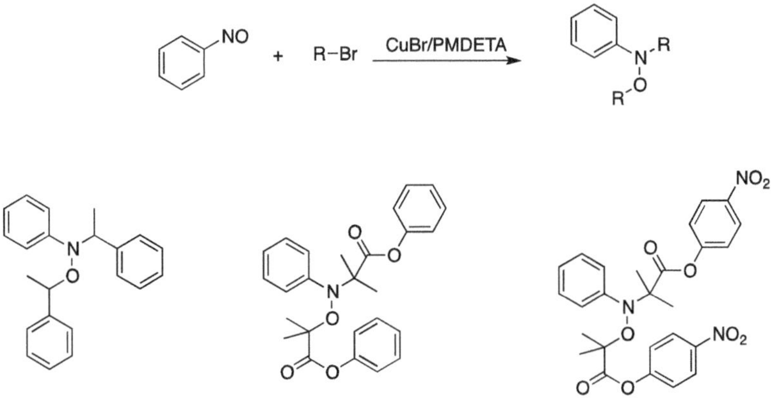

Figure 7.11 Synthesis of *N*-phenyl alkoxyamines from alkyl halides precursors and nitroso compounds.

"one pot" from commercially available compounds. Nevertheless, whatever the alkyl halide precursors, only moderate control at low conversion was obtained. The ^{1}H NMR analyses of the polymers presented vinylic protons that showed the occurrence of a disproportionation reaction that stopped the polymerization below 50% conversion.

Among the three alkoxyamines, the *para*-nitro derivative gave the best results due to the accelerated rate of dissociation of the initiating alkoxyamine that was already highlighted by Guillaneuf and co-workers.[46]

As the steric hindrance of the nitroxide has been shown to influence the rate of the intermolecular H-transfer, Studer and co-workers[55] tested a highly crowded nitroxide, 1-*tert*-butyl-3,3,5,5-tetraethyl-2-piperazinon-4-oxyl nitroxide, for the MMA polymerization in 2-methoxypropyl acetate using a microflow reactor. At 55 or 90 °C, the disproportionation stopped the polymerization at low to moderate conversion and thus prevented the successful NMP of methacrylate derivatives. In a similar approach, Bagryanskaya *et al.*[56] took advantage of their previous work on the mechanism of intramolecular H-transfer, where they found that one sterically hindered imidazoline nitroxide had a low k_{cD} value but also a low stability,[24] to prepare several imidazoline nitroxides *via* a 1,3-dipolar cycloaddition reaction. By an elegant synthetic pathway, they prepared imidazoline based nitroxides and the corresponding alkoxyamines with a spiro cyclic moiety (Figure 7.12). The fraction of disproportionation of such nitroxides was found to be in the range 1.3–3%. Therefore, they were considered as possible initiators for NMP of MMA. The bulk polymerization of MMA at 80 °C presented an increase of the M_n *vs.* conversion at low conversion followed by a plateauing of the M_n above 40% conversion when a high M_n (80 000 g mol^{-1}) was targeted (Figure 7.12). The dispersity also increased and reached 1.4. ESR analyses gave rather high livingness even if the reinitiation tests on both MMA and styrene were not convincing. The alkoxyamine is efficient for the NMP of styrene and then block copolymer

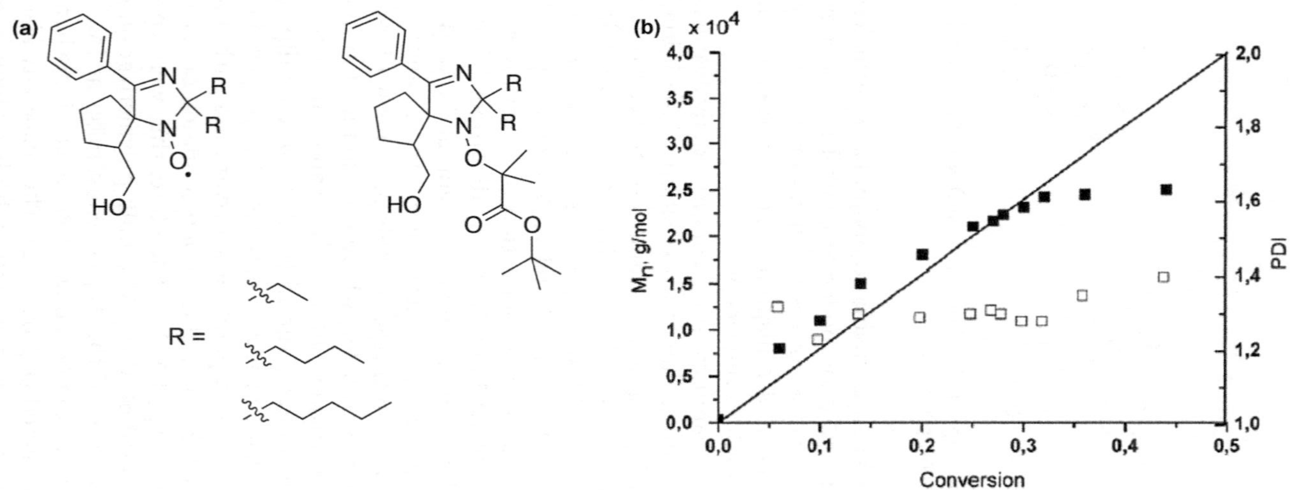

Figure 7.12 (a) Imidazoline-based nitroxides and alkoxyamines synthesized by Bagryanskaya *et al.*[56] (b) Evolution of M_n and $Đ$ *vs.* monomer conversion for NMP of MMA initiated with the ethyl-based alkoxyamine at 80 °C with monomer to initiator ratio = 800 : 1.
Reprinted with permission from ref. 56. Copyright (2014) Wiley.

PMMA-*b*-PS was obtained but with a rather high Đ and a very low styrene conversion (10%).

In conclusion, the better understanding of the disproportionation reaction and the kinetics of NMP allowed the design of new nitroxides able to induce partial or even good control of the methacrylate derivatives. The challenge is now to find a true "universal" nitroxide that will be able to control all classes of monomers.

7.4 The Copolymerization Approach

An alternative to significantly improve the control of methacrylic esters by NMP relies on their copolymerization with a small amount of an easy-to-control monomer such as styrene. Hawker and co-workers performed the copolymerization of MMA in the presence of an initial molar fraction of S (f_{S0}) ranging from 30 to 90 mol% using the styryl-TEMPO alkoxyamine.[57] The level of control was greatly improved for high initial amounts of S with a minimum of 50 mol% of S to reach a dispersity of 1.5. For lower values of f_{S0}, extensive occurrence of terminations by H-transfer led to a substantial increase of the dispersity (*e.g.*, Đ was 1.67 for $f_{S0} = 30$ mol%). A similar copolymerization approach was performed using the styryl-TIPNO alkoxyamine.[8] For a given f_{S0}, the control was better than the one obtained from styryl-TEMPO, especially for f_{S0} values lower than 60 mol%.

Six years later, the copolymerization approach by NMP between MMA and S was fully rationalized from a kinetics standpoint. The theoretical expression of the average activation–deactivation equilibrium constant, $\langle K \rangle$, and its product with the average propagation rate constant, $\langle k_p \rangle \langle K \rangle$, that is a direct function of the slope of ln[1/(1 − conv.)], were established by Charleux and co-workers.[58] Both the terminal model (TM) and the implicit penultimate unit effect model (IPUE) were considered for the propagation reaction $\langle k_p \rangle$. However, the activation–deactivation equilibrium constant was supposed to depend only on the terminal monomer unit. Using the TM, $\langle K \rangle$ can be described by eqn (7.2) and the product $\langle K \rangle \langle k_p \rangle$ by eqn (7.3).

$$\langle K \rangle = -\frac{(r_A f_A / k_{p,A}) + (r_B f_B / k_{p,B})}{(r_A f_A / k_{p,A} K_A) + (r_B f_B / k_{p,B} K_B)} \tag{7.2}$$

$$\langle k_p \rangle \langle K \rangle = \frac{r_A f_A^2 + 2 f_A f_B + r_B f_B^2}{(r_A f_A / k_{p,A} K_A) + (r_B f_B / k_{p,B} K_B)} \tag{7.3}$$

In eqn (7.2) and (7.3), r_A and r_B are the reactivity ratios ($r_A = k_{p,A}/k_{AB}$ and $r_B = k_{p,B}/k_{BA}$, where $k_{p,A}$ and $k_{p,B}$ are the homopropagation rate constants), k_{AB} and k_{BA} are the cross-propagation rate constants and f_A and f_B are the molar fractions of A and B in the initial comonomer mixture.

With the IPUE model, eqn (7.2) and (7.3) are still valid, by replacing $k_{p,A}$ and $k_{p,B}$ by eqn (7.4) and (7.5).

$$\overline{k_{p,A}} = k_{p,A} \frac{r_A f_A + f_B}{r_A f_A + (f_B/s_A)} \tag{7.4}$$

$$\overline{k_{p,B}} = k_{p,B} \frac{r_B f_B + f_A}{r_B f_B + (f_A/s_B)} \tag{7.5}$$

In eqn (7.4) and (7.5), s_A and s_B are the radical reactivity ratios ($s_A = k_{BAA}/k_{AAA}$ and $s_B = k_{ABB}/k_{BBB}$ where $k_{AAA} = k_{p,A}$ and $k_{BBB} = k_{p,B}$ are the homopropagation rate constants and the other rate constants, k_{BAA} and k_{ABB}, correspond to the propagation reactions).

From eqn (7.2) one can see that $\langle K \rangle$ not only depends on the individual activation–deactivation equilibrium constants (K_A and K_B) but also on the cross-propagation rate constants, which are expressed by the ratio of the homopropagation rate constant over the corresponding monomer reactivity ratio. Besides, it varies with the comonomer mixture composition (f_A and f_B). f_A and f_B change continuously during the copolymerization, and so does $\langle K \rangle$. This is important as a slow cross-propagation of the comonomer-based macroradical will ensure a longer lifetime, which will favour its efficient deactivation by the nitroxide. Consequently, a small amount of a suitable comonomer exhibiting a much smaller K than the one of MMA and a low k_p should induce a dramatic decrease of $\langle K \rangle$ and therefore drastically enhance the control of the polymerization. S, which satisfies the above-mentioned criteria, was predicted to be a suitable comonomer for the NMP of MMA (Figure 7.13). Above ~3 mol%, the value of $\langle K \rangle$ is closed to that for the homopolymerization of S. The system should therefore provide the same quality of control as for the SG1-mediated homopolymerization of S.

Expectations were fully met by performing the NMP of MMA at 90 °C in the presence of 2.2–8.8 mol% of S as a comonomer (Figure 7.14).[58,59] Fast and quantitative initiation was satisfied by using the BlocBuilder SG1-based alkoxyamine together with the addition of 10 mol% of free SG1 to reduce the occurrence of initial irreversible termination reactions. Without S, the polymerization rapidly stopped in agreement with expectations. However, only 2.2 mol% of S had a dramatic impact on the outcome of the copolymerization. It gave a rather linear evolution of the M_n with monomer conversion together with a lower $Đ$ (~1.4 at best). From 4.4 mol% of S, the copolymerization was nearly ideal with first order kinetics, linear evolution of M_n with conversion and decreasing dispersities reaching 1.3.

It was shown that the resulting P(MMA-*co*-S) copolymer ($f_{S0} = 0.088$) exhibited a high living chain fraction (~90%) and a vast majority of the chain-end structures displayed a MMA-S-SG1 sequence.[60] Well-defined diblock

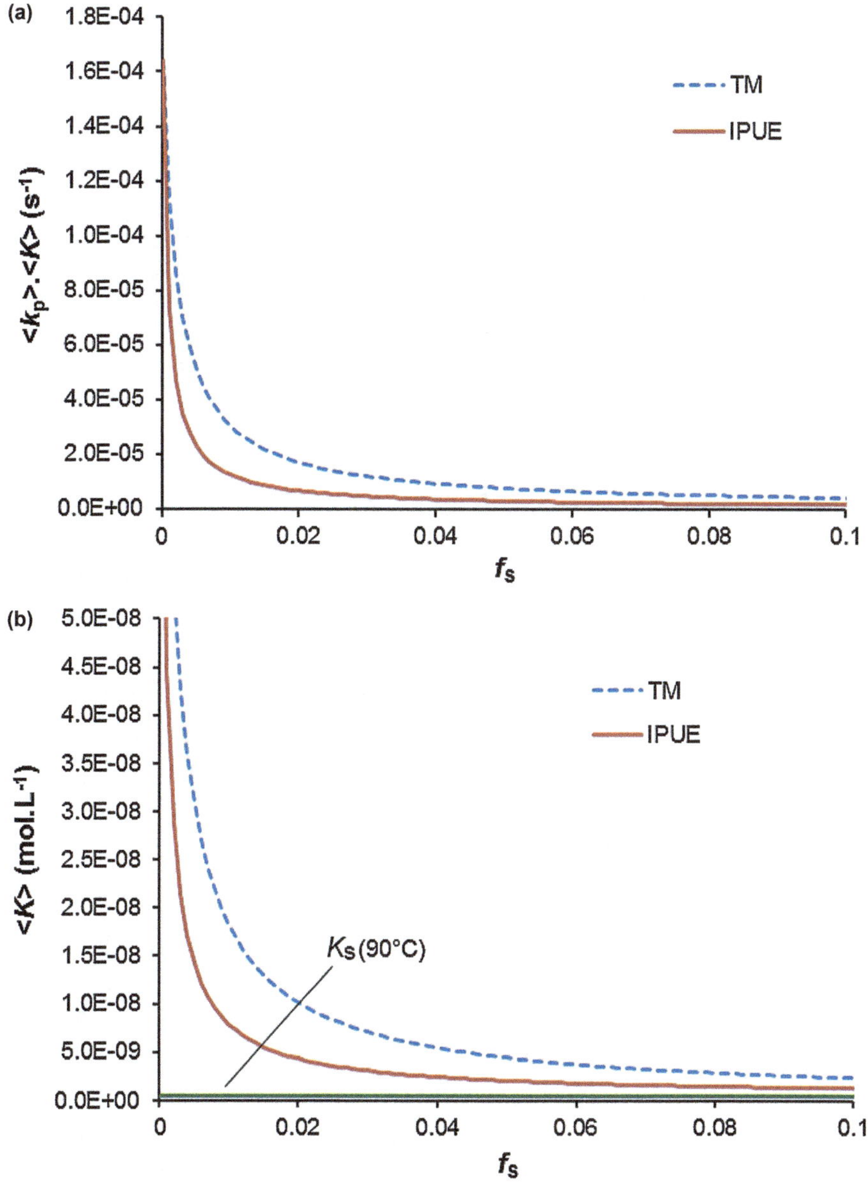

Figure 7.13 Predicted values of the average activation–deactivation equilibrium constant $\langle K \rangle$ (a) and of $\langle k_p \rangle \langle K \rangle$ (b) as a function of the molar fraction of styrene (S) in the copolymerization mixture (f_S), for the copolymerization of methyl methacrylate (MMA) and S at 90 °C, using either the terminal model (TM) or the implicit penultimate unit effect model (IPUE).
Reproduced with permission.[58] Copyright 2005, American Chemical Society.

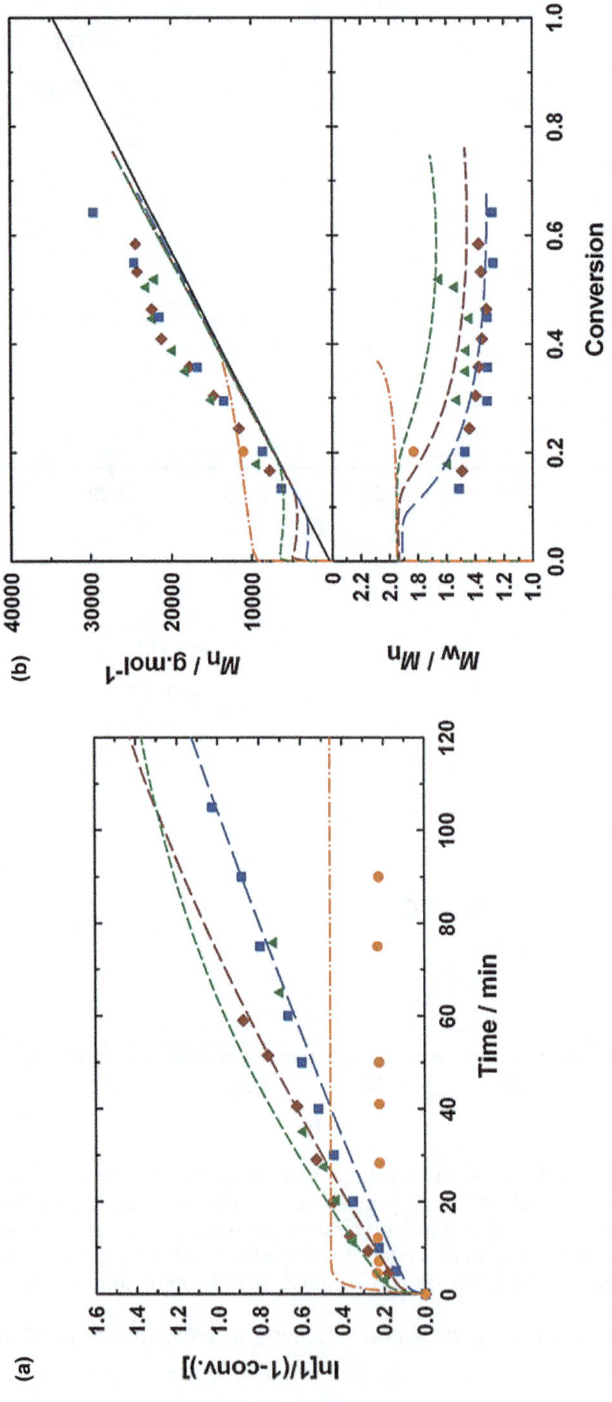

Figure 7.14 SG1-mediated bulk polymerization of methyl methacrylate (MMA) initiated by the BlocBuilder alkoxyamine at 90 °C with 10 mol% of free SG1 as a function of the initial amount of styrene (S): ■, $f_{S0} = 0.088$; ◆, $f_{S0} = 0.044$; ▲, $f_{S0} = 0.022$; ●, $f_{S0} = 0$. (a) ln[1/(1 − conv.)] *vs.* time; (b) number average molar mass, M_n, and dispersity, M_w/M_n, *vs.* conversion; the dotted lines correspond to the modelling studies and the full line represents the theoretical M_n.[59] Copyright 2009, American Chemical Society. Reproduced with permission.

copolymers were also reported using P(MMA-*co*-S)-SG1 macroinitiators. Due to the low dissociation rate constant of the resulting macroalkoxyamine formed during the copolymerization, the reaction can be conducted below 90 °C (typically in the 80–90 °C temperature range).

A comprehensive kinetic study of this copolymerization system by means of the PREDICI software confirmed the benefit of adding a small fraction of S during the NMP of MMA (see Figure 7.14) and also gave a better insight into the complex mechanism of nitroxide-mediated radical copolymerization.[59] It showed that an increase in S concentration caused an increase in the formation of S-S-SG1 macroalkoxyamines, which are less prone to be reinitiated under the experimental conditions. Therefore, it is important to find an appropriate balance for the S concentration to obtain the highest quality of control.

This copolymerization approach has been employed by various groups, not only with MMA[12,58–66] but also with other methacrylates, including methacrylic acid (MAA),[67,68] *n*-butyl methacrylate (nBMA),[69] benzylmethacrylate (BzMA),[70] (dimethylamino)ethyl methacrylate (DMAEMA),[71–73] poly(ethylene glycol) methyl ether methacrylate (MePEGMA)[12,62,68,74,75] and also a glucose-functionalized methacryloyl galactose (AcGalEMA).[76]

Apart from S, other styrenic derivatives can act as controlling comonomers, such as 4-styrene sulfonate (SS)[67] and pentafluorostyrene (PFS).[73] Maric and co-workers also used 9-(4-vinylbenzyl)-9H-carbazole (VBK).[64,65,71,74] For the copolymerization of MMA with VBK, using the BlocBuilder alkoxyamine, only 1 mol% VBK was enough to obtain a good control and livingness. This was assigned to the higher reactivity ratio for VBK compared to S.[64] MeOEGMA[74] and DMAEMA[71] were also successfully copolymerized with VBK.

To circumvent the inherent drawbacks of S and its derivatives (*e.g.*, high hydrophobicity except for SS, potential alteration of the well-known optical properties of PMMA, potential toxicity in the context of biomedical applications), alternative structures have been investigated. For instance, 2-vinyl pyridine (2VP)[72] was employed as a comonomer to control the NMP of DMAEMA. However, a moderate control was obtained ($Đ = 1.56$ for $f_{2VP} = 0.2$). Isoprene[77] had beneficial effects at 20 mol% for the NMP of MMA but *n*BA and isobornyl acrylate led to minimal improvement except at high concentrations (>30 mol%).[78,79] In fact, *N*,*N*-dimethylacrylamide (DMA)[63] and acrylonitrile (AN)[12,61,62] proved to be very efficient for the control of methacrylic esters by NMP. 10% DMA was enough to drastically improve the control of the NMP of MMA with dispersities reaching 1.4.[63] For AN, 2.2 to 8.8 mol% during the polymerization of MMA led to all the features of a controlled/living system (Figure 7.15).[61]

Interestingly, Nicolas and co-workers copolymerized OEGMA with AN in water-rich hydroalcoholic solutions (up to 75% water) due to the high water-solubility of AN. In addition, the polymerization temperature could be reduced down to 71 °C due to the accelerating effect of water,[62] which represents one of the lowest temperature ever reported for NMP. The resulting

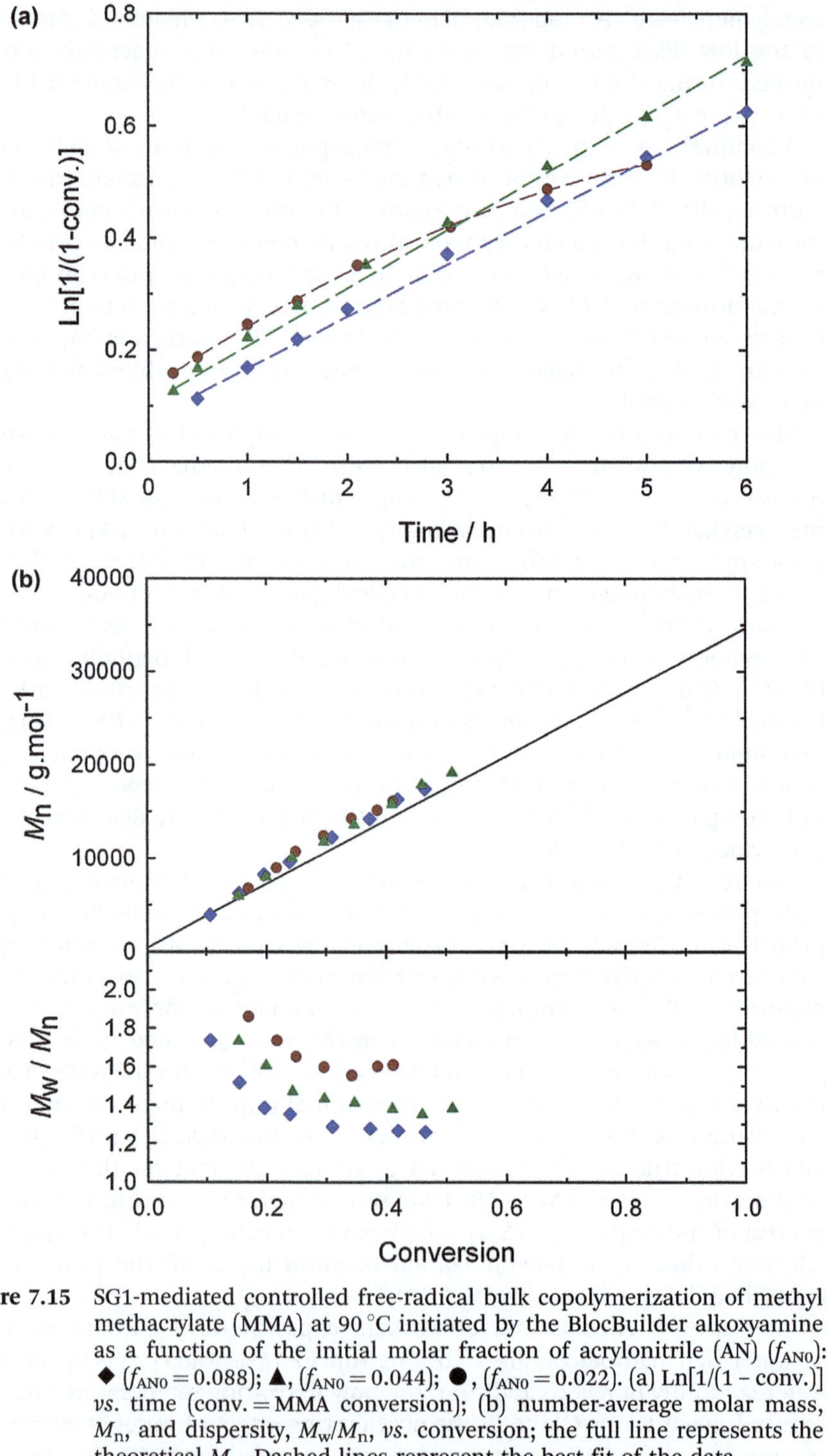

Figure 7.15 SG1-mediated controlled free-radical bulk copolymerization of methyl methacrylate (MMA) at 90 °C initiated by the BlocBuilder alkoxyamine as a function of the initial molar fraction of acrylonitrile (AN) (f_{AN0}): ◆ ($f_{AN0} = 0.088$); ▲, ($f_{AN0} = 0.044$); ●, ($f_{AN0} = 0.022$). (a) Ln[1/(1 − conv.)] *vs.* time (conv. = MMA conversion); (b) number-average molar mass, M_n, and dispersity, M_w/M_n, *vs.* conversion; the full line represents the theoretical M_n. Dashed lines represent the best fit of the data. Reproduced with permission.[61] Copyright 2010, Wiley.

PEG-based copolymers were not cytotoxic on various cell lines up to high concentrations (*i.e.*, 10 mg mL^{-1}). Also, by using NHS-functional SG1-based alkoxyamines, similar α-functional copolymers were prepared and successfully conjugated to amine-containing compounds such as a neuroprotective tripeptide and lysozyme.[12]

7.5 New NMP Alternatives to Control Methacrylic Esters

As an alternative strategy to traditional NMP, nitroxides and alkoxyamines can be prepared *in situ* in the polymerization medium from inexpensive readily and/or commercially available precursors. In that case, both the initiating radicals and the mediators are produced in a one-pot technique. This process is called *in situ* nitroxide-mediated polymerization (*in situ* NMP). This topic has been recently thoroughly discussed in the following review article.[80] Nitroso compounds represent an important family of alkyl radical scavengers and therefore have been widely applied in *in situ* NMP. The group of Matyjaszewski[81] investigated the polymerization of MMA at 80 °C using the AIBN/2-methyl-2-nitrosopropane (MNP) and AIBN/2,4,6-tri(*tert*-butyl)nitrosobenzene (TBNB) pair, and observed that the polymerization rate was governed by the concentration of nitroso compound for constant nitroso compound/initiator molar ratios. The controlled and living character of the polymerization was not investigated. A precursor of nitroso compounds, sodium nitrite ($NaNO_2$), was used by Detrembleur and co-workers[82] to control the polymerization at 80 °C of *tert*-butyl methacrylate (*t*BMA) in water. Reduction of $NaNO_2$ by an acidic solution of $FeSO_4$ is known to release NO, which is able to trap free radicals generated from a conventional initiator (AIBN or potassium persulfate), resulting in the formation of nitroso then nitroxides and finally parent alkoxyamines (Figure 7.16a). The simultaneous formation of NO and NO_2 that occurred during the reduction in the presence of *t*BMA produced α-nitro-ω-nitroso P*t*BMA in parallel.

It was shown that the molar mass was governed by the monomer over $NaNO_2$ molar ratio rather than by the amount of $FeSO_4$ (Figure 7.16b). Nevertheless the deviation of $\ln([M]_0/[M])$ *vs.* time from linearity, and the increase in dispersity when the monomer conversion was increased showed that side reaction occurred during the polymerization and only partial control was obtained at high conversion. The nitrite system is however rather complex because several species are formed *in situ* as a result of the addition of both the initiating and the propagating radicals onto different scavengers. Therefore, several nitroso compounds and thus nitroxides and alkoxyamines coexist in the polymerization medium. Detrembleur and co-workers[83] improved their system by performing the polymerization in organic medium by reaction at 25 °C MMA or *t*BMA with a mixture or NO/NO_2 to produce *in situ* the nitroso compounds. The further polymerization of

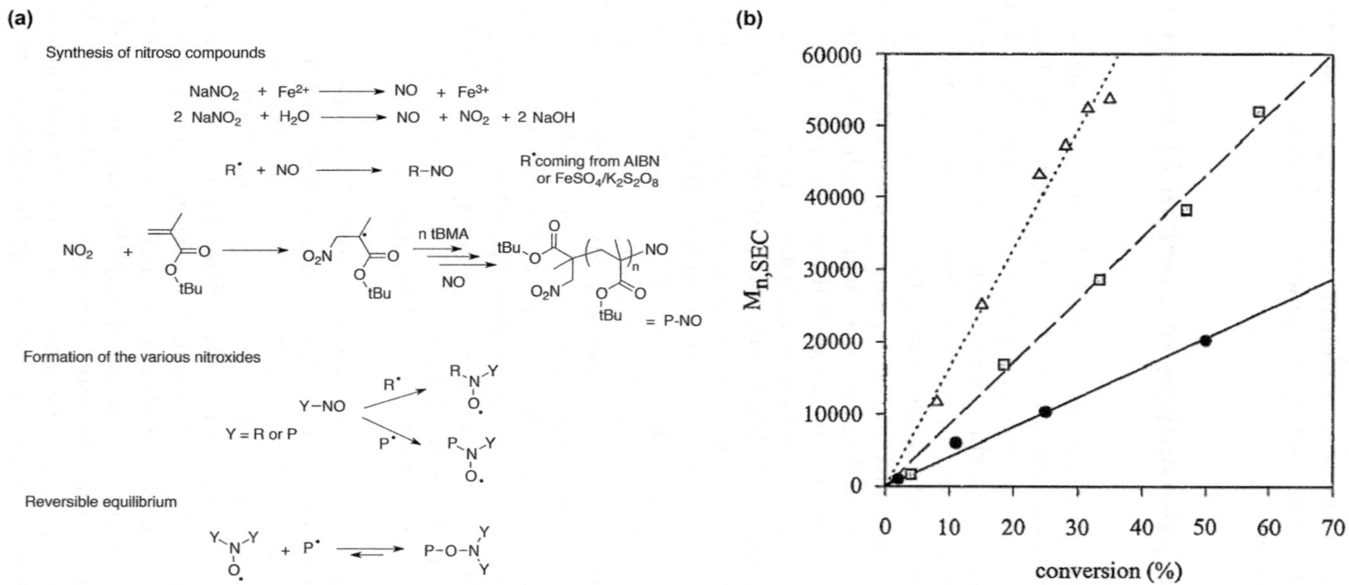

Figure 7.16 (a) Mechanism of the *in situ* NMP of *t*BMA in the presence of $NaNO_2$ and $FeSO_4$. (b) Evolution of the M_n *versus* *t*BMA conversion in water at 80 °C. The amount of $NaNO_2$ is changed, while keeping the concentration of the other components constant [× equiv. of $NaNO_2$, 1 equiv. of $FeSO_4$, 1 equiv. of $K_2S_2O_8$, 168 equiv. of *t*BMA, water/*t*BMA = 2.5]. (△) $NaNO_2$/ $K_2S_2O_8$ = 0.5/1 (□) $NaNO_2$/ $K_2S_2O_8$ = 1/1 (●) $NaNO_2$/ $K_2S_2O_8$ = 2/1.
Reprinted with permission from ref. 82. Copyright (2002) American Chemical Society.

MMA, *t*BMA and a mixture of HEMA/MMA (1 : 10) was then carried out at 60 °C using AIBN as radical initiator.

Recently Detrembleur and co-workers[84] made this system more straightforward by preparing and isolating the methyl 2-methyl-3-nitro-2-nitrosopropionate (NMMA) before the polymerization so that only one nitroso compound is present in the medium. Such a nitroso compound was then added to methacrylate derivatives in combination with V70 as low temperature azo initiator. The bulk polymerization at moderate temperature (40–50 °C) of methyl methacrylate (MMA), benzyl methacrylate (BzMA), trifluoroethyl methacrylate (TFMA) led to a good control of the polymerization combined with a high livingness (Figure 7.17). In particular, reinitiation experiments from low M_n PMMA macroinitiators were performed and led to very high M_n (280 000 and 82 000 g mol^{-1}) with still low dispersity (1.46 and 1.38, respectively).

Besides the *in situ* NMP, Junkers and Barner-Kowollik[85–87] developed a new system named enhanced spin capturing polymerization (ESCP) to control the polymerization of vinylic monomers based also on the use of radical scavenger and especially nitrones. The mechanism is based on the ability of the nitrones to limit the lifetime of a propagating radical and to induce termination by coupling between propagating macroradicals *via* two consecutive spin capturing reactions (Figure 7.18a). As the reactions are performed below the bond dissociation temperature of the generated alkoxyamine, the system enabled the prediction of the number-average molar mass, similarly to controlled polymerization techniques, but does not feature living characteristics since no reversible cleavage of the macroalkoxyamine is possible. This technique has been recently reviewed.[88]

ESCP has been employed for controlling the polymerization of S, *n*BA, ethylene and NIPAAm in either organic or aqueous solutions, and *via* thermally or photo-chemically initiated systems.[85,87,89,90]

Contrary to other monomers, the polymerization of MMA in the presence of nitrone (α-phenyl-*N*-*tert*-butylnitrone PBN) occurred but without any control over the molecular weight.[91] This was ascribed to the low rate of addition to PMMA macroradicals onto the nitrone functionality. Junkers and Barner-Kowollik[91] overcame this problem by the copolymerization approach that was originally proposed by Charleux and co-workers[42] (see above). With 5–10 mol% of styrene added into the monomer mixture, a good control was achieved. The average molecular weight of the samples varies between 57 000 g mol^{-1} for the uncontrolled reaction (featuring a nitrone concentration of zero) and close to 7000 g mol^{-1} for the highest PBN concentration of 0.4 mol L^{-1}, Figure 7.18b). The presence of the alkoxyamine functionality at the middle of the polymer chains was assessed by a NMP reinitiation procedure (bulk styrene polymerization at 110 °C, Figure 7.18c). Under such conditions, only chains carrying an intact alkoxyamine function can be reactivated and further chain extended. All other chains were dead and could not react again. A good shift of the molar

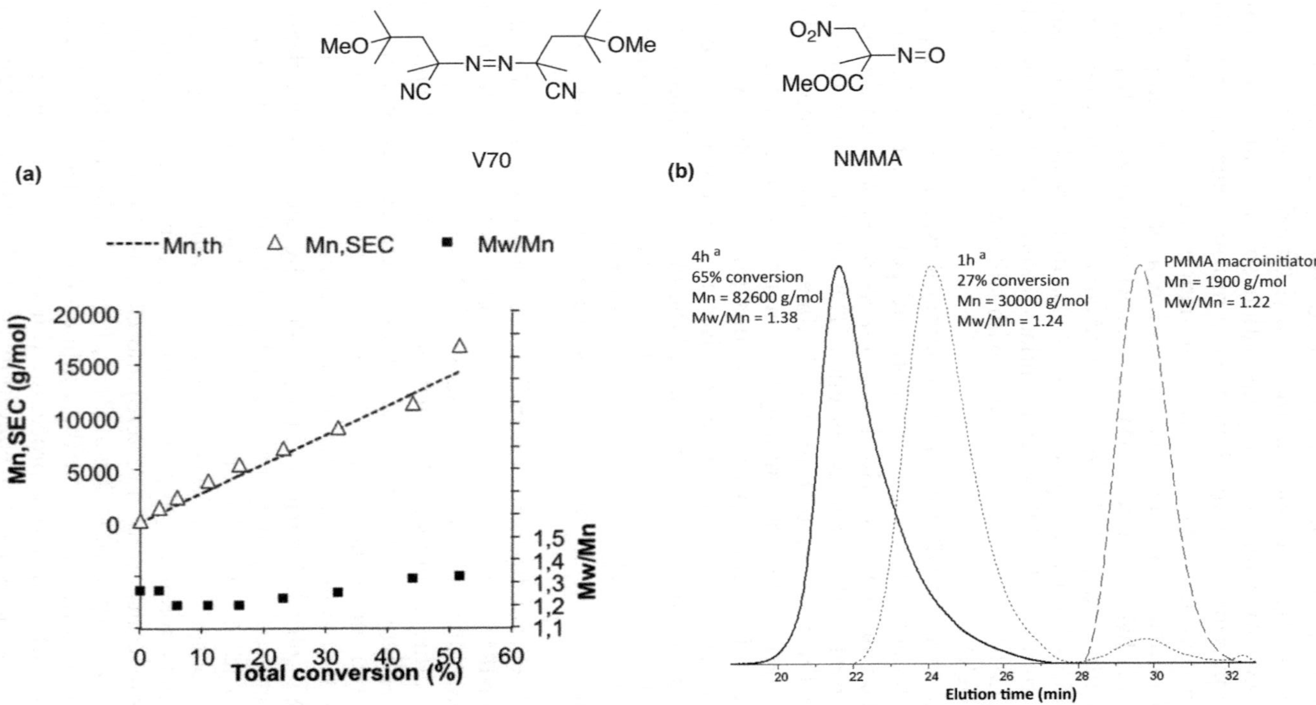

Figure 7.17 (a) Molar mass and dispersity evolution with monomer conversion for the bulk methyl methacrylate polymerization initiated by V70 in the presence of NMMA at 50 °C ([MMA]/[V70]/[NMMA] = 275/1/1). (b) Reinitiation experiments of PMMA macroinitiator ($M_n = 1900$ g mol^{-1}; $Đ = 1.22$) on bulk MMA at 50 °C ([MMA]/[PMMA] = 1776). Reprinted with permission from ref. 84. Copyright (2014) Royal Society of Chemistry.

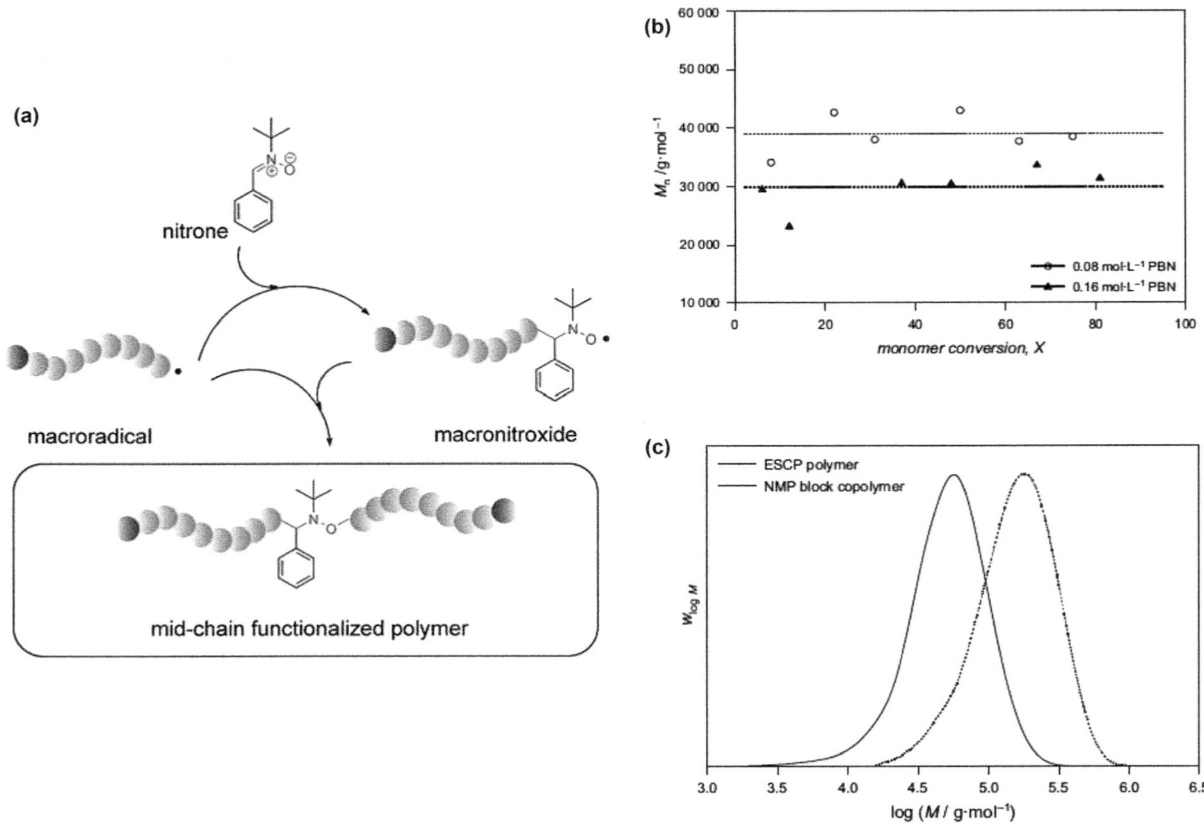

Figure 7.18 (a) Mechanism of the ESCP process. (b) Monomer to polymer conversion dependence of the number average molecular weight, M_n, in the bulk ESCP of MMA with 5 vol% styrene content at 60 °C, mediated with PBN and initiated by 0.04 mol L^{-1} AIBN. (c) Molecular weight distribution of polymer made *via* ESCP of MMA with 5 vol.% styrene as comonomer and the distribution of the polymer after reactivation of the alkoxyamine bonds at 110 °C and chain extension with pure styrene *via* NMP. Reprinted with permission from ref. 88 and 91. Copyright (2011) and (2010) Royal Society of Chemistry and Elsevier respectively.

mass distribution was observed and the preparation of a PMMA-*b*-PS-*b*-PMMA triblock copolymer was achieved even if a tailing was present on the low molecular weight side of the distribution, indicating a non-negligible proportion of dead chains during the NMP process.

7.6 PMMA-based Materials Prepared by NMP

Although homopolymerization of MMA cannot be controlled by the common nitroxides (*e.g.*, TEMPO, SG1), various materials based on PMMA are already prepared by NMP. The most representative example is perhaps the range of product developed by the ARKEMA company based on the use of the BlocBuilder alkoxyamine.[92] From this alkoxyamine, they produced at industrial scale (*i.e.*, up to a ton) various all-acrylic poly(methyl methacrylate)-*b*-poly(*n*-butyl acrylate)-*b*-poly(methyl methacrylate) (MAM) triblock copolymers under the trademark Nanostrength®.[92] By changing the volume fraction of the soft block the behaviour of these materials they could vary from though thermoplastic materials (20 mol% P*n*BA) to "soft acrylic" thermoplastic elastomers (80 mol% P*n*BA).

Although the control of the PMMA block was not effective, with final dispersities ranging between 2 and 3, the copolymers were shown to self-assemble without any phase separation, which was unexpected.[93] Also, it was shown that the interfacial curvature did not follow the phase diagram established for ideal monodisperse block copolymers. For example, lamellae are observed for MMA-rich compositions (65–55 vol% PMMA), whereas symmetric copolymers adopted a curved interface concave toward PMMA and formed cylinders or cylindrical micelles with a poor lattice order (Figure 7.19).[93]

This effect is consistent with the literature on polydisperse block copolymers,[94,95] and on the co-surfactant effect[96] and SCFT calculations.[97] Matyjaszewski and co-workers[98] obtained similar results with the stabilization of a particular morphology in the case of a polystyrene-*b*-poly(methyl acrylate) PS-*b*-PMA copolymer with a polydisperse PMA block. This was the consequence of the efficient alleviation of packing frustration that is facilitated by the balanced presence of small and large chains of PMA chains due to the high dispersity of this block. Nicolas and Charleux[99] managed to obtain these nanostructures directly inside latex particles by performing emulsion polymerization under NMP control, which could be very interesting from an industrial standpoint. "Onion-like" lamellar microphases inside latex particles were observed by AFM and TEM images of latex films. These nanostructures evolved toward more classical block copolymer lamellar morphologies upon re-dissolution and/or annealing of the latex films.

Two different applications were developed from these new all-acrylic materials. The first one is related to the increased toughness of thermoset and composite materials. Indeed, applications of epoxy-based thermosets

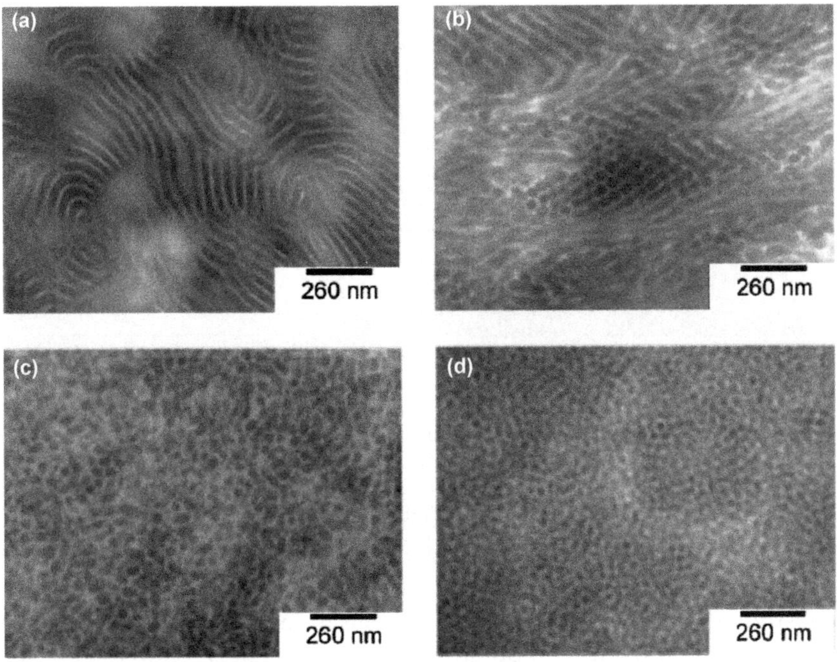

Figure 7.19 TEM micrographs of different Nanostrength® triblock copolymers showing lamellae for sample A [67 vol% PMMA: $M_n = 124\,000$ g mol^{-1}; $Đ$ (P*n*BA) = 1.5; $Đ$ (Nanostrength®) = 1.9]. (a) Cylinders for sample B [48 vol% PMMA: $M_n = 80\,000$ g mol^{-1}; $Đ$ (P*n*BA) = 1.9; $Đ$ (Nanostrength®) = 2.4]. (b) Cylindrical or spherical micelles [43 vol% PMMA: $M_n = 87\,000$ g mol^{-1}; $Đ$ (P*n*BA) = 1.4; $Đ$ (Nanostrength®) = 2.0]. (c) Spheres of PMMA [35 vol% PMMA: $M_n = 61\,000$ g mol^{-1}; $Đ$ (P*n*BA) = 1.5; $Đ$ (Nanostrength®) = 1.7]. (d) PMMA microdomains are stained with a 2 wt% PTA/benzyl alcohol aqueous solution and appear black.

are often limited by an inherent brittle behaviour of the matrix. The common solution to improve the toughness of such materials is to incorporate either conventional butadiene-based copolymers such as carboxyl-terminated butadiene-acrylonitrile copolymers (CTBN) or core-shell latex particles with methacrylate-based shells and a rubbery core.[100] To ensure a good dispersion into the matrix, a non-negligible quantity of particles has to be inserted, which has the disadvantage of reducing the effective stiffness of the final product and sometimes lowers the service temperature. An alternative approach to the toughening of epoxy network consisted of the addition of acrylic block copolymers, comprising rigid and rubbery block tethered together. Nanostrength® block copolymers were then inserted into various thermosets; *i.e.*, epoxy resin[101–103] and unsaturated polyester (UP).[104] In both cases, nanostructuration was observed.

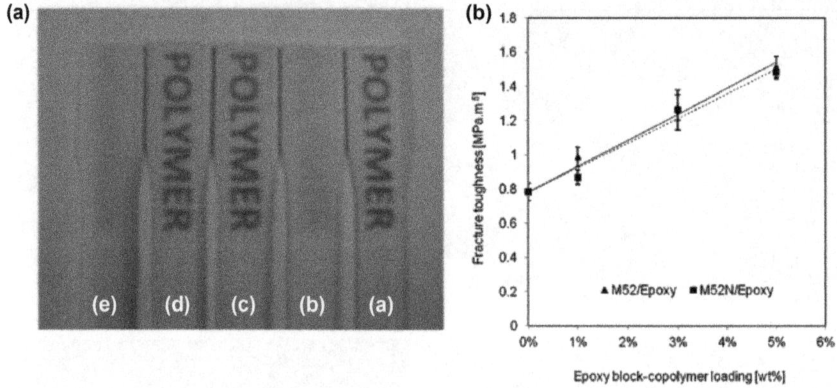

Figure 7.20 (a) Transparency of cured (a) neat epoxy and (b) 3 wt% M52, (c) 3 wt% M52N, (d) 5 wt% M52N and (e) 5 wt% M52 modified epoxy blends. (b) Fracture toughness of modified epoxy plotted against corresponding block-copolymer concentration.
Reprinted with permission from ref. 101. Copyright (2014) Wiley.

Nanostrength® modified epoxy resin (based on the diglycidylether of bisphenol A DGEBA in combination with a hardener) showed improved toughness, without significant reduction of the other desirable mechanical (*e.g.*, structural stiffness and strength) and thermal (*e.g.*, glass transition temperature) properties.[101] In the case of unsaturated polyester resins, the introduction of 5 wt% of the block copolymer led to an increase of about 50% of the critical stress intensity factor K_{Ic}.[104] It has to be noted that for a certain formulation of thermoset, a modified Nanostrength® formulation could be used, in which a polar dimethyl acrylamide functionality was randomly inserted into the PMMA block.[63,102–104] The insertion of such a polar fragment increased the solubility of the PMMA-rich block inside the matrix and led to a fully transparent thermoset (Figure 7.20).[101] Nevertheless, the use of pure PMMA gave an opaque thermoset due to macrophase separation during the curing of the thermoset but the final materials had the same mechanical properties in terms of toughness, T_g, tensile modulus, *etc.*[101,103]

Polymer composites reinforced with long fibres are widely used as high-performance materials in various industries, such as automotive, rail-industry, aerospace, sport and civil engineering. Since epoxy thermosets are usually used as matrix materials for glass, Kevlar and other fibre-reinforced composites, the introduction of Nanostrength® to increase the impact resistance has been tested. The first study was performed on Kevlar fibres embedded into DGEBA-based epoxy matrix and such composites were tested towards low-velocity impact.[105] A drop weight tower was used to conduct low-velocity impact tests on the composites to reproduce similar damage patterns on the specimen similar to that observed in industrial application (helmet). The composite without additives was perforated whereas low

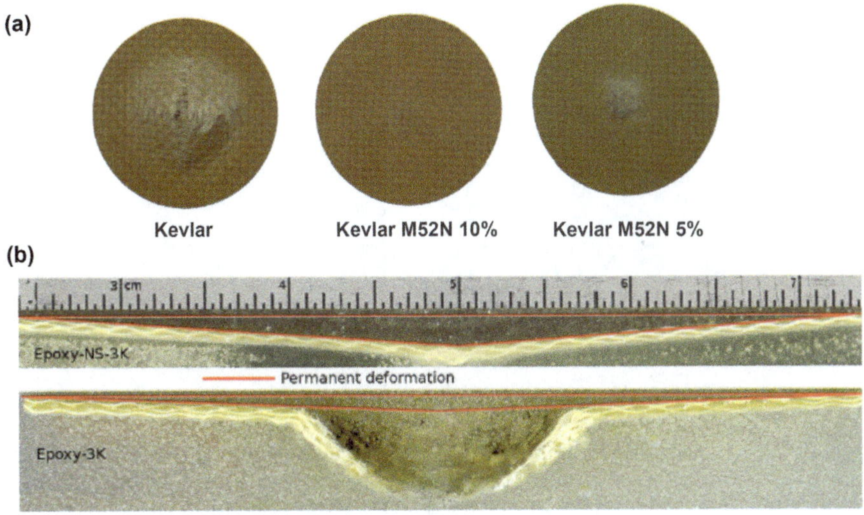

Figure 7.21 (a) Picture of the impact damage of the epoxy/Kevlar fibres with and without Nanostrength® (M52N). (b) Comparison between cross-section of the impacted plate.
Reprinted with permission from ref. 105. Copyright (2012) Elsevier.

damage was reported on the composites with various amounts of Nanostrength® (Figure 7.21).

The observation of the cross-section showed that the deformation (which corresponds to the impact energy) was spread on a larger part of the composite plate. Similar studies were also extended to basalt fibres[106] as well as glass fibres[107]/epoxy laminates. In both cases, Nanostrength® provided superior toughening as compared to reactive rubber flexibilizers or core shell toughening agents, without sacrificing T_g or modulus.

In a further study, Nanostrength® was also evaluated for increasing the low impact resistance of sandwich structures with Kevlar-reinforced facesheets.[108] Such lightweight structural materials (core made with polymer foam and skin by fibre-reinforced polymers) are used in a wide range of applications, including aerospace, automotive and marine structures. For such applications one of the main drawbacks is their relatively poor resistance to impact loading that could come from tool drops or runway debris. The introduction of Nanostrength® inside the epoxy matrix of the sandwich plates led to the absence of perforation of the facesheets at 16 J impact whereas it occurred without additive for impact at 12 J. The damage in the foam core was also not localised in the impact zone but was spread across a significant part of the structure, as already observed previously.

Acrylic glasses made from PMMA are very interesting materials due to their optical and lightweight properties and thus represent a good alternative to classic glass for many applications. Nevertheless, this material

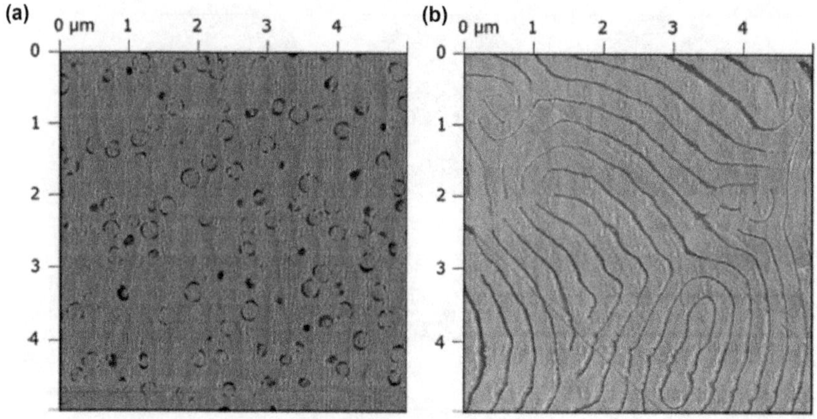

Figure 7.22 Tapping-mode atomic force microscopy, phase images from (a) a high impact cast PMMA (CI30) with 300 nm core shells rubber particles inserted in the matrix and (b) a cast PMMA ShieldUp sheet. Reprinted with permission from ref. 109. Copyright (2014) Elsevier.

suffers from the apparition of crazes that occurred during mechanical damages. Therefore, enhancement of the mechanical strength by the introduction of rubber phase inside the matrix is required. The introduction of small core–shell rubber particles, usually embedded in the PMMA matrix, led to the so-called high impact acrylic materials. In these products, the refractive index of rubber and PMMA phases are alike in the range −30 to 80 °C. Out of this range, a haze effect (lack of transparency) did not allow its use for automotive glazing when it would be a good way to decrease the overall weight of cars due to its lightness compared to common glass. Besides its role as a toughener for thermoset, the Nanostrength® triblock copolymer could be used to reinforce PMMA sheets as an alternative to the rubber-based particles. This product is commercially available from the Arkema Company under the trademark Altuglas ShieldUp®. An interesting feature of this product is the nanostructuration (lamellae of few tens nanometers) of the rubber phase of P*n*BA inside the PMMA matrix conversely to the particles that have diameters of about 300 nm (Figure 7.22).

The nanostructuration induced by the presence of the Nanostrength® copolymer avoided the transparency decrease at high temperature compared to the last generation of rubber-reinforced material (CI30), with a very low change of the haze even at 90 °C (Figure 7.23).

Besides the improved performance in transparency loss due to the introduction of the Nanostrength® triblock copolymer in the ShieldUp® sheets, the nanostructuration led to strong improvements in both impact and chemical (*e.g.*, isopropanol, acetone) resistance (Figure 7.24a and b). By varying the amount of the soft phase in the matrix and the molecular

(a)
(b)

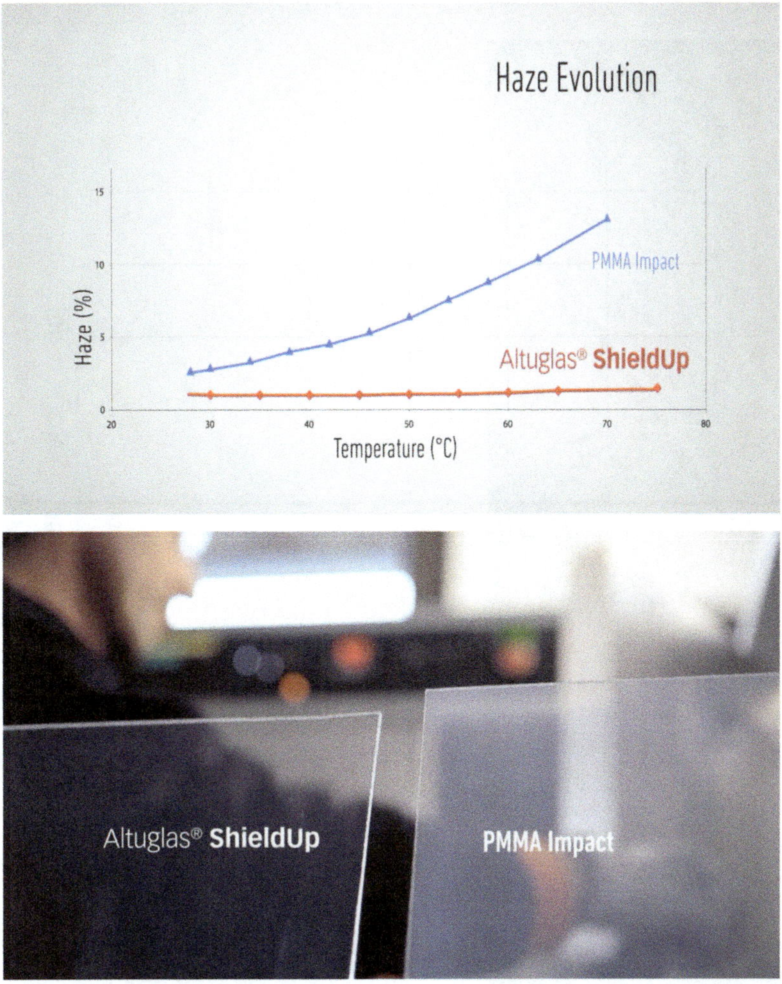

Figure 7.23 (a) Comparison of the haze evolution of high-impact cast PMMA and ShieldUp® with temperature. (b) High-impact cast PMMA and ShieldUp® heated to 90 °C. Courtesy of Arkema.

weight of the copolymer, it was possible to tune the mechanical properties of the ShieldUp® sheet (Figure 7.24c). This could extend the range of application of such materials. The Altuglas ShieldUp® Cast Sheet is now a good alternative for automotive glazing (*e.g.*, panoramic roof) since this material passed the ECE R43 test, which is a list of the main requirements needed for such applications (*e.g.*, resistances to impact, abrasion, humidity, chemicals and fire, weathering and optical properties) and has been used in the Renault Twizy and the Peugeot concept car Onyx.

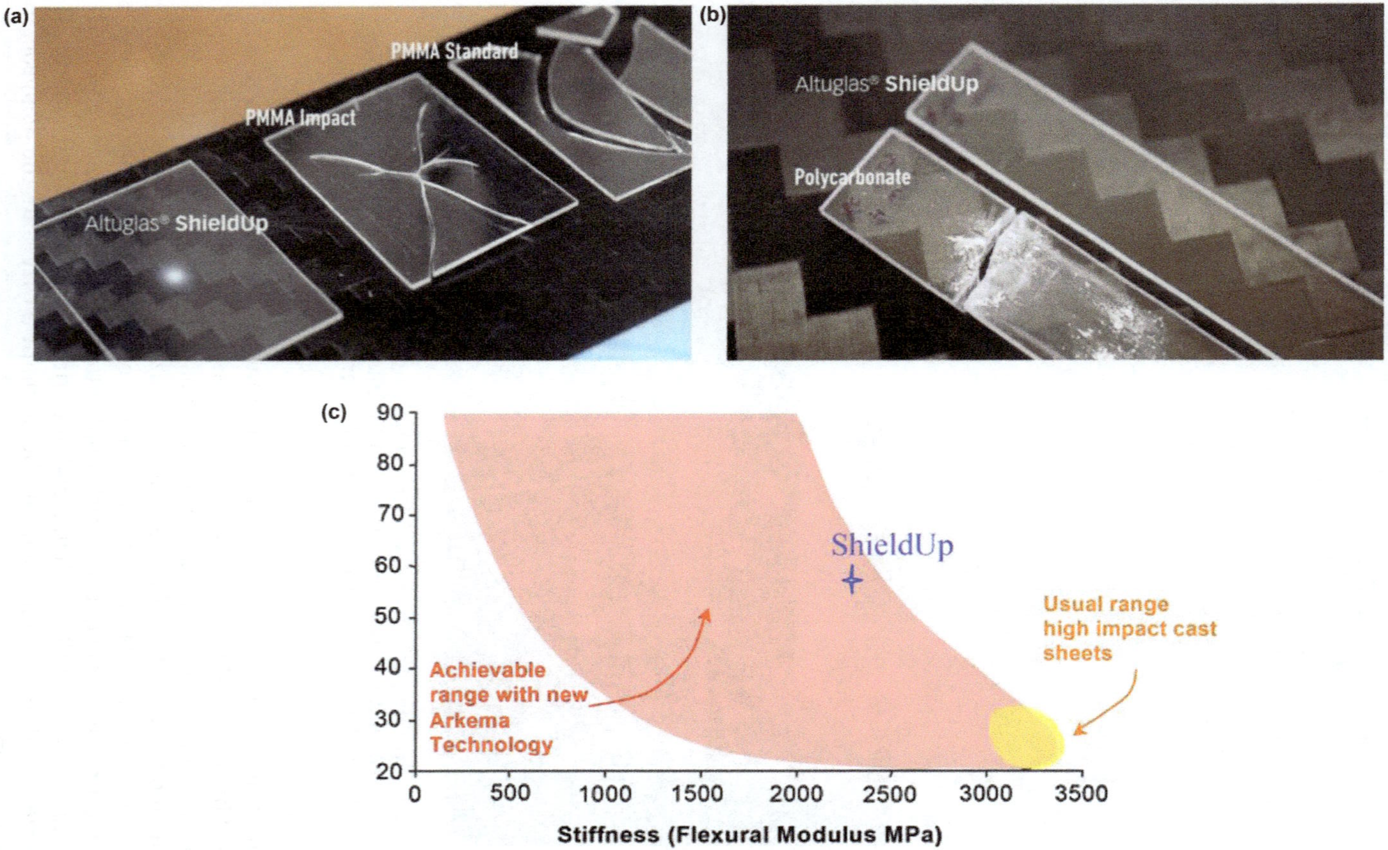

Figure 7.24 (a) Comparison of pure PMMA, high impact and ShieldUp® cast sheets after a ball drop test (3.1 kg). (b) Comparison of Polycarbonate and ShieldUp® cast sheet after acetone cleaning. (c) Impact strength *vs.* stiffness for cast PMMA sheets. Courtesy of Arkema.

7.7 Conclusion

This chapter first aimed at demonstrating the reason for the failure for the NMP of methacrylate derivatives. Due to a high propensity to undergo inter and/or intra hydrogen transfer reaction, the couple PMMA macroradical/nitroxide led in general to polymer chains with a terminal double bond.

The different strategies to control the polymerization of methacrylic esters by the NMP technique, and especially of MMA, have been detailed and discussed. Two main distinct strategies, each with pros and cons, have indeed emerged and hold great promise:

(i) The copolymerization approach that uses only a very small amount of a suitable comonomer is interesting for its simplicity and for the commercial availability of nitroxides and alkoxyamine initiators. Nevertheless, contamination of the resulting polymethacrylates by isolated comonomer units cannot be prevented;
(ii) The use of dedicated nitroxides and the corresponding alkoxyamines is interesting for the preparation of pure homopolymers of methacrylic esters. Nevertheless, such a controlling system is only very efficient for methacrylate derivatives and the combination with other monomer families is not yet totally efficient.

It has to be noted that *in situ* NMP has recently shown great promise to control the polymerization of MMA at rather low temperature and is now a credible alternative.

Concerning PMMA-based materials, the all-acrylic triblock copolymer Nanostrength® developed by ARKEMA is able to give nanostructured materials owing to the segregation of incompatible PMMA and P*n*BA blocks even if the polymerization of the PMMA outer block is not well controlled. Such self-assembly could be applied for the toughening of many thermosets but also of PMMA cast sheets.

References

1. James E. Mark, *Polymer Data Handbook*, Oxford University Press, 1999.
2. A.-V. Ruzette and L. Leibler, *Nat. Mater.*, 2005, **4**, 19–31.
3. R. Jérôme, P. Teyssié, B. Vuillemin, T. Zundel and C. Zune, *J. Polym. Sci., Part A: Polym. Chem.*, 1999, **37**, 1–10.
4. W. A. Braunecker and K. Matyjaszewski, *Prog. Polym. Sci.*, 2007, **32**, 93–146.
5. J. Nicolas, Y. Guillaneuf, C. Lefay, D. Bertin, D. Gigmes and B. Charleux, *Prog. Polym. Sci.*, 2013, **38**, 63–235.
6. M. K. Georges, R. P. N. Veregin, P. M. Kazmaier and G. K. Hamer, *Macromolecules*, 1993, **26**, 2987–2988.
7. M. K. Georges, R. P. N. Veregin, P. M. Kazmaier, G. K. Hamer and M. Saban, *Macromolecules*, 1994, **27**, 7228–7229.

8. D. Benoit, V. Chaplinski, R. Braslau and C. J. Hawker, *J. Am. Chem. Soc.*, 1999, **121**, 3904–3920.
9. D. Benoit, S. Grimaldi, S. Robin, J.-P. Finet, P. Tordo and Y. Gnanou, *J. Am. Chem. Soc.*, 2000, **122**, 5929–5939.
10. C. Lefay, B. Charleux, M. Save, C. Chassenieux, O. Guerret and S. Magnet, *Polymer*, 2006, **47**, 1935–1945.
11. E. Bloch, P. L. Llewellyn, T. Phan, D. Bertin and V. Hornebecq, *Chem. Mater.*, 2009, **21**, 48–55.
12. M. Chenal, C. Boursier, Y. Guillaneuf, M. Taverna, P. Couvreur and J. Nicolas, *Polym. Chem.*, 2011, **2**, 1523–1530.
13. L. Deng, P. T. Furuta, S. Garon, J. Li, D. Kavulak, M. E. Thompson and J. M. J. Fréchet, *Chem. Mater.*, 2006, **18**, 386–395.
14. R. Bouchet, S. Maria, R. Meziane, A. Aboulaich, L. Lienafa, J.-P. Bonnet, T. N. T. Phan, D. Bertin, D. Gigmes, D. Devaux, R. Denoyel and M. Armand, *Nat. Mater.*, 2013, **12**, 452–457.
15. G. Moad, A. G. Anderson, F. Ercole, H. J. Johnson, J. Krstina, C. L. Moad, E. Rizzardo, T. H. Spurling and S. H. Thang, *ACS Symp. Ser.*, 1998, **685**, 332–360.
16. C. Burguière, M. Dourges, B. Charleux and J. P. Vairon, *Macromolecules*, 1999, **12**, 3883–3890.
17. E. Guégain, Y. Guillaneuf and J. Nicolas, *Macromol. Rapid Commun.*, 2015, **28**, 25.
18. M. Souaille and H. Fischer, *Macromolecules*, 2001, **34**, 2830–2838.
19. I. Li, B. A. Howell, K. Matyjaszewski, T. Shigemoto, P. B. Smith and D. B. Priddy, *Macromolecules*, 1995, **28**, 6692–6693.
20. A. Ohno, Y. Tsujii and T. Fukuda, *Macromolecules*, 1997, **30**, 2503–2506.
21. G. S. Ananchenko and H. Fischer, *J. Polym. Sci., Part A: Polym. Chem.*, 2001, **39**, 3604–3621.
22. G. S. Ananchenko, M. Souaille, H. Fischer, C. Le Mercier and P. Tordo, *J. Polym. Sci., Part A: Polym. Chem.*, 2002, **40**, 3264–3283.
23. M. Edeleva, S. R. A. Marque, D. Bertin, D. Gigmes, Y. Guillaneuf, S. V. Morozov and E. G. Bagryanskaya, *J. Polym. Sci., Part A: Polym. Chem.*, 2008, **46**, 6828–6842.
24. M. V. Edeleva, I. A. Kirilyuk, D. P. Zubenko, I. F. Zhurko, S. R. A. Marque, D. Gigmes, Y. Guillaneuf and E. G. Bagryanskaya, *J. Polym. Sci., Part A: Polym. Chem.*, 2009, **47**, 6579–6595.
25. D. A. Parkhomenko, E. Bagryanskaya, S. R. A. Marque and D. Siri, *Phys. Chem. Chem. Phys.*, 2013, **15**, 13862–13871.
26. D. Benoit, Thèse de l'Université de Bordeaux-I, 1997.
27. H. Fischer and M. Souaille, *Chimia*, 2001, **55**, 109–113.
28. D. Gigmes, D. Bertin, C. Lefay and Y. Guillaneuf, *Macromol. Theory Simul.*, 2009, **18**, 402–419.
29. H. Fischer, *ACS Symp. Ser.*, 2003, **854**, 10–23.
30. Y. Guillaneuf, D. Gigmes, S. R. A. Marque, P. Tordo and D. Bertin, *Macromol. Chem. Phys.*, 2006, **207**, 1278–1288.

31. F. Chauvin, P. E. Dufils, D. Gigmes, Y. Guillaneuf, S. R. A. Marque, P. Tordo and D. Bertin, *Macromolecules*, 2006, **39**, 5238–5250.
32. D. Gigmes, J. Vinas, N. Chagneux, C. Lefay, T. N. T. Phan, T. Trimaille, P.-E. Dufils, Y. Guillaneuf, G. Carrot, F. Boue and D. Bertin, *ACS Symp. Ser.*, 2009, **1024**, 245–262.
33. Y. Guillaneuf, D. Bertin, P. Castignolles and B. Charleux, *Macromolecules*, 2005, **38**, 4638–4646.
34. J. Sobek, R. Martschke and H. Fischer, *J. Am. Chem. Soc.*, 2001, **123**, 2849–2857.
35. E. Beaudoin, D. Bertin, D. Gigmes, S. R. A. Marque, D. Siri and P. Tordo, *Eur. J. Org. Chem.*, 2006, 1755–1768.
36. D. Bertin, D. Gigmes, S. R. A. Marque and P. Tordo, *Chem. Soc. Rev.*, 2011, **40**, 2189–2198.
37. A. K. Nanda and K. Matyjaszewski, *Macromolecules*, 2003, **36**, 8222–8224.
38. D. Bertin, P. E. Dufils, I. Durand, D. Gigmes, B. Giovanetti, Y. Guillaneuf, S. R. A. Marque, T. Phan and P. Tordo, *Macromol. Chem. Phys.*, 2008, **209**, 220–224.
39. R. McHale, F. Aldabbagh and P. B. Zetterlund, *J. Polym. Sci., Part A: Polym. Chem.*, 2007, **45**, 2194–2203.
40. C. Dire, J. Belleney, J. Nicolas, D. Bertin, S. Magnet and B. Charleux, *J. Polym. Sci., Part A: Polym. Chem.*, 2008, **46**, 6333–6345.
41. M. Edeleva, S. R. A. Marque, K. Kabytaev, Y. Guillaneuf, D. Gigmes and E. Bagryanskaya, *J. Polym. Sci., Part A: Polym. Chem.*, 2013, **51**, 1323–1336.
42. B. Charleux, J. Nicolas and O. Guerret, *Macromolecules*, 2005, **38**, 5485–5492.
43. O. E. Ansong, S. Jansen, Y. Wei, G. Pomrink, S. Li and A. Patel, *Polym. Int.*, 2008, **57**, 863–871.
44. O. E. Ansong, S. Jansen, Y. Wei, G. Pomrink, H. Lu, A. Patel and S. X. Li, *Polym. Int.*, 2009, **58**, 54–65.
45. M. Steenbock, M. Klapper, K. Mullen, N. Pinhal and M. Hubrich, *Acta Polym.*, 1996, **47**, 276.
46. Y. Guillaneuf, D. Gigmes, S. R. A. Marque, P. Astolfi, L. Greci, P. Tordo and D. Bertin, *Macromolecules*, 2007, **40**, 3108–3114.
47. C. Berti, M. Colonna, L. Greci and L. Marchetti, *Tetrahedron*, 1975, **31**, 1745.
48. M. Edeleva, S. R. A. Marque, D. Bertin, D. Gigmes, Y. Guillaneuf and E. Bagryanskaya, *Polymers*, 2010, **2**, 364–377.
49. D. Bertin, D. Gigmes, S. Marque, S. Milardo, J. Peri and P. Tordo, *Collect. Czech. Chem. Commun.*, 2004, **69**, 2223–2238.
50. D. Bertin, D. Gigmes, R. Maurin and P. Tordo, *J. Polym. Sci., Part A: Polym.Chem.*, 2004, **42**, 3504–3515.
51. D. Gigmes, A. Gaudel-Siri, S. R. A. Marque, D. Bertin, P. Tordo, P. Astolfi, L. Greci and C. Rizzoli, *Helv. Chim. Acta*, 2006, **89**, 2312–2326.

52. P. Astolfi, L. Greci, P. Stipa, C. Rizzoli, C. Ysacco, M. Rollet, L. Autissier, A. Tardy, Y. Guillaneuf and D. Gigmes, *Polym. Chem.*, 2013, **4**, 3694–3704.
53. A. C. Greene and R. B. Grubbs, *Macromolecules*, 2009, **42**, 4388–4390.
54. A. C. Greene and R. B. Grubbs, *Macromolecules*, 2010, **43**, 10320–10325.
55. T. Fukuyama, Y. Kajihara, I. Ryu and A. Studer, *Synth.-Stuttgart*, 2012, **44**, 2555–2559.
56. M. V. Edeleva, D. A. Parkhomenko, D. A. Morozov, S. A. Dobrynin, D. G. Trofimov, B. Kanagatov, I. A. Kirilyuk and E. G. Bagryanskaya, *J. Polym. Sci., Part A: Polym. Chem.*, 2014, **52**, 929–943.
57. C. J. Hawker, E. Elce, J. L. Dao, W. Volksen, T. P. Russell and G. G. Barclay, *Macromolecules*, 1996, **29**, 2686–2688.
58. B. Charleux, J. Nicolas and O. Guerret, *Macromolecules*, 2005, **38**, 5485–5492.
59. J. Nicolas, L. Mueller, C. Dire, K. Matyjaszewski and B. Charleux, *Macromolecules*, 2009, **42**, 4470–4478.
60. J. Nicolas, C. Dire, L. Mueller, J. Belleney, B. Charleux, S. R. A. Marque, D. Bertin, S. Magnet and L. Couvreur, *Macromolecules*, 2006, **39**, 8274–8282.
61. J. Nicolas, S. Brusseau and B. Charleux, *J. Polym. Sci., Part A: Polym. Chem*, 2010, **48**, 34–47.
62. M. Chenal, S. Mura, C. Marchal, D. Gigmes, B. Charleux, E. Fattal, P. Couvreur and J. Nicolas, *Macromolecules*, 2010, **43**, 9291–9303.
63. T. N. T. Phan, S. Maiez-Tribut, J.-P. Pascault, A. Bonnet, P. Gerard, O. Guerret and D. Bertin, *Macromolecules*, 2007, **40**, 4516–4523.
64. B. H. Lessard, Y. Guillaneuf, M. Mathew, K. Liang, J.-L. Clement, D. Gigmes, R. A. Hutchinson and M. Marić, *Macromolecules*, 2013, **46**, 805–813.
65. B. Lessard, E. J. Y. Ling, M. S. T. Morin and M. Marić, *J. Polym. Sci., Part A: Polym. Chem*, 2011, **49**, 1033–1045.
66. Y. Miura, N. Nakamura, I. Taniguchi and A. Ichikawa, *Polymer*, 2003, **44**, 3461–3467.
67. S. Brusseau, J. Belleney, S. Magnet, L. Couvreur and B. Charleux, *Polym. Chem.*, 2010, **1**, 720–729.
68. C. Dire, J. Nicolas, S. Brusseau, B. Charleux, S. Magnet and L. Couvreur, *ACS Symp. Ser.*, 2009, **1024**, 303–318.
69. X. Qiao, M. Lansalot, E. Bourgeat-Lami and B. Charleux, *Macromolecules*, 2013, **46**, 4285–4295.
70. C. Zhang, B. Lessard and M. Maric, *Macromol. React. Eng.*, 2010, **4**, 415–423.
71. B. Lessard and M. Marić, *J. Polym. Sci., Part A: Polym. Chem*, 2011, **49**, 5270–5283.
72. C. Zhang and M. Maric, *J. Polym. Sci., Part A: Polym. Chem*, 2012, **50**, 4341–4357.
73. C. R. Becer, K. Kokado, C. Weber, A. Can, Y. Chujo and U. S. Schubert, *J. Polym. Sci., Part A: Polym. Chem*, 2010, **48**, 1278–1286.

74. B. H. Lessard, E. J. Y. Ling and M. Marić, *Macromolecules*, 2012, **45**, 1879–1891.
75. J. Nicolas, P. Couvreur and B. Charleux, *Macromolecules*, 2008, **41**, 3758–3761.
76. S. R. S. Ting, E.-H. Min, P. Escale, M. Save, L. Billon and M. H. Stenzel, *Macromolecules*, 2009, **42**, 9422–9434.
77. D. Benoit, E. Harth, P. Fox, R. M. Waymouth and C. J. Hawker, *Macromolecules*, 2000, **33**, 363–370.
78. P.-E. Dufils, Thèse de l'Université d'Aix-Marseille I, 2005.
79. J. Nicolas, PhD dissertation, Université Pierre et Marie Curie, France, 2005.
80. V. Sciannamea, R. Jerome and C. Detrembleur, *Chem. Rev.*, 2008, **108**, 1104–1126.
81. K. Matyjaszewski, S. Gaynor, D. Greszta, D. Mardare and T. Shigemoto, *Macromol. Symp.*, 1995, **98**, 73–89.
82. C. Detrembleur, P. Teyssie and R. Jerome, *Macromolecules*, 2002, **35**, 1611–1621.
83. C. Detrembleur, M. Claes and R. Jerome, *ACS Symp. Ser.*, 2003, **854**, 496–518.
84. C. Detrembleur, C. Jerome, J. De Winter, P. Gerbaux, J. L. Clement, Y. Guillaneuf and D. Gigmes, *Polym. Chem.*, 2014, **5**, 335–340.
85. T. Junkers, E. H. H. Wong, M. H. Stenzel and C. Barner-Kowollik, *Macromolecules*, 2009, **42**, 5027–5035.
86. E. H. H. Wong, M. H. Stenzel, T. Junkers and C. Barner-Kowollik, *J. Polym. Sci., Part A: Polym. Chem.*, 2009, **47**, 1098–1107.
87. E. H. H. Wong, T. Junkers and C. Barner-Kowollik, *J. Polym. Sci., Part A: Polym. Chem.*, 2008, **46**, 7273–7279.
88. E. H. H. Wong, T. Junkers and C. Barner-Kowollik, *Polym.Chem.*, 2011, **2**, 1008–1017.
89. C. Dommanget, C. Boisson, B. Charleux, F. D'Agosto, V. Monteil, F. Boisson, T. Junkers, C. Barner-Kowollik, Y. Guillaneuf and D. Gigmes, *Macromolecules*, 2013, **46**, 29–36.
90. T. Junkers, L. Zang, E. H. H. Wong, N. Dingenouts and C. Barner-Kowollik, *J. Polym. Sci., Part A: Polym. Chem.*, 2011, **49**, 4841–4850.
91. L. Zang, E. H. H. Wong, C. Barner-Kowollik and T. Junkers, *Polymer*, 2010, **51**, 3821–3825.
92. P. Gerard, L. Couvreur, S. Magnet, J. Ness and S. Schmidt, in *Controlled/Living Radical Polymerization: Progress in Raft, Dt, Nmp & Omrp*, ed. K. Matyjaszewsik, 2009, vol. 1024, pp. 361–373.
93. A.-V. Ruzette, S. Tence-Girault, L. Leibler, F. Chauvin, D. Bertin, O. Guerret and P. Gerard, *Macromolecules*, 2006, **39**, 5804–5814.
94. N. A. Lynd and M. A. Hillmyer, *Macromolecules*, 2005, **38**, 8803–8810.
95. D. Bendejacq, V. Ponsinet, M. Joanicot, Y. L. Loo and R. A. Register, *Macromolecules*, 2002, **35**, 6645–6649.
96. F. Court and T. Hashimoto, *Macromolecules*, 2001, **34**, 2536–2545.
97. S. W. Sides and G. H. Fredrickson, *J. Chem. Phys.*, 2004, **121**, 4974–4986.

98. J. Listak, W. Jakubowski, L. Mueller, A. Plichta, K. Matyjaszewski and M. R. Bockstaller, *Macromolecules*, 2008, **41**, 5919–5927.
99. J. Nicolas, A.-V. Ruzette, C. Farcet, P. Gerard, S. Magnet and B. Charleux, *Polymer*, 2007, **48**, 7029–7040.
100. Q. Guo, *Thermosets: Structure, Properties and Applications*, Woodhead Publishing, 2012.
101. M. T. Bashar, U. Sundararaj and P. Mertiny, *Polym. Eng. Sci.*, 2014, **54**, 1047–1055.
102. S. Maiez-Tribut, J. P. Pascault, E. R. Soule, J. Borrajo and R. J. J. Williams, *Macromolecules*, 2007, **40**, 1268–1273.
103. P. Gerard, N. Passade Boupat, T. Fine, L. Gervat and J. P. Pascault, *Macromol. Symp.*, 2007, **256**, 55–64.
104. E. Serrano, P. Gerard, F. Lortie, J. P. Pascault and D. Portinha, *Macromol. Mater. Eng.*, 2008, **293**, 820–827.
105. S. Denneulin, P. Viot, F. Leonardi and J. L. Lataillade, *Compos. Struct.*, 2012, **94**, 1471–1481.
106. M. T. Bashar, U. Sundararaj and P. Mertiny, *Polym. Test.*, 2013, **32**, 402–412.
107. R. M. Boumbimba, C. Froustey, P. Viot, J. M. Olive, F. Leonardi, P. Gerard and R. Inoubli, *Compos. Struct.*, 2014, **116**, 414–422.
108. K. R. Ramakrishnan, S. Guerard, P. Viot and K. Shankar, *Compos. Struct.*, 2014, **110**, 174–182.
109. A. Rubin, N. Rebutin, P. Gerard and C. Gauthier, *Mater. Lett.*, 2014, **135**, 184–187.

CHAPTER 8

Complex Macromolecular Architectures Prepared by NMP

DAVID RAYEROUX, GWÉNAËLLE BONZI, SÉBASTIEN MARIA AND DIDIER GIGMES*

Aix Marseille Université, CNRS, Institut de Chimie Radicalaire, UMR7273, 13397 Marseille cedex 20, France
*Email: didier.gigmes@univ-amu.fr

8.1 Introduction

Nowadays, synthetic polymers have prominent roles as materials in multiple sectors such as electronics, optics and medicine, among others. Hence, the impetus to conceive innovative polymer-based materials for specific applications is strong. The blossoming of polymer science through scientific research and discoveries has beneficially fostered a better understanding of the various issues behind the design of such materials. One of the major achievements of polymer science deals with the development of the reversible deactivation radical polymerization (RDRP) technique. Resting upon the settlement of a dynamic equilibrium between active and dormant species, RDRP provides appreciable tuning of macromolecular size, composition, structure and architecture. Among the most commonly studied techniques, such as atom transfer radical polymerization (ATRP)[1–3] and reversible addition-fragmentation chain transfer (RAFT),[4–6] nitroxide-mediated polymerization (NMP)[7–10] is of laudable simplicity. The achievements in the synthesis of nitroxides and alkoxyamines and their application to NMP alongside with the exhaustive exploration of the kinetics of NMP of the large families of monomers have been triggering utter zeal dedicated to

RSC Polymer Chemistry Series No. 19
Nitroxide Mediated Polymerization: From Fundamentals to Applications in Materials Science
Edited by Didier Gigmes

Published by the Royal Society of Chemistry, www.rsc.org

the fashion in a smart pattern of complex macromolecular architectures. Herein, a comprehensive though non-exhaustive overview of those accessible architectures provided by NMP is exposed. More particularly, linear and non-linear topologies afforded by NMP are highlighted.

8.2 General Considerations

In the context of complex macromolecular structures (Figure 8.1), linear topologies include diblock, triblock, random and gradient copolymers. Non-linear topologies group grafted, hyperbranched and star copolymers.

From a general point of view regarding complex macromolecular structures using exclusively or in part NMP, numerous strategies[11,12] may be employed among which successive NMP steps include introduction of an alkoxyamine moiety on a polymer chain that is further used as a macroinitiator, combination of polymerization techniques using a reactive functional group on a polymer chain-end for instance, using a dual initiator bearing an alkoxyamine function and another initiator, and macromolecular coupling where one block is previously made by NMP. Figure 8.2 gives an idea of the common strategies resorted to in diblock architecture synthesis.

The combination of NMP with other polymerization processes involves, for example, the use of a reactive functional group previously introduced at the alpha chain-end or a chemical modification of the nitroxide group to

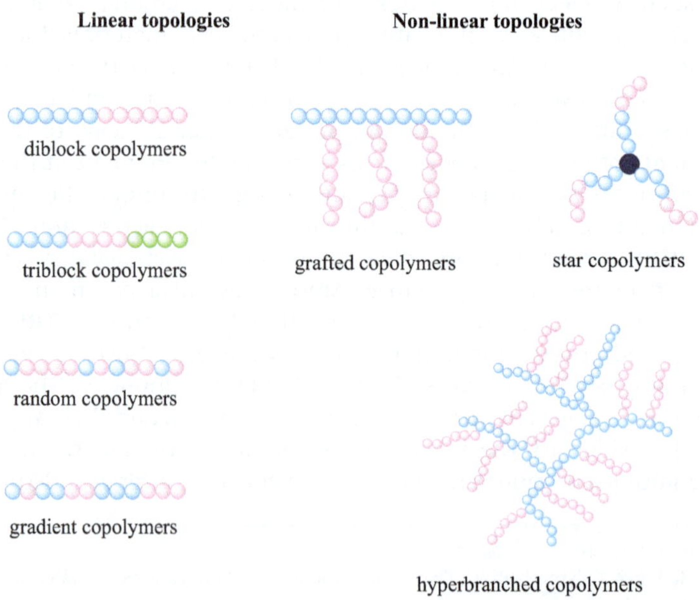

Figure 8.1 Linear and non-linear topologies in complex macromolecular architectures.

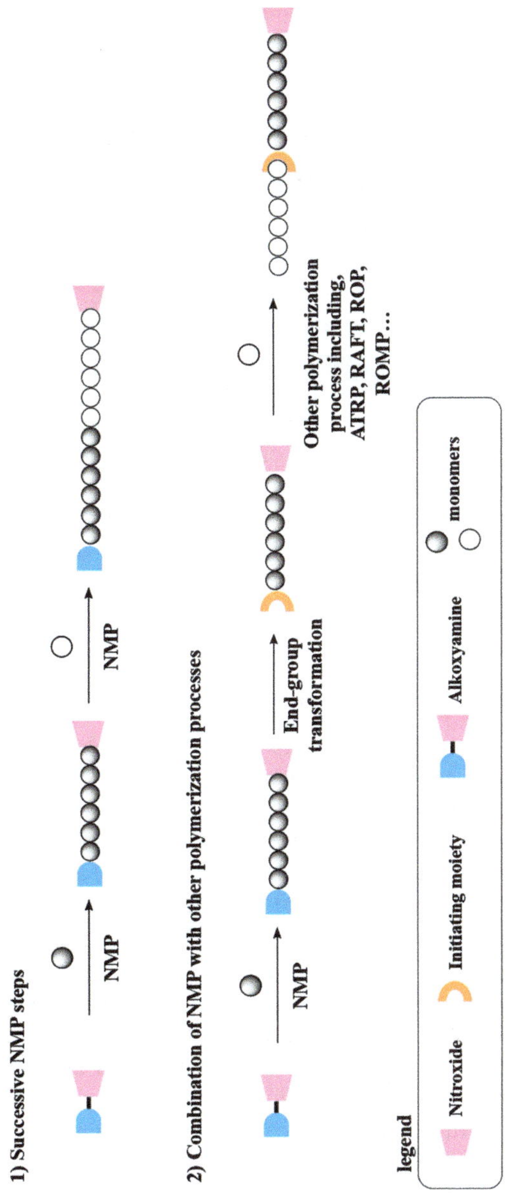

Figure 8.2 Examples of strategies employed in synthesis of complex macromolecular structures involving an NMP step.

transform it into a reactive initiating moiety. A small section of the chapter will later be assigned to the synthesis of complex linear macromolecular structures by the combination of NMP with other polymerization processes. In the following section, linear topologies of copolymers synthesized through either bimolecular or unimolecular NMP will be covered, with reference to the different families of monomers that make them up.

8.3 Linear Topologies by Unimolecular or Bimolecular Initiation

More commonly, linear topologies comprise AB-type copolymers as well as ABA-type and ABC-type copolymers. Various approaches leading to the synthesis of these linear structures have been adopted. More frequently, sequential monomer addition strategy through unimolecular or bimolecular initiation systems is chosen. Through this process, the polymerization of the first monomer (monomer A) is conducted, generating a macro-initiator meant for the successive chain extension with the second monomer (monomer B) and the third monomer (monomer A or C), depending on the case.

Along with the mastering of NMP of monomers of different chemical natures englobing styrenics,[13] acrylates,[14] methacrylates,[15] acrylamides,[16] acrylonitriles,[17] and dienes,[18] a large library of linear copolymers is accessible. In the following sections, the different attainable linear structures will be discussed through the categorization of their respective monomer units.

8.3.1 Styrene-based Copolymers

Styrene represents the simplest model involved in polymerization by NMP, catering for well-defined, highly-controlled and functional polymers, therefore ascertaining proper copolymerization with other monomers.[10] Adopting acyclic nitroxides like di-*tert*-butyl-nitroxide (DBNO),[19] 2,2,5-tri-methyl-4-phenyl-3-azahexane-3-nitroxide (TIPNO)[20,21] and *N*-*tert*-butyl-*N*-1-diethylphosphono-(2,2-dimethylpropyl)] (SG1)[22,23] or sterically-hindered (2,2,6,6-tetramethyl-1-piperidinyloxyl) (TEMPO) derivatives[24] provides for high polymerization rates at temperatures below 100 °C. Using these same nitroxides, equally good results are obtained for the NMP of styrene derivatives (substituted styrene) as in the case of *para*-substituted styrenes, which have behaviors similar to styrene.[25,26] This opens up remarkable possibilities for the combination of the different derivatives. For instance, through bimolecular initiation, Bertin *et al.*[27] conceived diblock copolymers based on chloromethylstyrene (CMSt) and styrene (St). Using benzoyl peroxide (BPO) as the initiator and TEMPO as the nitroxide, the synthesis of polychloromethylstyrene (PCMSt) macro-alkoxyamine of low molecular weight was carried out, thereafter enabling chain extension with styrene (Scheme 8.1). Though the successful synthesis was reported, the PCMSt-*b*-PSt copolymer presented a relatively high dispersity of 1.80 and comprised a marked

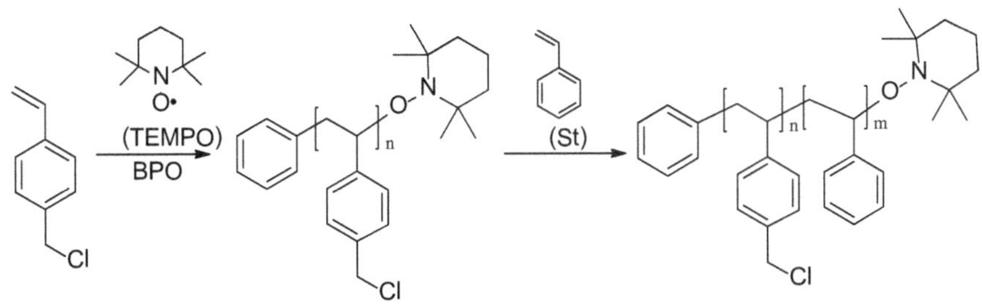

Scheme 8.1 Synthesis of PCMSt-*b*-PSt diblock copolymers.

proportion of residual non-blocking PCMSt homopolymer. Following the same reaction pathway, Yoshida *et al.*[28] designed diblock copolymers bearing individual 2-, 3-, and 4-chlorostyrenes (2-, 3-, 4-ClSt) and styrene moieties using 4-methoxy-2,2,6,6-tetramethylpiperidine-1-oxyl (MTEMPO) instead of TEMPO. The polymers obtained from 2- and 3-ClSt quantitatively acted as macro-initiators for the subsequent NMP of styrene giving the corresponding block copolymers, this being unsuccessful for poly(4-ClSt) macro-initiators. This was associated with the decreasing thermal stability of MTEMPO chain-end poly(ClSt)s according to the position of the Cl substituent (2- > 3- > 4-ClSt).

Moreover, the same group[29] was involved in the synthesis of diblock copolymers of *p*-bromostyrene (BrSt) and styrene with BPO as initiator in the presence of MTEMPO. Despite the fact that PBrSt-*b*-PSt diblock copolymers were obtained from the successive polymerization with styrene added onto the pBrSt macro-initiator, a significant proportion (around 10%) of dead PBrSt was reported. In addition, block copolymers of the same nature, but prepared in the reverse order, were elaborated. Indeed, these PSt-*b*-PBrSt diblock copolymers exhibited a relatively higher dispersity index value (1.34) as compared to the one of the former PBrSt-*b*-PSt diblock copolymers (1.14), suggesting lower control in the second strategy.

Diblock and triblock copolymers of styrene and N-(*p*-vinylbenzyl)phthalimide (VBP) were generated by Mariani *et al.*[30] as shown in Scheme 8.2.

In this strategy, in contrast with high molecular weight PSt ($DP_n > 100$), low molecular weight PSt ($DP_n = 38$) served as efficient macro-initiators for copolymerization with VBP to give rise to pure PSt-*b*-PVPB diblock copolymers, free from PVPB homopolymers formed as a result of thermal initiation polymerization. Ultimately, through extension with styrene, PSt-*b*-PVPB-*b*-PSt triblock copolymers were produced. Finally, the diblock and triblock copolymers were modified using hydrazine in order to form functional *p*-aminomethylstyrene (AMS) block copolymers (PSt-*b*-PAMS and PSt-*b*-PAMS-*b*-PSt). Later on, another group designed similar low-dispersity PSt-*b*-PAMS amphiphilic diblock copolymers, which were further treated with hydrogen chloride to yield cationic amphiphilic diblock copolymers with potential application as an emulsifier in emulsion polymerization.[31]

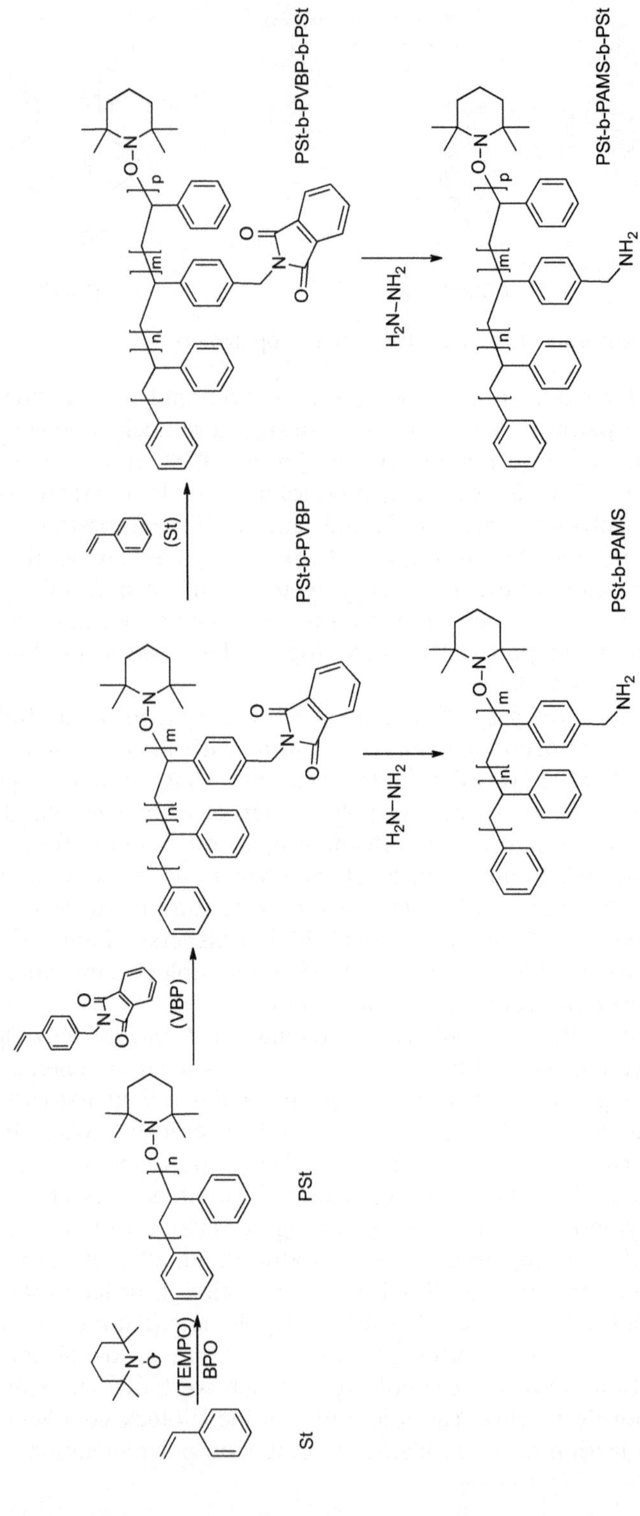

Scheme 8.2 Synthesis of styrene and amino-functional diblock copolymers.

Fukukawa *et al.*[32] showed particular interest in the synthesis and characterization of copolymers carrying St or 4-acetoxystyrene (AcOSt) and different silicon-containing styrenic units meant for direct patterning of nanostructures on a desired substrate using 2,2,5-trimethyl-3-(1-phenylethoxy)-4-phenyl-3-azahexane (α-H unimer) (Scheme 8.3). Through the optimization of parameters like polymerization time and temperature, solvent polarity, use of rate accelerants, the control of the polymerization with narrow dispersities below 1.25 and yields as high as 80% were achieved.

For their part, Okamura *et al.*[33] claimed the synthesis of random and block copolymers of St and neopentyl *p*-styrenesulfonate (nSS) of low polydispersity with AIBN/SG1 as the initiating system, followed by the ultimate transformation of sulfonate ester groups to sulfonic acid, hence affording polystyrene-*b*-poly(sulfonic acid) (PSt-*b*-PSSH) amphiphilic diblock copolymers (Scheme 8.4). The synthesis was also carried out in the reverse order, *i.e.* starting with the NMP of nSS to generate a macro-initiator allowing chain extension with styrene.

Other groups disclosed the synthesis of gradient and diblock copolymers based on styrene and 4-acetoxystyrene[34–36] using nitroxides such as TEMPO or DBNO (Scheme 8.5), followed by hydrolysis to produce polymers bearing polystyrene and poly(4-hydroxystyrene) (HS).

8.3.2 Acrylate-based Copolymers

The use of TEMPO hinders the synthesis of well-defined acrylate-based polymers and therefore copolymers, this being explained by the strong alkoxyamine C–ON bond formation inherent of the NMP of acrylates, for example *n*-butyl acrylate.[37–40] Cyclic nitroxides like PROXYL[40] and sterically-hindered TEMPO analogues[24,41] gave comparatively better results in terms of control and livingness of the polymer chains. Above all, acyclic nitroxides such as TIPNO,[20] SG1[23,42,43] and TIPNO derivatives[44,45] offered the best results. Hence, the synthesis of acrylate-based copolymers was directed towards the use of the classes of nitroxides mentioned above. Cameron *et al.*[46] polymerized BuA in the presence of an alkoxyamine initiator derived from the PROXYL nitroxide 2,2′,5-trimethyl-5′-phenylpyrrolidinyl-1-oxyl. Though the control was poor, the PBuA was able to act as a macro-initiator for the polymerization of styrene, giving PBuA-*b*-PSt diblock copolymer with a growth in M_n and a reduction of polydispersity with conversion. As for Nicolas *et al.*,[47] they successfully fulfilled the nitroxide-mediated emulsion polymerizations of BuA and styrene using a novel water-soluble SG1-based alkoxyamine as initiator, generating stable latex particles of controlled molar masses. This same group[48] also reported in another study the conception of triblock copolymer particle suspension by nitroxide-mediated emulsion polymerization. In this case, difunctional alkoxyamines based on the nitroxide SG1 were used as initiators.

Concerning copolymers synthesized by NMP from PBuA macro-alkoxyamine, diblock copolymers comprising 2-hydroxyethyl acrylate[49] and

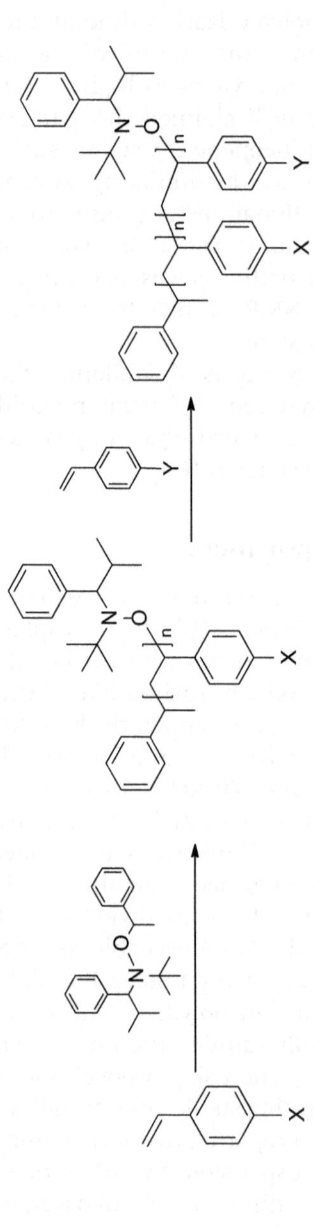

(X or Y = -H or -Si(CH$_3$)$_2$Si(CH$_3$)$_3$ or -CH(Si(CH$_3$)$_3$)$_2$ or Si(CH$_3$)$_2$(CH$_2$)$_2$OSi(CH$_3$)$_3$ or -OC(O)CH$_3$)

Scheme 8.3 Synthesis of silicon-containing styrenic diblock copolymers.

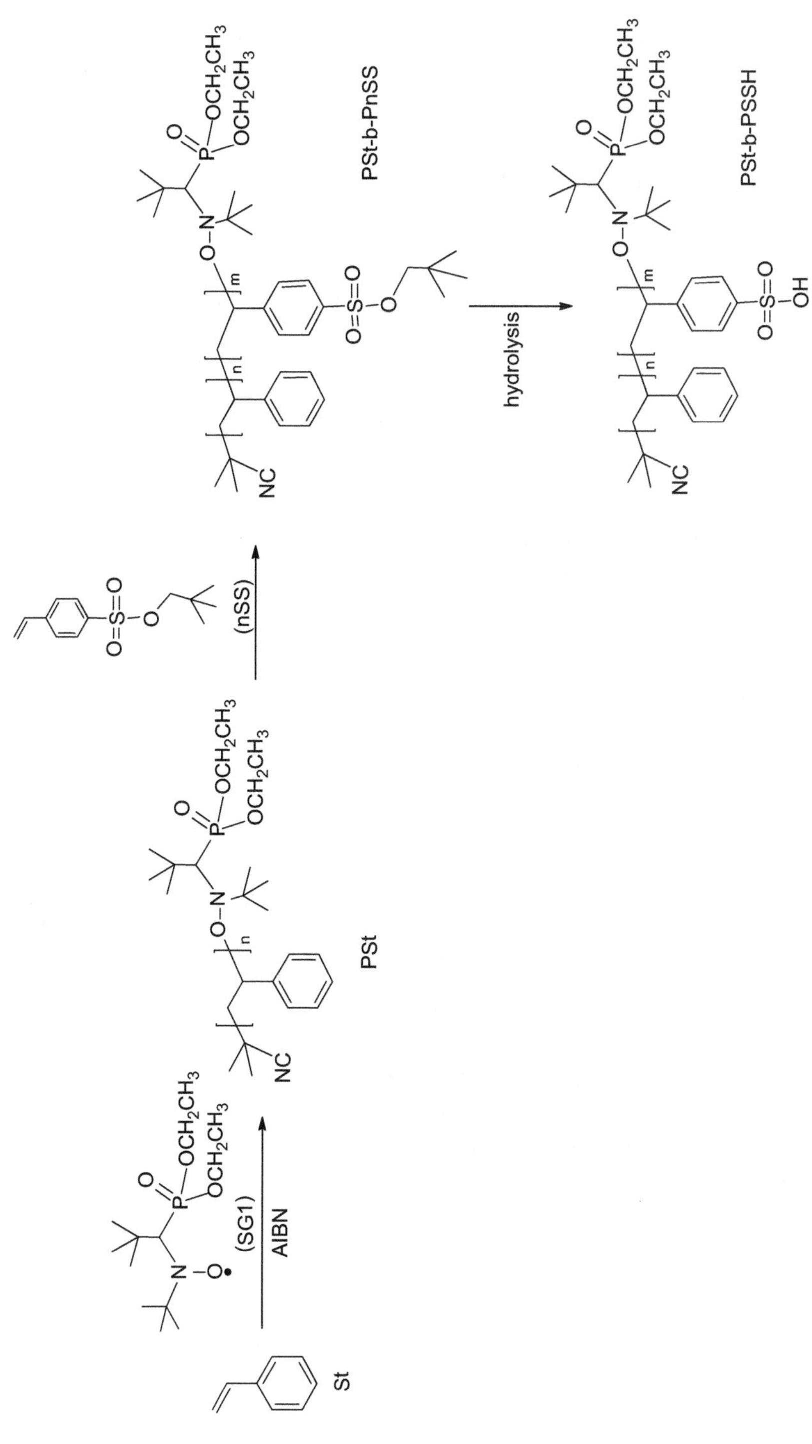

Scheme 8.4 Synthesis of PSt-*b*-PSSH amphiphilic diblock copolymers.

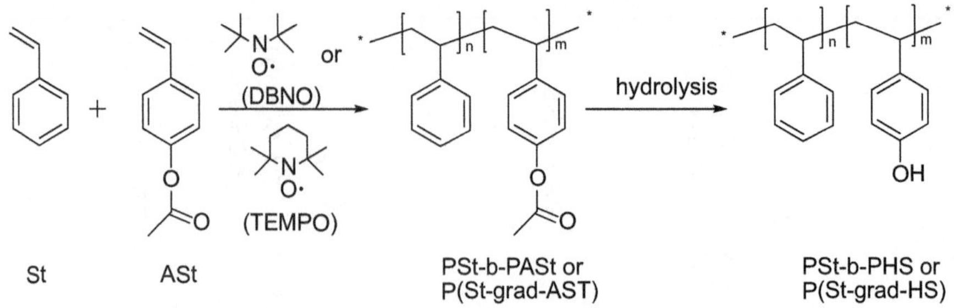

Scheme 8.5 Synthesis of amphiphilic diblock copolymers based on styrene and 4-acetoxystyrene.

2-(dimethylamino)ethyl acrylate[50] as the other segment were reported using SG1 as the nitroxide.

NMP of *tert*-butyl acrylate (tBuA) showed similar results to that of BuA using sterically hindered TEMPO or SG1. Diblock copolymers with appreciable polymerization control and living character were obtained. For example, the synthesis of diblock copolymers PSt-*b*-PtBuA was accomplished by Kuo *et al.*[51] using 4-oxo-TEMPO-capped polystyrene as the macroinitiator.

Among other acrylate-based copolymers noted in the literature, those based on poly(ethylene glycol) acrylate (PEGA)[52] are noteworthy. Chain extension *via* an initiator system SG1/(N-(2-methylpropyl)-*N*-(1-diethylphosphono-2,2-dimethylpropyl)-*O*-(2-carboxylprop-2-yl) hydroxylamine) (MAMA-SG1) of a comb-like poly(PEGA) (PPEGA) with styrene afforded PPEGA-*b*-PSt block copolymer with good architectural control and bearing a living character.

Recently, Handké *et al.*[53] showed that the NMP of *N*-acryloxysuccinimide (NAS) granted access to well-controlled polymers in the presence of MAMA-SG1 alkoxyamine as initiator with 5% free SG1 so as to contain a high propagation of NAS. In the same contribution, poly(*N*-acryloxysuccinimide-*co*-*N*-vinylpyrrolidone) [P(NAS-*co*-NVP)] random diblock copolymers were conceived using the same initiator system. PLA-*b*-P(NAS-*co*-NVP) block copolymer was then designed from NAS/NVP copolymerization with SG1-functionalized poly(D,L-lactide) (PLA-SG1) macro-alkoxyamine as the initiator. The copolymer found applications as a surface modifier for PLA diafiltration and drug delivery.

8.3.3 Methacrylate-based Copolymers

The syntheses of methacrylate-based copolymers represent a real challenge due to the ubiquitous β-hydrogen transfer from the propagating macroradicals to free nitroxides (*e.g.* TEMPO and other cyclic ones), drastically impairing the livingness of the chain-end functionality, leading to the formation of dead chains, hence impeding proper polymerization control.[54–57]

This side reaction was extensively investigated by Bagryanskaya *et al.*,[58,59] who showed its nature to be an intermolecular hydrogen transfer rather than an intramolecular hydrogen elimination and demonstrated its direct relationship with the nitroxide structure. With the omnipresence of this side reaction, the nitroxide is transformed to the corresponding hydroxylamine. However, this phenomenon did not deter several groups from synthesizing block copolymers, provided that the first block was well-controlled as in the case of the NMP of styrene followed by the polymerization with the methacrylate monomer. For instance, Lokaj *et al.*[60] brought their contribution through the synthesis of polystyrene-*b*-poly(2-(dimethylamino)ethyl methacrylate) amphiphilic diblock copolymers *via* TEMPO-mediated polymerization. Nonetheless, the authors reported that monomer conversion was somehow limited and did not increase with increased reaction times. Burguiere *et al.*[55] were implied in the synthesis and characterization of polystyrene-*b*-poly(*n*-butyl methacrylate) diblock copolymers using a TEMPO-capped polystyrene macro-initiator. MALDI-TOF MS analysis of the latter copolymer formed revealed a terminal unsaturation associated with the intermolecular hydrogen transfer side reaction described above. Although attempts were made in view of controlling the polymerization of methacrylates and its copolymerization, the issue remains a whole. Lately, prospects of better architectural control of the NMP of MMA were disclosed when sulfuric acid was used as an accelerant in the presence of several nitroxides, but further confirmation of the living character of the polymers formed is awaited.

8.3.4 Acrylamide-based Copolymers

SG1 nitroxide provides a satisfactory control in the NMP of acrylamide (AAm) and allows to generate associated diblock copolymers by chain extension.[61,62] Only a few works dedicated to associated copolymers of *N*-substituted acrylamides like *N*-isopropylacrylamide (NIPAAm) have been reported so far, this being largely explained by the shortcomings in terms of polymerization control in using classical nitroxides like TIPNO or TEMPO and also because of characterization issues encountered in order to evaluate their living character.[63] Nevertheless, though having high polydispersites, copolymers of NIPAAm and *N*-hydroxysuccinimide esters of methacrylic acid[64] using TIPNO-styryl based alkoxyamine were accounted. Still of high polydispersity values, poly(phenylene-ethynylene)-*b*-poly(*N*-isopropylacrylamide) (PPE-*b*-PNIPAAm) diblock copolymers were disclosed by Swager *et al.*[65] Poly(phenylene – ethynylene) (PPE) was end-capped with a TEMPO or DBNO followed by subsequent NMP of NIPAAm to yield the copolymer. Gibbons *et al.*[66] performed, for their part, the copolymerization of *N*-isopropylacrylamide (NIPAAm) and *N*-*tert*-butylacrylamide (tBAAm) *via* NMP with an AIBN/SG1 bimolecular system or a unimolecular polystyrene-SG1 macro-initiator in the presence of excess free SG1. Block copolymers of poly(*N*-isopropylacrylamide-stat-*N*-*tert*-butylacrylamide) [P(NIPAAm-stat-tBAAm)] with styrene were obtained by chain extension of either poly(NIPAAm-stat-tBAAm)-SG1 with styrene or PSt–SG1 with NIPAAm/tBAAm.

Some reports deal with disubstituted acrylamide-based copolymers, especially those of *N,N*-dimethylacrylamide (DMAAm) and *N,N*-ethylacrylamide (DEAAm). Concerning the former, in a publication by Fischer *et al.*,[67] amphiphilic block copolymers of DMAAm were synthesized using TEMPO-terminated poly(4-vinylpyridine) as a macro-initiator. In another paper,[68] the same group acknowledged the synthesis of diblock copolymers of *N*-hexadecyl-4-vinylpyridinium bromide and *N,N*-dimethylacrylamide through NMP using TEMPO as the nitroxide source. An example of DEAAm-based copolymer is one designed from poly(sodium acrylate)-based alkoxyamine macro-initiator to give rise to poly[(*N,N'*-diethylacrylamide)-*co*-(*N,N'*-methylenebisacrylamide)] amphiphilic copolymers in nitroxide-mediated controlled free-radical aqueous dispersion polymerization.

8.3.5 Acrylonitrile and Acrylic Acid-based Copolymers

The synthesis of block copolymers containing pure poly(acrylonitrile) segments in the backbone constitutes some ordeals; poor solubility in typical solvents or monomers and low blocking efficiency. Hence, few articles deal with the synthesis of block copolymers comprising poly(acrylonitrile) (PAN) segments. The great majority of the works treated the random copolymerization of acrylonitrile with styrene or butadiene accompanied by block copolymerization with other monomers, as in the case of polystyrene-*b*-poly(styrene-*co*-acrylonitrile) and poly(styrene-*co*-acrylonitrile)-*b*-polybutadiene-*b*-poly(styrene-*co*-acrylonitrile).[69,70] The preparation of block copolymers with acrylonitrile (AN) and *n*-butyl acrylate (BuA)[71] was carried out where chain extension from P(BuA)-SG1 macro-initiator offered better results regarding livingness as compared to PAN-SG1.

In the case of acrylic acid monomers, there was good control over polymerization and chain-end functionality[72,73] as far as moderate molar mass polymers are concerned when initiated by SG1-based alkoxyamine MONAMS (CH_3–OC(=O)–CH(CH_3)–SG1). Access to amphiphilic diblock copolymers, for example those consisting of poly(acrylic acid) and polystyrene as hydrophilic and hydrophobic moieties from SG1-capped polystyrene,[74] was granted.

8.3.6 Diene-based Copolymers

The NMP of conjugated diene monomers with TEMPO was pioneered by Georges *et al.*[38] and by Boutevin *et al.*[75,76] in the case of isoprene (IP) and butadiene (BD), respectively. In one particular study by Hong *et al.*,[77] side reactions, namely Diels–Alder dimerization and thermal auto-initiation, were reported to be present when TEMPO or 4-oxo-TEMPO was used with monomers of this nature, hence posing problems in subsequent block copolymerization. Recourse to TIPNO was urged fostering the synthesis of well-defined diene-based copolymers. Worth mentioning are block and random copolymers based on associations of 1,3-dienes with functionalized vinyl monomers, including styrene, acrylate and methacrylate derivatives using TIPNO.[78–82]

8.4 Linear Topologies by Combination of NMP with Other Polymerization Methods

In the scope of the synthesis of complex linear topologies, NMP may be combined with other polymerization methods like ionic or coordination polymerizations as well as other controlled radical polymerizations, such as ATRP or RAFT or even ring-opening polymerization (ROP). At moderate temperatures (typically below 80 °C), the nitroxide functionality is usually stable, therefore permitting this combination. In the case of the combination of NMP with conventional free radical polymerization (FRP), several alkoxyamines bearing an azoic group were synthesized.[83,84] An example of such an alkoxyamine is given in Scheme 8.6. This alkoxyamine was adopted, for example, in the synthesis of PSt-*b*-PMMA diblock copolymers overcoming the difficulties inherent of the NMP of MMA.

ATRP and NMP may be combined by using, for instance, a dual-headed initiator like an α-haloester-functionalized alkoxyamine. As an illustration, Tunca *et al.*[85] produced PMMA-*b*-PSt copolymers making use of such an alkoxyamine bearing a crown ether functionality as the initiator. The latter was involved in ATRP of MMA in the presence of the catalytic system CuCl/pentamethyldiethylenetriamine. The yielded PMMA was then implicated in the NMP of styrene to provide for well-defined PMMA-*b*-PSt with a crown ether per chain.

NMP may also be combined with ROP of specific monomers to produce the associated copolymers. As an example, heterobifunctional alkoxyamines carrying a hydroxyl group was used in the ROP of ε-caprolactone (εCL) and NMP of styrene, starting with ROP then NMP or *vice versa* or even in a one-step approach in which concomitant polymerization of εCL and styrene proceeds.

8.5 Non-linear Topologies

The creation of innovative complex non-linear topologies is a constant challenge in the field of polymers. Significant advances in the synthesis of these materials comprise grafted and comb polymers, star and hyperbranched copolymers or cyclic topologies. Though non-exhaustive, the following sections address the synthesis by NMP of these various families of non-linear polymers.

8.5.1 Grafted Copolymers

Grafted polymers are attractive complex macromolecular structures due to their modulable physicochemical properties and they may find potential applications as compatibilizers, thermoplastic elastomers, adhesives or

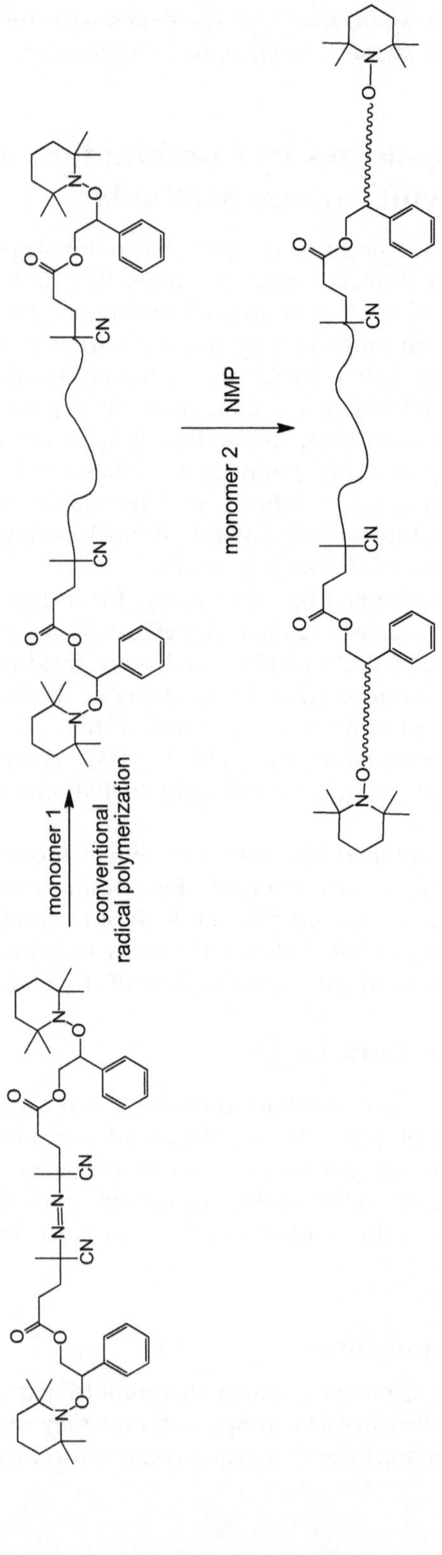

Scheme 8.6 A combination of NMP with FRP.

polymeric emulsifier. They are made of a backbone and pendant chains to which branches are attached. Grafted polymers are usually called "comb polymers" when the branches are composed of the same monomer units as the backbone. Different strategies can be performed to synthesize the suitable grafted polymer, traditionally called "grafting to", "grafting from" or "grafting through". Indeed, the backbone and the branches can be prepared separately and then the branches can be "grafted to" the backbone by a coupling reaction between the chain-ends of the branches and the reactive functions along the backbone. The "grafting from" strategy consists of the synthesis of a polymer bearing pendant groups able to initiate the polymerization of the branches from the backbone. Finally, it is also possible to obtain grafted polymers by the "grafting through" method, *i.e.* the polymerization of macromolecules with a polymerizable chain-end or macromonomers, alone or together with co-monomers. These three different strategies can be performed involving NMP alone or more generally in combination with another method of polymerization.

8.5.1.1 "Grafting to" Approach

This procedure offers the possibility of independently characterizing the backbone and the branches prior to the coupling reaction, which is the main advantage of this strategy. Nevertheless, due to steric hindrance, the density of branching is generally lower with the "grafting to" method compared to the others. Indeed, NMP in combination with another polymerization method for the design of grafted polymer by "grafting to" strategy was the object of different studies. One of the first examples is from Hadjichristidis *et al.*,[86] where *p*-chloromethylstyrene (CMS) was homopolymerized or copolymerized with styrene by NMP in bulk using the alkoxyamine 1-phenyl-1-TEMPO and a small amount of acetic anhydride as accelerator. Then poly(isoprene) or polystyrene were synthetized by anionic polymerization initiated by *sec*-BuLi in benzene and finally end-capped by 1,1-diphenylethylene (DPE). DPE-capped poly(isoprenylithium) or poly(styrenylithium) were successfully attached to the backbone by linking reaction with the chloromethyl groups to prepare, with a grafting efficiency near 100%, PSt-*g*-PI grafted copolymer and PSt-*g*-PSt comb copolymer, respectively (Scheme 8.7).

The same procedure was used for the synthesis of original architectures of grafted copolymers, where "V-shaped" and "Y-shaped" containing either poly(isoprenyllithium) or poly(styryllithium) were grafted onto PSt-*co*-PCMS made by NMP.[87] Similarly, a linear PCMS backbone was synthetized by a combination of TEMPO and AIBN, followed by grafting of poly(ethylene glycol) monomethyl ether (MPEG) using the Williamson etherification reaction.[88] Furthermore, comb-on-comb grafted copolymers were also obtained, starting from PSt-*co*-PCMS made by NMP mediated by TEMPO and using poly(isoprenyl lithium) synthesized by anionic polymerization for the branches.[89] The novelty of this work is the consecutive epoxidation of

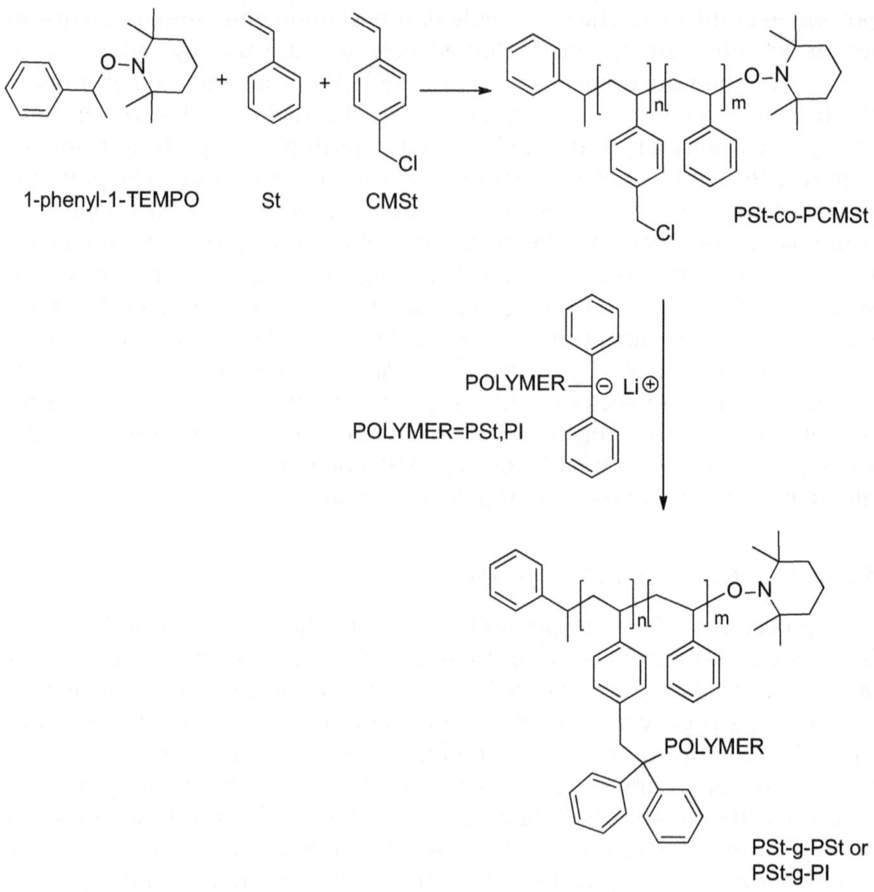

Scheme 8.7 "Grafting to" with combination of NMP and anionic polymerization.

the polyisoprene double-bond by performic acid to allow grafting of a second generation of branches by nucleophilic attack of poly(styryl lithium), poly(isoprenyl lithium), and poly(α-methyl styryl lithium) to design PSt-*g*-(PI-*g*-PSt) PSt-*g*-(PI-*g*-PI), and PSt-*g*-(PI-*g*-PαMS), respectively.

Chloro groups of PCMS were converted by azide substitution and the modified backbone has been clicked by Cu(I)-catalyzed 1,3-dipolar cycloaddition to α-yne-terminated polystyrenes bearing mesogenic side groups previously polymerized with an yne-functional ATRP initiator.[90] PSt-*co*-PCMS was also modified in azido copolymer, followed by grafting of propargylthiophene or -pyrrole by click chemistry and electropolymerized in the presence of pyrrole.[91]

The chloromethylstyrene monomer was also investigated by Hawker *et al.*[92] who copolymerized another styrene derivative bearing a succinimide moiety, *i.e.* *N*-oxysuccinimide 4-vinylbenzoate. Fréchet-type polyether dendrons were grafted through an amine group, giving an original hybrid dendritic–linear grafted copolymer, allowing an interesting study on the steric effects of the "grafting to" approach.

A recent example of "grafting to" is the work of García-Valdez *et al.*,[93] who developed a strategy in three steps to attach different polymers made by NMP to a chitosan (CTS) backbone. The –OH groups of the chitosan were first reacted under acidic conditions with glycidyl methacrylate, and then protonated amino groups were functionalized with sodium dodecylbenzenesulfonate (SDBS) to increase the solubility in organic solvents. Different SG1-terminated polymers were grafted onto the CTS backbone by radical addition on the methacrylate, a method that is similar to the intermolecular radical 1,2 addition of the BlocBuilder alkoxyamine on activated olefins.[93]

NMP has also been involved in the synthesis of dynamic grafted copolymers.[94] Poly(methacrylic ester) backbones bearing TEMPO-based alkoxyamine side groups were prepared by ATRP, and then polystyrene chains synthesized by NMP were grafted by radical exchange reaction between alkoxyamine side units and end-chains. Moreover by adding an excess of free alkoxyamine, the grafting became reversible. A similar strategy was employed recently to obtain reversible crosslinking.[95]

8.5.1.2 "Grafting From" Approach

NMP was involved in the design of grafted polymers by the "grafting from" approach, either for the synthesis of the backbone or branches. Several polymerization techniques initiated from a backbone synthesized by NMP were investigated. A good tandem method is to polymerize the backbone by NMP and then initiate ATRP from chlorine pendant groups. Since NMP is more active at higher temperature than ATRP, there is no risk of a simultaneous NMP process during ATRP of the branches.

For example, styrene and *p*-(4′-chloromethylbenzyloxymethyl)styrene were copolymerized by TEMPO-based alkoxyamines, and ATRP of methacrylates was initiated from chloromethyl groups.[96] Additionally, a dendrigraft copolymer was also made by three consecutive controlled radical polymerizations. Similarly, PSt-*co*-PCMS obtained by HTEMPO/BPO acted as a multi-initiator for ATRP of 2,2,3,3,4,4,5,5-octafluoropentyl acrylate.[97] NMP was also used in the synthesis of a block-graft copolymer where CROP of epichlorohydrin, initiated by –OH of HTEMPO, was controlled, followed by NMP of styrene and finally ATRP of MMA from chloromethyl moieties.[98]

NMP was also combined with RAFT as in the case of PSt-*co*-PCMS where chlorine was converted into a macro-RAFT agent capped with a fluorescent carbazole as the Z-group, followed by the RAFT polymerization of NIPAAM.[99] Functionalization by dithiocarbamates of PSt-*co*-PCMS was also used to polymerize tBuMA by photo-induced iniferter techniques to produce PSt-*g*-PtBuMA and PSt-*g*-PMAA.[100,101]

Recently, conductive grafted copolymers were elaborated *via* NMP for the synthesis of a PSt-*co*-PCMS backbone and oxidation polymerization of aniline[102] or thiophene[103] for the branches. In the case of aniline, after substitution of chlorine by azide, cycloaddition reaction was performed to introduce phenylamine groups, which are able to initiate, when oxidized, the polymerization of *p*-toluenesulfonic acid-doped polyaniline. Thiophene was also successfully polymerized from a PSt-*co*-PCMS backbone after replacement of chlorine by thiophene groups *via* Kumada cross-coupling reaction, followed by oxidation polymerization.

Many examples of the "grafting from" approach deal with NMP initiated from a backbone obtained by another method. The way to induce NMP polymerization from the main chain mostly consists of introducing TEMPO-based alkoxyamine as pendant groups, as first described by Hawker.[104] 1-hydroxy-2-phenyl-2-TEMPO was reacted with CMS in the presence of sodium hydride to give a monomer bearing pendant alkoxyamine, which was then copolymerized with styrene by Free-Radical Polymerization (FRP). Then, NMP was performed to obtain comb polystyrene.

TEMPO- or 2,2,5-trimethyl-4-(isopropyl)-3-azahexane-3-oxyl-based alkoxyamine functionalized methacrylate monomers were also described and copolymerized by FRP with styrene, methyl methacrylate, *n*-butyl methacrylate and *n*-butyl acrylate.[105] These functional macro-initiators were used for the NMP of styrene or *n*-butyl acrylate. Another methacrylate alkoxyamine monomer containing 4-benzyloxyl-TEMPO was polymerized together with MMA by FRP and PMMA-*g*-PSt was synthesized by NMP.

An alkoxyamine based on TEMPO with an acrylate function was also electropolymerized and polystyrene branches were made by NMP from the chemisorbed polyacrylate.[106]

Monomers containing alkoxyamine moieties were also included within polyolefin. Metallocene-catalyzed copolymerizations of TEMPO-[107] or TIPNO-based[108] alkoxyamine alkenes were achieved with propylene or 4-methylpentene and a mixture of ethylene/butadiene, respectively, followed by well-controlled NMP from the polyolefin backbone with different vinyl monomers.

An electron conductor polymer poly(phenyl acetylene) was also grafted by Rh-catalyzed polymerization of 1-(4-ethynylphenyl)-1-(2,2,6,6-tetramethyl-1-piperidinyloxyl)ethane and NMP of styrene.[109] Poly(*p*-phenylene terephthalamide)-*g*-PSt was also obtained starting from polycondensation of TEMPO-functionalized terephthaloyl chloride, and *p*-phenylenediamine.[110]

In some cases, instead of alkoxyamine, nitroxides were directly functionalized by a double bond and copolymerized. For example, TEMPO was attached to styrene (4-vinylbenzyl- TEMPO) and copolymerized by ATRP with MMA.[111] As a result, nitroxide was also transformed into pendant alkoxyamine from ethyl propionate stemming from the bromo-initiator of ATRP and the obtained macro-initiator was able to control the polymerization of styrene. Thus the control of the grafting occurred *via* a nitroxide immobilized on the backbone and the branches were reversibly homolytically disconnected from the backbone to propagate, called an insertion mechanism by Schmidt-Naake *et al.*[112] These authors described a TEMPO attached to an acrylate and its use in the FRP of styrene to synthesize the backbone. In this case, an initial consumption of radicals from AIBN by combination with the nitroxide, before the polymerization took place. Another interesting example of a nitroxide attached to a monomer involved an anionic ring opening homopolymerization of an epoxy monomer (4-glycidyloxy-TEMPO), followed by "grafting from" process by NMP of a styrene functionalized by carbazole (Scheme 8.8).[113]

Another example of a grafted copolymer entirely synthesized by NMP was reported: PSt-*co*-PCMS was obtained with 4-methoxy-TEMPO/BPO. After reaction of 4-hydroxy-TEMPO with chlorine, grafted controlled photopolymerization of MMA was achieved in the presence of a diazo compound as initiator and (4-*tert*-butylphenyl)diphenylsulfonium triflate as the photoacid generator.[114]

Another strategy to initiate NMP from a backbone is to modify the preformed polymer main chain, by attaching an alkoxyamine moiety on it instead of a nitroxide. For example, this approach was used to obtain a grafted polymer from polyolefins. In one of the first studies, isotactic PP was treated by γ-irradiation in air to produce PP-peroxides, enabling the initiation of NMP of styrene in the presence of TEMPO.[115] Moreover, isotactic poly(1-butene) was chemically modified by activation of alkane bonds with rhodium catalyst and reaction with a boron derivative. After various steps, 1-hydroxy-2-phenyl-2-TEMPO was bonded to the PP backbone and NMP of dipropyl(4-vinylphenyl)amine could occur.[116]

Furthermore, Saldìvar-Guerra's group studied polybutadiene modification by heating the monomer with a peroxide/nitroxide couple, only in the presence of 4-oxo-TEMPO, followed by grafting of PSt by NMP was performed.[117] Different elastomers were also tested, such as polyisoprene (PI), PSt-*b*-PI-*b*-PSt and PSt-*b*-PB-*b*-PSt (SBS) triblocks, EPDM rubber, polydimethylsiloxane (PDMS) or poly(ethylene glycol), and further grafted with PSt or poly(styrene-*co*-maleic anhydride).[118] The same group also grafted this last one in a controlled way by NMP from polydienes (PB and PI). Interestingly, the authors compared the grafting efficiency with different nitroxides in the presence of a peroxide initiator and it was observed that 4-oxo-TEMPO exhibited the best results compared to TEMPO or 4-hydroxy TEMPO, whereas SG1 was inefficient in these conditions, being prone to decomposition.[119]

In addition, PVC was also converted to a macro-initiator by successive Friedel–Crafts reaction with toluene, bromination of the arylated PVC and reaction with 1-hydroxy-TEMPO. After NMP polymerization, PVC-*g*-PSt was obtained.[120]

The grafting of poly(vinyl naphthalene) onto poly(vinyl alcohol)[121] and poly(sodium 2-acrylamido-2-methyl-1-propanesulfonate), *i.e.* a photoactive grafted amphiphilic polyelectrolyte, was reported.[122] Fenton's reagent generated a hydroxyl radical by abstraction of hydrogen atoms from the polymeric chains. The radicals formed along the backbone could capture 4-hydroxy-TEMPO and vinyl naphthalene is controlled by NMP (Scheme 8.9).

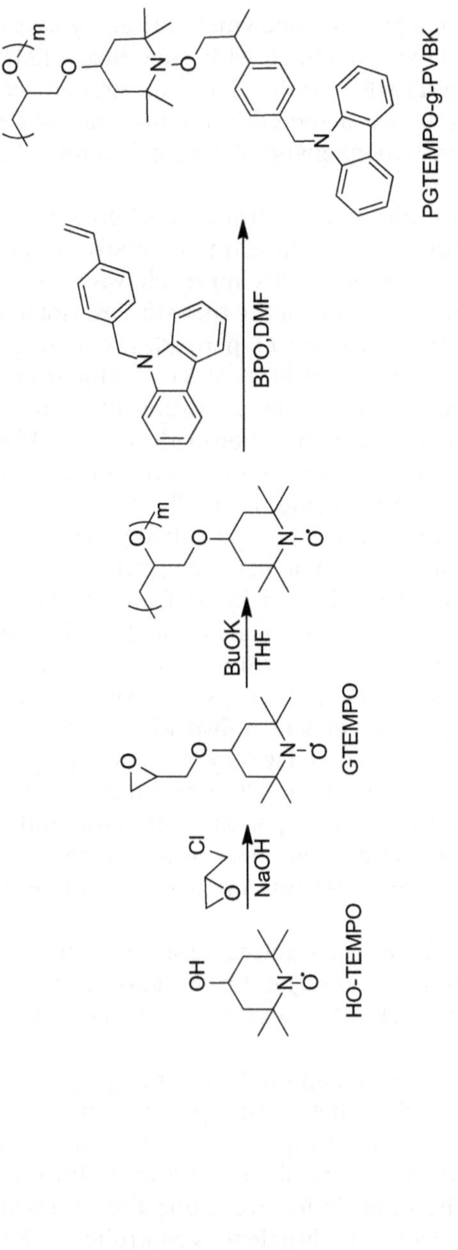

Scheme 8.8 Comb-shaped carbazole *via* combination of ROP and NMP *via* nitroxide-based monomer.

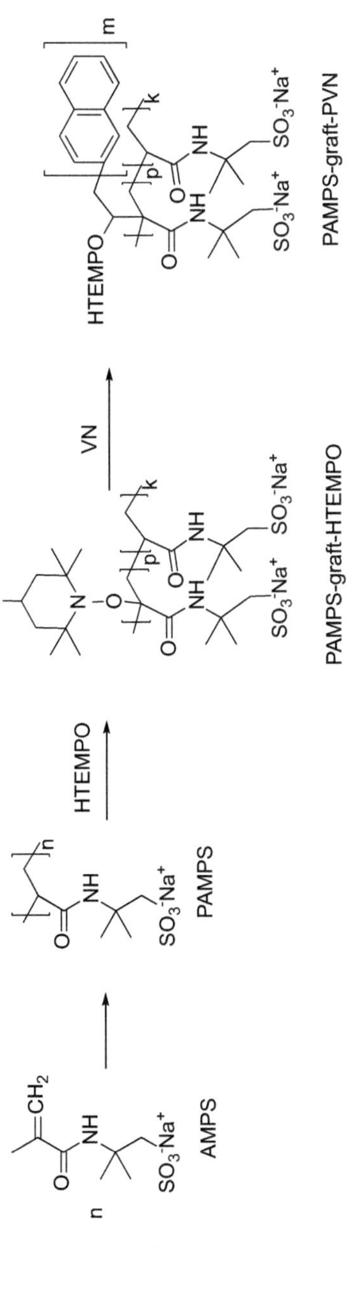

Scheme 8.9 Photoactive graft amphiphilic polyelectrolyte by "grafting from" NMP.

Syndiotactic PSt (sPSt) was also modified by a Friedel–Crafts acylation allowing for coupling of 1-hydroxy-TEMPO with chlorine.[123] NMP of styrene and *p*-methylstyrene (MS) was then performed and after bromination, ATRP was employed to obtain [sPSt-*g*-(PSt-*b*-PMS)]-*g*-PMMA.[124]

An intermediate approach consisted in the copolymerization with a functional monomer, easily transformable into a pendant alkoxyamine. It was achieved for polyethylene by metallocene catalyzed copolymerization of ethylene and *m*,*p*-methylstyrene. After bromination of methyl groups and Williamson etherification with 1-hydroxy-2-phenyl-2-TEMPO, PSt-*g*-PE could be obtained.[125]

Another example is the modification of carboxylated poly(arylene ether sulfone) by Barton esters (BE) under irradiation.[126] The BE-modified poly(arylene ether sulfone) reacted with styrene in the presence of TEMPO afforded controlled grafted polymerization of styrene. The same process was used to graft PSt from cellulose derivatives.[127] Another natural polymer, chitosan, was also successfully modified by different methods to introduce TEMPO and grafted with PSt,[128] poly(styrene-*co*-maleic anhydride),[129] poly(sodium 4-styrenesulfonate)[130] or poly(*n*-butyl acrylate).[131]

8.5.1.3 "Grafting Through" Approach

Typically, a polymer with a vinyl chain-end or macromonomer is copolymerized by NMP to obtain a grafted copolymer. For example, methacrylate terminated polycaprolactone, poly(D,L)lactide, poly(ethylene glycol) and acrylate terminated polyethylene were copolymerized in a controlled manner together with styrene, *p*-chloro-methylstyrene or 2-hydroxyethyl methacrylate in the presence of 1-phenyl-1-TEMPO.[132]

Methacrylate terminated PEGs (MePEGMA) were also well-controlled by MAMA-SG1, also called BlocBuilder MA, alkoxyamine initiator with a small amount of either styrene[133] or acrylonitrile.[134] By the same method, but from a MAMA-SG1 where the carboxylic acid was modified into a *N*-hydroxysuccinimidyl (NHS) ester, it was possible to a peptide/protein couple with P(MePEGMA-*co*-AN) derived from NMP (Scheme 8.10).[135]

8.5.2 Star Copolymers

Star polymers have gained considerable attention over the past decades due to their unique physical properties compared to their linear counterparts of equal molecular weight and monomer composition. Star topologies are *per se* the simplest branched polymers, which comprise a multifunctional core from which linear polymer chains diverge, offering a high extent of chain-end functional groups, which enable their application in cutting-edge materials for nanoelectronics, drug delivery, and tissue engineering, for instance. The main strategies utilized in RDRP for the synthesis of such architectures are the "core-first" and "arm-first" approaches,[136] while more recently, the use of tris-nitroxide was suggested for the elaboration of 3-arm

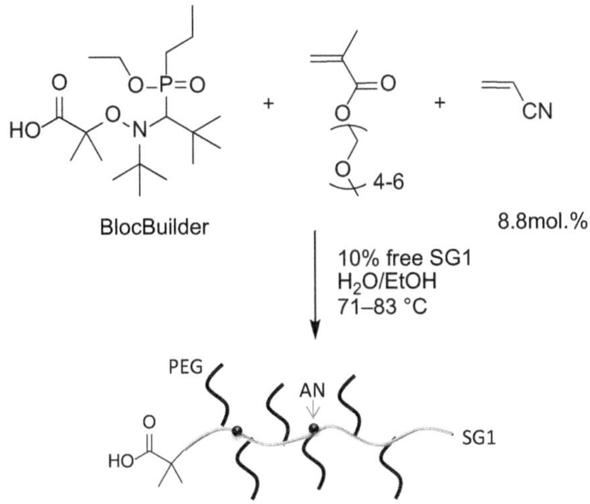

Scheme 8.10 Comb-shaped polymethacrylates with PEG side chains by "grafting through" NMP.

star PSt, though only 2-arm linear structures were produced experimentally.[137] In that case, the mechanism is associated with a "core-first" approach, combined with an "outside-in" approach.

8.5.2.1 "Core-first" Approach

The "core-first" approach involves the growth of arms directly from a multifunctional initiating core and allow for a great control of polymer architectures with a precise number of arms of equal length and a well-defined core structure.

In the context of NMP, this strategy implies the preparation of multifunctional alkoxyamines[136] as firstly described by Hawker *et al.*,[104] who reported the preparation of three initiating styrene-TEMPO groups generating well-defined PSt star polymers. Some other works are dedicated to a range of star polymers obtained by varying the number of arms (3, 6 or 12) using multifunctional TEMPO-based alkoxyamines[138–145] and by modulating the nature of the core. The latter was a dendrimer,[139] a porphyrin,[140] a cyclodextrin,[141] a mesitylene[144] or a polyhedral oligomeric siloxane fragment (POSS).[145] In another study, St, MA, isoprene and DMAAm based star polymers were synthesized by employing dendritic TIPNO-based alkoxyamine multifunctional macro-initiators to yield 6 or 12 arms.[146,147]

Further examples include the preparation of 8-arm based star polymers obtained after derivatization of resorcinarene with TEMPO or a derivatized SG1.[148]

An innovative approach was later established by Bertin *et al.*[149] to improve the preparation of multifunctional initiators by a facile one-step method. The intermolecular 1,2-radical addition onto different activated olefins (*n*-BA, AA, DMA, HEA and St) considerably simplified the preparation of multifunctional SG1-based alkoxyamines and resulted in well-defined 3- and 4-arm star polymers.

8.5.2.2 "Arm-first" Approach

The arm-first approach requires monofunctional linear polymer precursors with reactive end-groups able to couple with multifunctional coupling agents to generate star polymer topologies.

Hadjichristidis *et al.*[136] employed DVB as a coupling agent to build star-shape polymers composed of a cross-linked core and a shell made of linear preformed polymers. This method is advantageous for its simplicity but can suffer from some limitations including irregular number of arms leading to high polydispersities.

Several research groups[150–152] focused on the NMP of styrene derivatives and 4,4-divinylbiphenyl to produce microgels. Precise star polymer architectures were also achieved through the coupling of DVB with PSt-TEMPO[153,154] or various vinyl saccharides,[155] for example.

The potential of this technique to obtain very well-defined star polymers can be illustrated by the works of Hawker *et al.*,[156] who employed PSt-TIPNO and combinatorial methods to produce a range of 3-dimensional star polymer architectures towards the pursuit of new biologically active compounds or new catalytic sites. These complex topologies produced from a large range of block and random copolymers with different active end-groups were effective as multifunctional host/guest compounds for the encapsulation of various guest molecules as well as catalytic sites located in their core or in their periphery. From the perspective of reducing the typical issues related to highly-concentrated solutions of chromophore (aggregation, red-shift and fluorescence quenching), Harth *et al.*[157] described the synthesis of star polymers using fluorine- and thiophene-based divinyl derivatives to give rise to highly-photoluminescent macromolecules through site-isolated chromophore cores.

8.5.2.3 Star Topologies by Combination of NMP with Other Radical Polymerization Methods

Various CRP techniques may be combined with NMP to prepare complex star topologies bearing hetero-arms on their periphery, also named miktoarm star polymers. Typical examples are the combination of NMP with ROP, ATRP, or RAFT.

For example, ABC 3-miktoarm star terpolymers were synthesized by the combination of NMP, ATRP and ROP.[158] Within this goal, a trifunctional initiator carrying hydroxyl, bromide and TEMPO-based alkoxyamine, for successive ROP of ε-caprolactone, ATRP of methyl methacrylate, and NMP of styrene, was designed. ABC 3-miktoarms could also be afforded by the combination of NMP and ATRP from a bifunctional initiator, also bearing an alkyne group allowing grafting a third arm *via* click chemistry of a polymer with an azide end-chain.[159]

It is noteworthy that Miura *et al.*[160] achieved the preparation of arm-replaceable star-like nanogels by NMP of DVB initiated by an alkoxyamine-terminated PMMA previously prepared by ATRP.

8.5.3 Hyperbranched Copolymers

Hyperbranched copolymers may be described as those structures with densely branched structures consisting of a large proportion of end-groups. Fréchet and co-workers[161] devised a technique to prepare hyperbranched polymers from vinyl monomers. The strategy, coined self-condensing vinyl polymerization,[162] is grounded on the polymerization of inimers (INItiator + monoMER), in this case a styrene functionalized by a TEMPO-based alkoxyamine. The latter approach was applied to NMP to confer hyper-branched copolymers. Prospects for the design of alternate macro-molecular architectures like hyperbranched stars might be achieved, thanks to the bound nitroxide functionalities present at the chain-ends of the hyperbranched structure. Those living chain-ends were destined to participate in the initiation of the polymerization of vinyl monomers like styrene.

Indeed, a styrenic TIPNO-based alkoxyamine was copolymerized with styrene or *n*-butyl acrylate to afford branched polystyrene and for the first time branched poly(*n*-butyl acrylate), respectively, and this last one was an efficient hyperbranched macro-initiator for the polymerization of styrene.

Another method to target hyperbranched polymers is to react mono- and di-functional monomers through a controlled radical polymerization technique to contain complete reticulation or gelation. Hawker *et al.*[163] re-ported a procedure to prepare hyperbranched polymers by copolymerization of styrene and difunctional methacrylate (1,6-hexanediol dimethacrylate) mediated by TIPNO. These new hyperbranched copolymers were then tested as materials for imprint lithography.

8.5.4 Cyclic Polymers

Cyclic polymers have attracted increasing attention due to the distinctive properties arising from topological constrictions.[164–168] Intramolecular cyclization of α,α′-difunctional or α,ω-heterodifunctional linear polymer precursors are the two the main routes for the synthesis of such polymers. The first route is typically used for the coupling of a dicarbanionic polymer with a difunctional electrophile, while the second route allows for synthesis of telechelic polymers displaying functional end-groups and may be tailored to NMP.

The cyclisation of PSt was reported by Lepoittevin *et al.*,[169] who used α-carboxyl ω-hydroxy heterodifunctional PSt as a linear precursor,

4,4′-azobis(4-cyanovaleric acid) as the initiator, and ensured the control of the polymerization employing 4-hydroxy-TEMPO. The intramolecular cyclisation was favored by proceeding in dilute solutions using 1-methyl-2-chloropyridinium iodide as a catalyst to produce macrocyclic polystyrene exhibiting precise dimensions and narrow polydispersities. However, the final cyclic polymers contained a thermolabile alkoxyamine bond that was potentially dissociable and a hydrolysable chemically sensitive ester function: these limitations were overcome by combining NMP and post-functionalization in the synthesis of azide–alkyne telechelic linear polymers.[170]

The cyclization of these polymer precursors was performed in diluted medium through CuAAC to produce a scarcely separable mixture of linear and cyclic polymers,[140] thus alternative methods were proposed. One of them is the end-to-end cyclisation method in which cyclic polymers are prepared by ring-expansion polymerization (Scheme 8.11), as exampled by Ruehl *et al.*[171] who synthetized 5- and 17- membered cyclic alkoxyamines based on TIPNO to give macrocyclic polystyrene and reported, through ESR analyses, that the stability of the 5-membered ring prevented the cleavage of the alkoxyamine contrasting with the unstable 17-membered cyclic alkoxyamines.

A mixture of cyclic polymers wherein a lone alkoxyamine moiety was integrated onto the polymer chain and oligomeric cyclic polymers bearing several alkoxyamine bonds was produced. Similar conclusions were drawn when using an alkoxyamine tethered through azide/alkyne click chemistry and were ascribed to radical ring-crossover reactions, instigated from nitroxide exchange.[172] The same observations were made in the case of ring-expansion of cyclic alkoxyamines by both ATRP and Atom Transfer Nitroxide Radical Coupling (ATNRC).[173]

Maeda *et al.*[174] exploited this aspect by exploring the reversibility of the dynamic covalent bond characteristics of alkoxyamines and induced an appropriate external factor.

The stimuli chosen by Yamaguchi *et al.*[175] was the heating of a cyclic alkoxyamine compound to yield the corresponding polyalkoxyamine by intermolecular radical crossover reaction, also able to depolymerize by the same process, at very low concentrations.

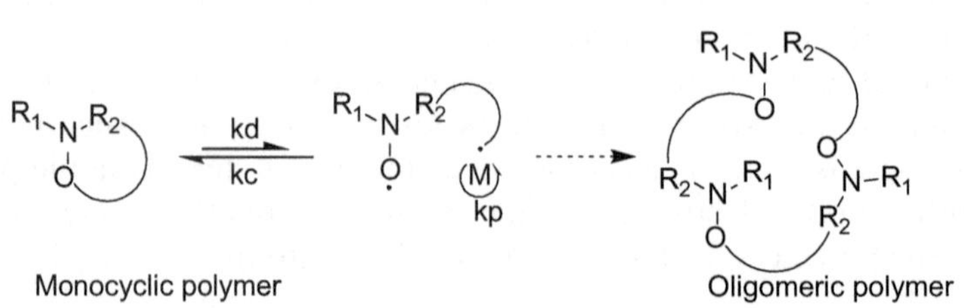

Scheme 8.11 Synthesis of macro-cycles by nitroxide-mediated ring-expansion polymerization.

8.6 Conclusion

Nitroxide-mediated polymerization as an RDRP technique represents a powerful method for the synthesis of copolymers. The accessible topologies may be linear (diblock, triblock, random, statistical, gradient) or non-linear (grafted, star, hyperbranched, cyclic). Even if the NMP technique is not applicable to all kinds of monomers, this technique is particularly convenient, non-toxic and offers different possibilities to introduce functional groups valuable to prepare a myriad of sophisticated composition and architecture polymer chains. Therefore, NMP is perfectly positioned to play a significant role in the synthesis of advanced polymer materials with a multitude of applications.

References

1. M. Kato, M. Kamigaito, M. Sawamoto and T. Higashimura, *Macromolecules*, 1995, **28**, 1721–1723.
2. J. S. Wang and K. Matyjaszewski, *J. Am. Chem. Soc.*, 1995, **117**, 5614–5615.
3. K. Matyjaszewski and J. Xia, *Chem. Rev.*, 2001, **101**, 2921–2990.
4. J. Chiefari, Y. K. Chong, F. Ercole, J. Krstina, J. Jeffery, T. P. T. Le, R. T. A. Mayadunne, G. F. Meijs, C. L. Moad, G. Moad, E. Rizzardo and S. H. Thang, *Macromolecules*, 1998, **31**, 5559–5562.
5. M. Destarac, D. Charmot, X. Franck and S. Z. Zard, *Macromol. Rapid Commun.*, 2000, **21**, 1035–1039.
6. G. Moad, E. Rizzardo and S. H. Thang, *Aust. J. Chem.*, 2005, **58**, 379–410.
7. D. H. Solomon, E. Rizzardo and P. Cacioli, CSIRO, Polymerization Process and Polymers Produced Thereby, *U.S. Pat.* 4581429, 1986.
8. M. K. Georges, R. P. N. Veregin, P. M. Kazmaier and G. K. Hamer, *Macromolecules*, 1993, **26**, 2987–2988.
9. D. H. Solomon, *J. Polym. Sci., Part A: Polym. Chem.*, 2005, **43**, 5748–5764.
10. C. J. Hawker, A. W. Bosman and E. Harth, *Chem. Rev.*, 2001, **101**, 3661–3688.
11. K. V. Bernaerts and F. E. Du Prez, *Prog. Polym. Sci.*, 2006, **31**, 671–722.
12. J. Nicolas, Y. Guillaneuf, C. Lefay, D. Bertin, D. Gigmes and B. Charleux, *Prog. Polym. Sci.*, 2013, **38**, 63–235.
13. S. Brinkmann-Rengel and N. Niessner, in *Controlled/Living Radical Polymerization: Progress in ATRP, NMP, and RAFT*, ed. K. Matyjaszewski, American Chemical Society, Washington DC, 1999, vol. 768, p. 394.
14. L. Hlalele and B. Klumperman, *Macromolecules*, 2011, **44**, 7100–7108.
15. B. Charleux, J. Nicolas and O. Guerret, *Macromolecules*, 2005, **38**, 5485–5492.
16. B. Grassl, G. Clisson, A. Khoukh and L. Billon, *Eur. Polym. J.*, 2007, **44**, 50–58.
17. D. Benoit, V. Chaplinski, R. Braslau and C. J. Hawker, *J. Am. Chem. Soc.*, 1999, **121**, 3904–3920.

18. D. Benoit, E. Harth, P. Fox, R. M. Waymouth and C. J. Hawker, *Macromolecules*, 2000, **33**, 363–370.
19. J.-M. Catala, F. Bubel and S. O. Hammouch, *Macromolecules*, 1995, **28**, 8441–8443.
20. D. Benoit, V. Chaplinski, R. Braslau and C. J. Hawker, *J. Am. Chem. Soc.*, 1999, **121**, 3904–3920.
21. D. Benoit, S. Grimaldi, J. P. Finet, P. Tordo, M. Fontanille and Y. Gnanou, *Polym. Prepr. (Am. Chem. Soc., Div. Polym. Chem.)*, 1997, **38**, 729–730.
22. S. Grimaldi, J. P. Finet, A. Zeghdaoui, P. Tordo, D. Benoit, Y. Gnanou, M. Fontanille, P. Nicol and J. F. Pierson, *Polym. Prepr. (Am. Chem. Soc., Div. Polym. Chem.)*, 1997, **38**, 651–652.
23. D. Benoit, S. Grimaldi, S. Robin, J.-P. Finet, P. Tordo and Y. Gnanou, *J. Am. Chem. Soc.*, 2000, **122**, 5929–5939.
24. C. A. Knoop and A. Studer, *J. Am. Chem. Soc.*, 2003, **125**, 16327–16333.
25. S. Butz, J. Hillermann, G. Schmidt-Naake, J. Kressler, R. Thomann, B. Heck and B. Stuehn, *Acta Polym.*, 1998, **49**, 693–699.
26. E. Bultz and T. P. Bender, *Macromolecules*, 2011, **44**, 3666–3669.
27. D. Bertin and B. Boutevin, *Polym. Bull.*, 1996, **37**, 337–344.
28. E. Yoshida and T. Fujii, *J. Polym. Sci., Part A: Polym. Chem.*, 1997, **35**, 2371–2378.
29. E. Yoshida, *J. Polym. Sci., Part A: Polym. Chem.*, 1996, **34**, 2937–2943.
30. M. Mariani, M. Lelli, K. Sparnacci and M. Laus, *J. Polym. Sci., Part A: Polym. Chem.*, 1999, **37**, 1237–1244.
31. K. Ohno, Y. Izu, Y. Tsujii, T. Fukuda and H. Kitano, *Eur. Polym. J.*, 2004, **40**, 81–88.
32. K.-i. Fukukawa, L. Zhu, P. Gopalan, M. Ueda and S. Yang, *Macromolecules*, 2004, **38**, 263–270.
33. H. Okamura, Y. Takatori, M. Tsunooka and M. Shirai, *Polymer*, 2002, **43**, 3155–3162.
34. J. Kim, M. M. Mok, R. W. Sandoval, D. J. Woo and J. M. Torkelson, *Macromolecules*, 2006, **39**, 6152–6160.
35. D. Woo, J. Kim, M.-H. Suh, H. Zhou, S. T. Nguyen, S.-H. Lee and J. M. Torkelson, *Polymer*, 2006, **47**, 3287–3291.
36. J. D. Quinn and R. A. Register, *J. Polym. Sci., Part B: Polym. Phys.*, 2009, **47**, 2106–2113.
37. M. K. Georges, J. L. Lukkarila and A. R. Szkurhan, *Macromolecules*, 2004, **37**, 1297–1303.
38. B. Keoshkerian, M. Georges, M. Quinlan, R. Veregin and B. Goodbrand, *Macromolecules*, 1998, **31**, 7559–7561.
39. A. Debuigne, T. Radhakrishnan and M. K. Georges, *Macromolecules*, 2006, **39**, 5359–5363.
40. D. Chan-Seng, A. Debuigne and M. K. Georges, *Eur. Polym. J.*, 2009, **45**, 211–216.
41. K. O. Siegenthaler and A. Studer, *Macromolecules*, 2006, **39**, 1347–1352.

42. C. Farcet, J. Nicolas and B. Charleux, *J. Polym. Sci., Part A: Polym. Chem.*, 2002, **40**, 4410–4420.
43. P. Lacroix-Desmazes, J.-F. Lutz, F. Chauvin, R. Severac and B. Boutevin, *Macromolecules*, 2001, **34**, 8866–8871.
44. L. Marx, Z. Eskandani and P. Hemery, *React. Funct. Polym.*, 2009, **69**, 306–318.
45. L. Marx and P. Hemery, *Polymer*, 2009, **50**, 2752–2761.
46. N. R. Cameron and A. J. Reid, *Macromolecules*, 2002, **35**, 9890–9895.
47. J. Nicolas, B. Charleux, O. Guerret and S. Magnet, *Angew. Chem., Int. Ed.*, 2004, **43**, 6186–6189.
48. J. Nicolas, A.-V. Ruzette, C. Farcet, P. Gerard, S. Magnet and B. Charleux, *Polymer*, 2007, **48**, 7029–7040.
49. K. Bian and M. F. Cunningham, *Macromolecules*, 2005, **38**, 695–701.
50. K. J. Bian and M. F. Cunningham, *J. Polym. Sci., Part A: Polym. Chem.*, 2006, **44**, 414–426.
51. K. H. Kuo, W. Y. Chiu and K. C. Cheng, *Polym. Int.*, 2008, **57**, 730–737.
52. B. Lessard and M. Maric, *Macromolecules*, 2008, **41**, 7870–7880.
53. N. Handké, T. Trimaille, E. Luciani, M. Rollet, T. Delair, B. Verrier, D. Bertin and D. Gigmes, *J. Polym. Sci., Part A: Polym. Chem.*, 2011, **49**, 1341–1350.
54. G. Moad, A. G. Anderson, F. Ercole, H. J. Johnson, J. Krstina, C. L. Moad, E. Rizzardo, T. H. Spurling and S. H. Thang, *ACS Symp. Ser.*, 1998, **685**, 332–360.
55. C. Burguiere, M. A. Dourges, B. Charleux and J. P. Vairon, *Macromolecules*, 1999, **32**, 3883–3890.
56. G. S. Ananchenko, M. Souaille, H. Fischer, C. Le Mercier and P. Tordo, *J. Polym. Sci., Part A: Polym. Chem.*, 2002, **40**, 3264–3283.
57. M. Souaille and H. Fischer, *Macromolecules*, 2001, **34**, 2830–2838.
58. M. V. Edeleva, I. A. Kirilyuk, D. P. Zubenko, I. F. Zhurko, S. R. A. Marque, D. Gigmes, Y. Guillaneuf and E. G. Bagryanskaya, *J. Polym. Sci., Part A: Polym. Chem.*, 2009, **47**, 6579–6595.
59. M. Edeleva, S. R. A. Marque, D. Bertin, D. Gigmes, Y. Guillaneuf, S. V. Morozov and E. G. Bagryanskaya, *J. Polym. Sci., Part A: Polym. Chem.*, 2008, **46**, 6828–6842.
60. J. Lokaj, P. Vlček and J. Kriz, *Macromolecules*, 1997, **30**, 7644–7646.
61. B. Grassl, G. Clisson, A. Khoukh and L. Billon, *Eur. Polym. J.*, 2007, **44**, 50–58.
62. J. Rigolini, B. Grassl, L. Billon, S. Reynaud and O. F. X. Donard, *J. Polym. Sci., Part A: Polym. Chem.*, 2009, **47**, 6919–6931.
63. F. Ganachaud, M. J. Monteiro, R. G. Gilbert, M. A. Dourges, S. H. Thang and E. Rizzardo, *Macromolecules*, 2000, **33**, 6738–6745.
64. E. N. Savariar and S. Thayumanavan, *J. Polym. Sci., Part A: Polym. Chem.*, 2004, **42**, 6340–6345.
65. K. Kuroda and T. M. Swager, *Macromolecules*, 2004, **37**, 716–724.
66. O. Gibbons, W. M. Carroll, F. Aldabbagh and B. Yamada, *J. Polym. Sci., Part A: Polym. Chem.*, 2006, **44**, 6410–6418.

67. A. Fischer, A. Brembilla and P. Lochon, *Eur. Polym. J.*, 2001, **37**, 33–37.
68. A. Fischer, A. Brembilla and P. Lochon, *Polymer*, 2001, **42**, 1441–1448.
69. T. Fukuda, T. Terauchi, A. Goto, Y. Tsujii, T. Miyamoto and Y. Shimizu, *Macromolecules*, 1996, **29**, 3050–3052.
70. M. Baumann, A. I. Roland, G. Schmidt-Naake and H. Fischer, *Macromol. Mater. Eng.*, 2000, **280–281**, 1–6.
71. C. Tang, T. Kowalewski and K. Matyjaszewski, *Macromolecules*, 2003, **36**, 1465–1473.
72. L. Couvreur, C. Lefay, J. Belleney, B. Charleux, O. Guerret and S. Magnet, *Macromolecules*, 2003, **36**, 8260–8267.
73. C. Lefay, J. Belleney, B. Charleux, O. Guerret and S. Magnet, *Macromol. Rapid Commun.*, 2004, **25**, 1215–1220.
74. G. Laruelle, J. Francois and L. Billon, *Macromol. Rapid Commun.*, 2004, **25**, 1839–1844.
75. J. L. Pradel, B. Ameduri and B. Boutevin, *Macromol. Chem. Phys.*, 1999, **200**, 2304–2308.
76. J. L. Pradel, B. Boutevin and B. Ameduri, *J. Polym. Sci., Part A: Polym. Chem.*, 2000, **38**, 3293–3302.
77. S. C. Hong, *Elastomers Compos.*, 2009, **44**, 55–62.
78. D. Benoit, E. Harth, P. Fox, R. M. Waymouth and C. J. Hawker, *Macromolecules*, 2000, **33**, 363–370.
79. C. Cheng, K. Qi, E. Khoshdel and K. L. Wooley, *J. Am. Chem. Soc.*, 2006, **128**, 6808–6809.
80. J. K. Wegrzyn, T. Stephan, R. Lau and R. B. Grubbs, *J. Polym. Sci., Part A: Polym. Chem.*, 2005, **43**, 2977–2984.
81. A. Sundararaman, T. Stephan and R. B. Grubbs, *J. Am. Chem. Soc.*, 2008, **130**, 12264–12265.
82. Y. Cai, K. B. Aubrecht and R. B. Grubbs, *J. Am. Chem. Soc.*, 2010, **133**, 1058–1065.
83. D. J. Gravert, A. Datta, P. Wentworth and K. D. Janda, *J. Am. Chem. Soc.*, 1998, **120**, 9481–9495.
84. I. Q. Li, B. A. Howell, M. T. Dineen, P. E. Kastl, J. W. Lyons, D. M. Meunier, P. B. Smith and D. B. Priddy, *Macromolecules*, 1997, **30**, 5195–5199.
85. O. Altintas, I. Yilmaz, G. Hizal and U. Tunca, *J. Polym. Sci., Part A: Polym. Chem.*, 2006, **44**, 3242–3249.
86. T. Tsoukatos, S. Pispas and N. Hadjichristidis, *Macromolecules*, 2000, **33**, 9504–9511.
87. F. Yu, J. He, X. Wang, G. Gao and Y. Yang, *J. Polym. Sci., Part A: Polym. Chem.*, 2007, **45**, 4013–4025.
88. D. Gromadzki, A. Jigounov, P. Stepanek and R. Makuska, *Eur. Polym. J.*, 2010, **46**, 804–813.
89. W. Sun, F. Yu, J. He, C. Zhang and Y. Yang, *J. Polym. Sci., Part A: Polym. Chem.*, 2008, **46**, 5518–5527.
90. Z. Ma, C. Zheng, Z. Shen, D. Liang and X. Fan, *J. Polym. Sci., Part A: Polym. Chem.*, 2012, **50**, 918–926.

91. B. Kiskan, B. Gacal, M. Asan, E. C. Gunaydin, I. Yilmaz and Y. Yagci, *Polym. Bull.*, 2011, **67**, 609–621.
92. A. Desai, N. Atkinson, F. Rivera, Jr., W. Devonport, I. Rees, S. E. Branz and C. J. Hawker, *J. Polym. Sci., Part A: Polym. Chem.*, 2000, **38**, 1033–1044.
93. O. Garcia-Valdez, S. George, R. Champagne-Hartley, E. Saldivar-Guerra, P. Champagne and M. F. Cunningham, *Polymer*, 2015, **67**, 139–147.
94. Y. Higaki, H. Otsuka and A. Takahara, *Macromolecules*, 2004, **37**, 1696–1701.
95. J. Su, Y. Amamoto, M. Nishihara, A. Takahara and H. Otsuka, *Polym. Chem.*, 2011, **2**, 2021–2026.
96. R. B. Grubbs, C. J. Hawker, J. Dao and J. M. J. Frechet, *Angew. Chem., Int. Ed. Engl.*, 1997, **36**, 270–272.
97. B. Liu, C. G. Yuan and C. P. Hu, *Macromol. Chem. Phys.*, 2001, **202**, 2504–2508.
98. M. A. Tasdelen, Y. Yagci, A. L. Demirel, T. Biedron and P. Kubisa, *Polym. Bull.*, 2007, **58**, 653–663.
99. S. Wang, Z. Cheng, J. Zhu, Z. Zhang and X. Zhu, *J. Polym. Sci., Part A: Polym. Chem.*, 2007, **45**, 5318–5328.
100. D. Gromadzki, S. Filippov, M. Netopilik, R. Makuska, A. Jigounov, J. Plestil, J. Horsky and P. Stepanek, *Eur. Polym. J.*, 2009, **45**, 1748–1758.
101. D. Gromadzki, R. Makuska, M. Netopilik, P. Holler, J. Lokaj, M. Janata and P. Stepanek, *Eur. Polym. J.*, 2007, **44**, 59–71.
102. M. Hatamzadeh and M. Jaymand, *RSC Adv.*, 2014, **4**, 28653–28663.
103. M. Hatamzadeh, M. Jaymand and B. Massoumi, *Polym. Int.*, 2014, **63**, 402–412.
104. C. J. Hawker, *Angew. Chem., Int. Ed. Engl.*, 1995, **34**, 1456–1459.
105. S. Flakus and G. Schmidt-Naake, *Macromol. Symp.*, 2009, **275–276**, 43–51.
106. S. Voccia, C. Jerome, C. Detrembleur, P. Leclere, R. Gouttebaron, M. Hecq, B. Gilbert, R. Lazzaroni and R. Jerome, *Chem. Mater.*, 2003, **15**, 923–927.
107. U. M. Stehling, E. E. Malmstrom, R. M. Waymouth and C. J. Hawker, *Macromolecules*, 1998, **31**, 4396–4398.
108. N. B. Bowden, M. Dankova, W. Wiyatno, C. J. Hawker and R. M. Waymouth, *Macromolecules*, 2002, **35**, 9246–9248.
109. Y. Miura and M. Okada, *Polymer*, 2004, **45**, 6539–6546.
110. S. Du, J. Zhang, Y. Guan and X. Wan, *Aust. J. Chem.*, 2014, **67**, 39–48.
111. F. J. Hua, B. Liu, C. P. Hu and Y. L. Yang, *J. Polym. Sci., Part A: Polym. Chem.*, 2002, **40**, 1876–1884.
112. M. Appelt and G. Schmidt-Naake, *Macromol. Mater. Eng.*, 2004, **289**, 245–253.
113. C. Chang, J. Zhu, Z. Zhang, N. Zhou, Z. Cheng and X. Zhu, *Polymer*, 2010, **51**, 1947–1953.
114. E. Yoshida, *Colloid Polym. Sci.*, 2011, **289**, 837–841.

115. Y. Miwa, K. Yamamoto, M. Sakaguchi and S. Shimada, *Macromolecules*, 2001, **34**, 2089–2094.
116. T. S. Jo, M. L. Yang, L. V. Brownell and C. Bae, *J. Polym. Sci., Part A: Polym. Chem.*, 2009, **47**, 4519–4531.
117. J. Bonilla-Cruz, E. Saldivar-Guerra, J. R. Torres-Lubian, R. Guerrero-Santos, B. Lopez-Carpy and G. Luna-Barcenas, *Macromol. Chem. Phys.*, 2008, **209**, 2268–2283.
118. E. Saldivar-Guerra, J. Bonilla-Cruz, B. Hernandez-Mireles and G. Ramirez-Manzanares, *Macromol. Symp.*, 2009, **283–84**, 110–119.
119. J. Bonilla-Cruz, C. Guerrero-Sanchez, U. S. Schubert and E. Saldivar-Guerra, *Eur. Polym. J.*, 2010, **46**, 298–312.
120. M. Abbasian and A. A. Entezami, *Polym. Adv. Technol.*, 2007, **18**, 306–312.
121. M. Rymarczyk-Machal, S. Zapotoczny and M. Nowakowska, *J. Polym. Sci., Part A: Polym. Chem.*, 2006, **44**, 2675–2683.
122. M. Rymarczyk-Machal, J. Szafraniec, S. Zapotoczny and M. Nowakowska, *Eur. Polym. J.*, 2014, **55**, 76–85.
123. M. Abbasian, *J. Elastomers Plast.*, 2011, **43**, 481–497.
124. M. Abbasian and S. E. Shoja, *Braz. J. Chem. Eng.*, 2012, **29**, 285–294.
125. S. Mohajery, S. Rahmani and A. A. Entezami, *Polym. Adv. Technol.*, 2008, **19**, 1528–1535.
126. W. H. Daly and T. S. Evenson, *Polymer*, 2000, **41**, 5063–5071.
127. W. H. Daly, T. S. Evenson, S. T. Iacono and R. W. Jones, *Macromol. Symp.*, 2001, **174**, 155–163.
128. D. B. Hua, W. C. Deng, J. Tang, J. X. Cheng and X. L. Zhu, *Int. J. Biol. Macromol.*, 2008, **43**, 43–47.
129. O. Garcia-Valdez, D. G. Ramirez-Wong, E. Saldivar-Guerra and G. Luna-Barcenas, *Macromol. Chem. Phys.*, 2013, **214**, 1396–1404.
130. J. L. Jiang, D. B. Hua, J. Jiang, J. Tang and X. L. Zhu, *Carbohydr. Polym.*, 2010, **81**, 358–364.
131. O. Garcia-Valdez, R. Champagne-Hartley, E. Saldivar-Guerra, P. Champagne and M. F. Cunningham, *Polym. Chem.*, 2015, **6**, 2827–2836.
132. C. J. Hawker, D. Mecerreyes, E. Elce, J. L. Dao, J. L. Hedrick, I. Barakat, P. Dubois, R. Jerome and W. Volksen, *Macromol. Chem. Phys.*, 1997, **198**, 155–166.
133. J. Nicolas, P. Couvreur and B. Charleux, *Macromolecules*, 2008, **41**, 3758–3761.
134. M. Chenal, S. Mura, C. Marchal, D. Gigmes, B. Charleux, E. Fattal, P. Couvreur and J. Nicolas, *Macromolecules*, 2010, **43**, 9291–9303.
135. M. Chenal, C. Boursier, Y. Guillaneuf, M. Taverna, P. Couvreur and J. Nicolas, *Polym. Chem.*, 2011, **2**, 1523–1530.
136. N. Hadjichristidis, H. Iatrou, M. Pitsikalis and J. Mays, *Prog. Polym. Sci.*, 2006, **31**, 1068–1132.
137. J. Li, Z. Zhang, X. Zhu, J. Zhu and Z. Cheng, *e-Polym.*, 2010, **145**, 1–11.
138. Y. Miura and Y. Yoshida, *Macromol. Chem. Phys.*, 2002, **203**, 879–888.

139. Y. Miura and H. Dote, *J. Polym. Sci., Part A: Polym. Chem.*, 2005, **43**, 3689–3700.
140. J. B. Beil and S. C. Zimmerman, *Macromolecules*, 2004, **37**, 778–787.
141. T. Kakuchi, A. Narumi, T. Matsuda, Y. Miura, N. Sugimoto, T. Satoh and H. Kaga, *Macromolecules*, 2003, **36**, 3914–3920.
142. P. Gopalan, Y. M. Zhang, X. F. Li, U. Wiesner and C. K. Ober, *Macromolecules*, 2003, **36**, 3357–3364.
143. Y. Miura and Y. Yoshida, *Polym. J.*, 2002, **34**, 748–754.
144. G. Chessa, A. Scrivanti, U. Matteoli and V. Castelvetro, *Polymer*, 2001, **42**, 9347–9353.
145. C. H. Lu, J. H. Wang, F. C. Chang and S. W. Kuo, *Macromol. Chem. Phys.*, 2010, **211**, 1339–1347.
146. J. Pyun, I. Rees, J. M. J. Frechet and C. J. Hawker, *Aust. J. Chem.*, 2003, **56**, 775–782.
147. Y. Miura and M. Dairoku, *J. Polym. Sci., Part A: Polym. Chem.*, 2007, **45**, 4364–4376.
148. S. Abraham, J. H. Choi, C. S. Ha and I. Kim, *J. Polym. Sci., Part A: Polym. Chem.*, 2007, **45**, 5559–5572.
149. P.-E. Dufils, N. Chagneux, D. Gigmes, T. Trimaille, S. R. A. Marque, D. Bertin and P. Tordo, *Polymer*, 2007, **48**, 5219–5225.
150. N. Ide and T. Fukuda, *Macromolecules*, 1997, **30**, 4268–4271.
151. N. Ide and T. Fukuda, *Macromolecules*, 1999, **32**, 95–99.
152. S. Abrol, P. A. Kambouris, M. G. Looney and D. H. Solomon, *Macromol. Rapid Commun.*, 1997, **18**, 755–760.
153. A. J. Pasquale and T. E. Long, *J. Polym. Sci., Part A: Polym. Chem.*, 2001, **39**, 216–223.
154. T. Tsoukatos, S. Pispas and N. Hadjichristidis, *J. Polym. Sci., Part A: Polym. Chem.*, 2001, **39**, 320–325.
155. H. Kaga, A. Narumi, T. Satoh, T. Matsuda, M. Sharfuddin and T. Kakuchi, *Macromol. Symp.*, 2002, **181**, 95–100.
156. A. W. Bosman, R. Vestberg, A. Heumann, J. M. J. Frechet and C. J. Hawker, *J. Am. Chem. Soc.*, 2003, **125**, 715–728.
157. C. T. Adkins and E. Harth, *Macromolecules*, 2008, **41**, 3472–3480.
158. T. He, D. Li, X. Sheng and B. Zhao, *Macromolecules*, 2004, **37**, 3128–3135.
159. O. Altintas, G. Hizal and U. Tunca, *J. Polym. Sci., Part A: Polym. Chem.*, 2006, **44**, 5699–5707.
160. Y. Miura, A. Narumi, S. Matsuya, Q. Duan, H. Kaga and T. Kakuch, *J. Polym. Sci., Part A: Polym. Chem.*, 2005, **43**, 4271–4979.
161. J. M. J. Frechet, M. Henmi, I. Gitsov, S. Aoshima, M. R. Leduc and R. B. Grubbs, *Science*, 1995, **269**, 1080–1083.
162. C. J. Hawker, J. M. J. Frechet, R. B. Grubbs and J. Dao, *J. Am. Chem. Soc.*, 1995, **117**, 10763–10764.
163. A. Khan, M. Malkoch, M. F. Montague and C. J. Hawker, *J. Polym. Sci., Part A: Polym. Chem.*, 2008, **46**, 6238–6254.
164. J. Roovers, *Macromolecules*, 1985, **18**, 1359–1361.

165. J. Roovers, *J. Polym. Sci., Part B: Polym. Phys.*, 1985, **23**, 1117–1126.
166. M. Schappacher and A. Deffieux, *Science*, 2008, **319**, 1512–1515.
167. S. Lecommandoux, R. Borsali, M. Schappacher, A. Deffieux, T. Narayaman and C. Rochas, *Macromolecules*, 2004, **37**, 1843–1848.
168. S. Cramail, M. Schappacher and A. Deffieux, *Macromol. Chem. Phys.*, 2000, **201**, 2328–2335.
169. B. Lepoittevin, X. Perrot, M. Masure and P. Hemery, *Macromolecules*, 2001, **34**, 425–429.
170. G. O'Bryan, N. Ningnuek and R. Braslau, *Polymer*, 2008, **49**, 5241–5248.
171. J. Ruehl, N. Ningnuek, T. Thongpaisanwong and R. Braslau, *J. Polym. Sci., Part A: Polym. Chem.*, 2008, **46**, 8049–8069.
172. A. Narumi, S. Zeidler, H. Barqawi, C. Enders and W. H. Binder, *J. Polym. Sci., Part A: Polym. Chem.*, 2010, **48**, 3402–3416.
173. R. Nicolay and K. Matyjaszewski, *Macromolecules*, 2011, **44**, 240–247.
174. T. Maeda, H. Otsuka and A. Takahara, *Prog. Polym. Sci.*, 2009, **34**, 581–604.
175. G. Yamaguchi, Y. Higaki, H. Otsuka and A. Takahara, *Macromolecules*, 2005, **38**, 6316–6320.

CHAPTER 9

NMP-derived Materials for Biomedical Applications

EVA HARTH[a] AND JULIEN NICOLAS*[b]

[a] Department of Chemistry, Vanderbilt University, Nashville, TN 37235, USA; [b] Institut Galien Paris-Sud, UMR CNRS 8612, Univ Paris-Sud, Faculté de Pharmacie, 5 rue Jean-Baptiste Clément, F-92296 Châtenay-Malabry cedex, France
*Email: julien.nicolas@u-psud.fr

9.1 Introduction

Nitroxide-mediated polymerization (NMP)[1] continues to be intensively investigated owing to its simplicity and versatility to produce not only high precision structures,[2–4] controlled in on–off polymerization methods initiated by low energy light,[5,6] but also by their ability to aid the preparation of biomaterials under various conditions. Polymer-based biomaterials have been recognized as ideal materials to study, detect or influence biological interactions. In this chapter, we particularly focus on the NMP technique to create biomaterials emphasizing glycopolymers, bioconjugates, nanoparticulate systems and polymer nanocomposites. While NMP has experienced a set-back in favor of fast developing ATRP[7] and RAFT[8] techniques, surpassing some of the early successes with NMP, developed SG1 and SG1-based alkoxyamine such as BlocBuilder[1] have led to increased research activity to further exemplify the versatility of using NMP, particularly in the preparation of biomaterials. Another contributing factor that has been limiting the use of NMP was the rather restricted commercial availability of SG1 and SG1-based alkoxyamines, which undermines the broader popularity of this technique.

RSC Polymer Chemistry Series No. 19
Nitroxide Mediated Polymerization: From Fundamentals to Applications in Materials Science
Edited by Didier Gigmes

Published by the Royal Society of Chemistry, www.rsc.org

Note that the choice of NMP is particularly relevant and appealing for the preparation of biomaterials due to: (i) its simplicity; (ii) the absence of sulfur-based compounds, metal-based catalysts or radical initiators; and (iii) the innocuousness of the employed nitroxide.[9]

The selection of biomaterials in this chapter will underline the ease of use of NMP in organic and aqueous environments,[10,11] at rather modest reaction temperatures[12] and fast reaction times, which are desired conditions when preparing biomacromolecules or model compounds towards an application in the biomedical field. The versatility of functional monomers[13–15] undergoing NMP to reach architectural diversity[16,17] and the precision of co- and block copolymers[18–20] with limited side reactions have led to numerous examples in the form of glycopolymers and synthetic polymers with polypeptides and related biomaterials. Nanocomposites with silica showed in various ways that their functionalization with SG1 or BlocBuilder accelerated the progress of these architectures as sensors, anticoagulants and polymer reinforcements. The preparation of selected polymeric nanoparticles for potential bioapplications[21,22] from NMP will conclude the chapter.

9.2 Glycopolymers

Together with DNA and proteins, sugar moieties are information-rich molecules in living organisms. They play a crucial role in life science during many recognition processes involving highly specific events in cell–cell recognition, cell–protein interactions and the targeting of hormones, antibodies and toxins. The density of information they can convey is the result of their structural diversification. These recognition processes are thought to proceed by specific non-covalent carbohydrate–protein interactions where the proteins are generically termed lectins; these are most frequently found on cell surfaces. Lectins are indeed able to decode the carbohydrate-encoded information from any variations in the oligosaccharide structure. Because the mechanism of carbohydrate–lectin interaction and the structure of the glycopolymers involved in this recognition process are still largely unknown, the synthesis of complex carbohydrate-based polymers has received increasing attention over the past few years and is emerging as a relevant strategy to investigate this kind of interaction.[23]

The use of controlled/living radical polymerization (CLRP) techniques offers flexible and promising synthetic routes to afford new glycopolymers with well-defined characteristics and complex architectures.[24] NMP has been employed as a relevant strategy to design carbohydrate-based polymers from the polymerization of carbohydrate-containing monomers (also termed glycomonomers). Various glycomonomers bearing different polymerizable groups have been used, such as styrenic, acrylate and methacrylate (Table 9.1).

In general, best results were obtained from protected glycomonomers. For instance, early attempts from *N*-(*p*-vinylbenzyl)-[*O*-β-D-galactopyranosyl-(1→4)]-D-gluconamide (VLA, **G1** with R = H) led to poor control by the

Table 9.1 Structure of glycomonomers polymerized by NMP.

	Glycomonomer	Ref.
G1	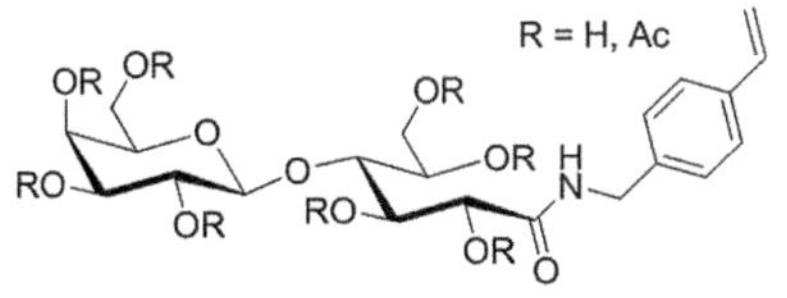	25–27
G2	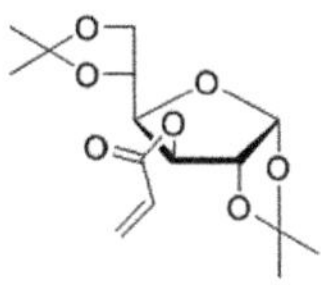	28
G3	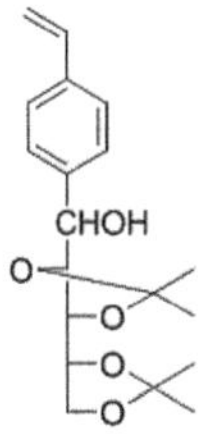	29, 30
G4	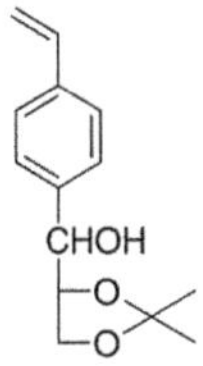	29, 30
G5	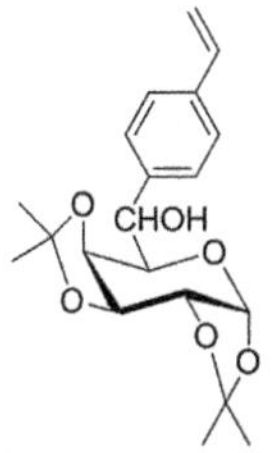	29, 30
G6	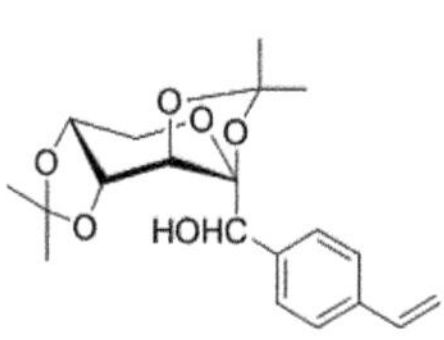	29, 30
G7	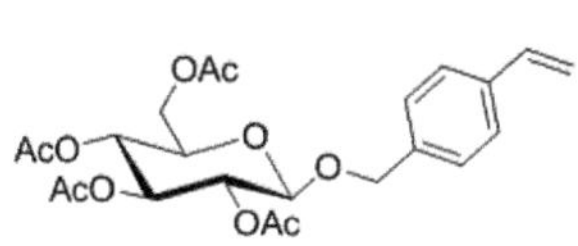	31–33

Table 9.1 (*Continued*)

	Glycomonomer	Ref.
G8	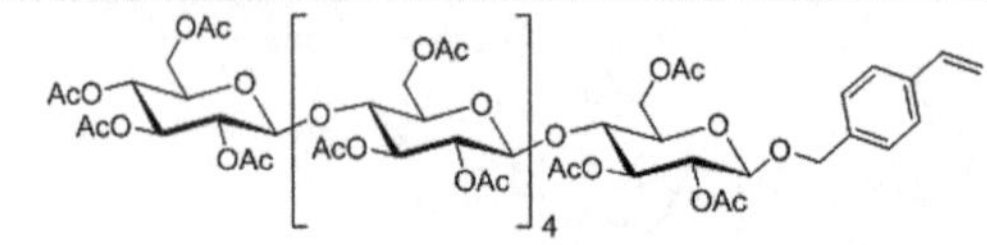	31, 33
G9	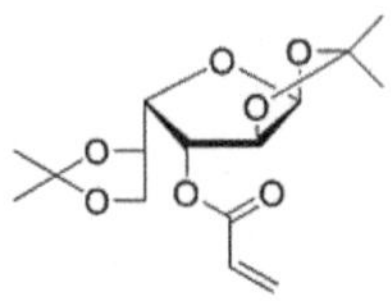	34
G10	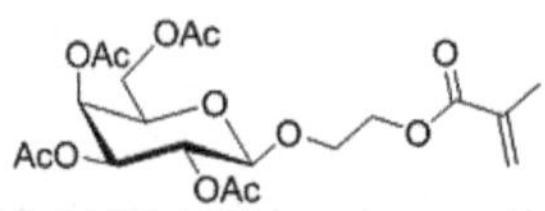	35

2-(benzoyloxy)-1-(phenylethyl)-DBNO (BS-DBNO) alkoxyamine in the presence of dicumyl peroxide in DMF at 90 °C, as opposed to its protected counterpart, (Ac-VLA, **G1** with R = Ac) which led to well-defined glycopolymers in the 2–40 kDa range under identical experimental conditions.[25–27] A lipophilic DBNO-based alkoxyamine, 1-(4-*N*,*N*-dioctadecylcarbamoylphenyl)-ethyl-DBNO, also yielded good control with Ac-VLA and the resulting glycolipid was further mixed with a phospholipid to form stable liposomes,[27] at the surface of which galactose residues specifically recognized RCA lectin. A similar pathway was also successfully applied to an acrylate-containing glycomonomer, 3-*O*-acryloyl-1,2:5,6-di-*O*-isopropylidene-α-D-glucofuranoside (AIpGlc, **G2**), in the presence of BS-DBNO in *p*-xylene at 100 °C.[28] Predicted molar masses and rather low dispersities were obtained, together with successful chain-extension to yield a well-defined PS-*b*-PAIpGlc block copolymer.

TEMPO-based molecular and macromolecular alkoxyamines were also shown to control the polymerization of other sugar-functionalized styrene (S) derivatives. For instance, the styryl-TIPNO alkoxyamine was employed for the bulk or solution polymerization at 130 °C of protected glycomonomers **G3–G6**.[29,30] Whereas bulk experiments from **G3–G5** yielded controlled structures of different targeted molar masses, the dispersity obtained from **G6** was unexpectedly high.[30] In diphenyl ether at 50 wt%, the solution polymerization of **G3** gave the lowest dispersity, even though the polymerization rate was lower. A small library of diblock copolymers was also prepared by chain extension with S from TEMPO-terminated glycopolymers obtained from **G3–G6**, even though dispersities significantly increased due to the presence of dead polymer chains. The protected glycomonomer **G1** was also controlled by a BPO-TEMPO alkoxyamine in 40 wt% DMF at 130 °C and gave low $Đ$ (<1.4) glycopolymers whose affinities towards lectins were shown to depend on the degree of polymerization; the higher the DP_n, the stronger the affinity.[26]

Access to well-defined diblock copolymers comprising a sugar block was more successful when protected glycomonomers were polymerized from a nitroxide end-capped polymer. For instance, polymerization of protected glycomonomers **G3**,[29] **G7**[31] and **G8**[31] from low molar mass TEMPO-terminated PS, followed by subsequent deprotection yielded low $Đ$ diblock copolymers. Interestingly, when divinylbenzene was used as a cross-linking agent upon chain extension with glycomonomer **G7** and **G8**, core-glyco-conjugated star-shaped PS architectures were obtained.[33] Nevertheless, a difunctional TEMPO-based alkoxyamine used for the polymerization of **G7** at 125 °C for 5 h followed by chain extension afforded triblock copolymers.

The efficiency of second-generation nitroxides such as SG1 and TIPNO makes them better candidates for the NMP of acrylate- and even methacrylate-based (under certain conditions) sugars. Hawker and co-workers successfully controlled the polymerization of glycomonomer **G9** in DMF at 105 °C with a TIPNO-based alkoxyamine functionalized with a lipophilic *N,N*-di(octadecyl) amine group.[34] The resulting lipo-glycopolymers were surface active and formed stable Langmuir films with accurate control over the molecular area. Interestingly, the copolymerization approach popularized by Charleux and co-workers,[1] which consists in copolymerizing methacrylic esters with a small amount of a suitable comonomer (*e.g.*, styrene, acrylonitrile), was applied to a sugar-based methacrylate monomer by Stenzel and co-workers.[35] Well-defined poly(2-(β-D-galactosyloxy)ethyl methacrylate-*co*-styrene)-*b*-polystyrene (P(GalEMA-*co*-S)-*b*-PS) amphiphilic block copolymers were obtained from SG1-mediated polymerization of glycomonomer **G10**, using either P(AcGalEMA-*co*-S)-SG1 or PS-SG1 macroinitiators, followed by deacetylation of AcGalEMA moieties (Figure 9.1). The self-assembling ability of PS-*b*-P(GalEMA-*co*-S) amphiphilic glycopolymers was then exploited to obtain micellar structures and honeycomb structured porous films of maintained biofunctionality.

Water-soluble, glucosylated poly(pentafluorostyrene) derivatives made by NMP were shown to be favorable coating materials for magnetic iron oxide nanoparticles.[36] The strategy relied on the design of a glucosylated, tetrafluorostyryl monomer, by thiol-para-fluorine "click" reaction, and its polymerization using the BlocBuilder alkoxyamine. These nanoparticles showed no cytotoxicity towards 3T3 cells and the application of fluorescein isothiocyanate (FITC)-labeled glycopolymer yielded fluorescent, superparamagnetic nanoparticles, which can be used for monitoring cell-carbohydrate interactions.

9.3 Bioconjugates

The combination of synthetic polymers with polypeptides has opened an avenue towards innovative materials with unprecedented properties

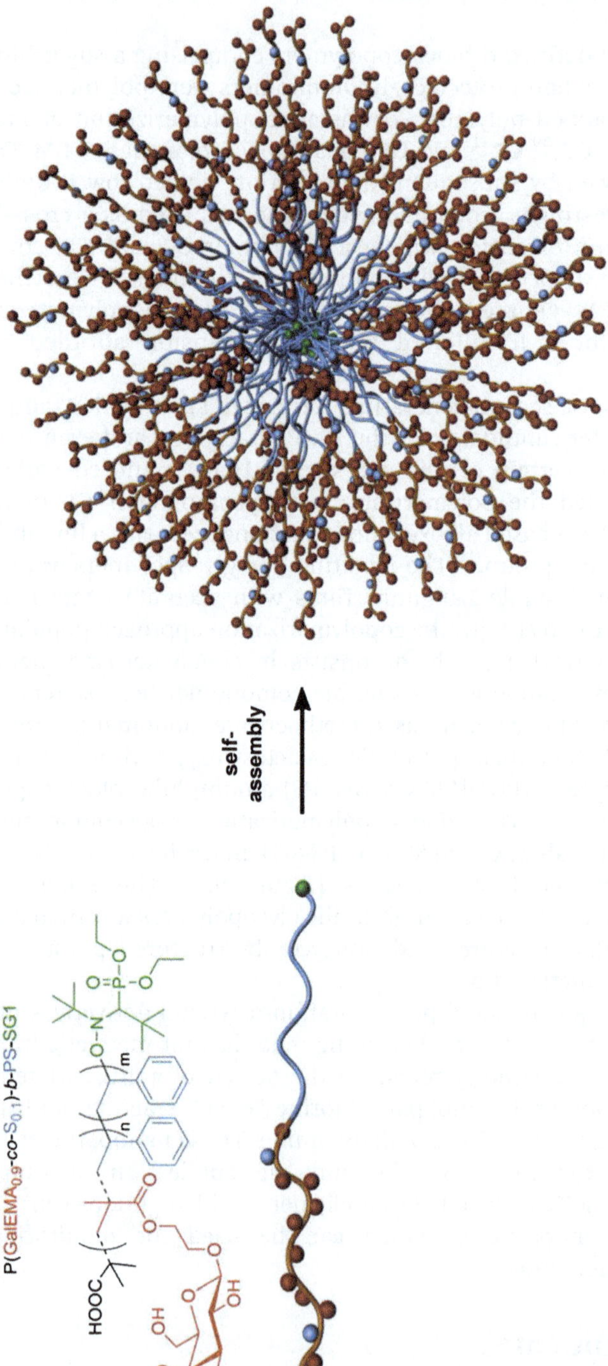

Figure 9.1 Self-assembly of amphiphilic poly(2-(β-D-galactosyloxy)ethyl methacrylate-*co*-styrene)-*b*-polystyrene (P(GalEMA-*co*-S)-*b*-PS) diblock copolymers obtained from NMP into carbohydrate-based micellar structures. Reproduced from ref. 1 with permission of Elsevier.

obtained in part because of the intrinsic (bio)functionalities of proteins and peptides. In this field, the most typical example is certainly the covalent linkage of poly(ethylene glycol) (PEG), a hydrophilic and flexible polymer, to poly(peptide)-based therapeutics, so-called PEGylation.[37,38] The resulting bioconjugates exhibit improved biodistribution and pharmacokinetics, better stability and solubility, reduced immunogenicity and longer plasma half-lives due to both reduced renal filtration and proteolysis, when compared to non-PEGylated analogs.[39]

The fast growth of CLRP techniques has brought about a significant renewal as new types of bioconjugates with unique properties have been successfully synthesized.[40–44] NMP has met some success with the preparation of polymer–protein/peptide bioconjugates. Different strategies have been used so far, including the conjugation of NMP-derived polymers to peptides attached to a solid support or to free peptides/proteins, the NMP from peptidic macroinitiators or the combination of NMP and NCA polymerization.

Conjugation to free peptides and proteins can be achieved by a combination of solid-phase peptide synthesis (SPPS) and surface-initiated NMP from the resin. This was shown by the derivatization of the protein transduction domain (PTD) of the human immunodeficiency virus (HIV) Tat protein[45] and the antimicrobial peptide tritrpticin (VRRFPWWWPFLRR),[46] by a TIPNO-based alkoxyamine for sequential NMP at 135 °C of *t*BA and MA (or S). Well-defined PTD-poly(acrylic acid)-*b*-poly(methyl acrylate) (PTD-PAA-*b*-PMA) and tritrpticin-poly(acrylic acid)-*b*-polystyrene (tritrpticin-PAA-*b*-PS) block copolymers were obtained. Interestingly, tritrpticin-PAA-*b*-PS micelles exhibited an enhancement of the antimicrobial activity against *Staphylococcus aureus* (*S. aureus*) and *Escherichia coli* (*E. coli*) when compared to free tritrpticin. An SG1-functionalized peptide (SG1-GGGWIKVAV) was also constructed by SPPS followed by the direct attachment of the BlocBuilder alkoxyamine to the terminal amine group of the peptide in the presence of PyBOP as the coupling agent (Figure 9.2).[47] Further surface-initiated NMP of S yielded the resulting peptide–PS bioconjugates following its cleavage from the support. The resin-supported SG1-GGGWIKVAV was also reacted with various vinyl-funtionalized peptides (either at its *C*-terminus or *N*-terminus), through intermolecular radical 1,2-addition.[48] This reaction proceeded without any coupling agent and represents an efficient peptide ligation strategy.

A simple approach for the conjugation of preformed polymer to free peptides and proteins (the so-called "*grafting to*" strategy) was recently reported. It relies on the design of α-functional comb-shaped polymethacrylates with PEG side chains from NMP of OEGMA with a few percent of acrylonitrile (to ensure good control due to the well-known inability of SG1 to control the homopolymerization of methacrylic esters) initiated by SG1-based alkoxyamines bearing an *N*-hydroxysuccinimidyl (NHS) moiety (Figure 9.3).[49] Importantly, the resulting copolymers showed no obvious cell toxicity on three representative cell lines (*i.e.*, J774, NIH3T3 and HUVEC)

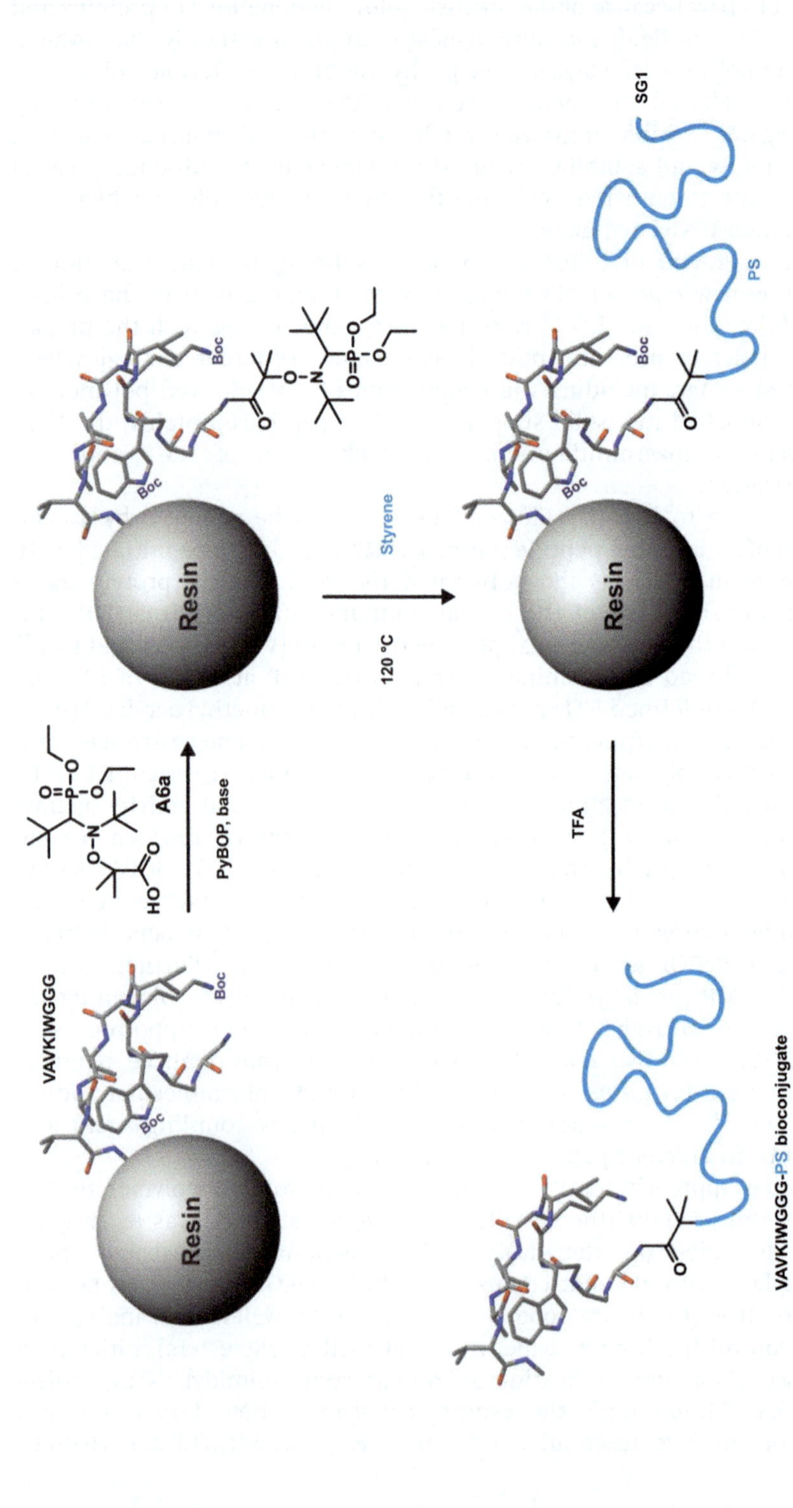

Figure 9.2 Synthesis of SG1-functionalized peptides as precursors for polymer–peptide conjugates. Reproduced from ref. 1 with permission of Elsevier.

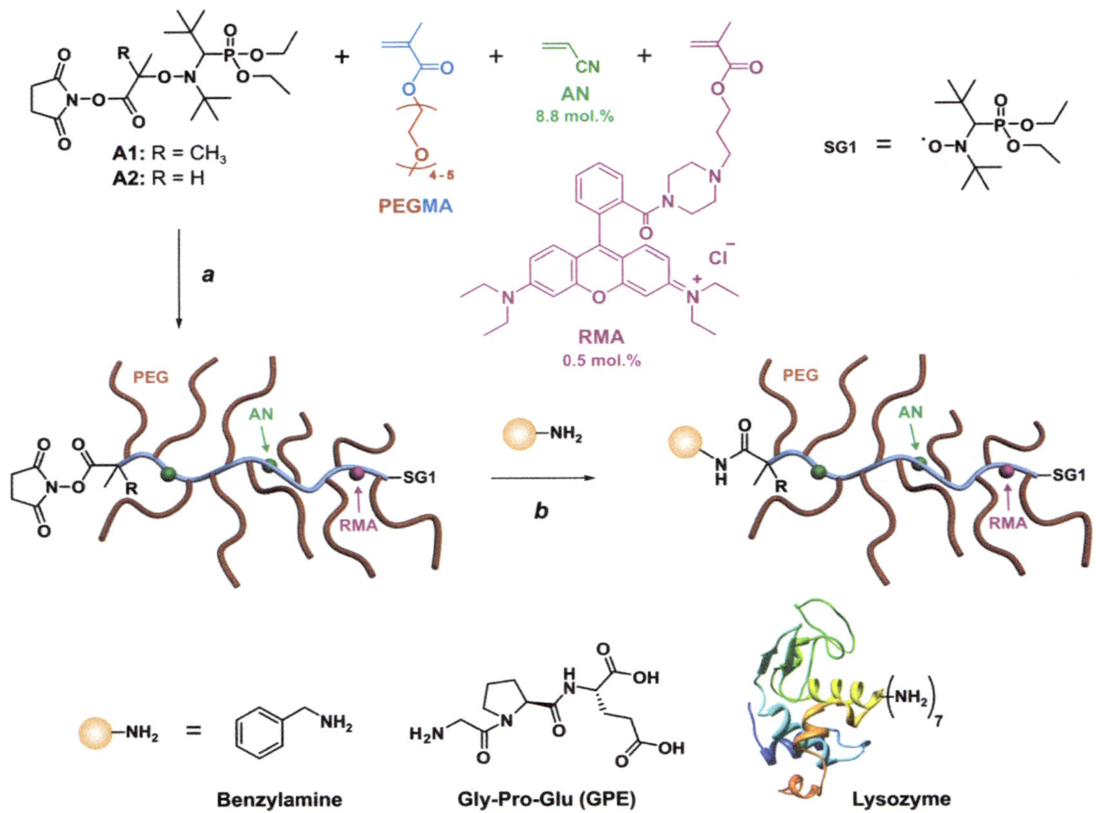

Figure 9.3 Design of α-functional, fluorescent comb-shaped polymethacrylates with PEG side chains for bioconjugation purposes. Reproduced from ref. 49 with permission of RSC Publishing.

even at very high doses, which is a prerequisite for biomedical applications. Conjugation reactions were demonstrated on a neuroprotective tripeptide and lysozyme as a model protein. Interestingly, the most efficient coupling to lysozyme was obtained from a less sterically hindered NHS moiety.

Conversely to the "*grafting to*" strategy, growing a polymer chain by NMP directly from peptide macroinitiators, usually termed "*grafting from*", is also a suitable method to prepare bioconjugates. A first approach consisted in the linkage of the nitroxide moiety (*e.g.*, TEMPO, TEMPO derivative) to an L-serine side chain *via* ether bond formation. The resulting compound was readily immobilized into various peptides using standard solution phase peptide chemistry.[50] Whereas a TEMPO-terminated peptide macroinitiator enabled a good control of the polymerization of S, the hindered TEMPO also allowed well-defined P*n*BA- or PNIPAAm-based bioconjugates to be obtained (Figure 9.4). With the latter, it was shown that the enzymatic reaction with chymotrypsin at 37 °C ($T >$ LCSTPNIPAAm) was totally suppressed, whereas at 22 °C ($T <$ LCSTPNIPAAm), the enzymatic reaction occurred and could be stopped by raising the temperature back to 37 °C. The synthetic pathway was further extended to a series of peptides with an alternating sequence of TEMPO-based alkoxyamine conjugated to lysine and glycine residues.[51] They were used as macroinitiators for NMP of S or NIPAAm to obtain peptide–polymer conjugates with well-defined linear peptide backbones and a defined number of polymeric side chains. Depending on their structures, their aggregation gave different topologies. For instance, peptide–PS conjugates led to honeycomb morphologies.

Another strategy to construct polymer–peptide bioconjugates is to combine the polymerization of NCA with NMP. NCA polymerization proceeds by a ring-opening mechanism and is initiated by nucleophiles or bases, such as primary amines or alkoxides.[52–56] Different alkoxyamines bearing suitable functionalities to initiate the ROP of NCA have been reported in the literature. Difunctional TIPNO-based initiators were used for the ROP of β-benzyl-L-glutamate-*N*-carboxyanhydrides (BLG-NCA), followed by the NMP of S in one-pot.[57] Well-defined high molar masses poly(γ-benzyl-L-glutamate)-*b*-polystyrene (PBLG-*b*-PS) bioconjugates (73–150 kDa) with low dispersities were obtained. A dual initiator containing a primary amine and a TIPNO group enabled the sequential ROP of BLG-NCA in DMF at 0 °C and NMP of S mediated by TIPNO at 120 °C.[58] Well-defined bioconjugates were obtained, either in a two-step process (*i.e.*, from a purified PBLG-TIPNO macroinitiator) or in a one-pot process, demonstrating the high compatibility of both polymerization techniques. 4-Amino-TEMPO enabled the preparation of block copolymers of PNVP and polypeptides (PNVP-*b*-PBLG, PNVP-*b*-PBLL and PNVP-*b*-PBLG-*b*-PBLL), with PBLL for poly(*tert*-butyloxycarbonyl-L-lysine) in a sequential fashion (Figure 9.5).[59] Well-defined polypeptides bearing TEMPO end-groups were first prepared, followed by the NMP of NVP in the presence of AIBN and acetic anhydride. Controlled molar masses from 23 to 97 kg mol^{-1} and narrow dispersities (typically 1.2) were obtained.

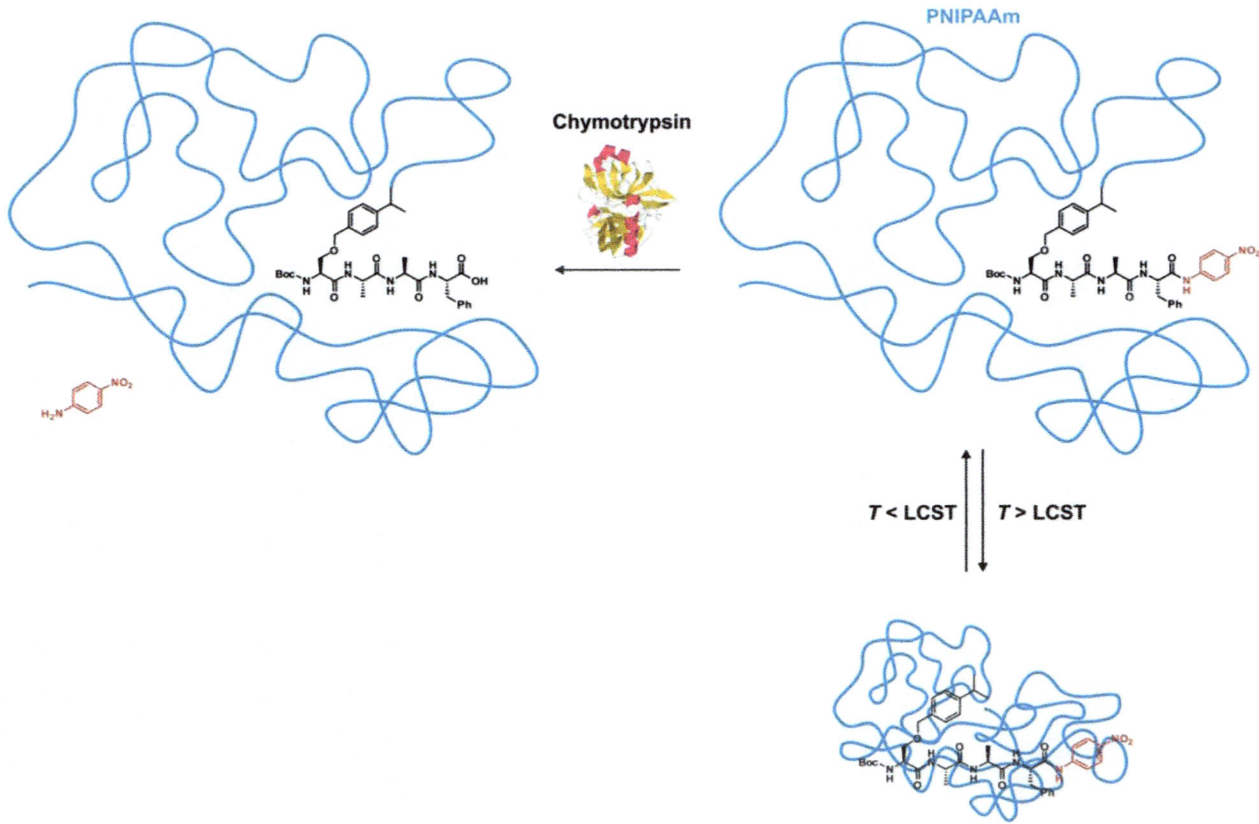

Figure 9.4 Reversible switching of the activity in PNIPAAm–peptide conjugates. Reproduced from ref. 1 with permission of Elsevier.

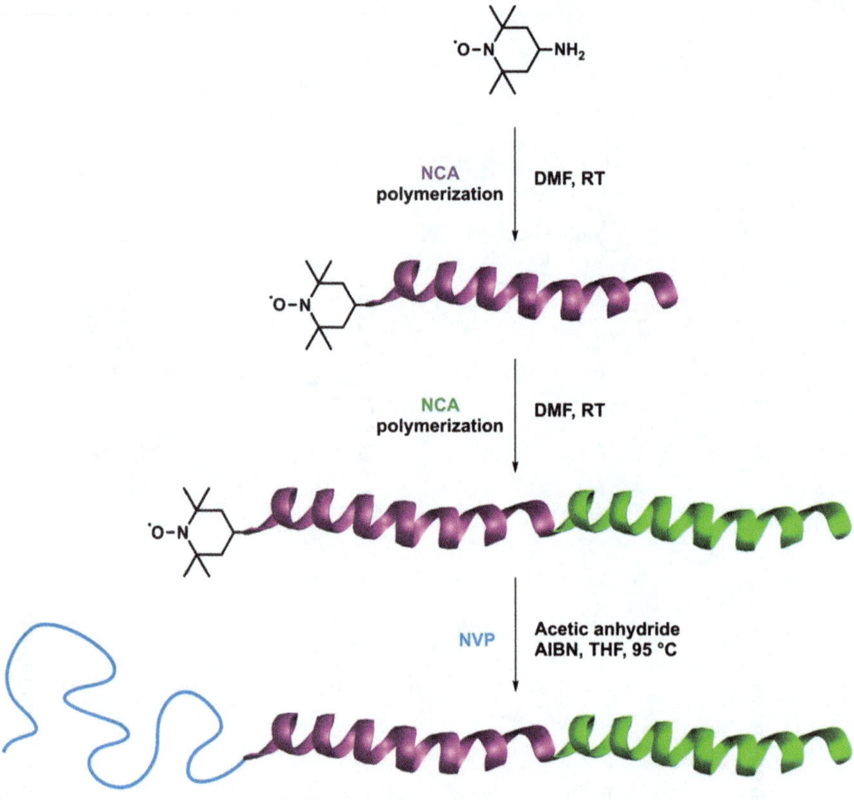

Figure 9.5 Synthesis of polypeptide–poly(*N*-vinyl pyrrolidone) bioconjugates by a combination of NCA polymerization and NMP from 4-amino-TEMPO. Reproduced from ref. 1 with permission of Elsevier.

9.4 Polymer Nanocomposites and Biocompatible Surfaces

One of the first implementations of NMP polymerization methods was in the preparation of polymer brushes in nanoscale patterns with different wettability *via* grafting-from approaches.[60] Polymer grafts from these types of surface functionalization can contribute to the prevention of nonspecific protein adsorptions in cardiovascular biomaterials and provide biomimetic cues to facilitate endothelial adhesion. The combination of non-fouling properties together with integrating biomimetic cell-adhesive end groups are therefore desirable features and are realized with surface-initiated polymerization of carboxybetaine methacrylamides from β-phoshonylated alkoxyamine initiators on silica wafers. Controlled charge densities are achieved *via* charge separating spacer alkyl groups in the carboxybetaine units as explored by Unsworth and co-workers.[61] The same concept and polymer structures have been employed by the same group to functionalize oxidized silicon nanoparticles (SiNP).[62] These modified nanoparticles were studied as model systems to understand how the polymer structure and function affect film hydration, absorbed protein conformation, and clot formation kinetics. The effect of the charge, hydration and end-groups that these films have on absorbed protein conformation were studied with three plasma proteins, lysozyme, bovine α-lactalbumin, and human serum albumin. In addition, it could be confirmed that the chain chemistry rather than the end group determines the film properties and clot formation (Figure 9.6).

Instead of using exclusively grafting-from approaches to functionalize and form polymer nanocomposite materials, a more practical way was facilitated with the development of SG1-based alkoxyamines such as BlocBuilder. Rather than growing chains from surfaces and particles, *N*-succidinimidyl ester entities on these initiators enable activated ester end groups on the α-end of the well-defined polymers. In this way, amine-functionalized silica particles could be grafted in the 0.1–0.2 chain nm^2 range, but with the expected lower density than compared from grafting-from approaches[63] as Charleux and co-workers demonstrated. In a similar approach, Gigmes and co-workers formed aminopropyl functionalized silica particles converted to an initiating species *via* the activated ester of the BlocBuilder alkoxyamine.[64] The optimization of the reaction parameters allowed for a high grafting density and avoided aggregation of particles during the PS growth to guarantee high quality particles for polymer reinforcement. Not only outer surfaces can be functionalized but also inner surfaces depending on the structure of the silica particle. Mesoporous silica was directly modified by a dimethylethoxysilane functional group as a derivative of the SG1-based initiator and polymerizations can be performed at the inner and outer surface of the particles. Billon and co-workers investigated a so-called *in situ* synthesis of acrylic/clay materials containing different weight-percentiles (1–4%) of raw and modified montmorillonite.[65] Also here, the SG1-based alkoxyamine was found to be a valuable initiator to realize materials with unusual thermo-mechanic properties. All these nanocomposites can be viewed also as model studies for further application to other polymers that are for instance thermo-responsive, such as poly(2-hydroxyethyl acrylate) or copolymers of HEA and HPA, as reported by Hoogenboom, and are suitable to be applied in biomedical applications.[66] A more complex polymer nanocomposite material was presented by Bourgeat-Lami[67] in which polymerization-induced self-assembly (PISA) was used to generate amphiphilic block copolymers nodules with morphologies of dumbbell-, daisy- or raspberry like structures randomly distributed around the central spheres of silica. Depending on the macroinitator concentration or the employed silica particle size, different morphologies are achieved (Figure 9.7). The first step in the synthesis encompasses the self-assembly of a water-soluble brush-type PEG-based macroalkoxyamine initiator (P(PEGMA$_{12}$-*co*-S$_1$)-SG1), composed of poly(ethylene glycol) methacrylate and a small amount of styrene, onto the surface of the silica particles. A following emulsion polymerization under

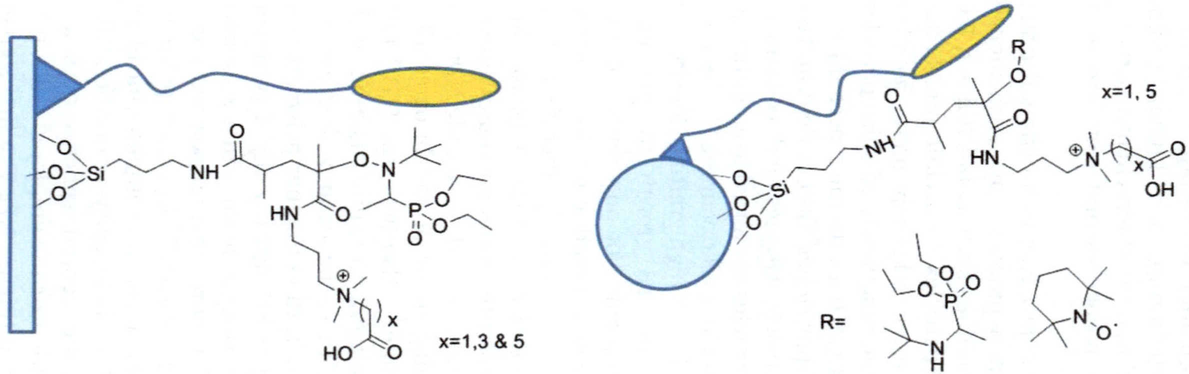

Figure 9.6 Combination of non-fouling properties with integrated biomimetic cell-adhesive end groups on silica surfaces and silica particles SiNP.

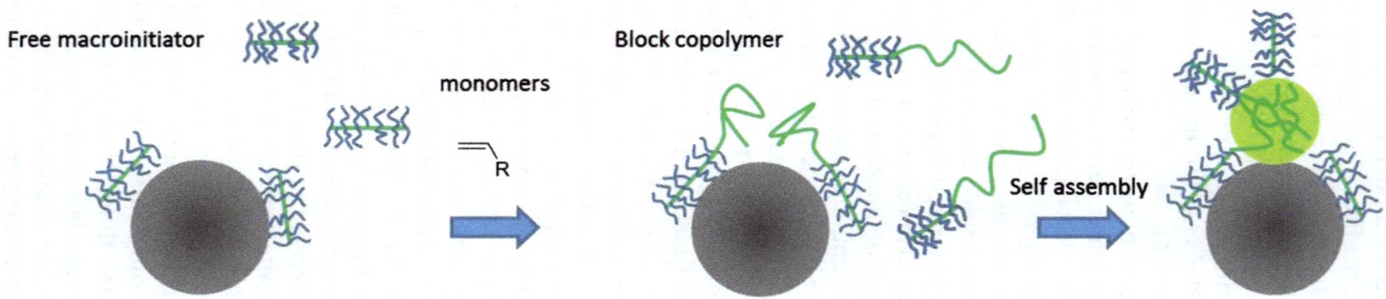

Figure 9.7 Design of polymer nanocomposite materials by polymerization-induced self-assembly (PISA).

mild conditions of *n*-butyl methacrylate with a small amount of styrene was initiated from these silica/macroinitiators and formed extraordinary multipod-like silica/polymer latexes.

9.5 Polymeric Nanoparticles

Polymeric nanoparticles gain importance as carrier systems for drugs,[68–70] imaging scaffolds or as component in polymer mixtures to change the rheology of materials. ATRP and RAFT polymerizations have been widely used for preparing star polymers and micellar systems as nanostructures for biomedical applications.[71] In this arena, NMP has been particularly investigated for particles derived from aqueous dispersion and emulsion polymerization techniques.[72,73] While acrylate-based polymers are non-degradable, their pH and thermoresponsive properties are attractive features for dual or triggered drug delivery approaches. For example, Charleux and co-workers[74] investigated the synthesis of micelles from diblock copolymers consisting of a poly(acrylic acid) (PAA) segment and an LCST-type poly(*N*,*N*-diethylacrylamide) (PDEAAm) block. Here, the water-solubility of the SG1-capped poly(sodium acrylate) macroalkoxyamine, prepared from the organosoluble alkoxyamine MONAMS, was critical for simultaneously synthesizing the double-responsive diblock copolymer and core-shell nanoparticles. The procedure proved to be more practical than comparable anionic polymerization methods because the stringent conditions and a stepwise procedure to form micellar structures were avoided. Furthermore, a third block can be inserted by adding styrene, with the purpose of stabilizing the *in situ* formed self-assembled structures by creating hydrophobic segregated nodules to produce core–shell–corona PS-PDEAAm-PAA crosslinked nanogels in a one-pot procedure. Indeed, a disassembly upon cooling is avoided and with a delayed incorporation of the styrene monomer, the responsiveness (Figure 9.8) of the nanoparticles can be tuned to work in favor for an application in the biomedical field.

It has to be emphasized that these nanostructures are only possible through the development of water-soluble SG1-based alkoxyamines to enable true miniemulsion and emulsion polymerization processes.[73] In another approach,[75] micelles from biohybrid polymers of poly(*n*-butyl acrylate) (P*n*BA) and polystyrene using a TIPNO and SG1-based bifunctional initiator in combination with a controlled polypeptide synthetic technique, such as *N*-carboxyanhydride (NCA) ring-opening polymerization, enabled the generation of well-defined block copolymers (Figure 9.9). In here, the peptide block, poly(L-glutamate) is used as the hydrophilic component and the formed superstructures (vesicles and micelles) have varying degrees of enzyme responsiveness when exposed to enzymes such as thermolysin and elastase. The gathered information around the enzyme responsiveness of these materials can be translated into a desired composition for the ideal interaction of target enzymes.

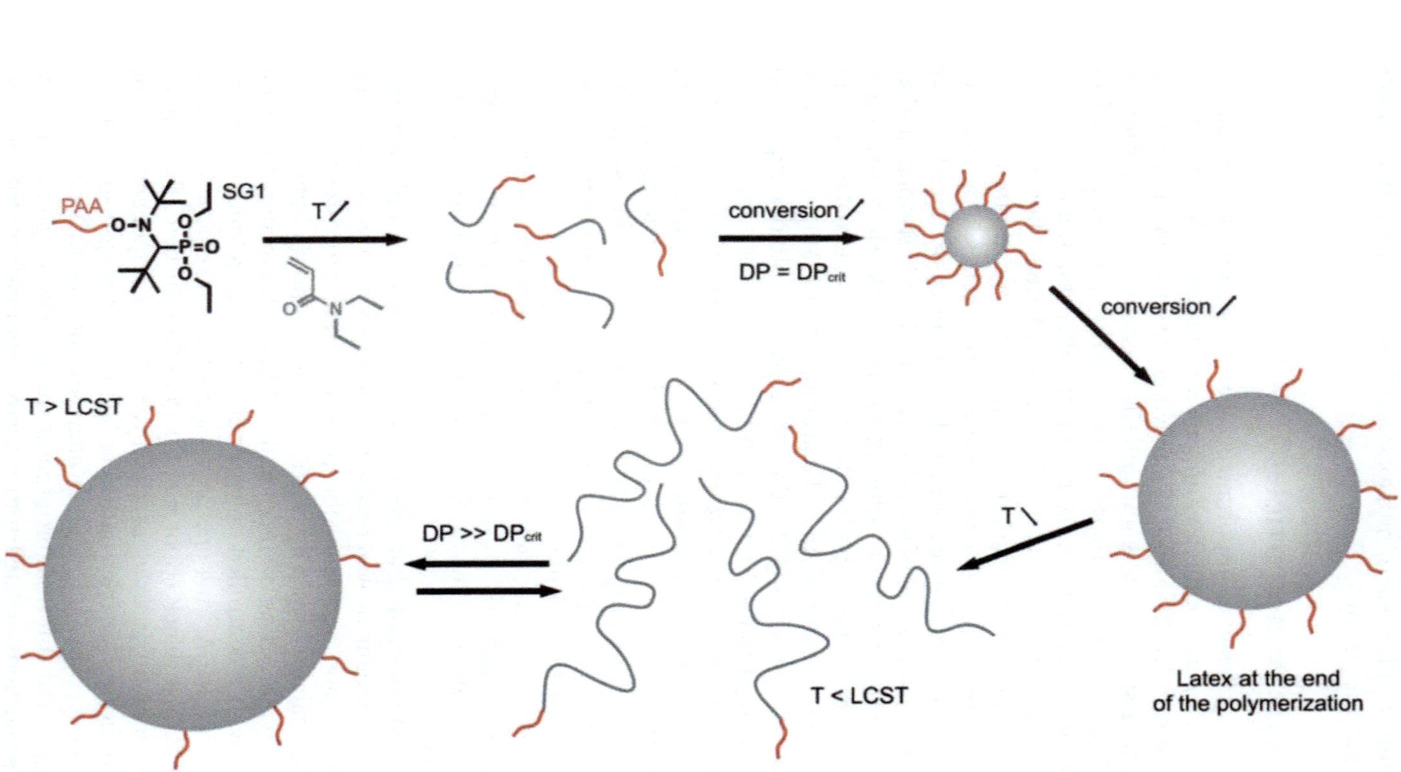

Figure 9.8 Synthesis of micelles *via* dispersion polymerization initiated by a water-soluble SG1-based macroalkoxyamine. Reproduced from ref. 74 with permission of RSC Publishing.

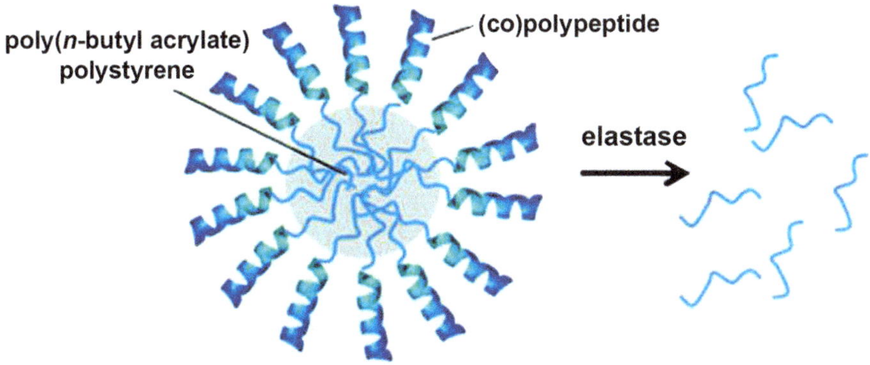

Figure 9.9 Enzyme-responsive polymeric micelles and vesicles. Reproduced from ref. 75 with permission of ACS.

NMP-derived polymer nanoparticles for potential biomedical applications can also be obtained by the formulation of preformed polymers in aqueous solutions, using a variety of methods (*e.g.*, nanoprecipitation, dialysis, *etc.*). In this field, PAA-based cross-linked block copolymer micelles represent an interesting class of materials.[76] Core-shell micelles were obtained from block copolymers designed by TIPNO-mediated polymerization *via* sequential polymerization of *t*BA and of a hydrophobic monomer (typically S derivatives or isoprene), followed by removal of the *tert*-butyl groups resulting in amphiphilic block copolymers able to self-assemble in aqueous solutions. The presence of many functionalities (*e.g.*, carboxylic acid, alkoxyamine, *etc.*) enabled their stabilization by cross-linking, their core- or/and shell-functionalization and their surface functionalization with bio-relevant molecules.[77–82]

More recently, a new class of polymer nanoparticles for drug delivery was developed. The methodology, termed 'drug-initiated', relies on the use of chemotherapeutics as initiating groups for subsequent CLRP of the desired monomer.[83,84] This was achieved by derivatization of the drug with CLRP moieties (*e.g.*, alkoxyamine for NMP,[83] chain transfer agent for RAFT[84,85]). In that case, the drug is covalently linked to the polymer which therefore provides all the benefits of the prodrug approach,[86] compared to traditional drug-loaded polymer nanoparticles (*e.g.*, no 'burst' release, high drug loading, *etc.*). Gemcitabine (Gem), a hydrophilic anticancer drug, was linked to an SG1-based alkoxyamine and served as initiator for the NMP of isoprene (Figure 9.10). The resulting well-defined Gem-PI conjugates led to well-defined and stable nanoparticles of 130–160 nm in diameter with narrow PSD and great colloidal stability. Drug loading ranged from 10.5 to 31.2 wt%, depending on the M_n. Interestingly, Gem-PI nanoparticles showed significant anticancer activity both *in vitro* on various cancer cell lines (*e.g.*, L1210, CCRF-CEM, MiaPaCa-2 and A549) and *in vivo* against a human pancreatic (MiaPaCa-2) carcinoma xenograft model in mice. This method was also applied to the RAFT polymerization of a methacrylate monomer based on

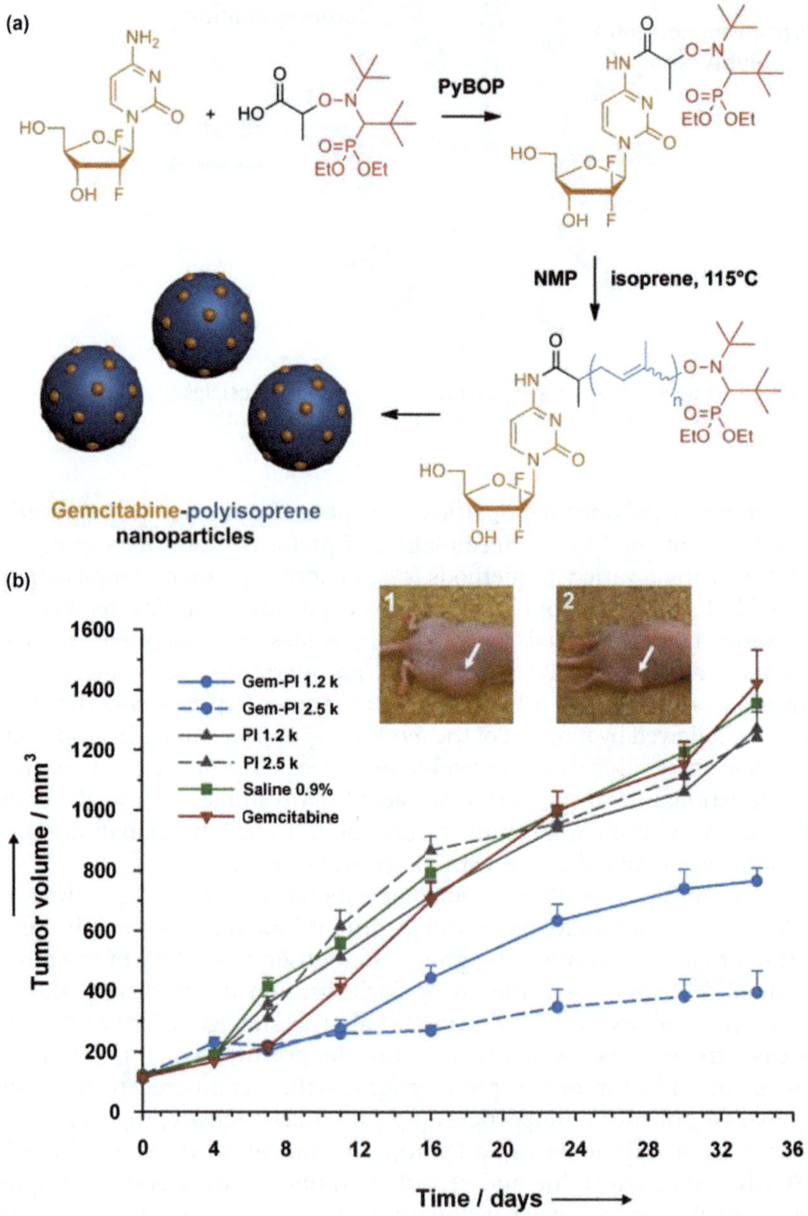

Figure 9.10 (a) Synthetic pathway for gemcitabine–polyisoprene (Gem–PI) conjugate nanoparticles by NMP. (b) *In vivo* anticancer activity of Gem–PI nanoparticles following intravenous treatment (on days 0, 4, 8 and 12) of mice bearing MiaPaCa-2 subcutaneous tumors: tumor progression as a function of time. White arrows indicate the position of the implanted tumor on the representative mouse at the end point for the Gem-treated group (inset 1) and the Gem–PI (2.5 kDa)-treated group (inset 2). Adapted from ref. 83 with permission of Wiley.

squalene, which is a natural isoprenoid widely distributed in nature.[84,85] Note that this strategy offers interesting features: (i) the efficiency of the conjugation should be quantitative as all the drugs should be retained at the chain ends; (ii) the purification of the conjugate should be facilitated since only the unreacted monomer has to be removed conversely to a preformed polymer; and (iii) the drug loading can adjusted simply by varying the polymer chain length.

9.6 Conclusion

We can conclude that NMP and the increased availability of related initiators allowed compliant polymerization conditions that were capitalized for the preparation of important classes of biomaterials. For example, well-defined glycopolymers were prepared from monomers carrying mono- or oligosaccharides for applications in which the interaction with biological matter relies on recognition mechanisms. In here, well-defined diblock copolymers comprising a sugar block were successfully prepared from protected glycopolymers using a homopolymer as macroinitator. Carbohydrate-based micellar structures were the result of the self-assembly of these block copolymers where the bioactive carbohydrate group was presented as surface functionality. In a similar fashion, peptide macroinitiators were applied to prepare block copolymers, in which the synthesis was performed with the peptide attached to the resin and cleaved in the last step to result a pure block copolymer. Simple grafting approaches were also employed to yield the desired peptide–polymer structure. Nanocomposite and surface modifications with non-fouling properties were accomplished with controlled charge densities, enabled by the controlled polymerization with Blocbuilder or TEMPO initiators. Activated esters deriving from the BlocBuilder alkoxyamine created silica initiating species to result in high grafting densities. The PISA method led to self-assembled PEG brushes that formed nodules with dumbbell-or raspberry structures on particle surfaces. Finally, the chapter ends with a presentation of nanoparticle structures enabled by water-soluble initiators through mini or emulsion polymerization techniques. Anticancer nanoparticles from self-assembled prodrugs were created by the 'drug-initiated' method in which a drug-based alkoxyamine served as initiator for the NMP of isoprene, leading to a well-defined amphiphilic drug–polymer conjugate with one drug positioned at the extremity of a polymer chain.

Overall, this chapter showed that NMP has gained incredibly in maturity and robustness over recent years and is perhaps the most suitable CLRP technique in terms of simplicity and innocuousness, especially when biomedical applications are envisioned.

References

1. J. Nicolas, Y. Guillaneuf, C. Lefay, D. Bertin, D. Gigmes and B. Charleux, *Prog. Polym. Sci.*, 2013, **38**, 63–235.

2. R. Kakuchi, M. Zamfir, J.-F. Lutz and P. Theato, *Macromol. Rapid Commun.*, 2012, **33**, 54–60.
3. S. Srichan, H. Mutlu, N. Badi and J.-F. Lutz, *Angew. Chem., Int. Ed.*, 2014, **53**, 9231–9235.
4. S. Srichan, H. Mutlu and J.-F. Lutz, *Eur. Polym. J.*, 2015, **62**, 338–346.
5. S. Telitel, F. Dumur, S. Telitel, O. Soppera, M. Lepeltier, Y. Guillaneuf, J. Poly, F. Morlet-Savary, P. Fioux, J.-P. Fouassier, D. Gigmes and J. Lalevee, *Polym. Chem.*, 2015, **6**, 613–624.
6. J. Morris, S. Telitel, K. E. Fairfull-Smith, S. E. Bottle, J. Lalevee, J.-L. Clement, Y. Guillaneuf and D. Gigmes, *Polym. Chem.*, 2015, **6**, 754–763.
7. N. V. Tsarevsky and K. Matyjaszewski, *Chem. Rev.*, 2007, **107**, 2270–2299.
8. G. Moad, E. Rizzardo and S. H. Thang, *Aust. J. Chem.*, 2009, **62**, 1402–1472.
9. M. Chenal, S. Mura, C. Marchal, D. Gigmes, B. Charleux, E. Fattal, P. Couvreur and J. Nicolas, *Macromolecules*, 2010, **43**, 9291–9303.
10. J. Rigolini, B. Grassl, L. Billon, S. Reynaud and O. F. X. Donard, *J. Polym. Sci., Part A: Polym. Chem.*, 2009, **47**, 6919–6931.
11. A. Darabi, A. R. Shirin-Abadi, P. G. Jessop and M. F. Cunningham, *Macromolecules*, 2015, **48**, 72–80.
12. C. Detrembleur, C. Jerome, J. De Winter, P. Gerbaux, J. L. Clement, Y. Guillaneuf and D. Gigmes, *Polym. Chem.*, 2014, **5**, 335–340.
13. M. Maric and V. Consolante, *J. Appl. Polym. Sci.*, 2013, **127**, 3645–3656.
14. X. Savelyeva and M. Marić, *J. Polym. Sci., Part A: Polym. Chem.*, 2014, **52**, 2011–2024.
15. M. Ghaemy and M. Bazzar, *J. Appl. Polym. Sci.*, 2010, **116**, 64–71.
16. A. Narumi, S. Zeidler, H. Barqawi, C. Enders and W. H. Binder, *J. Polym. Sci., Part A: Polym. Chem*, 2010, **48**, 3402–3416.
17. L. Petton, E. P. C. Mes, H. Van Der Wal, S. Claessens, F. Van Damme, S. Verbrugghe and F. E. Du Prez, *Polym. Chem.*, 2013, **4**, 4697–4709.
18. V. Delplace, S. Harrisson, A. Tardy, D. Gigmes, Y. Guillaneuf and J. Nicolas, *Macromol. Rapid Commun.*, 2014, **35**, 484–491.
19. B. Lessard, C. Tervo, S. De Wahl, F. J. Clerveaux, K. K. Tang, S. Yasmine, S. Andjelic, A. D'Alessandro and M. Maric, *Macromolecules*, 2010, **43**, 868–878.
20. X. Savelyeva, B. H. Lessard and M. Marić, *Macromol. React. Eng.*, 2012, **6**, 200–212.
21. N. Handké, T. Trimaille, E. Luciani, M. Rollet, T. Delair, B. Verrier, D. Bertin and D. Gigmes, *J. Polym. Sci., Part A: Polym. Chem*, 2011, **49**, 1341–1350.
22. X. Qiao, M. Lansalot, E. Bourgeat-Lami and B. Charleux, *Macromolecules*, 2013, **46**, 4285–4295.
23. D. A. Tirrell, *Nature*, 2004, **430**, 837.
24. V. Ladmiral, E. Melia and D. M. Haddleton, *Eur. Polym. J.*, 2004, **40**, 431–449.
25. K. Ohno, Y. Tsujii, T. Miyamoto, T. Fukuda, M. Goto, K. Kobayashi and T. Akaike, *Macromolecules*, 1998, **31**, 1064–1069.

26. Y. Miura, D. Koketsu and K. Kobayashi, *Polym. Adv. Technol.*, 2007, **18**, 647–651.
27. K. Ohno, T. Fukuda and H. Kitano, *Macromol. Chem. Phys.*, 1998, **199**, 2193–2197.
28. K. Ohno, Y. Izu, S. Yamamoto, T. Miyamoto and T. Fukuda, *Macromol. Chem. Phys.*, 1999, **200**, 1619–1625.
29. Y. Chen and G. Wulff, *Macromol. Chem. Phys.*, 2001, **202**, 3273–3278.
30. Y. Chen and G. Wulff, *Macromol. Chem. Phys.*, 2001, **202**, 3426–3431.
31. A. Narumi, T. Matsuda, H. Kaga, T. Satoh and T. Kakuchi, *Polym. J.*, 2001, **33**, 939–945.
32. A. Narumi, T. Matsuda, H. Kaga, T. Satoh and T. Kakuchi, *Polymer*, 2002, **43**, 4835–4840.
33. A. Narumi, T. Satoh, H. Kaga and T. Kakuchi, *Macromolecules*, 2002, **35**, 699–705.
34. H. Götz, E. Harth, S. M. Schiller, C. W. Frank, W. Knoll and C. J. Hawker, *J. Polym. Sci., Part A: Polym. Chem.*, 2002, **40**, 3379–3391.
35. S. R. S. Ting, E.-H. Min, P. Escale, M. Save, L. Billon and M. H. Stenzel, *Macromolecules*, 2009, **42**, 9422–9434.
36. K. Babiuch, R. Wyrwa, K. Wagner, T. Seemann, S. Hoeppener, C. R. Becer, R. Linke, M. Gottschaldt, J. R. Weisser, M. Schnabelrauch and U. S. Schubert, *Biomacromolecules*, 2011, **12**, 681–691.
37. F. M. Veronese, *Biomaterials*, 2001, **22**, 405–417.
38. F. M. Veronese and J. M. Harris, *Adv. Drug Delivery Rev.*, 2002, **54**, 453–456.
39. F. M. Veronese, C. Monfardini, P. Caliceti, O. Schiavon, M. D. Scrawen and D. Beer, *J. Controlled Release*, 1996, **40**, 199–209.
40. J. Nicolas, G. Mantovani and D. M. Haddleton, *Macromol. Rapid Commun.*, 2007, **28**, 1083–1111.
41. K. L. Heredia and H. D. Maynard, *Org. Biomol. Chem.*, 2007, **5**, 45–53.
42. B. Le Droumaguet and J. Nicolas, *Polym. Chem.*, 2010, **1**, 563–598.
43. H.-A. Klok, *J. Polym. Sci., Part A: Polym. Chem.*, 2005, **43**, 1–17.
44. H.-A. Klok, *Macromolecules*, 2009, **42**, 7990–8000.
45. M. L. Becker, J. Liu and K. L. Wooley, *Chem. Commun.*, 2003, 180–181.
46. M. L. Becker, J. Liu and K. L. Wooley, *Biomacromolecules*, 2005, **6**, 220–228.
47. T. Trimaille, K. Mabrouk, V. Monnier, L. Charles, D. Bertin and D. Gigmes, *Macromolecules*, 2010, **43**, 4864–4870.
48. T. Trimaille, L. Autissier, M. D. Rakotonirina, Y. Guillaneuf, C. Villard, D. Bertin, D. Gigmes and K. Mabrouk, *Chem. Commun.*, 2014, **50**, 2744–2747.
49. M. Chenal, C. Boursier, Y. Guillaneuf, M. Taverna, P. Couvreur and J. Nicolas, *Polym. Chem.*, 2011, **2**, 1523–1530.
50. K. Molawi and A. Studer, *Chem. Commun.*, 2007, 5173–5175.
51. M. Moller, C. Hentschel, L. Chi and A. Studer, *Org. Biomol. Chem.*, 2011, **9**, 2403–2412.
52. T. J. Deming, *Nature*, 1997, **390**, 386–389.
53. T. J. Deming, *J. Am. Chem. Soc.*, 1998, **120**, 4240–4241.

54. T. J. Deming, *Macromolecules*, 1999, **32**, 4500–4502.
55. T. J. Deming, *Adv. Drug Delivery Rev.*, 2002, **54**, 1145–1155.
56. T. J. Deming, *Adv. Polym. Sci.*, 2006, **202**, 1–18.
57. S. Steig, F. Cornelius, A. Heise, R. J. L. Knoop, G. J. M. Habraken, C. E. Koning and H. Menzel, *Macromol. Symp.*, 2007, **248**, 199–206.
58. R. J. I. Knoop, G. J. M. Habraken, N. Gogibus, S. Steig, H. Menzel, C. E. Koning and A. Heise, *J. Polym. Sci., Part A: Polym. Chem.*, 2008, **46**, 3068–3077.
59. A. Karatzas, P. Bilalis, H. Iatrou, M. Pitsikalis and N. Hadjichristidis, *React. Funct. Polym.*, 2009, **69**, 435–440.
60. M. Husseman, E. E. Malmstrom, M. McNamara, M. Mate, D. Mecerreyes, D. G. Benoit, J. L. Hedrick, P. Mansky, E. Huang, T. P. Russell and C. J. Hawker, *Macromolecules*, 1999, **32**, 1424–1431.
61. S. Abraham and L. D. Unsworth, *J. Polym. Sci., Part A: Polym. Chem.*, 2011, **49**, 1051–1060.
62. S. Abraham, A. So and L. D. Unsworth, *Biomacromolecules*, 2011, **12**, 3567–3580.
63. J. Parvole, L. Ahrens, H. Blas, J. Vinas, C. Boissiere, C. Sanchez, M. Save and B. Charleux, *J. Polym. Sci., Part A: Polym. Chem.*, 2010, **48**, 173–185.
64. C. Chevigny, J. Jestin, D. Gigmes, R. Schweins, E. Di-Cola, F. Dalmas, D. Bertin and F. Boué, *Macromolecules*, 2010, **43**, 4833–4837.
65. N. Cherifi, A. BenAboura, M. Save and L. Billon, *J. Polym. Sci., Part A: Polym. Chem.*, 2012, **50**, 3976–3985.
66. R. Hoogenboom, D. Popescu, W. Steinhauer, H. Keul and M. Möller, *Macromol. Rapid Commun.*, 2009, **30**, 2042–2048.
67. X. G. Qiao, P. Y. Dugas, B. Charleux, M. Lansalot and E. Bourgeat-Lami, *Macromolecules*, 2015, **48**, 545–556.
68. J. Nicolas, S. Mura, D. Brambilla, N. Mackiewicz and P. Couvreur, *Chem. Soc. Rev.*, 2013, **42**, 1147–1235.
69. C. Vauthier and P. Couvreur, *J. Biomed. Nanotechnol.*, 2007, **3**, 223–234.
70. O. C. Farokhzad and R. Langer, *ACS Nano*, 2009, **3**, 16–20.
71. M. Elsabahy and K. L. Wooley, *Chem. Soc. Rev.*, 2012, **41**, 2545–2561.
72. P. B. Zetterlund, Y. Kagawa and M. Okubo, *Chem. Rev.*, 2008, **108**, 3747–3794.
73. B. Charleux and J. Nicolas, *Polymer*, 2007, **48**, 5813–5833.
74. G. Delaittre, M. Save, M. Gaborieau, P. Castignolles, J. Rieger and B. Charleux, *Polym. Chem.*, 2012, **3**, 1526–1538.
75. G. J. M. Habraken, M. Peeters, P. D. Thornton, C. E. Koning and A. Heise, *Biomacromolecules*, 2011, **12**, 3761–3769.
76. R. K. O'Reilly, C. J. Hawker and K. L. Wooley, *Chem. Soc. Rev.*, 2006, **35**, 1068–1083.
77. R. K. O'Reilly, M. J. Joralemon, C. J. Hawker and K. L. Wooley, *J. Polym. Sci., Part A: Polym. Chem.*, 2006, **44**, 5203–5217.
78. N. S. Lee, G. Sun, W. L. Neumann, J. N. Freskos, J. J. Shieh, R. B. Dorshow and K. L. Wooley, *Adv. Mater.*, 2009, **21**, 1344–1348.
79. D. Pan, J. L. Turner and K. L. Wooley, *Chem. Commun.*, 2003, 2400–2401.

80. R. K. O'Reilly, M. J. Joralemon, K. L. Wooley and C. J. Hawker, *Chem. Mater.*, 2005, **17**, 5976–5988.
81. M. J. Joralemon, R. K. O'Reilly, C. J. Hawker and K. L. Wooley, *J. Am. Chem. Soc.*, 2005, **127**, 16892–16899.
82. R. K. O'Reilly, M. J. Joralemon, C. J. Hawker and K. L. Wooley, *Chem.-Eur. J.*, 2006, **12**, 6776–6786.
83. S. Harrisson, J. Nicolas, A. Maksimenko, D. T. Bui, J. Mougin and P. Couvreur, *Angew. Chem., Int. Ed.*, 2013, **52**, 1678–1682.
84. D. T. Bui, A. Maksimenko, D. Desmaele, S. Harrisson, C. Vauthier, P. Couvreur and J. Nicolas, *Biomacromolecules*, 2013, **14**, 2837–2847.
85. A. Maksimenko, D. T. Bui, D. Desmaële, P. Couvreur and J. Nicolas, *Chem. Mater.*, 2014, **26**, 3606–3609.
86. V. Delplace, P. Couvreur and J. Nicolas, *Polym. Chem.*, 2014, **5**, 1529–1544.

CHAPTER 10

Surface-initiated Nitroxide-mediated Polymerization

LAURENT BILLON,[a] MAUD SAVE[a] AND MICHAEL F. CUNNINGHAM*[b,c]

[a] Université de Pau et des Pays de l'Adour, CNRS UMR 5254, Equipe de Physique et Chimie des Polymères, Technopole Helioparc, 2, av. Pdt Angot 64 053 PAU cedex 09, France; [b] Department of Chemical Engineering, Queen's University, Kingston, Ontario, Canada K7L 3N6; [c] Department of Chemistry, Queen's University, Kingston, Ontario, Canada K7L 3N6
*Email: michael.cunningham@queensu.ca

10.1 Scope

While most of the research on nitroxide-mediated polymerization has been conducted in solution, bulk or heterogeneous dispersions, such as emulsion or miniemulsion polymerization, there is also considerable interest in conducting NMP from the surfaces of solid substrates in what has become known as surface-initiated nitroxide-mediated polymerization, or SI-NMP. Modification of surfaces has been an active research area for decades but advances in nanotechnology have focused attention on the development of new and innovative methods for modifying and even tailoring the individual chemical and physical nature of a wide variety of substrates. In nanoscale materials, the very high specific surface area ensures the surface properties play a dominant role in the overall behaviour and performance of the substance. The surfaces may be either flat or curved (*e.g.* nanoparticles). SI-NMP has been applied to a diverse range of substrates and applications including, for example, the formation of polymer brushes, and the graft modification of

RSC Polymer Chemistry Series No. 19
Nitroxide Mediated Polymerization: From Fundamentals to Applications in Materials Science
Edited by Didier Gigmes

Published by the Royal Society of Chemistry, www.rsc.org

nanoparticles (organic and inorganic) and renewable polymers to make new hybrid materials for electronics, optoelectronics, biomedicine, and advanced materials.

Before the advent of living radical polymerization, surface modification with polymers was synthetically challenging, usually involving multiple steps, and was often relatively inefficient and complex. While a "grafting to" approach, in which pre-formed polymer chains are covalently bound to the surface, became a well-developed technique, achieving the synthesis of high-density polymer brushes remained elusive because of steric hindrance restrictions. The "grafting from" approach using conventional free radical polymerization, in which polymer chains are grown from an initiator group bound to the surface, allows much higher graft densities to be achieved (so that the polymer chains are elongated well beyond their random coil conformation observed in bulk) but yields poor control of the molecular weight distribution. Furthermore, block copolymers cannot be prepared. Surface-initiated living radical polymerization does allow facile precision synthesis of polymer chains from surfaces, including block copolymers, at high chain densities.[1–5] Rigorous exclusion of air and moisture is not required as with anionic polymerization. NMP, which is the focus of this chapter, is also quite tolerant of many impurities. In addition, the only impurity in the final product is the nitroxide, which is bound to the chain end.

The chapter reviews recent research in SI-NMP with the objective of providing the reader with an understanding of the underlying principles, current capabilities, and the ongoing challenges of conducting NMP on different surfaces. The sections have been organized based on the substrate used, rather than application, as this structure allows for a more cohesive discussion of the critical issues and fundamental behaviour for each different substrate.

10.2 Introduction

While SI-NMP shares many common behaviours with NMP conducted in homogeneous environments, the presence of a surface can have a profound impact on the course of polymerization, including kinetics, chain length distribution, and chain end fidelity (livingness). One of the most significant differences is dramatically lower total chain concentrations resulting from having a comparatively small total surface area. Whereas total chain concentrations in homogeneous NMP systems are typically $\sim 10^{-2}$ M, in SI-NMP they can be orders of magnitude smaller. As importantly, the initiating and propagating sites are not uniformly distributed as with homogeneous environments, but rather in close proximity, possibly as little as several nanometers apart. This may also mean that the polymer chain ends remain constrained in closer proximity to each other than occurs in homogeneous NMP, with the possibility of increased irreversible termination. Thus, while the overall chain concentrations are usually quite low, the local chain concentrations can be much higher than in solution or bulk. A further practical consequence is that a "sacrificial" (or free or excess) alkoxyamine initiator or nitroxide often needs to be added to ensure that satisfactory control of the polymerization is maintained, since the total amount of bound nitroxide is usually insufficient to ensure the persistent radical effect is achieved. This not only consumes additional materials but also results in the formation of a significant amount of free (ungrafted) polymer contaminating the desired polymer/substrate grafted hybrid. Characterization of polymer-grafted SI-NMP materials often presents challenges, as conventional tools for determining critical properties such as molecular weight distribution and copolymer composition cannot always be applied. If the grafted chains cannot be effectively removed, which is often the case, investigators are required to make inferences about the grafted polymer, typically using instrumentation such as TGA (thermogravimetric analysis), XRD (X-ray diffraction), SANS (small angle neutron scattering) and contact angle measurements. The following sections describe recent progress made in SI-NMP on a variety of substrates, illustrating the characteristics detailed above and how they affect the synthesis and characterization of the grafted hybrid materials.

10.2.1 Polymer Brushes

The general features of polymer brushes, so-called tethered polymers, first gained real attention in the 1980s when reports based on the quantitative treatments of polymer brushes introduced theoretical approaches to study brush-conformations of coil-polymers.[6] Polymer brushes have become increasingly studied and most macromolecular and surface journals now routinely contain articles and reviews on polymer brushes, in particular exploiting achievements of controlled radical polymerization [or the more recent term reversible deactivation radical polymerization (RDRP)] techniques.[3–5,7,8] To have good control over the polymer mono-layer thickness and polymer structure, recent advances in stable free radical polymerization (SFRP) (*i.e.* NMP), *i.e.* strong decrease of terminations and chain transfer reactions, and a less constraining technique, have made the surface-initiation of well-defined and low dispersity ($Đ = M_w/M_n$) polymers viable.[9]

The synthesis of a dense film of polymer chains covalently bound to surfaces is an important field of research for its ability to control and tune the surface properties. The most straightforward techniques for elaborating well-defined polymer layers are the "grafting to" and "grafting from" processes (Figure 10.1).

If the "grafting to" approach does not usually give highly dense polymer brushes because chemisorption of the first fraction of chains hinders the diffusion of subsequent chains to the surface by way of forming a macromolecular barrier, this limitation can be overcome by using the approach named "grafting from" or "surface-initiated" polymerization. Indeed, this technique has been considered to result in higher grafting densities values due to a monolayer of small initiator molecules covalently attached to the solid surface. After activation, *i.e.* initiation of the polymerization, the chains

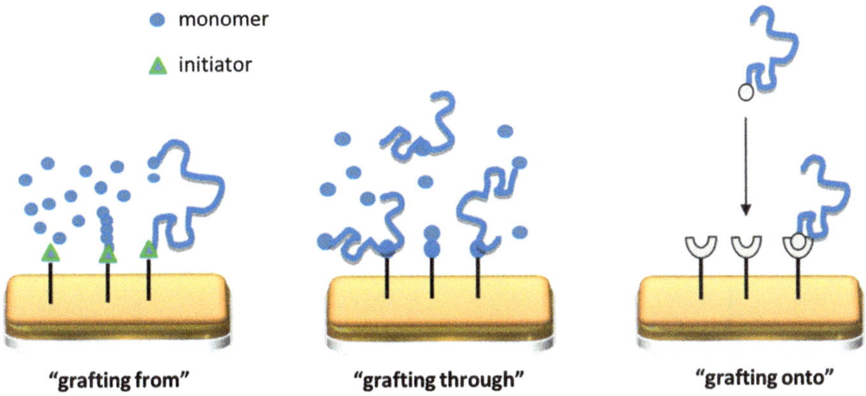

Figure 10.1 Schematic illustration of the attachment of polymers to surfaces.

grow simultaneously by controlled polymerization from the surface and then the only limit to propagation is the diffusion of small organic compounds, *i.e.* monomer molecules, to the growing active species.[10]

The structural properties of brushes are generally recognized as an array of coiled macromolecular chains attached to a surface and in sufficient proximity so that the unperturbed solution dimensions, *i.e.* in a good solvent, of the chains are altered. Such a situation causes an overlap of adjacent coil chains and significantly changes the conformational dimensions of individual polymer chains. Indeed, coiled polymer chains extend or alter their radius of gyration to avoid unfavorable interactions. Films composed of polymer chains that extend along a direction normal to the grafting surface can exhibit properties distinctly different from chains in solution, and thus make polymer brushes an interesting field for novel properties. These new properties were achieved with the "grafting from" procedure, so-called surface-initiated polymerization, in a process of advent and maturation, which can permit the targeting of high grafting densities ($\approx$1 chains nm^{-2}) and concentrated coil polymer brushes. Thus, polymer brushes can be considered as sensitive ultra-thin polymer films on a solid substrate. Indeed, effects of end-grafting on their structure and properties have been studied from mushrooms to concentrated brushes corresponding to very low to high grafting density values, respectively (Figure 10.2).

In the dry state, for high grafting density values, *i.e.* close to 1 coil chains nm^{-2}, the thickness value reaches about 40% of the fully extended chain length in the swollen state. Nevertheless, since the size of the unperturbed chain end-grafted on a repulsive surface is proportional to the square-root of chain length, this large value of thickness, close to 40%, means that the chains are already highly extended or stretched in their dry state. Such a phenomena was confirmed by neutron reflectivity measurements where a low swelling capacity of highly stretched polystyrene chains grown on a flat silicon surface using SI-NMP was described.[11] The control over the lateral density leads to brushes with a very high grafting density

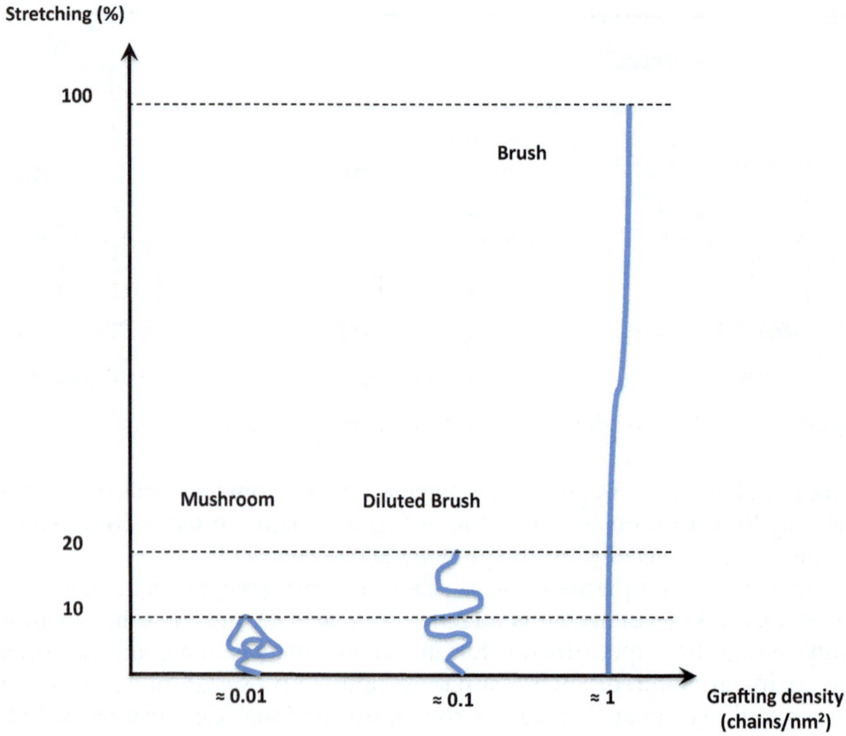

Figure 10.2 Macromolecular features of tethered polymer chains in solvent state *versus* the grafting density.[2]

(up to >1.0 chains nm^{-2}) which was evidenced by a specific feature of such brushes (high stretch) due to the controlled nature of the surface initiated polymerization leading to a very homogeneous growth of homopolymer or copolymer chains. The very homogeneous growth of the chains leads to homogeneous brushes with stretch or conformations controlled exclusively by the grafting density σ and thicknesses H tunable independently through the chain molecular weights ($H \approx \sigma$). At the same time by ellipsometry, the thickness was found to vary linearly with molecular weight for a constant σ in agreement with brush theory.[12,13] Cleaved brushes from silicon wafers showed a slightly higher dispersity and respective values of M_n 25% and 50% greater than the polymer simultaneously grown in the bulk that confirm the effect of confinement of the dormant and growing species on the surface polymerization kinetics.[14] This result seems to be in good agreement with the prediction resulting from the Monte Carlo simulations reported by Genzer regarding the effect of grafting density on the surface polymerization of polymer brushes.[15] The polymer brushes' macromolecular features have a direct impact on the physical properties of the tethered chains. Indeed, the glass transition temperature T_g of PMMA brushes with a constant grafting density was studied as a function of the film thickness H. If the thickness was also proportional to the M_n of the grafted chains, more interestingly for H below 50 nm, T_g increases sharply to 130 °C with decreasing L_d due to the restricted mobility of the end-grafted chains, *i.e.* surface effect.[16] Such behaviour was also observed on spherical nanoparticles for poly(butylacrylate) brushes.[17] For H larger than about 50 nm, T_g showed a constant value of about 120 °C which is already about 10 °C higher than the T_g of the bulk PMMA. In the same order, comparison of the results for a high-density PMMA brush and an equivalent spin-coated PMMA layer revealed differences in elastic and dielectric properties between them. In the glassy state, there was no appreciable difference, whereas in the molten state a more resistant behaviour against compression than the equivalent PMMA melt is observed.[18] The low compressibility of the molten high-density brushes was interpreted in terms of a rubber elasticity theory of a stretched polymer network with entanglements. This analysis suggested that the low compressibility is mainly attributed to a strain-hardening effect of the highly stretched entangled chains, and that there exists a considerable amount of entanglements among different grafted chains contributing to the elastic modulus.[18] A theoretical study of the polymer brush conformation when it is exposed to a molten and compatible polymer was described by Gay.[19] At large brush grafting densities σ and melt polymerization index P, the melt only penetrates the brush over a distance λ that is small compared to the brush thickness. For $P > N^{2/3}\ \sigma^{-2/3}$ (N is the polymerization degree of the grafted chains), the penetration length $\lambda \approx \sigma^{-1/3}\ N^{1/3}$ is associated with partial wetting or penetration. For $P < N^{2/3}\ \sigma^{-2/3}$, although wetting is complete Gay predicted that penetration is still partial: $\lambda \approx \sigma^{-1}\ N\ P^{-1}$. These differences of behaviour are of direct importance and interest for implications in interfacial rheology, wetting dynamics and also nanocomposites where the high grafting density and length of the polymer chains is not the "Holy Grail". Inoubli *et al.* validated this concept for silica particles with a mean radius of 11.5 nm and a layer thickness of 14.7 nm ($\sigma \approx 0.25$ poly(butylacrylate) chains nm^{-2}). From the SANS measurements, they calculated the hard sphere radius and a thickness of the impenetrable layer of about 11 nm. This implies that, at the percolation threshold (2.5 vol%), the polymer shell of the grafted silica particles presents a penetrable part of 3.7 nm thickness.[20] Such behaviour was also clearly illustrated by Green *et al.*[21] Using ultrasmall X-ray scattering (USAXS) and transmission electronic microscopy, they determined whether the polystyrene-grafted nanoparticles (NPs) were stable and dispersed uniformly, or were unstable and aggregated within the homologue matrix. From these measurements, they developed a phase diagram for NP miscibility with respect to σ, P, and N to determine the allophobic and autophobic transitions that correspond to the wetting and the drying of the graft polymer brush, which control NP stability in polymer matrices. On the basis of their interpretation, results imply that NPs curvature: (1) impacts the allophobic dewetting transition at lower graft densities where high curvature permits increased penetration of the matrix into the graft; and (2) appears to have less of an effect at higher graft densities where NPs are destabilized by the expulsion of the melt polymer.[21] Dewetting has also been observed in a series of mixed poly(*tert*-butylacrylate) (PtBA)/polystyrene (PS) brushes. A clear microphase separation is demonstrated, where the morphology of mixed brushes evolved from isolated, nearly spherical PS microdomains buried inside the PtBA matrix, to short PS cylindrical worm-like microdomains in the PtBA matrix, and to nearly bicontinuous nanostructures (Figure 10.3).[22]

In the swollen state, repulsive forces originating from the steric interaction between the solvent-swollen high-density PMMA brush and the AFM probe sphere were observed. The equilibrium thickness (L_e) of the solvent-swollen brushes was defined as the critical distance from the substrate surface beyond which no repulsive force was detectable. For the brush with $\sigma = 0.7$ chains nm^{-2}, the value of L_e/L_{cw} even approaches 0.9 (L_{cw} is the weight-average full length of the graft chain). This surprisingly large value certainly indicates that the graft chains in this brush are extended to an extraordinarily high extent, and compares with fully extended chains. The σ-dependency of L_e varies like $L_e/L_{cw} \propto \sigma^n$ with n increasing from about 1/3 to 1/2 with increasing σ from the "semi-dilute" to the brush regime (Figure 10.4).[23,24]

Neutron reflectivity measurements have also evidenced this specific feature of such brushes. In the case of very dense PS brushes (>1.0 chains nm^{-2}) in good solvent, the highest volume fraction (0.85) ever measured for polymer chains leads to a significant reduction of the swelling capacity of the layer and a noticeable deviation from classical brush scaling laws; a result somehow not completely surprising for such high volume fractions where scaling laws, applicable to semidilute polymer brushes in good solvent, have no reason to hold anymore.[11]

As described above, the macromolecular features of polymer brushes attached on planar surfaces have been largely studied by theoretical or experimental approaches. On the contrary, the structure of densely grafted polymer layers grown from the surfaces of spherical nanoparticles was not so examined over a broad range of graft densities and chain lengths N. At low enough molecular weight or at low enough grafting density, the radius of gyration of the chains, R_g, does not exceed the interchain spacing. Individual chains do not interact with adjacent chains, and thus they assume mushroom-like conformations on the surface (Figure 10.5A). Spherical brushes show similar behaviour in that a transition from mushrooms to semidilute polymer brushes (SDPB) to concentrated polymer brushes (CPB) is observed. There are two important differences between flat brushes and curved brushes that are worth emphasizing. The N scaling of the brush height assumes a different form than for a planar brush in the SDPB regime (Figure 10.5B). Second, in a spherical geometry, if the chains are long enough, then brushes initially in the CPB regime will transition to the SDPB regime at some distance, r_c, from the center of the nanoparticle (Figure 10.5C). Note that this transition arises because the packing constraints near the surface are relaxed by the increased volume available to chains as the radial distance away from

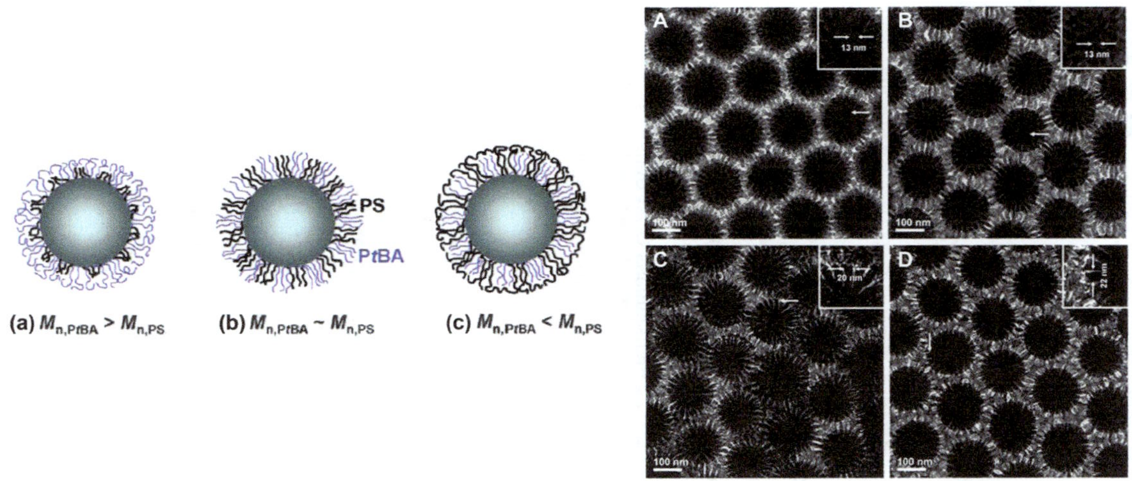

Figure 10.3 Schematic (left) and top-view TEM micrographs (right) illustration of phase morphologies of mixed PtBA/PS brushes with a fixed PtBA and varying PS molecular weights.[22]

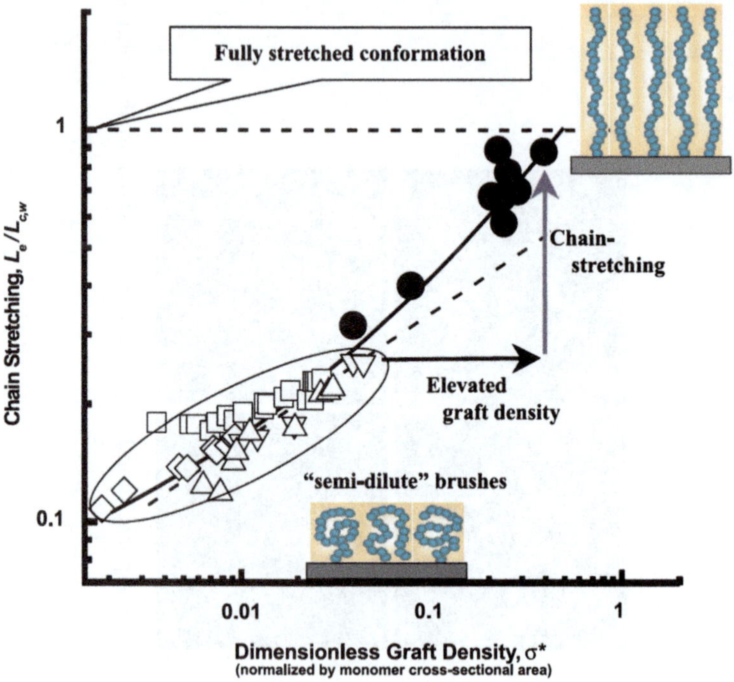

Figure 10.4 Chain stretching L_e/L_{cw} *versus* dimensionless graft density σ^*.[23,24]

a spherical surface increases. Kumar *et al.* showed by dynamic light scattering (DLS) experiments that the hydrodynamic thickness of the polymer layer, h, scales as $h \infty N^{4/5}$ for short chains and high grafting densities, that is, in the CPB regime, whereas $h \infty N^{3/5}$ for long enough chains in SDPB regime. The observations led to speculation that the brushes in the CPB regime do not show the $h \propto N$ dependence predicted for planar brushes for two reasons. First, because of the small radius particles values r_0 to which these chains are grafted, the ratio $h/r_0 \approx 0.5$, even for the shortest brushes considered. Second, if it was experimentally capable of pushing toward the limit $\sigma^* \rightarrow 1$ and investigating shorter brushes and/or larger particles, it might be able to access the flat brushlike linear scaling regime of the CPB more easily within the experimental range of the DLS technique. From these findings, the authors conclude that the result $h \propto N^{4/5}$ for the CPB is an intermediate crossover scaling, relevant to particles of finite curvature, analogous to the $h \propto N$ observed for concentrated flat brushes. Second, the scaling $h \propto N^{3/5}$ uniquely reflects the increased space available to the chain segments as one proceeds away from a curved grafting surface. Under these situations, the chains experience less packing frustration as compared to a planar brush, and the semidilute polymer brush shows scaling behaviour analogous to chains in good solvent, even though the chains are much more extended.[25]

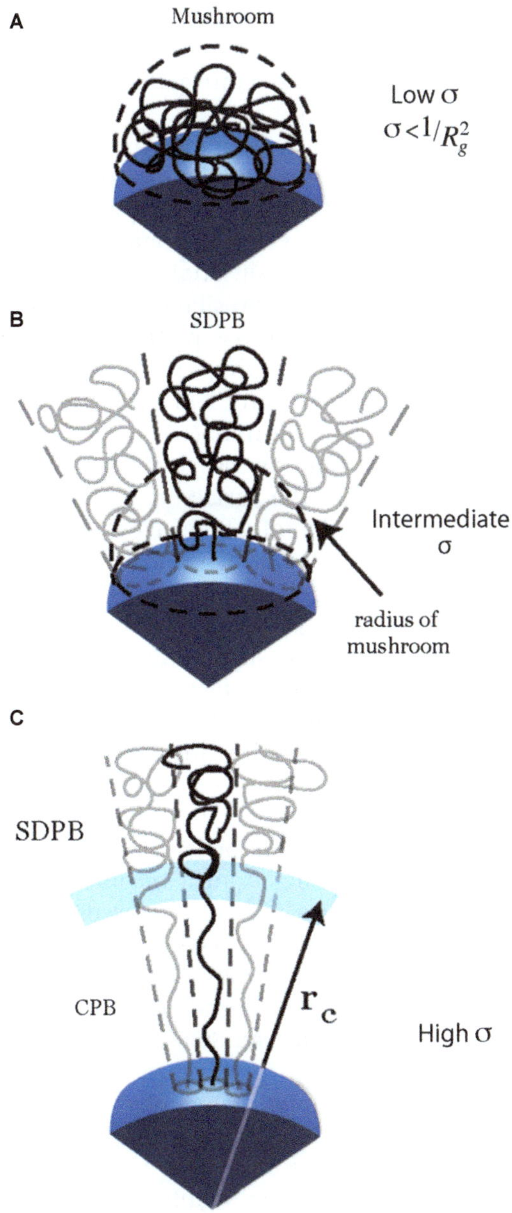

Figure 10.5 Schematic representation of the varied conformational features of polymer brushes attached to spherical interfaces.[25]

10.2.2 Controlling Systems

As detailed in previous chapters, nitroxide-mediated polymerization can be implemented either *via* a bi-component initiating system composed of an initiator and a stable control agent, *i.e.* nitroxide, or *via* a mono-component initiating system based on an alkoxyamine producing both initiator and control agent *via* a thermal dissociation/combination equilibrium. The various structures of controlling systems are gathered in Table 10.1, together with substrate and grafting density.

The first example of NMP initiated from inorganic surfaces was described in 1999 using a grafted alkoxyamine as a mono-component initiating system.[9] The TEMPO-styryl-based alkoxyamine was synthesized by reaction of a hydroxy functionalized alkoxyamine with 1-bromooct-8-ene in the presence of sodium hydride followed by hydrosilylation of the alkene-functionalized alkoxyamine with either dimethylchlorosilane or trichlorosilane. The covalent attachment of the alkoxyamine was performed by the triethylamine-catalyzed reaction of the chlorosilane groups with the surface silanol groups of silicon wafers or silica particles (Entry 1 in Table 10.1).

In order to implement surface-initiated polymerization *via* a bi-component system, a diazoic initiator carrying two trichlorosilane groups was designed to be grafted onto the surface of silica particles (Entry 2 in Table 10.1).[13,26] The polymerization of *n*-butyl acrylate was controlled by the addition of free SG1 nitroxide as a control agent. Surface initiation sites can also be created by atmospheric pressure plasma surface treatment with a surface density in active sites dependent on the plasma processing parameters (*i.e.* surface conditioning, plasma treatment time and power).[36] The growth of polystyrene chains was then controlled by NMP by adding TEMPO nitroxide in the reaction medium. The control of polymerization was confirmed by a linear increase of the film thickness *versus* time, *i.e. versus* monomer conversion.

In 2005, both Bourgeat-Lami *et al.*[28,37] and Billon *et al.*[29] implemented a method for surface functionalization of silica particles by creating the grafted alkoxyamine *in situ* starting from a bicomponent system. A first system consisted of grafting silica with (3-acryloxypropyl)trimethoxysilane (APTMS) to synthesize particles carrying acrylates functions. These particles were placed in the presence of *N,N′*-azobisisobutyronitrile (AIBN) initiator and free nitroxide in order to create the grafted alkoxyamine *in situ* by spin trapping the acryloxy radicals with the SG1 nitroxide acting as a radical scavenger (Entry 3, Table 10.1).[28,37] The analysis of the grafted silica by elemental analysis, thermogravimetric analysis and carbon solid-state nuclear magnetic resonance spectroscopy showed that unmodified APTMS molecules were still present at the surface of the particles.[28] Indeed, the addition of the AIBN radicals onto the double bond of the grafted APTMS molecule competes with direct spin trapping of the AIBN radicals with the SG1 radical or with termination of initiating radicals (cage effect). A variant system involved functionalization of silica particles with 3-methacryloxypropyl)trimethoxysilane, which were then further reacted with an excess of MONAMS alkoxyamine (Entry 4, Table 10.1).[29] For a given reaction time, the increase of the reaction temperature from 85 °C to 100 °C increased the grafting density of the alkoxyamine from 0.2 to 0.7 molecules nm^{-2}. The novel type of alkoxyamine, formed by the addition of the secondary radicals produced by dissociation of MONAMS onto the grafted methacrylic groups and trapped by SG1, is known to exhibit a higher dissociation rate than MONAMS.[38] Thus, the elevated temperature required for MONAMS dissociation might simultaneously produce a high concentration of radicals *via* the dissociation of the grafted alkoxyamine, hence inducing termination reactions between radicals. Solid state carbon NMR also highlighted the presence of residual vinylic carbons of the grafted monomer.[29] A more efficient pathway to design a silane-functionalized alkoxyamine by spin trapping of radicals was proposed by Inoubli *et al.* using the 2-methyl-2-[*N*-*tert*-butyl-*N*-(dimethoxyphosphoryl-2,2-dimethylpropyl)aminoxy]propionic acid (named BlocBuilder® or MAMA alkoxyamine) with (3-acryloxypropyl)-trimethoxysilane (APTMS) (Entry 5, Table 10.1).[30,39] The success of the intermolecular 1,2 addition of the BlocBuilder® alkoxyamine onto the APTMS relies on the high dissociation rate constant of the BlocBuilder® in comparison with the synthesized grafted alkoxyamine producing secondary carbon radicals by homolytic dissociation. At the low reaction temperature ($T = 80\,^{\circ}C$) or short reaction time (1 hour at 100 °C), the difference in reactivity between the two alkoxyamines prevents olefin polymerization *via* further dissociation of the secondary alkoxyamine.[40,41] The trimethoxysilyl-based APTMS-BlocBuilder alkoxyamine was subsequently grafted onto the silica particles by simple hydrolysis-condensation with the surface silanol groups. A high grafting density of 1.2 molecules nm^{-2} was observed for the alkoxyamine grafted silica particles.[30] As described in the early work of Husseman *et al.*,[9] the hydrosilylation of alkene functions is a suitable pathway to introduce alkoxysilane groups enabling the post-grafting of the alkoxyamine (Entries 6–7, Table 10.1). The alkene-functionalized alkoxyamine was first prepared by copper-catalyzed atom transfer radical addition (ATRA) involving the β-phosphonylated nitroxide radical (SG1) and a bromo-alkene derivative.[26,42] The effect of steric hindrance of the initiator on the grafting density was investigated by comparing the grafting density of the symmetrical diazoic initiator (0.8 molecules nm^{-2}, Entry 2, Table 10.1) with the one of the more hindered SG1-based alkoxyamine (0.5 molecules nm^{-2}, Entry 6, Table 10.1).[27] The grafting density of the alkoxyamine can be tuned by the initial amount of alkoxyamine (Entry 7, Table 10.1) but the grafting yield decreases from 50% for a grafting density of 0.35 molecules nm^{-2} to 6% for a grafting density of 0.60 molecules nm^{-2}.[31] ATRA also proved to be a suitable method to synthesize trimethoxysilyl-functionalized SG1-based alkoxyamine from 2-bromo-*N*-[3-(trimethoxysilyl) propyl] propanamide prepared directly by reacting (3-aminopropyl)trimethoxysilane with 2-bromopropionylbromide (Entry 8, Table 10.1).[32]

A two-step procedure involving the covalent coupling of the *N*-hydroxysuccinimide-based activated ester alkoxyamines with nanoparticles grafted with 3-aminopropyl triethoxysilane was proposed to covalently graft alkoxyamines (Entries 9–10, Table 10.1).[33,34] Such a grafting method offers the possibility to graft the BlocBuilder alkoxyamine bearing a tertiary alkyl moiety as initiating species, which gives better control of the polymerization

Table 10.1 List of grafted initiating and controlling system for surface-initiated NMP.

Entry	Functional initiator and controlling agent	Substrate (*Diameter of particle*, nm)	Initiator Grafting density (molecule · nm^{-2})
1[9]	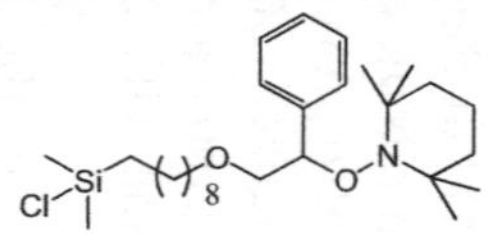	Silicon Wafer	—
2[26,27]	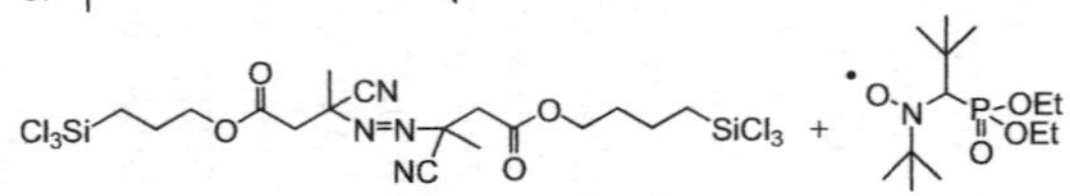	Silica (12 nm)	0.8
3[28]	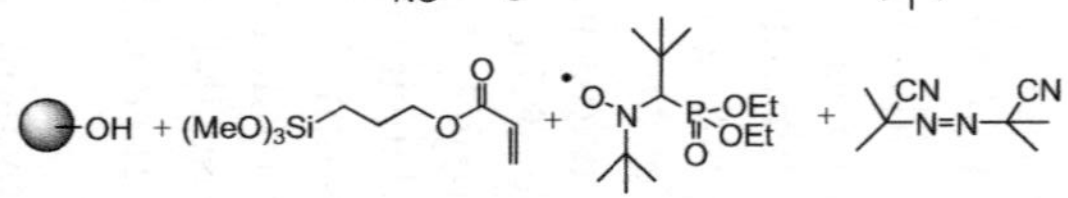	Silica (13 nm)	0.3–0.6
4[29]	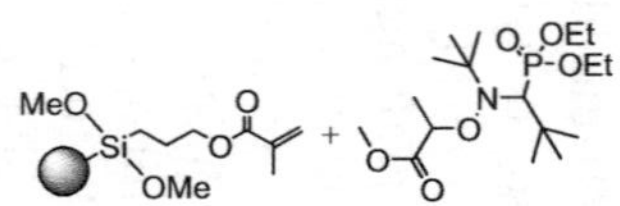	Silica (40 nm)	0.2–0.7
5[30]	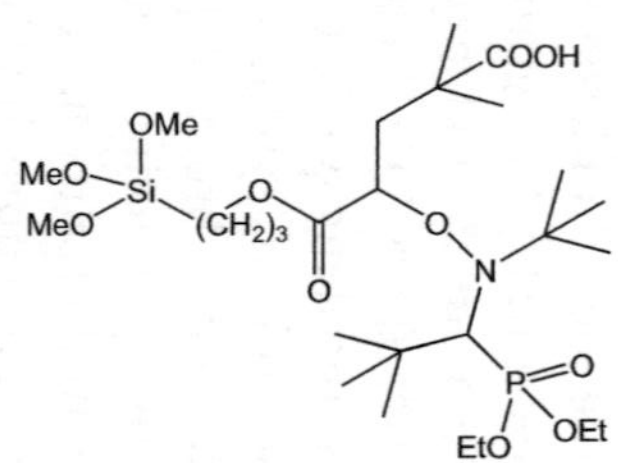	Silica (0.5–1 µm)	1.2
6[26,27]	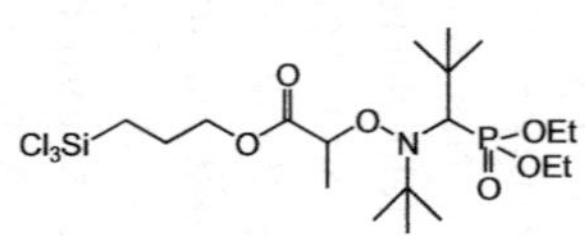	Silica (12 nm)	0.5

7[31]	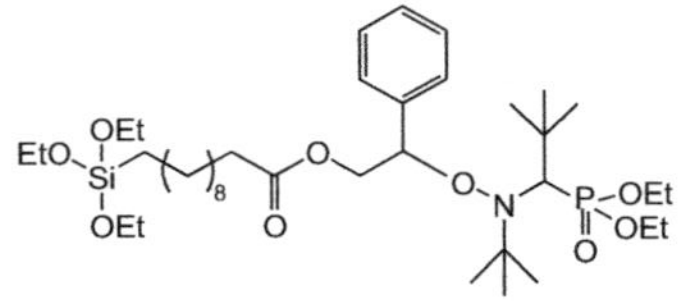	Silica (13 nm)	0.3–0.6
8[32]	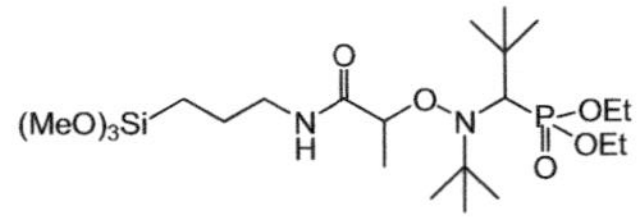	Silicon wafer	1.4
9[33]	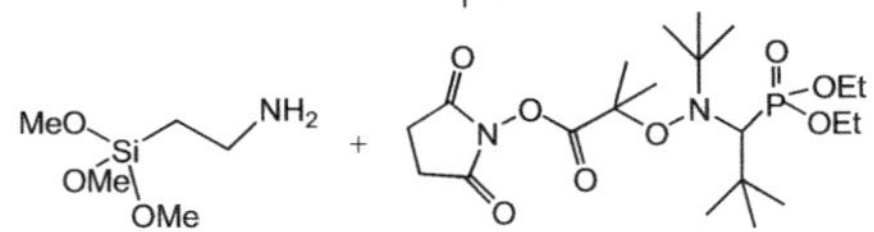	Silica	0.2
10[34]	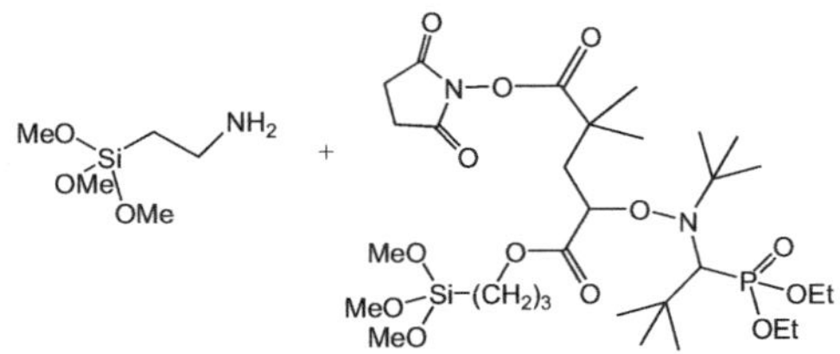	Silica (340 nm) Ordered mesoporous silica	1.9 0.2–0.5
11[35]	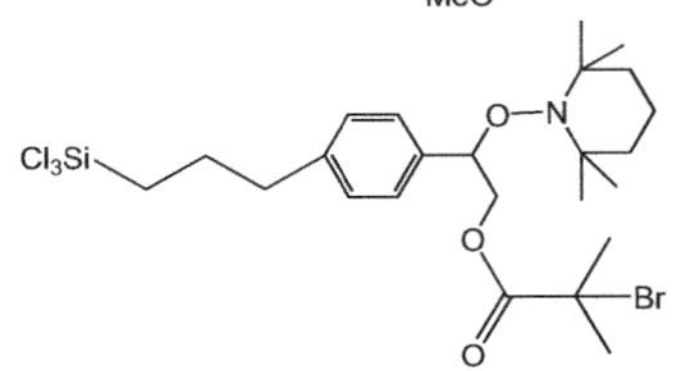	Silica (180 nm)	0.3
12[44,47]	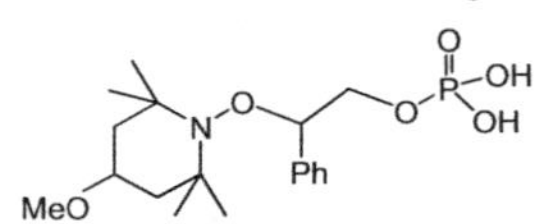	Magnetite (10 nm) Titanium oxide (15 nm)	0.1 0.2

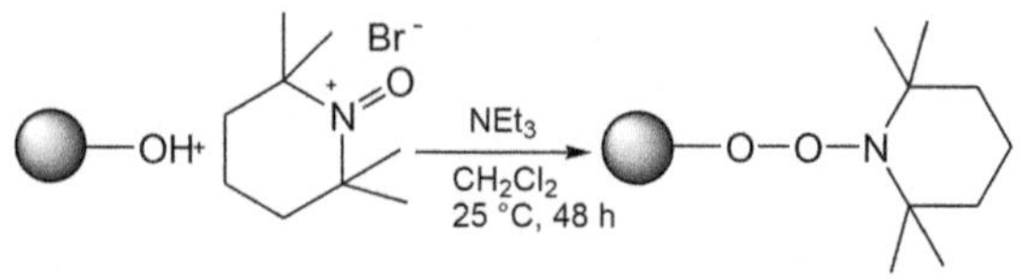

Scheme 10.1 Direct attachment of TEMPO nitroxide onto silica particles surface.[49]

reaction than the alkoxyamine bearing a secondary carbon moiety.[43] Surface modification of other types of inorganic nanoparticles (Fe_3O_4, TiO_2, CdSe) was performed *via* direct surface immobilization of TEMPO-based alkoxyamines carrying phosphoric acid or di-*n*-octyl phosphine oxide groups.[44–47] The design of cationic alkoxyamine offered the possibility to anchor alkoxyamines onto clay surface *via* cation exchange.[48]

Bonilla-Cruz *et al.* reported the direct immobilization of the TEMPO nitroxide onto silica particles followed by the copolymerization of styrene and maleic anhydride acid (Scheme 10.1). The oxoammonium salt of TEMPO is brought into contact with the oxide layer on the surface of the silica particle and reacts to covalently attach the TEMPO to the surface *via* the formation of a peroxide.[49] The direct attachment of TEMPO nitroxide was also possible by bonding the sodium salt of 4-hydroxy-TEMPO onto Merrifield resins with chloromethyl surface groups.[50]

Tandem initiating systems were interesting to promote the grafting of mixed polymer brushes composed of two chemically distinct homopolymers to produce microphase separation in the brush layer (Figure 10.3).[51] The mixed polymer brushes are grown from the grafted asymmetric difunctional initiator (Y-initiator, Entry 11, Table 10.1) by sequential surface-initiated atom transfer radical polymerization (SI-ATRP) (or ring opening polymerization, ROP) and nitroxide-mediated polymerization (SI-NMP). Such syntheses require the grafting of an alkoxyamine exhibiting a low dissociation rate constant (TEMPO-based alkoxyamine) to avoid its dissociation upon the first polymerization step carried out at 75 °C.[22,52]

As described in the following sections, the inorganic particles derivatized by the alkoxyamines or the tandem initiating systems were further involved in surface-initiated nitroxide-mediated polymerization to graft polymer chains in a controlled manner.

10.3 SI-NMP on Different Substrates

10.3.1 Silicon Wafers

The application of NMP for the preparation of polymer brushes by SIP was fueled and really started with the work of Hawker and co-workers, where polystyrene polymer brushes were prepared by using the 'grafting from' approach. Self-assembly of a TEMPO-based alkoxyamine onto a wafer from a solution led to covalent attachment of the initiator ('self-assembled monolayer', SAM). Subsequent SIP afforded the corresponding brushes. It was found that the addition of a free alkoxyamine was essential for successful control of polymerization at the surface. The sacrificial alkoxyamine provided a sufficiently high concentration of free nitroxide in the polymerization mixture, which was necessary to control the polymerization process. In agreement with a 'living' chain reaction, variation of the reaction time influenced the brush thickness and block copolymers could also be prepared. Compared to the 'grafting to' procedure, considerably thicker brushes (up to 80 nm) were obtained.[9] Ten years after the genesis, two major reviews of Klok *et al.* and Studer *et al.* reported polymer brushes *via* CRP and NMP, respectively.[3,4] A mature technique for controlling the SIP from planar substrates was then achieved.

Indeed, SI-NMP permits control of the polymerization and therefore control of the thickness of the brushes. TEMPO-based alkoxyamine and a bimolecular system AIBN-grafted initiator/counter radical SG1 were synthesized to control styrene and *n*-butyl acrylate polymerization, respectively. Both thermoplastic and elastomeric brushes were able to re-initiate SIP by thermo-cleaving of the alkoxyamine end-chain through a controlled behaviour characterized by ellipsometry with a linear evolution of the thickness with the molecular weight.[12,13] If chain extension was performed with the same initial monomer, block copolymers can be also synthesized. Poly(styrene-*b*-fluorinated styrene) and Poly(styrene-*b*-4-vinylpyridine) brushes were elaborated form a TEMPO-based alkoxyamine or *in situ* created alkoxyamine by trapping benzoyl peroxide radicals with hydroxy-TEMPO on a grafted methacrylate double bond.[53,54] Moreover, using three alkoxyamine chemical structures based on TEMPO, SG1 and TIPNO (Figure 10.6) and the surface pressure during Langmuir–Blodgett (LB) deposition, polystyrene brushes with grafting densities of 0.3–1.0 chains nm^{-2} and stretching values of 40–70% were obtained.

Whereas TEMPO- and TIPNO-based alkoxyamines 1 and 3 yielded relatively dense polystyrene brushes, the SG1-based alkoxyamine 2 yielded brushes with low grafting densities and stretching. This behaviour was associated to a consequence of the adsorption of the polar phosphonyl group at the air/water interface by LB.[14]

SIP combined with external physical or chemical processes have been used to pattern polymer brushes. Successful combination of photolithography with SI-NMP to provide continuous films in which there are discrete areas of hydrophilic and hydrophobic polymer brushes was reported by Hawker *et al.*[55] Following the same approach, biocompatible oligo(ethylene glycol)-containing polymer brushes were synthesized and tested for preventing nonspecific adhesion of cells.[53] Langmuir-Blodgett lithography was challenged by Studer *et al.* to control the spacial attachment of initiator to form regular stripes of polystyrene or poly(butylacrylate) brushes where the width and the periodicity were controlled.[56] Polymer layer growth by SI-NMP was achieved by atmospheric pressure hydrogen surface activation of a silicon wafer. A grafted polymer layer with lower surface roughness and a higher coverage was observed compared to free radical polymerization.[36] Gas-phase

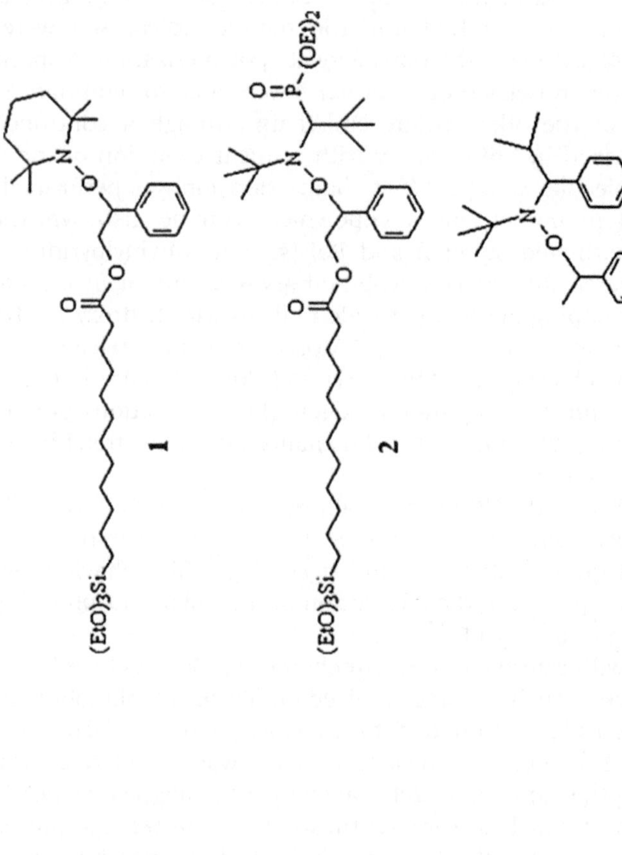

Figure 10.6 TEMPO (1), SG1 (2) and TIPNO (3)-based alkoxyamine structures.[14]

polymerization or "surface-initiated vapor deposition polymerization" was also achieved to create hydrophilic functional polymer brushes. Indeed, thermo-sensitive poly(*N*-isopropylacrylamide) brushes demonstrated a lower critical solution temperature at 32 °C, similar to the free chains.[57] SI-NMP of zwitterionic carboxybetaine methacrylamide brushes initiated by an SG1-based alkoxyamine, achieved by atom transfer radical addition (ATRA), was successfully demonstrated.[32] Photoreactive polymer brushes were also synthesized by SI-NMP to be transformed *via* photoclick trapping, *i.e.* photoreaction to surface-bound acyl radicals. Photolysis in the presence of persistent bio-conjugated nitroxide leads to site-specific streptavidin immobilization.[58] Classical PS brushes prepared by SI-NMP also showed either protein adsorption or protein-repellent properties depending on their brush thickness. This experiment was performed by AFM-based force spectroscopy.[59]

SI-NMP was not only used to create polymer brushes but also isolated polymer chains, *i.e.* low grafting density. In particular, a very low grafting density (0.005 chains nm^{-2}) was achieved by SI-NMP from silica planar substrates. AFM-based single molecule force spectroscopy has been also applied to study the detachment of isolated poly(acrylic acid) PAA chains pre-adsorbed onto a silicon nitride tip surface at varied pH. Comprehensive information about the molecular weight distribution of surface-grafted polymers was deduced by single molecule force spectroscopy and fitted the molecular weight distribution of free chains obtained by size exclusion chromatography.[60] The adhesion properties of polymers that bridge two surfaces were also studied. The measured interaction forces between the PAA chains and AFM tip depend on the grafting density of the polymer monolayers as well as on the contour length of the polymer chains. This study demonstrates that the stability of polymer bridges is determined by the adhesion strengths on both surfaces.[61]

10.3.2 Silica Particles (Dense and Mesoporous)

Surface modification of nanoparticles enhances their dispersibility and colloidal stability in various solvents or polymer matrices while they maintain their physical intrinsic characteristics. Among the numerous inorganic–organic hybrid materials, silica–polymer hybrid materials are the most commonly reported in the literature. Grafting of polymer brushes by SI-NMP has been widely dedicated to surface modification of spherical silica particles[13,17,20,22,26–31,33,37,39,62–68] and a few examples reported the grafting from ordered mesoporous silica particles exhibiting cylindrical organized mesopores of 2.5 to 10 nm.[34,69]

The first challenge was to control the macromolecular features and grafting densities of the polymer brushes in order to tune the dimension of the brush layer. A comparison of the polymerization of *n*-butyl acrylate from the surface of silica particles (13 nm diameter) in the presence of SG1 with a mono-component system and bi-component system was made by Parvole *et al.*[26] Although in both cases the control was maintained, in the case of the bi-component system it was difficult to predict the final number-average molecular weight (M_n) of the grafted chains. Indeed, after the decomposition of symmetrical diazoic initiator, free fragments of initiator can migrate in solution and change the free/grafted initiator ratio. On the contrary, the behaviour of the polymerization by the unimolecular system was in line with expectations, with experimental molecular weights close to theoretical ones.[26] The surface-initiated nitroxide-mediated polymerization from silica particles was mainly implemented in the presence of free alkoxyamine in order to improve the control of polymerization *via* persistent radical effect. It should be noticed that due to steric hindrance, the efficiency of the grafted alkoxyamines is generally below 40% providing polymer grafting density ranging between 0.05 and 0.6 polymer nm^{-2} for spherical silicas [26,28,30,31] and between 0.05 and 0.2 polymer nm^{-2} for ordered mesoporous silica.[34] The number-average molecular weight of polymer chains grafted onto spherical silica particles were in good accordance with the free chains [26,28,30,31] while the confined space of the mesopores of ordered mesoporous silica produced grafted chains with lower M_n in comparison with free chains.[34] The linear relationship between M_n of poly(*n*-butyl acrylate) brushes grafted from silica particles of 0.5 and 1 μm diameter and the thickness of the polymer shell was directly observed by transmission electron microscopy.[39] An increase of M_n from 27 to 91 kg mol^{-1} induces a shell thickness increase of 25 nm.[39] The analysis of polystyrene-grafted silica nanoparticles by small angle neutron scattering proved to be a powerful tool to show the increase of the polymer shell thickness with the increase of theoretical M_n for SI-NMP experiments.[63] The linear relationship was shown for polymerization carried out in the absence of free alkoxyamine initiator but only in the presence of free nitroxide as the control agent.[63] Of particular interest for controlled radical polymerization is the synthesis of block copolymers from the living chain end. Thus hybrid silica–poly(*n*-butyl acrylate)-*b*-polystyrene core–shell particles were synthesized by sequential surface-initiated nitroxide-mediated polymerization.[17] The grafted diblock copolymer shell showed two glass transitions characteristic of the soft P*n*BA block ($T_{g,PnBA} = -22$ °C) and the hard PS block ($T_{g,PS} = 83$ °C).

Mixed bushes of two distinct homopolymers chemically anchored from a dual grafted initiator have attracted interest because of their rich phase behaviour and responsive properties. Special attention was paid to the effect of different parameters on the self-assembled nanostructures (Figure 10.3). The relationship between the microphase separation and the overall grafting density,[22,65] the molecular weight of grafted P*t*BA and PS brushes[68] and the substrate curvature[70] was thoroughly investigated. The feature sizes of the patterns formed from microphase separation decrease with an increase of the grafting density[22] and with the decrease of average molecular weight.[68] The high grafting density asymmetric mixed P*t*BA/PS brushes (HDMB, $\sigma_{total} = 0.9$–1.2 chains nm^{-2}) are more stretched than the intermediate grafting density mixed brushes (IDMB, $\sigma_{total} = 0.6$–0.7 chains nm^{-2}).[22]

It should be noted that it is more difficult for mixed P*t*BA/PS brushes to undergo microphase separation at an overall grafting density of ~0.7–0.8 chains nm^{-2} in comparison with free P*t*BA-*b*-PS diblock copolymers.[68]

The properties of a polymer matrix can be enhanced by the incorporation of inorganic nanoparticles into a polymer matrix to design nanocomposite materials. The aggregation of the inorganic nanoparticles is detrimental to the improvement of the nanocomposite properties. Grafting polymer chains of controlled chain length and grafting density on the surface of the inorganic nanoparticles is a powerful method to control their spatial distribution inside the polymer matrix. The structure of dispersions of P*n*BA-grafted silica particles dispersed in P*n*BA matrix was first investigated by small angle neutron scattering (SANS), transmission electron microscopy (TEM) and rheological measurements.[20] SANS and TEM results showed the absence of nanoparticle aggregates.[20] Chevigny *et al.* analyzed the dispersion of PS-grafted silica particles, which were modified by surface-initiated nitroxide-mediated polymerization and subsequently dispersed inside a PS matrix.[64] The combination of scattering (SAXS/USAXS) and imaging (TEM) techniques showed that for intermediate polymer grafting density (0.2 chains nm^{-2}), one of the key parameters is the grafted to free chain molecular weight ratio ($R = M_{n,grafted}/M_{n,matrix}$). For R above 0.24, well-defined individual nanoparticles were dispersed in the PS matrix in the absence of any aggregation while compact and large aggregates were observed for R below 0.24.[64] The aggregation of the particles is associated with a significant collapse of the grafted chains, in agreement with the theoretical models describing the free energy as a combination of a mixing entropy term between the free and the grafted chains and an elastic term of deformation of the grafted brushes. Spectromechanical analyses of both silica–P*n*BA in P*n*BA and silica–PS in PS nanocomposites systems were performed for different nanoparticle contents. In the high frequency range, the perfect overlay of the elastic modulus (G') *versus* frequency curves for each silica content meant that the short characteristic times of the polymer matrix were not affected by the fillers.[20,64] However, in the low-frequency range, an increase of G' value was observed with increasing the inorganic nanoparticle content. The appearance of a longer relaxation time, distinct from that of the matrix, was ascribed to the particle diffusion [20] or to the contact between the grafted brushes for the individual particles.[64]

Core–shell silica particles of monodisperse 500 nm silica core grafted by adhesive P*n*BA brushes of controlled thickness can be self-assembled into colloidal crystals by simple dip coating of the core–shell dispersion.[39] The 2D micrometer-sized patterned array of the assembled particles which are in the same length range of the visible wavelength exhibited color diffraction due to the regular organization of the structured monolayer of particles. Surface-initiated NMP of 4-vinyl-triphenylamine from silica particles was carried out to design chemically and thermally stable core–shell particles.[67] The stability arises from the strong covalent bonding compared to the weak electrostatic interactions present in the usual *in situ* polymerization.

These conducting particles were combined with spiro-MeOTAD (2,2′,7,7′-tetrakis(*N,N*-di-*p*-methoxyphenylamine)-9,9′-spirobifluorene), a promising hole conductor, to improve conversion efficiency of solid-state dye-sensitized solar cells.[67]

10.3.3 Other Inorganic Surfaces

10.3.3.1 Clays

Polymer/clay nanocomposites, if they can be effectively prepared, offer the advantages of improved thermal properties for high temperature applications, barrier properties, reduced flammability and improved abrasion resistance and toughness. They have been explored for use in a range of polymer-based products and coatings. However, the improved properties obtained by incorporating clays are only realized if the clay platelets are well-dispersed, or exfoliated, in the polymer matrix. This has proven to be quite challenging for most desired polymers, which are typically hydrophobic and therefore incompatible with the hydrophilic clay surface. The use of surface-initiated living radical polymerization, and more specifically SI-NMP, provides a promising approach to address this challenge. Polymers grown from the surface of the clays can effectively overcome the attractive layers of the intercalated clay, resulting in partial or complete exfoliation to allow effective dispersion of the clay within the polymer matrix. The grafted polymer chains are chosen to be compatible with the intended host polymer matrix to improve dispersion quality and stability. Weimer *et al.*[71] were the first to report the use of SI-NMP. Using ion exchange, they tethered a TEMPO-based initiator to montmorillonite and then conducted the SI-NMP of styrene. Effectively complete delamination was achieved. Desorption of the polymer allowed molecular weight analysis; good livingness and control were preserved, even without the addition of excess (sacrificial) nitroxide.

A more detailed kinetic study of styrene SI-NMP from clay (laponite) was published by Konn et *al.*[48] The SG1 nitroxide (DEPN) was intercalated into the clay in the form of a quaternary ammonium salt alkoxyamine by ion exchange prior to polymerization. Excess DEPN alkoxyamine (styryl-DEPN) was added to improve control of the polymerization. The free and tethered polymer chains were collected separately, allowing comparison of the SI-NMP *versus* bulk NMP polymer properties. The bulk NMP M_n values agreed well with theoretical M_n values, while the tethered chains showed lower M_n although both populations had similar dispersities, indicating both the bulk NMP and SI-NMP exhibited comparable control. Dispersities were higher than typically observed for bulk NMP, a phenomenon often seen in surface-initiated polymerizations and likely due to low initiation efficiencies. Complete exfoliation of the laponite was observed, and the nanocomposite was found to be readily dispersible in a range of both polar and nonpolar monomers.

SI-NMP from montmorillonite was also reported by Shen *et al.*,[72] Micusik *et al.*[73] and Cherifi *et al.*[74] Micusik employed SG1 as the nitroxide in the

SI-NMP of MMA/*n*-BA mixtures. The authors undertook a different approach in preparation of the montmorillonite macroinitiator, first preparing a low molecular weight quaternary salt polymer (using vinylbenzyl trimethylammonium chloride) by SG1-mediated NMP, which was then ion exchanged into the montmorillonite. Polymerizations were conducted at 110 °C without using additional SG1. Complete exfoliation of the clay was observed, although some uncontrolled polymerization occurred. In addition to studying the bulk SI-NMP, Micusik further used the montmorillonite macroinitiator directly in miniemulsion polymerization to prepare colloidal dispersions of film forming particles containing clay. Colloidally stable latexes with relatively little coagulum were produced if a "masterbatch" approach was used in the miniemulsion polymerization; this procedure involved first preparing a masterbatch of montmorillonite with grafted polymer chains, and then dispersing this composite into monomer prior to miniemulsification. Attempts to directly disperse montmorillonite macroinitiator into monomer and then emulsify resulted in poor dispersion of the clay and unstable latexes. Cherifi *et al.* prepared organically modified clay by ion exchange with a cationic initiator in order to prepare polymer–clay materials by surface-initiated NMP of *n*BA and MMA in the presence of free alkoxyamine and free SG1 nitroxide.[74] The control of the gradient copolymer synthesis was not affected by the presence of clay except for a slight decrease of M_n observed by increasing the initiator fragments through the initiator-modified clay content. The enhancement of the thermal stability of P(BA-*grad*-MMA) by introduction of montmorillonite was observed by an increase of ~20 °C of both onset temperature taken at 10% of weight loss and maximum weight loss temperature.

The thermal degradation behaviour of magadiite/polystyrene nanocomposites has also been studied.[75] A TEMPO-based initiator was covalently bound to the clay surface using a silylated reaction. Exfoliation was observed, despite the low initiation efficiencies (~10% of the nitroxide groups formed chains) which is a common occurrence in any SI-NMP where access to the nitroxide groups may be restricted. The nanocomposite displayed significantly different thermal decomposition behaviour than magadiite/polystyrene physical mixtures. Whereas the mixtures decomposed to ash at high temperatures, the nanocomposites yielded a rigid, solid residue.

Billon *et al.*[76–78] reported the development of a new class of pigments based on nanocomposites of mica with a dye (alizarin) containing pigment. Initial efforts[77] used a copolymer consisting of a poly(*n*-butyl acrylate) block with a second block of styrene with an alizarin-bearing vinyl monomer synthesized by the authors. A cationic initiator was adsorbed onto the mica, and the SG1-mediated SI-NMP of *n*-butyl acrylate was then done to give a poly(*n*-butyl acrylate) macroinitiator for the subsequent polymerization of the second styrene-alizarin block (0.3 mol% of alizarin per mole of styrene). Some of the grafted polymer could be removed by toluene extraction but the authors reported a fraction of the graft polymer remained adsorbed on the mica surface. SEC analysis of the extracted polymers showed that initiation of the poly(*n*-butyl acrylate) block was highly efficient although the dispersities of the final block copolymer were high. While the free diblock copolymer P(BA-*b*-(S-A) was yellow in colour, the mica nanocomposite was unexpectedly pink, which was attributed to the dye colour being pH-sensitive and the mica surface being basic. A subsequent study used only styrene-alizarin copolymer bound to mica and further explored the relationship between the adsorption density and polymer chain length on the observed pigment colour. Well-controlled polymerizations were achieved. Oxyanions on the mica surface were shown to influence the observed colour. Variations in polymer chain adsorption density and molecular weight that resulted in the alizarin units being closer to the mica surface gave rise to larger effects on colour. A similar approach was used to prepare a hybrid pigment based on aluminum that had been coated with a thin layer of silica.[79] A silylation reaction was used to graft an SG1-based initiator to the aluminum platelets, prior to polymerization of either poly(*n*-butyl acrylate) or poly(*n*-butyl acrylate-*grad*-styrene) brushes containing a small amount of polymerizable dye. Perhaps counterintuitively, it was noted that superior colouration for the nanocomposites was obtained for more poorly controlled polymerizations, attributable to the interaction distance between the aluminum platelets and the dye as determined by the polymer chain length.

Fluoromica nanocomposites have been reported for potential applications in optoelectronics.[80,81] Fluoromica is a fluorine-containing semi-synthetic clay that is attracting interest for dye-based hybrid materials. However, it has poor solution processability due to limited swelling in solvents. Development of polymer-fluoromica nanocomposites may address this problem. Leone *et al.*[80] bound a TEMPO-based alkoxyamine to fluoromica using ion exchange with a cationic dye molecule followed by the SI-NMP of styrene. No sacrificial initiator was added, which is the likely explanation for the relatively poor control observed (dispersities >2). As other researchers have also observed, some of the tethered chains were removed by extraction following polymerization while a fraction of the chains did remain adsorbed on the fluoromica surface. Consistent with other SI-NMP of clays, low initiation efficiencies ($\sim$6%) were observed. Importantly, the solution processability was significantly improved, making solutions of the nanocomposite amenable to spin coating. A similar approach was then used to prepare fluoromica-based blue-emitting nanocomposites with possible application as polymer light-emitting devices (PLEDs).[81] In this case, a comonomer mixture consisting of styrene and styrene derivatives para-substituted with (oligo)fluorene moieties was used for the SI-NMP, yielding materials exhibiting blue photoluminescence.

10.3.3.2 Iron Oxide Particles

10.3.3.2.1 Fe_3O_4 (Magnetite). Magnetite nanoparticles are widely used in biomedical applications and research, including for example as antigen–antibody markers and in MRI (magnetic resonance imaging).

surface modification to allow tailoring of their interaction and dispersibility in different local environments, while preserving their magnetic properties, remains a topic of high interest. Takahara's group first reported the use of SI-NMP to graft polystyrene or poly(3-vinylpyridine) onto the surface of ~10 nm magnetite particles.[44,47] A 4-methoxy-TEMPO-containing alkoxyamine functionalized with phosphoric acid groups was designed to allow binding of the alkoxyamine to the nanoparticle surface through interaction with hydroxyls on the metal surface *via* chemisorption (Entry 12 in Table 10.1). Sacrificial initiator was added to improve control. Subsequent comparison of free polystyrene and grafted polystyrene revealed high monomer conversion and remarkable similarity in their molecular weight properties, despite the expectedly low initiation efficiencies (~15–30%). The 3-vinylpyridine polymerizations exhibited poorer control (dispersities ~1.5–1.7) compared to the styrene polymerizations (dispersities ~1.3). Graft densities were reported in the range of 0.12–0.20 chains nm^{-2}, reasonable values for SI-NMP (the alkoxyamine surface density was 0.73 chains nm^{-2}). The grafted magnetite nanoparticles could be readily dispersed in good solvents for polystyrene or poly(3-vinylpyridine), respectively. Protonation of the poly(3-vinylpyridine) grafted nanoparticles yielded stable aqueous dispersions. Binder *et al.*[82] grafted PNIPAM onto magnetite particles, also yielding a stimuli-responsive (thermally responsive) nanocomposite. Chen *et al.*[83] also used 4-hydroxy-TEMPO-mediated SI-NMP to polymerize 4-vinylpyridine onto 30 nm magnetite nanoparticles functionalized with the vinyl-containing 3-methacryloxypropyltrimethoxysilane (3-MPS). Benzoyl peroxide initiator, 4-hydroxy-TEMPO and monomer (4-vinylpyridine) were then added to start the polymerization, yielding covalently bound chains. The polymerizations exhibited good control (dispersities < 1.2) and molecular weights ranging from 2200 to 52 000 Da. Final particle diameters were ~120–150 nm, suggesting aggregates had formed. While the grafting was successful, magnetization of the grafted particles was substantially weaker than for the untreated magnetite.

10.3.3.2.2 γ-Fe_2O_3 (Maghemite). Robbes *et al.*[84,85] recently developed a methodology to conduct SG1-mediated SI-NMP from γ-Fe_2O_3 nanoparticles using a two-step process. Maintaining colloidal stability during the preparation of the macroalkoxyamine is especially challenging for this system, as the γ-Fe_2O_3 nanoparticles are synthesized in acidic aqueous media (and therefore possess positive charge surface). The silanization step to yield a surface suitable for attachment of a nitroxide requires a basic solution, and therefore the acidic dispersion must be transferred to a basic medium in polar solvent, without compromising colloidal stability. Free SG1 alkoxyamine (MAMA–NHS: 2-methyl-2-[*N*-*tert*-butyl-*N*-(1-diethoxyphosphoryl-2,2-dimethylpropyl)aminoxy]-*N*-propionyloxysuccinimide) was added for improved control. Dissolution of the maghemite particles yielded polymer chains for analysis. The M_n values for both the free and grafted polystyrene chains were in close agreement with theoretical values, although the dispersity for the grafted chains was somewhat higher. The initiation efficiency was surprisingly high for SI-NMP at approximately 70%. Extensive study was done on the grafted nanoparticle dispersions using SANS and SAXS to characterize the form factor and chain conformations. Further work more extensively examined the nanostructure of dispersions of the nanocomposites.[85]

10.3.3.3 Other Metal Oxides

Other metal oxide surfaces have been the subject of SI-NMP studies, and more are likely to be increasingly studied in the future as techniques and methodologies evolve. TiO_2 (titanium dioxide) is widely used industrially, valued for its high refractive index, which makes it an excellent opacifier, absorption of UV light and more recently for its photocatalytic activity. Using a similar approach to their work on magnetite, Takahara's group conducted SI-NMP of styrene from the rutile form of TiO_2.[44,47] The ~15 nm particles were first functionalized with 4-methoxy-TEMPO-based alkoxyamine with phosphoric acid groups by adsorption. The polymerization behaviour was similar to their results for magnetite. The polystyrene-grafted TiO_2 nanoparticles gave stable dispersions with high UV absorption in good solvents for polystyrene (tetrahydrofuran, ethyl acetate, toluene) but, not unexpectedly, unstable dispersions in poor or non-solvents such as acetone and methanol. Well-dispersed particles were believed to be not in the form of primary particles but rather small clusters (~90 nm) of a few aggregated particles. Solution-cast hybrid films of polystyrene-grafted TiO_2 and polystyrene displayed improved optical and wear resistance properties compared to films cast from ungrafted TiO_2 and polystyrene. P(*n*BA)–ZnO anisotropic core–shell building blocks were synthesized by SI-NMP to increase the solvent dispersion of the zinc oxide ZnO nanoparticles involved in the preparation of honeycomb porous hybrid films.[86]

CuO (cupric oxide) sheets functionalized with peroxide initiating groups were used in the TEMPO-mediated SI-NMP of a vinyl monomer derived from ionic liquids.[87] The grafted chains imparted tunable properties to the CuO surface, including wettability due to a reversible ion exchange procedure in which hydrophobic PF_6^- anions on the polymer were exchanged with hydrophilic Cl^- anions.

10.3.3.4 Metallic Compounds

Quantum dots [cadmium selenide (CdSe) nanoparticles] are sensitive to the presence of free radicals, which often causes quenching of the photoluminescence. While this is problematic for conventional free radical polymerization, it has been demonstrated that with the lower radical concentrations present in living radical polymerization, quenching may not be a serious issue. Sill and Emrick[45] used a ligand exchange procedure to covalently attach a TEMPO alkoxyamine to quantum dots, and then studied the SI-NMP of styrene or styrene–methyl methacrylate mixtures in both bulk and solution. Reasonably good control was achieved, although dispersities increased at higher molecular weight of the grafted chains. Importantly, there was not a significant blue shift in the fluorescence maxima of the polystyrene–CdSe nanocomposites compared to the TEMPO-functionalized CdSe, signifying the absence of appreciable degradation in the desired material properties. By contrast, the use of free radical initiators such as AIBN or benzoyl peroxide in previous studies to initiate styrene polymerization in the presence of CdSe nanoparticles caused degradation of the nanoparticles within minutes. Monodisperse ZnS (zinc sulfide) nanoparticles have also been the subject of SI-NMP.[88] A thin silica layer was first deposited on the ZnS surface, followed by functionalization with TEMPO and subsequent polymerization with styrene. A sacrificial initiator was added, giving well-controlled polymerization even at quite high polymer chain densities ($\sim$0.9 chains nm^{-2}). Highly ordered 2-dimensional arrays with colour resulting from the structure were observed. The chain length of the grafted polymer can be used to alter the spacing between adjacent particles in the array.

10.3.4 Organic Surfaces

Nitroxide-mediated polymerization has been initiated from the surfaces of organic substrates, although the range of substrates studied is limited compared to SI-NMP from inorganic substrates. Polystyrene surfaces have been the most commonly examined,[50,89,90] in addition to reports on MWCNTs (multi-walled carbon nanotubes),[91] graphene oxide[92] and chitosan.[93]

Mazurowski *et al.*[90] employed SI-NMP of styrene as a means of investigating the fundamental nature and structure of polymer brushes on curved surfaces of different diameters, and more specifically the nature of the "interphase" represented by the polymer chains grafted to the particle surface. Crosslinked polystyrene latex particles (of either $\sim$50 nm or $\sim$100 nm diameter) were chosen as the substrate, but were synthesized from deuterated monomers to provide contrast for SANS (small angle neutron scattering) and SAXS (small angle X-ray scattering) analyses between the particles and the grafted polystyrene chains. The TEMPO-mediated polymerizations yielded a range of graft molecular weights, with the observed shell thicknesses varying linearly with molecular weight.

Poor dispersibility in organic solvents represents a significant challenge to the effective utilization of MWCNTs in many desired applications, and hence numerous approaches have been explored to improve dispersibility by functionalizing their surface, including SI-NMP.[91] Nanotubes were first functionalized with TEMPO, followed by styrene grafting. Free TEMPO was not added to the polymerization, as is often done with SI-NMP. Molecular weight data was not reported, although TGA and FTIR evidence indicated that grafting did occur. Dispersibility of the grafted MWCNTs in $CHCl_3$, THF and *o*-dichlorobenzene was noticeably improved compared to the untreated

MWCNTs. Chain extension of the polystyrene chains with 4-vinylpyridine was also conducted to confirm the livingness of the grafted chains.

Graphenes are comprised of two-dimensional structures of sp^2 hybridized carbon atoms resembling sheets, with a thickness of one atom. Graphene oxides, which can bear functional groups (*e.g.* epoxy, hydroxyl and carboxyl) are attracting much attention as of late for their potential use in photonic, electronic and biomedical applications. Garcia-Valdez *et al.*[92] functionalized graphene oxide with TEMPO, which then was used to initiate the polymerization of polyisoprene and polystyrene. The functionalization step (attachment of TEMPO to the graphene oxide) resulted in some exfoliation of the graphene oxide, and polymerization appeared to result in further exfoliation, although the authors note that complete exfoliation was not observed, and that many of the TEMPO initiating sites were likely not readily accessible during polymerization. As the grafted chains could not be cleaved, no molecular weight data was reported. Confirmation of grafting was provided using a variety of characterization techniques, including TGA, FT-IR, Raman, XPS, XRD, and TEM.

Crosslinked polystyrene microspheres (Merrifield resins)[50] and nanospheres[89] were used to demonstrate the effectiveness of SI-NMP on polymer particle substrates. Merrifield resins are relatively large porous particles (0.05–015 mm) containing chloromethyl groups that are widely used in solid-phase synthesis. Bian and Cunningham[50] bound TEMPO to the resins and grafted several polymers (including random copolymers and block copolymers) from the resin surface. With sacrificial initiator added, well-controlled polymerizations were observed. Subsequent work by Bian[89] used much smaller particles (60 nm diameter prior to functionalization), and used SG1 instead of TEMPO. Atom transfer radical addition proved to be the most efficient route to bind the SG1 to the particle surface through chloromethyl groups, achieving loadings of $\sim$1 mmol SG1 g particles^{-1}. The pH-sensitive monomer 2-(dimethylamino)ethyl acrylate (DMAEA) was readily grafted from the particle surface, giving a core–shell morphology (Figure 10.7). Amphiphilic block copolymers of DMAEA with styrene and *n*-butyl acrylate were also grafted, with excess added SG1 required to achieve good control and livingness. Typical increases in diameter upon grafting were $\sim$5–20 nm for molecular weights varying from $\sim$5–25 kDa.

Natural polymers are becoming the subject of more graft modification studies for their possible applications in composite materials incorporating a renewable component or as alternative substitutes for traditional petroleum-based polymers. Typically, the properties of the natural polymers are not inherently suitable and therefore modification is required. For example, many natural polymers are hydrophilic and not compatible with hydrophobic materials. Lefay *et al.*[94] functionalized the amino group of chitosan with an acrylamide moiety followed by intermolecular 1,2 radical addition of the SG1-based BlocBuilder alkoxyamine, which then enabled the heterogeneous graft polymerization of sodium 4-styrene sulfonate or methyl methacrylate (with a small fraction of acrylonitrile for improved control of

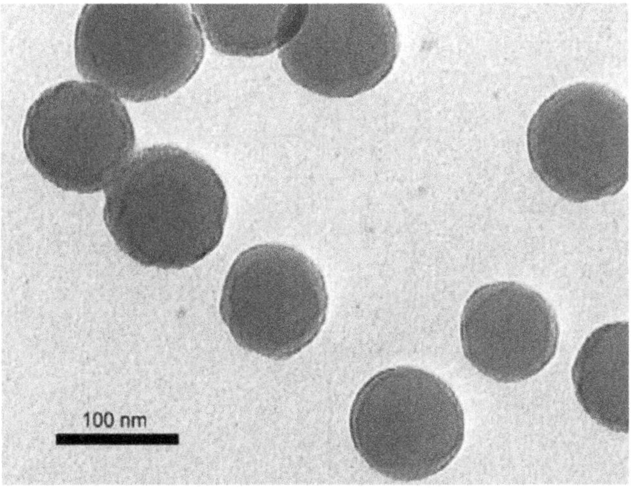

Figure 10.7 TEM of PS latex particles functionalized with SG1 and then grafted with poly(S-*b*-DMAEA).[50]

the polymerization).[94] TGA and solid-state ^{13}C CP-MAS NMR spectroscopy confirmed the presence of grafting. Garcia-Valdez *et al.*[95] modified chitosan in supercritical carbon dioxide *via* a 4-hydroxy-TEMPO mediated process where the grafted polymer was a styrene-maleic anhydride alternating co-polymer, and subsequently grafted polystyrene and *n*-butyl acrylate in homogeneous solution using either TIPNO (2,2,5-trimethyl-4-phenyl-3-aza-hexane-3-nitroxide) or SG1 [*N*-*tert*-butyl-*N*-(1-diethylphosphono-2,2-dimethyl-propyl) nitroxide] as the mediating nitroxides.[93] The SG1 and TIPNO nitroxides are more versatile and allow a wider selection of monomers to be grafted as well as requiring lower reaction temperatures compared to TEMPO, which may be a concern for some natural polymers where exposure to temperatures >120 °C for several hours may lead to degradation. Natural polymers may differ from inorganic substrates in offering more than one possible site to attach the nitroxide. With chitosan, either the amino[91] or primary hydroxyl[93,95] may be selected. For some applications, preservation of the amino functionality may be desired. Garcia-Valdez' approach of functionalizing the primary hydroxyl allowed the amino groups to be reversibly complexed with SDBS (sodium dodecyl benzene sulfonate), which imparted solubility in DMSO to the chitosan macroinitiator, thereby allowing the SI-NMP to proceed under homogeneous conditions, which is likely to improve the effectiveness and uniformity of the grafting procedure (Scheme 10.2).

10.4 Perspective on Future Research Opportunities

SI-NMP has proven to be a robust and comparatively facile synthetic approach for custom tailoring of a variety of organic and inorganic surfaces

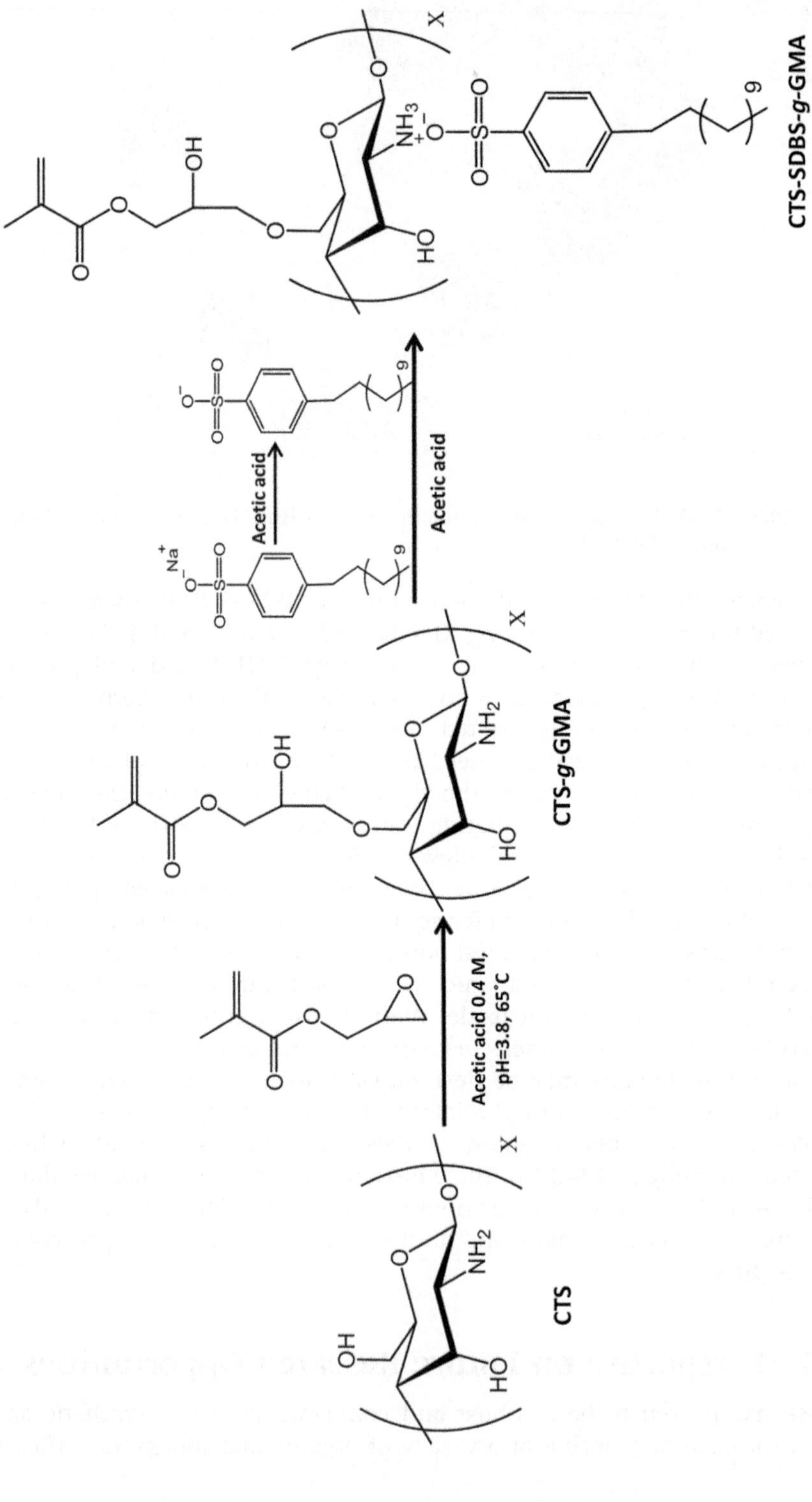

Scheme 10.2 Schematic illustration of the synthesis of CTS-SDBS-*g*-GMA.[93]

with polymers of controlled chain length distribution and composition. Yet despite the considerable advances made in recent years, a number of challenges remain to be adequately addressed, with greater opportunity for more widespread adoption of SI-NMP in commercial products if these issues are resolved. Efficient synthetic routes to functionalize initiating groups to surfaces can be problematic. For some substrates and with some nitroxides, efficient routes have been developed. However, in many cases the procedures are multi-step, time consuming, expensive and require considerable volumes of solvent. Development of simpler, lower cost routes employing more environmentally benign methods would be a significant breakthrough for facilitating commercial use of SI-NMP. Accessibility of the initiating groups on various surfaces can also be a severe challenge, leading to low initiator utilization (often ~10–30%) and therefore increasing costs and reducing economic viability. Improved design of the structure of initiating groups to improve accessibility and initiation efficiency is an important objective. A commonly used strategy in SI-NMP is the use of free or sacrificial initiator or nitroxide to improve control of the polymerization, necessitated by the very low polymer chain concentrations characteristic of SI-NMP. While effective, this poses two problems. First is the additional cost of the additional initiator or nitroxide, and second is the additional required step to separate the free polymer from the polymer-grafted substrate. In cases where complete removal of ungrafted polymer is essential or desirable for end-use properties, this may be a particularly difficult and expensive task for materials such as clays, where the ungrafted polymer is strongly adsorbed, or very low specific surfaces of microplatelets for pigments. Difficulty in characterizing the polymer-grafted hybrid materials also remains an issue. While several characterization techniques have been applied to SI-NMP studies and much can be learned about the structure of the grafted materials, it is often difficult to accurately determine the molecular weight distribution of the grafted polymer. If cleavable bonds to the substrate are present, effective molecular weight characterization can usually be conducted. However, for most applications, highly stable linkages between the substrate and polymer chains will be desirable. The commonly used approach of adding sacrificial initiator and assuming the grafted polymer has the same chain length distribution as the free polymer is only sometimes reliable. A generic approach that would facilitate accurate molecular weight characterization of the grafted polymer would be a valuable contribution. TGA is widely used to determine the ratio of substrate to grafted polymer, and works well for some substrates. However, for organic substrates, including polymer nanoparticles and natural polymers, it is often not possible to distinguish between the substrate and graft polymer in the thermograms.

Recent years have seen a surge in activity in SI-NMP, spurred by the use of nitroxides such as SG1 and TIPNO that offer greater versatility in the range of usable monomers compared to TEMPO and its derivatives. A solid foundation of the fundamental principles has now been established, and the opportunity now exists to expand research on SI-NMP to new areas and applications, and to promote commercial development.

Acknowledgements

Financial support from the Ontario Research Chairs program (MFC) is gratefully acknowledged.

References

1. L. Ghannam, J. Parvole, G. Laruelle, J. Francois and L. Billon, *Polym. Int.*, 2006, **55**, 1199.
2. Y. Tsujii, K. Ohno, S. Yamamoto, A. Goto and T. Fukuda, *Surface-Initiated Polymerization I*, ed. R. Jordan, Springer Berlin Heidelberg, Germany, 2006, vol. 197, pp. 1–45.
3. R. Barbey, L. Lavanant, D. Paripovic, N. Schüwer, C. Sugnaux, S. Tugulu and H.-A. Klok, *Chem. Rev.*, 2009, **109**, 5437.
4. M. K. Brinks and A. Studer, *Macromol. Rapid Commun.*, 2009, **30**, 1043.
5. A. Olivier, F. Meyer, J.-M. Raquez, P. Damman and P. Dubois, *Prog. Polym. Sci.*, 2012, **37**, 157.
6. P. G. de Gennes, *Macromolecules*, 1980, **13**, 1069.
7. B. Radhakrishnan, R. Ranjan and W. J. Brittain, *Soft Matter*, 2006, **2**, 386.
8. *Polymer Brushes: Synthesis: Synthesis and Characterization, Applications*, ed. R. Advincula, W. J. Brittain, K. C. Caster and J. Rühe, Wiley-VCH Verlag GmbH & Co. KGaA, Weinheim, 2004.
9. M. Husseman, E. E. Malmström, M. McNamara, M. Mate, D. Mecerreyes, D. Benoit, J. L. Hedrick, P. Mansky, E. Huang, T. P. Russell and C. J. Hawker, *Macromolecules*, 1999, **32**, 1424.
10. A. Ulman, *Chem. Rev.*, 1996, **96**, 1533.
11. C. Devaux, F. Cousin, E. Beyou and J.-P. Chapel, *Macromolecules*, 2005, **38**, 4296.
12. C. Devaux, J. P. Chapel, E. Beyou and Ph. Chaumont, *Eur. Phys. J. E: Soft Matter Biol. Phys.*, 2002, **7**, 345.
13. J. Parvole, J.-P. Montfort, G. Reiter, O. Borisov and L. Billon, *Polymer*, 2006, **47**, 972.
14. R.-V. Ostaci, C. Celle, G. Seytre, E. Beyou, J.-P. Chapel and E. Drockenmuller, *J. Polym. Sci., Part A: Polym. Chem.*, 2008, **46**, 3367.
15. M. R. Tomlinson and J. Genzer, *Polymer*, 2008, **49**, 4837.
16. S. Yamamoto, Y. Tsujii and T. Fukuda, *Macromolecules*, 2002, **35**, 6077.
17. G. Laruelle, J. Parvole, J. Francois and L. Billon, *Polymer*, 2004, **45**, 5013.
18. K. Urayama, S. Yamamoto, Y. Tsujii, T. Fukuda and D. Neher, *Macromolecules*, 2002, **35**, 9459.
19. C. Gay, *Macromolecules*, 1997, **30**, 5939.

20. R. Inoubli, S. Dagréou, A. Lapp, L. Billon and J. Peyrelasse, *Langmuir*, 2006, **22**, 6683.
21. D. Sunday, J. llavsky and D. L. Green, *Macromolecules*, 2012, **45**, 4007.
22. X. Jiang, B. Zhao, G. Zhong, N. Jin, J. M. Horton, L. Zhu, R. S. Hafner and T. P. Lodge, *Macromolecules*, 2010, **43**, 8209.
23. S. Yamamoto, M. Ejaz, Y. Tsujii, M. Matsumoto and T. Fukuda, *Macromolecules*, 2000, **33**, 5602.
24. S. Yamamoto, M. Ejaz, Y. Tsujii and T. Fukuda, *Macromolecules*, 2000, **33**, 5608.
25. D. Dukes, Y. Li, S. Lewis, B. Benicewicz, L. Schadler and S. K. Kumar, *Macromolecules*, 2010, **43**, 1564.
26. J. Parvole, J. P. Montfort and L. Billon, *Macromol. Chem. Phys.*, 2004, **205**, 1369.
27. J. Parvole, G. Laruelle, C. Guimon, J. Francois and L. Billon, *Macromol. Rapid Commun.*, 2003, **24**, 1074.
28. C. Bartholome, E. Beyou, E. Bourgeat-Lami, P. Chaumont, F. Lefebvre and N. Zydowicz, *Macromolecules*, 2005, **38**, 1099.
29. R. Inoubli, S. Dagreou, A. Khoukh, F. Roby, J. Peyrelasse and L. Billon, *Polymer*, 2005, **46**, 2486.
30. R. Inoubli, S. Dagreou, M.-H. Delville, A. Lapp, J. Peyrelasse and L. Billon, *Soft Matter*, 2007, **3**, 1014.
31. C. Bartholome, E. Beyou, E. Bourgeat-Lami, P. Chaumont and N. Zydowicz, *Macromolecules*, 2003, **36**, 7946.
32. S. Abraham and L. D. Unsworth, *J. Polym. Sci., Part A: Polym. Chem.*, 2011, **49**, 1051.
33. C. D. Chevigny Gigmes, D. Bertin, J. Jestin and F. Boué, *Soft Matter*, 2009, **5**, 3741.
34. H. Blas, M. Save, C. Boissière, C. Sanchez and B. Charleux, *Macromolecules*, 2011, **44**, 2577.
35. D. Li, X. Sheng and B. Zhao, *J. Am. Chem. Soc.*, 2005, **127**, 6248.
36. G. T. Lewis and Y. Cohen, *Langmuir*, 2008, **24**, 13102.
37. C. Bartholome, E. Beyou, E. Bourgeat-Lami, P. Chaumont and N. Zydowicz, *Polymer*, 2005, **46**, 8502.
38. J. Nicolas, Y. Guillaneuf, C. Lefay, D. Bertin, D. Gigmes and B. Charleux, *Prog. Polym. Sci.*, 2013, **38**, 63.
39. C. Deleuze, M. H. Delville, V. Pellerin, C. Derail and L. Billon, *Macromolecules*, 2009, **42**, 5303.
40. D. Gigmes, P. E. Dufils, D. Gle, D. Bertin, C. Lefay and Y. Guillaneuf, *Polym. Chem.*, 2011, **2**, 1624.
41. J. Nicolas, B. Charleux, O. Guerret and S. Magnet, *Macromolecules*, 2005, **38**, 9963.
42. E. Beyou, J. Humbert and P. Chaumont, *e-Polym.*, 2003, **3**, 249.
43. F. Chauvin, P. E. Dufils, D. Gigmes, Y. Guillaneuf, S. R. A. Marque, P. Tordo and D. Bertin, *Macromolecules*, 2006, **39**, 5238.

44. R. Matsuno, K. Yamamoto, H. Otsuka and A. Takahara, *Macromolecules*, 2004, **37**, 2203.
45. K. Sill and T. Emrick, *Chem. Mater.*, 2004, **16**, 1240.
46. R. Matsuno, H. Otsuka and A. Takahara, *Soft Matter*, 2006, **2**, 415.
47. M. Kobayashi, R. Matsuno, H. Otsuka and A. Takahara, *Sci. Technol. Adv. Mater.*, 2006, **7**, 17.
48. C. Konn, F. Morel, E. Beyou, P. Chaumont and V. Cedex, *Macromolecules*, 2007, **40**, 7464.
49. J. Bonilla-Cruz, T. Lara-Ceniceros, E. Saldivar-Guerra and E. Jimenez-Regalado, *Macromol. Rapid Commun.*, 2007, **28**, 1397.
50. K. Bian and M. F. Cunningham, *J. Polym. Sci., Part A: Polym. Chem.*, 2005, **43**, 2145.
51. B. Zhao and L. Zhu, *Macromolecules*, 2009, **42**, 9369.
52. W. Li, C. Bao, R. A. E. Wright and B. Zhao, *RSC Adv.*, 2014, **4**, 18772.
53. L. Andruzzi, W. Senaratne, A. Hexemer, E. D. Sheets, B. Ilic, E. J. Kramer, B. Baird and C. K. Ober, *Langmuir*, 2005, **21**, 2495.
54. S. Zhou, Y. Xu and S. Fang, *Polym.-Plast. Technol. Eng.*, 2012, **51**, 1377.
55. M. Husemann, M. Morrison, D. Benoit, J. Frommer, M. C. Mate, W. D. Hinsberg, J. L. Hedrick and C. J. Hawker, *J. Am. Chem. Soc.*, 2000, **122**, 1844.
56. M. K. Brinks, M. Hirtz, L. Chi, H. Fuchs and A. Studer, *Angew. Chem., Int. Ed.*, 2007, **46**, 5231.
57. J. Li, X. Chen and Y.-C. Chang, *Langmuir*, 2005, **21**, 9562.
58. A. Mardyukov, Y. Li, A. Dickschat, A. H. Schaefer and A. Studer, *Langmuir*, 2013, **29**, 6369.
59. C. Hentschel, H. Wagner, J. Smiatek, A. Heuer, H. Fuchs, X. Zhang, A. Studer and L. Chi, *Langmuir*, 2013, **29**, 1850.
60. L. Sonnenberg, J. Parvole, O. Borisov, L. Billon, H. E. Gaub and M. Seitz, *Macromolecules*, 2006, **39**, 281.
61. L. Sonnenberg, J. Parvole, F. Kuhner, L. Billon and H. E. Gaub, *Langmuir*, 2007, **23**, 6660.
62. B. Zhao, *Polymer*, 2003, **44**, 4079.
63. C. Chevigny, D. Gigmes, D. Bertin, R. Schweins, J. Jestin and F. Boue, *Polym. Chem.*, 2011, **2**, 567.
64. C. Chevigny, F. Dalmas, E. Di Cola, D. Gigmes, D. Bertin, F. Boue and J. Jestin, *Macromolecules*, 2011, **44**, 122.
65. C. Bao, S. Tang, J. M. Horton, X. Jiang, P. Tang, F. Qiu, L. Zhu and B. Zhao, *Macromolecules*, 2012, **45**, 8027.
66. D. Cimen and T. Caykara, *J. Appl. Polym. Sci.*, 2013, **129**, 383.
67. W. Yuan, H. Zhao, H. Hu, S. Wang and G. L. Baker, *ACS Appl. Mater. Interfaces*, 2013, **5**, 4155.
68. C. Bao, S. Tang, R. A. E. Wright, P. Tang, F. Qiu, L. Zhu and B. Zhao, *Macromolecules*, 2014, **47**, 6824.

69. M. Lenarda, G. Chessa, E. Moretti, S. Polizzi, L. Storaro and A. Talon, *J. Mater. Sci.*, 2006, **41**, 6305.
70. J. M. Horton, S. Tang, C. Bao, P. Tang, F. Qiu, L. Zhu and B. Zhao, *ACS Macro Lett.*, 2012, **1**, 1061.
71. M. W. Weimer, H. Chen, E. P. Giannelis and D. Y. Sogah, *J. Am. Chem. Soc.*, 1999, **121**, 1615.
72. Y. Shen, Y. Wang, J. Chen, H. Li, Z. Li and C. Li, *J. Appl. Polym. Sci.*, 2010, **118**, 1198.
73. M. Micusik, A. Bonnefond, M. Paulis and J. R. Leiza, *Eur. Polym. J.*, 2012, **48**, 896.
74. N. Cherifi, A. BenAboura, M. Save and L. Billon, *J. Polym. Sci., Part A: Polym. Chem.*, 2012, **50**, 3976.
75. H. Yukutake, M. Kobayashi, H. Otsuka and A. Takahara, *Polymer*, 2009, **41**, 555.
76. L. Ghannam, M. Bacou, H. Garay, M. Shanahan, J. Francois and L. Billon, *Polymer*, 2004, **45**, 7035.
77. L. Ghannam, H. Garay, M. Shanahan, J. Francois and L. Billon, *Chem. Mater.*, 2005, **17**, 3837.
78. L. Ghannam, H. Garay and L. Billon, *Macromolecules*, 2008, **41**, 7374.
79. M. Joubert, A. Khoukh, J.-F. Tranchant, F. Morvan and L. Billon, *Macromol. Chem. Phys.*, 2009, **210**, 1544.
80. G. Leone, U. Giovanella, F. Bertini, W. Porzio, F. Meinardi, C. Bottaa and G. Riccia, *J. Mater. Chem. C*, 2013, **1**, 1450.
81. G. Leone, U. Giovanella, F. Bertini, S. Hoseinkhani, W. Porzio, G. Ricci, C. Botta and F. Galeotti, *J. Mater. Chem. C*, 2013, **1**, 6585.
82. W. H. Binder, D. Gloger, H. Weinstabl, G. Allmaier and E. Pittenauer, *Macromolecules*, 2007, **40**, 3097.
83. J. Chen, H. Li, L. Zhang, Q. Zhou and Y. Li, *J. Macromol. Sci., Pure Appl. Chem.*, 2011, **48**, 241.
84. A.-S. Robbes, F. Cousin, F. Meneau, C. Chevigny, D. Gigmes, J. Fresnais, R. Schweins and J. Jestin, *Soft Matter*, 2012, **8**, 3407.
85. A.-S. Robbes, F. Cousin, F. Meneau, F. Dalmas, R. Schweins, D. Gigmes and J. Jestin, *Macromolecules*, 2012, **45**, 9220.
86. C. Deleuze, C. Derail, M. H. Delville and L. Billon, *Soft Matter*, 2012, **8**, 8559.
87. S. Long, F. Wan, W. Yang, H. Guo, X. He, J. Ren and J. Gao, *J. Appl. Polym. Sci.*, 2013, **128**, 2687.
88. V. Ladmiral, T. Morinaga, K. Ohno, T. Fukuda and Y. Tsujii, *Eur. Polym. J.*, 2009, **45**, 2788.
89. K. Bian and M. F. Cunningham, *Polymer*, 2006, **47**, 5744.
90. M. Mazurowski, K. Sondergeld, J. Elbert, C. J. Kim, J. Li, H. Frielinghaus, M. Gallei, B. Stuehn and M. Rehahn, *Macromol. Chem. Phys.*, 2013, **214**, 1094.
91. X.-D. Zhao, X.-H. Fan, X.-F. Chen, C.-P. Chai and Q.-F. Zhou, *J. Polym. Sci., Part A: Polym. Chem.*, 2006, **44**, 4656.

92. O. Garcia-Valdez, R. Ledezma-Rodríguez, E. Saldívar-Guerra, L. Yate, S. Moya and R. F. Ziolo, *Polymer*, 2014, **55**, 2347.
93. O. García-Valdez, R. Champagne-Hartley, E. Saldívar-Guerra, P. Champagne and M. F. Cunningham, *Polym. Chem.*, 2015, **6**, 2827–2836.
94. C. Lefay, Y. Guillaneuf, G. Moreira, J. J. Thevarajah, P. Castignolles, F. Ziarelli, E. Bloch, M. Major, L. Charles, M. Gaborieau, D. Bertina and D. Gigmes, *Polym. Chem.*, 2013, **4**, 322.
95. O. Garcia-Valdez, D. G. Ramirez-Wong, E. Saldivar-Guerra and G. Luna-Barcenas, *Macromol. Chem. Phys.*, 2013, **214**, 1396.

CHAPTER 11

Novel Materials: From Nanoporous Materials to Micro-Electronics

BENOÎT H. LESSARD

Department of Chemical and Biological Engineering, University of Ottawa, Ottawa, Ontario, Canada
Email: benoit.lessard@uottawa.ca

11.1 Introduction

Nitroxide mediated polymerization (NMP) is a reversible-deactivation radical polymerization technique (RDRP[1]) that facilitates the synthesis of homogeneous, well-defined and complex macromolecular architectures.[2–4] These polymers with well-controlled features allow for final materials with predictable and tunable properties. For example, NMP can be used for the synthesis of block copolymers with narrow molecular weight distributions, containing two distinct glass transitions originating from each block.[5] While the simple blending of two homopolymers could give similar properties, the use of a block copolymer would prevent phase separation, which can be essential to the application requirements. The ability to synthesize homogeneous polymers with high chain-end fidelity allows for subsequent coupling reactions for grafting-to procedures. This functional chain end chemistry can facilitate the functionalization of surfaces or to promote improved dispersion of inorganic nanoparticles. The use of NMP has been found to enhance, augment and improve numerous current and next

RSC Polymer Chemistry Series No. 19
Nitroxide Mediated Polymerization: From Fundamentals to Applications in Materials Science
Edited by Didier Gigmes

Published by the Royal Society of Chemistry, www.rsc.org

generation applications by introducing well-defined polymers with predictable and tunable properties and function.

NMP is an industrially robust technique that is inherently simple, often only requiring a single unimolecular initiator with a monomer to perform the polymerization followed by a simple precipitation/filtration to workup the final product.[2–4] In addition, the use of NMP for some biological and electronic applications is preferred over other RDRP techniques, such as atom transfer-radical polymerization (ATRP), due to the avoidance of transition metal catalysts that can still be present in the final polymers.[2–4] The reduction in metal contamination in the final polymer when using ATRP as well as the other short falls of respective RDRP techniques have been addressed over the years and has been the subject of many recent reviews.[3,6–10] The NMP process is governed by the reversible termination of propagating chains with a nitroxide based stable free radical (SFR). The controlled pseudo-living polymerization takes place when an appropriate equilibrium is established between the dormant, SFR-terminated polymers chains and the active, propagating chains with unbound SFRs. Traditionally, the major deficiency of NMP was the range of monomers which could be polymerized in a controlled manner.[2–4] First generation polymerization systems based on 2,2,6,6-tetramethyl-1-piperidinyloxy (TEMPO,[11,12] Figure 11.1) SFR required

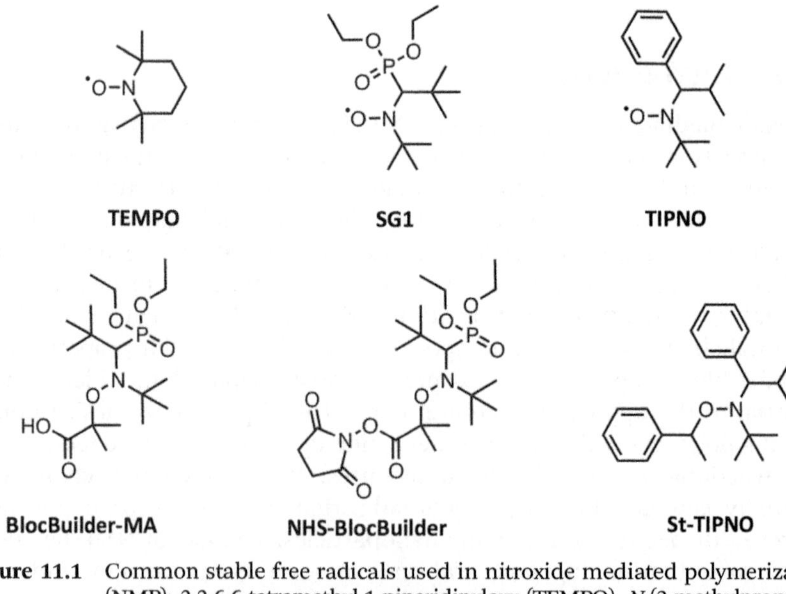

Figure 11.1 Common stable free radicals used in nitroxide mediated polymerization (NMP): 2,2,6,6-tetramethyl-1-piperidinyloxy (TEMPO), *N*-(2-methylpropyl)-*N*-(1-diethyl-phosphono-2,2-dimethylpropyl)-*N*-oxyl) (SG1), 2,2,5-trimethyl-4-phenyl-3-azahexane-3-nitroxide (TIPNO), (*N*-(2-methylpropyl)-*N*-(1-diethylphosphono-2,2-dimethylpropyl)-*O*-(2-carboxylprop-2-yl)hydroxylamine (BlocBuilder-MA), BlocBuilder-MA bearing a succinimidyl ester group (NHS-BlocBuilder) and 2,2,5-trimethyl-3-(1-phenylethoxy)-4-phenyl-3-azahexane (*st*-TIPNO).

elevated polymerization temperature (125 °C) which meant several families of monomers such as acrylates and acrylamides could not be targeted. In the past 15 years the development of second generation initiators based on *N*-(2-methylpropyl)-*N*-(1-diethyl-phosphono-2,2-dimethylpropyl)-*N*-oxyl) (SG1,[13,14] Figure 11.1) and 2,2,5-trimethyl-4-phenyl-3-azahexane-3-nitroxide (TIPNO,[15] Figure 11.1) have allowed for a lower polymerization temperature (80–100 °C) resulting in a controlled homopolymerization of previously unattainable monomers. Currently, (*N*-(2-methylpropyl)-*N*-(1-diethylphosphono-2,2-dimethylpropyl)-*O*-(2-carboxylprop-2-yl)hydroxylamine available from Arkema under the trade name BlocBuilder-MA and 2,2,5-trimethyl-3-(1-phenylethoxy)-4-phenyl-3-azahexane (*st*-TIPNO) available from Sigma-Aldrich are two prevalent unimolecular alkoxyamine initiators based on SG1 and TIPNO, respectively (Figure 11.1). Several great reviews have discussed the advances in NMP over the past two decades and can be found elsewhere.[2–4]

Since its first demonstrations, NMP has proven to be a significant tool available for enhancing the function and performance of numerous more complex applications, such as organic electronic devices, nanoporous membranes and several other emerging technologies and next generation applications. This chapter will briefly discuss some of these applications and demonstrate the benefits presented by the implementation of NMP.

11.2 Photoresists/Imprint Lithography

The use of photonic-based data transfer applications is growing due to their very broad bandwidth with low transmission losses.[16,17] The next generation of photonics will demand higher densities of interconnections, smaller feature sizes and even the possibility for their integration into flexible substrates. Polymer based waveguides have the potential to displace conventional silicon based technologies due to the reduced processing costs, flexibility, chemical tunability and compatibility with other printed circuit board manufacturing processes, such as lamination.[16–18]

Polymer waveguides are often fabricated in a two-step process that includes the polymer processing followed by the patterning or etching of the polymer film. The processing of the polymers for waveguides is performed by similar techniques used with other polymer systems for electronic applications. The key is for the resulting polymer film to be thin with a uniform thickness while avoiding the formation of bubbles or discolourations due to striations.[17] Such common techniques are spin-coating, doctor blade coating, spray coating and lamination.[19] Once the polymer film has been cast, the patterning of the optical waveguides can be accomplished by photoresist-based patterning, direct lithographic patterning, and even soft lithography.[17] As an example, direct lithography can be used when the waveguide polymer film includes a UV cross-linking agent. The cast polymer film is exposed to UV light through a mask resulting in selective crosslinking of the exposed polymer film followed by the dissolution and removal of the

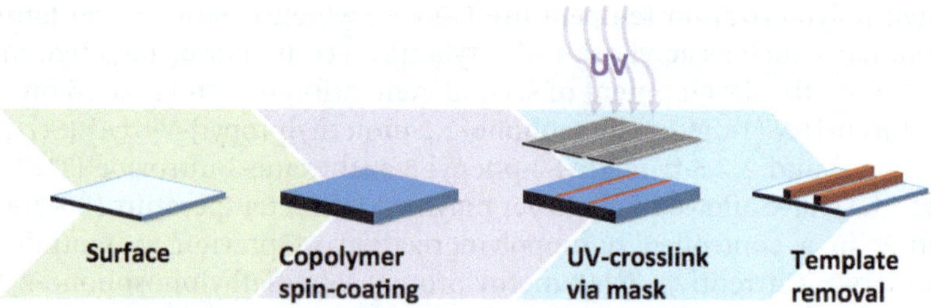

Figure 11.2 Graphical representation of direct lithography using a UV-crosslinking method for the fabrication of a polymer photoresist.

rest of the polymer film (Figure 11.2). The resulting cross-linked pattern can be used as an optical waveguide or can undergo further modification, depending on the desired application.

11.2.1 Photoresists Using Copolymers Synthesized by NMP

As previously mentioned, the use of NMP can result in homopolymers and copolymers that are characterized by having a narrow molecular weight distribution compared to polymers synthesized by conventional free radical polymerization. This improved homogeneity between the polymer chains can result in improved control over the final materials microstructure and ultimately the final application of the resulting polymer. Hawker and coworkers were interested in synthesizing polymers for use as the matrix for deep-UV wavelength (220–260 nm) waveguides that are amenable to aqueous base microlithographic fabrication technologies.[20,21] The authors studied the use of TEMPO to synthesize poly(4-hydroxystyrene) homopolymers due to their preferred absorption characteristics in the UV range.[20] It was identified that the UV absorption and the dissolution kinetics were strongly affected by the molecular weight and the dispersity of the poly(4-hydroxystyrene) homopolymers, which could easily be tuned using NMP.[20]

Maric and coworkers have used NMP to synthesize well-defined copolymers for application in photoresists. The authors targeted the synthesis of narrowly disperse glycidyl methacrylate (GMA) and pentafluorostyrene (PFS) random copolymers (Figure 11.3).[22] Fluorinated polymers are considered to be one of the best candidates due to their low optical loss at near-infrared wavelengths[23,24] and GMA contains a reactive epoxy group that can be used for selective crosslinking. The authors often use BlocBuilder-MA (Figure 11.1), which is a unimolecular alkoxyamine initiator based on the SG1 stable free radical and is commercially available through Arkema.[25–27] Previously, the authors found that to obtain a controlled copolymerization of GMA it was necessary to protect the carboxylic group of BlocBuilder-MA prior to copolymerization, to avoid detrimental reaction with the epoxy in the GMA monomer.[28,29] Wang *et al.* therefore used BlocBuilder-MA bearing a

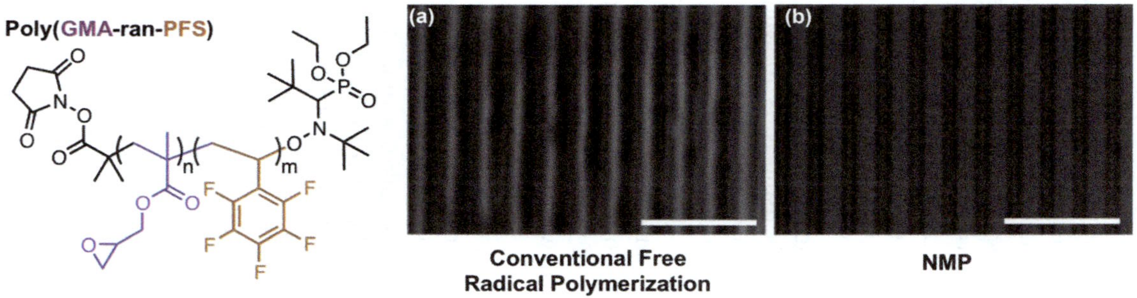

Figure 11.3 Glycidyl methacrylate (GMA) and pentafluorostyrene (PFS) random copolymers synthesized using BlocBuilder-MA bearing a succinimidyl ester group (left). SEM images of single-mode channel waveguides fabricated by direct UV lithographic patterning using poly(GMA-*ran*-PFS) random copolymers, which were synthesized (a) by conventional radical polymerization and (b) by NMP. The images of the resulting photoresists were taken at 4000× magnification and the scale bars represent 10 μm.
Adapted from ref. 22 with permission from John Wiley & Sons.

succinimidyl ester group (NHS-BlocBuilder, Figure 11.1) to synthesize a series of narrowly disperse GMA and PFS random copolymers (Figure 11.3).[22] The resulting copolymers were characterized by having a relatively narrow molecular weight distribution, $M_w/M_n < 1.35$, compared to the copolymers that were synthesized by conventional radical polymerization ($M_w/M_n > 3.2$). Both sets of polymers were used to fabricate the core structure of single-mode channel waveguides by direct UV lithographic patterning (Figure 11.3).[22] When fabricating the polymer films, the authors added a commercially available photoacid generator to the copolymer mixture so that upon UV-irradiation the generated acid would activate the epoxy groups in the GMA, resulting in localized crosslinking. The lithographic patters fabricated using the copolymers synthesized by NMP resulted in a much greater pattern resolution, where the pattern was characterized by having a line width roughness (LWR) of about 0.16 μm. The LWR of the copolymers synthesized by conventional radical polymerization was about 0.5 μm under similar copolymer compositions (Figure 11.3).[22] The same group followed up with a subsequent study using NMP to synthesize a series of well-defined 5-methacryloyloxy-2,6-norboranecarbolactone (NLAM) containing copolymers for use in 193 nm photoresists.[30] While the authors did not report the use of the copolymers in photoresists, they did meet all the necessary requirements, such as narrow molecular weight ($M_w/M_n < 1.30$) and minimal absorbance in the 193 nm range.[30] In these examples, the use of NMP benefitted the fabrication of polymer optical waveguides by simply producing a starting material that was characterized by having a narrow molecular weight distribution.

Recently Morris *et al.*[31] and Telitel *et al.*[32] explored the use of novel photosensitive alkoxyamines to fabricate nano patterned polymer surfaces. The authors began by spin coating and crosslinking a thin film of acrylic polymers containing the photosensitive alkoxyamine. The resulting acrylic film bearing nitroxide moieties at the free-surface is then coated with a second thin film of monomer and treated with UV light through a chromium mask.[32] The UV exposure facilitates the surface-initiated photo NMP of the second block of polymers, resulting in nanofeatured acrylic surfaces.[31,32] Similarly, Sarav *et al.* used imprint lithography of functional photopolymer resins to create controlled nanoscale patterns of a cross-linked photopolymer containing embedded initiator groups.[33] Functionalized polymer brushes consisting of polystyrene and poly(*N*,*N*-dimethyl acrylamide) were grown from these patterned layers by TIPNO-based NMP initiator. Fluorescent chain-end functionalization of the brush layer was accomplished by nitroxide radical exchange with a pyrene functionalized TIPNO molecule.[33]

Other lithographic techniques have also benefited from the use of NMP, such as negative-tone electron beam resists using C60-containing copolymers,[34] atomic force microscopy (AFM) scratch lithography of polymer brush coatings grown by surface initiated NMP (SI-NMP)[35] and even site-selective Langmuir–Blodgett lithography by SI-NMP using a TEMPO based initiator.[36]

11.3 Polymer Templating and Nanoporous Membranes

A variety of next generation biological, optical and even electronic applications are reliant on material with nano-sized features.[37] Well-defined block copolymers containing two or more covalently bound, discrete and chemically different polymer blocks are known to self-assemble on a nano-scale.[38,39] The resulting thermodynamically favourable nano-phase separation is consistently and reproducibly observed when the block copolymers are either in the bulk or dissolved in a solution. The manipulation and formation of three-dimensional networks from these block copolymers is dependent on several factors, such as the block copolymer composition, molecular weight and ratio of the respective blocks. To be able to tune and predict these nano-morphologies, the synthesis of well-defined homogeneous block copolymers with narrow molecular weight distribution is essential. The development of controlled radical polymerization techniques has facilitated the straightforward synthesis of well-defined block copolymers that were traditionally only available by true living polymerization techniques, such as ionic polymerization. Once the desired self-assembled structure is obtained, one of the polymer blocks can be preferentially removed while leaving the second block intact, resulting in a nanoporous matrix.[38] Block copolymer self-assembly has therefore been exploited for a variety of applications, such as nano porous membranes and organic templates for imprint lithography.[39] Nanoprorous membranes and nanoporous surfaces could find applications in such fields as chromatography, pharmaceutical purification and filtration, high surface area catalysis, microelectronics and others.

Typically, thin films of block copolymers are processed in a similar fashion to other polymers for electronic or photoresist applications, such as spin coating or blade coating (Figure 11.4). In some cases the substrate has undergone a treatment to help selectively orient the phase separation of the block copolymers and to induce long-range order (Figure 11.4). The phase separation of block copolymers is thermodynamically stable and therefore an annealing step is often included to eliminate the kinetically trapped morphology and induce this desired phase separation. Once the block copolymer film has self-assembled onto the surface, an etching step is

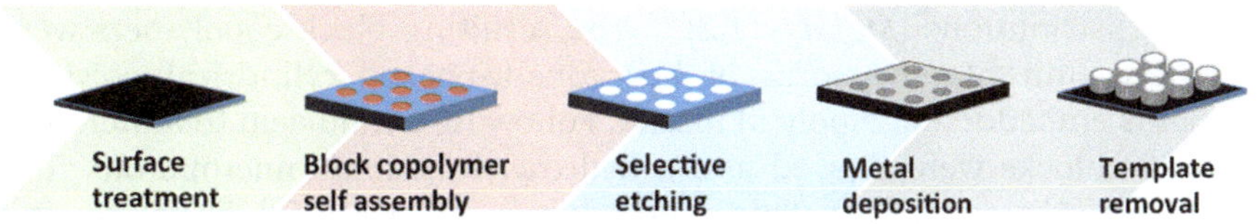

Figure 11.4 Graphical representation of a general fabrication procedure for nano-featured substrates by using a template made from block copolymer self-assembly.

employed to selectively remove one of the blocks. This etching step is generally executed using an acid or UV-light (Figure 11.4). The preferential removal of one of the blocks results in a nanoporous film. At this point the film can be used as is for high surface area applications, such as catalysis and filtration, or the film can be further processed. For example, these porous membranes can be filled with metal, by thermal evaporation followed by complete dissolution and removal of the polymer matrix (Figure 11.4). These additional steps would reveal nano-featured metal objects. This process is known as chemical templating using block copolymers and can lead to metal nanofeatured objects without the need for high precision lasers, which are very expensive on the nanosized feature scale.[39]

11.3.1 Nanoporous Thin Films

It has been reported by several groups that poly(styrene)-*block*-poly(ethylene oxide) [poly(S)-*block*-poly(EO)] block copolymers are ideal candidates for nanoporous membranes and templates due to their inherent desire to self-assemble in films with long range lateral order and where the poly(EO) cylinders are oriented normal to the surface.[40–42] When working with bulk samples of block copolymers, harsh conditions can be employed to degrade the poly(EO) block. For example Mao *et al.* used hydroiodic acid to degrade the poly(EO) block in a bulk sample.[43] However, the use of a strong acid is not amenable to thin films. To overcome the use of harsh conditions, Zhang *et al.* synthesized a poly(EO)-*block*-poly(S) block copolymer that had a triphenylmethyl (trityl) ether linkage between the poly(EO) and poly(S) blocks.[41] The authors first synthesized a TEMPO-derived trityl ether group, which was coupled to a commercially available poly(EO) group.[41] The group homopolymerized styrene, using this poly(EO) macroinitiator to give the resulting block copolymer, which self-assembled in a thin film. The films were then treated with a relatively mild Lewis acid under ambient conditions resulting in the selective removal of the poly(EO) block.[41] Hawker and coworkers also converted commercially available poly(EO) into NMP macroinitiators by first modifying the poly(EO) chain ends with azide groups. Subsequently, the azide functional polymers were subjected to Cu-catalyzed azide–alkyne cycloaddition reaction with an alkyne functional TEMPO initiator (Figure 11.5).[42] Mono functional or bifunctional macroinitiators were used to homopolymerize styrene resulting in either poly(EO)-*block*-poly(S) or poly(S)-*block*-poly(EO)-*block*-poly(S) block copolymers with narrow molecular weight distributions ($M_w/M_n < 1.2$).[42] The resulting block copolymers were cast into thin film networks of hexagonally-packed cylindrical poly(EO) domains embedded in a poly(S) matrix. Following a mild acid treatment the poly(EO) blocks were cleaved and a hydroxyl functional microporous film remained (Figure 11.5).[42] This method proved successful in fabricating nanoporous thin films with long-range order (Figure 11.5).

Mugemana *et al.* took a different approach to building nanoporous membranes from poly(EO)-*block*-poly(S) block copolymers; the authors

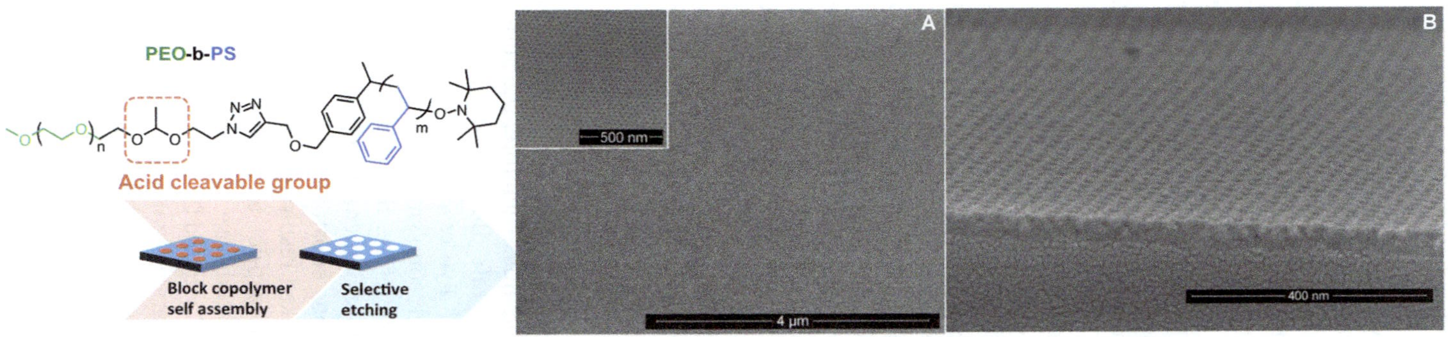

Figure 11.5 Microporous thin films made from acid cleaving and washing of a thin film of self-assembled poly(ethylene oxide)-*block*-poly(styrene) [poly(EO)-*block*-poly(S)] synthesized by NMP. Upon acid cleaving the minor poly(EO) phase is removed resulting in a microporous poly(S) thin films as seen in the SEM images in A top view and B side view.
Adapted from ref. 42 with permission from The Royal Society of Chemistry.

synthesized metallo-supramolecular block copolymers. Mugemana *et al.* synthesized end functional terpyridine poly(S) homopolymers by NMP and coupled them to other terpyridine functional poly(EO) homopolymers by forming nickel–terpyridine complexes.[44] The resulting block copolymers self-assembled into a thin film with hexagonal morphology and were treated with a competing ligand. These ligands reacted with the poly(S) blocks resulting in the dissolution of now free poly(EO) blocks, which could easily be removed.[44] The pores of the resulting poly(S) nanoprorous membrane were therefore coated with terpyridine groups, which could be used for further heterogeneous surface reactions.[44]

When attempting to template a surface using block copolymer self-assembly it is important to take into account the interaction between the surfaces and the individual polymer blocks. Preferential interactions between the surface and one of the blocks could result in undesired block copolymer phase separation. A method to avoid this non-uniform interaction is by functionalizing the surface with a coating that will interact with both segments of the block copolymer equally. For example, Shin *et al.* utilized NMP to synthesize poly(styrene-*ran*-methyl methacrylate) [poly(S-*ran*-MMA)] random copolymers with benzyl alcohol functional groups that allowed the tethering of the copolymers to the surface of a silicon wafer.[45] Once treated the authors used the surfaces for the self-assembly of poly(S)-*block*-poly(MMA) block copolymers and obtained vertical oriented cylinder morphology.[45] This controlled morphology facilitated the long-range templated growth of gold and cadmium nano-dots, which was not possible without surface treatment. Other researchers have also used surface initiated NMP as a method to template a surface. For example, Xu *et al.* used resist-patterned silicon wafer and selectively grew polymers by anchoring a functional NMP initiator to the oxide covered microdomains of the surface.[46] the authors also demonstrated the use of surface initiated ATRP to grow polymers in the H-terminated silicon microdomains. Therefore, the authors used both NMP and ATRP to simultaneously graft different polymers to different microdomains of the silicon wafer.[46] Ryu *et al.* were interested in applying long-range order templating to block copolymer self-assembly while avoiding the surface tethering reactions. Therefore, the authors synthesized terpolymers of MMA, styrene and benzocyclobutene using *st*-TIPNO, which was subsequently spun cast onto various substrates, such as gold or silicon.[47] After casting, the films were crosslinked resulting in a universal surface treatment. Regardless of the substrate used prior to the terpolymer crosslinking treatment, the resulting surfaces would have similar block copolymer templating characteristics.[47] The authors illustrated that the self-assembly of poly(S)-*block*-poly(MMA) block copolymers varied significantly between untreated gold and silicon surfaces, however, the self-assembly of the same block copolymers on either surface that was treated with this cross-linked terpolymer resulted in identical self-assembled morphology.[47] This approach represents a simple technique to modify the effect of a substrate on the self-assembly of block copolymers without the need for surface reactions or chemical modification.

These examples illustrate that NMP is versatile and can be used to synthesize well defined polymers for the fabrication of nanoporous thin films. The technique can be used to synthesize the block copolymers themselves that will self-assemble into the three-dimensional networks and be used to modify a substrate surface for long range block copolymer self-assembly.

11.3.2 Breath Figure (BF) Technique

The breath figure (BF) technique is a reproducible method used to obtain well defined, microporous arrays. The BF technique is based on the principle of a solution evaporating in the presence of wet air, which results in the polymers self-assembling around the drying water droplets (Figure 11.6). Several research groups have used this technique for the self-assembly of polymers into microporous membranes. For example, Bolognesi *et al.* synthesized poly(fluorine)-*block*-poly(styrene) [poly(FO)-*block*-poly(S)] microporous arrays where there assembly was facilitated by the BF technique.[48] The authors first synthesized a TEMPO functionalized tetrahydropyranyl as a

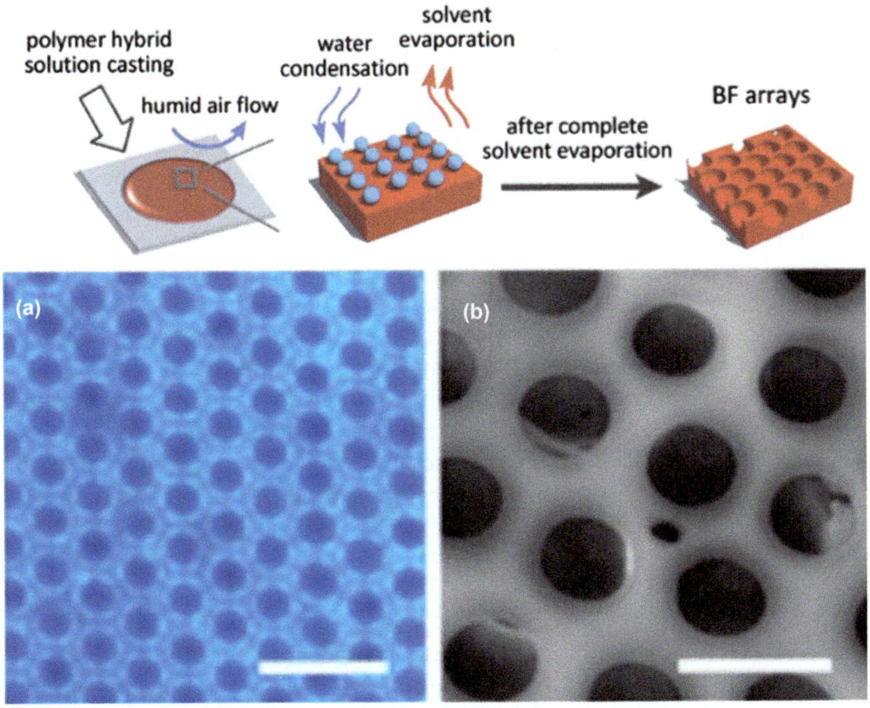

Figure 11.6 Graphical representation of the breath figure (BF) technique (top) as well as florescent and SEM images of the resulting honeycomb matrices (bottom).
Adapted from ref. 51 with permission from The Royal Society of Chemistry.

hydrophilic macroinitiator. This π-conjugate polymer was subsequently used to initiate the polymerization of styrene resulting in a well-defined poly(FO)-*block*-poly(S) block copolymer. These amphiphilic block copolymers were shown to self-assemble using the BF technique (Figure 11.6).[48] The use of the BF technique on the poly(FO)-*block*-poly(S) block copolymers resulted in a fluorescent honeycomb structure where the average hole diameter was roughly 2.2 μm.[48] To further study this BF technique, Galeotti *et al.*[49] synthesized primary amine functional poly(S) homopolymers by NMP and coupled them to fluorescamine, a known fluorescent tag common in peptide labeling, as well as coupling boron-dipyrromethene (BODIPY) functional TEMPO to the end of poly(S) chains.[49] The authors used these fluorescent tags to demonstrate that even with just one hydrophilic group, such as the amine, the polymer will self-assemble into honeycomb structures with the greatest concentration of hydrophilic groups at the pore surface.[49] The same group has also polymerized styrene[50] and a mixture of styrene and a styrenic π-conjugated oligo(fluorene) side-chain[51] in the presence of fluoromica silicate layers. The authors accomplished this by first employing a cationic exchange between the silicate layers and amine functional TEMPO molecules followed by a polymerization of the styrenic monomers in the presence of this hybrid TEMPO-containing surface.[50,51] The authors hypothesize that this technique results in polymer intercalation that effectively reduce π-stacking, chain–chain interactions, twists and bends, and interfacial effects, leading to a significant difference in the electronic/optoelectronic properties, and improved optical, thermal and chemical stability of the materials.[51] When employing the BF technique, the authors were able to obtain a well-organized honeycomb network with these hybrid fluorescent materials (Figure 11.6).[51]

Recently, Chen *et al.* designed polysaccharide-based linear amphiphilic copolymers by NMP. The authors used NHS-BlocBuilder to synthesize a dextran based macroinitiator, which was then used to polymerize styrene or a mixture of MMA and styrene.[52] The authors employed the BF technique to generate honeycomb structures, which experienced nanophase separation within the pores, resulting in a hierarchically structured network.[52] These examples illustrate that NMP can facilitate the controlled placement of hydrophilic groups and ultimately tune the properties of a candidate material for the BF technique.

11.3.3 Inorganic Templating

Nanofeatured metal or inorganic structures are desired for high temperature and electronic applications. For these applications, block copolymers can be used as a template for the nanofeatured design of inorganic materials (Figure 11.4). For example, when treating silicon containing polymers with an oxygen plasma, the silicon is oxidized to silicon oxide, which has very high thermal and mechanical stability, making it an ideal candidate for a dielectric insulator for microchip applications.[53] Fukukawa *et al.* used block copolymer self-assembly of silicon-containing block copolymers to make silicon oxide hierarchical nanostructures. The authors synthesized a series of silicon containing styrenic polymers and block copolymers by NMP with narrow molecular weight distribution ($M_w/M_n < 1.2$) and exposed the final materials to an oxygen plasma. It was determined that if the silicon content in the block copolymer was elevated enough then the block copolymer self-assembly would remain intact before and after the oxygen treating. However, at low silicon content the self-assembled morphology would deteriorate with the oxygen plasma treatment.[53] Regardless, the authors illustrated that this strategy was effective in producing silicon oxide nanofeatured arrays.

Hadziioannou and coworkers synthesized block copolymers using a TEMPO functional poly(2,5-dioctyloxy-*p*-phenylene vinylene) [poly(DpPV)] that was initially prepared by a Siegrist polycondensation. This macro-initiator was subsequently used for the homopolymerization of styrene, resulting in rigid-flexible block copolymers (Figure 11.7).[54] The rigid poly(DpPV) block and the flexible poly(S) block self-assembled into honeycomb structures when drop cast from solution.[54] The resulting micro-structures were used as a template for the deposition of aluminum cups onto the surface. The authors estimated that the method resulted in roughly, 6.5×10^6 cups of aluminum per cm^2. This procedure constitutes a simple method for preparing patterned electrodes or photolithography masks with relatively large features compared to modern techniques like e-beam lithography (Figure 11.7).[54]

Bloch *et al.* used poly(S)-*block*-poly(EO)-*block*-poly(S) block copolymers for the self assembly of inorganic particles. BlocBuilder-MA initiator was first coupled to terminal-vinyl functionalized poly(EO) homopolymers by performing an intramolecular addition.[55] The resulting poly(EO) SG1 functional macroinitiators were employed to homopolymerize styrene resulting in poly(S)-*block*-poly(EO)-*block*-poly(S) block copolymers. The block copolymers were subsequently used to facilitate the templated growth of mesoporous silica. Bloch *et al.* determined that by changing parameter such as the poly(S) and poly(EO) length the micropore volume as well as the mesopore size could be tuned.[55] In a similar fashion, Quémener *et al.* prepared poly(styrene-*ran*-acrylonitrile)-*block*-poly(EO)-*block*-poly(styrene-*ran*-acrylonitrile) [poly(S-*ran*-AN)-*block*-poly(EO)-*block*-poly(S-*ran*-AN)] block copolymers and found that when spin coating the resulting polymers from different solvent mixtures, they would obtain free-standing nanoparticle films.[56] For example, when using a 1 : 1 volume ratio of *N,N*-dimethylformamide (DMF):toluene the authors obtained 50 nm micelle-like particles in the dried films. However if the authors changed the solvent ratio to 1 : 3 volume ratio of DMF:toluene they obtained interconnected worm-like micelles that the authors called spider web films.[56] Gromadzki *et al.* synthesized a series of poly(S)-*block*-poly(S-*ran*-AN) block copolymers by NMP and found that, even though styrene was present in the second block, the entire block copolymer still phase separated into lamellar morphologies before and after 1,3-dipolar cycloaddition modification of the acetonitrile.[57] Hermans *et al.* utilized a

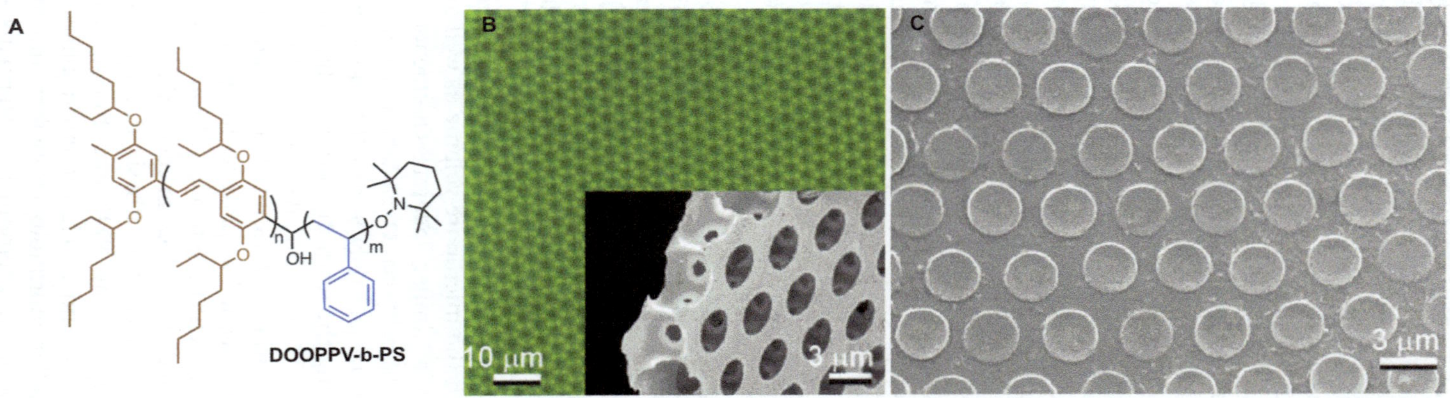

Figure 11.7 (A) Chemical structure of rod–coil poly(*p*-phenylenevinylene)-*block*-poly(styrene) block copolymer [poly(DpPV)-*block*-poly(S)] and (B) Fluorescence microscopy and SEM image inset of block copolymer template and C) SEM image of the resulting aluminum cups after template removal.
Adapted from ref. 54 with permission from John Wiley & Sons.

hydroxyl functional TIPNO-based initiator to homopolymerize *N,N*-dimethyl acrylamide (DMA) followed by a ring opening polymerization of polylactide [poly(LA)] to give a poly(DMA)-*block*-poly(LA) block copolymer.[58] The authors solubilized a mixture of the resulting block copolymers with either methyl silsesquioxanes (MSSQ) or organosilicates, which are both low-dielectric-constant materials. The mixtures were subsequently spun cast into thin films. MSSQ have shown great potential for next generation electronic technologies, however the ability to control the porosity of these materials is essential.[58] Hermans *et al.* found that when controlling the solution conditions it was possible to increase the range of accessible morphologies from discrete and minimally interconnected pores to packed microporous nanoparticle domains. The change in morphology resulted in films with significantly different dielectric constants.[58]

The versatility of NMP facilitates the synthesis of block copolymers with narrow molecular weight distributions and varying chemical functionality between blocks. The resulting polymers can be used to template inorganic nanoparticles as well as template the growth of inorganic nano-objects. These well-defined inorganic nano-objects will provide significant opportunities for novel inorganic materials for next generation electronic applications.

11.4 Proton Exchange Membrane and Ionic Polymers

11.4.1 Styrene Sulfonate Containing Polymers

Proton-exchange membranes or polymer electrolyte membranes (PEMs) are an integral part of the modern fuel cell. PEMs allow the transfer of protons while preventing the passage of reactive gases such as hydrogen and oxygen. Therefore an effective PEM should have high proton conductivity, good mechanical strength, and high chemical stability. Developed in the 1960s by DuPont, The most common PEM material is commercially known as Nafion® and is a tetrafluoroethylene based polymer with pendant sulfonic acid groups (SO_3H). The proton conductivity comes from these sulfonic acid groups. Several have proposed that the conductivity of Nafion® is due to the clustering of SO_3H groups resulting in nm-wide channels across the membrane.[59,60] Holdcroft and coworkers wanted to study the effect of morphology on conductivity by synthesizing SO_3H-containing copolymers.[61] The authors therefore synthesized poly(styrene)-*graft*-poly(sodium styrene sulfonate) [poly(S)-*graft*-poly(SSNa)] graft copolymers by NMP (Figure 11.8). This was accomplished by synthesizing poly(SSNa) homopolymers using TEMPO followed by the reaction with divinylbenzene to make poly(SSNa) macromonomers. These poly(SSNa) macromonomers could then be copolymerized with styrene to make graft copolymers with variable amounts of SO_3H groups[61,62] or used to synthesize poly(SSNa) star polymers in a one-pot TEMPO controlled synthesis.[63] As a comparison, simple poly(S)-*ran*-poly(SSNa) random copolymers were also synthesized. After acidification, the copolymers were used to fabricate PEMs and the resulting conductivities

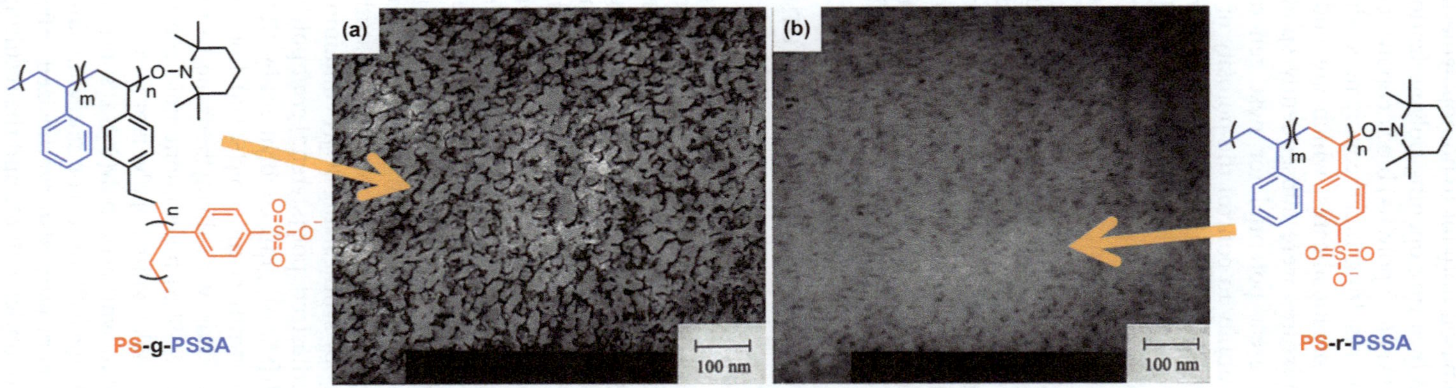

Figure 11.8 TEM micrographs of copolymer membranes: (a) poly(styrene)-*graft*-poly(styrene sulfonate) (poly(S)-*graft*-poly(S)SA); (b) poly(styrene)-*ran*-poly(styrene sulfonate) (poly(S)-*ran*-SSA). In both cases, the styrene sulfonic acid content is roughly 12.0%.
Adapted with permission from ref. 61 Copyright 2001 American Chemical Society.

were measured by impedance spectroscopy. The authors identified that PEMs made using poly(S)-*graft*-poly(SSA) [poly(styrene)-*graft*-poly(styrene sulfonate)] were characterized as having a conductivity an order of magnitude greater than PEMs made using poly(S)-*ran*-poly(SSA) or Nafion®.[61] The authors attribute this significant increase in proton conductivity to the morphology of the films (Figure 11.8).[61] The authors further identified that varying the length and density of ionic graft chains on the polymer would vary the degree of phase separation, resulting in a change in the membrane's proton conductivity as well as water content uptake.[62] Due to the versatile nature of NMP, the authors could then explore the use of different polymer backbones, such as using acrylonitrile instead of styrene, resulting in poly(acrylonitrile)-*graft*-poly(sodium styrene sulfonate) [poly(AN)-*graft*-poly(SSNa)] graft copolymers.[64] Using NMP, the authors could also study the effect of separating the poly(SSNa) blocks from the graft polymer backbone by first synthesizing poly(S)-*block*-poly(SSNa) block copolymers followed by the reaction with divinylbenzene to ultimately give [poly(S)-*graft*-(poly(S)-*block*-poly(SSNa)] graft block copolymers.[65] Recently, Li *et al.* synthesized a series of poly(CMS)-*block*-poly(S) by NMP and blended with poly(2,6-(dimethyl-1,4-phenylene oxide) to fabricate anion exchange membranes.[66] The benzyl chloride groups in the blended films were subsequently converted to quaternary ammonium groups by reaction with trimethylamine. The ionic conductivities of the membranes showed an increase in ionic conductivity and water uptake after either annealing in the presence of water and THF or high-temperature annealing in the presence of water.[66]

In a working fuel cell the temperatures can be elevated and the need for thermal stability in the PEMs is crucial. In attempts to increase the thermal stability of the SO_3H containing polymers, Holmberg *et al.* explored the use of poly(vinylidene fluoride) [poly(VDF)]. For example, the authors irradiated films of poly(VDF) with an electron beam resulting in the formation of free radicals that were immediately quenched with TEMPO.[67] The resulting TEMPO functionalized poly(VDF) homopolymers were used as the polymer initiators for the NMP grafting-from homopolymerization of styrene. These styrene grafts were subsequently sulfonated using chlorosulfonic acid in 1,2-dichloroethane at ambient temperature for 24 h giving rise to poly(VDF)-*graft*-poly(SSA) copolymers.[67] The resulting PEMs were found to operate in a fuel cell for 930 h at 70 °C without any drop in current density.[67] The authors also determined that substituting the chain ends with a maleimide derivative prior to sulfonation also enhanced the thermo-oxidative stability of the PEMs. Sawada *et al.* also showed that this method for grafting polymers using an irradiation/TEMPO coupling procedure was also possible when starting with films of poly(ethylene-*co*-tetrafluoroethylene) [poly(ETFE)] films.[68]

11.4.2 Polymer Ionic Liquids

Ionic liquids are thermally stable, molten salts with negligible vapour pressure and have even been advertised as a potential "green solvent".[69,70]

Polymerizing an ionic liquid containing monomers (PILs) allows for the combination of the ionic liquid properties with the desirable polymer mechanical properties. PILs are showing promise as potential candidates for next generation ion conducting membranes, polymer electrolytes and other conductive electrochemical applications.[70] NMP has been utilized to incorporate ionic liquid-containing monomers into a copolymer. For example, Long and coworkers have reported the controlled homopolymerization of *N*-vinylmidazole (N-Vim) as well as a styrenic derivative, 1-(4-vinylbenzyl)-imidazole (VBIm) using BlocBuilder-MA as the initiator.[71] While these monomers are not an ionic liquid in this form, a relatively straightforward alkylation and anion exchange can be performed to give an ionic polymer counterpart. It has also been reported that the conductivity of PILs decreases with increasing glass transition temperature (T_g).[72] Long and coworkers have explored the use of imidazole-based ABA triblock copolymers to reduce the T_g of the PILs block while keeping the mechanical property of the overall polymer high enough to maintain mechanical stability. The authors used a difunctional alkoxyamine NMP initiator based on SG1[73] for the synthesis of am imidazole-based ABA triblock copolymer.[74,75] This imidazole-based ABA triblock copolymer achieved a microphase-separated morphology with an ion-conducting central block separating into one phase while the poly(S) peripheral blocks would separate into a second phase (Figure 11.9). Long and coworkers could therefore increase the conductivity of the polymer by independently reduce the T_g of the central conducting block while keeping the T_g of the matrix elevated. The authors used the copolymerization of butyl acrylate into the central block to reduce the T_g and increase the conductivity while the peripheral blocks were poly(styrene) to give mechanical stability to the conductive matrix (Figure 11.9).[74] The same technique was further employed for the synthesis of phosphonium ionic liquid ABA tripolymers.[76] The authors also applied these ABA PILs to the fabrication of electromagnetic actuators.[75] Evaluation of electromechanical actuators using an imidazole-based ABA triblock copolymer revealed device curvatures over twice those of devices made using Nafion® (Figure 11.9).

Stancik *et al.* synthesized a series of ionic liquid-containing poly(styrene)-*block*-poly(4-vinylbenzyl alkylimidazolium chloride) block copolymers by NMP. Using *st*-TIPNO (Figure 11.1), the authors first synthesized a poly(styrene) macroinitiator, which was subsequently used to initiate a fresh batch of chloromethyl styrene (CMS).[77,78] The final poly(S)-*block*-poly(CMS) diblock copolymers were reacted with 1-methylimidazole in *N,N*-dimethylacetamide resulting in the clean substitution of the chloride functionality with the imidazole.[77] The authors used small angle neutron scattering (SANS) and dynamic light scattering (DLS) to study the effect of the block copolymer poly(S) length on the size of the micelles which resulted from the self-assembly in various solutions.[77] Following the same approach as Stancik *et al.*, Weber *et al.* also synthesized ionic liquid containing poly(styrene)-*block*-poly(4-vinylbenzyl-(alkylimidazolium bis(trifluoromethanesulfonyl)imide) (poly(S)-*block*-poly(VBn-(alkyl)ImTFSI); alkyl = methyl, *n*-butyl or hexyl) block copolymers with relatively

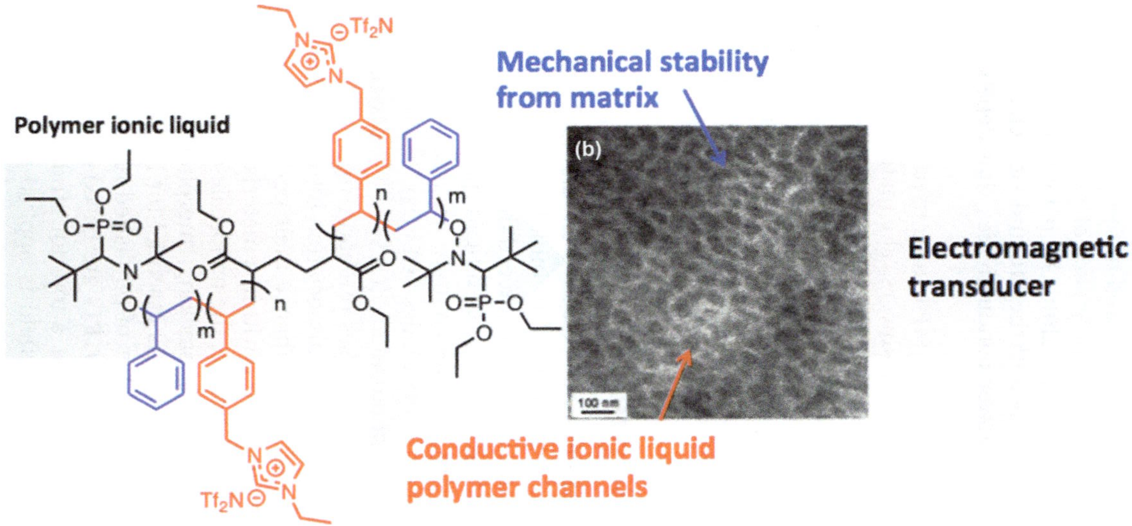

Figure 11.9 Chemical structure of imidazole-based ABA triblock copolymer (left) with resulting AFM figure of block copolymer self-assembly (right).
Adapted with permission from ref. 75 Copyright 2012 American Chemical Society.

narrow molecular weight distribution ($M_w/M_n \approx 1.3$) using TIPNO.[79] The authors correlated the block copolymer self-assembly to electrochemical impedance spectroscopy measurements and identified the strong relationship between morphology and conductivity. The potential for cylindrical domains to suffer from morphological defects and grain boundaries that decrease macroscopic connectivity resulted in an order of magnitude decrease of impede conductivity compared to the lamellar morphology.[79] The same group again studied homopolymers of poly(VBn-(alkyl)ImTFSI) and found that the hydrophilic PILs exhibit high ionic conductivities that depend on the relative humidity and the PIL and side chain length while the hydrophobic PILs exhibit lower conductivities that depend predominantly on the homopolymer T_g.[78]

These examples illustrate that NMP can be used to synthesize conductive materials for use in PEMs or transducers. The materials are fabricated from either sulfonic acid or ionic liquid containing block copolymers. In these examples, the use of NMP facilitates the independent tuning of the conductivity and mechanical properties while controlling the block copolymer morphology. In several cases the well-defined molecular architectures with optimized conductivities were found to outperform commercially available material such as Nafion®.

11.5 Organic Electronics

Due to the increase in demand for active circuits and circuitry for smart cards carrying personal information, to building entry cards, it is believed that the need for organic circuitry will exceed that for silicon electronics.[80] In addition, over the past two decades, significant progress has been made in the performance of relatively more complex organic optoelectronic devices, such as organic light-emitting diodes (OLEDs),[81,82] thin film transistors (OTFT),[83–86] and organic photovoltaic (OPV)[87,88–90] devices. In all organic optoelectronic devices, the processes of charge injection, charge transport, charge recombination and charge dissociation into holes and electrons, and charge collection are critical to enhance their performance. In most cases, these phenomena are experienced within a material on the nanoscale and at the materials interfaces. Therefore, the nanoscale control of structure, size, sequence, conformation, order, stacking, morphology and surface interaction of the functional polymers is essential for highly efficient electronic materials and devices.[91] A number of published reviews provide deeper insight into organic electronics.[87] OTFTs are based on field-effect controlled current and can also be viewed as a tool for studying charge transport in solid materials. Once an electric field is applied through the dielectric, a current will flow from the source electrode to the drain electrode allowing for basic logic-gate operations. OTFTs are essential components for numerous novel organic electronics, such as flexible OLEDs, flash memory and energy storage devices.[92] In an OLED, the balance of injection of charge carriers from the anode and the cathode are intimately connected to the overall light output and device efficiency, which can be enhanced by engineering the polymer morphology and the electrode–polymer interfaces.[93,94] Similarly to an OLED, in an OPV, good control over the polymer/polymer or polymer/small molecule morphology can be the difference between record breaking efficiency and a non-functional device.[95,96]

11.5.1 Organic Photovoltaic (OPV) Devices

Commercial solar cells are almost exclusively made with silicon or silicon derived wafers, which in most cases represents roughly 50% of the total module cost due to the numerous steps involved, such as silicon purification, crystal growth and wafering, cell processing, and cell encapsulation.[97] In addition to the already high costs, the rapidly escalating demand for ultra-pure silicon wafers is creating a supply shortage, which can increase the cost as well as impede the market expansion of these solar panels.[98] Arguably, one of the holy grails of the 21st century in the field of new materials is the development of inexpensive and highly efficient OPV devices. The potential for polymers to be used for organic electronics, such as OPV devices, is appealing due to the low potential fabrication costs, the relative ease of production, as well as the thin and flexible nature of the final product, which can result in new applications, such as wearable electronics and painted OPVs. While a couple of companies are beginning to attempt to commercialize these products, the majority of the work is still in the research and development stages and a great deal of improvement is still necessary.[96]

11.5.1.1 Introduction to the Working Principles of OPV Devices

When solar energy, in the form of a photon, hits an inorganic solar cell, such as a typical silicon based device, a free hole/electron pair is created. The electron and hole are then capable of separating under ambient conditions due to the binding energy being lower than the thermal energy at room temperature ($kT < 5$ meV).[99] The free electron, present in a conduction band, and the hole, residing in a valence band, can then both easily diffuse to their respective electrodes and create electricity. However, in contrast to inorganic cells, when an organic cell comprising of small molecules or polymers is struck by photons, a relatively small hole bound to an electron pair is formed. This neutral electron/hole pair, known as an exciton, has a much higher binding energy, approximately 10 times larger than the electron/hole pair produced in an inorganic cell ($kT = 5$–15 meV), which is much greater than thermal energy at room temperature.[100] While the use of heat or a magnetic field could induce the dissociation of the photogenerated excitons, the more practical approach is to use two organic compounds with different and complimentary highest molecular orbitals (HOMO) and lowest molecular orbitals (LUMO). Once the exciton reaches the interface between these two organic molecules, it becomes

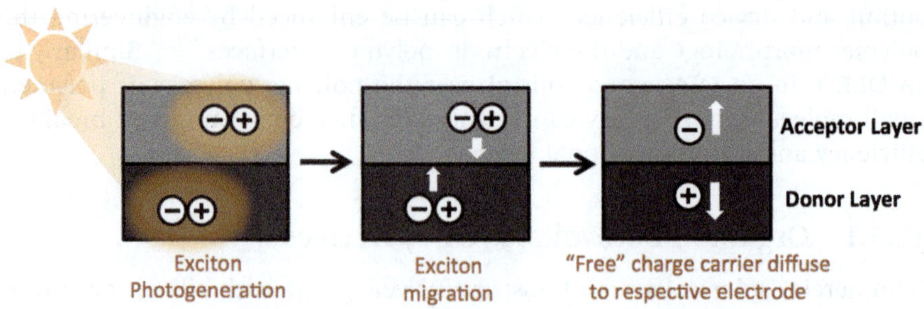

Figure 11.10 Exciton formation in organic photovoltaic (OPV) device in either the donor molecule or the acceptor molecule will result in charge separation at the interface and the formation of free charge carriers.

energetically favourable for the excitons to dissociate into free holes and electrons, which are subsequently able to travel to the respective electrodes (Figure 11.10).

The first successful organic solar cell was reported in 1986 by Cheng W. Tang, from the Kodak Research laboratories in Rochester, New York, and was constructed by depositing two layers of organic small molecules: an electron acceptor molecule (large electron affinity) and an electron donor molecule (low ionization potential).[101] This device configuration is now known as the planar bilayer configuration, or planar heterojunction (PHJ), OPV device.[101] Depending on the nature of the organic compounds in which the excitons are being formed, the excitons have very limited time to reach these interfaces before recombination of the free charge occurs. The excitons' lifetimes have been reported to be on the order of pico to nano seconds[100,102] and for some donor/acceptor cases using fullerene blends, the lifetime can drop to as low as femto seconds.[103] As a consequence, the photocurrent of an organic solar cell is directly proportional to the distance in which the exciton can travel and reach the donor/acceptor interface before recombination. This length is known as the characteristic length (L_D). The original bilayer device therefore only extracts photocurrent from the excitons, which are generated within this L_D. While this L_D is very small, several research groups have been able to obtain significant increases in photovoltaic efficiency by optimizing the bilayer thicknesses as well as the HOMO/LUMO levels of the organic compounds.[89] In parallel, researchers have attempted to improve the photovoltaic efficiency by changing the device configurations, such as by using a bulk hetero junction (BHJ) OPV device structure. The BHJ OPV structures are often based on blends of electron-donating polymers and electron-acceptor soluble fullerene derivatives (Figure 11.11). The BHJ OPV devices therefore possess an increased amount of interfaces between the donor and acceptor layer, resulting in greater probability for photogenerated exciton dissociation into free charge. However, for increased photocurrent generation to be observed, the appropriate morphology plays a crucial role in avoiding charge recombination and facilitating charge transport to the respective electrodes.[104] A great number of

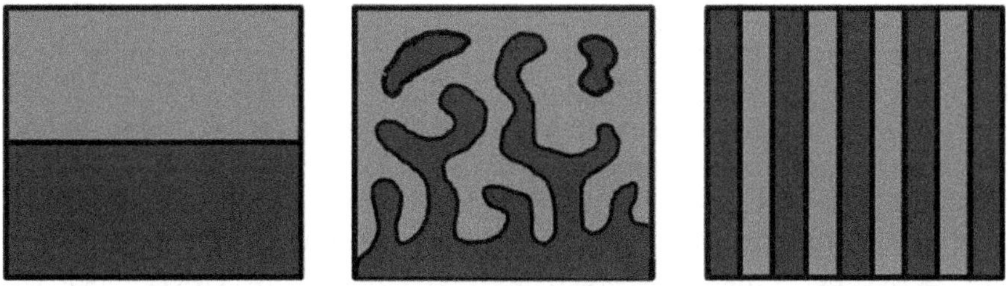

Figure 11.11 Graphical representation of polymer heterojuntion morphologies.

variables can alter the morphology and ultimately affect the BHJ OPV performance, such as choice of materials, solvents, donor : acceptor weight ratio, solution processing methods and parameters, and post deposition heating annealing steps.[104–106] In the past decade, researchers have begun to explore the potential of using block copolymers to induce morphology organization of the donor and acceptor polymers and small molecules on the nanoscale (Figure 11.11).[107,108] This increased order of the donor and acceptor molecules can increase the opportunity for charge carriers to reach the electrodes, reducing the recombination rate and resulting in more efficient BHJ OPV devices. The use of donor–acceptor block copolymers can increase the interfaces, which is inherent with blending the two molecules while benefiting from the order and well defined channels for charge collection.[107,108]

11.5.1.2 Coil–Coil Block Copolymer Synthesized by NMP

Hadziioannou and coworkers were among the first to investigate the use of donor–acceptor block copolymer using NMP.[109–111] The group utilized benzyl peroxide (BPO) and TEMPO to synthesize block copolymers by sequential homopolymerization of styrenic derivatives of polyphenylenevinylene [poly(PV)] and oxidiazole based monomer: 2-(4-(*tert*-butyl)phenyl)-5-(4′-vinyl-[1,1′-biphenyl]-4-yl)-1,3,4-oxadiazole.[109–111] The researchers reported some preliminary devices by sandwiching the poly(PV)-*block*-poly(PBD) block copolymer in between ITO and aluminum and exciting the sample with a laser 350 nm (40 mW cm^{-2}).[109] The device made from the block copolymers was found to be 120 times more efficient then a device fabricated using a simple blend of two poly(PBD) and poly(PV) containing homopolymers.[109] Thelakkat and coworkers also explored the use of st-TIPNO (Figure 11.1) to polymerize a vinyl triphenylamine monomers [poly(vTPA)], resulting in homopolymers having a $M_n = 15–20$ kg mol^{-1} and $M_w/M_n \approx 1.2$. The poly(vTPA) macroinitiators were subsequently used for the homopolymerization of a fresh batch of acrylate functional perylene bisimide monomer (PerAcr, Figure 11.12), resulting in a series of donor/acceptor block copolymers with $M_n = 17–37\,\text{kg mol}^{-1}$ and $M_w/M_n \approx 1.5–2.0$.[112] The authors examined films cast from the resulting block copolymers by TEM and observed phase separation on the nanometer scale in the form of nanowires of about 40–60 nm thick and 1–2 μm long. According to the authors, this large aspect ratio should favor a better percolation of charge transport and ultimately result in a better device. The differences in morphology, photophysical properties, electrochemical behavior and BHJ OPV device performance between poly(vTPA)-*block*-poly(PerAcr) block copolymers and simple blends of poly(vTPA) and poly(PerAcr) homopolymers have been the focus of several studies.[113–115] In addition to the observed micro phase separation, the devices made from poly(vTPA)-*block*-poly(PerAcr) block copolymers exhibited an increase in power conversion efficiency (PCE) from 0.007% to 0.07% compared to similar devices made from a blend of the respective homopolymers.[115] Using steady-state and transient spectroscopies, King *et al.* studied the effect of degree of polymerization of each block of a poly(vTPA)-*block*-poly(PerAcr) block copolymer and determined that donor–acceptor block copolymers can be used to control both the nanomorphology and the charge-separation efficiency in organic solar cells.[116] Therefore, when optimizing the chain length and donor–acceptor composition the authors obtained a peak PCE ≈ 0.11% for BHJ OPV devices made from poly(vTPA)-*block*-poly(PerAcr) block copolymers, which was roughly an order of magnitude greater than the devices fabricated from a blend of the respective homopolymers.[116] Thelekkat and coworkers have also explored various synthetic routes and chemical modifications to this preliminary block copolymer in attempts to increase the device PCE. For example, the authors initially synthesized an alkyne-containing homopolymer by NMP followed by the coupling of azide functional perylene bismide groups by performing a copper-catalyzed azide–alkyne reaction.[117] In addition, the group has also explored the use of methoxy groups on the triphenyl moiety (vDMTPA, Figure 11.12) as well as the use of a tetrabenzyl monomer (vDMTPD, Figure 11.12), instead of the basic vTPA monomer. The methoxy groups and the tetrabenzyl groups were found to increase the solubility and raising the HOMO levels of the electron donor block of the donor–acceptor block copolymer (Figure 11.12).[118,119] The same researchers studied the charge transport properties of these different triphenyl containing homopolymers in OTFTs,[120] which measures the charge mobility parallel to a surface as well as in a diode sandwich configuration,[121] which measures the charge mobility perpendicular to a surface. In addition to triphenyl containing homopolymers they also built OTFTs using poly(PerAcr) homopolymers.[122,123] These OTFT studies identified that homopolymers of vDMTPD [poly(vDMTPD)] had a much great charge mobility compared to the other triphenylamine containing homopolymers, such as poly(vTPA) and poly(vDMTPA) and therefore justified their incorporation into a donor–acceptor block copolymers for use in BHJ OPV devices.[118,124] Following these findings, the authors built a series of BHJ OPV devices using poly(vTPA)-*block*-poly(PerAcr), poly(vDMTPA)-*block*-poly(PerAcr) and poly(vDMTPD)-*block*-poly(PerAcr) block copolymers as

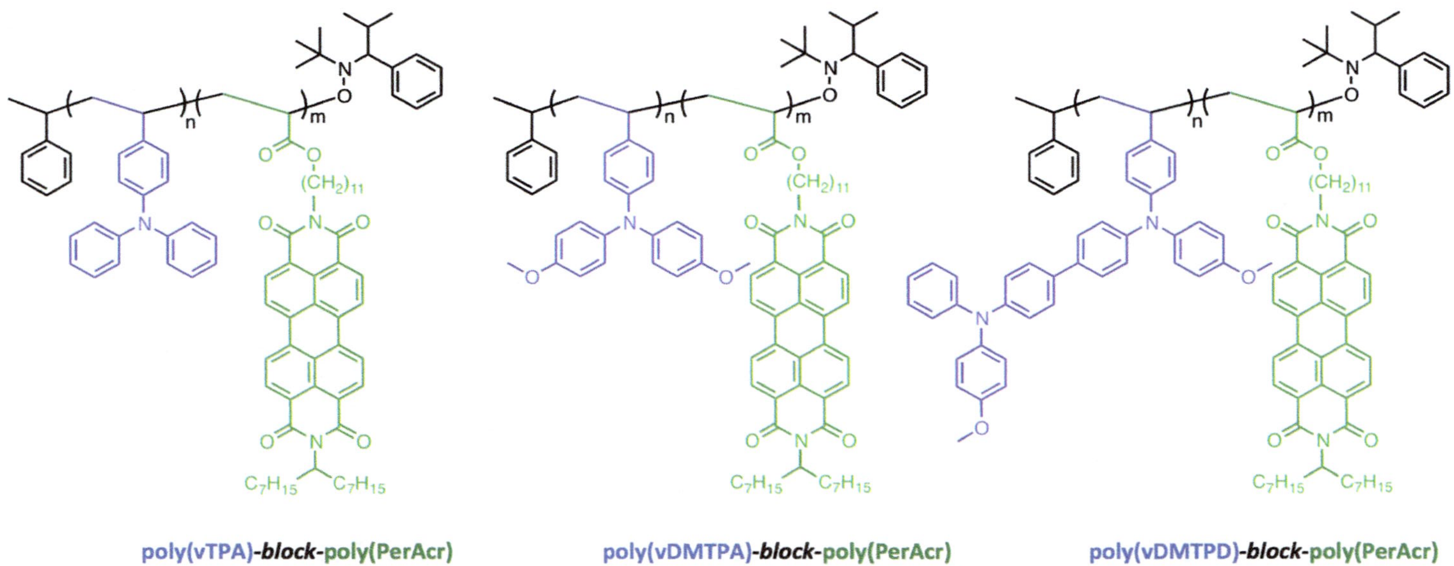

Figure 11.12 Chemical structure of coil–coil triphenylamine and acrylate functional perylene bisimide (PerAcr) containing block copolymer.

the active layer (Figure 11.12). In all cases, the block copolymers were synthesized by NMP using *st*-TIPNO resulting in well-defined block copolymers. The methoxy functional vTPA containing block copolymer poly(vDMTPA)-*block*-poly(PerAcr) experienced the greatest increase in PCE ≈ 0.32% compared to the poly(vTPA)-*block*-poly(PerAcr).[118,124] This increase in efficiency was credited to the raised HOMO level as well as the increase in charge mobility compared to poly(vTPA).[118,124] When examining the morphology of the poly(vDMTPD)-*block*-poly(PerAcr) block copolymers it was apparent that the phase separation was not ideal due to the short poly(vDMTPD) chains.[124] The researchers proceeded to attempt the synthesis of a larger poly(vDMTPD) block, however concluded that this was not possible due to irreversible chain termination reactions which reduced the living fraction.[119,125]

11.5.1.3 Rod–Coil Block Copolymers Synthesized by NMP

Increasing the conjugation in a polymer is known to have two effects: it increases charge transport in the final material and reduces the solubility of the final polymer. In fact, conjugated polymers are often insoluble due to their rigid backbone, which has a tendency to stack and form crystalline rod-like morphologies. Over the years, several researchers have focused on simultaneously increasing the conjugation and solubility of polymers for organic electronic device applications. Poly(3HT) is characterized by having good light absorption properties, high charge transport properties and adequate solubility in several organic solvents. These desired attributes have made poly(3HT) an ideal candidate for a variety of applications, including BHJ OPV devices. Poly(3HT) is one of the most well studied conjugated polymers and is the subject of several reviews.[126–129] Often blended with small molecule fullerene derivatives, the use of poly(3HT) in BHJ OPV devices has resulted in power conversion efficiencies (PCE) as high as 3–5%.[88,130] Grignard metatheses (GRIM) developed by McCullough and coworkers is currently one of the most common methods used to synthesize regioregular poly(3HT).[127,131] GRIM can also be combined with other synthetic techniques to facilitate the incorporation of several other functional monomers and to facilitate the synthesis of well-defined block copolymers. For example, after chemical substitutions, rigid conjugated polymers such as poly(3HT) can be used as macroinitiators for the polymerization of inherently flexible vinyl-based monomers resulting in rod–coil block copolymers. These rod–coil block copolymers have been reported to exhibit very different self-assembly behaviors compared to the traditional coil–coil block copolymer self-assembly model as reported by Gaussian chain statistics and could be very interesting candidates for BHJ OPV devices.[107,132,133]

NMP can be used to combine both poly(3HT) and vinyl-based polymers and numerous examples can be found in the litterature.[134] For example, Richard *et al.* synthesized a poly(3HT) containing TIPNO macroinitiator by first synthesizing a bromine end functionalized poly(3HT) by GRIM. The authors proceeded to couple this poly(3HT) homopolymer to a TIPNO-based

Grignard reagent.[135] The resulting poly(3HT) macroinitiator was used to initiate the homopolymerization chloromethyl styrene (CMS), giving a rod–coil block copolymer.[135] The same authors also performed similar chemistry to obtain well-defined dioctylpoly(thienylenevinylene)-*block*-poly(chloromethyl styrene) low band gap rod–coil conjugated block copolymers.[136] In a similar fashion, Britze *et al.*[137] and Kaul *et al.*[138] both synthesized poly(3HT)-poly(S) block copolymers by either combining GRIM with azide/alkyne coupling chemistry or by employing Kumada catalyst-transfer polycondensation (KCTP) for the preparation of a TIPNO functional poly(3HT) macroinitiator prior to the NMP synthesis of poly(S).[137,138] Yuan *et al.* followed a similar procedure for synthesizing a poly(3HT) functional TIPNO macroinitiators, which were subsequently used to copolymerize zinc dimethacrylate (ZnMa). The resulting poly(3HT)-*block*-poly(ZnMa) block copolymers were crosslinked and used to stabilize zinc oxide (ZnO)/poly(3HT) hybrid OPV devices.[139] The authors identified that the use of the crosslinked block copolymer reduced the deterioration of the PCE performance over time.[139]

Emrick and coworkers synthesized donor–acceptor block copolymers, composed of poly(3HT) homopolymers that were synthesized by GRIM followed by the palladium-catalyzed Stille and Heck coupling using a bromine functionalized TIPNO based alkoxyamine (Figure 11.13).[135,140,141] The authors subsequently used the poly(3HT) macroinitiator to homopolymerize a perylene diimide acrylate, resulting in donor–acceptor block copolymers characterized by having a narrow molecular weight distribution of 1.2–1.4.[140] In addition to displaying microphase separation, these copolymers showed efficient photoluminescence quenching in the solid state, indicative of charge separation, and were used to produce a BHJ OPV device with a PCE of 0.49% (Figure 11.13).[140] Sommer *et al.* synthesized a similar poly(3HT)-*block*-poly(PerAcr) block copolymer by first synthesizing an iodo-functionalized poly(3HT) homopolymer followed by the coupling of a bromine-functionalized TIPNO-based alkoxyamine prior to its use as a macroinitiator for the homopolymerization of PerAcr.[142] The resulting block copolymers were used to fabricate OTFTs[143] and OPV devices.[144] The champion poly(3HT)-*block*-poly(PerAcr) block copolymer BHJ OPV devices were characterized by having a PCE of 0.20%, which the authors suggest was a result of unsuitable block copolymer orientation in the nanomorphology.[144] When introduced into an OTFT, the n-type and p-type blocks of the poly(3HT)-*block*-poly(PerAcr) self-assemble into two-dimensional lattices of π–π stacks and main chain polymer lamellae, respectively, over a broad composition range. Different annealing conditions as well as varying block copolymer lengths allowed the OTFT polarity to be changed from either p-type to n-type or p-type to ambipolar, respectively.[143] The same research group has also explored the use of a triphenyl containing second generation dendrons as macroinitiatiors for the NMP homopolymerization of PerArc monomer, however their use in devices was not reported.[145] Lohwasser *et al.* explored the use of combining copper-catalyzed azide-alkyne coupling

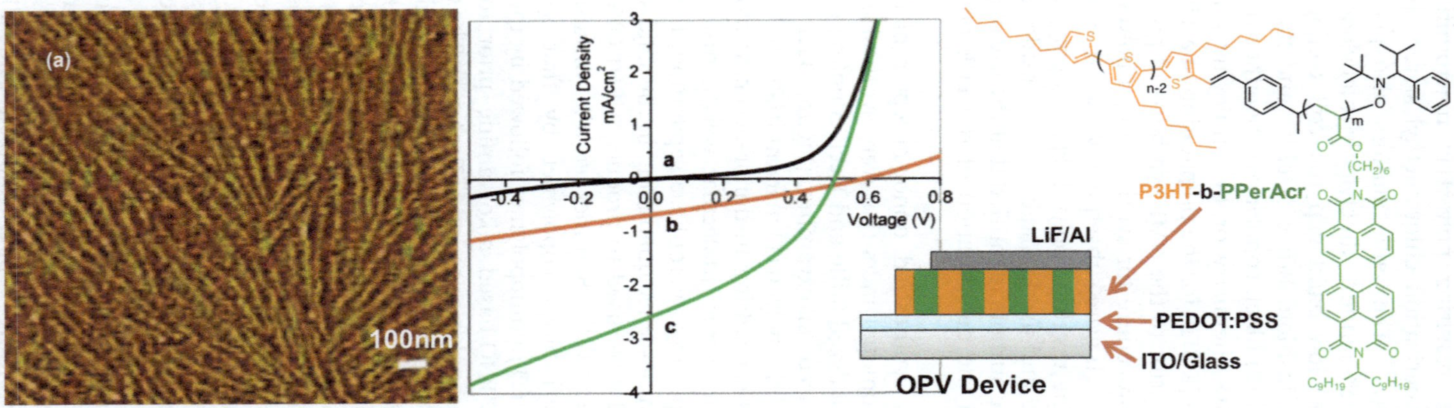

Figure 11.13 AFM image of the rod-like morphology of poly(3HT)-poly(PerAcr) block copolymer (left). Device characteristics using poly(3HT)-poly(PerAcr) block copolymer (green) or a blend of the respective homopolymers (red, middle). Chemical structure of poly(3HT)-poly(PerAcr) block copolymer and graphical representation of device structure (right).
Adapted with permission from ref. 140 Copyright 2009 American Chemical Society.

reaction and NMP to synthesize well defined poly(3HT)-*block*-poly(4VP)[146] and poly(3HT)-*block*-poly(PerAcr) block copolymers resulting in both lamellar and cylindrical morphology depending on the size of the poly(PerAcr) block.[147] In addition to poly(3HT), Hadziioannou and coworkers have also published several studies exploring the use of another conjugated block, namely, poly(*p*-phenylenevinylene) (PPV) synthesized *via* the Siegrist polycondensation followed by the nucleophilic attack of a Grignard reagent functionalized TEMPO alkoxyamine[148] and TIPNO alkoxyamine[149–151] to the aldehyde group on the poly(PV) polymer.[152] These examples illustrate that the combination of conjugated polymerization techniques, such as GRIM, and NMP can be used to synthesize well-defined, donor–acceptor rod–coil block copolymers that can be utilized in BHJ OPV devices.

As mentioned earlier, in a typical BHJ OPV device, poly(3HT) is often paired with a fullerene (C_{60}) derivatives, namely, [6,6]-penyl-C_{61}-butyric-acid-methyl-ester ($PC_{61}BM$).[126,130] When making poly(3HT):$PC_{61}BM$ BHJ OPV devices, the poly(3HT) and $PC_{61}BM$ are solubilized together in an organic solvent and spun cast to make a thin film that can be used as the active layer of the BHJ OPV device.[126,130] This physical blend of molecules results in a kinetically trapped morphology unlike block copolymer self-assembly, which is a result of a thermodynamically stable morphology. Therefore, researchers have attempted to take advantage of block copolymer self-assembly by coupling the C_{60} molecule to the block copolymer itself. For example, researchers have synthesized chloromethyl styrene (CMS) containing homopolymers by NMP or block copolymers by combining NMP and GRIM followed by the coupling of the C_{60} to the CMS. The coupling can either be accomplished by directly using atom transfer radical addition,[148,153] or by first performing an azidation on the CMS containing polymer followed by heating in the presence of C_{60} (Figure 11.14).[149,154,155] Gholamkhass *et al.* synthesized a TEMPO functionalized poly(3HT) macroinitiator by Suzuki coupling the brominated poly(3HT) followed by the NMP copolymerization of CMS and styrene, resulting in poly(3HT)-*graft*-poly(CMS-*ran*-styrene) graft block copolymers with varying amounts of CMS/styrene chains depending on the degree of bromination.[155] After an azidation on the CMS, C_{60} was coupled to the graft copolymers (Figure 11.14).[155] The authors introduced these graft copolymers as the active material in BHJ OPV devices, and while the films experienced a micro-bicontinuous morphology with features as small as 5 nm, the photovoltaic efficiencies were significantly reduced compared to poly(3HT):$PC_{61}BM$ blended devices.[155] The authors concluded that the aggregation of the less soluble fullerene group during film formation restricts the self-organization of the poly(3HT) polymer, resulting in a reduction in its domain size, lowering crystallinity, charge mobility, and PV activity.[155] The same authors then proceeded to study the use of the poly(3HT) graft azide functional graft copolymers to perform a solid state reaction with the blended $PC_{61}BM$ prior to being used in an OPV.[156] The authors studied this technique as a potential route for stabilizing the morphology of the poly(3HT):$PC_{61}BM$ active layer in a BHJ OPV device.

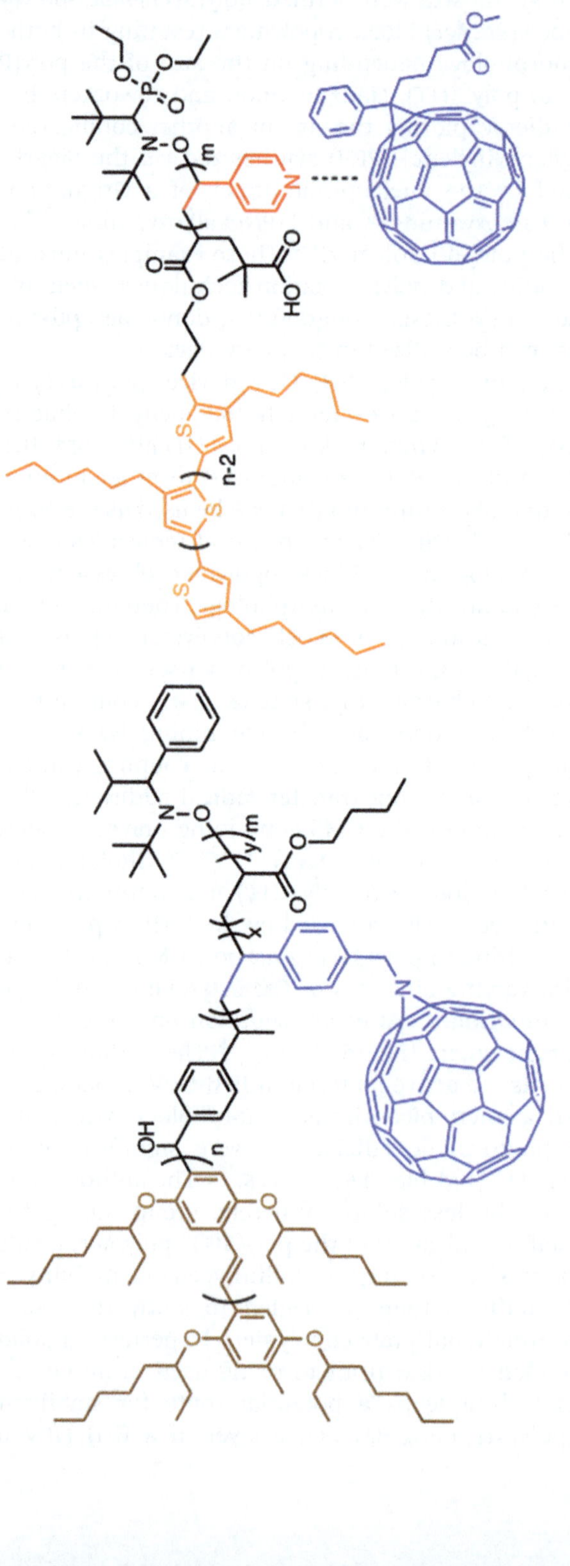

PPV-b-(PCMS-C_{60})

P3HT-b-(P4VP-complex-$PC_{61}BM$)

Figure 11.14 Chemical structure of rod–coil [6,6]-penyl-C_{61}-butyric-acid-methyl-ester ($PC_{61}BM$)-containing block copolymers where poly-(*p*-phenylenevinylene) (PPV, left) or poly(3-hexylthiophene) (poly(3HT), right) are converted to nitroxide mediated polymerization conjugated macroinitiators. In both cases, the second block is coupled to $PC_{61}BM$ through covalent bonding (left) or charge transfer complexation (right).

The OPV devices prepared from these $PC_{61}BM$ fused graft copolymer layers exhibited PCE ≈ 1.85% which fell to 0.93% over 3 h at 150 °C, whereas native poly(3HT)/$PC_{61}BM$ devices initially displayed a PCE ≈ 2.5%, which dropped to 0.5% over the same period.[156] Gholamkhass *et al.* go on to show that a combination of phase stabilization using this graft copolymer and replacement of thermally labile PEDOT:poly(SS) with gold nano-dots can substantially improve device stability in accelerated aging tests.[157] Zhou *et al.* synthesized poly(triazole) functionalized poly(3HT) graft copolymers by first coupling a TEMPO derived macroinitiator throughout a poly(3HT) chain followed by "graft from" NMP of a triazole containing monomer.[158] The authors proceeded to blend the poly(triazole)-*graft*-poly(3HT) copolymers with a triazole functionalized C_{60} and observed microscopic blending, before and after annealing, while mixtures of poly(3HT) with triazole functionalized C_{60} exhibited micron-sized phase-segregated morphologies before and after annealing.[158] Heuken *et al.* also investigated the use of C_{60}, however they focused on the use of Bingel conditions to covalently bind C_{60} to propargyloxy styrene-containing copolymers[159] and poly(3HT)-containing block copolymers.[160] In all these examples, the C_{60} molecule was coupled to the polymer using a multi-step covalent bonding route. Unfortunately, no device data is available and therefore no conclusion on the best synthetic route can be drawn. Regardless, these findings illustrate that the covalent bonding of C_{60} molecules to rod–coil donor–acceptor block copolymers is facilitated by the use of NMP.

Instead of using covalent binding, another method of manipulating the self-assembly of C_{60} in a polymer matrix is to use the complexation reaction between 4-vinyl pyridine (4VP) and C_{60}.[161] Laiho *et al.* identified that when blending C_{60} molecules with block copolymers of poly(S)-poly(4VP), the C_{60} molecules would preferentially self-assemble in the poly(4VP) domain of the resulting film.[161] While it is impossible to synthesize poly(4VP) using NMP in the presence of C_{60},[162] it can however be preferentially incorporated into the poly(4VP) domain of the polymer matrix after the synthesis is complete.[161] Several studies have successfully combined NMP and GRIM to synthesize 4VP or even 2-vinyl pyridine (2VP)-containing block copolymers such as, poly(3HT)-*block*-poly(4VP)[163] and poly(3HT)-*block*-poly(2VP).[146] For example, Renaud *et al.* synthesized a poly(3HT) macroinititaor by first coupling an acrylate to the poly(3HT) followed by a coupling of BlocBuilder-MA *via* an intermolecular radical 1,2-addition.[164] The now SG1 functional poly(3HT) macroinitiator was subsequently used to initiate the homopolymerization of 4VP, resulting in well-defined poly(3HT)-*block*-poly(4VP) block copolymers. The poly(3HT)-*block*-poly(4VP) block copolymers were incorporated into poly(3HT):$PC_{61}BM$ blends that were used as the active layer in BHJ OPV devices.[164] The addition of as little as 4 wt% of poly(3HT)-*block*-poly(4VP) block copolymers in the poly(3HT):$PC_{61}BM$ blend resulted in a significant increase in overall PCE ≈ 3.4% compared to ≈ 2.7% for the simple poly(3HT):$PC_{61}BM$ devices. When a thermal annealing step was included the PCE of the poly(3HT)-*block*-poly(4VP)-containing device rose to as high as ≈4.3% (Figure 11.14).[164] The authors conclude that the increase in performance suggests the formation of a well-optimized nanoscale structure that allows for a more efficient exciton dissociation and charge transport, by lowering the charge recombination and/or charge trapping. In addition to C_{60}, poly(4VP) has also been found to complex to inorganic nanoparticles. For example, Maria *et al.* synthesized a series of tetrabenzyl-*block*-poly(4VP) block copolymers by NMP, which were found to micro-phase separate into lamellar morphologies as determined by TEM.[165] When the authors blended CdSe:Te nanoparticles with these block copolymers, the hybrid material again experienced micro-phase separation into lamellar morphologies, however, the CdSe:Te nanoparticles were exclusively found in the poly(4VP) phase.[165]

This ability to combine the synthesis of conjugated polymers and vinylic polymers synthesized by NMP gives rise to intricate rod–coil nanomorphologies and unique BHJ OPV devices properties. While the chemistry and synthesis are well established, as evidenced by the numerous examples, the overall BHJ OPV device performances are still relatively low. This suggests that exploiting the benefits of block copolymer phase separation is not straightforward and that a great deal of research is still needed.

11.5.2 Polymer Organic Light-emitting Diodes (PLEDs or Polymer OLEDs)

For years, the most common lighting source was the incandescent light bulb, which is very inefficient. Only 5% of the energy driven into the bulb is converted to light, while the rest is discharged as heat.[166] While energy-saving compact fluorescent lamps (CFLs) are about 20% efficient, they come at an environmental cost since they contain mercury, which is highly toxic and a cumulative pollutant.[166,167] CFLs are expected to be replaced by more efficient and environmentally sustainable solid state lighting technologies, such as inorganic light-emitting diodes (LEDs).[167] The problems associated with LEDs, such as heat management, lifetime, colour rendering, cost and efficiency are all being addressed and currently the technology is already displacing CFLs in several commercial lighting applications.[167] White organic light-emitting diodes (OLEDs) are considered the most promising technology for next generation solid-state lighting due to their high-energy efficiency, compatibility with flexible substrates, eye-friendly diffusive warm light and ultrathin form factor.[168,169] In addition, due to the tunability of organic molecules and the negligible heat production, OLEDs are void of several problems that plague LEDs, such as colour rendering and heat management. While some startup companies are beginning to commercialize white OLEDs for lighting, such as aerelight™ by OTI Lumionics Inc. (Figure 11.15), the majority of the applications of OLEDs have been in display technologies. In 1997, companies began to commercialize OLEDs for use in cellular phone and television displays and currently the majority of

Figure 11.15 White organic light-emitting diode (OLED) desk lamp. Reproduced with permission from aerelight™ and OTI Lumionics Inc.

international electronic companies, such as Sony, Ritdisplay, Eastman Kodak, LG Display, Epson, Samsung SDI, Matsushita and Sharp Corporation, are also exploring the use of this technology.[170]

11.5.2.1 Introduction to the Working Principles of OLEDs

In an OLED device, holes and electrons are injected from the anode and the cathode, respectively, and combine to make excitons. Relaxation of the electron of this exciton to the ground state releases energy. This released energy can be in the form of radiative emission where a portion may even be in the visible spectrum. The emission wavelength of this relaxation corresponds to the difference between the HOMO and LUMO levels of the organic molecule in which the resulting excitons are formed. Quantum mechanics dictates that the combination of a hole and an electron with two possible half integer spins will result in the formation of excitons with singlet states and triplet states in a ratio of 1 : 3, respectively.[171] The emissive relaxation from singlets results in fluorescence, while the emission from triplets corresponds to longer time scale phosphorescence. OLED devices based on fluorescence only benefit from the singlets as the triplets tend to decay through non-emissive pathways. On the other hand, phosphorescent-based OLED devices take advantage of the higher proportion of emissive triplets. The use of phosphorescent molecules, also known as phosphors, results in intersystem crossing, which leads to the conversion of excited singlets into excited triplets. This transfer facilitates all of the generated excitons to phosphoresce, resulting in the potential to reach 100% internal quantum efficiency.[172,173] Phosphors are molecules containing heavy atoms, such as iridium or platinum, and are often introduced in OLEDs as a dopant in a host material due to the potential for triplet quenching between neighboring molecules.[174,175]

In 1990, Burroughes *et al.* published the first report of a conjugated polymer based OLED (PLED) using poly(*p*-phenylene vinylene) [poly(PV)].[176] Since then numerous researchers at academic as well as commercial and industrial institutions have explored the use of organic molecules and polymers in PLEDs. Several groups have attempted to take advantage of the phosphorescence by introducing phosphors into PLEDs. Similarly to OPV devices, there are several routes that can be used to accomplish the desired morphology: blending small molecule phosphors into a polymer matrix, covalently binding the phosphor into the main-chain of the polymer or even pendant to the polymer itself.[174] Simple blending of the polymer and the phosphor can result in poor device efficiencies due to aggregation of the dopants resulting in concentration quenching and is therefore less desirable.[174,177] Several good review articles on OLEDs and PLEDs have been published recently and cover all aspects of device engineering and device photophysics.[166,174,178]

11.5.2.2 The Use of NMP for the Synthesis of Active Polymers for PLEDs

As mentioned, the use of NMP can facilitate the synthesis of polymers with narrow molecular weight distributions, tunable molecular compositions, and materials with nano-sized features that can be implemented into organic electronic devices. In 2004, Fréchet and coworkers investigated the use of solution processable block copolymers with a controlled amount of phosphor dopant. The block copolymer served as a host for the dopants, which were covalently bound pendant to the polymer chain, to reduce aggregation.[179] By using block copolymers the authors could intentionally separate different emitters in both phase and independently tune the host of each dopant.[179] The authors, synthesized terpolymers of oxidiazole containing monomers (2-phenyl-5-(4-vinylphenyl)-1,3,4-oxadiazole, 3OXA), 3-vinyltriphenylamine (v3TPA) and a diketonate moieties-containing monomer that were subsequently metalated with a platinum ligand. The resulting block copolymers were solution processed to give simple layer near-white OLED devices with an external quantum efficiency (EQE) of up to 4.6%.[179] The researchers followed up with an extensive study, varying not only the copolymer composition but also using novel oxadiazole- and triphenylamine-containing monomers. For example, the novel monomers were similar to 3OXA or v3TPA except substituents such as *tert*-butyl groups (*t*-OXA) were added to the peripheral aromatic groups and the vinyl-containing groups were changed from styrenic to acrylic.[82] In this study, the researchers used

NMP to synthesize a library of random copolymers and block copolymers with narrow molecular weight distributions and tunable molecular weights. By controlling the copolymer composition, the authors could accurately tune the ratio of hole transport and electron transport properties of the resulting phosphor-containing copolymers. This allowed the researchers to ultimately maximize the charge transport properties and the device efficiency. Optimization of these copolymers resulted in white OLEDs with an EQE of 4.9% and green OLEDs with an EQE of 10.5%.[82] Ma *et al.* synthesized random and block copolymers containing v3TPA and *t*-OXA, which were doped with platinum or iridium containing complexes prior to being used in a single layer OLED (the dopants were not bound to the polymers).[180] The authors identified that the nano-phase separation observed for the diblock copolymers was unchanged by the addition of the dopant. In addition, they found that, compared to the use of random copolymers, the use of block copolymers resulted in a 2–3 times increase in hole–electron recombination, resulting in increased device efficiency.[180] The authors also identified that the phase separation did in fact increase the dopant aggregation causing increased triplet–triplet annihilation (resulting in non-emissive decay and a decrease in device efficiency). Poulsen *et al.* synthesized, using NMP, a series of poly(TPA-*ran*-BLUE)-*block*-poly(*t*-OXA-*ran*-RED) containing block copolymers, where "BLUE" and "RED" are green–blue and orange–red emissive pendant styryl heteroleptic Ir(III) complexes (Figure 11.16). This copolymerization approach eliminated the need for post-coupling reactions and gave rise to block copolymer systems that can deliver site isolation of the two emitters (Figure 11.16).[181] The authors explored the effect of block copolymer chain length on the micro phase separation and the resulting device EQE of the white OLED. Poulsen *et al.* observed an increase in EQE when the block lengths got larger, which corresponded to more discrete microphase separation and therefore better site isolation of the corresponding emitters (Figure 11.16).[181] Similarly, Dumur *et al.* synthesized a series of iridium containing well defined random copolymers by copolymerizing cationic iridium(III) complex-based monomers using BlocBuilder-MA. The resulting copolymers were used as electrophosphorescent emitters in OLEDs, when doped into poly(*N*-vinylcarbazole) [poly(VK)] or alone as a single layer.[182,183]

Tsuchiya *et al.* also utilized NMP to synthesize a series of homopolymers and block copolymers from v3TPA and 4-vinylbenzyl 4-[5-(4-*t*-butylphenyl)-1,3,4-oxadiazole- 2-yl]benzoate or 4-vinylbenzyl 4-[5-(3,5-bis(trifluoromethyl) phenyl)-1,3,4-oxadiazole-2-yl]benzoate, e prior to their use as host materials for an iridium complex dopant.[184] Similar to Frechet and coworkers,[82] Tsuchiya *et al.* observed an EQE of 10.1% when using the oxidiazole/triphenylamine-based block copolymer, which was significantly greater than that observed for the device fabricated using a simple blend of the two respective homopolymers.[184] The same group followed with the synthesis of a series of 2-(4-*t*-butylphenyl)-5-(4-vinylbiphenyl-4′-yl)-1,3,4-oxadiazole (PBD) and 2,7-dimethoxy-*N*-(4-vinylphenyl)carbazole (DMK) containing homopolymers, random copolymers and block copolymers by NMP, which

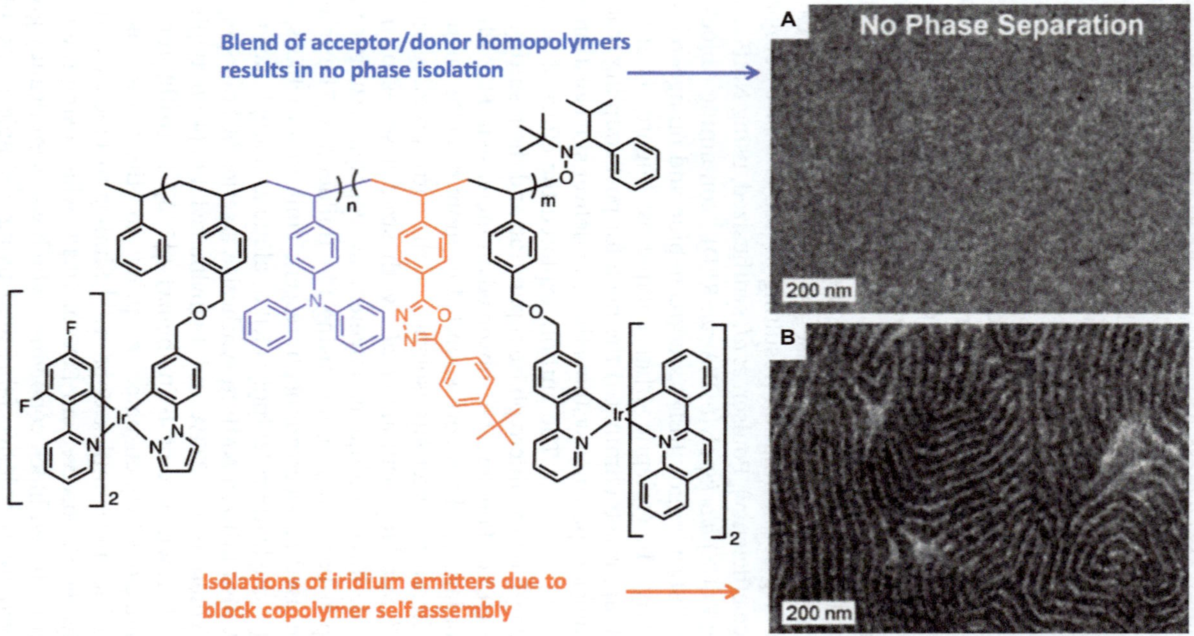

Figure 11.16 Chemical structure (left) of oxidiazole/triphenylamine-based block copolymer used in organic light-emitting diodes (OLEDs). A) and B) are AFM images of a blend of oxidiazole/triphenylamine homopolymers and an oxidiazole/triphenylamine block copolymer, respectively, where the iridium complexes are phase isolated.
Adapted from ref. 181 with permission from John Wiley & Sons.

were utilized as the host material for iridium complex dopants as the emitting layer of polymer OLEDs.[185] Similarly to the observations for the PBD/v3TPA-based block copolymer study,[184] the authors observed a significant increase in EQE≈6.3% for the poly(PBD)-*block*-poly(DMK) block copolymer compared to the OLEDs fabricated using the respective poly(PBD-*ran*-DMK) random copolymers.[185] In addition, the devices made from poly(PBD)-*block*-poly(DMK) block copolymer resulted in a significant increase in lifetime compared to the devices fabricated from simple blends of respective poly(PBD) and poly(DMK) homopolymers.[185] The authors also observed an increase in OLED efficiency due to a significant modification of the overall morphology when 12 wt% of the oxidiazole/carbazole-based block copolymer was added to the blend of respective poly(PBD) and carbazole homopolymers.[186] NMP has also been employed to synthesize well-defined poly(2-(*N*-carbazolyl)ethyl acrylate) [poly(VAK)] homopolymers, which are carbazole pendant acrylic polymers.[187] Poly(VAK) homopolymers were subsequently used as effective polymer hosts for blue and blue-green emitting phosphor-based PLEDs and were found to outperform the well-known and structurally similar, poly(*N*-vinylcarbazole) [poly(VK)].[188]

In a similar fashion to the rod–coil block copolymers synthesized for use in OPV devices, Tao *et al.* synthesized a series of poly(PV)-*block*-poly(3OXA) block copolymers by combining a Siegrist polycondensation followed by an azide–alkyne coupling reaction to make poly(PV) functionalized TIPNO macroinitiators that were used to homopolymerize oxidiazole containing monomers. The resulting poly(PV)-*block*-poly(3OXA) block copolymers were introduced as the active layer in a simple single layer OLED and were found to give devices with significantly greater EQE compared to devices made of pure poly(PV) and those made from blends of poly(PV) and poly(3OXA) homopolymers.[189]

In all these examples, the PLED devices were fabricated using a single emissive polymer layer. NMP can also be used to fabricate more complex hybrid, patterned and even multi layer PLED based devices. For example, Chang *et al.* employed TIPNO to synthesize functional oligomers containing triphenylamine groups paired with a photolabile ortho-nitrobenzyl carboxylate groups.[190] The authors then applied these polymers to ITO glass substrates that were used to build PLEDs.[190] When the working devices were exposed to UV light (365 nm) through a patterned photomask and developed in chloroform, the triphenylamine-containing nitrobenzyl group was cleaved and a positive image was obtained. The same technique was used to fabricate patterned PLEDs.[190] Park *et al.* synthesized triphenylamine and chloro styrene functional poly(vTPA-*ran*-CMS) copolymers with narrow molecular weight by *st*-TIPNO.[191] The Cl groups on the CMS monomers were converted to N_3 groups, which produce nitrene radicals under UV-irradiation. Therefore, cast films of the now N_3 functional poly(vTPA-*ran*-N_3MS) random copolymers could be cross-linked under UV-light prior to depositing additional solution processed layers.[191] Park *et al.* used the cross-linked vTPA containing layer as a hole transport/electron blocking layer in a Ir(ppy)3-based multilayer solution processed PLED. The resulting orthogonally processed multilayer solution processed PLEDs were characterized by having an EQE of as high as 11.8%.[191] In addition, the authors also made devices with a patterned mask during the UV-crosslinking, resulting in pattered PLEDs.[191] Kozma *et al.* synthesized well-defined styrene and styrenic perylene diimide containing copolymers using TEMPO. The authors blended these styrenic copolymers with green-emitting conjugated polymers into films. The polymer films were subsequently combined with commercially available inorganic blue LEDs to convert the blue light into white light.[192] Compared to simple blends of poly(styrene) with perylene diimide dyes, the pendant functional copolymer efficiently suppressed the quenching processes due to aggregation states of perylene diimides, causing a relative increase in efficiency.[192]

These studies illustrate that NMP is a versatile tool for synthesizing well-defined random copolymers, block copolymers and phosphor-containing polymers that can improve the active layer morphology and improve overall device efficiency. In addition, NMP can introduce novel functionality that can lead to interesting processing techniques and novel applications.

11.5.3 Memory Device Applications

With the unavoidable necessity for digitalization and miniaturization of our things, the demand for data storage, programmable options and digital memory is high. Digital memory is ubiquitous, from personal computers, smart phones and tablets, to remote controls, thermostats, and other household products. In all these examples, the need for increased capacity, reduced power consumption and improved performance is growing.[193–195] Polymer based memory devices can reach all these targets and are particularly promising due to the potential for low cost processing *via* roll-to-roll printing, spray-coating, ink jet printing and other well established industrial techniques.[193–195]

11.5.3.1 Introduction to the Working Principles of Memory Device Applications

Electrical memory is inherently fast compared to other memory storage devices, such as hard drives, floppy disks, DVDs or CDs. These other techniques require a translation from data stored by physical, optical, magnetic or other means to electrical signals prior to being used by the processing system.[195] This necessary conversion in signal increases the delay and the size of the overall system, prompting the interest in the use of electrical memory. Electrical memory has several different modes of operation. For example, electrical memory devices can either store the data as long as power is supplied to the device but lose it the moment power is withdrawn (volatile memory) or the memory can remain even when the power is removed (nonvolatile memory). In addition, their operation can further be subdivided into more specific categories, such as write-once read-many times (WORM) memory,[196] dynamic or static random access memory (DRAM or SRAM), flash memory,[197] and others. WORM memory is non-volatile and can only be written once but read often, and it is not feasible to modify the stored data. WORM memories can be used to store databases, replace the function of conventional DVDs and even be used as electronic RFID tags.[195,196] DRAM is volatile memory that can be used to read, write, and erase, but must have its electrical states refreshed. DRAM is often used in computers as the memory. SRAM is similar to DRAM except for some significant data remanence, which implies the device will eventually lose all the information making it a volatile memory but it will take longer than DRAM. Flash is a non-volatile memory that is readable, re-writable and is based on the same working principle as organic thin film transistors (OTFTs).[194,197,198] A better understanding of the differences between these methods and how they operate can be found in great detail in several recent reviews.[193–195]

11.5.3.2 Polymer by NMP for Memory Device Application

While NMP has not been extensively used to synthesize polymers for applications in memory devices, there are a few examples in the literature. For example, Kwon *et al.* showed that an oxadiazole-containing brush polymer, poly-((5-phenyl-1,3,4-oxadiazol-2-yl-[1,10-biphenyl]carboxyloxy-(*n*-nonyl acrylate), synthesized using *st*-TIPNO could be used for memory applications.[199] Interestingly, the authors found that the resulting device would exhibit volatile or non-volatile memory behavior based on the packing of the oxidiazole molecules within the polymer film, which could be controlled with subsequent annealing steps.[199] Fang *et al.* synthesized a series of random copolymers containing pendant electron-donating 9-(4-vinylphenyl)carbazole (VBK) and electron-accepting 2-phenyl-5-(4-vinylphenyl)-1,3,4-oxadiazole (4OXA) or 2-(4-vinylbiphenyl)-5-(4-phenyl)-1,3,4-oxadiazole (BOXD) with various copolymer ratios by NMP.[200] The authors found that when building simple anode/polymer/cathode devices, the electrical switching behavior could be tuned through the donor/acceptor ratio. The authors reported that the memory device based on these copolymers exhibited SRAM behavior. The slightly lower HOMO level of 4OXA moieties compared to that of BOXD led to polymer devices that could store charge for a longer period of time. The same authors followed with the synthesis of a series of poly(BOXD-*ran*-vTPA) random copolymers with varying copolymer ratios by NMP.[201] The donor/acceptor ratios and the addition of cyano side groups on the vTPA were found to have significant effects on the memory characteristics of the respective devices. Under the same device configuration, the authors found the devices exhibited SRAM or DRAM behavior. The authors identified that when the copolymers with BOXD and the cyano functional vTPA monomers were used, the resulting memory devices exhibited nonvolatile behavior due to the preferred interaction of Al atoms with the cyano group. These examples illustrate that the memory switching characteristics of different electrical memory devices can be tuned by using different donor–acceptor polymers and that NMP can be used to incorporate these different pendant monomers into polymers in a controlled manner.

11.6 Materials Property Enhancement and Industrial Implementation

Over a decade ago it was estimated that controlled radical polymerization would be a $20 billion market, replacing many application where conventional free radical polymerization already had a firm footing.[202] However in the last 15 years, the industrial implementation of controlled radical polymerization has been relatively slow for a technology with such academic interest.[3,202,203] It is the opinion of some that controlled radical polymerizations, such as NMP, will penetrate the market by providing more sophisticated microstructures for more niche applications. In fact, the majority of the examples of industrial NMP implementation have been for the synthesis of material property enhancement additives.[204] This section focuses on a few examples of companies that publicize the use of NMP to make their products, however for a look at more commercial products made by other CRP methods the reader is encouraged to consult several good reviews that have been published recently.[202,203]

Ciba Specialty Chemicals from Switzerland, acquired by BASF in 2010, was a company that developed, manufactured and marketed products present in many items we use every day, such as pigments and dyes, antioxidants and light stabilizers, UV absorbers, polymeric flocculants, photolatent polymerization catalysts, and even components or additives for the making of mobile phone screens, DVDs, transparent packaging materials and personal care products.[204] The company has a long list of commercial products, including the use of TEMPO-based oligomer antioxidants and polymer light stabilizers such as Ciba® TINUVIN® 770, TINUVIN® 103 and CHIMASSORB® 944. In addition, the company focused on the synthesis of a variety of hindered alkoxyamine stable free radicals (SFRs) that facilitated the controlled NMP synthesis of styrenics and even acrylates at elevated temperatures (125–135 °C).[205–211]

The most documented commercial use of NMP by Ciba Specialty Chemicals is for the synthesis of block copolymers for the stabilization of colour pigments (Figure 11.17).[212–216] The authors identified that the most effective polymeric pigment dispersants are block copolymers with a pigment stabilizing block and a soluble dispersing polymeric block. This marriage of both functionalities allows for the solubilizing of the pigments in the solvent while reducing pigment agglomeration. These dispersants were also found to improve the rheology of the pigment concentrates, their stability, and their colouring properties (Figure 11.17). Auschra *et al.* synthesized a series of pigment dispersing poly(butyl

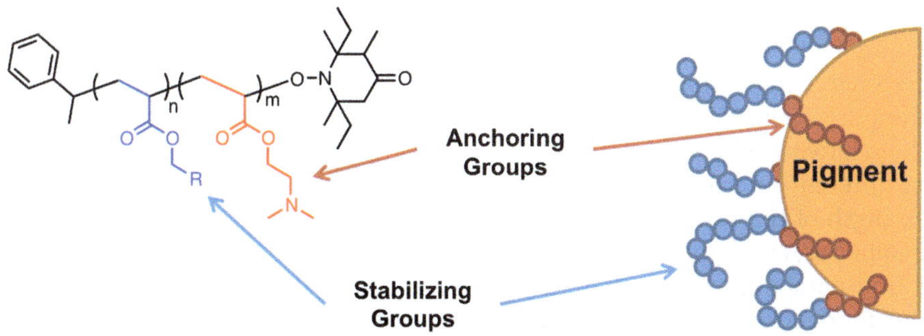

Figure 11.17 Chemical representation of commercialized block copolymer used as pigment stabilizer synthesized by Ciba Specialty Chemicals (now part of BASF).

acrylate)-*block*-poly(dimethylaminoethylacrylate) [poly(BA)-*block*-poly(DMAEA)] block copolymers.[212] Due to the ranging particle size of the various pigments, the authors found that by tuning the ratio of poly(BA) to poly-(DMAEA) block length, they could optimize the performance of both opaque pigments as well as transparent pigments.[212] Ciba Specialty Chemicals has also patented the use of NMP to synthesize a variety of vinyl chloride-containing block copolymers under pressure at low temperatures and even propargyl functional block copolymers.[217,218]

Arkema Inc., previously known as Atofina, a spin-off company from Total, has invested a great deal of resources in making NMP alkoxyamine initiators[25,219,220] as well as well-defined block copolymers by NMP.[221,222] Publishing with Arkema Inc. researchers, Schierholz *et al.* synthesized a series of acrylamide-based amphiphilic block copolymers using SG1,[223] while Poivet *et al.* exemplified their use as adhesive properties under aqueous environments.[224] Arkema Inc. has also protected the use of acid functional block copolymers with tunable segment length, resulting in varying adhesive properties and thermoplastic composition.[225]

Maiez-Tribut *et al.* studied the use of amphiphilic block copolymers containing a random copolymer block of methyl methacrylate (MMA) and *N,N*-dimethylacrylamide (DMA), synthesized by using BlocBuilder-MA.[226] This poly(DMA-*ran*-MMA) was used as the miscible block in the amphiphilic block copolymers used to generate nanostructured phases in epoxy networks.[226] The authors found that modifying the DMA/MMA composition in the block copolymers resulted in a change in block copolymer miscibility in the epoxy as well as a modification in the resulting material opacity. The possibility of varying the DMA fraction in the MMA-DMA random block gives the possibility to tailor these new block copolymers for their use in a variety of impact strength modifiers for epoxy thermosets.[227,228] Another interesting note is that while the homopolymerization of MMA is problematic using SG1, the copolymerization of MMA using DMA resulted in a controlled copolymerization to high conversion.[229] A poly(*n*-butyl acrylate) macroinitiator was also used to initiate a fresh batch of DMA and MMA to make AB and ABA block copolymers.[226,229,230] These ABA block copolymers are currently commercially available from Arkema inc. under the trade name Nanostrength®. Gerard *et al.* published the use of Nanostrength®, and found that the relative size of the ABA blocks as well as the DMA *vs.* MMA ratio resulted in different micro phase separation and ultimately varying material impact strength of the resulting blends made from liquid epoxy, diglycidyl ether of bisphenol A and poly propylene oxide triamine mixtures.[230] The same group found parallel benefits when using similar DMA/MMA block copolymers in the fabrication of rubber-toughened unsaturated polyester resins.[231] Following similar polymerization methods, Jullian *et al.* synthesized a series of diblocks and triblocks of poly(butyl acrylate) and poly(styrene) for use in pressure sensitive adhesives.[232] These examples illustrate some of the interests Arkema has in well-defined block copolymers by NMP for adhesives and mechanical property enhancements for a variety of resins.

Marić and coworkers synthesized a telechelic poly(ethylene-*ran*-butylene) [poly(EB)] SG1-based macroinitiator that was used to copolymerize styrene and maleic anhydride.[233] The resulting ABA block copolymers were blended with nylon-6 at elevated temperatures, causing coupling reactions between the nylon-6 and the maleic anhydride functional groups. The resulting resin experienced a significant reduction in particle size of the poly(EB) phase, which was expected to give better impact strength to the nylon-6 matrix.[233] The same research group used NMP to synthesize styrene/acrylonitrile copolymers, which were subsequently functionalized with primary amines at the chain ends and used for reactive blending with maleic anhydride functional poly(ethylene).[234] The resulting blends were found to be thermally stable compared to the non-reactively blended mixtures of polymers.[234] These studies illustrate that NMP can be used to introduce functional groups into polymers that facilitate the reactive blending of commercial resins, resulting in improved material properties.

Recently, Coatex SAS, a subsidiary company of Arkema Inc., filed for a series of patents based on acrylic and methacrylic random copolymers and block copolymers for applications in cosmetic formulations,[235] hair treatment products,[236] paper coatings for inkjet printing,[237] and calcium carbonate hydraulic grinding aids.[238,239] This interest in the use of well-defined copolymers and block copolymers exemplifies that the potential for commercial application made by NMP is growing.

11.7 Conclusion and Perspective

Nitroxide mediated polymerization (NMP) is a robust and industrially relevant technique. Often only requiring a single unimolecular initiator for the synthesis of pseudo-living well-defined homopolymers and copolymers. In this chapter, the use of NMP was explored in the context of improving several emerging applications, such as organic electronics, nanoporous templating, photoresists and even membranes for fuel cells. In these examples, the application enhancements were related to the ability of NMP to reduce the dispersity, to increase chain end homogeneity, to synthesize block copolymers that could self-assemble, to tether the polymers to surfaces and to incorporate novel functional monomers. While the improvements due to the use of NMP were significant, several factors need to be in line for a new technology to displace a current technology. It is therefore always very difficult to hypothesize what target must be reached before a material can become a commercial success.

As the interest in NMP grows and as the number of applications that benefit from its use increases, the commercial successes will also expand. In this chapter, a short discussion on the current industrial implementations identified that a few major companies are producing commercially available products that are synthesized by NMP. These applications are currently in adhesives, coatings, and paints. These are well-established markets and therefore it is relatively simple to identify the cost to benefit analysis of introducing new products. However, for the majority of the applications in this chapter, such as organic electronics, the markets are niche, poorly defined or non-existent. As these emerging technologies become larger, well-understood markets it will become easier for companies to accurately identify the value of NMP based products and the use of NMP-based processes, resulting in further industrial adoption of the technique.

References

1. A. Jenkins and R. Jones, *Pure Appl. Chem.*, 2010, **82**, 483–491.
2. C. J. Hawker, A. Bosman and E. Harth, *Chem. Rev.*, 2001, **101**, 3661–3688.
3. J. Nicolas, Y. Guillaneuf, C. Lefay, D. Bertin, D. Gigmes and B. Charleux, *Prog. Polym. Sci.*, 2012, **38**, 63–235.
4. R. Grubbs, *Polym. Rev.*, 2011, **51**, 104–137.
5. B. H. Lessard and M. Maric, *Macromolecules*, 2008, **41**, 7870–7880.
6. Y. Wang, Y. Zhang, B. Parker and K. Matyjaszewski, *Macromolecules*, 2011, **44**, 4022–4025.
7. D. Siegwart and J. Oh, *Prog. Polym. Sci.*, 2012, **37**, 18–37.
8. M. Cunningham, *Prog. Polym. Sci.*, 2008, **33**, 365–398.
9. J. Oh, *J. Polym. Sci. Pol. Chem.*, 2008, **46**, 6983–7001.
10. M. Stenzel, *Chem. Commun.*, 2008, **30**, 3486–3503.
11. M. Georges, R. Veregin and P. Kazmaier, *Macromolecules*, 1993, **26**, 2987–2988.
12. M. K. Georges, G. K. Hamer and N. A. Listigovers, *Macromolecules*, 1998, **31**, 9087–9089.
13. D. Benoit, S. Grimaldi, S. Robin, J.-P. Finet, P. Tordo and Y. Gnanou, *J. Am. Chem. Soc.*, 2000, **122**, 5929–5939.
14. S. Grimaldi, J.-P. Finet, F. Le Moigne, A. Zeghdaoui, P. Tordo, D. Benoit, M. Fontanille and Y. Gnanou, *Macromolecules*, 2000, **33**, 1141–1147.

15. D. Benoit, V. Chaplinski, R. Braslau and C. J Hawker, *J. Am. Chem. Soc.*, 1999, **121**, 3904–3920.
16. S. S. Zakariyah, P. P. Conway, D. A. Hutt, D. R. Selviah, K. Wang, J. Rygate, J. Calver and W. Kandulski, *J. Lightwave Technol.*, 2011, **29**, 3566–3576.
17. H. Ma, A. Jen and L. R. Dalton, *Adv. Mater.*, 2002, **14**, 1339–1365.
18. L. Dalton, in *Polymers for Photonics Applications I; Advances in Polymer Science*; Springer, Berlin Heidelberg: Berlin, Heidelberg, 2002, vol. 158, pp. 1–86.
19. F. C. Krebs, *Sol. Energy Mater. Sol. Cells*, 2009, **93**, 394–412.
20. G. G. Barclay, C. J. Hawker, H. Ito, A. Orellana, P. R. L. Malenfant and R. F. Sinta, *Macromolecules*, 1998, **31**, 1024–1031.
21. G. G. Barclay, C. J. Hawker, H. Ito, A. Orellana, R. F. Sinta and P. R. Malenfant, *Proc. SPIE 2724, Advances in Resist Technology and Processing XIII* 1996, 249–260.
22. Z. J. Wang and M. Maric, *J. Polym. Sci. Pol. Chem.*, 2013, **51**, 2970–2978.
23. Y. Wan, X. Fei, Z. Shi, J. Hu, X. Zhang, L. Zhao, C. Chen, Z. Cui and D. Zhang, *J. Polym. Sci. Pol. Chem.*, 2011, **49**, 762–769.
24. J. P. Kim, J. W. Kang, J. J. Kim and J. S. Lee, *Polymer*, 2003, **44**, 4189–4195.
25. J.-L. Couturier, O. Guerret, D. Bertin, D. Gigmes, S. R. A. Marque, P. Tordo and P.-E. Dufils, Alkoxyamines derived from beta-phosphorylated nitroxides, and use thereof for preparing polymerized or non-polymerized mono-or polyalkoxyamines, US 2005/0065119 A1, 2004-03-24 2004.
26. J. Nicolas, B. Charleux, O. Guerret and S. P. Magnet, *Macromolecules*, 2004, **37**, 4453–4463.
27. B. H. Lessard, X. Savelyeva and M. Maric, *Polymer*, 2012, **53**, 5649–5656.
28. A. Moayeri, B. H. Lessard and M. Maric, *Polym. Chem.*, 2011, **2**, 2084–2092.
29. B. H. Lessard and M. Maric, *J. Polym. Sci. Pol. Chem.*, 2009, **47**, 2574–2588.
30. Z. Wang and M. Maric, *Polymers*, 2014, **6**, 565–582.
31. J. Morris, S. Telitel, K. E. Fairfull-Smith, S. E. Bottle, J. Lalevée, J.-L. Clement, Y. Guillaneuf and D. Gigmes, *Polym. Chem.*, 2015, **6**, 754–763.
32. S. Telitel, S. Telitel, J. Bosson, A. Spangenberg, J. Lalevée, F. Morlet-Savary, J.-L. Clement, Y. Guillaneuf, D. Gigmes and O. Soppera, *Adv. Mater. Interfaces*, 2014, **1**, 1400067.
33. S. B. Jhaveri, M. Beinhoff, C. J. Hawker, K. R. Carter and D. Y. Sogah, *ACS nano*, 2008, **2**, 719–727.
34. H. Okamura, D. C. Forman and C. K. Ober, *Polym. Bull.*, 2014, **71**, 2395–2405.
35. M. Hirtz, M. K. Brinks, S. Miele, A. Studer, H. Fuchs and L. Chi, *Small*, 2009, **5**, 919–923.
36. M. K. Brinks, M. Hirtz, L. Chi, H. Fuchs and A. Studer, *Angew. Chem., Int. Ed.*, 2007, **46**, 5231–5233.

37. J. Kim, S. Yang, Y. Lee and Y. Kim, *Prog. Polym. Sci.*, 2010, 1325–1349.
38. C. J. Hawker and T. P. Russell, *MRS Bull.*, 2011, **30**, 952–966.
39. C. G. Gamys, J.-M. Schumers, C. Mugemana, C.-A. Fustin and J.-F. Gohy, *Macromol. Rapid Commun.*, 2013, **34**, 962–982.
40. S. H. Kim, M. J. Misner, T. Xu, M. Kimura and T. P. Russell, *Adv. Mater.*, 2004, **16**, 226–231.
41. M. Zhang, L. Yang, S. Yurt, M. J. Misner, J.-T. Chen, E. B. Coughlin, D. Venkataraman and T. P. Russell, *Adv. Mater.*, 2007, **19**, 1571–1576.
42. K. Satoh, J. E. Poelma, L. M. Campos, B. Stahl and C. J. Hawker, *Polym. Chem.*, 2012, **3**, 1890–1898.
43. H. Mao and M. A. Hillmyer, *Macromolecules*, 2005, **38**, 4038–4039.
44. C. Mugemana, J.-F. Gohy and C.-A. Fustin, *Langmuir*, 2012, **28**, 3018–3023.
45. K. Shin, K. A. Leach, J. T. Goldbach, D. H. Kim, J. Y. Jho, M. Tuominen, C. J. Hawker and T. P. Russell, *Nano Lett.*, 2002, **2**, 933–936.
46. F. J. Xu, Y. Song, Z. P. Cheng, X. L. Zhu, C. X. Zhu, E. T. Kang and K. G. Neoh, *Macromolecules*, 2005, **38**, 6254–6258.
47. D. Y. Ryu, K. Shin, E. Drockenmuller, C. J. Hawker and T. P. Russell, *Science*, 2005, **308**, 236–239.
48. A. Bolognesi, F. Galeotti, U. Giovanella, F. Bertini and S. Yunus, *Langmuir*, 2009, **25**, 5333–5338.
49. F. Galeotti, V. Calabrese, M. Cavazzini, S. Quici, C. Poleunis, S. Yunus and A. Bolognesi, *Chem. Mater*, 2010, **22**, 2764–2769.
50. G. Leone, U. Giovanella, F. Bertini, W. Porzio, F. Meinardi, C. Botta and G. Ricci, *J. Mater. Chem.*, 2013, **1**, 1450–1460.
51. G. Leone, U. Giovanella, F. Bertini, S. Hoseinkhani, W. Porzio, G. Ricci, C. Botta and F. Galeotti, *J. Mater. Chem. C*, 2013, **1**, 6585–6593.
52. S. Chen, M. Alves, M. Save and L. Billon, *Polym. Chem.*, 2014, **5**, 5310–5319.
53. K.-I. Fukukawa, L. Zhu, P. Gopalan, M. Ueda and S. Yang, *Macromolecules*, 2005, **38**, 263–270.
54. B. de Boer, U. Stalmach, H. Nijland and G. Hadziioannou, *Adv. Mater.*, 2000, **12**, 1581–1583.
55. E. Bloch, T. Phan, D. Bertin, P. Llewellyn and V. Hornebecq, *Microporous and Mesoporous Mater.*, 2008, **112**, 612–620.
56. D. Quemener, G. Bonniol, T. N. T. Phan, D. Gigmes, D. Bertin and A. Deratani, *Macromolecules*, 2010, **43**, 5060–5065.
57. D. Gromadzki, J. Lokaj, P. Cernoch, O. Diat, F. Nallet and P. Stepanek, *Eur. Polym. J.*, 2008, **44**, 189–199.
58. T. M. Hermans, J. Choi, B. G. G. Lohmeijer, G. Dubois, R. C. Pratt, H.-C. Kim, R. M. Waymouth and J. L. Hedrick, *Angew. Chem. Int. Ed.*, 2006, **45**, 6648–6652.
59. T. D. Gierke, G. E. Munn and F. C. Wilson, *J. Polym. Sci., Polym. Phys. Ed.*, 1981, **19**, 1687–1704.
60. W. Y. Hsu and T. D. Gierke, *Macromolecules*, 1982, **15**, 101–105.
61. J. Ding, C. Chuy and S. Holdcroft, *Chem. Mater.*, 2001, **13**, 2231–2233.

62. J. Ding, C. Chuy and S. Holdcroft, *Macromolecules*, 2002, **35**, 1348–1355.
63. J. Ding and S. Holdcroft, *Aust. J. Chem.*, 2012, **65**, 1117–1123.
64. C. Chuya, J. Ding, E. Swanson, S. Holdcroft, J. Horsfall and K. V. Lovell, *J. Electrochem. Soc.*, 2003, **150**, E271.
65. J. Ding, Q. Tang and S. Holdcroft, *Aust. J. Chem.*, 2002, **55**, 461–466.
66. Y. Li, A. C. Jackson, F. L. Beyer and D. M. Knauss, *Macromolecules*, 2014, **47**, 6757–6767.
67. S. Holmberg, P. Holmlund, R. Nicolas, C.-E. Wilén, T. Kallio, G. Sundholm and F. Sundholm, *Macromolecules*, 2004, **37**, 9909–9915.
68. S.-I. Sawada, A. Suzuki, T. Terai and Y. Maekawa, *Radiat. Phys. Chem.*, 2010, **79**, 471–478.
69. P. G. Jessop, D. J. Heldebrant, X. Li, C. A. Eckert and C. L. Liotta, *Nature*, 2005, **436**, 1102.
70. J. Lu, F. Yan and J. Texter, *Prog. Polym. Sci.*, 2009, **34**, 431–448.
71. M. D. Green, M. H. J. Allen, J. M. Dennis, D. Salas-de la Cruz, R. Gao, K. I. Winey and T. E. Long, *Eur. Polym. J.*, 2011, **47**, 486–496.
72. S. Lascaud, M. Perrier, A. Vallee, S. Besner, J. Prud'homme and M. Armand, *Macromolecules*, 1994, **27**, 7469–7477.
73. B. D. Mather, M. B. Baker, F. L. Beyer, M. A. G. Berg, M. D. Green and T. E. Long, *Macromolecules*, 2007, **40**, 6834–6845.
74. M. D. Green, J.-H. Choi, K. I. Winey and T. E. Long, *Macromolecules*, 2012, **45**, 4749–4757.
75. M. D. Green, D. Wang, S. T. Hemp, J.-H. Choi, K. I. Winey, J. R. Heflin and T. E. Long, *Polymer*, 2012, **53**, 3677–3686.
76. S. Cheng, F. L. Beyer, B. D. Mather, R. B. Moore and T. E. Long, *Macromolecules*, 2011, **44**, 6509–6517.
77. C. M. Stancik, A. R. Lavoie, J. Schütz, P. A. Achurra, P. Lindner, A. P. Gast and R. M. Waymouth, *Langmuir*, 2004, **20**, 596–605.
78. R. L. Weber, Y. Ye, S. M. Banik, Y. A. Elabd, M. A. Hickner and M. K. Mahanthappa, *J. Polym. Sci., Part B: Polym. Phys.*, 2011, **49**, 1287–1296.
79. R. L. Weber, Y. Ye, A. L. Schmitt, S. M. Banik, Y. A. Elabd and M. K. Mahanthappa, *Macromolecules*, 2011, **44**, 5727–5735.
80. S. Forrest, P. Burrows and M. Thompson, *IEEE Spectrum*, 2000, **37**, 29–34.
81. C.-L. Ho, H. Li and W.-Y. Wong, *J. Organomet. Chem.*, 2014, **751**, 261–285.
82. L. Deng, P. Furuta, S. Garon, J. Li, D. Kavulak, M. E. Thompson and J. M. J. Fréchet, *Chem. Mater.*, 2006, **18**, 386–395.
83. O. Knopfmacher, M. L. Hammock, A. L. Appleton, G. Schwartz, J. Mei, T. Lei, J. Pei and Z. Bao, *Nat. Commun.*, 2014, **5**, 1–9.
84. L. Pan, A. Chortos, G. Yu, Y. Wang, S. Isaacson, R. Allen, Y. Shi, R. Dauskardt and Z. Bao, *Nat. Commun.*, 2014, **5**, 1–8.
85. B. Sun, W. Hong, H. Aziz, N. M. Abukhdeir and Y. Li, *J. Mater. Chem. C*, 2013, **1**, 4423–4426.
86. Y. Li, P. Sonar, L. Murphy and W. Hong, *Energy Environ. Sci.*, 2013, **6**, 1684–1710.

87. N. Kaur, M. Singh, D. Pathak, T. Wagner and J. M. Nunzi, *Synth. Met.*, 2014, **190**, 20–26.
88. B. H. Lessard, J. D. Dang, T. M. Grant, D. Gao, D. S. Seferos and T. P. Bender, *ACS Appl. Mater. Interfaces*, 2014, **6**, 15040–15051.
89. K. Cnops, B. P. Rand, D. Cheyns, B. Verreet, M. A. Empl and P. Heremans, *Nat. Commun.*, 2014, **5**, 1–6.
90. L. Murphy, W. Hong, H. Aziz and Y. Li, *Sol. Energy Mater. Sol. Cells*, 2013, **114**, 71–81.
91. B. Kumar, B. K. Kaushik and Y. S. Negi, *J. Mater. Sci.: Mater. Electron.*, 2013, **25**, 1–30.
92. Y.-C. Lai, Y.-C. Huang, T.-Y. Lin, Y.-X. Wang, C.-Y. Chang, Y. Li, T.-Y. Lin, B.-W. Ye, Y.-P. Hsieh, W.-F. Su, Y.-J. Yang and Y.-F. Chen, *NPG Asia Mater.*, 2014, **6**, 1–7.
93. H. Ma, M. S. Liu and A. K.-Y. Jen, *Polym. Int.*, 2009, **58**, 594–619.
94. E. L. Ratcliff, B. Zacher and N. R. Armstrong, *J. Phys. Chem. Lett.*, 2011, **2**, 1337–1350.
95. X. Yang and A. Uddin, *Renewable Sustainable Energy Rev.*, 2014, **30**, 324–336.
96. A. J. Heeger, *Adv. Mater.*, 2014, **26**, 10–28.
97. M. Green, *J. Materi. Sci.: Mater. Electron.*, 2007, **18**, 15–19.
98. P. Woditsch and W. Koch, *Sol. Energy Mater. Sol. Cells*, 2002, **72**, 11–26.
99. R. Bube, *Photoelectronic Properties of Semiconductors*, Cambridge University Press, New York, 1992.
100. S. Sun, Z. Fan, Y. Wang and J. Haliburton, *J. Mater. Sci.*, 2005, **40**, 1429–1443.
101. C. Tang, *Appl. Phys. Lett.*, 1986, **48**, 183–185.
102. S. Veenstra, J. Loos and J. Kroon, *Progress in Photovoltaics: Research and Applications*, 2007, **15**, 727–740.
103. L. Smilowitz, N. Sariciftci, R. Wu, C. Gettinger, A. Heeger and F. Wudl, *Phys. Rev. B*, 1993, **47**, 13835–13842.
104. X. Yang and J. Loos, *Macromolecules*, 2007, **40**, 1353–1362.
105. O. A. Abdulrazzaq, V. Saini, S. Bourdo, E. Dervishi and A. S. Biris, *Part. Sci. Technol.*, 2013, **31**, 427–442.
106. P. Hudhomme, *EPJ Photovoltaics*, 2013, **4**, 1–11.
107. R. Segalman, B. McCulloch, S. Kirmayer and J. Urban, *Macromolecules*, 2009, **42**, 9205–9216.
108. S. Darling, *Energy Environ. Sci.*, 2009, **2**, 1266–1273.
109. A. Hilberer, M. Moroni, R. E. Gill, H.-J. Brouwer, V. V. Krasnikov, T.-A. Pham, G. G. Malliaras, S. Veenstra, M. P. L. Werts, P. F. van Hutten and G. Hadziioannou, *Macromol. Symp.*, 1997, 99–109.
110. M. Moroni, A. Hilberer and G. Hadziioannou, *Macromol. Rapid Commun.*, 1996, **17**, 693–702.
111. L. Boiteau, M. Moroni, A. Hilberer, M. Werts, B. de Boer and G. Hadziioannou, *Macromolecules*, 2002, **35**, 1543–1548.
112. S. M. Lindner and M. Thelakkat, *Macromolecules*, 2004, **37**, 8832–8835.

113. S. M. Lindner, N. Kaufmann and M. Thelakkat, *Org. Electron.*, 2007, **8**, 69–75.
114. S. M. Lindner and M. Thelakkat, *Macromol. Chem. Phys.*, 2006, **207**, 2084–2092.
115. S. M. Lindner, S. Huttner, A. Chiche, M. Thelakkat and G. Krausch, *Angew. Chem., Int. Ed.*, 2006, **45**, 3364–3368.
116. S. King, M. Sommer, S. Huettner, M. Thelakkat and S. A. Haque, *J. Mater. Chem.*, 2009, **19**, 5436–5441.
117. A. Lang, A. Neubig, M. Sommer and M. Thelakkat, *Macromolecules*, 2010, **43**, 7001–7010.
118. M. Sommer and M. Thelakkat, *EPJ Appl. Phys.*, 2006, **36**, 245–249.
119. M. Sommer, S. Huettner and M. Thelakkat, *J. Mater. Chem.*, 2010, **20**, 10788–10797.
120. S. Huettner, M. Sommer, U. Steiner and M. Thelakkat, *Appl. Phys. Lett.*, 2010, **96**, 073503-1–3.
121. E. M. Barea, G. Garcia-Belmonte, M. Sommer, S. Huettner, H. J. Bolink and M. Thelakkat, *Thin Solid Films*, 2010, **518**, 3351–3354.
122. S. Huttner, M. Sommer and M. Thelakkat, *Appl. Phys. Lett.*, 2008, **92**, 093302-1–3.
123. M. Sommer, S. Huttner, S. Wunder and M. Thelakkat, *Adv. Mater.*, 2008, **20**, 2523–2527.
124. M. Sommer, S. M. Lindner and M. Thelakkat, *Adv. Funct. Mater.*, 2007, **17**, 1493–1500.
125. M. Sommer, S. Huettner and M. Thelakkat, *Adv. Polym. Sci.*, 2010, **228**, 123–153.
126. J. Hou; X. Guo, in Organic Solar Cells*; Green Energy and Technology*, Springer London, London, 2012, pp. 17–42.
127. R. D. McCullough, *Adv. Mater.*, 1998, **10**, 93–116.
128. M. Brinkmann, *J. Polym. Sci., Part B: Polym. Phys.*, 2011, **49**, 1218–1233.
129. T. Higashihara and M. Ueda, *Macromol. Res.*, 2013, **21**, 257–271.
130. M. T. Dang, L. Hirsch and G. Wantz, *Adv. Mater.*, 2011, **23**, 3597–3602.
131. R. McCullough and S. Tristram-Nagle, *J. Chem. Soc.*, 1993, **115**, 4910–4911.
132. F. Bates and G. Fredrickson, *Physics Today*, 1999, **52**, 32–38.
133. B. Olsen and R. Segalman, *Mater. Sci. Eng., R*, 2008, **62**, 37–66.
134. M. C. Iovu, R. Zhang, J. R. Cooper, D.-M. Smilgies, A. E. Javier, E. E. Sheina, T. Kowalewski and R. D. McCullough, *Macromol. Rapid Commun.*, 2007, **28**, 1816–1824.
135. F. Richard, C. Brochon, N. Leclerc, D. Eckhardt, T. Heiser and G. Hadziioannou, *Macromol. Rapid Commun.*, 2008, **29**, 885–891.
136. K. van de Wetering, C. Brochon, C. Ngov and G. Hadziioannou, *Macromolecules*, 2006, **39**, 4289–4297.
137. A. Britze, V. Moellmann, G. Grundmeier, H. Luftmann and D. Kuckling, *Macromol. Chem. Phys.*, 2011, **212**, 679–690.

138. E. Kaul, V. Senkovskyy, R. Tkachov, V. Bocharova, H. Komber, M. Stamm and A. Kiriy, *Macromolecules*, 2010, **43**, 77–81.
139. K. Yuan, F. Li, L. Chen and Y. Chen, *Thin Solid Films*, 2012, **520**, 6299-6306.
140. Q. Zhang, A. Cirpan, T. P. Russell and T. Emrick, *Macromolecules*, 2009, **42**, 1079–1082.
141. C. Brochon, F. Richard, N. Leclerc and P. Leveque, *Polym. Prepr. (Am. Chem. Soc., Div. Polym. Chem.)*, 2008, **49**, 391-392.
142. M. Sommer, A. Lang and M. Thelakkat, *Angew. Chem., Int. Ed.*, 2008, **47**, 7901–7904.
143. S. Huettner, M. Sommer, J. Hodgkiss, P. Kohn, T. Thurn-Albrecht, R. H. Friend, U. Steiner and M. Thelakkat, *ACS Nano*, 2011, **5**, 3506–3515.
144. S. Huettner, J. M. Hodgkiss, M. Sommer, R. H. Friend, U. Steiner and M. Thelakkat, *J. Phys. Chem. B*, 2012, **116**, 10070–10078.
145. A. S. Lang, F. R. Kogler, M. Sommer, U. Wiesner and M. Thelakkat, *Macromol. Rapid Commun.*, 2009, **30**, 1243–1248.
146. R. H. Lohwasser and M. Thelakkat, *Macromolecules*, 2012, **45**, 3070–3077.
147. R. H. Lohwasser, G. Gupta, P. Kohn, M. Sommer, A. S. Lang, T. Thurn-Albrecht and M. Thelakkat, *Macromolecules*, 2013, **46**, 4403–4410.
148. U. Stalmach, B. de Boer, C. Videlot, P. F. van Hutten and G. Hadziioannou, *J. Am. Chem. Soc.*, 2000, **122**, 5464–5472.
149. M. H. van der Veen, B. de Boer, U. Stalmach, K. I. van de Wetering and G. Hadziioannou, *Macromolecules*, 2004, **37**, 3673–3684.
150. U. Stalmach, B. de Boer, A. D. Post, P. F. van Hutten and G. Hadziioannou, *Angew. Chem.*, 2001, **113**, 442–444.
151. S. Barrau, T. Heiser, F. Richard, C. Brochon, C. Ngov, K. van de Wetering, G. Hadziioannou, D. V. Anokhin and D. A. Ivanov, *Macromolecules*, 2008, **41**, 2701–2710.
152. T. Heiser, G. Adamopoulos, M. Brinkmann, U. Giovanella, S. Ould-Saad, C. Brochon, K. van de Wetering and G. Hadziioannou, *Thin Solid Films*, 2006, **511–512**, 219–223.
153. G. Adamopoulos, T. Heiser, U. Giovanella, S. Ould-Saad, K. van de Wetering, C. Brochon, T. Zorba, K. Paraskevopoulos and G. Hadziioannou, *Thin Solid Films*, 2006, **511–512**, 371–376.
154. X. Chen, X. Han, G. Vamvounis and S. Holdcroft, *Polym. Prepr. (Am. Chem. Soc., Div. Polym. Chem.)*, 2007, **48**, 102–103.
155. B. Gholamkhass, T. J. Peckham and S. Holdcroft, *Polym. Chem.*, 2010, **1**, 708–719.
156. B. Gholamkhass and S. Holdcroft, *Chem. Mater*, 2010, **22**, 5371–5376.
157. B. Gholamkhass and S. Holdcroft, *Sol. Energy Mater. Sol. Cells*, 2011, **95**, 3106–3113.
158. Z. Zhou, X. Chen and S. Holdcroft, *J. Am. Chem. Soc.*, 2008, **130**, 11711–11718.
159. M. Heuken, H. Komber and B. Voit, *Macromol. Chem. Phys.*, 2012, **213**, 97–107.

160. M. Heuken, H. Komber, T. Erdmann, V. Senkoyskyy, A. Kiriy and B. Voit, *Macromolecules*, 2012, **45**, 4101–4114.
161. A. Laiho, R. H. A. Ras, S. Valkama, J. Ruokolainen, R. Osterbacka and O. Ikkala, *Macromolecules*, 2006, **39**, 7648–7653.
162. B. H. Lessard and M. Maric, *e-Polym.*, 2012, **63**, 1–14.
163. S. Mougnier, C. Brochon and E. Cloutet, *J. Polym. Sci. Pol. Chem.*, 2012, **50**, 2463–2470.
164. C. Renaud, S.-J. Mougnier, E. Pavlopoulou, C. Brochon, G. Fleury, D. Deribew, G. Portale, E. Cloutet, S. Chambon, L. Vignau and G. Hadziioannou, *Adv. Mater.*, 2012, **24**, 2196–2201.
165. S. B. Maria, A. S. Susha, M. Sommer, D. V. Talapin, A. L. Rogach and M. Thelakkat, *Macromolecules*, 2008, **41**, 6081–6088.
166. G. M. Farinola and R. Ragni, *Chem. Soc. Rev.*, 2011, **40**, 3467–3482.
167. C. J. Humphreys, *MRS Bull.*, 2011, **33**, 459–470.
168. Y.-L. Chang, S. Yin, Z. Wang, M. G. Helander, J. Qiu, L. Chai, Z. Liu, G. D. Scholes and Z. Lu, *Adv. Funct. Mater.*, 2013, **23**, 705–712.
169. M. Eritt, C. May, K. Leo, M. Toerker and C. Radehaus, *Thin Solid Films*, 2010, **518**, 3042–3045.
170. L. Xiao, Z. Chen, B. Qu, J. Luo, S. Kong, Q. Gong and J. Kido, *Adv. Mater.*, 2010, **23**, 926–952.
171. M. A. Baldo, D. F. O'Brien and S. R. Forrest, *Phys. Rev. B*, 1999, **60**, 14422–14428.
172. Y. Kawamura, K. Goushi, J. Brooks, J. J. Brown, H. Sasabe and C. Adachi, *Appl. Phys. Lett.*, 2005, **86**, 071104.
173. C. Adachi, M. A. Baldo, M. E. Thompson and S. R. Forrest, *J. Appl. Phys.*, 2001, **90**, 5048–5051.
174. S. Gong, C. Yang and J. Qin, *Chem. Soc. Rev.*, 2012, **41**, 4797–4807.
175. B. Geffroy, P. le Roy and C. Prat, *Polym. Int.*, 2006, **55**, 572–582.
176. J. H. Burroughes, D. Bradley, A. R. Brown and R. N. Marks, *Nature*, 1990, **347**, 539–548.
177. X. Chen, J.-L. Liao, Y. Liang, M. O. Ahmed, H.-E. Tseng and S.-A. Chen, *J. Am. Chem. Soc*, 2003, **125**, 636–637.
178. N. Tessler, Y. Preezant, N. Rappaport and Y. Roichman, *Adv. Mater.*, 2009, **21**, 2741–2761.
179. P. T. Furuta, L. Deng, S. Garon, M. E. Thompson and J. M. J. Fréchet, *J. Am. Chem. Soc.*, 2004, **126**, 15388–15389.
180. B. Ma, B. J. Kim, L. Deng, D. A. Poulsen, M. E. Thompson and J. M. J. Fréchet, *Macromolecules*, 2007, **40**, 8156–8161.
181. D. Poulsen, B. Kim, B. Ma, C. Zonte and J. Frechet, *Adv. Mater.*, 2009, **22**, 77–82.
182. F. Dumur, Y. Guillaneuf, A. Guerlin, G. Wantz, D. Bertin, F. Miomandre, G. Clavier, D. Gigmes and C. R. Mayer, *Macromol. Chem. Phys.*, 2011, **212**, 1616–1628.
183. F. Dumur, Y. Guillaneuf, A. Guerlin, G. Wantz, D. Bertin, F. Miomandre, G. Clavier, C. R. Mayer and D. Gigmes, *Polym. Prepr. (Am. Chem. Soc., Div. Polym. Chem.)*, 2011, **52**, 783–784.

184. K. Tsuchiya, H. Kasuga and A. Kawakami, *J. Polym. Sci.,: Part A: Polym. Chem.*, 2010, **48**, 1461–1468.
185. K. Tsuchiya, K. Sakaguchi, H. Kasuga, A. Kawakami, H. Taka, H. Kita and K. Ogino, *Polymer*, 2010, **51**, 616–622.
186. K. Tsuchiya, K. Sakaguchi, A. Kawakami, H. Taka, H. Kita, T. Shimomura and K. Ogino, *Synth. Met.*, 2010, **160**, 1679–1682.
187. X. Savelyeva and B. Lessard, *Macromol. React. Eng.*, 2012, **6**, 200–212.
188. B. H. Lessard, L. Beouch, F. Goubard, M. Marić, D. Gigmes and F. Dumur, *Org. Electron.*, 2015, **17**, 377–385.
189. Y. Tao, B. Ma and R. A. Segalman, *Macromolecules*, 2008, **41**, 7152–7159.
190. C.-C. Chang and M.-K. Leung, *Chem. Mater*, 2008, **20**, 5816–5821.
191. J. Park, C. Lee, J. Jung, H. Kang, K.-H. Kim, B. Ma and B. J. Kim, *Adv. Funct. Mater.*, 2014, **24**, 7588–7596.
192. E. Kozma, W. Mróz and F. Galeotti, *Dyes Pigm.*, 2015, **114**, 138–143.
193. J. C. Scott and L. D. Bozano, *Adv. Mater.*, 2007, **19**, 1452–1463.
194. W. L. Leong, N. Mathews, B. Tan, S. Vaidyanathan, F. Doetz and S. Mhaisalkar, *J. Mater. Chem.*, 2011, **21**, 5203–5214.
195. Q.-D. Ling, D.-J. Liaw, C. Zhu, D. S.-H. Chan, E.-T. Kang and K.-G. Neoh, *Prog. Polym. Sci.*, 2008, **33**, 917–978.
196. S. Möller, C. Perlov, W. Jackson, C. Taussig and S. R. Forrest, *Nature*, 2003, **426**, 166–169.
197. J. Y. Ouyang, C. W. Chu, C. R. Szmanda, L. P. Ma and Y. Yang, *Nat. Mater.*, 2004, **3**, 918–922.
198. H.-C. Chang, C.-L. Liu and W.-C. Chen, *ACS Appl. Mater. Interfaces*, 2013, **5**, 13180–13187.
199. W. Kwon, B. Ahn, D. M. Kim, Y.-G. Ko, S. G. Hahm, Y. Kim, H. Kim and M. Ree, *J. Phys. Chem. C*, 2011, **115**, 19355–19363.
200. Y.-K. Fang, C.-L. Liu and W.-C. Chen, *J. Mater. Chem.*, 2011, **21**, 4778–4786.
201. Y.-K. Fang, C.-L. Liu, G.-Y. Yang, P.-C. Chen and W.-C. Chen, *Macromolecules*, 2011, **44**, 2604–2612.
202. M. Destarac, *Macromol. React. Eng.*, 2010, **4**, 165–179.
203. P. Nesvadba, *Radical Polymerization in Industry*, John Wiley & Sons, Ltd, Chichester, UK, 2012.
204. W. Rutsch and M. A. Cech, *CHIMIA*, 2007, **61**, 33–41.
205. A. Kramer and P. Nesvadba, 1-alkoxy-polyalkyl-peridine derivatives and their use as polymerization regulators, 7402639, 2008.
206. P. Nesvadba, A. Kramer and L. Bugnon, Alkoxyamines containing a radically polymerizable group, 7619050, 2009.
207. P. Nesvadba, L. Bugnon and R. Sift, *J. Polym. Sci. Pol. Chem.*, 2004, **42**, 3332–3341.
208. L. A. Engelbrecht and P. Nesvadba, Alkoxyamines containing unsaturated groups, 7812100, 2010.
209. P. Nesvadba, A. Kramer and M. O. Zink, Heterocyclic alkoxyamines as regulators in controlled radical polymerization processes. US 6479608 B1, 2002-12-11 2002.

210. P. Nesvadba, A. Steinmann and W. Stauffer, Polymerizable compositions containing alkoxyamine compounds derived from nitroso- or nitrone compounds. US 6262260 B1, 2001-07-17 2001.
211. M. O. Zink, A. Kramer and P. Nesvadba, *Macromolecules*, 2000.
212. C. Auschra, E. Eckstein, A. Muhlebach, M. O. Zink and F. Rime, *Prog. Org. Coat.*, 2002, **45**, 83–93.
213. A. T. Termaten, M. van der Sluis, F. Pirrung and C. Auschra, *Polym. Prepr. (Am. Chem. Soc., Div. Polym. Chem.)*, 2008, **49**, 353–354.
214. C. Auschra, E. Eckstein and R. Knischka, *Eur. Coat. J.*, 2004, **26**, 1–9.
215. C. Auschra, E. Eckstein, M. O. Zink, Pigment compositions with modified block copolymer dispersants, US 6849679 B2, 2005-02-01 2005.
216. C. Auschra, E. Eckstein and R. Knischka, *Eur. Coat. J.*, 2005, **156**, 162–163.
217. R. Pfaendner, T. Wannemacher and D. Braun, (Co)polymerization of vinyl chloride in the presence of a stable nitroxyl radical, 6844405, 2005.
218. L. A. Engelbrecht and A. Noordam, Polymers obtained via nitroxyl radical polymerization, 7858694, 2010.
219. J.-P. Gillet, O. Guerret and J.-P. Lascombe, Method for preparing nitroxides, US 6538141 B1, 2003-03-25 2003.
220. O. Guerret, J.-L. Couturier and C. LeMercier, Method for preparing β-phosphorus nitroxide radicals. US 7126021 B2, 2006-10-24 2006.
221. O. Guerret, Method for preparing a polymeric material including a multiblock copolymer prepared by controlled free-radical polymerisation, US 2010/0168338 A1, July 1, 2010.
222. P. Gerard, L. Couvreur, S. Magnet and J. Ness, *Polym. Prepr. Polym. Prepr. (Am. Chem. Soc., Div. Polym. Chem.)*, 2009, **49**, 56–57.
223. K. Schierholz, M. Givehchi, P. Fabre, F. Nallet, E. Papon, O. Guerret and Y. Gnanou, *Macromolecules*, 2003, **36**, 5995–5999.
224. S. Poivet, P. Fabre, F. Nallet, K. Schierholz, G. Abraham, E. Papon, Y. Gnanou, R. Ober, O. Guerret and N. E. El-Bounia, *Eur. Phys. J. E*, 2006, **20**, 273–287.
225. N. P. Boupat, O. Guerret, S. Magnet and P. Gerard, Adjustable block copolymer having acid functional groups and adhesive and thermoplastic composition containing it, US 2008/0214712 A1, September 4, 2008.
226. S. Maiez-Tribut, J. P. Pascault, E. R. Soulé, J. Borrajo and R. J. J. Williams, *Macromolecules*, 2007, **40**, 1268–1273.
227. F. Court, L. Leibler, J.-P. Pascault and S. Ritzenthaler, Thermoset Materials with Improved Impact Resistance, US 6894113, 2005-03-17 2005.
228. O. Guerret, P. Gerard, J.-P. Pascault and A. Bonnet, Thermoset Materials with improved impact resistance, US 2008/0051511 A1, February 28, 2008.
229. T. Phan, S. Maiez-Tribut, J.-P. Pascault, A. Bonnet, P. Gerard, O. Guerret and D. Bertin, *Macromolecules*, 2007, **40**, 4516–4523.

230. P. Gerard, N. P. Boupat, T. Fine, L. Gervat and J.-P. Pascault, *Macromol. Symp.*, 2007, **256**, 55–64.
231. E. Serrano, P. Gerard, F. Lortie, J.-P. Pascault and D. Portinha, *Macromolecular Mater. Eng.*, 2008, **293**, 820–827.
232. N. Jullian, L. Rubatat, P. Gerard, J. Peyrelasse and C. Derail, *J. Rheol.*, 2011, **55**, 379–400.
233. B. H. Lessard, C. Aumand-Bourque, R. Chaudury, D. Gomez, A. Haroon, N. Ibrahimian, S. Mackay, M.-C. Noel, R. Patel, S. Sitaram, S. Valla, B. White and M. Marić, *Int. Polym. Process.*, 2011, **XXVI**, 197–204.
234. K. J. Oxby and M. Maric, *Macromol. React. Eng.*, 2013, **8**, 160–169.
235. J.-M. Suau, O. Guerret, R. Souzy and Y. Kensicher, Use of acrylic comb copolymers as a colour developing agent in cosmetic compositions, US 2012/0276032 A1, November 1, 2012.
236. J.-M. Suau and O. Guerret, Comb polymers for the hair, US 8658144 B2, February 25 2014.
237. M. Guillot, O. Guerret and F. Dupont, Weakly anionic polymers, uses thereof, US2013/0130046, May 23, 2013.
238. J.-M. Suau, J. Mongoin, Y. Kensicher and O. Guerret, Use in a hydraulic binder composition of a dry-ground calcium carbonate with a copolymer of (meth) acrylic acid and an alkoxy or hydroxy polyalkyleneglycol group, US 2011/0319526 A1, December 29, 2011.
239. J.-M. Suau, J. Mongoin, D. Ruhlmann and O. Guerret, Use in a paint of a dry-ground calcium carbonate with a copolymer of (meth)acrylic acid with an alkoxy or hydroxy polyalkyleneglycol group, US 2010/0111809 A1, May 6, 2010.

Subject Index